T0358753

Financial Mathematics

Chapman & Hall/CRC Financial Mathematics Series

Aims and scope:
The field of financial mathematics forms an ever-expanding slice of the financial sector. This series aims to capture new developments and summarize what is known over the whole spectrum of this field. It will include a broad range of textbooks, reference works and handbooks that are meant to appeal to both academics and practitioners. The inclusion of numerical code and concrete real-world examples is highly encouraged.

Series Editors
M.A.H. Dempster
Centre for Financial Research
Department of Pure Mathematics and Statistics
University of Cambridge

Dilip B. Madan
Robert H. Smith School of Business
University of Maryland

Rama Cont
Department of Mathematics
Imperial College

Robert A. Jarrow
Lynch Professor of Investment Management
Johnson Graduate School of Management
Cornell University

Handbook of Financial Risk Management
Thierry Roncalli
Optional Processes
Stochastic Calculus and Applications
Mohamed Abdelghani, Alexander Melnikov
Machine Learning for Factor Investing
Guillaume Coqueret and Tony Guida
Malliavin Calculus in Finance
Theory and Practice
Elisa Alos, David Garcia Lorite
Risk Measures and Insurance Solvency Benchmarks
Fixed-Probability Levels in Renewal Risk Models
Vsevolod K. Malinovskii

Giuseppe Campolieti
Professor, Department of Mathematics,
Wilfrid Laurier University,
Waterloo, Ontario

Roman N. Makarov
Associate Professor, Department of Mathematics,
Wilfrid Laurier University,
Waterloo, Ontario

For more information about this series please visit: https://www.crcpress.com/Chapman-and-HallCRC-Financial-Mathematics-Series/book-series/CHFINANCMTH

Financial Mathematics
A Comprehensive Treatment
in Discrete Time
Second Edition

by

Giuseppe Campolieti
Professor, Department of Mathematics,
Wilfrid Laurier University,
Waterloo, Ontario

Roman N. Makarov
Associate Professor, Department of Mathematics,
Wilfrid Laurier University,
Waterloo, Ontario

CRC Press
Taylor & Francis Group
Boca Raton London New York

CRC Press is an imprint of the
Taylor & Francis Group, an **informa** business
A CHAPMAN & HALL BOOK

Second edition published 2021
by CRC Press
6000 Broken Sound Parkway NW, Suite 300, Boca Raton, FL 33487-2742

and by CRC Press
2 Park Square, Milton Park, Abingdon, Oxon, OX14 4RN

© 2021 Taylor & Francis Group, LLC

First edition published by CRC Press 2014

CRC Press is an imprint of Taylor & Francis Group, LLC

Reasonable efforts have been made to publish reliable data and information, but the author and publisher cannot assume responsibility for the validity of all materials or the consequences of their use. The authors and publishers have attempted to trace the copyright holders of all material reproduced in this publication and apologize to copyright holders if permission to publish in this form has not been obtained. If any copyright material has not been acknowledged please write and let us know so we may rectify in any future reprint.

Except as permitted under U.S. Copyright Law, no part of this book may be reprinted, reproduced, transmitted, or utilized in any form by any electronic, mechanical, or other means, now known or hereafter invented, including photocopying, microfilming, and recording, or in any information storage or retrieval system, without written permission from the publishers.

For permission to photocopy or use material electronically from this work, access www.copyright.com or contact the Copyright Clearance Center, Inc. (CCC), 222 Rosewood Drive, Danvers, MA 01923, 978-750-8400. For works that are not available on CCC please contact mpkbookspermissions@tandf.co.uk

Trademark notice: Product or corporate names may be trademarks or registered trademarks and are used only for identification and explanation without intent to infringe.

ISBN: 978-1-138-58787-8 (hbk)
ISBN: 978-1-032-02307-6 (pbk)
ISBN: 978-0-429-50366-5 (ebk)

Typeset in CMR10 font
by KnowledgeWorks Global Ltd.

To our students

Contents

List of Figures

List of Tables

Preface

Objectives and Audience

This book has evolved from financial mathematics courses that the authors have developed and taught mainly within the bachelor's and master's programs in financial mathematics at Wilfrid Laurier University over the past 15+ years. The material has been tested and refined through years of classroom teaching experience. This volume is a revision and extension of Parts I and II of the first edition, "Financial Mathematics: A Comprehensive Treatment," published in 2014. The first edition combines both discrete-time and continuous-time asset pricing theory, with various models and applications into a single volume consisting of Parts I-IV. Part I covers topics at various levels that include some simple mathematics of compounding, basic fixed income securities and annuities, an introduction to portfolio theory and management, a primer on derivative securities, the concept of arbitrage, replication and no-arbitrage pricing of financial securities. Part II is a more formal self-contained and rigorous mathematical development of financial derivative (asset) pricing theory and stochastic processes in the discrete-time setting. In particular, it begins with a general single-period economic model where the key concepts of arbitrage, replication, market completeness, and pricing and hedging of financial derivatives are formulated. Further chapters lay down the foundation for stochastic processes, martingales, martingale measures for pricing derivatives in a multi-period binomial model as well as in a more general multi-asset discrete-time setting. Part II also covers the two important fundamental theorems of asset pricing in discrete time.

Parts III and IV of the first edition constitute material for senior undergraduate and master's level graduate courses in continuous-time asset pricing theory. The third part provides a comprehensive coverage of stochastic (Itô) calculus for Brownian motion in one and many dimensions, no-arbitrage pricing in the Black–Scholes–Merton framework for single and mutli-asset European options, cross-currency options, American options as well as other path-dependent options. Various option pricing formulae are derived using different techniques. Part III also presents risk-neutral asset pricing theory within a mathematically rigorous framework that incorporates equivalent martingale measures and change of numéraire methods for pricing with various option pricing applications. Part III also tackles interest-rate modelling and pricing of fixed-income derivatives, as well as alternative asset pricing models, including the local volatility model and solvable state-dependent volatility (e.g., the CEV diffusion) models, stochastic volatility models, jump-diffusion and pure jump processes and variance gamma models. Finally, Part IV of the first edition covers several computational techniques for pricing financial derivatives, as well as a comprehensive chapter on simulation and Monte Carlo methods.

In comparison with Parts I and II of the first edition, this volume contains new material in all chapters, as well as a new chapter on elementary probability theory in Appendix A and an expanded set of problems throughout each chapter, including answers to all problems in Appendix C. In total, this volume has almost 200 pages worth of new material. As the title suggests, this book is a comprehensive, self-contained, and unified treatment of the main theory and application of mathematical methods behind modern-day financial mathematics in the discrete-time setting. In writing this book, the authors have really strived to create a single source that can be used as a complete standard university textbook for several interrelated courses in financial mathematics at the undergraduate as well as graduate levels. As such, the authors have aimed to introduce both the financial theory and the relevant mathematical methods in a mathematically rigorous yet student-friendly and engaging style, that includes an abundance of examples, problem exercises, and fully worked-out solutions.

In contrast to most published manuscripts on the subject of financial mathematics, this book presents multiple problem-solving approaches. It hence bridges together related comprehensive techniques for pricing different types of financial derivatives. The book contains a rather complete and in-depth comprehensive coverage of discrete-time financial models that form the foundation of financial derivative pricing theory. This book also provides a self-contained introduction to discrete-time stochastic calculus and martingale theory, which is an important cornerstone in quantitative finance. The material in many of the chapters is presented at a level that is mainly accessible to undergraduate students of mathematics, finance, actuarial science, economics, and other related quantitative fields. The textbook covers a breadth of material, from beginner to more advanced levels, that is required, i.e., absolutely essential, in the core curriculum courses on financial mathematics currently taught at second-year, intermediate and senior undergraduate levels, as well as master's graduate levels, at many universities worldwide.

The book has the following key features:

- comprehensive treatment covering a complete undergraduate program in discrete-time financial mathematics;

- student-friendly presentation with numerous fully worked out examples and exercise problems with answers in every chapter;

- in-depth coverage of discrete-time theory and methodology;

- mathematically rigorous and consistent, yet simple, style that bridges various basic and more advanced concepts and techniques;

- judicious balance of financial theory, mathematical, and computational methods.

Guide to Material

This book is divided into two main parts with each part consisting of several chapters. There are a total of eight chapters, and every chapter (with the exception of Appendix A on elementary probability theory) ends with a comprehensive and exhaustive set of exercises of varying difficulty with answers available in Appendix C.

Part I is an introduction to pricing and management of financial securities. This part has four chapters. Chapter 1 introduces the reader to time value of money, compounding interest, and the basic concepts of fixed income markets. Chapter 2 introduces basic derivative securities and the concept of arbitrage. Chapter 3 covers standard theoretical topics of portfolio management and only requires some very basic linear algebra and optimization. Chapter 4 presents more formal definitions and gives a thorough discussion on basic options theory, including payoff replication, hedging, put-call parity relations, forward and futures contracts, swaps, American options, and other contracts.

Part II is devoted to discrete-time financial modelling. Chapters 5-8 of Part II can be

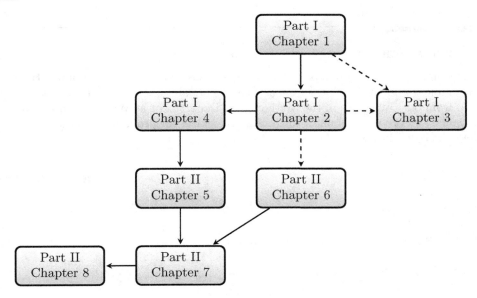

FIGURE: Guide to material.

considered as a complete course on discrete-time asset pricing. Part II introduces the main financial concepts in risk-neutral pricing theory. Chapter 5 covers the financial and formal mathematical underpinnings of the single-period (Arrow–Debreu) economic model. Chapter 6 lays down the foundation for stochastic processes in discrete time, which is then essential for Chapter 7. The latter chapter covers the multi-period binomial market model and is considered the centrepiece of discrete-time financial derivative pricing. All the important concepts for pricing and hedging standard European, as well as path-dependent derivatives within this model, are presented. A general multi-asset, multi-period, discrete-time model is covered in Chapter 8. This chapter also presents the two fundamental theorems of asset pricing and equivalent martingale measures for discrete-time derivative pricing.

The inter-relationship among the different chapters is summarized in the figure, which represents a flow chart of the material in the textbook. Each solid arrow indicates a strong connection between the material in the respective chapters, i.e., when a chapter is viewed as a prerequisite for the other. A dashed arrow indicates that a chapter is relevant but not necessarily a prerequisite for the other. Finally, the table below is a reference guide for instructors. It displays two courses for which this book can be adopted as a required textbook. The relevant chapters for each course and the basic prerequisites are indicated in the table.

Course	Chapters	Prerequisites
Introduction to Financial Mathematics	1-4, App. 1	Calculus, Linear Algebra, Elementary Probability Theory
Discrete-Time Derivative Pricing	2, 4-8	Calculus, Linear Algebra, Probability Theory

Acknowledgements

We would like to thank all our past and current undergraduate and graduate students for their valuable comments, feedback, and advice. We are also grateful for the feedback we have received from our colleagues in the Mathematics Department at Wilfrid Laurier University. In particular, we are very thankful to Dr. Adam Metlzer for providing several examples and problems included in Part I of this book.

Giuseppe Campolieti and Roman N. Makarov

Waterloo, Ontario
October 2020

Part I

Introduction to Pricing and Management of Financial Securities

1

Mathematics of Compounding

1.1 Interest and Return

1.1.1 Amount Function and Return

A rational investor prefers a dollar in his or her pocket today to a dollar in the pocket one year from now. If an investor lends money to a borrower, the investor expects to be compensated for the use of the money. *Interest* is compensation that a borrower of capital pays to a lender of capital for its use. If an initial amount P grows to an amount V over time, then the difference $I = V - P$ is *interest*. This situation is illustrated in Figure 1.1. For investments, it is often called interest earned, but it goes by other names such as return on investment or coupon payment. We assume that a nonnegative amount, and usually a positive amount, of interest is paid. The other crucial assumption is that there is no risk involved in this operation. So in this chapter, we only deal with risk-free financial instruments. For example, the capital is deposited in a risk-free bank account or invested in government bonds (here, we neglect the possibility of government default on financial obligations).

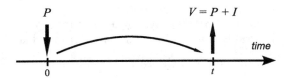

FIGURE 1.1: A time diagram for a simple cash flow. The interest is earned at time t. The accumulated value V is a sum of the principal P and interest I.

There are various explanations for the existence of interest, including the following.

The time value of money. Generally, people prefer to have money now rather than the same amount of money at a later point in time.

Inflationary expectations. The actual cost of the same amount changes in time.

Alternative investments. A lender of capital no longer has the option of immediately using the money invested. Interest compensates a lender for this loss of choices.

The following notation will be used.

P denotes the initial capital borrowed or invested. It is called the *principal* or the *present value* of capital.

$V(t)$ denotes the accumulated value at time $t \geqslant 0$, called the *value function*. Its initial value is equal to the principal, $V(0) = P$.

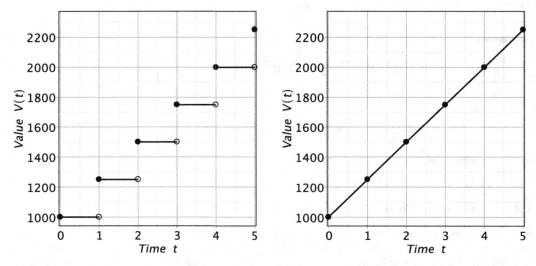

(a) Interest is paid at the end of each year.   (b) Interest is paid continuously such that the amount of interest earned over a period of time is proportional to the length of the period.

FIGURE 1.2: An investment of $1000 grows by a constant amount of $250 each year for five years.

t is the time measured in years or, more generally, in time periods; typically, one period is one year, although it can be one month, one week, one day, etc. In practice, the duration of one period relates to the frequency at which interest is earned.

Example 1.1. An investment of $1000 grows by a constant amount of $250 each year for five years. What does the graph of the value function $V(t)$ look like if

(a) interest is only paid at the end of each year;

(b) interest is paid continuously so that the amount of interest earned over a period of time is proportional to the length of the period.

Solution. In case (a), the value function is a piecewise step function with jumps at the end of every year, i.e., $V(t) = V(t-1) + 250$ for all $t \geqslant 1$. In case (b), the value function is a linear function of time t, i.e., $V(t) = V(0) + ct$ for some parameter c. Since $V(1) = V(0) + 250$, the parameter c is equal to 250, and, therefore, $V(t) = 1000 + 250t$. The plots of the value functions are given in Figure 1.2. $\qquad\square$

Typically, the amount an investment is worth at time t is proportional to the principal P deposited at time 0. Let $a(t)$ denote the accumulated value for principal $1. This function is called the *accumulation function*. So, $1 invested at time 0 grows to $a(t)$ at time t. Then, the accumulated value for principal P is given by

$$V(t) = P\,a(t). \tag{1.1}$$

Consider an investment with the value function $V(t)$, $t \geqslant 0$. The *total return* on the investment is the ratio of the amount received at the end of a period to the amount invested at the beginning of the period,

$$\text{total return} = \frac{\text{amount received}}{\text{amount invested}}.$$

Thus, the total return for the time interval $[s, t]$ with $0 \leqslant s < t$, denoted $R_{[s,t]}$, on an investment commencing at time s and terminating at time t is

$$R_{[s,t]} = \frac{V(t)}{V(s)} \, . \tag{1.2}$$

Note that the total return is a function of a time period rather than a function of an instantaneous time moment.

The *rate of return* or *rate of interest* is the ratio of the amount of interest earned during the period to the investment value at the beginning of the period;

$$\text{rate of return} = \frac{\text{interest earned}}{\text{amount invested}} \, .$$

The aggregate rate of return for the time interval $[s, t]$, denoted $r_{[s,t]}$, is given by

$$r_{[s,t]} = \frac{V(t) - V(s)}{V(s)} = \frac{V(t)}{V(s)} - 1 \, . \tag{1.3}$$

It can be expressed in terms of the accumulation function. For example, for the time interval of t periods from the date of the investment we have

$$r_{[0,t]} = \frac{V(t) - V(0)}{V(0)} = \frac{P \, a(t) - P}{P} = a(t) - 1 \, .$$

It is clear that the two notions are related by

$$R = 1 + r \, .$$

Equations (1.2) and (1.3) can be rewritten as

$$V(t) = R_{[s,t]} \, V(s) = (1 + r_{[s,t]}) \, V(s) \, . \tag{1.4}$$

Note that upper- and lowercase letters, such as R and r, are used for total returns and rates of return, respectively. Often, for simplicity, the term *return* is used for both notions. Let n be a positive integer. The interval $[n - 1, n]$ is the nth year (or the nth period, in general). Denote by R_n and r_n the total return and rate of return during the nth year from the start of investment, respectively. That is, we have

$$R_n := R_{[n-1,n]} = \frac{V(n)}{V(n-1)} \quad \text{and} \quad r_n := r_{[n-1,n]} = \frac{V(n) - V(n-1)}{V(n-1)} \tag{1.5}$$

for $n = 1, 2, 3, \dots$. The accumulated value $V(n)$ can be now written as follows:

$$\frac{V(n) - V(n-1)}{V(n-1)} = r$$
$$V(n) - V(n-1) = rV(n-1)$$
$$V(n) = (1 + r)V(n-1), \quad \text{with } r = r_n \, . \tag{1.6}$$

For two nonnegative integers n and m with $n < m$, the amount $V(m)$ accumulated by the end of period m can be written in two different ways:

$$V(m) = (1 + r_{[n,m]}) \, V(n)$$

and

$$V(m) = (1 + r_m) V(m-1) = (1 + r_{m-1}) (1 + r_m) V(m-2) = \cdots$$
$$= (1 + r_{n+1}) \cdots (1 + r_{m-1}) (1 + r_m) V(n). \tag{1.7}$$

Therefore, one-period returns and aggregate returns are related as follows:

$$1 + r_{[n,m]} = (1 + r_{n+1}) (1 + r_{n+2}) \cdots (1 + r_m), \tag{1.8}$$
$$R_{[n,m]} = R_{n+1} R_{n+2} \cdots R_m. \tag{1.9}$$

Example 1.2 (The rate of return). Joe invested \$200 for three years in an account with the accumulation function $a(t) = 0.1t^2 + 1$. Find annual returns r_1, r_2, and r_3.

Solution. The accumulated values and returns at the end of years 1, 2, and 3 are, respectively,

$$V(1) = 200 \cdot (0.1 \cdot 1^2 + 1) = 220 \implies r_1 = \frac{V(1) - V(0)}{V(0)} = \frac{220 - 200}{200} = \frac{1}{10} = 10\%,$$

$$V(2) = 200 \cdot (0.1 \cdot 2^2 + 1) = 280 \implies r_2 = \frac{V(2) - V(1)}{V(1)} = \frac{280 - 220}{220} = \frac{3}{11} \cong 27.27\%,$$

$$V(3) = 200 \cdot (0.1 \cdot 3^2 + 1) = 380 \implies r_3 = \frac{V(3) - V(2)}{V(2)} = \frac{380 - 280}{280} = \frac{5}{14} \cong 35.71\%.$$

$\square$

Example 1.3 (The accumulated value). Let $V(3) = 1000$ and $R_n = \frac{3n+2}{2n+2}$, $n = 1, 2, 3, \ldots$. Find $V(6)$.

Solution. By using (1.4) and (1.9), we obtain

$$V(6) = V(3) R_4 R_5 R_6 = 1000 \cdot \frac{7}{5} \cdot \frac{17}{12} \cdot \frac{10}{7} \cong \$2833.33. \qquad \square$$

1.1.2   Simple Interest

For simple interest, the rate of return $r_{[0,t]}$ is proportional to time t measured in years. That is, there exists a positive constant r, called a *simple interest rate* (per year), so that $r_{[0,t]} = rt$ for all $t \geqslant 0$. In particular, the rate of return for year 1 is $r_1 = r_{[0,1]} = r$. The interest $I(t)$ earned on the original principal P during the time period of length t is then calculated by the simple formula

$$I(t) = r_{[0,t]} P = rt P.$$

The accumulated value after t years will be

$$V(t) = P + I(t) = P + rtP = (1 + rt)P, \quad t \geqslant 0. \tag{1.10}$$

Clearly, the accumulated value $V(t)$ is a linear function of time t (see Figure 1.2b).

To find the initial capital (the principal) whose accumulated value at time t is given, we invert the formula (1.10) to obtain

$$P = V(0) = \frac{V(t)}{1 + rt} = (1 + rt)^{-1} V(t). \tag{1.11}$$

This number is called the *present* or *discounted value* of the amount $V(t)$, and $(1 + rt)^{-1}$ is a *discount factor at a simple interest rate* r. Formula (1.11) allows us to find the original principal P invested at time 0.

Example 1.4. How long will it take $3000 to earn $60 interest at 6%?

Solution. We have $P = 3000$, $I = 60$, $r = 0.06$, and then

$$t = \frac{I}{Pr} = \frac{60}{3000 \cdot 0.06} = \frac{1}{3} \text{ years} = 4 \text{ months.}$$ □

Example 1.5. Treasury bills (T-bills) are popular short-term securities issued by the Federal Government of Canada with maturities of 1, 3, 6, or 12 months. T-bills are issued in different denominations, or face values. The face value of a T-bill is the amount the government guarantees it will pay on the maturity date. There is no interest stated on a T-bill. Instead, to determine its purchase price, you need to discount the face value to the date of sale at an interest rate that is determined by market conditions.

Suppose that a six-month T-bill with a face value of $25,000 is purchased by an investor who wishes to yield 3.80%. What price is paid?

Solution. We have $V = 25{,}000$, $t = \frac{6}{12} = \frac{1}{2}$, and $r = 0.038$. Therefore, the investor will pay

$$P = V \cdot (1 + rt)^{-1} = 25{,}000 \cdot \left(1 + 0.038 \cdot \frac{1}{2}\right)^{-1} \cong \$24{,}533.86$$

for the T-bill and receive $25,000 in six months. □

Let us study the return on an account earning interest at rate r. The distinction between the interest rate and the rate of return is that the former refers to a period of one year (i.e., interest per annum) and is independent of the actual duration of investment, whereas the latter reflects both the interest rate and the length of time the investment is held. For all $0 \leqslant t < s$, we have

$$r_{[t,s]} = \frac{V(s) - V(t)}{V(t)} = \frac{(1 + rs)P - (1 + rt)P}{(1 + rt)P} = \frac{(s - t)r}{1 + rt}.$$

In particular, for the first year, we have $r_1 = r_{[0,1]} = r$. The rate of return during the nth year from the date of investment is

$$r_n = r_{[n-1,n]} = \frac{r}{1 + (n - 1)r}.$$

Clearly, these rates form a decreasing sequence that converges to zero as n approaches ∞. In other words, if the investment earns simple interest at the same rate every year, the effective annual rate of return is decreasing. To guarantee a constant annual return, the interest should be reinvested, as is demonstrated in the next section.

1.1.3 Periodic Compound Interest

Assume that the interest earned at a constant rate $i > 0$ is automatically *reinvested*, i.e., it is added to the investment periodically (e.g., annually, semi-annually, quarterly, weekly, daily). In this situation, we are dealing with *periodic compounding*. The interest is said to be *compounded* or *converted*.

Example 1.6. Determine the compound interest earned on $1000 for one year at an annual rate of 8% compounded quarterly and compare it with the simple interest earned on the same amount for one year at 8% per annum.

Solution. Since the compounding period is one quarter, the interest rate per period is equal to $\frac{1}{4} \cdot 8\% = 2\%$. In the case of simple interest, the amount of interest earned during each quarter is fixed and equal to $1000 \cdot 0.02 = \$20$. In the case of compounded interest, the interest earned during one quarter depends on the amount invested at the beginning of the period. The interest on the investment of V dollars over one quarter is $I = 0.02 \cdot V$. Results of our calculations are given in Table 1.1, where we also compare the accumulated values calculated using simple interest and compound interest, respectively.

TABLE 1.1: Calculation of compound interest

	Compound Interest		Simple Interest
At the End of	Interest Earned	Accumulated Value	Accumulated Value
Quarter 1	$1000.00 \cdot 0.02 = \$20.00$	\$1020.00	\$1020.00
Quarter 2	$1020.00 \cdot 0.02 = \$20.40$	\$1040.40	\$1040.00
Quarter 3	$1040.40 \cdot 0.02 = \$20.81$	\$1061.21	\$1060.00
Quarter 4	$1061.21 \cdot 0.02 = \$21.22$	\$1082.43	\$1080.00

In summary, the compound interest earned on \$1000 for one year at 8% compounded quarterly is \$82.43, whereas the simple interest on \$1000 for one year at 8% is equal to $1000 \cdot 0.08 \cdot 1 = \$80$. $\qquad\square$

Let δt denote the time between two consecutive interest conversions. Assume that there are m interest conversion periods per year, and thus $\delta t = \frac{1}{m}$. The interest earned during one period is

$$I = V\, i\, \delta t = V\, \frac{i}{m}\,,$$

where V is the amount invested at the beginning of the period, and i is the annual interest rate. Thus, the interest rate per period of δt is equal to $\frac{i}{m}$. Denote this one-period rate by j. Let us find the accumulated value $V(n\,\delta t)$ at the end of period $n = 1, 2, \ldots$. In the beginning, we have $V(0) = P$. By compounding the interest at the end of each period, we obtain the following accumulated values:

at the end of the 1st period $\quad V(\delta t) = V(0) + jV(0) = (1+j)V(0) = (1+j)P$,
at the end of the 2nd period $\quad V(2\,\delta t) = V(\delta t) + jV(\delta t) = (1+j)V(\delta t) = (1+j)^2 P$,
at the end of the 3rd period $\quad V(3\,\delta t) = (1+j)V(2\,\delta t) = (1+j)^3 P$,

$$\vdots$$

at the end of the nth period $\quad V(n\,\delta t) = (1+j)V((n-1)\,\delta t) = (1+j)^n P$.

Therefore, the accumulated value at time $t = n\,\delta t$ is

$$V(t) = \left(1 + \frac{i}{m}\right)^n P\,. \tag{1.12}$$

The quantity m is called the *frequency of compounding*. Commonly used frequencies of compounding are $m = 1$ for annual compounding, $m = 2$ for semi-annual compounding, $m = 4$ for quarterly compounding, $m = 12$ for monthly compounding, $m = 52$ for weekly compounding, and $m = 365$ for daily compounding. The length of time between two consecutive interest calculations is called the *interest conversion period* or just *interest period*. Let $i^{(m)}$ denote a *nominal* interest rate that is compounded (convertible) m times per year. Note that $i^{(m)}$ is stated as an annual rate of interest. The interest rate compounded annually is simply denoted by i, i.e., $i^{(1)} \equiv i$. The interest rate per period of $\delta t = 1/m$ years is equal to

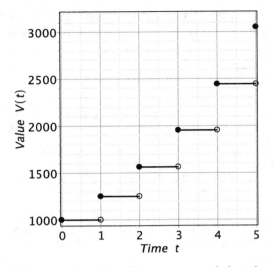

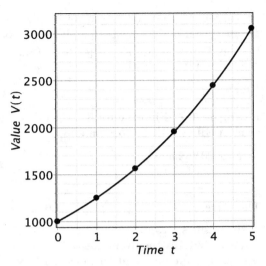

(a) Interest is paid and then compounded at the end of each year.

(b) Interest is paid and then compounded continuously such that the amount function grows exponentially.

FIGURE 1.3: An investment of \$1000 grows at a rate of 25% per year for five years.

$i^{(m)}/m$. In what follows, we denote such a periodic rate by j. The accumulated value of \$1 at the end of m interest conversion periods (i.e., at the end of one year) is

$$\left(1 + \frac{i^{(m)}}{m}\right)^m .$$

Thus, after t years the principal P grows to

$$V(t) = \left(1 + \frac{i^{(m)}}{m}\right)^{mt} P . \tag{1.13}$$

Here, mt gives the cumulative number of periods. The factor $\left(1 + \frac{i^{(m)}}{m}\right)^m$ is called the *accumulation factor* over a one-year period, and it is, in fact, the accumulated value of \$1 at the end of one year with m as the compounding frequency.

Note that the annual nominal rate is meaningless unless we specify the frequency m. In particular, under interest compounded at the end of every year, the accumulation function is $V(t) = (1+i)^t P$ (see Figure 1.3). At the same nominal rate the accumulated value depends on this frequency: it increases with increasing m.

To find the return on a deposit attracting interest compounded periodically, we use the general formula $r_{[s,t]} = \frac{V(t)-V(s)}{V(s)}$ and readily arrive at:

$$r_{[s,t]} = \frac{V(t) - V(s)}{V(s)} = \frac{V(t)}{V(s)} - 1 = \left(1 + \frac{i^{(m)}}{m}\right)^{(t-s)m} - 1 \quad \text{for } 0 \leqslant s < t .$$

For the nth year the rate of return is

$$r_n = r_{[n-1,n]} = \left(1 + \frac{i^{(m)}}{m}\right)^m - 1 .$$

As is seen, the annual rate of return r_n does not depend on n and is a constant quantity for any given m. It is clear that the return on a deposit subject to periodic compounding is not additive, i.e., $R_{[t_1,t_2]} + R_{[t_2,t_3]} \neq R_{[t_1,t_3]}$ and $r_{[t_1,t_2]} + r_{[t_2,t_3]} \neq r_{[t_1,t_3]}$ for $t_1 < t_2 < t_3$ in general. For example, take $m = 1$, then $r_{[0,1]} = r_{[1,2]} = i$ and

$$r_{[0,2]} = (1+i)^2 - 1 = 2i + i^2 \neq 2i = r_{[0,1]} + r_{[1,2]}.$$

Example 1.7. Find the total return and the rate of return over two years under quarterly compounding with 8% per annum.

Solution. We have

$$R_{[0,2]} = \left(1 + \frac{0.08}{4}\right)^{2 \cdot 4} = 1.02^8 = 1.17166.$$

Therefore, $r_{[0,2]} = R_{[0,2]} - 1 = 17.166\%$.   □

In business transactions, it is frequently necessary to determine what principal P now will accumulate at a given interest rate to a specified amount $V(t)$ at a specified future date t. From the fundamental formula (1.13), we obtain

$$P = \frac{V(t)}{(1 + i^{(m)}/m)^{mt}} = \left(1 + \frac{i^{(m)}}{m}\right)^{-mt} V(t). \tag{1.14}$$

P is called the *discounted value or present value* of $V(t)$. The process of determining P from $V(t)$ is called *discounting*. Similarly, we can find the value $V(s)$ of an investment at any intermediate date s given the value $V(t)$ at some fixed future time t so that $0 < s < t$:

$$V(s) = \frac{V(t)}{(1 + i^{(m)}/m)^{(t-s)m}}.$$

1.1.4   Continuous Compound Interest

Consider a fixed nominal rate of interest r compounded at different frequencies m. Let us find the limiting value of the accumulated value of $\$1$ over one year as $m \to \infty$:

$$\lim_{m \to \infty} \left(1 + \frac{r}{m}\right)^m = \left(\lim_{m \to \infty} \left(1 + \frac{r}{m}\right)^{\frac{m}{r}}\right)^r$$

(denote r/m by x, which converges to 0 as $m \to \infty$)

$$= \left(\lim_{x \searrow 0} (1 + x)^{\frac{1}{x}}\right)^r = e^r.$$

[Note: The number e $= 2.71828\ldots$ is the base of the natural logarithm; it is occasionally called Euler's number.] Therefore, the accumulated value of principal P compounded continuously at rate r over a period of t years (note that t may be fractional) is given by

$$V(t) = \lim_{m \to \infty} \left(1 + \frac{r}{m}\right)^{mt} P = \left[\lim_{m \to \infty} \left(1 + \frac{r}{m}\right)^m\right]^t P = e^{rt} P. \tag{1.15}$$

We denote the continuously compounded interest rate by r rather than by $i^{(\infty)}$. The discounted value P for given $V(t)$, r, and t is

$$P = e^{-rt} V(t). \tag{1.16}$$

Being given the annual rate of return $i = \frac{V(1) - V(0)}{V(0)}$, one can find the continuously compounded rate r as follows: $r = \ln(1 + i)$.

Example 1.8. Calculate the accumulated value of $P = \$1000$ at the end of five years if the rate is 4% and interest is compounded

(a) yearly; (b) quarterly; (c) continuously.

Solution. Calculate the accumulated value $V = V(5)$ for each scenario:

(a) $V = 1000 \cdot (1 + 0.04)^5 \cong \$1216.65$;

(b) $V = 1000 \cdot \left(1 + \frac{0.04}{4}\right)^{5 \cdot 4} = 1000 \cdot 1.01^{20} \cong \$1220.19$;

(c) $V = 1000 \cdot e^{0.04 \cdot 5} \cong \$1221.40$.

As we can see, the accumulated value V increases with the frequency m. $\square$

1.1.5 Equivalent Rates

Two nominal compound interest rates are said to be *equivalent* if they yield the same accumulated value at the end of one year, and hence at the end of any number of years. For example, the rate i compounded annually and rate $i^{(m)}$ compounded m per year are equivalent if

$$1 + i = \left(1 + \frac{i^{(m)}}{m}\right)^m.$$

Example 1.9. Determine what rate $i^{(4)}$ is equivalent to (a) $i^{(12)} = 6\%$, (b) $i^{(2)} = 6\%$.

Solution.

(a) Equate the two accumulation factors:

$$(1 + 0.06/12)^{12} = (1 + i^{(4)}/4)^4,$$

and then solve for the rate $i^{(4)}$:

$$i^{(4)} = 4 \cdot ((1 + 0.06/12)^{12/4} - 1) = 4 \cdot (1.005^3 - 1) \cong 6.030\%.$$

(b) We have $(1 + 0.06/2)^2 = (1 + i^{(4)}/4)^4$, and hence

$$i^{(4)} = 4 \cdot ((1 + 0.06/2)^{2/4} - 1) = 4 \cdot (\sqrt{1.03} - 1) \cong 5.956\%. \qquad \square$$

For a given nominal rate $i^{(m)}$ compounded m times per year or for a rate r compounded continuously, we define the corresponding annual *effective* rate, i, to be that rate which, if compounded annually, produces the same interest. In other words, i is the annual rate of interest that is equivalent to $i^{(m)}$ or r. Note that equivalent compound rates have the same annual effective rate. To determine i, we compare the accumulated values of $\$1$ at the end of one year; thus

$$i = \begin{cases} \left(1 + \frac{i^{(m)}}{m}\right)^m - 1 & \text{for a rate compounded } m \text{ times per year,} \\ e^r - 1 & \text{for a rate compounded continuously.} \end{cases} \tag{1.17}$$

Example 1.10. You wish to invest a sum of money for a number of years and have narrowed your choices to the following three investments:

A at the interest rate $i^{(2)} = 10.35\%$;

B at the interest rate $i^{(12)} = 10.15\%$;

C at the interest rate $i^{(4)} = 10.25\%$.

Which investment should you choose?

Solution. To decide which investment is best, you need to calculate the equivalent rate compounded with the same frequency (e.g., annually) for each investment. So, let us calculate the annual effective rate of interest for each investment:

A: $i = \left(1 + \frac{0.1035}{2}\right)^2 - 1 \cong 10.6178\%$;

B: $i = \left(1 + \frac{0.1015}{12}\right)^{12} - 1 \cong 10.6358\%$;

C: $i = \left(1 + \frac{0.1025}{4}\right)^4 - 1 \cong 10.6508\%$.

It turns out that investment **C** has the highest annual effective rate, so you should choose this investment.   □

Example 1.11. Determine what rates $i^{(2)}$ and $i^{(12)}$ are equivalent to $r = 6\%$.

Solution. We equate the annual accumulation factors at each interest rate to obtain

$$(1 + i^{(2)}/2)^2 = e^{0.06} \qquad\qquad (1 + i^{(12)}/12)^{12} = e^{0.06}$$
$$i^{(2)}/2 = e^{0.06/2} - 1 \qquad\qquad i^{(12)}/12 = e^{0.06/12} - 1$$
$$i^{(2)} \cong 6.091\% \qquad\qquad i^{(12)} \cong 6.015\%$$

□

We can generalize the result presented above and state that for rates of interest that are all equivalent we have

$$i^{(1)} > i^{(2)} > i^{(4)} > i^{(12)} > i^{(365)} > i^{(\infty)}, \tag{1.18}$$

where $i^{(1)} \equiv i$ is the effective annual rate, and $i^{(\infty)} \equiv r$ is an equivalent rate compounded continuously.

Proposition 1.1. *For a fixed annual effective rate i, the compound interest rates $i^{(m)}$, where $m = 1, 2, 3, \ldots$, all equivalent to i, form a strictly decreasing sequence that converges to the rate $r = \ln(1 + i)$ as $m \to \infty$.*

Proof. The interest rate $i^{(m)}$ is equivalent to the annual effective rate i iff $(1 + i^{(m)}/m)^m = 1 + i$ holds. This defines the interest rate $i(m) \equiv i^{(m)}$ as a function of $m > 0$:

$$i(m) = \left((1 + i)^{1/m} - 1\right) m = \left(e^{r/m} - 1\right) m.$$

Its derivative w.r.t. m is

$$i'(m) = i(m)/m - (r/m)e^{r/m} = i(m)/m - (1 + i(m)/m) \ln(1 + i(m)/m).$$

Let us prove that the derivative $i'(m)$ is strictly negative for all $m > 0$.

- To show that $(1 + i(m)/m) \ln(1 + i(m)/m) > i(m)/m$ holds for all $m > 0$, it suffices to prove that the inequality $\ln(1 + x) > x/(1 + x)$ holds for all $x > 0$.

- Notice that if two functions $f(x)$ and $g(x)$ with integrable derivatives are such that $f(0) \geqslant g(0)$ and $f'(x) > g'(x)$ holds for all $x > 0$, then $f(x) > g(x)$ is valid for all $x > 0$. Indeed, we have

$$f(x) = f(0) + \int_0^x f'(t) \, dt > g(0) + \int_0^x g'(t) \, dt = g(x).$$

- Let $f(x) := \ln(1 + x)$ and $g(x) := x/(1 + x) = 1 - (1 + x)^{-1}$. Their derivatives are $f'(x) = (1 + x)^{-1}$ and $g'(x) = (1 + x)^{-2}$, respectively.

- Clearly, $(1+x)^{-1} > (1+x)^{-2}$ holds for all $x > 0$. Since $f(0) = g(0) = 0$, we obtain that $\ln(1 + x) > x/(1 + x)$ holds for $x > 0$.

Using l'Hôpital's rule (LHR) gives

$$\lim_{m \to \infty} i^{(m)} = \lim_{m \to \infty} \frac{e^{r/m} - 1}{1/m} = \lim_{t \to 0} \frac{e^{rt} - 1}{t} \stackrel{\text{(LHR)}}{=} \lim_{t \to 0} re^{rt} = r \lim_{t \to 0} e^{rt} = r. \qquad \square$$

Proposition 1.2. *The accumulation function* $a(m) = \left(1 + \frac{r}{m}\right)^m$ *for a fixed nominal interest rate r compounded with frequency m over a one-year period is a strictly increasing function of m that converges to e^r as $m \to \infty$.*

Proof. Differentiate the function $a(m)$ w.r.t. m to obtain

$$A'(m) = (1 + r/m)^m \left(\ln(1 + r/m) - \frac{r/m}{1 + r/m}\right).$$

Using the inequality $\ln(1 + x) > x/(1 + x)$ with $x = r/m > 0$ gives that $A'(m) > 0$. Therefore, $\{a(m)\}_{m \geqslant 1}$ is a strictly increasing sequence that converges to e^r as $m \to \infty$. $\square$

As follows from the above proposition, continuous compounding produces higher accumulated value than periodic compounding with any frequency m, that is, $\left(1 + \frac{r}{m}\right)^m < e^r$ for all $m \geqslant 1$.

Example 1.12. Consider an investment at: (a) simple interest rate 6%, (b) annual compound interest rate 6%, and (c) continuous compound interest rate 6%. Find the time that is necessary to double the original principal.

Solution. For each investment, we find time t so that the respective accumulation function $a(t)$ equals 2, and thus $V(t) = P \cdot a(t) = 2P$.

(a) For simple interest: $1 + 0.06t = 2$, hence $t = 1/0.06 \cong 16.667$ years.

(b) For interest compounded annually: $(1 + 0.06)^t = 2$, hence $t = \frac{\ln 2}{\ln 1.06} \cong 11.896$ years.

(c) For interest compounded continuously: $e^{0.06t} = 2$, hence $t = \frac{\ln 2}{0.06} \cong 11.5525$ years. $\square$

Let us summarize the above three methods of calculating interest. Let r be the nominal annual rate. The *accumulation function* $a(t) = \frac{V(t)}{V(0)} = R_{[0,t]}$ has the following form:

$$a(t) = \begin{cases} 1 + rt & \text{for a rate of simple interest,} \\ \left(1 + \frac{r}{m}\right)^{mt} & \text{for a rate of interest compounded } m \text{ times per year,} \\ e^{rt} & \text{for a rate of interest compounded continuously.} \end{cases} \qquad (1.19)$$

In what follows, we will also use the *discounting function* $d(t) = \frac{V(0)}{V(t)} = \frac{1}{a(t)}$ taking the form:

$$d(t) = \begin{cases} (1 + rt)^{-1} & \text{for a rate of simple interest,} \\ \left(1 + \frac{r}{m}\right)^{-mt} & \text{for a rate of interest compounded } m \text{ times per year,} \\ e^{-rt} & \text{for a rate of interest compounded continuously.} \end{cases} \qquad (1.20)$$

1.1.6   Continuously Varying Interest Rates

Consider the case of continuously compounded interest but with a rate that is changing in time. Let T be the time horizon of investment; let $r(t)$ denote the *instantaneous* interest rate at time $t \in [0, T]$. Consider a time interval $[t, t + \delta t]$ with small $\delta t > 0$. Suppose that the rate $r(t)$ is approximately constant on the interval $[t, t + \delta t]$. Then the amount $V(t)$ invested at time t will grow to $V(t + \delta t) = V(t)e^{r(t)\,\delta t}$ at time $t + \delta t$.

Let the principal P be invested at time 0. To determine $V(t)$ in terms of the principal and the instantaneous interest rate function $r(t)$ for $0 \leqslant t \leqslant T$, we split the time interval $[0, T]$ in N equal subintervals of length $\delta t = T/N$ and apply the above approach. For times $t_k = k\,\delta t$ with $k = 0, 1, \ldots, N$, we obtain

$$V(t_1) = V(t_0)e^{r(t_0)\,\delta t} = Pe^{r(t_0)\,\delta t}$$

$$V(t_2) = V(t_1)e^{r(t_1)\,\delta t} = Pe^{r(t_0)\,\delta t}\,e^{r(t_1)\,\delta t} = Pe^{(r(t_0)+r(t_1))\,\delta t}$$

$$V(t_3) = V(t_2)e^{r(t_2)\,\delta t} = Pe^{(r(t_0)+r(t_1))\,\delta t}\,e^{r(t_2)\,\delta t} = Pe^{(r(t_0)+r(t_1)+r(t_2))\,\delta t}$$

$$\vdots$$

$$V(t_N) = P\prod_{k=0}^{N-1} e^{r(t_k)\,\delta t} = Pe^{(r(t_0)+r(t_1)+\cdots+r(t_{N-1}))\,\delta t} = P\exp\left(\sum_{k=0}^{N-1} r(t_k)\,\delta t\right).$$

Recognizing the Riemann sum in the last equation with the time step $\delta t = T/N$ and taking the limit as $N \to \infty$, we obtain

$$\lim_{N \to \infty} \sum_{k=0}^{N-1} r(t_k)\,\delta t = \int_0^T r(t)\,dt\,.$$

Therefore, in the limiting case as $N \to \infty$, the value at the maturity time T is given by

$$V(T) = P\exp\left(\int_0^T r(t)\,dt\right). \tag{1.21}$$

If $r(t)$ is constant and equal to r for all $t \in [0, T]$, then (1.21) reduces to the usual formula for the accumulated value under continuous compounding:

$$V(T) = P\exp\left(\int_0^T r\,dt\right) = P\exp(rT)\,.$$

Let $y(T)$ denote the average of the instantaneous interest rates from time 0 to time T:

$$y(T) := \frac{1}{T}\int_0^T r(t)\,dt\,.$$

Then, (1.21) takes a familiar form:

$$V(T) = P\,e^{y(T)T}.$$

The interest rate $y(T)$ is called the *yield rate* or *spot rate* for maturity T. In finance, the instantaneous interest rate $r(t)$ is often called the *short rate*.

It is possible to derive (1.21) in a different way. Consider an investment with principal $V(0) = P$ and accumulated value $V(t)$ at time $t > 0$. The nominal rate of interest compounded m times per year and evaluated at time t is given by

$$i^{(m)}(t) = \frac{V(t + \frac{1}{m}) - V(t)}{\frac{1}{m}V(t)}\,.$$

Let $\delta t = \frac{1}{m}$ denote the length of one period. As $m \to \infty$ and $\delta t \to 0$, we have

$$r(t) := \lim_{m \to \infty} i^{(m)}(t) = \lim_{\delta t \to 0} \frac{V(t + \delta t) - V(t)}{\delta t\, V(t)} = \frac{V'(t)}{V(t)} = \frac{\mathrm{d}\ln V(t)}{\mathrm{d}t}. \tag{1.22}$$

Thus, $r(t)$ is a measure of the relative instantaneous rate of growth at time t. In actuarial science, it is also called the *force of interest*. Integrating $r(t) = \frac{\mathrm{d}\ln V(t)}{\mathrm{d}t}$ from 0 to T gives

$$\int_0^T r(t)\,\mathrm{d}t = \int_0^T \frac{\mathrm{d}\ln V(t)}{\mathrm{d}t}\,\mathrm{d}t = \ln V(T) - \ln V(0) = \ln\left(\frac{V(T)}{V(0)}\right),$$

and then (1.21) follows by exponentiation.

Example 1.13. Calculate the accumulated value of \$1000 at the end of four years if $r(t) = 0.05 + 0.1t$.

Solution. The accumulation function is

$$a(T) = \exp\left(\int_0^T r(t)\,\mathrm{d}t\right) = \exp\left(\int_0^T (0.05 + 0.1t)\,\mathrm{d}t\right) = e^{0.05T + 0.05T^2}$$

for every $T \geqslant 0$. Thus, after four years we have

$$a(4) = e^{0.05 \cdot 4 + 0.05 \cdot 4^2} = e \cong 2.71828182$$

and $V(4) = P\,a(4) = 1000\,e \cong \$2{,}718.28$. ◻

Example 1.14. Find the yield rate $y(t)$ for $0 \leqslant t \leqslant 4$ if the instantaneous interest rate is

$$r(t) = \begin{cases} 0.02 & \text{for } 0 \leqslant t < 1, \\ 0.025 & \text{for } 1 \leqslant t < 2, \\ 0.025 + 0.005(t-2) & \text{for } 2 \leqslant t \leqslant 4. \end{cases}$$

Solution. The instantaneous interest rate is a piecewise function. Thus, the yield rate is also a piecewise function. Find its values on the intervals $[0,1)$, $[1,2)$, and $[2,4]$, respectively:

for $t \in [0,1)$, $\quad y(t) = \dfrac{1}{t}\displaystyle\int_0^t 0.02\,\mathrm{d}s = \dfrac{0.02t}{t} = 0.02;$

for $t \in [1,2)$, $\quad y(t) = \dfrac{1}{t}\left(\displaystyle\int_0^1 0.02\,\mathrm{d}s + \int_1^t 0.025\,\mathrm{d}s\right) = \dfrac{0.02 + 0.025(t-1)}{t};$

for $t \in [2,4]$, $\quad y(t) = \dfrac{1}{t}\left(\displaystyle\int_0^1 0.02\,\mathrm{d}s + \int_1^2 0.025\,\mathrm{d}s + \int_2^t (0.025 + 0.005(s-2))\,\mathrm{d}s\right)$

$$= \frac{0.02 + 0.025 + 0.025(t-2) + 0.0025(t-2)^2}{t}$$

$$= \frac{0.02 + 0.025(t-1) + 0.0025(t-2)^2}{t}.$$

It is easy to show that $y(t)$ is a continuous function. ◻

1.2 Time Value of Money and Cash Flows

1.2.1 Equations of Value

All financial decisions take into account the basic idea that money has its time value. The mathematics of finance deals with *dated values*. In order to compare or combine different amounts of money, we place the amounts at the same point in time, called the *focal date*. In general, we compare dated values by the definition of equivalence described below.

Assume the periodic compounding of interest. Amount $\$X$ due on a given date is *equivalent* to $\$Y$ due n interest conversion periods later, if $Y = X(1 + j)^n$ or, equivalently, $X = Y(1+j)^{-n}$, where $j = i^{(m)}/m$ is a given compound interest rate per period. The time diagram in Figure 1.4 illustrates dated values that are equivalent to a given amount X.

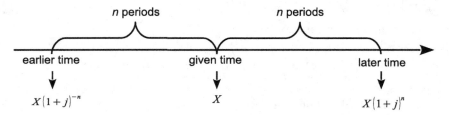

FIGURE 1.4: Equivalence at a given compound interest rate.

Example 1.15. A debt of $5000 is due at the end of three years. Determine an equivalent debt due at the end of (a) three months, (b) three years, and nine months. Assume quarterly compounding with $i^{(4)} = 12\%$.

Solution. We arrange the data in a time diagram below where one period is three months.

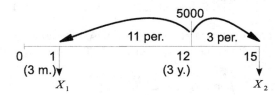

By the definition of equivalence:

(a) $X_1 = \$5000 \cdot (1 + 0.12/4)^{-11} = \$5000 \cdot 1.03^{-11} \cong \$3612.11$;

(b) $X_2 = \$5000 \cdot (1 + 0.12/4)^3 = \$5000 \cdot 1.03^3 \cong \$5463.64$. $\qquad\square$

We say that two sets of payments are equivalent at a given interest rate if the dated values of the sets, on any common date, are equal. An equation stating that the dated values of two sets of payments are equal is called an *equation of value*. For example, two payments X_1 and X_2 due at the end of periods n_1 and n_2, respectively, are equivalent to a single payment due now if

$$Y = X_1(1+j)^{-n_1} + X_2(1+j)^{-n_2}.$$

Example 1.16. A debt of \$5000 is due at the end of five years. It is proposed that \$$X$ be paid now and with another \$$X$ paid in 10 years' time to liquidate the debt. Calculate the value of X if the effective interest rate is 12% for the first six years and 8% for the next four years.

Solution. The value of the first payment at time $t = 5$ is $X \cdot 1.12^5$. The value of the second payment at time $t = 5$ is $X \cdot 1.08^{-4} \cdot 1.12^{-1}$. We equate the sum of these two values and the value of the debt of \$5000 at $t = 5$ to obtain

$$5000 = X \cdot 1.12^5 + X \cdot 1.08^{-4} \cdot 1.12^{-1} = X \cdot (1.76234 + 0.656277) = 2.41862\,X \,.$$

Hence, $X = \$5000/2.41862 \cong \$2067.30$. □

Example 1.17. Suppose you can buy a house for \$84,000 cash (option 1) or for three payments of \$50,000 now, \$20,000 in one year, and \$20,000 in two years (option 2). If money is worth $i^{(12)} = 9\%$, which option is more profitable?

Solution. The present value of option 1 is \$84,000. Using monthly compounding, the present value of option 2 is

$$\$50{,}000 + \$20{,}000 \cdot (1 + 0.09/12)^{-12} + \$20{,}000 \cdot (1 + 0.09/12)^{-24}$$
$$\cong \$50{,}000 + \$18{,}284.76 + \$16{,}716.63 = \$85{,}001.39.$$

Clearly, option 1 is better than option 2. □

Often, we need to find an equivalent rate of return or solve for the interest rate. Consider a couple of problems on this topic.

Example 1.18. If you file your taxes late, the Canada Revenue Agency charges you an immediate 5% late penalty, as well as 5% compounded daily on your outstanding balance. Express your total penalty as a single rate (compounded daily) assuming you pay 26 weeks late.

Solution. If you pay n days late, every dollar 708 originally owed swells to

$$1.05 \cdot (1 + 0.05/365)^n \,.$$

Therefore, on a per dollar basis your total debt after $n = 26 \cdot 7 = 182$ days is

$$1.05 \cdot (1 + 0.05/365)^{182}.$$

Had you been charged the single rate $i^{(365)}$, your total debt would have been $(1 + i^{(365)}/365)^{182}$. Equating and solving for $i^{(365)}$ we get

$$1.05 \cdot (1 + 0.05/365)^{182} = (1 + i^{(365)}/365)^{182} \,,$$

which can be rearranged to find

$$i^{(365)} = 365 \cdot \left(1.05^{1/182} \cdot (1 + 0.05/365) - 1 \right) \cong 0.14787 \,.$$

Therefore your effective penalty rate is approximately 14.79%. □

Remark. If you pay n days late, then your penalty rate is

$$i_n^{(365)} = 365 \cdot \left(1.05^{1/n}(1 + 0.05/365) - 1 \right) \,.$$

Observe that $\lim_{n \to \infty} i_n^{(365)} = 0.05$. Thus, if you pay extremely late, the initial 5% penalty is dwarfed by the compound interest. Of course, this does not mean that it is a good idea to pay extremely late.

Remark. Suppose you are offered a Guaranteed Investment Certificate (GIC) that promises 4% compounded annually with an upfront service fee of 1% of the amount invested. If invested for n years, the effective rate that you earn on the GIC can be determined by solving $0.99 \cdot (1 + r)^n = \left(1 + i^{(n)}\right)^n$, which yields $i^{(n)} = 0.99^{1/n} \cdot (1 + r) - 1$. It is easy to prove (do this as an exercise) that $i^{(n)}$ is less than r and converges to r as $n \to \infty$.

Example 1.19. You have access to an account that pays $100r\%$ per annum, compounded annually. You are given two options to repay a debt. Option A is an immediate payment of $30,000; option B is the payment of $16,000 made at the end of each of the next two years. For what values of r would you choose option B?

Solution. The present value of B, in thousands of dollars, is

$$P_B(r) = \frac{16}{1 + r} + \frac{16}{(1 + r)^2} \,.$$

The present value of option A, in thousands of dollars, is $P_A(r) = 30$. It is only sensible to choose option B provided $P_B(r) < P_A(r)$, which occurs if and only if (iff)

$$30 > \frac{16}{1 + r} + \frac{16}{(1 + r)^2} \quad \Longleftrightarrow \quad 30(1 + r)^2 > 16(1 + r) + 16 \,,$$

which occurs iff $30r^2 + 44r - 2 > 0$. The zeros of the quadratic function in the left-hand side are -1.510794 and 0.044127, and since the leading coefficient of the quadratic function is positive, we should, therefore, choose option B if r exceeds 0.044127. In summary, we choose option A if the interest rate is less than (approximately) 4.4127% and choose option B otherwise.   □

1.2.2   Deterministic Cash Flows and Their Net Present Values

From a broad point of view, an investment can be defined as a sequence of expenditures and receipts spanning a time period. Let all expenses and incomes be denominated in cash. The series of cash payments made over a time period is called a *cash flow stream*. Suppose that the total time interval $[0, T]$ is divided into n subintervals. Let the payments of $C_0, C_1, \ldots, C_n$ dollars be respectively received on the dates $t_0, t_1, \ldots, t_n$ so that $0 = t_0 < t_1 < t_2 < \ldots < t_n = T$. The following table represents such a scenario.

Dates	t_0	t_1	...	t_{n-1}	t_n
Cash Flows	C_0	C_1	...	C_{n-1}	C_n

A cash flow stream can also be represented as a pair $(\mathbf{C}, \mathbf{T})$ of two vectors:

$$\mathbf{C} = [C_0, C_1, \ldots, C_n] \quad \text{and} \quad \mathbf{T} = [t_0, t_1, \ldots, t_n] \,.$$

Every coin has two sides. If one party holds the cash flow stream $\mathbf{C}$, then the opposite party holds the stream $-\mathbf{C} = [-C_0, -C_1, \ldots, -C_n]$. Here, we assume that if $C_k > 0$, then the payment of C_k dollars is received at time t_k (an inflow of cash); if $C_k < 0$, then the payment of $|C_k|$ dollars is made to the counter-party at time t_k (an outflow of cash).

To visualize a cash flow stream $(\mathbf{C}, \mathbf{T})$, one can also use a time diagram which is constructed as follows. First, draw a horizontal line, which represents time increasing from the present (denoted by 0) as we are moving from left to right. After that, draw short vertical lines that start on the horizontal line. Those that go up represent cash coming in (positive cash flows, or receipts), while those that go down represent cash going out (negative cash flows, or disbursements). In general, without knowing the sign of C_k we cannot, in the time diagram, correctly indicate whether C_k should be above or below the time line. The cash flow stream for compounding is given in Figure 1.5. Here, the time is measured in periods; $j = i^{(m)}/m$ is the interest rate for one period. This rate is assumed to be constant. The cash flow stream for discounting is represented in Figure 1.6.

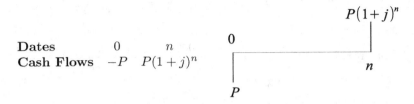

Dates	0	n
Cash Flows	$-P$	$P(1+j)^n$

FIGURE 1.5: The cash flow stream for compounding

Dates	0	n
Cash Flows	$-V(1+j)^{-n}$	V

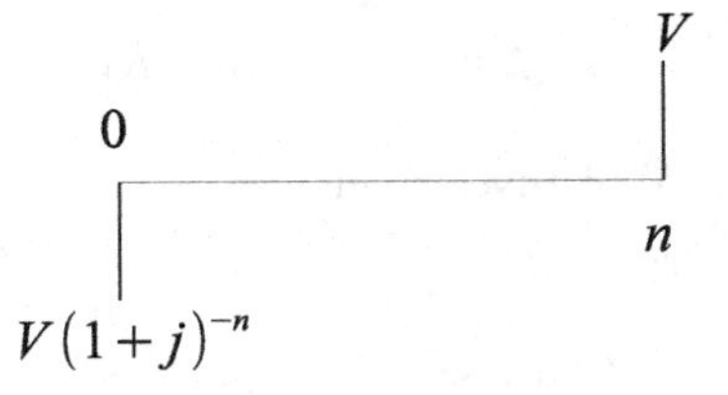

FIGURE 1.6: The cash flow stream for discounting

The *net present value* (NPV), of an investment is the difference between the present value of the cash inflows ($C_k > 0$) and the present value of the cash outflows ($C_k < 0$), that is,

$$\text{NPV}(\mathbf{C}) = C_0 + d(t_1)C_1 + d(t_2)C_2 + \cdots + d(t_n)C_n, \tag{1.23}$$

where $d(t)$ is the discounting function given by (1.20).

For example, let us assume that the lengths of the periods (t_{k-1}, t_k), $k = 1, 2, \ldots, n$, are equal, and interest is periodically compounded at interest rate j. Then

$$\text{NPV}(\mathbf{C}) = C_0 + C_1(1+j)^{-1} + C_2(1+j)^{-2} + \cdots + C_n(1+j)^{-n}.$$

The interest rate used is known as the cost of capital and can be considered as the cost of borrowing money by the business or the rate of return that an investor may obtain if the money is invested with security. The NPV can be used to make a business decision. If NPV ≥ 0, we can conclude that the rate of return from the cash inflows is greater than or equal to the cost of the cash outflows and the project has economic merit and should proceed. If NPV < 0, the project should not proceed. When attempting to choose between two investments with the same levels of risk, the investor generally chooses the one with the higher net present value. If both investments have the same net present value and the same time interval, then the investor is said to be indifferent between the investments.

Example 1.20 (Calculating the NPV). An investor is considering two investments with the following annual cash flows:

Years	0	1	2	3
Cash Flows (Investment 1)	$-\$13{,}000$	$\$5000$	$\$6000$	$\$7000$
Cash Flows (Investment 2)	$-\$13{,}000$	$\$7000$	$\$4800$	$\$6000$

Which is the better investment if the prevailing annual compound interest rate is (a) $i =$ 4.5%; (b) $i = 9$%?

Solution.
(a) At $i = 4.5$%, we have

$$\text{NPV of Investment 1} = -13{,}000 + \frac{5000}{1 + 0.045} + \frac{6000}{(1 + 0.045)^2} + \frac{7000}{(1 + 0.045)^3}$$

$$\cong \$3413.14,$$

$$\text{NPV of Investment 2} = -13{,}000 + \frac{7000}{1 + 0.045} + \frac{4800}{(1 + 0.045)^2} + \frac{6000}{(1 + 0.045)^3}$$

$$\cong \$3351.85.$$

Thus, at $i = 4.5$%, Investment 1 is a better choice.
(b) At $i = 9$%, we have

$$\text{NPV of Investment 1} = -13{,}000 + \frac{5000}{1 + 0.09} + \frac{6000}{(1 + 0.09)^2} + \frac{7000}{(1 + 0.09)^3}$$

$$\cong \$2042.52,$$

$$\text{NPV of Investment 2} = -13{,}000 + \frac{7000}{1 + 0.09} + \frac{4800}{(1 + 0.09)^2} + \frac{6000}{(1 + 0.09)^3}$$

$$\cong \$2095.18.$$

Thus, at $i = 9$%, Investment 2 is a better choice.   □

Example 1.21. Consider the two investments from Example 1.18. For what values of the interest rate i compounded annually, the NPV of investment 1 exceeds the NPV of investment 2?

Solution. Let us write down the NPVs as functions of $x = (1 + i)^{-1}$:

$$\text{NPV}_1(x) = -13{,}000 + 5000x + 6000x^2 + 7000x^3,$$

$$\text{NPV}_2(x) = -13{,}000 + 7000x + 4800x^2 + 6000x^3.$$

Solve the inequality $\text{NPV}_1(x) > \text{NPV}_2(x)$ for x. We are interested in positive solutions. So, we have:

$$-13 + 5x + 6x^2 + 7x^3 > -13 + 7x + 4.8x^2 + 6x^3$$

$$x^3 + 1.2x^2 - 2x > 0$$

$$x^2 + 1.2x - 2 > 0 \quad \text{(since } x > 0\text{)}.$$

The quadratic function $x^2 + 1.2x - 2$ has two zeros:

$$x_1 = \frac{-1.2 - \sqrt{9.44}}{2} \cong -2.13622915 \quad \text{and} \quad x_2 = \frac{-1.2 + \sqrt{9.44}}{2} \cong 0.93622915.$$

Thus, the set of positive solutions is (x_2, ∞). Solving $x(i) > x_2$ for i yields $i < 1/x_2 - 1$. Therefore, the NPV of investment 1 exceeds the NPV of investment 2 iff the rate i is (approximately) less than 6.8115%.   □

The NPV is defined as the present value of the cash flow stream. In general, the dated value $V(t)$ of the stream $(\mathbf{C}, \mathbf{T})$ with n payments is given by

$$V(t) = \sum_{k=1}^{n} a(t - t_k)C_k = \sum_{k=1}^{n} d(t_k - t)C_k$$

for $t \geqslant 0$.

1.3 Annuities

One of the main types of cash flow streams is an *annuity*, which is defined as a sequence of periodic payments, usually equal, made at equal intervals of time. There are many examples of annuities in the financial world: mortgage payments on a home, car loan payments, payments on rent, dividends, payments on instalment purchases, etc. We will be using the following standard terminology.

- The time between successive payments of an annuity is called the *payment interval.*

- The time from the beginning of the first payment interval to the end of the last payment interval is called the *term* of an annuity.

- The payments of an *ordinary annuity* (also known as an *annuity immediate*) are made at the end of the payment intervals. When the payments are made at the beginning of the payment intervals, the annuity is called an *annuity due.*

- When the payment interval and interest conversion period coincide, the annuity is called a *simple annuity*; otherwise, it is a *general annuity.*

We define the *accumulated value* of an annuity as the equivalent dated value of the set of payments due at the end of the term. Similarly, the *discounted value* of an annuity is defined as the equivalent dated value of the set of payments due at the beginning of the term. Throughout this section, we shall use the following notations.

C denotes the periodic payment of the annuity.

n is the number of interest compounding periods during the term of an annuity. In the case of a simple annuity, n equals the total number of payments.

j denotes interest rate per conversion period (assume $j > 0$).

V_A is the future (accumulated) value of an annuity at the end of the term.

P_A is the present (discounted) value of an annuity.

1.3.1 Simple Annuities

1.3.1.1 Ordinary Annuities

The present value P_A of an ordinary simple annuity is defined as the equivalent dated value of the set of payments due at the beginning of the term, i.e., *one period before the first payment*. The accumulated value V_A of an ordinary simple annuity is defined as the equivalent dated value of the set of payments due at the end of the term, i.e., *on the date of the last payment*. Figure 1.7 contains a time diagram for an ordinary simple annuity. The cash flow streams for the repayment of a loan amounting to P_A and for the accumulation of a fund amounting to V_A are given in Tables 1.2 and 1.3, respectively.

Now we develop a formula for the accumulated value V_A of an ordinary simple annuity of n payments of \$$C$ each using the sum of a geometric progression. Let $V(k)$ denote the accumulated value at the end of period k. Calculate the value $V(3)$ at the end of 3 periods as follows.

Period	Investment	Interest	Balance at the end of the period
1	C	0	$C = V(1)$
2	C	$jV(1)$	$C + (1+j)V(1) = C + (1+j)C = V(2)$
3	C	$jV(2)$	$C + (1+j)V(2) = C + (1+j)C + (1+j)^2 C = V(3)$

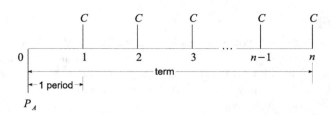

FIGURE 1.7: The time diagram for an ordinary simple annuity.

TABLE 1.2: Cash flow streams for the repayment of a loan amounting to P_A.

Dates	0	1	...	$n-1$	n
Cash Flows	P_A	$-C$	...	$-C$	$-C$

TABLE 1.3: Cash flow streams for the accumulation of a fund amounting to V_A.

Dates	0	1	...	$n-1$	n
Cash Flows	0	$-C$	...	$-C$	$V_A - C$

At the end of n periods, we have

$$V_A = V(n) = C + (1+j)C + (1+j)^2C + (1+j)^3C + \cdots + (1+j)^{n-1}C. \qquad (1.24)$$

Recall the formula for the sum of a geometric progression. For $a > 0$, $q \neq 1$, and $m = 1, 2, \ldots$, we have

$$a + aq + aq^2 + \cdots + aq^m = a \sum_{k=0}^{m} q^k = a \cdot \frac{q^{m+1} - 1}{q - 1}.$$

Applying this formula to a geometric progression in (1.24) with $m = n-1$, whose first term is $a = C$ and common ratio is $q = 1+j$, we obtain the accumulated value V_A of an ordinary simple annuity of n payments of $\$C$ each:

$$V_A = C \frac{(1+j)^n - 1}{j}. \qquad (1.25)$$

To calculate the periodic payment, we solve equation (1.25) for C and obtain

$$C = V_A \frac{j}{(1+j)^n - 1}.$$

It is possible to derive the discounted value P_A in several ways. First, let us consider the amortization of a loan by regular payments of $\$C$. Calculate the value $V(2)$ at the end of 2 periods as follows.

Period	Payment	Interest	Balance at the end of the period
0			$P_A = V(0)$
1	C	$jV(0)$	$(1+j)V(0) - C = (1+j)P_A - C = V(1)$
2	C	$jV(1)$	$(1+j)V(1) - C = (1+j)^2P_A - (1+j)C - C = V(2)$

At the end of n periods, we have

$$V(n) = (1+j)^n P_A \underbrace{-C - (1+j)^1 C - (1+j)^2 C - \ldots - (1+j)^{n-1} C}_{=-V_A} = (1+j)^n P_A - V_A = 0.$$

Therefore,

$$V_A = (1+j)^n P_A. \tag{1.26}$$

Note that this relation also follows directly from the fact that P_A and V_A are both dated values of the same set of payments and thus they are equivalent. Therefore, the basic formula for the discounted value P_A of an ordinary simple annuity of n payments of $\$C$ each is

$$P_A = (1+j)^{-n} V_A = C(1+j)^{-n} \frac{(1+j)^n - 1}{j} = C \frac{1 - (1+j)^{-n}}{j}. \tag{1.27}$$

It is convenient to use the following standard notation:

$$a_{\overline{n}|j} = \frac{1 - (1+j)^{-n}}{j}.$$

This number is called an *annuity symbol* (read "a angle n at j"). It is the present value of an ordinary simple annuity of n payments of \$1 each. That is,

$$a_{\overline{n}|j} = (1+j)^{-1} + (1+j)^{-2} + \cdots + (1+j)^{-n},$$

where $(1+j)^{-1}$ is the present value of the first payment of \$1, $(1+j)^{-2}$ is the present value of the second payment of \$1, and so forth. The expressions for P_A and V_A take the following concise forms:

$$P_A = C a_{\overline{n}|j}, \qquad V_A = C a_{\overline{n}|j} (1+j)^n = C \frac{(1+j)^n - 1}{j}. \tag{1.28}$$

The formulae (1.25) and (1.27) can also be derived by using the fact that the NPV of the respective cash flow stream is zero:

$$\text{NPV}(P_A, \underbrace{-C, \ldots, -C}_{n \text{ payments}}) = P_A - C d(1) - C d(2) - \ldots - C d(n) = 0$$

$$\implies P_A = C \sum_{k=1}^{n} (1+j)^{-k} = C \frac{1 - (1+j)^{-n}}{j}$$

$$\text{NPV}(0, \underbrace{-C, \ldots, -C}_{n \text{ payments}} + V_A) = -C d(1) - C d(2) - \ldots - C d(n) + V_A d(n) = 0$$

$$\implies V_A = \frac{C}{d(n)} \sum_{k=1}^{n} (1+j)^{-k} = C(1+j)^n \frac{1 - (1+j)^{-n}}{j} = C \frac{(1+j)^n - 1}{j}.$$

Here, $d(k) = (1+j)^{-k}$ is the discounting function.

Example 1.22. You make deposits of \$1500 every six months starting today into a fund that pays interest at $i^{(2)} = 7\%$. How much do you have in your fund immediately after the 30th deposit?

Solution. We have $C = 1500$, $n = 30$, and the interest rate is $j = 0.07/2 = 0.035$ per half year. Since we have been asked to determine the accumulated value on the date of the 30th deposit, we deal with an ordinary simple annuity. Therefore,

$$V_A = 1500 \cdot \frac{1.035^{30} - 1}{0.035} \cong \$77{,}434.02. \qquad \square$$

Example 1.23. It is estimated that a machine will need replacing 10 years from now at a cost of $80,000. How much must be put aside each year to provide that amount of money if the company's savings earn interest at an 8% annual effective rate?

Solution. Assume that an amount of $$C$ is deposited at the end of each of 10 years. So we deal with an ordinary simple annuity. The accumulated value is

$$V_A = C \, \frac{1.08^{10} - 1}{0.08} = 80,000 \implies C = 80,000 \cdot \frac{0.08}{1.08^{10} - 1} \cong \$5522.36 \,. \qquad \square$$

Example 1.24. A used car is purchased for $2000 down and $200 a month for six years. Interest is at $i^{(12)} = 10\%$.

(a) Determine the price of the car.

(b) Assuming no payments are missed, what single payment at the end of two years will completely pay off the debt?

Solution.

(a) We have that the price P of the car is a sum of the down payment and the present value P_A of a simple annuity paying $200 a month:

$$P = 2000 + P_A = 2000 + 200 \cdot \frac{1 - 1.008333^{-72}}{0.008333} \cong \$12,795.73 \,.$$

(b) The value of the debt at the end of two years is given by the sum of one regular monthly payment and the present value of $48 = 4 \cdot 12$ payments left:

$$200 + 200 \cdot \frac{1 - 1.008333^{-48}}{0.008333} \cong \$8085.63 \,. \qquad \square$$

1.3.1.2   Annuities Due

An *annuity due* is an annuity whose periodic payments are due at the beginning of each payment interval. The term of an annuity due starts at the time of the first payment and ends one payment period after the date of the last payment. The diagram in Figure 1.8 shows the simple case of an annuity due with n payments when the payment intervals and interest periods coincide.

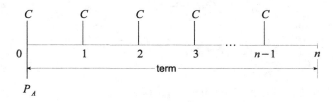

FIGURE 1.8: The time diagram for a simple annuity due.

Annuities due (and deferred annuities presented below) can be handled using the concept of an equation of value. The accumulated value of the payments on the date of the last nth payment (which is time $n - 1$) is $C \, a_{\overline{n}|j}(1 + j)^n$. We then accumulate this amount for one interest period to obtain the accumulated value V_A of an annuity due:

$$V_A = C \, a_{\overline{m}|j} \, (1 + j)^{n+1} \,. \qquad (1.29)$$

The discounted value of the payments one interest period before the first payment is $C a_{\overline{n}|j}$. We then accumulate this amount for one interest period to determine the present value P_A of an annuity due:

$$P_A = C\, a_{\overline{n}|j}\, (1 + j)\,. \tag{1.30}$$

Example 1.25. Determine the discounted value and the accumulated value of $400 payable semi-annually at the beginning of each half-year over 10 years if interest is 8% per year payable semi-annually.

Solution. We deal with an annuity due. There are $n = 20$ payment periods. The interest rate over one period of six months is $0.08/2 = 0.04$. Therefore,

$$V_A = 400 \cdot \left(\frac{1.04^{20} - 1}{0.04}\right) \cdot (1 + 0.04) = 400 \cdot 30.9692 \cong \$12{,}387.68\,,$$

$$P_A = 400 \cdot \left(\frac{1 - 1.04^{-20}}{0.04}\right) \cdot (1 + 0.04) = 400 \cdot 14.1339 \cong \$5653.58\,. \qquad \square$$

1.3.1.3 Deferred Annuities

A *deferred annuity* is an annuity whose first payment is due later than the end of the first interest period. Thus, an ordinary deferred annuity is an ordinary simple annuity whose term is deferred for, say, k periods. The time diagram (Figure 1.9) depicts this case.

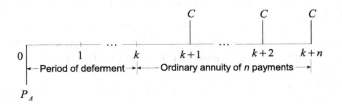

FIGURE 1.9: The time diagram for a deferred annuity.

Deferred annuities can be handled using the concept of an equation of value. The period of deferment is k periods, and the first payment of the ordinary annuity is at time $k + 1$. Hence, the discounted value one period before the first payment (time k) is $C\, a_{\overline{n}|j}$; the discount factor for k periods is $(1 + j)^{-k}$. The present value P_A of a deferred annuity is

$$P_A = C\, a_{\overline{n}|j}\, (1 + j)^{-k}\,. \tag{1.31}$$

Example 1.26. A used car sells for $9550. Brent wishes to pay for it in 18 monthly instalments, the first due in three months from the day of purchase. If 12% compounded monthly is charged, determine the size of the monthly payment.

Solution. The discounted value of the deferred annuity is

$$9550 = C \cdot \frac{1 - 1.01^{-18}}{0.01} \cdot 1.01^{-2} = 16.0751579\,C\,.$$

Hence, the monthly payment is $C = \frac{9550}{16.0751579} \cong \$594.08\,.$ $\qquad \square$

1.3.2   Determining the Term of an Annuity

In some problems, the accumulated value V_A or the discounted value P_A, the periodic payment C, and the rate j are specified. This leaves the number of payments n to be determined. Note that n can be found by using logarithms:

$$V_A = C\,\frac{(1+j)^n - 1}{j} \implies (1+j)^n = \frac{jV_A}{C} + 1 \implies n = \frac{\ln(jV_A/C + 1)}{\ln(1+j)}$$

$$P_A = C\,\frac{1 - (1+j)^{-n}}{j} \implies (1+j)^{-n} = 1 - \frac{jP_A}{C} \implies n = -\frac{\ln(1 - jP_A/C)}{\ln(1+j)}.$$

Usually, being given V_A or P_A, C, and j, we cannot find an integer number of periods n for the annuity. It is necessary to make the concluding payment differ from C in order to have equivalence. There are two ways, as follows.

> **Procedure 1.** The last payment is increased by a sum that will make the payments equivalent to the accumulated value V_A or the discounted value P_A. This increase is sometimes referred to as a *balloon* payment.

> **Procedure 2.** A smaller concluding payment is made one period after the last full payment. The smaller concluding payment is sometimes referred to as a *drop* payment.

Example 1.27. A debt of \$4000 bears interest at $i^{(2)} = 8\%$. It is to be repaid by semi-annual payments of \$400. Determine the number of full payments needed and the final payment. Use both procedures.

Solution. We have $P_A = 4000$, $C = 400$, $j = i^{(2)}/2 = 0.08/2 = 0.04$, and we want to calculate n. Substituting all data in equation (1.27), we obtain:

$$\frac{1 - (1+0.04)^{-n}}{0.04} = 10 \implies -n\ln(1.04) = \ln(0.6) \implies n = 13.024.$$

> **Procedure 1.** There will be 12 regular payments and a final payment. Let X be the balloon payment that will be added to the last regular payment to make the payments equivalent to the discounted value $P_A = 4000$. Using time 0 as the focal date, we obtain the following equation of value:
>
> $$400 \cdot \frac{1 - (1+0.04)^{-13}}{0.04} + X\,(1+0.04)^{-13} = 4000$$
>
> $$\implies X = 1.04^{13} \cdot \left(4000 - 400\,\frac{1 - 1.04^{-13}}{0.04} \right)$$
>
> $$\implies X = 5.7409 \cdot 1.04^{13} \cong \$9.56.$$

So the last, 13th payment, will be $400 + 9.56 = \$409.56$.

> **Procedure 2.** There will be 13 full payments and a final smaller payment. Let Y be the size of a smaller concluding payment (the drop payment). Using time 0 as the focal date, we obtain the following equation of value:
>
> $$400 \cdot \frac{1 - 1.04^{-13}}{0.04} + Y\,1.04^{-14} = 4000 \implies Y = 5.7409 \cdot 1.04^{14} \cong \$9.94. \qquad \square$$

1.3.3 General Annuities

Let us consider annuities for which payments are made either more or less frequently than interest is compounded. Such a series of payments is called a *general annuity*. One way to solve general annuity problems is to replace the given interest rate by an equivalent rate for which the interest compounding period is the same as the payment period. Another approach used in solving a general annuity problem is to replace the given payment by equivalent payments made on the stated interest conversion dates. As a result, for both these approaches, a general annuity problem is transformed into a simple annuity problem.

Example 1.28 (Interest is compounded more frequently than payments are made). Joe deposits $200 at the beginning of each year into a bank account that earns interest at $i^{(4)} = 6\%$. How much money will be in his bank account at the end of five years?

Solution. First, determine the rate j compounded annually that is equivalent to $i^{(4)} = 6\%$:

$$1 + j = (1 + 0.06/4)^4 \implies j = 1.015^4 - 1 \implies j = 6.136355\%$$

Second, calculate the accumulated value V_A of an ordinary annuity due with $C = \$200$, $n = 5$, and $j = 6.136355\%$:

$$V_A = 200 \left(\frac{(1+j)^6 - 1}{j} - 1 \right) = 200 \cdot 5.99931 \cong \$1199.86. \qquad \square$$

Example 1.29 (Payments are made more frequently than interest is compounded).

A car is purchased by paying $2000 down and then $300 each quarter for three years. If the interest on the loan was $i^{(2)} = 9.2\%$, what did the car sell for?

Solution. First, determine the rate j per quarter equivalent to $0.092/2 = 4.6\%$ per half-year:

$$(1 + j)^4 = (1 + 0.092/2)^2 \implies 1 + j = \sqrt{1.046} \implies j = 2.27414\%.$$

Second, calculate the discounted value P_A of an ordinary simple annuity with $C = \$300$, $n = 3 \cdot 4 = 12$, and $j = 2.27414\%$:

$$P_A = 300 \frac{1 - (1+j)^{-12}}{j} = 300 \cdot 10.39946 \cong \$3119.84.$$

The sale price of the car is $P = \$2000 + P_A = \$5119.84.$ $\qquad \square$

Example 1.30. Suppose the following.

- You have access to an account paying $100r\%$ compounded annually for the indefinite future.

- Your salary will grow by $100g\%$ per year for as long as you work, and your salary is paid in a lump sum at the end of each year and you just got paid.

- You plan to retire N years from now.

- You expect to live for M years after retirement.

- Upon retirement, you would like an annual income equal to $100p\%$ of your final salary, adjusted for inflation.

- Inflation is expected to remain stable at $100i\%$ for the indefinite future.

- You will save a fixed percentage $100q\%$ of your salary each year until retirement.

Here, r, g, p, i, and q are positive real parameters; N and M are positive integer parameters. What is the minimum value of q that will allow you to achieve your retirement goals?

Solution. Suppose you currently earn S (just paid), so that next year you will earn $S(1+g)$, the year after $S(1+g)^2$, etc. You are depositing the fraction q of this each year for the next N years and will earn a return of r each year. Thus, the moment you retire (i.e., the moment you make your final deposit), you will have

$$qS(1+g)(1+r)^{N-1} + qS(1+g)^2(1+r)^{N-2} + \cdots + qS(1+g)^{N-1}(1+r) + qS(1+g)^N.$$

This is the same as

$$qS(1+g)^N \left[\frac{(1+r)^{N-1}}{(1+g)^{N-1}} + \frac{(1+r)^{N-2}}{(1+g)^{N-2}} + \cdots + \frac{(1+r)}{(1+g)} + 1\right].$$

If we let $1+h = \frac{1+r}{1+g}$ (so that $h = \frac{1+r}{1+g} - 1$), we can recognize the term in square brackets as the accumulated amount

$$s_{\overline{N}|h} := \frac{(1+h)^N - 1}{h} = (1+h)^N\, a_{\overline{N}|h}.$$

Thus, upon retirement, you will have

$$qS(1+g)^N s_{\overline{N}|h}.$$

Exactly one year after your final deposit, you will withdraw $pX(1+i)$, where $X = S(1+g)^N$ is your final salary (remember that you are withdrawing fraction p of your final salary and that you are adjusting that final salary for inflation). The year after you will withdraw $pX(1+i)^2$, etc. As you plan to make M withdrawals, you will need at least

$$\frac{pX(1+i)}{1+r} + \frac{pX(1+i)^2}{(1+r)^2} + \cdots + \frac{pX(1+i)^M}{(1+r)^M} = pS(1+g)^N a_{\overline{j}|M}$$

in the account at date N, where $\frac{1}{1+j} = \frac{1+i}{1+r}$ (so that $j = \frac{1+r}{1+i} - 1$). Thus we must ensure that $qS(1+g)^N s_{\overline{h}|N} \geqslant pS(1+g)^N a_{\overline{j}|M}$ holds. Equivalently, we have

$$q \geqslant p \cdot \frac{a_{\overline{j}|M}}{s_{\overline{h}|N}}.$$

For example, if $r = 5\%$, $g = 3\%$, $i = 2\%$, $p = 50\%$, $N = 30$ years, and $M = 20$ years, then q should be no less than 18.61%. If you want to receive 75% of your final salary (i.e., $p = 75\%$), then you need to save about 27.91% of your income. $\square$

1.3.4   Perpetuities

A *perpetuity* is an annuity whose payments begin on a fixed date and continue forever. Examples of perpetuities are the series of interest payments from a sum of money invested permanently at a certain interest rate, or a scholarship paid from an endowment on a perpetual basis. We shall discuss an *ordinary simple perpetuity*, that is, when a lump sum is invested and a series of level periodic payments is made, with the first payment made at the end of the first interest period and payments continuing forever. It is meaningless to speak about the accumulated value of a perpetuity. The discounted value, however, is well

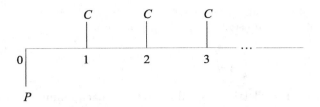

FIGURE 1.10: The time diagram for a perpetuity.

defined as the equivalent dated value of the set of payments at the beginning of the term of the perpetuity.

Let j denote the interest rate per period. The discounted value P of an ordinary simple perpetuity must be equivalent to the set of payments C, as shown in Figure 1.10.

From the equation of value we get

$$P = C(1+j)^{-1} + C(1+j)^{-2} + C(1+j)^{-3} + \ldots . \tag{1.32}$$

We know that the sum of an infinite geometric progression can be expressed as

$$a + aq + aq^2 + aq^3 + \ldots = \frac{a}{1-q} \quad \text{if } -1 < q < 1.$$

The expression in (1.32) is a geometric progression with $a = C(1+j)^{-1}$ and $q = (1+j)^{-1}$. Clearly, $0 < q < 1$ for $j > 0$, hence

$$P = \frac{C(1+j)^{-1}}{1-(1+j)^{-1}} = \frac{C}{(1+j)-1} = \frac{C}{j}.$$

Alternatively, it is evident that P will perpetually provide $C = Pj$ as an interest payment on the invested capital P at the end of each interest period as long as it remains invested at rate j per period.

Example 1.31. How much money is needed to establish a scholarship fund paying \$1500 annually if the fund will earn interest at $i = 6\%$ and the first payment will be made (a) at the end of the first year, (b) immediately, (c) five years from now?

Solution.

(a) We have an ordinary simple perpetuity with $C = 1500$ and $j = 0.06$; therefore, we calculate $P = \frac{\$1500}{0.06} = \$25{,}000$.

(b) We have a simple perpetuity due and $P = \frac{C}{j} + C$. So, we calculate $P = \$25{,}000 + \$1500 = \$26{,}500$.

(c) We have a simple perpetuity deferred $k = 4$ periods for which $P = \frac{C}{j}(1+j)^{-k}$ (we set the focal date four years from now). Therefore, $P = \$25{,}000 \cdot 1.06^{-4} \cong \$19{,}802.34$. □

1.3.5 Continuous Annuities

Consider a general annuity in which the payments are made more frequently than interest is compounded. Let m payments be made throughout each of n interest periods. Interest is periodically compounded at interest rate j. Suppose that each payment equals $\$\frac{1}{m}$ for a total of \$1 per period. The present value of such an annuity is

$$\sum_{k=1}^{nm} \frac{1}{m}(1+j)^{-\frac{k}{m}} = \frac{1}{m}\, a_{\overline{nm}|j_m}\,,$$

where $j_m = (1+j)^{\frac{1}{m}}-1$. By letting the value of m approach infinity, we obtain a continuous annuity in which payments are made continuously for a total of \$1 per period. The present value of such an annuity, denoted $\bar{a}_{\overline{n}|j}$, is

$$\bar{a}_{\overline{n}|j} = \lim_{m\to\infty}\frac{1}{m}a_{\overline{nm}|j_m} = \lim_{m\to\infty}\sum_{k=1}^{nm}\frac{1}{m}(1+j)^{-\frac{k}{m}} = \lim_{h\to 0}\sum_{k=1}^{nm}h(1+j)^{-hk}\quad(\text{where } h=\frac{1}{m})$$

(note that we deal with a limit of Riemann sums)

$$= \int_0^n (1+j)^{-t}\,\mathrm{d}t = \frac{-(1+j)^{-t}}{\ln(1+j)}\Big|_0^n = \frac{1-(1+j)^{-n}}{\ln(1+j)} = \frac{1-(1+j)^{-n}}{j_\infty}\,, \tag{1.33}$$

where $j_\infty = \ln(1+j)$ is an equivalent interest rate compounded continuously. Here, one unit of time is one period. Hence, j_∞ is a continuous compound rate per period. Equation (1.33) takes the following concise form:

$$\bar{a}_{\overline{n}|j} = \frac{j}{j_\infty}a_{\overline{n}|j}\,.$$

If the payment is \$$C$ per period payable continuously, the present value P_A and accumulated value V_A are, respectively,

$$P_A = C\bar{a}_{\overline{n}|j}\quad\text{and}\quad V_A = C\bar{a}_{\overline{n}|j}(1+j)^n\,.$$

Similarly, we can derive the present value of an annuity in which payments are made continuously, and interest is compounded continuously at rate j_∞:

$$\bar{a}_{\overline{n}|j_\infty} = \lim_{m\to\infty}\sum_{k=1}^{nm}\frac{1}{m}\mathrm{e}^{-j_\infty\frac{k}{m}}$$

$$= \int_0^n \mathrm{e}^{-j_\infty t}\,\mathrm{d}t = -\frac{\mathrm{e}^{-j_\infty t}}{j_\infty}\Big|_0^n = \frac{1-\mathrm{e}^{-j_\infty n}}{j_\infty}\,. \tag{1.34}$$

Alternatively, using the equivalency $1+j = \mathrm{e}^{-j_\infty}$, we can derive (1.34) directly from (1.33).

Lastly, let us consider a general case with payments made continuously at a varying rate. Let $c(t)$ be the intensity of payments at time t. The payment made throughout an infinitesimally small time interval $[t, t+\mathrm{d}t]$ is $c(t)\,\mathrm{d}t$. The present value of this payment is $(1+j)^{-t}c(t)\,\mathrm{d}t$ in the case of periodic compounding. To find the present value P_A of such an annuity, we only need to sum up the present values of all payments. Since there are infinitely many of such infinitesimally small payments, the summation becomes an integral from 0 to n:

$$P_A = \int_0^n c(t)(1+j)^{-t}\,\mathrm{d}t\,.$$

If interest is compounded continuously at a constant rate j_∞, then

$$P_A = \int_0^n c(t)\mathrm{e}^{-j_\infty t}\,\mathrm{d}t\,.$$

Finally, if interest is compounded continuously at a varying rate $j_\infty(t)$, we have

$$P_A = \int_0^n c(t)\mathrm{e}^{-\int_0^t j_\infty(u)\,\mathrm{d}u}\,\mathrm{d}t\,.$$

1.4 Bonds

1.4.1 Introduction and Terminology

Bonds are a method used to borrow money from a large number of investors to raise funds for financing long-term debts. From the investor's point of view, bonds provide steady periodic interest payments along with returning the amount borrowed at some point in the future. A bond is a written contract between the issuer (borrower) and the investor (lender) that specifies the following.

- The *face value*, or the *denomination*, denoted F, of the bond (usually is a multiple of 100).

- The *redemption date*, or *maturity date*, denoted T, is the date on which the loan will be repaid. In most cases, the *redemption value*, denoted W, of a bond is the same as the face value (i.e., $W = F$), and in such cases we say the bond is *redeemed at par*.

- The *bond rate*, or *coupon rate*, denoted c, is the rate at which the bond pays interest on its face value at equal time intervals until the maturity date. For example, an 8% bond with semi-annual coupons has $c = 0.04$ or, equivalently, $c^{(2)} = 0.08$.

The following notations will be used in calculating bond prices:

P denotes the purchase price of a bond (its present value).

C is the amount of the coupon; it is a fraction of the face value of the bond, i.e., $C = Fc$, where c is the periodic coupon rate. If the coupons are paid m times per year, then we say that the bond pays interest at the nominal rate $c^{(m)}$; the periodic coupon rate is $c = c^{(m)}/m$.

n is the total number of coupon/interest payment periods; we assume here that the bond rate and interest rate have the same conversion periods.

j denotes the yield rate per interest period, often called the yield to maturity, i.e., the interest rate earned by the investor, assuming that the bond is held until it matures. If the interest is compounded with the frequency m at the annual nominal rate $i^{(m)}$, then $j = i^{(m)}/m$.

When a bond is sold, its present value is reported as a percent of its face value. For example, the price of a \$1000 bond may be reported as 98, which means that the market value of the bond is $\$1000 \cdot (98/100) = \$980$. A price of 100 means that the value of the bond is equal to its face value.

There are two main examples of bonds: (1) *saving bonds* (such as Canada Savings Bonds) can be cashed in at any time before the redemption date, and the bondholder will receive the full face value plus accrued interest; (2) *marketable bonds* (such as corporate or government bonds) do not allow the bondholder to cash the bond in before maturity. Marketable bonds can be sold on the bond market, where current interest rates will influence the price you receive.

1.4.2 Zero-Coupon Bonds

The simplest example of a bond is a *zero-coupon bond*, which pays no coupons but instead just returns the investor an amount equal to the face value of the bond on the date of maturity. In this case, the coupon rate c is zero. Why would anyone buy such a bond? The point is that the investor pays an amount that is smaller than the face value of the bond. Given the interest rate, the present value of such a bond can be easily computed. Suppose that a bond with face value F dollars is maturing in T years, and the annual effective rate is i; then the present value (i.e., the purchase price) of the bond is

$$P = (1+i)^{-T}F.$$

In reality, the opposite happens: bonds are freely traded, and their prices are determined by the markets, whereas the interest rates are implied by the bond prices. If P is the market price of a zero-coupon bond, then the implied annual effective rate is then

$$i = \left(\frac{F}{P}\right)^{1/T} - 1.$$

For simplicity, let us consider a bond whose face value is equal to one unit of the home (domestic) currency, i.e., $F = 1$. Denote the purchase price of a unit bond at time $t \leqslant T$ by $Z(t,T)$, where times t and T are measured in years. In particular, $Z(0,T) = P$ is the present value of the bond at time $t = 0$, and $Z(T,T) = 1$ is equal to the face value of $1. Let us summarize pricing formulae for a unit zero-coupon bond.

- If interest is compounded annually at rate i, then

$$Z(t,T) = (1+i)^{-(T-t)}.$$

- Using periodic compounding with frequency m, we have

$$Z(t,T) = \left(1 + \frac{i^{(m)}}{m}\right)^{-m(T-t)}.$$

- In the case of continuous compounding, we obtain

$$Z(t,T) = e^{-(T-t)r}.$$

- If the instantaneous interest rate is a function of time, i.e., the interest rate at time $s \in [t,T]$ is $r(s)$, then
$$Z(t,T) = e^{-\int_t^T r(s)\,ds}.$$

Note that the instantaneous interest rate can be readily obtained by differentiating the log-price of a bond:
$$r(t) = \frac{\partial \ln Z(t,T)}{\partial t} = \frac{1}{Z(t,T)}\frac{\partial Z(t,T)}{\partial t}.$$

The rate $r(t)$ is sometimes referred to as the *short rate*.

For comparing different investment opportunities, we can use the *yield rate* or *spot rate*, denoted $y(t,T)$, which is defined implicitly as the yield to maturity of a zero-coupon bond:

$$Z(t,T) = e^{-y(t,T)\,(T-t)}.$$

The spot rate can be viewed as an average interest rate for the time interval $[t,T]$. For time-varying interest rates, we have $y(t,T) = \frac{1}{T-t}\int_t^T r(s)\,ds$. Typically, the yield rate is considered as a function of the time to maturity $T - t$, so we can also use the notation $y(\tau) := y(t,T)$ where $\tau = T - t$.

1.4.3   Coupon Bonds

The buyer of a coupon bond will receive two types of payments, namely, a coupon payment Fc at the end of each interest period (with the length of $\delta t = \frac{1}{m}$ years) and the redemption value W on the redemption date. Thus, a coupon bond can be viewed as a cash flow stream of the form

$$[-P, \underbrace{Fc, Fc, \dots, Fc}_{n \text{ coupon payments}} +W].$$

It is depicted in Figure 1.11.

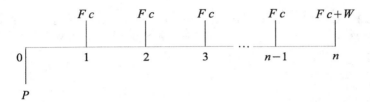

FIGURE 1.11: The time diagram for a coupon bond.

First, let us assume that the interest is compounded periodically at rate j, and the coupons are paid at the interest conversion dates. To determine the price paid to buy the bond, we discount each cash flow to the date of sale at rate j. The purchase price P is the sum of the discounted value of all coupons and the discounted value of the redemption value:

$$
\begin{aligned}
P &= \frac{Fc}{1+j} + \frac{Fc}{(1+j)^2} + \cdots + \frac{Fc}{(1+j)^{n-1}} + \frac{Fc+W}{(1+j)^n} \\
&= Fc \left(\frac{1}{1+j} + \frac{1}{(1+j)^2} + \cdots + \frac{1}{(1+j)^n} \right) + \frac{W}{(1+j)^n} \qquad (1.35) \\
&= Fca_{\overline{n}|j} + W(1+j)^{-n}. \qquad (1.36)
\end{aligned}
$$

The formula (1.36) can be simplified by eliminating the factor $(1+j)^{-n}$ as follows:

$$P = Fca_{\overline{n}|j} + W(1 - ja_{\overline{n}|j}) = W + (Fc - Wj)a_{\overline{n}|j}. \qquad (1.37)$$

Hence, if $P > W$ (the purchase price of a bond exceeds its redemption value), the bond is said to have been purchased *at a premium*. The size of the premium is given by

$$\text{Premium} = P - W = (Fc - Wj)a_{\overline{n}|j}.$$

That is, a premium occurs when $Fc > Wj$. For par value bonds (i.e., $W = F$) the bond is purchased at a premium when $c > j$. Similarly, if $P < W$ (the purchase price is less than the redemption value) the bond is said to have been purchased *at a discount*. The size of the discount is

$$\text{Discount} = W - P = (Wj - Fc)a_{\overline{n}|j}.$$

That is, a discount occurs when $Fc < Wj$. In other words, each coupon is less than the interest desired by the investor. For par value bonds (i.e., $W = F$) the bond is purchased at a discount when $j > c$. Formula (1.37) is computationally more efficient than (1.36) since it requires only the calculation of $a_{\overline{n}|j}$. It also tells us whether the bond is purchased at a premium or at a discount.

Suppose a bond is redeemable at par (that is, $W = F$). Then its price is equal to the face value (i.e., $P = F$) iff the coupon rate and yield rate coincide (i.e., $c = j$). In this case, the bond is said to be *purchased at par*. The proof is straightforward. Assume $W = F$ in (1.36), then

$$P = F \iff ca_{\overline{n}|j} + (1+j)^{-n} = 1 \iff c = \frac{1 - (1+j)^{-n}}{a_{\overline{n}|j}} = j \,.$$

Alternatively, setting $W = F$ in (1.37) gives

$$P = F \iff (Fc - Fj)a_{\overline{n}|j} = 0 \iff c = j \,.$$

Example 1.32. A \$1000 bond that pays interest at $c^{(2)} = 8\%$ is redeemable at par at the end of five years. Determine the purchase price to yield an investor 10% compounded semi-annually.

Solution. We know that:

the redemption value is $W = F = \$1000$;

the coupon rate is $c = 0.08/2 = 0.04$, and the coupon is hence $C = Fc = \$40$;

the number of coupon payments is $n = 5 \cdot 2 = 10$;

the yield rate is $j = 0.1/2 = 0.05$.

The purchase price P to yield $i^{(2)} = 10\%$ is the sum of the discounted values of coupons and of the redemption value:

$$P = 40 \cdot \frac{1 - (1 + 0.05)^{-10}}{0.05} + F(1 + 0.05)^{-10} \cong 308.87 + 613.91 = \$922.78 \,.$$

Since the buyer is buying the bond for less than the redemption value, that is, $P < F$, we say that the bond is purchased at a discount. □

Example 1.33. A corporation issues 15-year bonds redeemable at par. Under the contract, interest payments will be made at the rate $c^{(2)} = 10\%$. The bonds are priced to yield 8% per annum compounded monthly. What is the issue price of a \$1000 bond?

Solution. We have $F = W = \$1000$, $c = 0.1/2 = 0.05$, $n = 15 \cdot 2 = 30$. Calculate a half-year rate j equivalent to $i^{(12)} = 8\%$:

$$(1+j)^2 = \left(1 + \frac{0.08}{12}\right)^{12} \implies j = \left(1 + \frac{0.08}{12}\right)^6 - 1 \implies j = 0.040672622 \,.$$

Now the purchase price is

$$P = W + (Fc - Wj)a_{\overline{30}|j} = 1000 + (50 - 40.672622)a_{\overline{30}|j}$$
$$= 1000 + 9.327378 \cdot 17.15168 \cong \$1159.98 \,. \qquad \square$$

1.4.4 Serial Bonds, Strip Bonds, and Callable Bonds

A set of bonds issued at the same time but having different maturity dates is called *serial bonds*. Serial bonds can be thought of as several bonds covered under one bond contract. The present value of the entire issue of the bonds is just the sum of the present values of the individual bonds.

Example 1.34. To finance an expansion of services, the City of Waterloo issued $30,000,000 of serial bonds on March 15, 2013. Bond interest at 4% per annum is payable half yearly on March 15 and September 15, and the contract provides for redemption as follows: $10,000,000 of the issue to be redeemed March 15, 2018; $10,000,000—March 15, 2023; $10,000,000—March 15, 2028. Calculate the purchase price of the issue to the public to yield $i^{(2)} = 4\%$ on those bonds redeemable in five years and $i^{(2)} = 5\%$ on the remaining bonds.

Solution. This issue can be viewed as a sequence of three coupon bonds redeemable at par. The face value of each bond is $F = \$10,000,000$. The coupon rate is $c = 0.02$; each coupon is $C = \$200,000$. Let us find the present values.

1. The bond is redeemed March 15, 2018. There are $n = 5 \cdot 2 = 10$ coupon payments. The yield rate $j = 0.04/2 = 0.02$ is equal to the coupon rate c. Therefore, the bond is purchased at par. The purchase price is $P_1 = \$10,000,000$.

2. The bond is redeemed March 15, 2023. There are $n = 10 \cdot 2 = 20$ coupons. The yield rate is $j = 0.05/2 = 0.025 > c$; the bond is purchased at a discount. The purchase price is
$$P_2 = 10,000,000 + (200,000 - 10,000,000 \cdot 0.025)a_{\overline{20}|0.025} \cong \$9,220,541.89 \,.$$

3. The bond is redeemed March 15, 2028. There are $n = 15 \cdot 2 = 30$ coupons. The yield rate is $j = 0.025$. The purchase price is
$$P_3 = 10,000,000 + (200,000 - 10,000,000 \cdot 0.025)a_{\overline{30}|0.025} \cong \$8,953,485.37 \,.$$

The purchase price of the whole issue is $P = P_1 + P_2 + P_3 = \$28,174,027.26$. $\qquad\square$

Some investors may separate coupons from the principal of coupon bonds, so that different investors may receive the principal (i.e., the redemption value) and each of the coupon payments. The coupons and remainder are sold separately. This creates a supply of new zero-coupon bonds. This method of creating zero-coupon bonds is known as stripping and the contracts are known as *strip bonds*. **STRIPS** stands for **S**eparate **T**rading of **R**egistered **I**nterest and **P**rincipal **S**ecurities. Recall that the discounted value of the redemption value is $W(1+j)^{-n}$. Each coupon has the discounted value $Fc(1+j)^{-k}$, where k is the number of interest conversion periods from now to the time the coupon is paid. Thus, the present value of coupons is $Fca_{\overline{n}|j}$.

Example 1.35. Investor **A** buys a $2,500 10-year bond paying interest at $c^{(2)} = 6\%$, redeemable at 105, to yield $i^{(2)} = 7\%$. She sells the coupons to investor **B**, who wishes to yield $i^{(2)} = 6.25\%$, and she sells the strip bond to investor **C**, who wishes to yield $i^{(4)} = 6.5\%$. What profit does investor **A** make?

Solution. The face value is $F = \$2500$. The redemption value is $W = \$2500 \cdot 1.05 = \$2625$. Investor **A** (with the yield rate $j = 7\%/2 = 3.5\%$) pays
$$P_{\mathbf{A}} = 2625 + (2500 \cdot 0.03 - 2635 \cdot 0.035) \cdot a_{\overline{20}|0.035} \cong \$2385.17 \,.$$

Investor **B** (with the yield rate $j = 6.25\%/2 = 3.125\%$) pays for the coupons
$$P_{\mathbf{B}} = 2500 \cdot 0.03 \cdot a_{\overline{20}|0.03125} \cong \$1103.02 \,.$$

Investor **C** pays for the strip bond
$$P_{\mathbf{C}} = 2625 \cdot \left(1 + \frac{0.065}{4}\right)^{-40} \cong \$1377.55 \,.$$

The profit of investor **A** is

$$P_{\mathbf{B}} + P_{\mathbf{C}} - P_{\mathbf{A}} = \$1103.02 + \$1377.55 - \$2385.17 = \$95.40 \,. \qquad \square$$

A *callable bond* is a bond that allows the issuer to pay off the loan (or a fraction of the loan) at any of a set of designated *call dates*. If interest rates decline, a bond issuer would call a bond early, pay off the old issue, and replace it with a new series of bonds with a lower coupon rate. A bond is said to have a *European option* if it has a single call date prior to maturity. A bond has an *American option* if it is callable at *any* date following the lockout period (the period before the first call date). When an investor calculates the price of a callable bond, the investor must determine a price that will guarantee the desired yield regardless of the call date. *Putable* (or *extendible*, or *retractable*) bonds are bonds that allow the bond owner (not the issuer) to redeem the bond at a time other than the stated redemption date.

Example 1.36. A \$5000 callable bond matures on September 1, 2023, at par. It is callable on September 1, 2018 at \$5250. Interest on the bond is $c^{(2)} = 6\%$. Calculate the price on September 1, 2013, to yield an investor $i^{(2)} = 5\%$.

Solution.

Scenario 1: The bond is called prior to maturity. The present value is

$$P_1 = 5250 + (5000 \cdot 0.03 - 5250 \cdot 0.025) \cdot a_{\overline{10}|0.025} \cong \$5414.10 \,.$$

Scenario 2: The bond is not called prior to maturity. The present value is

$$P_2 = 5000 + 5000 \cdot (0.03 - 0.025) \cdot a_{\overline{20}|0.025} \cong \$5389.73 \,.$$

To guarantee that the buyer will yield at least $i^{(2)} = 5\%$, the purchase price must be the smallest of the two prices:

$$P = \min\{P_1, P_2\} = \min\{\$5414.10, \$5389.73\} = \$5389.73 \,. \qquad \square$$

1.5 Yield Rates

1.5.1 Internal Rate of Return and Evaluation Criteria

The interest rate that produces a zero NPV of some cash flow stream $[C_0, C_1, \ldots, C_n]$ is called the *internal rate of return* (IRR) or simply the *rate of return*. Its calculation is another financial tool that can be used to determine whether or not to proceed with a project. For the case with equal time periods and when the interest is periodically compounded, the internal rate of return j_{int} solves the equation $\text{NPV}(j_{\text{int}}) = 0$, where

$$\text{NPV}(j) = C_0 + C_1(1+j)^{-1} + C_2(1+j)^{-2} + \cdots + C_n(1+j)^{-n} = \sum_{k=0}^{n} C_k d^k$$

with $d := (1+j)^{-1}$. Note that multiple solutions are possible.

Let j_{cc} be the cost of capital (the risk-free interest rate), and let j_{int} be the internal rate of return.

- If $j_{cc} < j_{int}$, then $\text{NPV}(j_{cc}) > 0$ and the project will return a profit.

- If $j_{cc} = j_{int}$, then $\text{NPV}(j_{cc}) = 0$.

- If $j_{cc} > j_{int}$, then $\text{NPV}(j_{cc}) < 0$ and the project will not return the required rate of return j_{cc}.

Example 1.37 (Multiple Solutions). Find the rate of return for

(a) an investment with the principal of 100 that yields returns of 70 at the end of each of two periods, i.e., find the rate of return for the cash flow stream $\mathbf{C} = [-100, 70, 70]$;

(b) the cash flow stream $\mathbf{C} = [a\,b, -a - b, 1]$ with $a, b > 0$.

Solution.

(a) The rate of the return j solves the equation

$$100 = (1 + j)^{-1}70 + (1 + j)^{-2}70\,.$$

Denoting $d = (1 + j)^{-1}$ and solving the quadratic equation $70d^2 + 70d - 100 = 0$ for d, we obtain $d = \frac{-7 \pm \sqrt{329}}{14}$. Since $70 + 70 > 100$ we are looking for a positive rate j solving the equation. Since $j > 0$ implies that $0 < d < 1$, we obtain the solution $d = 0.795597$. Thus, the internal rate of return is $j = \frac{1}{d} - 1 \cong 25.69\%$.

(b) The net present value is given by

$$\text{NPV} = ab - (a + b)d + d^2 = (d - a)(d - b),$$

where $d = (1+j)^{-1}$. The NPV is zero if $d = a$ or $d = b$, and it is not uniquely determined when $a \neq b$. Furthermore, the NPV of the cash flow stream $-\mathbf{C}$ has the same zeros. It is therefore not clear how to use the internal rate to determine which of the two cash flows is preferable. □

The equation $\text{NPV}(j_{int}) = 0$ can be solved analytically (for short cash flow streams) or numerically. As is shown in the previous example, this equation may have multiple solutions. However, in some special cases, the internal rate j_{int} is uniquely determined by the cash flows.

Proposition 1.3. *If $C_0 > 0$ and $C_k < 0$ for all $k = 1, 2, \ldots, n$ (or $C_0 < 0$ and $C_k > 0$ for all $k = 1, 2, \ldots, n$), then the internal rate j_{int} is uniquely determined. In this case, the rate is nonnegative, i.e., the discount factor $d = (1 + j_{int})^{-1} \leqslant 1$, iff*

$$|C_0| \leqslant \sum_{k=1}^{n} |C_k|\,. \tag{1.38}$$

Proof. Define $P(d) := C_0 + C_1 d + C_2 d^2 + \cdots + C_n d^n$. Then, in the first case with $C_0 > 0$,

$$P(d) = |C_0| - (|C_1|d + |C_2|d^2 + \cdots + |C_n|d^n)\,.$$

Clearly, $P(d)$ is a strictly decreasing, continuous function for $d \geqslant 0$ so that $P(0) > 0$ and $\lim_{d \to \infty} P(d) = -\infty$. By the Intermediate Value Theorem (IVT), $P(d)$ has a unique zero d_0 in $(0, \infty)$. Therefore, $j_{int} > -1$ exists and is unique. It is given by $j_{int} = d_0^{-1} - 1$. If $|C_0| \leqslant \sum_{k=1}^{n} |C_k|$ holds, then $P(1) \leqslant 0$. Hence, by the IVT, $P(d)$ has a unique zero $d_0 \in (0, 1]$. Solving $d_0 = (1 + j)^{-1}$ for j gives us the rate $j_{int} \geqslant 0$. If the condition (1.38) does not hold,

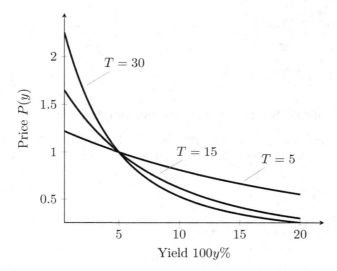

FIGURE 1.12: The coupon bond price as a function of the yield rate compounded annually for different maturities T (a par value bond with the face value of $F = 1$ and annual coupons of $C = 0.05$ each).

then $P(1) > 0$ and hence $P(d)$ does not have zeros in the interval $(0, 1]$. Thus, the rate $j_{\text{int}} \geqslant 0$ does not exist.

Similarly, in the second case with $C_0 < 0$, we have

$$P(d) = -|C_0| + (|C_1|d + |C_2|d^2 + \cdots + |C_n|d^n).$$

The function $P(d)$ is monotonically increasing and continuous so that $P(0) < 0$ and $\lim_{d \to \infty} P(d) = \infty$. If the condition (1.38) holds, then $P(1) > 0$, and hence $P(d)$ has a unique zero $d_0 \in (0, 1]$. Again, solving $d_0 = (1 + j_{\text{int}})^{-1}$ yields the rate $j_{\text{int}} \geqslant 0$. If (1.38) is not satisfied, $P(1) < 0$ and hence the solution to $P(d_0) = 0$ is greater than 1. Thus, $j_{\text{int}} = d_0^{-1} - 1 \in (-1, 0)$.

$\square$

1.5.2   Determining Yield Rates for Bonds

One of the most important properties of a coupon bond is the relation between its price and the yield rate. Bonds are freely traded, and the markets determine their prices. Therefore, the yield rate of a bond is implied by the purchase price. Clearly, the higher the yield rate, the lower the bond price. This property is illustrated in Figure 1.12, where the price of a coupon bond is plotted as a function of the yield rate for different maturities. Let us formally prove this fact.

Theorem 1.4 (The Price-Yield Theorem). *The price and yield of a coupon bond move in opposite directions.*

Proof. Let us think of P as a function of the periodic yield rate j. Let us differentiate (1.35) with respect to j:

$$\frac{\mathrm{d}P}{\mathrm{d}j} = Fc\left(-\frac{1}{(1 + j)^2} - \frac{2}{(1 + j)^3} - \cdots - \frac{n}{(1 + j)^{n+1}}\right) - \frac{nW}{(1 + j)^{n-1}}.$$

The derivative is negative, and hence P is a decreasing function of j. $\square$

In practice, the market price of a bond is often given without stating the yield rate. The yield of the bond reflects expectations of future rates. Bonds of different maturities can have different yields implied by the market. The investor is interested in determining the actual rate of return on investment. Based on the yield rate, the investor can decide whether a particular bond is an attractive investment or can find out which of several bonds available is the best investment. Let us consider the problem of determining what rate of return a bond gives to the buyer when bought for a specified market price.

To find the yield rate y for a given market price B, we need to solve $P(y) = B$ for the rate y. This problem can be expressed as a root-finding problem where we need to find a zero of the function $f(y) = P(y) - B$.

The yield rate can be found *analytically* in simple cases for zero-coupon bonds as well as for bonds with up to four periodic coupons by using Cardano's formula for cubic equations or Ferrari's method for quartic equations.

In more complex cases, the yield rate can be computed *numerically* using one of the following methods.

- *The method of averages.* The yield is obtained by dividing the average interest payment per period by the average amount invested. The accuracy of this method is low, and hence it is often used to find a first approximation of the yield.

- *The method of linear interpolation.* On a given interval that contains the yield rate, the bond price function is approximated by a linear function. The yield rate is then calculated by solving a linear equation.

- *The bisection method.* The main idea of this iterative method is to construct a nested sequence of intervals that contain the yield rate. At every step, the current interval is halved, and the half containing the yield is selected.

- *Newton's method* has the highest accuracy and speed. It is also an iterative method that requires a derivative of the bond price function.

1.5.2.1 Zero-Coupon Bonds

The purchase price of a unit zero-coupon bond defines the implied yield rate. In the case of continuous compounding, the annual yield rate $y = y(t, T)$ is implicitly given by

$$Z(t, T) = e^{-y(T-t)}.$$

Solve this equation for y to obtain

$$y = -\frac{\ln Z(t, T)}{T - t}.$$

The yield rate obtained is a measure of the return for a single payment made at maturity to the holder of a zero-coupon bond. The yield rate is a function of time to maturity. Recall that we use the notation $y(t, T)$ to denote the yield rate at time t for a zero-coupon bond maturing at time T (time is measured in years). If we plot the yield rates of zero-coupon bonds with different maturities, we obtain the *yield curve*, which is discussed in Subsection 1.5.4.

1.5.2.2 Coupon Bonds

A direct method for determining the yield rate for a coupon bond with the purchase price B involves the solution of the nonlinear algebraic equation $B = Fc\,a_{\overline{n}|j} + W(1 + j)^{-n}$ for the periodic rate j. A typical question is whether this equation has a solution j for a given a purchase price. The following theorem settles this issue.

Theorem 1.5 (The Yield Existence Theorem). *If the purchase price $B > 0$ of a coupon bond satisfies the inequality $B < nFc + W$, then there exists a unique yield rate $j > 0$ such that $B = Fca_{\overline{n}|j} + W(1+j)^{-n}$ holds.*

Proof. We know that $P(j) := Fca_{\overline{n}|j} + W(1+j)^{-n}$ is a continuous decreasing function of $j \in \mathbb{R}_+$. Furthermore,

$$\lim_{j \to \infty} P(j) = 0 \quad \text{and} \quad \lim_{j \searrow 0} P(j) = nFc + W.$$

Hence, by the Intermediate Value Theorem, there is a positive solution j to the equation $B = Fca_{\overline{n}|j} + W(1+j)^{-n}$.

Alternatively, we can use Proposition 1.4 to prove the existence of the yield. The rate j is, in fact, the IRR for the following cash flow stream:

$$C_0 = -B, \quad C_1 = C_2 = \ldots = C_{n-1} = Fc, \quad C_n = Fc + W.$$

That is, the bondholder purchases the bond for B dollars to receive n coupons of Fc each and the redemption value W. These cash flows satisfy conditions of Proposition 1.4. Therefore, a unique yield $j > 0$ exists iff $B < nFc + W$. $\square$

Note that the yield rate y compounded annually can be expressed in terms of the periodic yield rate $j = i^{(m)}/m$ as follows:

$$1 + y = (1+j)^m \iff y = (1+j)^m - 1.$$

Here, m is the number of coupon (or interest) periods per year. For the yield rate y compounded continuously, we have

$$e^y = (1+j)^m \iff y = m\ln(1+j).$$

Example 1.38. Consider a bond that matures in exactly five years and pays semi-annual coupons of 3% on a face value of \$1000 each. The price $P(y)$ of this bond is a function of the yield rate y compounded semi-annually. Sketch the inverse function P^{-1} of this bond as a function of the purchase price.

Solution. The bond price is given by

$$P(y) = W + (Fc - W(y/2))a_{\overline{n}|y/2} = 1000 + (30 - 500y)a_{\overline{10}|y/2}.$$

The inverse function P^{-1} can be obtained by solving $P(y) = B$ with a given purchase price B for the yield y. Since $P(y)$ is a strictly decreasing function, the inverse function P^{-1} is also strictly decreasing. As the yield y approaches 0, the bond price $P(y)$ approaches \$1300. Additionally, $\lim_{y \to \infty} P(y) = 0$. Thus, the domain of $P^{-1}(B)$ is the interval $[0, 1300]$. The range of $P^{-1}(B)$ is the interval $[0, \infty)$. The vertical axis is an asymptote. When $y = 6\%$, the price $P(y)$ is equal to \$1000. Thus, $P^{-1}(0.06) = 1000$. The plot of $y = P^{-1}(B)$ is presented in Figure 1.13. $\square$

1.5.3 Approximation Methods

1.5.3.1 The Method of Averages

The method of averages calculates an approximate value of the yield rate j as a ratio of the average interest payment to the average amount invested. If n is the number of interest

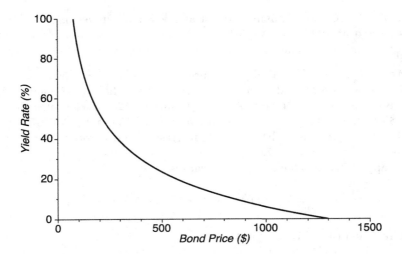

FIGURE 1.13: The yield rate as a function of the coupon bond price in Example 1.38.

periods from the date of sale until the redemption date, then the average interest payment is

$$\frac{nFc + W - P}{n},$$

and the average amount invested is $\frac{W+P}{2}$. The approximate value of j is then given by the ratio

$$j \approx \frac{\text{average interest payment}}{\text{average amount invested}} = \frac{\frac{nFc+W-P}{n}}{\frac{W+P}{2}}. \tag{1.39}$$

To derive the method of averages, we first solve the equation $P = W + (Fc - Wj)a_{\overline{n}|j}$ for j treating the annuity symbol $a_{\overline{n}|j}$ as a constant. This yields

$$j = \frac{Fc}{W} - \frac{P-W}{a_{\overline{n}|j}\,W}. \tag{1.40}$$

We have the following series expansion:

$$(a_{\overline{n}|j})^{-1} = \frac{j}{1 - (1-j)^{-n}} = \frac{1}{n} + \frac{1}{2}\frac{n+1}{n}j + \mathcal{O}(j^2).$$

Plug it in (1.40) to obtain the following approximation:

$$j \approx \frac{Fc}{W} - \frac{P-W}{W}\left(\frac{1}{n} + \frac{1}{2}\frac{n+1}{n}j\right).$$

Solving this equation for j gives

$$j \approx \frac{Fc + \frac{W-P}{n}}{\frac{P+W}{2} + \frac{P-W}{2n}}, \tag{1.41}$$

which can also be used to calculate an approximation for j. If n is large, then $\frac{P-W}{2n}$ is small in comparison with $\frac{P+W}{2}$. Setting $\frac{P-W}{2n} = 0$ in (1.41) leads to equation (1.39).

Example 1.39. A \$500 bond, paying interest at $c^{(2)} = 9.5\%$, redeemable at par on August 15, 2018, is quoted at 109.50 on August 15, 2006. Compute an approximate value of the yield rate compounded semi-annually.

Solution. We have $F = W = \$500$. The purchase price is $P = \$500 \cdot 1.095 = \$547.50$ since the bond is sold on a bond interest date. If the buyer holds the bond until maturity, she will receive 24 coupon payments of $\$500 \cdot 0.095/2 = \$23.75$ each plus the redemption payment of \$500. In total, the buyer will receive $24 \cdot \$23.75 + \$500 = \$1070$. The net gain $\$1070 - \$547.50 = \$522.50$ is realized over 24 interest periods, so that the average interest per period is $\$522.50/24 = \$21.77$. The average amount invested is $(\$547.50 + \$500)/2 = \$523.75$. The approximate value of the yield rate is

$$ j \approx \frac{21.77}{523.75} \cong 0.0416 = 4.16\% \implies i^{(2)} \approx 8.32\% \,. \qquad \square $$

1.5.3.2   The Method of Interpolation

The method of interpolation for computing yield rates is based on a linear approximation of the function $P^{-1}(B)$. Let us recall the idea of linear interpolation. Given two points $(a, f(a))$ and $(b, f(b))$ and a number c such that $a < c < b$, how can one estimate $f(c)$? Without any other information, we simply join the two given points with a straight line segment, $y = g(x)$, and then approximate $f(c)$ with the value of $g(c)$. The line $y = g(x)$ passes through the points $(a, f(a))$ and $(b, f(b))$ and has the slope $\frac{f(b)-f(a)}{b-a}$; therefore its equation is given by

$$ \frac{g(x) - f(a)}{x - a} = \frac{f(b) - f(a)}{b - a} \iff g(x) - f(a) = \frac{f(b) - f(a)}{b - a}(x - a) \,. $$

With $x = c$, where $a < c < b$, we have $f(c) \approx g(c)$ and hence

$$ \frac{f(c) - f(a)}{f(b) - f(a)} \approx \frac{c - a}{b - a} \iff f(c) \approx \frac{f(b) - f(a)}{b - a}(c - a) + f(a) \,, \qquad (1.42) $$

or in the alternative form,

$$ f(c) \approx g(c) = \left(\frac{b - c}{b - a} \right) f(a) + \left(\frac{c - a}{b - a} \right) f(b) \,. $$

The linear interpolation methods is used to solve $P(j) = B$ for j as follows (we can also directly solve for $i^{(m)}$). First, calculate two adjacent rates, j_1 and j_2, so that the market price of the bond lies between the prices determined by those two rates:

$$ P(j_1) > B > P(j_2) \text{ with } j_1 < j < j_2 \,. $$

To find two rates bracketing the yield rate, one can use the method of averages. Second, determine the actual yield rate using linear interpolation between the two adjacent rates. In particular, use (1.42) with $f = P$, $a = j_1$, $b = j_2$, $c = j$ to obtain

$$ \frac{B - P(j_1)}{P(j_2) - P(j_1)} \approx \frac{j - j_1}{j_2 - j_1} \iff j \approx j_1 + \left(\frac{B - P(j_1)}{P(j_2) - P(j_1)} \right)(j_2 - j_1). $$

Example 1.40. Compute the yield rate as in the previous example by the method of interpolation.

Solution. By the method of averages we determined that the yield rate is approximately $j = 4.16\%$. Now we compute the purchase prices to yield $j = 4\%$ and $j = 4.75\%$ compounded semi-annually. To yield $j = 4\%$ the purchase price is

$$P = 500 + (23.75 - 20)a_{\overline{24}|0.04} \cong \$557.18.$$

If the bond interest rate is $c = 9.5\%/2 = 4.75\%$, and the yield rate $j = 4.75\%$, the purchase price is equal to the redemption value: $P = 500$. So the market purchase price $B = \$547.50$ lies between $P(j_1 = 4\%) = \$557.18$ and $P(j_2 = 4.75\%) = \$500$. Assuming that the purchase price P is a linear function of the yield rate j, we obtain

$$\frac{j - 4.75}{4 - 4.75} = \frac{547.50 - 500}{557.18 - 500}$$

with solution $j = \frac{547.50-500}{557.18-500}(4 - 4.75) + 4.75 = 4.75 - 0.75\frac{47.5}{57.18} = 4.127\%$. So the approximate yield rate is $i^{(2)} = 2j = 8.254\%$. The bond price for this yield is

$$P = 500 + (23.75 - 20.635)a_{\overline{24}|0.04127} \cong \$546.88.$$

It is close to the actual market price of $\$547.50$. □

Example 1.41. An investor purchases a $1000 bond, paying interest at $c^{(2)} = 8\%$ and redeemable at par in 10 years. The bond is priced to yield $i^{(2)} = 7\%$ if held to maturity. After holding the bond for three years, it is sold to yield the new holder $i^{(2)} = 6\%$. Calculate the actual investor's yield rate $i^{(2)}$ over the three-year investment period.

Solution. Using $Fc = 1000 \cdot 0.08/2 = 40$, $W = 1000$, $n = 10 \cdot 2 = 20$, and $j = 0.035$, we find the purchase price:

$$\text{Purchase Price} = 40\,a_{\overline{20}|0.035} + 1000(1.035)^{-20} \cong \$1071.06.$$

After three years the selling price can be calculated using $Fc = 40$, $W = 1000$, $n = 7 \cdot 2 = 14$, and $j = 0.03$:

$$\text{Selling Price} = 40\,a_{\overline{14}|0.03} + 1000(1.03)^{-14} \cong \$1112.96.$$

Now consider the three-year investment. The investor pays $\$1071.06$ for the bond, receives a $400 coupon payment every six months for three years and finally gets $\$1112.96$ when the bond is sold. The equation of value at rate j per half-year is

$$1071.06 = 40\,a_{\overline{6}|j} + 1112.96(1 + j)^{-6}.$$

Using the method of averages, an approximate value of the yield rate j is

$$j \cong \frac{6 \cdot 40 + 1112.96 - 1071.06}{(1112.96 + 1071.06)/2} = 0.043 \text{ or } i^{(2)} = 8.6\%.$$

By linear interpolation we obtain a more accurate answer. The bond prices for the two yields of 8% and 9% are

$$P(i^{(2)} = 8\%) = 40\,a_{\overline{6}|0.04} + 1112.96(1.04)^{-6} \cong \$1089.27,$$
$$P(i^{(2)} = 9\%) = 40\,a_{\overline{6}|0.045} + 1112.96(1.045)^{-6} \cong \$1060.95.$$

Hence for the yield rate $i^{(2)}$ we use interpolation,

$$\frac{i^{(2)} - 8}{9 - 8} = \frac{1071.06 - 1089.27}{1060.95 - 1089.27} = \frac{18.21}{28.32} = 0.643$$

giving the yield rate $i^{(2)} = 8.643\%$. □

1.5.3.3 Numerical Methods

Although the method of interpolation provides a better approximation of the yield rate in comparison with the method of averages, the result is still inaccurate. The approximation error can be significantly reduced by employing a numerical algorithm such as the bisection method or Newton's method for solving nonlinear equations.

First, let us consider the *bisection method* for solving an equation $f(x) = 0$. To start the method, we need to bracket a zero of the function $f(x)$. That is, we need to find an interval $[a_0, b_0]$ that contains a zero of f. If the values $f(a_0)$ and $f(b_0)$ have opposite signs, i.e., $f(a_0)f(b_0) < 0$, and f is a continuous function on $[a_0, b_0]$, then by the Intermediate Value Theorem there exists a zero $x^* \in (a_0, b_0)$ such that $f(x^*) = 0$. Note that the interval may contain several zeros, and the bisection method pinpoints one of them.

The main idea is as follows: given a bracketed root, halve the interval while continue bracketing the root. As a result, we construct a sequence of nested intervals $\{[a_n, b_n]\}_{n \geqslant 0}$ so that $f(a_n)f(b_n) < 0$ for every $n \geqslant 0$. Thus, every bracket contains x^*. The length of the interval $[a_n, b_n]$ converges to zero, as $n \to \infty$, and hence the sequences of endpoints $\{a_n\}$ and $\{b_n\}$ converge to x^*, as $n \to \infty$. Indeed, by construction, $a_n < b_n$ for every $n \geqslant 0$ and the length of the nth bracket is

$$b_n - a_n = \frac{b_0 - a_0}{2^n} \to 0, \quad \text{as } n \to \infty.$$

Therefore, the sequence of midpoints $\{c_n = \frac{a_n + b_n}{2}\}$ converges to x^* as well. At every step, the midpoint c_n can be used as an approximation to x^*. Since the length of the nth bracket is $(b_0 - a_0)/2^n$ and the distance between x^* and c_n is not greater than $(b_n - a_n)/2$, the error $\epsilon_n = |c_n - x^*|$ does not exceed $\frac{b_0 - a_0}{2^{n+1}}$.

We can apply the bisection method to the problem of computing a yield rate for a coupon bond. Let $P(y)$ be the bond price as a function of the yield y, and B be the market price. Since $P(y)$ is a strictly decreasing function of y, the interval $[a_0, b_0]$ brackets a zero of the function $f(y) := P(y) - B$ (and it is a unique zero) iff $f(a_0) > 0$ and $f(b_0) < 0$ hold. Find the middle point $c_0 = \frac{a_0 + b_0}{2}$ and calculate $P(c_0)$. If $P(c_0) > B$, then the yield rate is greater than c_0 and hence the next bracket is $[a_1, b_1] = [c_0, b_0]$. If $P(c_0) < B$, then the yield rate is less than c_0 and hence the next bracket is $[a_1, b_1] = [a_0, c_0]$. We repeat this process as many times as necessary to find an approximate value of the yield.

The Newton iterative rule for finding a zero of a continuously differentiable function $f = f(x)$ takes the following form:

$$x_{k+1} = x_k - \frac{f(x_k)}{f'(x_k)}, \quad k = 0, 1, 2, \ldots. \tag{1.43}$$

If initial approximation x_0 is sufficiently close to a zero of the function f, then the sequence $\{x_k\}_{k \geqslant 0}$ produced by (1.43) converges to a solution of $f(x) = 0$.

To justify Newton's method, let us write the first-degree Taylor polynomial of f at x_k to approximate $f(x_{k+1})$:

$$f(x_{k+1}) = f(x_k) + f'(x_k)(x_{k+1} - x_k) + o(x_{k+1} - x_k).$$

Assuming that x_{k+1} is close to a zero of f and hence $f(x_{k+1}) \approx 0$, we have the following approximation:

$$f(x_k) + f'(x_k)(x_{k+1} - x_k) = 0.$$

Solving it for x_{k+1} yields the iterative rule (1.43).

To determine the yield rate for a given market price B of a coupon bond, we need to solve the equation

$$W + (Fc - Wj)a_{\overline{n}|j} = B$$

for the periodic rate j. Having found the rate j, we can calculate the annual yield y. Define the function $P(j) := W + (Fc - Wj)a_{\overline{n}\rceil j}$ and rewrite the above equation as $P(j) - B = 0$. According to Theorem 1.5, we have

$$P(0) = nFc + W, \quad \lim_{j \to \infty} P(j) = 0, \quad \text{and} \quad P'(j) < 0.$$

Thus, for any given price $B \in (0, nFc + W)$, there exists a unique yield rate $j \in (0, \infty)$ such that $B = P(j)$.

Newton's method can be applied to find a zero of the function $f(j) := P(j) - B$. An initial approximation of the rate j can be calculated with the method of averages. Such an approximation can be used as a starting value j_0 within Newton's method. It is worth noting that

$$\frac{da_{\overline{n}\rceil j}}{dj} = \frac{n(1+j)^{-n-1}}{j} - \frac{a_{\overline{n}\rceil j}}{j}.$$

Hence, the derivative of the bond value function P w.r.t. j is

$$P'(j) = (Fc - Wj)\frac{n(1+j)^{-n-1}}{j} - Fc\frac{a_{\overline{n}\rceil j}}{j}.$$

Let us describe the method of solving for the corresponding yield to maturity. Newton's method begins with an initial guess j_0 and recursively computes approximate rates j_n by

$$j_{n+1} = j_n - \frac{f(j_n)}{f'(j_n)} = j_n - \frac{P(j_n) - B}{P'(j_n)}, \quad n = 0, 1, \ldots. \tag{1.44}$$

The rate of convergence of Newton's method is quadratic since $P'(j) \neq 0$ for all $j > 0$. So the approximation j_n obtained after n iterations is very close to the desired solution.

Newton's method can be applied to a more general bond with varying coupons. Let us find the annual yield y by using Newton's iterative rule

$$y_{n+1} = y_n - \frac{P(y_n) - B}{P'(y_n)}, \quad n = 0, 1, \ldots. \tag{1.45}$$

Consider the following three cases.

- The yield y is compounded continuously, the coupons are paid at times t_k with $k = 1, 2, \ldots, n$:

$$P(y) = \sum_{k=1}^{n} C_k e^{-y t_k} + W e^{-y t_n},$$

$$P'(y) = -\sum_{k=1}^{n} C_k t_k e^{-y t_k} - W t_n e^{-y t_n}.$$

- The yield y is compounded annually, and the bond pays annual coupons:

$$P(y) = \sum_{k=1}^{n} C_k (1+y)^{-k} + W (1+y)^{-n},$$

$$P'(y) = -\sum_{k=1}^{n} C_k k (1+y)^{-k-1} - W n (1+y)^{-n-1}.$$

- The yield y is compounded semi-annually, and the bond pays semi-annual coupons:

$$P(y) = \sum_{k=1}^{n} C_k (1 + y/2)^{-k} + W (1 + y/2)^{-n},$$

$$P'(y) = -\sum_{k=1}^{n} C_k (k/2) (1 + y/2)^{-k-1} - W (n/2) (1 + y/2)^{-n-1}.$$

To compute the yield y in each of the above cases, we apply the iterative rule (1.45) with $P(y_n)$ and $P'(y_n)$ calculated by the corresponding formulae.

Example 1.42. The price of a two-year par-value bond paying a semi-annual coupon of 4% (i.e., $c^{(2)} = 8\%$) on a face value of $1000 is $1078.16.

(a) Find the exact value of the yield (as a continuously compounded rate) on the bond.

(b) Find an approximate value of the yield on the bond using

    (i) the method of averages;
    (ii) the bisection method;
    (iii) Newton's method.

Solution.

(a) The continuously compounded yield rate y solves the equation

$$P(y) = B \iff 40\,e^{-0.5y} + 40\,e^{-y} + 40\,e^{-1.5y} + 1040\,e^{-2.0y} = 1078.16\,.$$

Denote $x = e^{-y/2}$ to obtain an equivalent quartic equation:

$$1040x^4 + 40x^3 + 40x^2 + 40x = 1078.16\,.$$

There are two real solutions: $x_1 = -1.01863$ and $x_2 = 0.980875$. Since x has to be positive, we have a unique yield rate:

$$e^{-y/2} = x_2 \implies y = -2\ln x_2 = -2\ln 0.98087 \cong 3.8621\%.$$

(b) The method of averages gives the following approximate value of a periodic yield rate compounded semi-annually:

$$j \approx \frac{\text{average interest payment}}{\text{average amount invested}} = \frac{40 + \frac{1000 - 1078.16}{4}}{\frac{1000 + 1078.16}{2}} \cong 0.01969\,.$$

Thus, we can find an equivalent yield rate compounded continuously as follows:

$$e^{y/2} = 1 + j \implies y \approx 2\ln 1.01969 \cong 3.8998\%.$$

Now, find an approximate yield by the bisection method. The interval $[3.75\%, 4\%]$ brackets a zero of the function $f(y) = P(y) - B$. Indeed, $f(3.75\%) \cong 2.291 > 0$ and $f(4.0\%) \cong -2.808 < 0$. Perform five iterations of the bisection method.

Iteration, n	Left endpoint, a_n	Right endpoint, b_n	Midpoint, c_n	$f(c_n)$
0	3.5000%	4.0000%	3.7500%	$2.290
1	3.7500%	4.0000%	3.8750%	$-$0.263
2	3.7500%	3.8750%	3.8125%	$1.013
3	3.8125%	3.8750%	3.8438%	$0.375
4	3.8438%	3.8750%	3.8594%	$0.056
5	3.8594%	3.8750%	3.8672%	$-$0.103

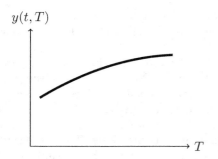

$y(t,T)$

T

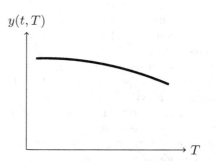

$y(t,T)$

T

(a) *Normal shape*: long bonds have higher yields to compensate investors for the uncertainty.

(b) *Inverted shape*: short bonds are more preferable since interest rates are expected to fall.

FIGURE 1.14: Normal and inverted shapes of a yield curve.

After five iterations, an approximate yield is $y \approx c_5 = 3.8672\%$.

Lastly, we apply Newton's method to find a zero of the function $f(y) = P(y) - B$. Let us use the rate $y_0 = 3.9\%$, provided by the method of averages, as a starting point. Perform two iterations of Newton's method $y_{n+1} = y_n - f(y_n)/f'(y_n)$ with $n = 0, 1$.

Iteration, n	y_n	$f(y_n)$	$f'(y_n)$	y_{n+1}
0	3.9000%	$-\$0.772283$	-2038.6005	3.86212%
1	3.8649%	$-\$0.000286$	-2040.1092	3.86213%

After two iterations, an approximate yield is $y \approx y_2 = 3.86213\%$. □

1.5.4 The Yield Curve

For many reasons (credit risk, liquidity, etc.) cash flows occurring at different points in time need to be discounted using different rates. Yield curves are often used to determine what rates should be used. Such a curve is a graph of the yield-to-maturity of a zero-coupon bond plotted against tenor (the number of years to maturity). The value of the spot (yield) rate is calculated based on the assumption that the investor holds the bond until redemption and that the yield-to-maturity is the same yield rate used in pricing the bond. The market price reflects today's expectation of interest rates. The effective interest implied by the market prices determines the yield of the bond. Bonds of different maturities can have different yields implied by the market.

The yield curve is an instrument used for analysis of the bond market. The shape of the yield curve reflects the market's expectation of where interest rates are heading. As the date of maturity of a bond approaches, there is less and less uncertainty left. The principal of the bond is known and will be paid on the redemption date, and the interest rate is unlikely to move much in a short period. A bond of longer maturity will be exposed to more uncertainty (i.e., it is riskier) than one of short maturity. We can, therefore, expect long-dated bonds to have higher yields to compensate investors for this additional risk. So the yield curve is upward sloping: $y(t, T_1) \leqslant y(t, T_2)$ for $T_1 < T_2$. It is a so-called *normal yield curve* with upward sloping. See Figure 1.14 (a). The market expects that interest rates are likely to rise in the future. However, sometimes shorter maturities have higher yields so that the yield curve is known as downward sloping. The market predicts that the interest rates will fall. In such situations, the yield curve is said to be *inverted*. See Figure 1.14 (b).

There are well-established markets in the major government bonds so an investor need not hold the bond until maturity. Instead, the investor can sell it on the market. A normal yield curve highlights the fact that if you purchase a bond, hold on to the investment for a long period, and then sell it, it is likely that your selling yield will be lower than your purchase yield.

So far, we discussed the pricing of a coupon bond in the situation where the desired yield rate is given and is constant during the lifetime of the bond. Now we consider the general case where the yield rate varies with time and is implicitly defined by values of zero-coupon bonds. First, note that a coupon payment of $\$C$ due at time t gives the same cash flow as that of C unit zero-coupon bonds maturing at time t. Therefore, a coupon bond can be viewed as a portfolio of zero-coupon bonds with different maturity times. The present value of the coupon paid at time t is equal to $CZ(0,t)$. The present value of the redemption value paid at time T is $WZ(0,T)$. Therefore, the purchase price P of a coupon bond (i.e., its value at time 0) can be represented as a weighted sum of unit zero-coupon bonds' values:

$$P = CZ(0, \delta t) + CZ(0, 2\delta t) + \cdots + CZ(0, (n-1)\delta t) + (C+W)Z(0,T)$$
$$= \sum_{k=1}^{n} CZ(0, k\delta t) + WZ(0,T), \tag{1.46}$$

where δt is the length of the coupon/interest payment period and $T = n\delta t$ is the maturity time.

For each zero-coupon bond with maturity time t we can find the annual yield rate $y(t)$ implied by the bond price $Z(0,t)$ as follows:

$$Z(0,t) = e^{-y(t)t} \implies y(t) = -\frac{\ln Z(0,t)}{t}.$$

The formula (1.46) takes the form

$$P = \sum_{k=1}^{n} e^{-y_k k\delta t} C + e^{-y_n T} W, \tag{1.47}$$

where $y_k = y(k\delta t)$ is the yield rate (at time 0) for zero-coupon bond maturing at time $t = k\delta t$.

Some bonds have variable coupon rates. If it is the case, then (1.47) takes the form:

$$P = \sum_{k=1}^{n} e^{-y_k k\delta t} C_k + e^{-y_n T} W, \tag{1.48}$$

where C_k is the coupon paid at time $t = k\delta t$.

Let us consider an arbitrary stream of cash flows. Let $t_1, t_2, \ldots, t_n$ be the payments dates so that $0 < t_1 < t_2 < \ldots < t_n$, with respective cash flows $C_1, C_2, \ldots, C_n$. To find the present (time-0) value of this cash flow stream using the yield curve, we follow the following procedure.

Step 1. Determine the annual yield rates $y(t_1), y(t_2), \ldots, y(t_n)$ that correspond to the dates $t_1, t_2, \ldots, t_n$.

Step 2. Find the present value of the unit zero-coupon bond with maturity time t_k using

$$Z(0, t_k) = e^{-y(t_k)t_k} \text{ (for continuously compounded yield rates) or}$$
$$Z(0, t_k) = (1 + y(t_k))^{-t_k} \text{ (for annually compounded yield rates)}$$

for each $k = 1, 2, \ldots, n$.

Step 3. Evaluate the present value of the stream by simply discounting all the future cash flows back to time 0:

$$P = C_1 Z(0, t_1) + C_2 Z(0, t_2) + \cdots + C_n Z(0, t_n). \tag{1.49}$$

Example 1.43. Suppose that the continuously compounded yield curve on a zero-coupon bond with t years to maturity is

$$y(t) = 0.05 - 0.03 \left(1 - e^{-t}\right)/t, \quad t \geqslant 0.$$

How much would I pay for a five-year bond that pays an annual coupon of 3% on a face value of $100?

Solution. The cash flows that I receive from the coupon bond (assuming zero credit risk) are no different from the cash flows that I would receive from a portfolio of (i) one zero-coupon bond with face value $3 maturing in one year, (ii) one zero-coupon bond with face value $3 maturing in two years, ..., (v) one zero-coupon bond with face value $103 maturing in five years. Therefore I should pay the same amount for the coupon bond as I would for the portfolio. The cost of the portfolio is

$$P = 3e^{-y(1)\cdot 1} + 3e^{-y(2)\cdot 2} + 3e^{-y(3)\cdot 3} + 3e^{-y(4)\cdot 4} + 103e^{-y(5)\cdot 5}$$

$$\cong 3e^{-0.031} + 3e^{-2\cdot 0.037} + 3e^{-3\cdot 0.04} + 3e^{-4\cdot 0.043} + 103e^{-5\cdot 0.044}$$

$$\cong \$93.52.$$

Therefore I should be willing to pay $93.52 for the bond. □

Example 1.44. Assume we have the interest rate structure as given below. Assume annual compounding.

Time to maturity	Spot rate
1 year	1.50%
2 years	1.50%
3 years	1.75%
4 years	2.00%
5 years	3.00%

Calculate the price of a five-year par value $1000 bond if it is

(a) a zero-coupon bond;

(b) a bond with annual coupons of 4% on the face value.

Solution.
(a) The price of a zero-coupon bond is equal to the discounted value of the redemption value paid at maturity. To determine the price, we discount the value $W = 1000$ using the five-year spot rate of 4%:
$$P = 1000 \cdot 1.03^{-5} \cong \$862.61.$$

(b) To determine the price of the coupon bond, we use (1.49). The amount of each coupon payment is $C = 1000 \cdot 0.04 = \$40$. The cash flows received from the coupon bond are no different from the cash flows received from a portfolio of four zero-coupon bonds with face value $40 maturing in one, two, three, and four years, respectively, and one zero-coupon bond with face value $1040 maturing in five years. Using the spot rates from the table gives

$$P = 40 \cdot 1.015^{-1} + 40 \cdot 1.015^{-2} + 40 \cdot 1.0175^{-3} + 40 \cdot 1.02^{-4} + 1040 \cdot 1.03^{-5}$$

$$\cong \$1050.27. \qquad \square$$

Example 1.45. The continuously compounded yield curve on a zero-coupon bond with t years to maturity is given by

$$y(t) = 0.05 - 0.03\,\mathrm{e}^{-0.5t}, \quad t \geqslant 0.$$

Show that the price of the bond from Example 1.42 is indeed $1,078.16.

Solution. There are four coupon payments of $40 each paid semi-annually and the redemption amount of $1000 paid at the end of year 2. Find the present value of each cash flow:

Time t in years	Rate $y(t)$	Payment in dollars	Present value in dollars
0.5	2.66360%	$40.00	$39.471
1.0	3.18041%	$40.00	$38.748
1.5	3.58290%	$40.00	$37.907
2.0	3.89636%	$1040.00	$962.033

The purchase price B of the bond is a sum of the present values and is equal to

$$B = 39.471 + 38.748 + 37.907 + 962.033 \cong \$1078.16. \qquad \square$$

If we only know the actual yields for a very small number of tenors, we can fit a curve to the data to determine what the yield on an arbitrary tenor should be. For example, Figure 1.15 illustrates actual yield curves using data from Government of Canada bonds as of January 28, 2013. Using the quadratic fit we would estimate that the yield on a 15-year bond issued by the Government of Canada should be $y(15) = 2.43938$, or approximately 2.44% compounded semi-annually.

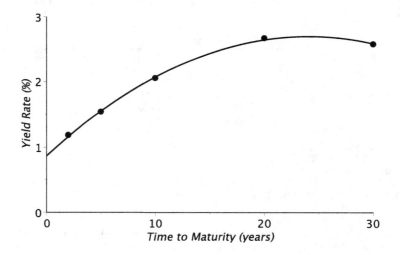

FIGURE 1.15: Yields of several Government of Canada zero-coupon bonds (with maturities 2, 5, 10, 20, and 30 years) and a quadratic fit $y(t) = 0.87008 + 0.15202t - 0.00316t^2$ to the data.

A popular method for fitting yield curves used in practice is a so-called "Nelson–Siegel" curve given by

$$y(t) = \alpha_0 + \alpha_1 \left(\frac{1 - \mathrm{e}^{-t/\beta}}{t/\beta} \right) + \alpha_2 \left(\frac{1 - \mathrm{e}^{-t/\beta}}{t/\beta} - \mathrm{e}^{-t/\beta} \right),$$

where $\alpha_0, \alpha_1, \alpha_2, \beta$ are parameters to be calibrated to the data (for example, using the method of least squares or another curve fitting technique).

Example 1.46. Evaluate $\lim_{t \to \infty} y(t)$ and $\lim_{t \to 0} y(t)$ for a Nelson–Siegel curve. Use your answers to interpret the parameters α_0 and α_1.

Solution. We use l'Hôpital's rule to obtain

$$\lim_{t \to 0} \frac{1 - e^{-t/\beta}}{t/\beta} = \lim_{t \to 0} e^{-t/\beta} = 1,$$

and, therefore,

$$\lim_{t \to 0} \left(\frac{1 - e^{-t/\beta}}{t/\beta} - e^{-t/\beta} \right) = 1 - 1 = 0.$$

Thus, $y(0) = \alpha_0 + \alpha_1$, and we can interpret the sum of these parameters as the so-called "short rate" at time 0. It is obvious that

$$\lim_{t \to \infty} \left(\frac{1 - e^{-t/\beta}}{t/\beta} \right) = 0 = \lim_{t \to \infty} \left(\frac{1 - e^{-t/\beta}}{t/\beta} - e^{-t/\beta} \right),$$

and thus $\lim_{t \to \infty} y(t) = \alpha_0$. So we can interpret α_0 as the long rate, and $-\alpha_1$ represents how steep the yield curve is, in the sense that it measures the differential cost of long-term borrowing relative to short-term borrowing. □

1.6 Yield Risk and Duration

When market yields change, it impacts prices of fixed-income instruments. We say that bond prices are subject to the *yield risk*. The sensitivity of the bond value w.r.t. the yield rate y is measured by the derivative of the purchase price P w.r.t. y. Since the derivative $\frac{\partial P}{\partial y}$ is negative, the price P is a decreasing function of the yield y.

Let us look at changes in the bond value relative to yield rate changes for coupon bonds with different maturities. As we see in Figure 1.12, the prices of bonds with long maturities are more sensitive to yield rate changes than those of short-maturity bonds. However, maturity itself does not give a complete quantitative measure of yield rate sensitivity. A better measure is a so-called *duration*.

The *duration* of a fixed-income instrument is a weighted average of the times that payments are made. The *duration* of a stream of cash flows received at times $t_1, t_2, \ldots, t_n$ is

$$\text{Duration} = \frac{\text{PV}_1 \cdot t_1 + \text{PV}_2 \cdot t_2 + \cdots + \text{PV}_n \cdot t_n}{\text{PV}}, \tag{1.50}$$

where PV_k is the present value of the kth payment with $k = 1, 2, \ldots, n$, and PV is the present value of the whole cash stream given by $\text{PV} = \text{PV}_1 + \text{PV}_2 + \cdots + \text{PV}_n$. The duration is measured in units of time, e.g., in years.

The duration can be written as a weighted sum of payment times:

$$\text{Duration} = \sum_{k=1}^{n} w_k \cdot t_k,$$

where $w_k = \text{PV}_k / \text{PV}$. The percentage of the kth payment in the whole stream is $100 w_k \%$.

Assume that all cash flows are nonnegative (or they are all non-positive). In this case, all weights w_k are nonnegative, as well. Additionally, the sum of weights is one:

$$\sum w_k = \sum (\text{PV}_k/\text{PV}) = \left(\sum \text{PV}_k\right)/\text{PV} = \text{PV}/\text{PV} = 1.$$

Therefore, the duration cannot be less than the earliest payment time and cannot exceed the latest time:

$$\min t_k \leqslant \text{Duration} \leqslant \max t_k$$

If the cash stream contains only a single payment made at time T, e.g., it is a zero-coupon bond with maturity T, then its duration is equal to T.

Let us find the duration of a coupon bond with n semi-annual coupons assuming that the interest is compounded continuously at rate r. In the literature it is called the *Fisher–Weil* duration. Let $c = c^{(2)}/2$ be the coupon rate per half a year. So, the first $n-1$ cash flows are $C_k = Fc$ with $k = 1, 2, \ldots, n-1$, where F is the face value. The nth cash flow is $C_n = Fc + W$ where W is the redemption value. Thus, the duration D is equal to

$$D = \frac{1}{P}\left(\sum_{k=1}^{n} t_k C_k e^{-r t_k}\right) = \frac{1}{P}\left(Fc\sum_{k=1}^{n}\frac{k}{2}e^{-rk/2} + W\frac{n}{2}e^{-rn/2}\right)$$

$$= \frac{1}{P}\left(\frac{Fc}{2}\sum_{k=1}^{n}k(1+j)^{-k} + W\frac{n}{2}(1+j)^{-n}\right), \tag{1.51}$$

where $t_k = \frac{k}{2}$ are payment dates, P is the bond purchase price, and $j = e^{r/2} - 1$ is an equivalent rate for a half-year period. The bond price is given by

$$P = \sum_{k=1}^{n} C_k e^{-r t_k} = \sum_{k=1}^{n} Fc e^{-rk/2} + W e^{-rn/2} = Fca_{\overline{n}|j} + W(1+j)^{-n}. \tag{1.52}$$

Using the formula

$$\sum_{k=1}^{n} k\,x^k = \frac{x^{n+1}(xn - n - 1) + x}{(x-1)^2}, \tag{1.53}$$

we can simplify (1.51) as follows:

$$D = \frac{\frac{Fc}{2}\left(\frac{1+j-(1+j(n+1))(1+j)^{-n}}{j^2}\right) + W\frac{n}{2}(1+j)^{-n}}{Fca_{\overline{n}|j} + W(1+j)^{-n}}. \tag{1.54}$$

Consider a bit more general situation, where a par-value bond pays m coupons per year. Let the size of each coupon be Fc, and the interest be compounded periodically at rate $j = i^{(m)}/m$. The duration D is given by (1.50) with $C_1 = C_2 = \cdots = C_{n-1} = Fc$, $C_n = F + Fc$, and $t_k = \frac{k}{m}$. So, we have

$$D = \frac{\sum_{k=1}^{n} Fc\frac{k}{m}(1+j)^{-k} + F\frac{n}{m}(1+j)^{-n}}{\sum_{k=1}^{n} Fc(1+j)^{-k} + F(1+j)^{-n}}$$

$$= \frac{c\sum_{k=1}^{n} k(1+j)^{-k} + n(1+j)^{-n}}{m\left(ca_{\overline{n}|j} + (1+j)^{-n}\right)}.$$

It is called the *Macaulay duration*. As we can see, the duration of a par-value bond does not depend on the face value. Again, the above formula can be simplified, with the help of (1.53), as follows:

$$D = \frac{c(1+j - (1+j(n+1))(1+j)^{-n}) + nj^2(1+j)^{-n}}{mj^2\left(ca_{\overline{n}|j} + (1+j)^{-n}\right)}. \tag{1.55}$$

Example 1.47. Calculate the Macaulay duration for each of the following two par-value bonds with semi-annual coupons.

(A) $c^{(2)} = 8\%$, $T = 20$ years;

(B) $c^{(2)} = 7\%$, $T = 15$ years.

Assume that the yield to maturity is $i^{(2)} = 6\%$.

Solution. The bond durations are:

$$D_A = \frac{0.04 \cdot \sum_{k=1}^{40} k \cdot 1.03^{-k} + 40 \cdot 1.03^{-40}}{2 \cdot \left(0.04 \cdot \sum_{k=1}^{40} 1.03^{-k} + 1.03^{-40}\right)},$$

$$D_B = \frac{0.035 \cdot \sum_{k=1}^{30} k \cdot 1.03^{-k} + 30 \cdot 1.03^{-40}}{2 \cdot \left(0.035 \cdot \sum_{k=1}^{30} 1.03^{-k} + 1.03^{-30}\right)}.$$

Calculate each summation in the above two expressions:

$$\sum_{k=1}^{40} 1.03^{-k} = \frac{1.03^{-41} - 1.03^{-1}}{1.03^{-1} - 1} = 23.11477197,$$

$$\sum_{k=1}^{40} k \cdot 1.03^{-k} = \frac{(40 \cdot 1.03^{-1} - 41) \cdot 1.03^{-41} + 1.03^{-1}}{(1.03^{-1} - 1)^2} = 384.8647177,$$

$$\sum_{k=1}^{30} 1.03^{-k} = \frac{1.03^{-31} - 1.03^{-1}}{1.03^{-1} - 1} = 19.60044135,$$

$$\sum_{k=1}^{30} k \cdot 1.03^{-k} = \frac{(30 \cdot 1.03^{-1} - 31) \cdot 1.03^{-31} + 1.03^{-1}}{(1.03^{-1} - 1)^2} = 260.9617255.$$

Now, we can find the value P and the duration D of each bond:

$$P_A = 100 \cdot (0.04 \cdot 23.11477197 + 1.03^{-40}) \cong \$123.1147 \quad \text{(for } F = \$100\text{)},$$

$$P_B = 100 \cdot (0.04 \cdot 19.60044135 + 1.03^{-30}) \cong \$109.8002 \quad \text{(for } F = \$100\text{)},$$

$$D_A = \frac{0.04 \cdot 384.8647177 + 40 \cdot 1.03^{-40}}{2 \cdot (0.04 \cdot 23.11477197 + 1.03^{-40})} \cong 11.2321,$$

$$D_B = \frac{0.035 \cdot 260.9617255 + 30 \cdot 1.03^{-30}}{2 \cdot (0.04 \cdot 19.60044135 + 1.03^{-30})} \cong 19.7874.$$

$\square$

Let y be the yield rate compounded continuously. The present value of a stream of n cash flows is

$$P = \sum_{k=1}^{n} C_k e^{-y t_k}.$$

Differentiate P w.r.t. y to obtain

$$\frac{\partial P}{\partial y} = -\sum_{k=1}^{n} C_k t_k e^{-y t_k}.$$

As we can see, $\frac{\partial P}{\partial y} = -DP$, where D is the Fisher–Weil duration. Therefore, the change in value due to a change in yield y is proportional to duration:

$$\Delta P \approx \frac{\partial P}{\partial y}\, \Delta y = -DP\,\Delta y. \tag{1.56}$$

That is, for every 1% increase (or decrease) in the yield, the percent decrease (or increase) in the price is approximately equal to $100D\%$. Hence, the duration is a measure of sensitivity of bond values w.r.t. changes in yield: the greater the duration—the larger the yield risk. Since, $\frac{\partial P}{\partial y} = -DP$, the slope of the line tangent to the price-yield curve at P is $-DP$. Notice that the price-yield curve becomes steeper as the duration increases (e.g., if the maturity increases). See Figure 1.12.

Let y be the yield rate compounded m times per year. The present value of a stream of n cash flows is

$$P = \sum_{k=1}^{n} C_k\,(1 + y/m)^{-t_k\,m}.$$

Differentiate P w.r.t. y to obtain

$$\frac{\partial P}{\partial y} = -\sum_{k=1}^{n} C_k\, t_k\,(1 + y/m)^{-t_k\,m-1}$$

$$= -\frac{1}{1 + y/m}\sum_{k=1}^{n} C_k\, t_k\,(1 + y/m)^{-t_k\,m} = -\frac{D}{1 + y/m}\,P,$$

where D is the Macaulay duration. Therefore, the change in value due to a change in the yield y is

$$\Delta P \approx \frac{\partial P}{\partial y}\,\Delta y = -\frac{D}{1 + y/m}\,P\,\Delta y. \tag{1.57}$$

The quantity $\dfrac{D}{1 + y/m}$ is called the *modified duration*. As $m \to \infty$, the modified duration converges to the Fisher–Weil duration.

Example 1.48. Estimate the changes in value due to an increase in yield by 3% for bonds from Example 1.47.

Solution. The yield changes from $i^{(2)} = 6\%$ to $i^{(2)} = 9\%$. According to (1.57), the changes in value for bonds A and B are, respectively,

$$\Delta P_A \approx -\frac{D_A}{1 + 0.03} \cdot 0.03 \cdot P_A \cong -\$40.28 \quad (\text{per } \$100),$$

$$\Delta P_B \approx -\frac{D_B}{1 + 0.03} \cdot 0.03 \cdot P_B \cong -\$31.30 \quad (\text{per } \$100).$$

$\square$

Equations (1.56) and (1.57) tell us that the magnitude of the change ΔP is directly proportional to $|\Delta y|$ and does not depend on the sign of Δy. To derive the approximations (1.56) and (1.57), we have used the linearization of the bond value function. Applying the second order Taylor approximation to the function $P(y)$ gives

$$\Delta P \approx \frac{\partial P}{\partial y}\,\Delta y + \frac{1}{2}\frac{\partial^2 P}{\partial y^2}\,(\Delta y)^2,$$

where we assume that the interest is compounded continuously. Since the bond price is a monotonically decreasing, convex function of the yield, the magnitude of the change, $|\Delta P|$, due to an increase in yield by $100\Delta y\%$ is less than the magnitude of the change due to a decrease in yield of the same magnitude. In other words,

$$P(y) - P(y + \Delta y) < P(y - \Delta y) - P(y)$$

holds for any $y > 0$ and $\Delta y > 0$ so that $y - \Delta y > 0$.

1.6.1 Immunization

Immunization is the method of structuring a bond portfolio to protect ("immunize") against interest rate changes (i.e., protect against the yield risk). Suppose that you face a series of cash obligations and you wish to form a portfolio of bonds that you will use to pay these obligations as they arise. One solution is to purchase several ZCBs (such as Treasury bonds) whose maturities and face values exactly match the individual obligations. However, perfect matching is usually not possible. Secondly, you may prefer to purchase corporate coupon bonds with higher yields. The desired bond portfolio can be found by solving two equations:

$$\begin{cases} \text{Present Value of Bonds} = \text{Present Value of Obligations,} \\ \text{Duration of Bonds} = \text{Duration of Obligations.} \end{cases} \tag{1.58}$$

To find the present value of a portfolio of bonds, we need to sum up present values of individual bonds. Let us calculate the duration of a bond portfolio. First, consider just two bonds, denoted A and B. Assume that their coupons are paid on the same dates, $t_1, t_2, \ldots, t_n$. Let PV_k^A and PV_k^B with $k = 1, 2, \ldots, n$ denote the present value of the kth payment for the bonds A and B, respectively. The durations of the bonds A and B are, respectively,

$$D^A = \frac{\sum_{k=1}^{n} t_k \, \text{PV}_k^A}{\text{PV}^A}, \qquad D^B = \frac{\sum_{k=1}^{n} t_k \, \text{PV}_k^B}{\text{PV}^B}.$$

The duration of a combined portfolio of bonds A and B is

$$\begin{aligned} D &= \frac{\sum_{k=1}^{n} t_k \left(\text{PV}_k^A + \text{PV}_k^B\right)}{\text{PV}^A + \text{PV}^B} \\ &= \frac{\sum_{k=1}^{n} t_k \, \text{PV}_k^A}{\text{PV}^A + \text{PV}^B} + \frac{\sum_{k=1}^{n} t_k \, \text{PV}_k^B}{\text{PV}^A + \text{PV}^B} \\ &= \left(\frac{\text{PV}^A}{\text{PV}^A + \text{PV}^B}\right) D^A + \left(\frac{\text{PV}^B}{\text{PV}^A + \text{PV}^B}\right) D^B. \end{aligned}$$

The weights $\frac{\text{PV}^A}{\text{PV}^A + \text{PV}^B}$ and $\frac{\text{PV}^B}{\text{PV}^A + \text{PV}^B}$, respectively represent the percentage compositions of the bonds A and B in the portfolio. So, the duration of a bond portfolio is a weighted sum of durations of individual bonds.

In general, the duration of a portfolio of ℓ bonds with respective present values P_k and durations D_k with $k = 1, 2, \ldots, \ell$ (all calculated on the same date) is

$$D = w_1 D_1 + w_2 D_2 + \cdots + w_\ell D_\ell, \quad \text{where} \quad w_k = \frac{P_k}{P_1 + P_2 + \cdots + P_\ell}. \tag{1.59}$$

If all P_k are positive, then all the weights are positive as well. Their sum is one. Thus, the duration of a bond portfolio is bounded by the minimum and maximum durations of individual bonds included in the portfolio:

$$\min D_k \leqslant D \leqslant \max D_k.$$

Example 1.49. The X Corporation must pay $1 million in 10 years. It wishes to invest money now that will be sufficient to meet this obligation. Suppose that the yield rate is $i^{(2)} = 9\%$. The X Corporation is planning to select from the three par-value bonds with semi-annual coupons. Find a bond portfolio using the immunization method.

	Coupon rate	Maturity
Bond 1	6%	30
Bond 2	11%	10
Bond 3	9%	20

Solution. The present value of the obligation is

$$PV = 10^6 \cdot \left(1 + \frac{0.09}{2}\right)^{-10 \cdot 2} \cong \$414{,}642.86.$$

Since the obligation is a single amount, its duration is 10 years. Calculate prices and durations of the three bonds:

$$P_1 = \$69.04 \qquad P_2 = \$113.01, \qquad P_3 = \$100.00,$$
$$D_1 = 11.4448, \qquad D_2 = 6.5355, \qquad D_3 = 9.6148,$$

where the purchase prices are calculated per $100. The corporation needs at least two bonds to form a portfolio whose duration matches the duration of the obligation. Since the duration of a bond portfolio is always between $\min D_k$ and $\max D_k$, we can use bonds of types 1 and 2 or bonds of types 2 and 3, but we cannot use bonds of types 2 and 3 only since both D_2 and D_3 are less than 10. Let us use bonds of types 1 and 2. First, find the amounts V_1 and V_2 invested in bonds of types 1 and 2, respectively, by solving the following system.

$$\begin{cases} V_1 + V_2 = PV \\ D_1 V_1 + D_2 V_2 = D\, PV \end{cases} \iff \begin{cases} V_1 + V_2 = 414{,}642.86, \\ 11.4448\, V_1 + 6.5355\, V_2 = 10 \cdot 414{,}642.86, \end{cases}$$

$$\iff \begin{cases} V_1 = 292{,}614.06, \\ V_2 = 122{,}028.80, \end{cases}$$

Second, find the number of bonds of each type (assuming that the face value of each bond is $100):

$$\begin{cases} \text{No. of bonds of type } 1 = {}^{V_1}\!/_{P_1} \cong 4238.33, \\ \text{No. of bonds of type } 2 = {}^{V_2}\!/_{P_2} \cong 1079.81. \end{cases}$$

□

Example 1.50. Study the sensitivity of the bond portfolio constructed in Example 1.49 to changes in the yield rate. Consider two scenarios: $\Delta y = 1\%$ and $\Delta y = -1\%$.

Solution. We have that $\Delta V_\Pi \approx -\dfrac{D_\Pi V_\Pi}{1 + y/2}\, \Delta y$, where V_Π is the bond portfolio value and D_Π is its duration. By construction, $V_\Pi = \$414{,}642.86$ and $D_\Pi = 10$. Thus, $\Delta V_\Pi \approx \mp\$39{,}678.74$ if $\Delta y = \pm 1\%$.

□

1.7 Exercises

Exercise 1.1. Let r_k and $r_{[0,k]}$ denote, respectively, the rate of return for year k and the aggregate rate of return for the first k years with $k = 1, 2, \ldots$.

(a) Show that the aggregate rate for the first two years, equals $r_1 + r_2 + r_1 r_2$.

(b) Show that the aggregate rate for the first $k + 1$ years equals $r_{[0,k]} + r_{k+1} + r_{[0,k]} r_{k+1}$.

Exercise 1.2. The rates of return for years 1, 2, and 3 are $r_1 = 3\%$, $r_2 = 2\%$, and $r_3 = 3\%$, respectively.

(a) Find the aggregate rate of return, $r_{[0,3]}$, for the three-year period.

(b) If \$3900 is invested at the beginning of year 1, what is the accumulated value at the end of year 3?

(c) Find the equivalent rate compounded annually.

Exercise 1.3. Show that the return on a deposit subject to periodic compounding is not additive, i.e., in general, $R_{[t_1,t_2]} + R_{[t_2,t_3]} \neq R_{[t_1,t_3]}$ for $t_1 < t_2 < t_3$.

Exercise 1.4. Prove that $V(t) = P(1 + i)^t$ with $P, i > 0$ is an increasing, convex function of time $t \geqslant 0$.

Exercise 1.5. (a) Find the effective annual rate when the nominal rate of 10% is compounded (i) semi-annually, (ii) monthly, (iii) continuously.

(b) If the effective annual rate is 10%, find the nominal rate compounded (i) semi-annually, (ii) monthly, (iii) continuously.

Exercise 1.6. Show that

(a) the internal rate of return corresponding to a nominal rate $i^{(m)}$ compounded m times a year does not depend on the number of years it is invested;

(b) the internal rate of return corresponding to a simple interest investment rate of r does depend on the number of years it is invested.

Exercise 1.7. Prove that for a fixed nominal interest rate i, the accumulation function $A(m) = (1 + \frac{i}{m})^m$ for the interest rate compounded m times a year is a strictly increasing function of $m \geqslant 1$. [*Hint*: Use the inequality $\ln(1 + x) > \frac{x}{1+x}$ for $x > 0$.]

Exercise 1.8. Let interest be paid at a nominal rate of 15% per year. How long will it take for a principal to double if the interest is

(a) compounded annually;

(b) compounded quarterly;

(c) compounded continuously?

Exercise 1.9. Assume a fixed yearly compounding rate $i > 0$. Find a formula that gives the number of years required to increase initial funds by n times for $n = 2, 3, 4, \ldots$.

Exercise 1.10. You are considering three different investments:

(a) one paying 7.1% compounded annually;

(b) one paying 7% compounded quarterly;

(c) one paying 6.9% compounded continuously.

Which investment has the highest effective annual rate of return?

Exercise 1.11. The interest rate compounded continuously is equal to 3.5% for the first two years, 2% for the next three years, and 2.75% for the last three years.

(a) Find the yield curve $y(t)$ for $0 \leqslant t \leqslant 8$.

(b) If $3500 is invested at the beginning of year one, what is the accumulated value at time 7.5 years?

(c) The accumulated value of an investment at time four years is $5134.88. Find the present value at the beginning of year one.

(d) An investment is made at the beginning of year one. The interest earned on the investment from time 4 to 7.5 (in years) is equal to $372.956. Find the original principal.

Exercise 1.12. Construct a time diagram that represents the following cash flow stream and find the internal rate of return.

$$\frac{\quad 0 \qquad 1 \qquad 2 \quad}{\$1000 \quad -\$500 \quad -\$600}$$

Exercise 1.13. A firm can buy a piece of equipment for $200,000 paid immediately, or in three instalments: $70,000 paid now, $70,000 in one year, and $70,000 in two years time. Which option is better if money can be invested at a nominal rate of 6% compounded monthly?

Exercise 1.14. Which of the following projects should a firm choose if each proposal costs $50,000 and the cost of capital is $i^{(1)} = 8\%$? The projects provide the following end-of-year net cash flows:

Years	1	2	3	4	5
Project A	$20,000	$10,000	$5000	$10,000	$20,000
Project B	$20,000	$20,000	$10,000	$10,000	$5000

Exercise 1.15. The instantaneous rate of return is given by

$$r(t) = 0.01 + 0.02\sqrt{t} + 0.001t, \quad t \geqslant 0.$$

(a) Find the yield curve $y(T)$ for $T \geqslant 0$.

(b) If $2200 is invested at time 0, what is the accumulated value at time $t = 2.5$?

(c) If the accumulated value at time $t = 2.5$ is $5000, what is the present value of the investment?

(d) The cash flows $[C_1, C_2, C_3] = [\$500, -\$500, -\$500]$ are made at times $[T_1, T_2, T_3] = [0.25, 0.5, 1.5]$ (in years), respectively. Determine the present value of the stream of cash flows.

Exercise 1.16. Using the yield curve $y(T) = 0.04 - 0.015e^{-0.9T}$ with $T \geqslant 0$, determine:

(a) the purchase price of a zero-coupon bond with a face value of $100 that matures after 8.5 years;

(b) the present value and the duration of cash flows $[C_1, \ C_2, \ C_3] = [\$200, \$350, \$250]$ at times $[T_1, T_2, T_3] = [1.25, 2.75, 3.5]$ (in years).

Exercise 1.17. Find the closed-form expression for

$$S_n = \sum_{k=1}^{n} x^{k-1} = 1 + x + x^2 + \cdots + x^{n-1}$$

by first showing that $xS_n - S_n = x^n - 1$ and then solving for S_n.

Exercise 1.18. A used car sells for $9550. You wish to pay for it in 18 monthly instalments, the first due on the day of purchase. If 12% compounded monthly is charged, determine the size of the monthly payment.

Exercise 1.19. A couple is thinking of buying a new house by amortizing a loan. They can afford to pay $1000 a month for the first 10 years and $1200 a month for the next 10 years. How much can they borrow, if the current interest rate is $i^{(12)} = 6\%$?

Exercise 1.20. A company is able to borrow money at $i^{(4)} = 9\frac{1}{2}\%$. It is considering the purchase of a machine costing $110,000, which will save the company $5000 at the end of every quarter for seven years and may be sold for $14,000 at the end of the term. Should the company borrow the money to buy the machine?

Exercise 1.21. Find the yield curve $y(t)$ and the present value of a zero-coupon bond $Z(0, T) = e^{-y(T)T}$ if the instantaneous interest rate is

$$r(t) = \frac{1}{1+t} a + \frac{t}{1+t} b$$

with $a, b > 0$.

Exercise 1.22. (a) Sketch the purchase price P for this bond as a function of the yield rate y compounded semi-annually.

(b) Sketch the inverse of the pricing function, P^{-1}, for this bond.
   Be sure to label important points, intercepts, and asymptotes.

Exercise 1.23. The short rate is given by $r(t) = 0.03 + 0.02t - 0.01t^2$ for $t \in [0, 3]$.

(a) Find and plot the yield curve $y(T)$ for $T \in [0, 3]$.

(b) Find the price of a unit ZCB, $Z(0, T)$, for $T \in [0, 3]$.

(c) Find the present value of an ordinary annuity paying $1000 annually for three years.

Exercise 1.24. A $1,000 callable bond pays interest at $c^{(2)} = 8\%$ and matures at 105 in 20 years. It may be called at 110 at the end of year 5, or at 107.5 at the end of year 15.

(a) Determine the price of the bond to yield at least $i^{(2)} = 7\%$.

(b) Estimate, using first the method of averages and then linear interpolation, the yield rate earned by the buyer if the bond is called at the end of year 5.

Exercise 1.25. On September 15, 2013, you were given a set of yield rates for various terms to maturity, generated using pricing data for Government of Canada bonds and treasury bills. Use the yield rates from the table below to price a coupon bond (issued on September 15, 2013) with the face value $F = \$1000$. The bond is redeemable at par at the end of five years and it pays five annual coupons at a rate of 2.5%.

Time t	1	2	3	4	5
Rate $y(t)$	2.395%	2.487%	2.623%	2.758%	2.882%

Exercise 1.26. Tom and Jack purchased bonds on the same day. Both bonds are redeemable at par and have a yield of 6% compounded annually and a face value of $10,000.

(a) If Tom's bond pays annual coupons at the rate of 8% and it has 15 years to maturity, then how much does he pay for it?

(b) If Jack's bond has 10 years to maturity and he pays $11,487.75 for it, then what is the annual coupon rate?

Exercise 1.27. Jack invests $2000 in Bond A and $2000 in Bond B. On the same day Tom invests $2000 in Bond C and $8000 in Bond D. At the end of six months Bond A is worth $2050, Bond B is worth $2100, Bond C is worth $2040, and Bond D is worth $8360. Confirm that Jack has a higher rate of return on each of his bonds than Tom, but Tom has a higher rate of return on his total portfolio.

Exercise 1.28. Consider a coupon bond. Which would be larger: the price increase associated with a fall in yield by 1%, or the price decrease associated with a rise in yield by 1%?

Exercise 1.29. A par value bond matures in six years, pays a semi-annual coupon at a nominal rate of 4.3% on a face value of $500 and is currently trading at $457.70. Prove that its yield is at least 5% compounded semi-annually.

Exercise 1.30. A par value bond matures in six years, pays a semi-annual coupon at a nominal rate of 1.4% on a face value of $500 and is currently trading at $385.53. Prove that its yield does not exceed 7% compounded semi-annually.

Exercise 1.31. Six-month and one-year zero-coupon bonds with a face value of $100 are trading at $99.50 and $98.51, respectively. A 1.5-year bond with a semi-annual coupon at the rate $c^{(2)} = 5\%$ and a face value of $1000 is trading at $1014.80. What are the six-month, one-year and 1.5-year zero-coupon yields?

Exercise 1.32. A two-year bond with an annual coupon of 10% and a face value of $100 trades at $113.40. A two-year bond with an annual coupon of 2% and a face value of $100 trades at $98.02. Find the corresponding yields, expressed as annual rates with continuous compounding, for one-year and two-year zero-coupon bonds. [*Hint*: Set up an appropriate system of two equations in two unknowns.]

Exercise 1.33. Consider a bond market consisting of the following three bonds. Bond A is a zero-coupon bond with a face value of $80, maturing one year from now. Bond B is a zero-coupon bond with a face value of $100, maturing two years from now. Bond C matures two years from now and pays an annual coupon of 4% on a face value of $1000.

(a) If Bond A is trading at $75, what is its continuously compounded yield to maturity?

(b) If the continuously compounded yield on Bond A is 6.45% and Bond C is trading at $900, what is the (continuously compounded) yield on Bond B?

Exercise 1.34. A $1000 bond paying interest at the rate $c^{(2)} = 6\%$ matures at par on August 1, 2013. If this bond is quoted at 92 on August 1, 2011, determine the yield rate compounded semi-annually by

(a) using the method of averages;

(b) using the method of interpolation.

Exercise 1.35. Consider a two-year $2500 par value bond that pays semi-annual coupons at the rate $c^{(2)} = 7\%$. Suppose that the bond was purchased for $2644.46.

(a) Use the method of averages to approximate the effective yield rate, $i^{(2)}$, compounded semi-annually.

(b) Perform three iterations of the bisection method and find the midpoint $c(3)$ to approximate the yield rate $i^{(2)}$. To create the initial bracket $[a(0), b(0)]$, use the rate $i^{(2)}$ from part (a) rounded off to two decimals as follows:

$$a(0) = i^{(2)} - 0.005 \quad \text{and} \quad b(0) = i^{(2)} + 0.005.$$

Exercise 1.36. Consider a $1000 par value bond that pays seven semi-annual coupons at a nominal rate of 7% compounded semi-annually. Suppose that the bond was purchased for $1094.38.

(a) Use the method of averages to approximate the effective yield rate, $i^{(2)}$, compounded semi-annually.

(b) Use the result in part (a) to complete one iteration of Newton's method to calculate a more accurate approximation of the yield rate $i^{(2)}$.

Exercise 1.37. The Government of Ontario has an outstanding bond that matures in exactly six years and pays a semi-annual coupon at a rate of 4.2% on a face value of $1000. The current market price of the bond is $1028.14.

(a) Use the method of averages to estimate the bond's yield (expressed as an annual rate with semi-annual compounding).

(b) Determine whether or not your answer from part (a) is an overestimate or underestimate.

(c) What happens to the interval $(2\%, 4.2\%)$ after two iterations of the bisection method?

(d) Use two iterations of Newton's method to refine the initial guess of 4.2% compounded semi-annually.

Exercise 1.38. Consider a two-year $4000 bond that is redeemable at par and pays semi-annual coupons at $c^{(2)} = 8\%$. Determine the purchase price and the duration of the bond if the yield rate equals 3.5% and is

(a) compounded annually;

(b) compounded semi-annually;

(c) compounded continuously.

Exercise 1.39. Using the yield curve $y(T) = 0.04 - 0.035e^{-0.8T}$ with $T \geqslant 0$, determine the purchase price and the duration of a two-year $1500 bond that is redeemable at par and pays semi-annual coupons at $c^{(2)} = 4\%$.

Exercise 1.40. The zero-coupon yield curve is

$$y(T) = 0.055 - 0.025 \cdot \left(\frac{1 - e^{-T/2}}{T} \right), \quad T \geqslant 0.$$

What is the duration of a portfolio that consists of one five-year ZCB, one 10-year ZCB, and one 30-year ZCB? All bonds have the same face value.

Exercise 1.41. Stream A consists of $1 to be received at the end of each of the next 10 years. Stream B consists of $100 to be received at the end of each of the next 10 years. Which is larger—the duration of A or the duration of B?

Exercise 1.42. A portfolio consists of one of each of the following bonds

Bond	1	2	3	4
Duration (years)	0.7	3.2	5.1	8.7
Price ($)	98.10	101.50	112.30	90.25

(a) Find the duration of the portfolio.

(b) Estimate the percentage change in the value of each bond assuming the continuously compounded yield decreases by 0.15%, and use the results to estimate the change in the value of the whole portfolio.

(c) Estimate the percentage change in the value of the portfolio assuming the continuously compounded yield decreases by 0.15% by using the portfolio's duration.

Exercise 1.43. Consider the following zero-coupon bonds each of which is redeemable at par and has a yield rate of 3.2% compounded continuously.

Bond	Face value ($)	Maturity (years)
A	5000	10
B	4000	15
C	6000	20
D	9000	30

(a) Determine the purchase price of each bond.

(b) Determine the present value and the duration of the portfolio of bonds A, B, C, and D.

(c) Estimate the absolute and relative changes in the portfolio value if the yield decreases by 0.1%.

Exercise 1.44. Consider the following bonds each of which is redeemable at par, pays semi-annual coupons, and has a yield rate of 6.8% compounded semi-annually.

Bond	Coupon rate	Face value	Maturity	Price	Duration
A	7%	$3500	3 years	$3518.71	2.758 years
B	4%	$3500	10 years	$2797.25	8.097 years
C	7%	$1000	6 years	$1009.72	5.006 years

Consider a bond portfolio that contains four bonds A, four bonds B, and three bonds C.

(a) Determine the present value and duration of the bond portfolio.

(b) Use the duration to estimate the absolute and relative change in the portfolio value if the yield increases by 0.5%.

Exercise 1.45. Suppose that a corporation is obligated to pay a debt of $265,000 in nine years. In order to pay the debt, the corporation plans to create a portfolio consisting of the following two par-value bonds that pay semi-annual coupons.

Bond	Coupon rate	Face value	Maturity	Price	Duration
A	3%	$1000	10 years	$933.96	8.658 years
B	4%	$1500	15 years	$1534.06	11.469 years

(a) Determine the present value of the debt if the current market yield is 3.8% compounded semi-annually.

(b) Determine how much should be invested in each of the bonds using the immunization method.

(c) Using the result in part (b), determine how many shares of each of Bond A and Bond B that need to be purchased.

Exercise 1.46. Consider cash flows $\{C_j\}_{1 \leqslant j \leqslant n}$ received at times $\{t_j\}_{1 \leqslant j \leqslant n}$.

(a) Prove that if we double each cash flow, then the duration of the stream remains unchanged. That is, prove that the duration of this stream is equal to that of the stream $\{2C_j\}_{1 \leqslant j \leqslant n}$.

(b) Prove the duration of the stream $\{-C_j\}_{1 \leqslant j \leqslant n}$ is the same as that for the original one.

Exercise 1.47. Consider four coupon bonds as given in the table below. Assume that all bonds have the same redemption and face values, the same frequency of coupons, and that the yield rate is the same for all maturities.

Bond	Coupon rate (%)	Time to maturity (years)
A	0	25
B	10	20
C	15	20
D	10	25

Rank the bonds in order of increasing duration.

Exercise 1.48. Prove the formula (1.59) for the duration of a portfolio of several bonds.

2

Primer on Pricing Risky Securities

2.1 Stocks and Stock Price Models

2.1.1 Underlying Assets and Derivative Securities

Let us start with a primary classification of financial instruments. First, we differentiate between underlying assets (or securities) and financial derivative contracts. *Financial security* is a legal claim to some future benefit. Bonds, bank deposits, common stocks, and the like are all examples of financial securities.

In contrast, a general *financial contract* links nominally two or more parties. Such a contract specifies conditions under which payments or payoffs are to be made between the parties. The main example is a *derivative contract* whose payoff depends on values of other financial variables such as stock and bond prices, market indices, and exchange rates all called *underlying assets*. Examples include forward contracts, futures, swaps, and options. Derivatives are discussed in more detail in Chapter 4 and later chapters.

In Chapter 1, we discussed various types of financial assets and investments, such as bank accounts, bonds, annuities, mortgages, and other types of loans. They all have one thing in common. It is the assumption that all future payments are guaranteed. However, in any investment portfolio, where a certain amount is invested today, and future cash flows are returned over time, there is always a possibility that not all future payments will actually be received or the amount of future payments is not certain. For most investments, future cash flows are contingent upon the financial position of the company issuing the investment. On the other hand, there are many financial assets, such as stocks, foreign currencies, or commodities, whose future market values are unpredictable. They depend on the choices and decisions made by a great number of agents acting under conditions of uncertainty. Such assets can be viewed as "risky" assets compared with "risk-free" assets with certain cash flows.

There are several types of risk involved in the purchase and sale of financial assets. These include economic uncertainty, interest rate sensitivity, company failure, company management problems, competition with other companies, and governmental rulings that may negatively affect the company. It is reasonable to assume that future prices of risky assets depend on random factors. That is, the asset values can be treated as random variables.

A corporation that needs funds for its development or expansion may issue *stock* to investors. Stock represents ownership of a corporation. Owners of *common stock* have voting rights and are entitled to the earnings of the company after all obligations are paid. If an investor purchases shares of stock in the company, then such an investor assumes a large amount of risk in return for the possibility of growth of the company and a corresponding increase in the value of the shares. It is essential to realize that the corporation receives its money when the stock shares are issued. Any trading after that point takes place between the shareholders and investors wishing to purchase the stock and does not directly represent a profit or loss to the company. Investors who have purchased stock may receive dividends periodically (usually quarterly). Thus, the investor may profit by an increase in the value of the stock or by receipt of dividends. The price that an investor is willing to pay for a share of common stock is based upon the investor's expectations regarding dividends and the future price of the stock.

In order to buy or sell shares of stocks, the investor uses an investment firm registered with the appropriate governmental agencies. The fee or commission charged for the transaction is an important consideration. The person making the transactions for the firm is called a broker. If an investor believes that a stock is going to decrease in value, then the investor may borrow shares from the broker (if such a transaction is possible), and then sell the stock. Such a financial operation is called *short selling*. If the stock decreases in value, then the investor may purchase an equivalent number of shares at the lower price and use those shares to repay the loan.

Market *indices* are used to compute "average" prices for groups of assets. A stock market index attempts to mirror the performance of the stock portfolio it represents through the use of one number, the index. Indices may represent the performance of all stocks in an exchange or a smaller group of stocks, such as an industrial or technological sector of the market. Examples of major U.S. stock indices are Dow Jones, S&P 500 (Standard and Poor's), and NASDAQ. The S&P/TSX index is the Canadian equivalent of the S&P 500 in the United States. There are international stock indices such as DAX and Euro Stoxx 50.

2.1.2   Basic Assumptions for Asset Price Models

In practice, financial mathematicians generally make the following assumptions when dealing with asset price models.

Not moving the market. Our actions do not affect the market prices. In other words, we can buy or sell any amount of assets without affecting their prices. Clearly, this is not true for a free market, since an increasing demand moves market prices up.

Liquidity. At any time we can buy or sell as much as we wish at the market price without being forced to wait until a counter-party can be found. A liquid asset can be sold rapidly, any time within market hours. Cash is the most liquid asset. A market is said to be liquid when there are ready and willing buyers and sellers at all times.

Shorting. We can have a negative amount of an asset by selling assets we do not hold. In this case we say that a *short position* is taken or that the asset is *shorted*. A short position in bonds means that the investor borrows cash and the interest rate is determined by the bond prices. A short position in stock means that the investor borrows the stock, sells it, and uses the proceeds to make other investments. The opposite of going "short," i.e., holding an asset, is sometimes called being "long" in the asset.

Fractional quantities. We can purchase fractional quantities of an asset. It is a reasonable assumption when the size of a typical financial transaction is sufficiently larger than the smallest unit one can hold.

No transactions costs. We can buy and sell assets without paying any additional fees. In the market, one of the typical ways to collect transaction costs is that buy and sell prices differ slightly. Such a difference in prices is called a *bid–ask spread*. The size of the bid–ask spread is closely related to liquidity. For a very liquid asset, the bid–ask spread tends to be quite small.

Stochastic prices. The future prices of financial assets are uncertain. Thus, we deal with *stochastic asset price models*. All factors that can affect the outcome of an economy are commonly called risk factors. We assume that all prices are equilibrium prices. We are not concerned with modeling the mechanism by which prices equilibrate.

Model assumptions. At the time we introduce asset price models, some assumptions on the models will be stated.

Basic Components of a Financial Model:

- **The time horizon** $T > 0$ is a date at which trading activity stops.

- **Trading dates** are calendar dates between the initial time $t = 0$ and time horizon $t = T$ at which trading are allowed to take place.

- **A state of the world,** denoted by ω, characterizes the "real-world" state, which is relevant to the economic environment we wish to model. Such outcomes are also called *market scenarios*. The set of all feasible scenarios or states of the world, denoted by Ω, is called the *state space*.

- **Tradable base assets** are underlying assets available for trading. The price of any derivative instrument depends on the current price or history of prices of the corresponding underlying asset.

- **Trading rules,** such as the allowance of short selling, the presence of taxes and transaction costs, should be specified.

Financial models can be categorized as either of two general types: continuous-time or discrete-time. For a continuous-time model, the allowable time moments and trading dates form a continuous interval $[0, T]$ with specified initial and final calendar times 0 and T, respectively. For a discrete-time model, assets are only observed at a finite set of calendar dates $\{0, 1, \ldots, T\}$. The discrete-time models can be further subdivided into *single-period* models with only two relevant trading dates consisting of the current date $t = 0$ and the maturity date $t = T$ and multi-period models, where trading takes place at several dates. Financial models can also be categorized with respect to the state space Ω, which can be finite or infinite.

Consider some financial asset/security denoted by A or S. The price of asset A at time t is denoted by A_t (or $A(t)$ in some cases). The price is assumed to be positive for all times t. The collection of asset prices indexed by time, $\{A_t\}_{t \in \mathbf{T}}$, is called the *price process* of asset A. Here, we have $\mathbf{T} = \{0, T\}$ for a single-period model, $\mathbf{T} = \{0, 1, \ldots, T\}$ for a multi-period discrete-time model, and $\mathbf{T} = [0, T]$ for a continuous-time model. At the initial time $t = 0$, the price A_0 is known to all investors. In general, the future values of any asset are uncertain. From the mathematical point of view, the prices A_t, $0 \leqslant t \leqslant T$, are positive random variables on the state space Ω. That is, for any $t \in \mathbf{T}$, A_t is a function from Ω to $(0, \infty)$. Therefore, the price process $\{A_t\}_{t \in \mathbf{T}}$ is a sequence of random variables indexed by time. Such a series of uncertain variates is called a *random* or *stochastic process*.

The notation $A_t(\omega)$ is used to denote the price at time t given that the market follows scenario $\omega \in \Omega$. Fixing ω gives a particular realization of the asset price process for the given scenario. Such a realization $A_t(\omega)$ considered as a function of time $t \in \mathbf{T}$ is called a *sample asset price path* or simply a *sample path*.

2.2 Basic Price Models

2.2.1 A Single-Period Binomial Model

Consider an investor who is faced with an economy whose future state is not known for certain. Let us construct the simplest possible model that describes such an economy.

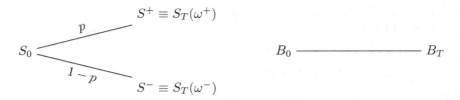

FIGURE 2.1: A schematic representation of a single-period model with two scenarios. B is a risk-free asset. S is a risky asset with two time-T prices, $S^{\pm} = S_T(\omega^{\pm})$.

Suppose that there are only two dates/times: a current date labelled 0 and a future (maturity) date labelled T. Since the economy is uncertain, there should be at least two future states, each corresponding to one of two possible scenarios or outcomes. Let these two states in our model be denoted by ω^- and ω^+. For example, these states can respectively represent "bad" news and "good" news about the economy. So our first financial model is a *single-period* model with the binomial state space $\Omega = \{\omega^-, \omega^+\}$.

The wealth V of the investment in underlying assets is a function of time: $V = V_t$ with $t \in \{0, T\}$. It is reasonable to assume that the *initial wealth* V_0 of the investment is known at time $t = 0$. However, the *terminal wealth* V_T at the maturity date is uncertain and is a function of state $\omega \in \Omega$. There are two states; hence V_T can take on two possible values: $V_T(\omega^-)$ and $V_T(\omega^+)$.

To further develop our model, we first need to quantify the chances of finding our economy in either of the possible states of Ω. Second, we need to specify what assets are available to form an investment portfolio. Assume that state ω^+ occurs with probability $p \in (0, 1)$. Thus, state ω^- occurs with probability $1 - p$. So the terminal wealth V_T can be viewed as a random variable on the finite probability space $(\Omega, \mathcal{F}, \mathbb{P})$ with Ω as a scenario set and $\mathbb{P}$ as a probability distribution function defined on the collection of events $\mathcal{F} = \{\emptyset, \{\omega^-\}, \{\omega^+\}, \Omega\}$ as follows:

$$\mathbb{P}(E) = \begin{cases} 0 & E = \emptyset, \\ p & E = \{\omega^+\}, \\ 1 - p & E = \{\omega^-\}, \\ 1 & E = \Omega = \{\omega^-, \omega^+\}. \end{cases}$$

We note that $\mathbb{P}(E)$ represents the *real-world probability* of event E. We will see shortly that the fair price of a derivative contract does *not* depend on the real-world probabilities p and $1 - p$.

We consider the case with only two tradable base assets, namely, a risk-free bond B and a risky stock S. Let S_t and B_t, $t \in \{0, T\}$, be the respective time-t prices of the stock and the bond. The initial prices S_0 and B_0 are positive constants. The terminal prices S_T and B_T are positive random variables on the probability space $(\Omega, \mathcal{F}, \mathbb{P})$. The bond is a risk-free asset iff the variable B_T is certain, i.e., $B_T(\omega^+) = B_T(\omega^-) \equiv B_T$. The stock is a risky asset iff the variable S_T is uncertain, i.e., $S_T(\omega^+) \neq S_T(\omega^-)$. This is depicted in Figure 2.1, where for convenience we take $S^+ \equiv S_T(\omega^+) > S_T(\omega^-) \equiv S^-$.

A *static portfolio* in the bond and stock is a pair of real numbers $(x, y) \in \mathbb{R}^2$ that represents the positions (fixed in time) of the investment in the stock and bond:

$$x = \text{Number of shares of the stock} \quad \text{and} \quad y = \text{Number of units of the bond.}$$

If $x > 0$ (or $y > 0$) then the position in the stock (or bond) is said to be *long*. If $x < 0$ (or

$y < 0$) then the position in the stock (or bond) is said to be *short*. The time-t value of a portfolio (x, y), denoted by $\Pi_t^{(x,y)}$, is given by

$$\Pi_t^{(x,y)} = xS_t + yB_t, \quad t \in \{0, T\}.$$

Note that in this economic model the investor takes on positions (x, y) in the two base assets at initial time $t = 0$ and holds these positions until the end of the period at time $t = T$. The initial value $\Pi_0 = \Pi_0^{(x,y)}$ is constant, while the terminal value Π_T is a generally nonconstant random variable on $(\Omega, \mathcal{F}, \mathbb{P})$. Note that the position x represents the number of units held in the risky asset. Hence, Π_T is constant iff $x = 0$, in which case the portfolio contains only positions in the bond.

Before proceeding further with the discussion of this two-state single period economy, we note the basic model assumptions as follows.

Divisibility. Positions in the base assets, x and y, may have noninteger value.

Liquidity. There are no bounds on x and y. That is, any asset can be bought or sold on demand at the market price in arbitrary quantities.

Short Selling. The positions x and y may be negative. In this case we say that a *short position* in the asset is taken or that the asset is *shorted* (otherwise, we say that an investor has a *long position* in the asset). A short position in the bond means that the investor borrows cash and the interest rate is determined by the bond prices. A short position in the stock means that the investor borrows shares of the stock, sells them, and uses the proceeds to make other investments.

Solvency. The portfolio value must be nonnegative at all times, $\Pi_t \geqslant 0$ for $t \in \{0, T\}$. A portfolio satisfying this condition is said to be *admissible*.

Clearly, there are many possible ways to form an investment portfolio. While the initial wealth of such portfolios is the same, the terminal values may have different distributions. Different investments can be compared by calculating their returns. *Return on investment* is the ratio of the terminal value of the investment to the initial value of the investment. The total return on the portfolio (x, y) from time 0 to time T, denoted by R_Π, can be expressed as a weighted sum of returns on the base assets:

$$R_\Pi = \frac{\Pi_T^{(x,y)}}{\Pi_0^{(x,y)}} = \frac{xS_T + yB_T}{xS_0 + yB_0} = \underbrace{\frac{xS_0}{xS_0 + yB_0}}_{\equiv w_S} \underbrace{\frac{S_T}{S_0}}_{\equiv R_S} + \underbrace{\frac{yB_0}{xS_0 + yB_0}}_{\equiv w_B} \underbrace{\frac{B_T}{B_0}}_{\equiv R_B}$$

$$= w_S R_S + w_B R_B, \tag{2.1}$$

where $R_S := \frac{S_T}{S_0}$ and $R_B := \frac{B_T}{B_0}$ are the total returns on the stock and bond, respectively. Note that the return on the bond is certain, i.e., $R_B = R_B(\omega^\pm)$ is a constant. The weights w_S and w_V, with the property $w_S + w_B = 1$, correspond to the respective fractions of the initial wealth invested in the stock and bond. Since the sum of the weights is one, the rate of return on the portfolio (x, y) is a weighted sum of the respective rates of return on the stock and bond:

$$r_\Pi = R_\Pi - 1 = w_S R_S + w_B R_B - 1 = w_S(r_S + 1) + w_B(r_B + 1) - 1$$

$$= w_S r_S + w_B r_B. \tag{2.2}$$

As noted above, it is convenient in what follows to denote the stock price at maturity in the two possible states as $S^{\pm} \equiv S_T(\omega^{\pm})$ and let $S^+ > S^-$. The return on the stock is a random variable whose values on the two states are denoted by $r_S^{\pm} \equiv r_S(\omega^{\pm})$, where $r_S^{\pm} = S^{\pm}/S_0 - 1$ and $S^{\pm} = S_0(1 + r_S^{\pm})$. From (2.2), we see that the rate of return on the portfolio, r_{Π}, is a random variable with two possible values: $r_{\Pi}(\omega^+) = w_S r_S^+ + w_B r_B$ for the outcome ω^+ and $r_{\Pi}(\omega^-) = w_S r_S^- + w_B r_B$ for the outcome ω^-. Hence, the return on any portfolio without short selling (with positive x and y) always falls in between the largest and smallest possible returns on the stock and bond:

$$\min\{r_B, r_S^-, r_S^+\} \leqslant r_{\Pi} \leqslant \max\{r_B, r_S^-, r_S^+\}.$$

The terminal value of a portfolio in base assets is a function of the form $\Pi_T \colon \Omega \to \mathbb{R}$. Any such function is called a *payoff* function (or payoff for short). A nonnegative payoff is called a *claim*. A payoff is also a random variable defined on the probability space $(\Omega, \mathcal{F}, \mathbb{P})$. Every contract C in our economic model can be specified by the terminal payoff C_T and by the initial market price C_0 at which the contract is traded. Since a linear combination of payoffs is again a payoff function, we can construct a new contract as a portfolio of existing contracts.

A typical financial problem is the evaluation of the fair initial price of a contract with a given terminal payoff. Such an evaluation problem can be done by constructing a portfolio whose terminal value is equivalent to the payoff of the contract. This portfolio is said to *replicate* the payoff. Being given a replicating portfolio, the fair initial price of the contract is simply equal to the initial cost of setting up the portfolio. An important question that arises, and which we now answer, is as follows.

How do we *replicate* the payoff of a specified target financial contract with a portfolio in the two base assets B and S? In other words, how do we form an investment portfolio (x, y) whose terminal value coincides with the payoff of the contract in every state?

The claims that we wish to replicate are generally contingent (derivative) claims with uncertain payoff dependent on the outcome. In the two-state economy, any payoff C_T has two possible values, $C^+ \equiv C_T(\omega^+)$ and $C^- \equiv C_T(\omega^-)$. For a contingent payoff, $C^- \neq C^+$ holds; otherwise, when $C^- = C^+$, the payoff is said to be *deterministic*. To replicate a claim C_T with a portfolio in B and S, we must form a so-called *replicating portfolio* (x, y) such that $\Pi_T^{(x,y)}(\omega) = C_T(\omega)$ for both outcomes $\omega = \omega^+$ and $\omega = \omega^-$. The replication is equivalent to solving a system of two linear equations:

$$\begin{cases} xS^+ + yB_T = C^+, \\ xS^- + yB_T = C^-. \end{cases}$$

Since $S^+ \neq S^-$, this system has a unique solution:

$$x = \frac{C^+ - C^-}{S^+ - S^-}, \quad y = \frac{C^- S^+ - C^+ S^-}{B_T(S^+ - S^-)}. \tag{2.3}$$

This solution states that, given an arbitrary claim with payoff values $C_T(\omega^{\pm}) = C^{\pm}$, we can form a unique replicating portfolio (x, y) with x, y given by (2.3) where $\Pi_T^{(x,y)}(\omega^{\pm}) = C^{\pm}$. We can rewrite (2.3) in terms of the initial prices S_0 and B_0, the return on the bond, where $B_T = (1 + r_B)B_0$, and the return on the stock in the two states, where $S^{\pm} = (1 + r_S^{\pm})S_0$, as follows:

$$x = \frac{C^+ - C^-}{S_0(r_S^+ - r_S^-)}, \quad y = \frac{C^-(1 + r_S^+) - C^+(1 + r_S^-)}{B_0(1 + r_B)(r_S^+ - r_S^-)}. \tag{2.4}$$

As we now see, and as discussed later in Section 2.3 and more formally and mathematically in-depth in later chapters, replication is the key to fair pricing and valuation of derivative contracts. By replicating the exact payoff structure of a target contract, by means of a portfolio in tradable assets, we are arriving at the fair price of the contract, which is given by the initial cost of setting up the replicating portfolio: $C_0 = \Pi_0^{(x,y)}$. In particular, by substituting the above values for x, y, we can represent the initial value of the replicating portfolio, and hence the fair price C_0 of the derivative contract, in the following equivalent ways:

$$C_0 = \Pi_0^{(x,y)} = xS_0 + yB_0 = \frac{C^+ - C^-}{S^+ - S^-}S_0 + \frac{C^-S^+ - C^+S^-}{B_T(S^+ - S^-)}B_0$$

$$= \frac{B_0}{B_T}\left[\left(\frac{(B_T/B_0)S_0 - S^-}{S^+ - S^-}\right)C^+ + \left(\frac{S^+ - (B_T/B_0)S_0}{S^+ - S^-}\right)C^-\right]$$

$$= \frac{1}{1+r_B}\left[\left(\frac{(1+r_B)S_0 - S^-}{S^+ - S^-}\right)C^+ + \left(\frac{S^+ - (1+r_B)S_0}{S^+ - S^-}\right)C^-\right]$$

$$= \frac{1}{1+r_B}\left[\left(\frac{(1+r_B)S_0 - (1+r_S^-)S_0}{(r_S^+ - r_S^-)S_0}\right)C^+ + \left(\frac{(1+r_S^+)S_0 - (1+r_B)S_0}{(r_S^+ - r_S^-)S_0}\right)C^-\right]$$

$$= \frac{1}{1+r_B}\left[\left(\frac{r_B - r_S^-}{r_S^+ - r_S^-}\right)C^+ + \left(\frac{r_S^+ - r_B}{r_S^+ - r_S^-}\right)C^-\right]. \tag{2.5}$$

It is clear from (2.5) that the value of the replicating portfolio, and hence the initial fair price of the derivative contract, does not depend on the real-world probabilities of the outcomes $\omega^\pm$. Later we shall formally introduce the notions of arbitrage and risk-neutral probabilities, which will bring a more complete meaning to the result encapsulated in (2.5).

Example 2.1. Suppose $B_0 = \$100$, $B_T = \$110$, $S_0 = \$100$, $S^+ = \$120$, $S^- = \$90$.

(a) Determine a portfolio (x, y) whose value at time T is given by $\Pi_T(\omega^+) = \$930$ and $\Pi_T(\omega^-) = \$780$.

(b) Find the initial value Π_0 of the portfolio constructed in (a) and the rate of return, r, on the portfolio.

(c) Assuming that $\mathbb{P}(\omega^+) = \frac{2}{5}$, find the expected rate of return $E[r]$ and the expected terminal value $E[\Pi_T]$.

Solution.

(a) Apply (2.3) with $C^+ = \Pi_T(\omega^+) = \$930$, $C^- = \Pi_T(\omega^-) = \$780$ to obtain:

$$x = \frac{930 - 780}{120 - 90} = 5, \quad y = \frac{780 \cdot 120 - 930 \cdot 90}{110 \cdot (120 - 90)} = 3.$$

(b) From (2.5) we have $\Pi_0 = xS_0 + yB_0 = 5 \cdot 100 + 3 \cdot 100 = \$800$. Equation (2.2) gives $r(\omega^\pm) = \frac{\Pi_T(\omega^\pm) - \Pi_0}{\Pi_0}$, hence

$$r(\omega^+) = (930 - 800)/800 = 0.1625 = 16.25\%,$$
$$r(\omega^-) = (780 - 800)/800 = -0.025 = -2.5\%.$$

(c) The real-world expected return on the portfolio given that $\mathbb{P}(\omega^+) = p = \frac{2}{5}$ and $\mathbb{P}(\omega^-) = 1 - p = \frac{3}{5}$ is

$$E[r] = pr(\omega^+) + (1-p)r(\omega^-) = \frac{2}{5} \cdot 0.1625 + \frac{3}{5} \cdot (-0.025) = 0.05 = 5\%.$$

The expected terminal value is

$$\mathrm{E}[\Pi_T] = p\Pi_T(\omega^+) + (1-p)\Pi_T(\omega^-) = \frac{2}{5}\cdot 930 + \frac{3}{5}\cdot 780 = \$840\,.$$

This number is consistent with the fact that the expected return on the portfolio is

$$\mathrm{E}[r] = \mathrm{E}[\Pi_T/\Pi_0 - 1] = \mathrm{E}[\Pi_T]/\Pi_0 - 1,\ \text{ giving } \mathrm{E}[\Pi_T] = \Pi_0(1+\mathrm{E}[r])\,.$$

As a check: $840 = 800\cdot(1+0.05)$. $\square$

When dealing with a discrete-time financial model, we are interested in the fundamental characteristics of the model, such as whether or not the following statements hold:

(1) There exists a replicating portfolio for every derivative contract or claim.

(2) Every replicating portfolio is unique for a given claim.

(3) If there are different replicating portfolios for a given claim, then they all have the same initial cost.

(4) The initial cost of a portfolio replicating a *positive* claim C_T (i.e., $C_T \geqslant 0$ and $C_T(\omega) > 0$ for at least one $\omega \in \Omega$) is necessarily positive.

In the above simplest example of a two-state model with two base assets, we have already seen that Statements 1 and 2 hold; hence Statement 3 is irrelevant. Statement 4 can be guaranteed once we impose the extra condition that $r_S^- < r_B < r_S^+$, which is equivalent to requiring that there is no *arbitrage* in the model. The concept of arbitrage is discussed in detail in Section 2.3.

The replication method allows for constructing a portfolio that perfectly matches a given payoff. If someone writes and sells a derivative contract (and, as a result, takes a short position in the contract) and then uses the proceed to form an additional portfolio in the base assets that replicates the derivative's payoff, then the resulting portfolio is risk-free since the total payoff is zero. Indeed, if C_T is the payoff of the derivative, and Π_T is the value of the replicating portfolio, then the terminal value of the investments combining the derivative (a short position) and the replicating portfolio is $-C_T + \Pi_T = 0$.

In practice, it may be not always possible to eliminate the risk associated with a derivative contract. Let us discuss a more general approach to the minimization of such a risk. Consider an investor who has written a contract with payoff C_T. The investor takes a short position in the contract. Additionally, the investor includes x shares of the risky stock to *hedge* the short position in the contract. To finance the transactions, the investor adds y units of the risk-free bond. The time-t value of the *hedging* portfolio is

$$V_t = xS_t + yB_t - C_t\,.$$

What is the optimal value of x that minimizes the risk associated with this portfolio measured as $\mathrm{Var}(V_T)$? Let us find the hedging portfolio by minimizing $\mathrm{Var}(V_T)$ subject to $V_0 = 0$.

Calculate the variance of V_T, which is a sum of two random variables, namely, $x S_T$ and $-C_T$ and a non-stochastic quantity $y B_T$ as follows:

$$\begin{aligned}
\mathrm{Var}(V_T) &= \mathrm{Var}(x S_T + y B_T - C_T) \\
&= \mathrm{Var}(x S_T - C_T) = \mathrm{Var}(x S_T) - 2\,\mathrm{Cov}(x S_T, C_T) + \mathrm{Var}(C_T) \\
&= x^2\,\mathrm{Var}(S_T) - 2x\,\mathrm{Cov}(S_T, C_T) + \mathrm{Var}(C_T).
\end{aligned}$$

Differentiate the above expression w.r.t. x:

$$\frac{d}{dx} \text{Var}(V_T) = 2x \, \text{Var}(S_T) - 2x \, \text{Cov}(S_T, C_T).$$

The derivative is zero iff $x = \dfrac{\text{Cov}(S_T, C_T)}{\text{Var}(S_T)}$. Since $\frac{d^2}{dx^2} \text{Var}(V_T) = 2 \, \text{Var}(S_T) > 0$, the extreme point is a point of minimum. The minimum value of $\text{Var}(V_T)$ is equal to

$$\left(\frac{\text{Cov}(S_T, C_T)}{\text{Var}(S_T)}\right)^2 \text{Var}(S_T) - 2\left(\frac{\text{Cov}(S_T, C_T)}{\text{Var}(S_T)}\right) \text{Cov}(S_T, C_T) + \text{Var}(C_T)$$
$$= \text{Var}(C_T)\left[1 - \text{Corr}^2(S_T, C_T)\right].$$

For the single-period binomial model, we can show that

$$\text{Var}(S_T) = (S^+ - S^-)^2 p(1-p), \, \text{Var}(C_T) = (C^+ - C^-)^2 p(1-p),$$
$$\text{Cov}(S_T, C_T) = (S^+ - S^-)(C^+ - C^-)p(1-p), \quad \text{and} \quad \text{Corr}(S_T, C_T) = 1.$$

Therefore, the position x in the risky stock is $x = \dfrac{C^+ - C^-}{S^+ - S^-}$. It is exactly the same value as the number of stock shares in the replicating portfolio! Additionally, by taking such a position in the risky stock, we can entirely eliminate the risk since the variance of the resulting portfolio is zero. The condition $V_0 = 0$ implies that $y = \dfrac{C_0 - xS_0}{B_0}$.

2.2.1.1 A Single-Period Trinomial Model

As was demonstrated previously, every derivative can be replicated/hedged in the single-period binomial model. Let us modify this model by including an additional state of economy. Consider a market model with **three states** of the world, i.e., $\Omega = \{\omega^-, \omega^0, \omega^+\}$. We still have two assets, B and S. The risk-free asset B has a rate of return r_B. The risky asset S admits three possible cash flows at time T:

$$S_T(\omega) = \begin{cases} S_0 u & \text{if } \omega = \omega^+ \text{ with probability } p^+, \\ S_0 m & \text{if } \omega = \omega^0 \text{ with probability } p^0, \\ S_0 d & \text{if } \omega = \omega^- \text{ with probability } p^-, \end{cases}$$

where $S_0 > 0$ is the initial price and $0 < d < m < u$ are price factors. The probabilities are positive and sum up to one:

$$p^+, p^0, p^- > 0 \quad \text{and} \quad p^+ + p^0 + p^- = 1.$$

Figure 2.2 depicts this model. Let us see if we can construct a replicating portfolio for every payoff.

Example 2.2. Consider a trinomial model with $B_0 = 100$, $B_T = 110$, $S_0 = 100$, $d = 0.9$, $m = 1$, $u = 1.2$. Find a portfolio replicating the payoff

(a) $C_T(\omega^-) = 0$, $C_T(\omega^0) = 0$, $C_T(\omega^+) = 20$;

(b) $C_T(\omega^-) = 13$, $C_T(\omega^0) = 12$, $C_T(\omega^+) = 10$.

Solution. To find a replicating portfolio (x, y), we need to solve $xS_T + yB_T = C_T$ for x and y. In fact, it is a system of three equations since the model has three states.

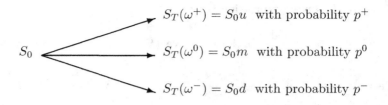

$$S_T(\omega^+) = S_0 u \ \text{ with probability } p^+$$

$$S_0 \qquad S_T(\omega^0) = S_0 m \ \text{ with probability } p^0$$

$$S_T(\omega^-) = S_0 d \ \text{ with probability } p^-$$

FIGURE 2.2: A schematic representation of the stock dynamics in a single-period trinomial model.

(a) The linear system
$$\begin{cases} 120x + 110y = 20 \\ 100x + 110y = 0 \\ 90x + 110y = 0 \end{cases}$$

does not have a solution since the rank of the coefficient matrix is not equal to the rank of the augmented matrix:

$$\text{rank} \begin{bmatrix} 120 & 110 \\ 100 & 110 \\ 90 & 110 \end{bmatrix} = 2 \neq 3 = \text{rank} \begin{bmatrix} 120 & 110 & 20 \\ 100 & 110 & 0 \\ 90 & 110 & 0 \end{bmatrix}.$$

Therefore, the payoff cannot be replicated.

(b) The system
$$\begin{cases} 120x + 110y = 10 \\ 100x + 110y = 12 \\ 90x + 110y = 13 \end{cases}$$

has a unique solution $x = -\frac{1}{10}$ and $y = \frac{2}{10}$. Therefore, the portfolio $(x, y) = (-\frac{1}{10}, \frac{1}{5})$ replicates the payoff. The initial value is $\Pi_0 = -\frac{1}{10} \cdot 100 + \frac{1}{5} \cdot 100 = 10$.

$\square$

As we can see, there exist contracts whose payoff cannot be replicated in the trinomial model. We say that the trinomial model is an *incomplete* asset price model. A model is said to be *complete* if every payoff can be replicated. That is, for every payoff C_T, there is a portfolio in base assets such that $\Pi_T = C_T$ for all market scenarios. The single-period binomial model is an example of a complete market model. Can the trinomial model be turned into a complete market model? Are there other complete and incomplete single-period models? These questions will be discussed in Chapter 5.

2.2.2 A Discrete-Time Model with a Finite Number of States

Let us enrich our single-period model by increasing the number of observation periods and adding more states. We assume that trading can take place at discretely monitored times $t = 0, 1, 2, \ldots, T$. That is, the time horizon T is an integer, and time is measured in periods. One observation period may correspond to one year, one week, one day, or even one second. The state space Ω is finite and contains M states of the world $\omega^1, \omega^2, \ldots, \omega^M$. Consider an asset such as stock ($A = S$) or bond ($A = B$) with the price A_t monitored at times $t = 0, 1, 2, \ldots, T$. Assume that $A_t > 0$ for all t. At the initial time $t = 0$, the current price A_0 is assumed to be known. The future prices A_t for $t > 0$ remain uncertain until information about the market state is revealed. So, mathematically, for every fixed t, price A_t is a positive random variable on a finite probability space.

2.2.2.1 Asset Returns

Consider an asset A (or a portfolio of assets) with the price process $\{A_t\}_{t=0,1,\ldots,T}$. The total return and the rate of return on the asset A over a time interval $[s, t]$ with $0 \leqslant s < t$, respectively denoted by $R^A_{[s,t]}$ and $r^A_{[s,t]}$, are random variables defined by

$$R^A_{[s,t]} := \frac{A_t}{A_s} \quad \text{and} \quad r^A_{[s,t]} := \frac{A_t - A_s}{A_s},$$

respectively. They are related by $R^A = 1 + r^A$. Often, for simplicity, the term *return* is used for both notions. Returns on risky assets are uncertain, and, therefore, they are functions of $\omega \in \Omega$, e.g., $R^A_{[s,t]}(\omega) = \frac{A_t(\omega)}{A_s(\omega)}$. For returns on asset A over a single period $[t-1, t]$ where $t = 1, 2, \ldots, T$, we use notations $R^A_t := R^A_{[t-1,t]}$ and $r^A_t := r^A_{[t-1,t]}$. For simplicity and when the context is clear, we will omit the superscript A and will denote the total return by R and the rate of return by r.

Now the asset prices can be written in terms of single period returns on the stock. For every $t = 1, 2, \ldots, T$, we have that

$$r_t = R_t - 1 = \frac{A_t - A_{t-1}}{A_{t-1}} = \frac{A_t}{A_{t-1}} - 1 \implies A_t = R_t A_{t-1} \quad \text{and} \quad A_t = (1 + r_t) A_{t-1}.$$

By applying successively this rule to $A_{t-1}, A_{t-2}, \ldots, A_1$, we obtain

$$A_t = (1 + r_t) A_{t-1} = (1 + r_{t-1})(1 + r_t) A_{t-2} = \cdots$$
$$= (1 + r_1)(1 + r_2) \cdots (1 + r_t) A_0.$$

Equivalently, we have $A_t = R_1 R_2 \cdots R_t A_0$ for $t = 0, 1, \ldots, T$. Therefore, the dynamics of asset prices can also be described by single-period asset returns and initial price A_0. Note that the aggregate returns $r_{[s,t]} = \frac{A_t - A_s}{A_s}$ and $R_{[s,t]} = \frac{A_t}{A_s}$ on asset A from time s to time t with $0 \leqslant s < t \leqslant T$ respectively satisfy

$$1 + r_{[s,t]} = (1 + r_{s+1})(1 + r_{s+2}) \cdots (1 + r_t) \quad \text{and} \quad R_{[s,t]} = R_{s+1} R_{s+2} \cdots R_t. \tag{2.6}$$

Example 2.3. Consider a market that assumes three possible scenarios: $\Omega = \{\omega^1, \omega^2, \omega^3\}$. Suppose that stock S with $S_0 = 100$ takes on the following values over a two-period interval:

Scenario, ω	S_1	S_2
ω^1 (boom)	120	150
ω^2 (stability)	105	100
ω^3 (recession)	80	60

Find the returns r_1, r_2 and compare them with $r_{[0,2]}$.

Solution. The one-period returns $r_1 = \frac{S_1 - S_0}{S_0}$ and $r_2 = \frac{S_2 - S_1}{S_1}$ and the two-period return $r_{[0,2]} = \frac{S_2 - S_0}{S_0}$ have the following values:

Scenario	r_1	r_2	$r_{[0,2]}$
ω^1	20%	25%	50%
ω^2	5%	−4.76%	0%
ω^3	−20%	−25%	−40%

As is seen from the table above, the returns satisfy (2.6), which takes the following form for a two-period model: $1 + r_{[0,2]} = (1 + r_1)(1 + r_2)$. $\qquad\qquad\qquad\qquad\qquad\qquad\qquad\qquad$ □

The returns on a risky asset are random variables. Expected returns can be calculated when the scenario probabilities are given. Suppose that the probabilities $p_k := \mathbb{P}(\omega^k)$ are known for every $\omega^k \in \Omega$. Suppose that the probabilities p_k are strictly positive and sum up to one:

$$p_1, p_2, \ldots, p_M > 0, \quad p_1 + p_2 + \cdots + p_M = 1.$$

For any given scenario ω^k, $k = 1, 2, \ldots, M$, the return $r_{[s,t]}(\omega^k)$ is known. The expected return for the period $[s,t]$ with $0 \leqslant s < t \leqslant T$ can be calculated as follows:

$$\mathrm{E}[r_{[s,t]}] = r_{[s,t]}(\omega^1) \cdot p_1 + r_{[s,t]}(\omega^2) \cdot p_2 + \cdots + r_{[s,t]}(\omega^M) \cdot p_M. \tag{2.7}$$

If the one-period returns $r_{s+1}, r_{s+2}, \ldots, r_t$ are *independent* random variables, then the expected aggregate return can be expressed in terms of expected one-period returns as follows:

$$1 + \mathrm{E}[r_{[s,t]}] = \left(1 + \mathrm{E}[r_{s+1}]\right)\left(1 + \mathrm{E}[r_{s+2}]\right) \cdots \left(1 + \mathrm{E}[r_t]\right).$$

Suppose that the one-period returns r_t, $t \geqslant 1$, are independent and identically distributed (i.i.d.) random variables with the common expected value of a one-period return, $r_A := \mathrm{E}[r_1]$. Then, we obtain

$$1 + \mathrm{E}\big[r_{[s,t]}\big] = \left(1 + \mathrm{E}[r_1]\right)^{t-s} = (1 + r_A)^{t-s}.$$

Since $A_t = (1 + r_{[0,t]})\, A_0$ and A_0 is certain, the expected asset price at time t is

$$\mathrm{E}[A_t] = (1 + \mathrm{E}[r_{[0,t]}])\, A_0 = (1 + r_A)^t\, A_0. \tag{2.8}$$

Note that this expression is very similar to the formula of the accumulated value under periodic compounding of interest. In practice, it may be difficult to estimate probability distributions of returns. What can be easily computed from historical data is an average return over a certain period. As a result, one can estimate expected values of future cash flows.

The *log-return* on asset A over a time interval $[s,t]$, denoted $L^A_{[s,t]}$, is given by

$$L^A_{[s,t]} := \ln\left(\frac{A_t}{A_s}\right) = \ln R^A_{[s,t]} = \ln(1 + r^A_{[s,t]}).$$

A single-period log-return, denoted L^A_t, is

$$L^A_t := L^A_{[t-1,t]}, \quad t = 1, 2, \ldots, T.$$

Since $R^A_{[s,t]} = R^A_{s+1}\, R^A_{s+2} \cdots R^A_t$, we obtain that the log-returns are additive:

$$L^A_{[s,t]} = L^A_{s+1} + L^A_{s+2} + \cdots + L^A_t.$$

The bond prices B_t, $t \geqslant 0$, are nonrandom (deterministic). In other words, they do not depend on the world state: $B_t(\omega) = B_t$ for all $\omega \in \Omega$ and all $t \geqslant 0$. The returns on the bond $r_t = \frac{B_t - B_{t-1}}{B_{t-1}}$ with $t \geqslant 1$ are deterministic as well. We can express the bond price B_t in terms of the initial price B_0 and one-period returns as follows:

$$B_t = B_0(1 + r_1)(1 + r_2) \cdots (1 + r_t) \quad \text{for } t \geqslant 1.$$

Assuming that the one-period returns r_t all have the same constant value r_B, we arrive at a formula that is analogous to (2.8):

$$B_t = B_0(1 + r_B)^t. \tag{2.9}$$

Equations (2.8) and (2.9) enlighten us on how to compare performances of a risky asset and a risk-free asset—we can simply compare the expected one-period returns on assets of interest.

In summary, to construct a discrete-time financial model we need to know the initial price and the probability distribution of one-period returns for each base asset. For a model with a finite number of states, all returns are discrete random variables defined on a common probability space. Usually, we can distinguish one underlying whose returns are certain, e.g., a risk-free bond. A more detailed analysis of single-period models and multi-period models will be respectively done in later chapters. In the next section, we present the binomial tree model, which is the simplest example of a multi-period model.

2.2.3   Introducing the Binomial Tree Model

In this subsection, we introduce a discrete-time model with one risky stock S and one risk-free asset B such as a bond (or a bank account). Assume that there are T periods, and the time is $t = 0, 1, 2, \ldots, T$. The one-period return on the risk-free asset is denoted by $r \equiv r_B > 0$. In fact, r is a risk-free interest rate compounded periodically. The bond prices are $B_t = B_0(1+r)^t$, $t \geqslant 1$. Assume that the one-period returns on the stock, $R_t \equiv R_t^S$ with $t = 1, 2, \ldots, T$, are i.i.d. random variables such that

$$R_t = \begin{cases} u & \text{with probability } p, \\ d & \text{with probability } 1 - p, \end{cases} \tag{2.10}$$

where the factors d and u satisfy the condition $0 < d < u$, and $p \in (0,1)$ is the real-world probability that the stock price moves up. Notice that the average one-period return on the stock is given by

$$\mathrm{E}[R_t] = p\,u + (1 - p)\,d\,.$$

Since $R_t = S_t/S_{t-1}$, the stock price S_t at time t is expressed in terms of the price S_{t-1} at the previous time moment and the return R_t as follows:

$$S_t = R_t\,S_{t-1} = \begin{cases} S_{t-1}u & \text{with probability } p, \\ S_{t-1}d & \text{with probability } 1 - p. \end{cases} \tag{2.11}$$

In other words, at each time t the stock price S_t can move up by a factor u or down by a factor d (relative to S_{t-1}). The inequality $d > 0$ guarantees the positiveness of stock prices, i.e., $S_t > 0$ for all $t \geqslant 1$, provided that $S_0 > 0$. Equivalently, we may work with rates of return, $r_t = R_t - 1$. Equations (2.10)–(2.11) take the form

$$S_t = (1 + r_t)\,S_{t-1}, \quad \text{where} \quad r_t = \begin{cases} u - 1 & \text{with probability } p, \\ d - 1 & \text{with probability } 1 - p. \end{cases}$$

Iterating equation (2.11) allows us to express the stock price S_n (with $n \geqslant 1$) in terms of the returns $R_1, R_2, \ldots, R_n$ as follows:

$$S_1 = S_0 R_1,$$
$$S_2 = S_1 R_2 = S_0 R_1 R_2,$$
$$S_3 = S_2 R_3 = S_0 R_1 R_2 R_3,$$
$$\vdots$$
$$S_n = S_0 R_1 R_2 \cdots R_n. \tag{2.12}$$

Our goal is to find the probability distribution of S_n for any $n \geqslant 1$. We start with representing (2.11) in the following form:

$$S_t = u^{X_t} d^{1-X_t} S_{t-1}, \quad \text{where} \quad X_t = \begin{cases} 1 & \text{with probability } p, \\ 0 & \text{with probability } 1 - p. \end{cases} \tag{2.13}$$

Indeed, if $X_t = 1$, then $S_t = u^1 d^0 S_{t-1} = u S_{t-1}$; if $X_t = 0$, then $S_t = u^0 d^1 S_{t-1} = d S_{t-1}$;

Notice that X_t is a Bernoulli random variable. For each t, the return R_t can be written as a function of X_t: $R_t = u^{X_t} d^{1-X_t}$. Equivalently, we can solve the above equation for X_t to obtain $X_t = \frac{\ln R_t - \ln d}{\ln u - \ln d}$. If the returns $\{R_t\}$ are mutually independent, the variables $\{X_t\}$ are mutually independent as well. Therefore, under our assumptions about stock returns, $\{X_t\}_{t \geqslant 1}$ is a collection of i.i.d. Bernoulli random variables with probability p of success.

Iterating equation (2.13) gives

$$S_1 = S_0 u^{X_1} d^{1-X_1},$$
$$S_2 = S_1 u^{X_2} d^{1-X_2} = S_0 u^{X_1+X_2} d^{2-(X_1+X_2)},$$
$$S_3 = S_2 u^{X_3} d^{1-X_3} = S_0 u^{X_1+X_2+X_3} d^{3-(X_1+X_2+X_3)},$$
$$\vdots$$
$$S_n = S_0 u^{X_1+X_2+\cdots+X_n} d^{n-(X_1+X_2+\cdots+X_n)}. \tag{2.14}$$

Comparing (2.12) and (2.14) gives us the following formula for the aggregate return $R_{[0,n]}$:

$$R_{[0,n]} = R_1 R_2 \cdots R_n = u^{X_1+X_2+\cdots+X_n} d^{n-(X_1+X_2+\cdots+X_n)}.$$

The expression (2.14) gives the stock price at time $t = n$ in terms of its initial price and the cumulative sum $X_1 + X_2 + \cdots + X_n$. Let us denote the latter by Y_n. Recall that a sum of n i.i.d. Bernoulli random variables can be interpreted as the number of successes in n independent trials with probability p of success on each trial. It has the binomial distribution; the probability mass function of $Y_n \sim Bin(n, p)$ is

$$b(k; n, p) = \mathbb{P}(Y_n = k) = \binom{n}{k} p^k (1-p)^{n-k}, \quad k = 0, 1, 2, \ldots, n.$$

Thus, at time $t = n$ we have the following formula for stock prices:

$$S_n = S_0 u^{Y_n} d^{n-Y_n}, \quad \text{where} \quad Y_n \sim Bin(n, p).$$

Since the range of Y_n is the set $\{0, 1, \ldots, n\}$, the stock price S_n can only take on a value in the set

$$\left\{ S_{n,k} := S_0 u^k d^{n-k} \; : \; k = 0, 1, \ldots, n \right\}.$$

By the equivalence of the events $\{S_n = S_{n,k}\}$ and $\{Y_n = k\}$, the probability distribution of S_n is then

$$\mathbb{P}(S_n = S_{n,k}) = \mathbb{P}(Y_n = k) = \binom{n}{k}p^k(1-p)^{n-k}, \quad k = 0, 1, \ldots, n. \tag{2.15}$$

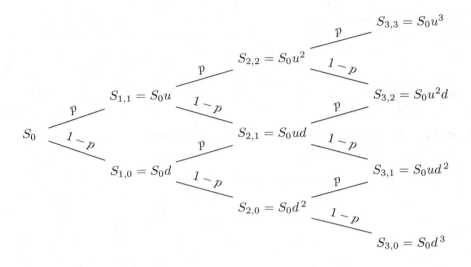

FIGURE 2.3: A schematic representation of the binomial lattice with three periods.

Equation (2.15) tells us that the stock price S_n at time $t = n$ can admit $n+1$ different values, i.e., the values $S_{n,k}$ with $k = 0, 1, \ldots, n$. There are $\binom{n}{k}$ different scenarios (n-step price paths) with exactly k upward and $n - k$ downward price moves that produce the same (terminal) stock price $S_{n,k}$ at time n. The set Ω of all the possible scenarios can be compactly represented by the n-step *recombining binomial tree* or *binomial lattice* (see Figure 2.3). Each n-step path or scenario leading to $S_{n,k}$ has equal probability $p^k(1-p)^{n-k}$ of occurring. Every such path starts at S_0, moves up a total of k times and down a total of $n - k$ times. Since there are $\binom{n}{k}$ paths leading to the same value $S_{n,k}$, then summing this over values $k = 0, 1, \ldots, n$ must give the total number of all possible paths, which is 2^n. This calculation, of course, corresponds to the binomial formula:

$$2^n = (1+1)^n = \sum_{k=0}^{n}\binom{n}{k} = \sum_{k=0}^{n}\left\{\begin{array}{l}\text{Number of scenarios with } k \text{ upward} \\ \text{and } n - k \text{ downward price moves}\end{array}\right\}.$$

A recombining binomial tree with n periods has 2^n distinct paths of length n. Each of them represents a possible state of the world in the model. For example, a three-period binomial tree has eight market scenarios listed in the table below.

Scenario, ω	Path in the tree	$\mathbb{P}(\omega)$	S_1	S_2	S_3
ω^1	Down-Down-Down	$(1-p)^3$	$S_{1,0} = S_0 d$	$S_{2,0} = S_0 d^2$	$S_{3,0} = S_0 d^3$
ω^2	Down-Down-Up	$p(1-p)^2$	$S_{1,0} = S_0 d$	$S_{2,0} = S_0 d^2$	$S_{3,1} = S_0 u d^2$
ω^3	Down-Up-Down	$p(1-p)^2$	$S_{1,0} = S_0 d$	$S_{2,1} = S_0 u d$	$S_{3,1} = S_0 u d^2$
ω^4	Down-Up-Up	$p^2(1-p)$	$S_{1,0} = S_0 d$	$S_{2,1} = S_0 u d$	$S_{3,2} = S_0 u^2 d$
ω^5	Up-Down-Down	$p(1-p)^2$	$S_{1,1} = S_0 u$	$S_{2,1} = S_0 u d$	$S_{3,1} = S_0 u d^2$
ω^6	Up-Down-Up	$p^2(1-p)$	$S_{1,1} = S_0 u$	$S_{2,1} = S_0 u d$	$S_{3,2} = S_0 u^2 d$
ω^7	Up-Up-Down	$p(1-p)^2$	$S_{1,1} = S_0 u$	$S_{2,2} = S_0 u^2$	$S_{3,2} = S_0 u^2 d$
ω^8	Up-Up-Up	p^3	$S_{1,1} = S_0 u$	$S_{2,2} = S_0 u^2$	$S_{3,3} = S_0 u^3$

Example 2.4. Construct a three-period binomial tree model with $S_0 = \$100$, $u = 1.2$, $d = 0.8$, $p = 60\%$. After that, find the probability distribution of S_3 and compute $\mathrm{E}[S_3]$.

Solution. First, let us construct the three-period binomial tree.

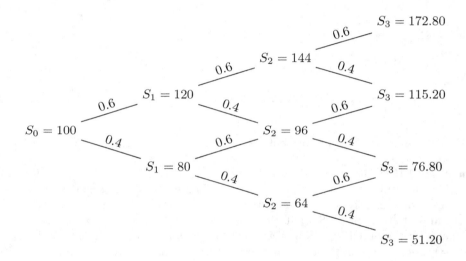

The probability distribution of S_3 is as follows:

k	$S_{3,k} = S_0\, u^k\, d^{\,3-k}$	$p_{S_3}(S_{3,k}) = \mathbb{P}(S_3 = S_{3,k}) = \binom{3}{k} u^k d^{3-k}$
0	$100 \cdot 1.2^0 \cdot 0.8^3 = 51.20$	$\binom{3}{0} \cdot 0.6^0 \cdot 0.4^3 = 21.6\%$
1	$100 \cdot 1.2^1 \cdot 0.8^2 = 76.80$	$\binom{3}{1} \cdot 0.6^1 \cdot 0.4^2 = 43.2\%$
2	$100 \cdot 1.2^2 \cdot 0.8^1 = 115.20$	$\binom{3}{2} \cdot 0.6^2 \cdot 0.4^1 = 28.8\%$
3	$100 \cdot 1.2^3 \cdot 0.8^0 = 172.80$	$\binom{3}{3} \cdot 0.6^3 \cdot 0.4^0 = 6.4\%$

We can find the expected value of S_3 by using its PMF:

$$E[S_3] = \sum_s s\, p_{S_3}(s) = \sum_{k=0}^{3} S_{3,k}\, p_{S_3}(S_{3,k}) = \sum_{k=0}^{3} \left(S_0\, u^k\, d^{3-k}\right) \binom{3}{k} u^k\, d^{3-k}$$

$$= 51.20 \cdot 0.216 + 76.80 \cdot 0.432 + 115.20 \cdot 0.288 + 172.80 \cdot 0.064 \cong \$112.49.$$

Alternatively, we can use the property that $E[S_3] = S_0\, (E[R_S])^3$ where

$$E[R_S] = u\,p + d\,(1-p) = 1.2 \cdot 0.6 + 0.8 \cdot 0.4 = 1.04.$$

Thus, $E[S_3] = 100 \cdot (1.04)^3 = 112.4864 \cong \$112.49.$ □

For the sake of comparison, let us consider a two-step binomial tree model under three different assumptions on the probability distributions of returns.

Example 2.5. Let $S_0 = \$100$. Find the distributions of prices S_1 and S_2 assuming one of the following.

(a) The returns R_1 and R_2 are i.i.d. and $R_t = \begin{cases} 1.1 & \text{with prob. } 1/2, \\ 0.9 & \text{with prob. } 1/2. \end{cases}$

(b) The returns R_1 and R_2 are independent, but not identically distributed:

$$R_1 = \begin{cases} 1.1 & \text{with prob. } 1/2, \\ 0.9 & \text{with prob. } 1/2; \end{cases} \qquad R_2 = \begin{cases} 1.15 & \text{with prob. } 1/3, \\ 0.9 & \text{with prob. } 2/3. \end{cases}$$

(c) The returns R_1 and R_2 are not independent and not identically distributed:

$$R_1 = \begin{cases} 1.1 & \text{with prob. } 1/2, \\ 0.9 & \text{with prob. } 1/2; \end{cases}$$

$$R_2|\{R_1 = 1.1\} = \begin{cases} 1.15 & \text{with prob. } 1/3, \\ 0.9 & \text{with prob. } 2/3; \end{cases}$$

$$R_2|\{R_1 = 0.9\} = \begin{cases} 1.1 & \text{with prob. } 1/4, \\ 0.85 & \text{with prob. } 3/4. \end{cases}$$

Here, $R_2|\{R_1 = a\}$ represents values of R_2 given that $R_1 = a$.

Solution. Since the probability distribution of R_1 is the same for all three cases, the distribution of S_1 is the same as well:

$$S_1 = S_0\, R_1 = \begin{cases} 100 \cdot 1.1 & \text{with prob. } 1/2 \\ 100 \cdot 0.9 & \text{with prob. } 1/2 \end{cases} = \begin{cases} 110 & \text{with prob. } 1/2 \\ 90 & \text{with prob. } 1/2. \end{cases}$$

The stock price at the end of period 2 is $S_2 = S_1\, R_2$. So we find the probability distribution of S_2 for each of the three cases using on the values of R_2 and S_1.

(a) We have a two-period binomial tree model discussed just above with parameters $u = 1.1$, $d = 0.9$, and $p = 1/2$. Applying (2.15) gives

$$S_2 = S_{2,2} = 100 \cdot 1.1^2 = 121 \qquad \text{with probability } \binom{2}{0} 0.5^2 = 0.25,$$

$$S_2 = S_{2,1} = 100 \cdot 1.1 \cdot 0.9 = 99 \qquad \text{with probability } \binom{2}{1} 0.5^2 = 0.5,$$

$$S_2 = S_{2,0} = 100 \cdot 0.9^2 = 81 \qquad \text{with probability } \binom{2}{2} 0.5^2 = 0.25.$$

(b) Given that $S_1 = 110$, the price $S_2 = 110\,R_2$ takes one of two values: $110 \cdot 1.15 = 126.5$ or $110 \cdot 0.9 = 99$. If $S_1 = 90$, then $S_2 = 90\,R_2$ is either $90 \cdot 1.15 = 103.5$ or $90 \cdot 0.9 = 81$. So there are four distinct values for S_2. We can compute the probabilities by using the independence of the returns R_1 and R_2, i.e., using that $P(R_1 = x, R_2 = y) = \mathbb{P}(R_1 = x) \cdot \mathbb{P}(R_2 = y)$ for all x and y (hence, S_1 and R_2 are independent random variables):

$$\mathbb{P}(S_2 = 126.5) = \mathbb{P}(R_1 = 1.1, R_2 = 1.15) = \mathbb{P}(R_1 = 1.1) \cdot \mathbb{P}(R_2 = 1.15) = \frac{1}{2} \cdot \frac{1}{3} = \frac{1}{6},$$

$$\mathbb{P}(S_2 = 103.5) = \mathbb{P}(R_1 = 0.9, R_2 = 1.15) = \mathbb{P}(R_1 = 0.9) \cdot \mathbb{P}(R_2 = 1.15) = \frac{1}{2} \cdot \frac{1}{3} = \frac{1}{6},$$

$$\mathbb{P}(S_2 = 99) = \mathbb{P}(R_1 = 1.1, R_2 = 0.9) = \mathbb{P}(R_1 = 1.1) \cdot \mathbb{P}(R_2 = 0.9) = \frac{1}{2} \cdot \frac{2}{3} = \frac{1}{3},$$

$$\mathbb{P}(S_2 = 81) = \mathbb{P}(R_1 = 0.9, R_2 = 0.9) = \mathbb{P}(R_1 = 0.9) \cdot \mathbb{P}(R_2 = 0.9) = \frac{1}{2} \cdot \frac{2}{3} = \frac{1}{3}.$$

(c) If $R_1 = 1.1$ (and hence $S_1 = 110$), then $S_2 = 110 \cdot R_2$ takes on the value $110 \cdot 1.15 = 126.5$ or $110 \cdot 0.9 = 99$. If $R_1 = 1.1$ (and hence $S_1 = 90$), then $S_2 = 90 \cdot R_2$ is either $90 \cdot 1.1 = 99$ or $90 \cdot 0.85 = 76.5$. The probability distribution of S_2 is given by conditioning on R_1:

$$\mathbb{P}(S_2 = 126.5) = \mathbb{P}(R_1 = 1.1, R_2 = 1.15)$$
$$= \mathbb{P}(R_2 = 1.15 \mid R_1 = 1.1) \cdot \mathbb{P}(R_1 = 1.1) = \frac{1}{3} \cdot \frac{1}{2} = \frac{1}{6},$$
$$\mathbb{P}(S_2 = 99) = \mathbb{P}(R_1 = 0.9, R_2 = 1.1) + \mathbb{P}(R_1 = 1.1, R_2 = 0.9)$$
$$= \mathbb{P}(R_2 = 1.1 \mid R_1 = 0.9) \cdot \mathbb{P}(R_1 = 0.9)$$
$$+ \mathbb{P}(R_2 = 0.9 \mid R_1 = 1.1) \cdot \mathbb{P}(R_1 = 1.1) = \frac{1}{4} \cdot \frac{1}{2} + \frac{2}{3} \cdot \frac{1}{2} = \frac{11}{24},$$
$$\mathbb{P}(S_2 = 76.5) = \mathbb{P}(R_1 = 0.9, R_2 = 0.85)$$
$$= \mathbb{P}(R_2 = 0.85 \mid R_1 = 0.9) \cdot \mathbb{P}(R_1 = 0.9) = \frac{3}{4} \cdot \frac{1}{2} = \frac{3}{8}. \qquad \square$$

Example 2.6. Suppose that stock prices follow a binomial tree with $S_0 = \$100$. Find the factors d and u if (a) $S_1 \in \{\$120, \$90\}$; (b) $\min_\omega S_2(\omega) = \$64$ and $\max_\omega S_2(\omega) = \$121$. These are simple examples of calibration of a two-period binomial model.

Solution.

(a) In one period, the stock prices are $S^+ = S_0 u = 100u = 120$ and $S^- = S_0 d = 100d = 90$. Solve these equations for d and u to obtain $d = 0.9$ and $u = 1.2$.

(b) In two periods:

$$\min_\omega S(2, \omega) = S_0 \min_{0 \leqslant k \leqslant 2} u^k d^{2-k} = S_0 \min\{d^2, ud, u^2\} = S_0 d^2 = 100 d^2$$

and

$$\max_\omega S(2, \omega) = S_0 \max_{0 \leqslant k \leqslant 2} u^k d^{2-k} = S_0 \max\{d^2, ud, u^2\} = S_0 u^2 = 100 u^2.$$

Hence, $u = \sqrt{1.21} = 1.1$ and $d = \sqrt{0.64} = 0.8$. $\qquad \square$

2.2.4   Recursive Construction of a Binomial Tree

In this subsection we discuss a recursive construction of a binomial tree with asset prices and probabilities. This method can be used for constructing a binomial tree model on a spreadsheet.

A binomial tree with T periods can be described by two upper-triangular matrices with dimensions $(T+1) \times (T+1)$. The first matrix contains asset prices, whereas the other one consists of respective probabilities:

$$\begin{bmatrix} S_{0,0} & 0 & 0 & \cdots & 0 \\ S_{1,0} & S_{1,1} & 0 & \cdots & 0 \\ S_{2,0} & S_{2,1} & S_{2,2} & \cdots & 0 \\ \vdots & \vdots & \vdots & \ddots & \vdots \\ S_{T,0} & S_{T,1} & S_{T,2} & \cdots & S_{T,T} \end{bmatrix} \text{ and } \begin{bmatrix} P_{0,0} & 0 & 0 & \cdots & 0 \\ P_{1,0} & P_{1,1} & 0 & \cdots & 0 \\ P_{2,0} & P_{2,1} & P_{2,2} & \cdots & 0 \\ \vdots & \vdots & \vdots & \ddots & \vdots \\ P_{T,0} & P_{T,1} & P_{T,2} & \cdots & P_{T,T} \end{bmatrix}$$

where $S_{n,k} := S_0 u^k d^{n-k}$ and $P_{n,k} = \binom{n}{k} p^k (1-p)^{n-k}$ with $0 \leqslant k \leqslant n \leqslant T$. If $n = 0$, then $S_{0,0} = S_0$ and $P_{0,0} = 1$. For any $n = 1, 2, \ldots, T$, we have that $\mathbb{P}(S_n = S_{n,k}) = P_{n,k}$.

Stock prices satisfy the following recurrence:

$$S_{n+1,k} = S_{n,k}\, d \ \text{ for } k = 0, 1, \ldots, n, \ \text{ and } \ S_{n+1,n+1} = S_{n,n}\, u.$$

Alternatively, we can write

$$S_{n+1,k+1} = S_{n,k}\, u \ \text{ for } k = 0, 1, \ldots, n, \ \text{ and } \ S_{n+1,0} = S_{n,0}\, d.$$

Using the above relations, we can build the matrix with asset prices row by row.

The probabilities also satisfy a recurrence, where $P_{n+1,k}$ is expressed in terms of $P_{n,k-1}$ and $P_{n,k}$. This recursive relation, which is stated in the proposition below, can be used to iteratively construct the matrix with probabilities.

Proposition 2.1. *Let us denote* $P_{n,k} = \dbinom{n}{k} p^k (1-p)^{n-k}$ *for* $k = 0, 1, \ldots, n$ *and* $P_{n,k} = 0$ *for all* $k < 0$ *and* $k > n$. *Then, the following recurrence holds for all* $n \geqslant 0$ *and* $k \geqslant 0$:

$$P_{n+1,k} = p\, P_{n,k-1} + (1-p)\, P_{n,k}. \tag{2.16}$$

Proof. A combinatorial proof of this proposition is left as an exercise for the reader. Equation (2.16) can also be derived using a probabilistic argument. The node $(n+1, k)$ of a binomial tree can only be reached by descending from the node (n, k) with probability $1-p$ of by ascending from the node $(n, k-1)$ with probability p. Therefore, the probability $P_{n+1,k}$ of being at the node $(n+1, k)$ is a weighted sum of probabilities to be at the nodes (n, k) and $(n, k-1)$, respectively. $\square$

For the two boundary cases, where, respectively, $k = 0$ and $k = n+1$, the formula (2.16) has the form:

$$P_{n+1,0} = (1-p)\, P_{n,0}, \quad P_{n+1,n+1} = p\, P_{n,n}.$$

2.2.5 Self-Financing Investment Strategies in the Binomial Model

Here we give a brief introduction to the concept of a self-financing trading strategy within the binomial model. Recall that a static portfolio (x, y) in the two base assets is an investment consisting of fixed x positions in the stock (the risky asset) and fixed y positions in the bond or money market account (the risk-free asset). For such a portfolio there is no trading, or re-balancing, of the positions over all periods. In contrast, in the binomial model trading is allowed to take place at the beginning of each period so that the positions held in the two base assets are generally not static over time. That is, within each period the positions are allowed to be *re-balanced* at the beginning of the period and are subsequently held fixed until the end of the period. The model allows for trading at the discrete times $t = 0, 1, \ldots, T$.

Imagine that an investor begins by setting up a portfolio (x_0, y_0) at time $t = 0$ and holds it until the end of the first period. At time $t = 1$, trading is allowed and the investor re-balances the portfolio positions to (x_1, y_1), and this new portfolio is held to the end of the second period at time $t = 2$. At time $t = 2$, trading takes place and the investor again re-balances the positions to (x_2, y_2) and holds this portfolio until time $t = 3$, and so on until maturity time $t = T$. The sequence of portfolios $(x_0, y_0), (x_1, y_1), \ldots, (x_{T-1}, y_{T-1})$ forms what is called a two-asset *trading*, or *investment*, or *portfolio strategy* in the binomial model with maturity time T.

At each time t, *after* re-balancing, the portfolio value is

$$\Pi_t = x_t S_t + y_t B_t.$$

During the period from time t to $t + 1$ the portfolio positions are held static and the new portfolio value, *before* re-balancing, at time $t + 1$ is

$$\Pi_{t+1} = x_t S_{t+1} + y_t B_{t+1}.$$

Note that the change in value for the one period is due solely to the change in prices of the base assets:

$$\Pi_{t+1} - \Pi_t = x_t(S_{t+1} - S_t) + y_t(B_{t+1} - B_t). \tag{2.17}$$

The first term on the right of this equation gives the change in portfolio value due to the change in the share price of the stock. Assuming the number of shares x_t is positive (i.e., long the stock), then $x_t(S_{t+1} - S_t)$ corresponds to a gain (or loss) if the share price increases (or decreases) whereas the opposite is true if x_t is negative (short the stock). The second term gives the change in value due to the bond component. If $y_t > 0$, then $y_t(B_{t+1} - B_t) > 0$ is a gain or earning based on the interest paid by the bond or a bank account during the period. If $y_t < 0$, the position in the bond is negative and this gives a loss corresponding to borrowing money from a bank account. At time $t + 1$, the portfolio value *after re-balancing* must be

$$\Pi_{t+1} = x_{t+1} S_{t+1} + y_{t+1} B_{t+1}.$$

At this point, we bring in the important notion of self-financing within the above strategy. So far, we allowed for the positions to be re-balanced at each time step without any restriction. Of course, any decision made by the investor about when to change the asset positions and what assets to sell or to buy is only based on the historical information about the market currently available. Generally, the investor may alter the positions at any time by selling some assets and investing the proceeds in others. However, we now enforce the condition that there cannot be any consumption or injection of funds within the strategy at any time past initial time $t_0 = 0$. In other words, after initially setting up the port-folio, we only allow for self-financed strategies. Hence, at each time t, the portfolio value before re-balancing the positions must be the same as its value after re-balancing. The above two expressions for Π_{t+1} must, therefore, be equal, and this gives rise to the so-called *self-financing condition (s.f.c.)*:

$$S_{t+1}(x_{t+1} - x_t) + B_{t+1}(y_{t+1} - y_t) = 0 \tag{2.18}$$

for all $0 \leqslant t \leqslant T - 1$. The first term corresponds to the cost (at time $t + 1$) of re-balancing in the stock with share price S_{t+1} and position change $\delta x_t := x_{t+1} - x_t$. The second term gives the re-balancing cost (at time $t + 1$) for the bond holdings where the position change is $\delta y_t := y_{t+1} - y_t$. Note that (i) $\delta x_t < 0$ iff $\delta y_t > 0$ and (ii) $\delta x_t > 0$ iff $\delta y_t < 0$. In case (i) re-balancing involves selling $|\delta x_t|$ stock shares at price S_{t+1} and investing the proceeds in buying δy_t bonds at price B_{t+1}. In case (ii) re-balancing involves selling $|\delta y_t|$ bonds at price B_{t+1} and investing this amount in buying δx_t stock shares at price S_{t+1}.

Let us denote the one-period changes in the portfolio value and asset prices by

$$\delta\Pi_t := \Pi_{t+1} - \Pi_t, \quad \delta S_t := S_{t+1} - S_t, \quad \delta B_t := B_{t+1} - B_t,$$

respectively. Since $S_{t+1} = S_t + \delta S_t$ and $B_{t+1} = B_t + \delta B_t$, the above s.f.c. takes the form

$$(S_t + \delta S_t)\,\delta x_t + (B_t + \delta B_t)\,\delta y_t = 0. \tag{2.19}$$

In later chapters we shall see that this form is similar to the s.f.c. for continuous-time modeling of positions and prices. If we now compute the change in the re-balanced portfolio values from time t to $t + 1$, we see that the s.f.c. is equivalent to the statement that this change is due only to the changes in the asset prices:

$$\begin{aligned}
\delta\Pi_t &= x_{t+1}S_{t+1} + y_{t+1}B_{t+1} - (x_t S_t + y_t B_t) \\
&= (x_t + \delta x_t)(S_t + \delta S_t) + (y_t + \delta y_t)(B_t + \delta B_t) - (x_t S_t + y_t B_t) \\
&= x_t\,\delta S_t + y_t\,\delta B_t + (S_t + \delta S_t)\,\delta x_t + (B_t + \delta B_t)\,\delta y_t \\
&= x_t\,\delta S_t + y_t\,\delta B_t .
\end{aligned}$$

This recovers (2.17) above and follows by enforcing the s.f.c. in (2.19) in the last expression.

We can write the s.f.c. in terms of the one-period ($t \to t + 1$) returns, where $S_{t+1} = (1 + r_{t+1}^S)\,S_t$ and $B_{t+1} = (1 + r_B)\,B_t$. Given the positions x_t, y_t at time t, then by (2.17)

$$\Pi_{t+1} = x_t\,(1 + r_{t+1}^S)\,S_t + y_t\,(1 + r_B)\,B_t$$

and the s.f.c. takes the form

$$S_t(1 + r_{t+1}^S)(x_{t+1} - x_t) + B_t(1 + r_B)(y_{t+1} - y_t) = 0. \tag{2.20}$$

In particular, assume the rate of return r_{t+1}^S is one of two known values r_S^+ or r_S^-, i.e., two scenarios are possible for one period. Then, given knowledge of S_t, B_t, r_B, x_t, and y_t, the above s.f.c. gives a linear relation between the position in the stock and that in the bond at time $t + 1$ for either ($+$ or $-$) stock return scenarios:

$$S_t\,(1 + r_S^\pm)\,(x_{t+1}^\pm - x_t) + B_t\,(1 + r_B)\,(y_{t+1}^\pm - y_t) = 0, \tag{2.21}$$

where the time $t + 1$ portfolio (x_{t+1}, y_{t+1}) can be explicitly denoted by (x_{t+1}^+, y_{t+1}^+) in case $r_{t+1}^S = r_S^+$ and by (x_{t+1}^-, y_{t+1}^-) for $r_{t+1}^S = r_S^-$. Note that if we are given another independent linear relation between the positions at time $t+1$, then the portfolio (x_{t+1}, y_{t+1}) is uniquely given. The example below describes such a situation where we impose an added independent condition on the positions besides the s.f.c..

Example 2.7. Consider a two-period binomial model with $r_B = 10\%$, $r_S^- = -10\%$, $r_S^+ = 20\%$, $S_0 = \$100$, and $B_0 = \$10$. Construct a self-financing strategy with an initial value of $\$1000$ such that 50% of the wealth is always invested in risk-free bonds.

Solution. We solve this problem by applying the above formulae for a single period from time 0 to time 1. Initially, $\Pi_0 = 1000 = x_0 S_0 + y_0 B_0$, such that $y_0 B_0 = 0.5 \cdot 1000 = 500$. Hence, $y_0 = \frac{500}{B_0} = 50$ units of the bond and $x_0 = \frac{500}{S_0} = 5$ stock shares, i.e., $(x_0, y_0) = (5, 50)$. At time 1, $S_1 = S_0(1 + r_S)$ and $\Pi_1 = x_1 S_1 + y_1 B_1 = x_1 S_0(1 + r_S) + y_1 B_0(1 + r_B)$. Since our strategy is constrained to have 50% equally invested in the stock and the bond, then $x_1 S_0(1 + r_S) = y_1 B_0(1 + r_B)$. Combining this relation with the above one-period s.f.c. in (2.21) (with $t = 0$) gives us two linear equations in the two unknowns x_1, y_1:

$$\begin{aligned}
S_0(1 + r_S)x_1 + B_0(1 + r_B)y_1 &= \Pi_1, \\
S_0(1 + r_S)x_1 - B_0(1 + r_B)y_1 &= 0,
\end{aligned}$$

where $\Pi_1 = x_0 S_0(1 + r_S) + y_0 B_0(1 + r_B)$. The solution is:

$$x_1 = \frac{\Pi_1}{2S_0(1 + r_S)}, \quad y_1 = \frac{\Pi_1}{2B_0(1 + r_B)}.$$

(a) For $r_S = r_S^+ = 0.2$, $S_0(1 + r_S) = 120$, $B_0(1 + r_B) = 11$, $\Pi_1 = 5 \cdot 120 + 50 \cdot 110 = 1150$, giving the portfolio $(x_1, y_1) = (\frac{115}{24}, \frac{575}{11}) \equiv (x_1^+, y_1^+)$. In this case, $\delta x \equiv x_1 - x_0 = \frac{115}{24} - 5 = -\frac{5}{24}$ and $\delta y \equiv y_1 - y_0 = \frac{575}{11} - 50 = \frac{25}{11}$. Hence, at time $t = 1$, re-balancing involves selling $\frac{5}{24}$ stock shares at price $S_1 = \$120$ and investing the proceeds in buying $\frac{25}{11}$ bonds at price $B_1 = \$110$.

(b) For $r_S = r_S^- = -0.1$, $S_0(1 + r_S) = 90$, $\Pi_1 = 5 \cdot 90 + 50 \cdot 110 = 1000$, giving the portfolio $(x_1, y_1) = (\frac{50}{9}, \frac{500}{11}) \equiv (x_1^-, y_1^-)$. Since $\delta x = \frac{50}{9} - 5 = \frac{5}{9}$ and $\delta y = \frac{500}{11} - 50 = -\frac{50}{11}$, we re-balance by buying $\frac{5}{9}$ stock shares at price $S_1 = \$90$ and finance this by selling off $\frac{50}{11}$ bonds at price $B_1 = \$110$.

Finally, note that in both cases we have the s.f.c. $\delta x\, S_1 + \delta y\, B_1 = 0$. $\qquad\square$

Let us end this subsection with the definition of an admissible strategy.

Definition 2.1. An *admissible strategy* is a self-financing strategy with nonnegative values for all dates from time zero until maturity.

At any date, the holder of an admissible strategy will have no potential liabilities which he or she would not able to honour. In particular, since the terminal value will be nonnegative in any state of the world, the liquidation of the terminal portfolio will not result in a loss.

2.2.6   Log-Normal Pricing Model

The binomial tree model has apparent disadvantages as a discrete-time and discrete-price model. We shall remove these restrictions by passing to the continuous-time limit from the binomial tree model. As a result, we will obtain a continuous model of the stock price $S(T)$. Although $S(T)$ is uncertain at time 0, we assume that the mathematical expectation μ and variance σ^2 of the annual log-return on the stock over the time interval $[0, T]$ are given by

$$\mu := \frac{1}{T} \operatorname{E}\left[\ln \frac{S(T)}{S(0)}\right] \quad \text{and} \quad \sigma^2 := \frac{1}{T} \operatorname{Var}\left(\ln \frac{S(T)}{S(0)}\right)$$

Note that μ and σ^2 can be estimated from historical observations $\{\widehat{S}(k\delta)\}$ as follows:

$$\mu \approx \frac{1}{N\delta} \sum_{k=1}^{N} \ln \frac{\widehat{S}(k\delta)}{\widehat{S}((k-1)\delta)},$$

$$\sigma^2 \approx \frac{1}{N\delta} \sum_{k=1}^{N} \left(\ln \frac{\widehat{S}(k\delta)}{\widehat{S}((k-1)\delta)} - \frac{1}{N} \sum_{k=1}^{N} \ln \frac{\widehat{S}(k\delta)}{\widehat{S}((k-1)\delta)}\right)^2.$$

For each $N \geqslant 1$, consider an N-period binomial tree model with the trading dates

$$0, \delta_N, 2\delta_N, \ldots, N\delta_N = T$$

spaced uniformly in $[0, T]$ with the time step $\delta_N := \frac{T}{N}$. Let $\{S_t^{(N)}; t = 0, 1, \ldots, N\}$ denote the stock price process in the N-period recombining binomial tree model; let the initial price be $S_0 \equiv S(0)$. Recall that the stock prices are given by a product of the initial stock price and single-period returns on the stock S,

$$S_t^{(N)} = S_0 \prod_{k=1}^{t} R_k^{(N)} .$$

The returns $R_k^{(N)}$, $k = 1, 2, \ldots, N$, are i.i.d. random variables having the common probability distribution

$$R_k^{(N)} = \begin{cases} u_N & \text{with probability } p_N, \\ d_N & \text{with probability } 1 - p_N. \end{cases}$$

The next step is to parametrize the binomial tree model. The upward and downward factors, u_N and d_N, can be obtained by matching the first two moments of the stock price returns. In the binomial model, the aggregate log-returns on the stock, $L_{[0,N]}^{(N)} := \ln \frac{S_N^{(N)}}{S_0}$, have the following expectation and variance:

$$\mathrm{E}\left[L_{[0,N]}^{(N)}\right] = N \, \mathrm{E}\left[L_1^{(N)}\right] = N \left(\ln(u_N)p_N + \ln(d_N)(1 - p_N)\right), \tag{2.22}$$

$$\mathrm{Var}\left[L_{[0,N]}^{(N)}\right] = N \, \mathrm{Var}\left(L_1^{(N)}\right) = N \left(\mathrm{E}\left[(L_1^{(N)})^2\right] - \mathrm{E}\left[L_1^{(N)}\right]^2\right)$$

$$= N \left(\ln(u_N)^2 p_N + \ln(d_N)^2(1 - p_N) - (\ln(u_N)p_N + \ln(d_N)(1 - p_N))^2\right)$$

$$= N \left(p_N(1 - p_N)(\ln(u_N/d_N))^2\right) . \tag{2.23}$$

Here, we use the fact that $L_{[0,N]}^{(N)} = L_1^{(N)} + \cdots + L_N^{(N)}$ where $\{L_k^{(N)}\}$ are i.i.d. single-period log-returns. Equating the respective moments of the log-returns $\ln \frac{S(T)}{S(0)}$ and $L_{[0,N]}^{(N)}$ gives us two equations:

$$\ln(u_N)p_N + \ln(d_N)(1 - p_N) = \mu \, \delta_N , \tag{2.24}$$
$$\ln(u_N)^2 p_N + \ln(d_N)^2(1 - p_N) = \sigma^2 \delta_N + \mu^2 \, \delta_N^2 . \tag{2.25}$$

Note that there are three unknowns, namely, u_N, d_N, and the probability p_N. Hence, we need a third equation. A convenient choice is the symmetry condition

$$u_N \cdot d_N = 1. \tag{2.26}$$

The following solution is the most commonly used in binomial models:

$$u_N = e^{\sigma \sqrt{\delta_N}}, \tag{2.27}$$
$$d_N = e^{-\sigma \sqrt{\delta_N}}, \tag{2.28}$$
$$p_N = \frac{1}{2} + \frac{1}{2}\frac{\mu}{\sigma}\sqrt{\delta_N}. \tag{2.29}$$

Under this parametrization, the log-returns on the stock have the following properties:

$$\mathrm{E}\left[L_{[0,N]}^{(N)}\right] = N \mu \, \delta_N = \mu T, \tag{2.30}$$

$$\mathrm{Var}\left[L_{[0,N]}^{(N)}\right] = N \sigma^2 \, \delta_N + N\mu^2 \, \delta_N^2 = \sigma^2 T + (\mu T)^2/N . \tag{2.31}$$

That is, the solution (2.27)–(2.29) satisfies (2.22)–(2.23) up to $\mathcal{O}(\delta_N)$. In other words, the binomial tree models we constructed here conserve the expected log-return on the stock and asymptotically (as $N \to \infty$) conserve the variance of the log-return.

The binomial price $S_N^{(N)}$ at the end of the Nth period is a discrete random variable taking on a value in the set $\{S_{N,k} = S_0 u_N^k d_N^{N-k} \; ; \; k = 0, 1, \ldots, N\}$. As the number of periods N increases to ∞, the length δ_N of one period approaches zero. Moreover, the density of the points $S_{N,k}$ increases and their range expands. Let us find the limiting distribution of the binomial prices. At the maturity date T, we define the limiting asset price:

$$S(T) := \lim_{N \to \infty} S_N^{(N)} = \lim_{N \to \infty} S(0) \exp\left(L_{[0,N]}^{(N)}\right).$$

To understand the distribution of the limiting price $S(T)$, we take a closer look at the distribution of the log-return $L_{[0,N]}^{(N)} = L_1^{(N)} + L_2^{(N)} + \cdots + L_N^{(N)}$. The one-step log-returns $L_k^{(N)}$, $k = 1, 2, \ldots, N$, are i.i.d. random variables. For each value of N, we introduce a sequence of i.i.d. Bernoulli random variables $X_k^{(N)}$, $k = 1, 2, \ldots, N$, having the following two-point probability distribution:

$$X_k^{(N)} = \begin{cases} 1 & \text{with probability } p_N, \\ 0 & \text{with probability } 1 - p_N. \end{cases}$$

We have that $X_k^{(N)} = 1$ and $1 - X_k^{(N)} = 0$ with probability p_N, and $X_k^{(N)} = 0$ and $1 - X_k^{(N)} = 1$ with probability $1 - p_N$. Therefore, we can express the log-return $L_k^{(N)}$ in terms of $X_k^{(N)}$ as follows:

$$L_k^{(N)} = \ln(u_N) X_k^{(N)} + \ln(d_N)\left(1 - X_k^{(N)}\right) = \ln\left(\frac{u_N}{d_N}\right) X_k^{(N)} + \ln(d_N).$$

As a result, the aggregate log-return is given by

$$L_{[0,N]}^{(N)} = \sum_{k=1}^{N} L_k^{(N)} = \ln\left(\frac{u_N}{d_N}\right) \sum_{k=1}^{N} X_k^{(N)} + N \ln(d_N).$$

Denote $Y_N := \sum_{k=1}^{N} X_k^{(N)}$. A sum Y_N of N i.i.d. Bernoulli variables has the binomial probability distribution: $Y_N \sim Bin(N, p_N)$. By the de Moivre–Laplace Theorem provided below, for large N, the distribution of Y_N is approximately normal with mean $N p_N$ and variance $N p_N (1 - p_N)$. Therefore, when N is large, the distribution of the log-return $L_{[0,N]}^{(N)} = \ln(u_N/d_N) Y_N + N \ln(d_N)$ is also approximately normal with mean μT and variance $\sigma^2 T$ (see formulae (2.30) and (2.31)). That is, in the limiting case, the probability distribution of $\lim_{N \to \infty} L_{[0,N]}^{(N)}$ is $Norm(\mu T, \sigma^2 T)$. Therefore, the limiting stock price $S(T) = \lim_{N \to \infty} S(0) \exp(L_{[0,N]}^{(N)})$ has the *log-normal probability distribution* and admits the following representation:

$$S(T) = S(0)\, e^{\mu T + \sigma \sqrt{T} Z}, \quad \text{where} \quad Z \sim Norm(0, 1). \tag{2.32}$$

The parameter μ is called the *drift parameter*; σ is the *volatility parameter*.

Here and below, $Norm(a, b^2)$ denotes the normal probability distribution with mean a and variance b^2. The cumulative distribution function (CDF) of the standard normal distribution $Norm(0, 1)$, denoted $\mathcal{N}$ (or Φ in some other texts), is

$$\mathcal{N}(z) := \frac{1}{\sqrt{2\pi}} \int_{-\infty}^{z} e^{-x^2/2} \, dx. \tag{2.33}$$

If Z is a standard normal variate, then for any real a and b, the random variable $a + bZ$ has the $Norm(a, b^2)$ probability distribution. Hence, the CDF F of $Norm(a, b^2)$ is

$$F(z) = \mathbb{P}(a + bZ \leqslant z) = \mathbb{P}\left(Z \leqslant \frac{z - a}{b}\right) = \mathcal{N}\left(\frac{z - a}{b}\right) = \frac{1}{\sqrt{2\pi}b} \int_{-\infty}^{z} e^{-\frac{(x-a)^2}{2b^2}} \, dx.$$

Rigorous justification of (2.32) is based on the following theorem, which is stated without a proof.

Theorem 2.2 (De Moivre–Laplace). *Consider a sequence $\{p_n\}_{n\geqslant 1}$ in $(0, 1)$ that converges to $p \in (0, 1)$ as $n \to \infty$. Let $\{Y_n\}_{n\geqslant 1}$ be a sequence of independent binomial random variables with $Y_n \sim Bin(n, p_n)$. Then the sequence of rescaled (normalized) random variables*

$$Y_n^* := \frac{Y_n - \mathrm{E}[Y_n]}{\sqrt{\mathrm{Var}(Y_n)}} = \frac{Y_n - np_n}{\sqrt{np_n(1 - p_n)}}$$

converges weakly (in distribution) to a standard normal variable.

Recall that a sequence of random variables $\{Y_n\}_{n\geqslant 1}$ converges in distribution to another random variable Y, denoted $Y_n \overset{d}{\to} Y$, as $n \to \infty$, if the CDF of Y_n converges to the CDF of Y, as $n \to \infty$, that is, for almost all $x \in \mathbb{R}$, we have $F_{Y_n}(x) \to F_Y(x)$, as $n \to \infty$.

Proposition 2.3. *Suppose that a sequence $\{Y_n\}_{n\geqslant 1}$ converges weakly to a standard normal random variable: $Y_n \overset{d}{\to} Norm(0, 1)$, as $n \to \infty$. Consider two converging sequences of real numbers: $a_n \to a$ and $b_n \to b \neq 0$, as $n \to \infty$. Then,*

$$a_n + b_n Y_n \overset{d}{\to} Norm(a, b^2), \quad as \ n \to \infty.$$

We have a sequence of binomial random variables $Y_N \sim Bin(N, p_N)$, where the probability $p_N = \frac{1}{2} + \frac{1}{2}\frac{\mu}{\sigma}\sqrt{\delta_N} \to \frac{1}{2}$, as $N \to \infty$. Therefore, by Theorem 2.2,

$$\frac{Y_N - \mathrm{E}[Y_N]}{\sqrt{\mathrm{Var}(Y_N)}} = \frac{Y_N - Np_N}{\sqrt{Np_N(1 - p_N)}} \overset{d}{\to} Norm(0, 1), \quad as \ N \to \infty.$$

On the other hand, for the log-returns $L_N \equiv L_{[0,N]}^{(N)}$, we have the identity

$$\frac{L_N - \mathrm{E}[L_N]}{\sqrt{\mathrm{Var}(L_N)}} = \frac{Y_N - \mathrm{E}[Y_N]}{\sqrt{\mathrm{Var}(Y_N)}} \tag{2.34}$$

the proof of which is left as an exercise for the reader. Thus, $L_N^* := \frac{L_N - \mathrm{E}[L_N]}{\sqrt{\mathrm{Var}(L_N)}} \to Norm(0, 1)$, as $N \to \infty$. Since we can express L_N in terms of L_N^*,

$$L_N = \mathrm{E}[L_N] + L_N^* \sqrt{\mathrm{Var}(L_N)} = \mu T + L_N^* \sqrt{\sigma^2 T + (\mu T)^2/N},$$

and $\sqrt{\sigma^2 T + (\mu T)^2/N} \to \sigma\sqrt{T}$, as $N \to \infty$, then, by Proposition 2.3, we have

$$L_N \overset{d}{\to} Norm(\mu T, \sigma^2 T), \quad as \ N \to \infty.$$

To summarize, in the binomial tree model, the stock price $S_N^{(N)}$ is a **discrete** random variable, which is a function of a binomial random variable. In the log-normal model, the stock price $S(T)$ is a **continuous** random variable having the log-normal probability distribution. As seen in Figure 2.4, the shape of the probability distribution of binomial prices is close to that of log-normal prices.

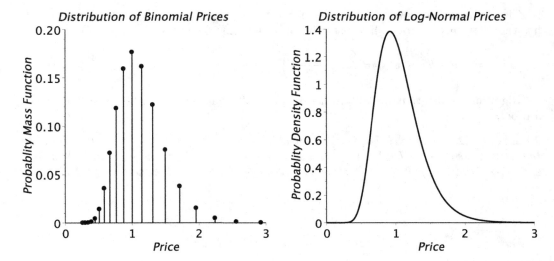

FIGURE 2.4: The probability distributions of asset prices in the binomial tree model (a) and log-normal model (b). The initial price is $S_0 = 1$; the time to maturity is $T = 1$; the binomial tree model has $N = 20$ periods; the model parameters are $\mu = 1\%$ and $\sigma = 30\%$.

Example 2.8 (The log-normal distribution). Find the cumulative distribution function (CDF) and the probability density function (PDF) of the log-normal price

$$S(T) = S_0 e^{\mu T + \sigma \sqrt{T} Z} \text{ with } T > 0,$$

where $Z \sim Norm(0, 1)$.

Solution. First, obtain the CDF F of $S(T)$:

$$
\begin{aligned}
F(s) = \mathbb{P}(S(T) \leqslant s) &= \mathbb{P}\left(S_0 e^{\mu T + \sigma \sqrt{T} Z} \leqslant s \right) \\
&= \mathbb{P}\left(\mu T + \sigma \sqrt{T} Z \leqslant \ln(s/S_0) \right) = \mathbb{P}\left(Z \leqslant \frac{\ln(s/S_0) - \mu T}{\sigma \sqrt{T}} \right) \\
&= \mathcal{N}\left(\frac{\ln(s/S_0) - \mu T}{\sigma \sqrt{T}} \right) \quad \text{for} \quad s > 0,
\end{aligned}
$$

where $\mathcal{N}$ is the *standard normal* CDF defined by (2.33): $\mathcal{N}(z) = \frac{1}{\sqrt{2\pi}} \int_{-\infty}^{z} e^{-x^2/2} \, \mathrm{d}x$. The standard normal PDF is given by $\mathcal{N}'(z) = \frac{1}{\sqrt{2\pi}} e^{-z^2/2}$. Differentiating the CDF $F(s)$ w.r.t. s gives us the PDF f of the log-normal price:

$$
\begin{aligned}
f(s) = F'(s) &= \mathcal{N}'\left(\frac{\ln(s/S_0) - \mu T}{\sigma \sqrt{T}} \right) \cdot \frac{\mathrm{d}}{\mathrm{d}s}\left(\frac{\ln(s/S_0) - \mu T}{\sigma \sqrt{T}} \right) \\
&= \frac{1}{s \sigma \sqrt{2\pi T}} e^{-(\ln(s/S_0) - \mu T)^2/(2\sigma^2 T)}, \quad s > 0. \qquad \square
\end{aligned}
$$

Example 2.9. Calculate $\mathrm{E}[S(T)]$, $\mathrm{E}[S^a(T)]$ and $\mathrm{Var}(S(T))$ for the log-normal price

$$S(T) = S_0 e^{\mu T + \sigma \sqrt{T} Z} \text{ where } T > 0 \text{ and } Z \sim Norm(0, 1).$$

Solution. Recall that the MGF of a normal random variable $X \sim Norm(a, b^2)$ is

$$M_X(u) = e^{au + (bu)^2/2}.$$

The log-normal price $S(T)$ is the exponentiation of

$$X = \ln S_0 + \mu T + \sigma\sqrt{T}Z \sim \text{Norm}(\ln S_0 + \mu T, \sigma^2 T).$$

Thus, we can compute the expected value of $S(T)$ as well as the mathematical expectation powers of $S(T)$ by employing the normal MGF:

$$\mathrm{E}[S(T)] = \mathrm{E}[e^X] = M_X(1) = e^{\ln S_0 + \mu T + \sigma^2 T/2} = S_0 e^{\mu T + \sigma^2 T/2},$$

$$\mathrm{E}[S^a(T)] = \mathrm{E}[e^{aX}] = M_X(a) = e^{a\ln S_0 + a\mu T + a^2\sigma^2 T/2} = S_0^a e^{a\mu T + a^2\sigma^2 T/2},$$

$$\mathrm{Var}(S_T) = \mathrm{E}[S^2(T)] - \mathrm{E}[S(T)]^2 = S_0^2 e^{2\mu T + 4\sigma^2 T/2} - S_0^2 e^{2\mu T + 2\sigma^2 T/2}$$

$$= S_0^2 e^{2\mu T + \sigma^2 T}\left(e^{\sigma^2} - 1\right). \qquad \square$$

Example 2.10. Let $\mu = 5\%$, $\sigma = 20\%$, and $S_0 = 100$. Calculate the probabilities

$$\text{(a) } \mathbb{P}(S(1) > 110) \quad \text{and} \quad \text{(b) } \mathbb{P}(90 \leqslant S(1) < 110)$$

for the log-normal price

$$S(T) = S_0 e^{\mu T + \sigma\sqrt{T}Z} \text{ where } T > 0 \text{ and } Z \sim \text{Norm}(0,1).$$

Solution. Let $F(s)$ denote the CDF of $S(T)$.

$$\text{(a)} \quad \mathbb{P}(S(1) > 110) = 1 - \mathbb{P}(S(1) \leqslant 110) = 1 - F(110) = 1 - \mathcal{N}\left(\frac{\ln(110/100) - 0.05}{0.2}\right)$$

$$\cong 1 - \mathcal{N}(0.2266) \cong 1 - 0.58961 = 41.039\%;$$

$$\text{(b)} \quad \mathbb{P}(90 \leqslant S(1) < 110) = \mathbb{P}(S(1) < 110) - \mathbb{P}(S(1) < 90) = F(110) - F(90)$$

$$= \mathcal{N}\left(\frac{\ln(110/100) - 0.05}{0.2}\right) - \mathcal{N}\left(\frac{\ln(90/100) - 0.05}{0.2}\right)$$

$$\cong \mathcal{N}(0.2266) - \mathcal{N}(-0.7768) \cong 0.58961 - 0.21864 = 37.097\%. \qquad \square$$

2.3 Arbitrage and Risk-Neutral Pricing

An *arbitrage opportunity* arises when someone can buy an asset at a low price to immediately sell it for a higher price. For example, such a combination of matching deals can be done by taking advantage of an asset price difference between two or more markets. Both buying in one market and selling on the other must occur simultaneously to avoid exposure to any type of market risk. In practice, such simultaneous transactions are only possible with assets and financial products traded electronically. The prices should not change before both transactions are complete; the cost of transport, storage, transaction, or insurance should not eliminate the arbitrage opportunity. In other words, arbitrage is a risk-free opportunity of gaining money.

A trader who engages in arbitrage is called an *arbitrageur*. Arbitrage opportunities are often hard to come by, due to transaction costs, the costs involved with finding an arbitrage opportunity, and the number of people who are also looking for such opportunities. Arbitrage profits are generally short-lived, as the buying and selling of assets will change the price of those assets in such a way as to eliminate that arbitrage opportunity. This situation is particularly the case in an efficient market.

Arbitrageurs can often be found in currency markets. Such financial markets have the advantage of being quite liquid, so we do not take the risk of acquiring an asset that may take some time to sell. The transaction costs are minimal for large currency transactions. Since foreign currency markets are an ideal environment for arbitrageurs, arbitrage opportunities tend to be very limited, as any discrepancies in exchange rates tend to be corrected quite quickly by investors trying to exploit those differences.

While arbitrage opportunities may exist in financial markets, in what follows we assume that the financial models we deal with do not allow for arbitrage. All asset prices are equilibrium prices and all arbitrage opportunities are eliminated. We are going to develop a non-arbitrage pricing theory. In a market model that admits arbitrage, wealth can be created from nothing. Thus, it is reasonable to assume that the financial model of consideration does not admit arbitrage opportunities. Let us start with a basic definition of arbitrage without reference to any model.

Definition 2.2. An *arbitrage opportunity* is a trading strategy that costs nothing to begin with (i.e., zero initial capital is used to set up) and has no chance of incurring any loss, but has a nonzero chance of making a gain.

This is reminiscent of a free lottery ticket. One of the fundamental properties of a good mathematical model for a financial market is that it does not allow for arbitrage.

2.3.1   The Law of One Price

As is mentioned in Section 2.2, replication is a key to pricing derivatives. Indeed, the following theorem states that if the future values of any two assets (or two portfolios of base assets) at some time are equal to each other in all possible market scenarios, then the present prices of these assets must be the same as well. Therefore, being given a derivative security which can be replicated by a portfolio or trading strategy in base assets (it means that the future values of the derivative and the portfolio are equal in all market states), the initial price of the derivative has to be equal to the initial value of the replicating portfolio in the absence of arbitrage.

Theorem 2.4 (Law of One Price). *Assume that the market is arbitrage free. Let there be two assets X and Y whose respective initial prices are X_0 and Y_0. Suppose at some time $T > 0$ the prices of X and Y are equal in all states of the world: $X_T(\omega) = Y_T(\omega)$, for all $\omega \in \Omega$. Then $X_0 = Y_0$.*

Proof. We shall show that if $X_0 \neq Y_0$, then there exists an arbitrage. Without loss of generality, we suppose that $X_0 > Y_0$ (if $X_0 < Y_0$ then we may relabel X and Y). Let us construct an arbitrage portfolio in these two assets. Starting with \$0, we first borrow and sell one unit of X and realize \$$X_0$. We then buy one unit of Y, costing us \$$Y_0$. Both transactions give us a positive amount \$$(X_0 - Y_0)$, which we can keep in cash or invest in a risk-free asset. So at time zero, we have a portfolio of one unit of Y, negative one unit of X, and the cash amount of \$$(X_0 - Y_0)$. The initial value of this portfolio is zero:

$$\Pi_0 = -X_0 + Y_0 + (X_0 - Y_0) = 0.$$

Note that setting up this portfolio requires no initial investment.

At time T, we sell the unit of Y to obtain \$$Y_T$. We buy and return the unit of X. This costs \$$X_T$. Since $X_T = Y_T$, the net cost of these two trades is zero. However, we still have the positive cash amount $X_0 - Y_0$ (and possible interest earned), and hence we have exhibited an arbitrage opportunity:

$$\Pi_T = -X_T + Y_T + X_0 - Y_0 = X_0 - Y_0 > 0.$$

Therefore, to eliminate the arbitrage, we must have $X_0 = Y_0$.   □

In this proof we have assumed that there are no transaction costs in carrying out the trades required, short sells are allowed, and the assets involved can be bought and sold at any time at will.

2.3.2 A First Look at Arbitrage in the Single-Period Binomial Model

In the single-period binomial model of a financial market considered in the previous section, an arbitrage strategy simply reduces to an arbitrage portfolio. Consider a portfolio (x, y) in stock S and bond B with initial value $\Pi_0 = \Pi_0^{(x,y)} = xS_0 + yB_0$ and terminal values $\Pi_T(\omega^\pm) = \Pi_T^{(x,y)}(\omega^\pm) = xS_T(\omega^\pm) + yB_T \equiv xS^\pm + yB_T$. The above definition of arbitrage hence implies that (x, y) will be an arbitrage portfolio when the following conditions are met:

(a) $\Pi_0 = 0$,

(b) $\Pi_T(\omega^\pm) \geqslant 0$, and $\Pi_T(\omega^+) > 0$ or $\Pi_T(\omega^-) > 0$.

Note that condition (b) can be stated using probabilities:

(b) $\mathbb{P}(\Pi_T \geqslant 0) = 1$ and $\mathbb{P}(\Pi_T > 0) > 0$.

Hence, the single-period binomial model admits an arbitrage iff there exists a portfolio (x, y) satisfying conditions (a) and (b). As Theorem 2.5 shows, there is no such arbitrage portfolio (x, y) when the return on the bond falls strictly in between the higher and lower returns on the stock.

Theorem 2.5 (Arbitrage: Single-period binomial model). *The single-period binomial model admits no arbitrage iff* $r_S^- < r_B < r_S^+$.

Proof. First, note that condition $r_S^- < r_B < r_S^+$ is equivalent to $S^- < (1 + r_B)S_0 < S^+$. So we can formulate our proof using either returns or prices.

- Consider any zero cost, nontrivial portfolio (x, y), i.e., a portfolio such that $\Pi_0^{(x,y)} = xS_0 + yB_0 = 0$ and $(x, y) \neq (0, 0)$. This implies that the portfolio (x, y) has to be of the form $(x, -xS_0/B_0)$ with $x > 0$ or $(-yB_0/S_0, y)$ with $y > 0$. The first portfolio type corresponds to buying the stock and borrowing money while the second is a portfolio short in stock and positively invested in a bond.

- For a portfolio of the form $(x, -xS_0/B_0)$, we have: $\Pi_T^{(x,y)}(\omega^\pm) = x[S^\pm - (1 + r_B)S_0]$. In the worst case scenario $\Pi_T^{(x,y)}(\omega^-) = x[S^- - (1 + r_B)S_0]$. Hence, $S^- \geqslant (1 + r_B)S_0$ implies that

$$\Pi_T^{(x,y)}(\omega^\pm) \geqslant 0 \quad \text{and} \quad \Pi_T(\omega^+) = x[S^+ - (1 + r_B)S_0] > 0.$$

Therefore, such a portfolio is an arbitrage unless $S^- < (1 + r_B)S_0$.

- Similarly, for the second type we have $\Pi_T^{(x,y)}(\omega^\pm) = y(B_0/S_0)[(1 + r_B)S_0 - S^\pm]$. In the best case scenario $\Pi_T^{(x,y)}(\omega^+) = y(B_0/S_0)[(1 + r_B)S_0 - S^+]$. Hence, $S^+ \leqslant (1 + r_B)S_0$ implies that

$$\Pi_T^{(x,y)}(\omega^\pm) \geqslant 0 \quad \text{and} \quad \Pi_T^{(x,y)}(\omega^-) = y(B_0/S_0)[(1 + r_B)S_0 - S^-] > 0.$$

The portfolio is an arbitrage unless $S^+ > (1 + r_B)S_0$.

Hence, by combining the two cases, we have shown that there is no arbitrage iff

$$S^- < (1 + r_B)S_0 < S^+. \qquad \qquad \square$$

Note that the result of Theorem 2.5 can be formulated in terms of asset values: there is no arbitrage iff $\frac{S^-}{S_0} < \frac{B_T}{B_0} < \frac{S^+}{S_0}$. If market prices do not allow for profitable arbitrage, then the prices are said to constitute an arbitrage-free market. Later we shall see that the assumption that there is no arbitrage is used in quantitative finance to calculate unique (no-arbitrage) prices for derivatives that can be replicated.

Example 2.11. Consider a single-period binomial model with two base assets so that $S_0 = B_0 = \$100$ and $r_B = 5\%$. Find an arbitrage portfolio if the stock returns are

(a) $r_S^- = -5\%$ and $r_S^+ = 5\%$;

(b) $r_S^- = 5\%$ and $r_S^+ = 10\%$.

Solution. Clearly, since the condition $r_S^- < r_B < r_S^+$ is not satisfied in both cases, there exist arbitrage opportunities. Due to the condition $S_0 = B_0$, any portfolio with initial cost zero has the form $(x, y) = (c, -c)$ with some $c \in \mathbb{R}$.

(a) The bond outperforms the stock, so we should sell short the stock and buy the bond. Form the portfolio $(x, y) = (-1, 1)$. At maturity, we have

$$\Pi_T = x\,S_0\,(1 + r_S) + y\,B_0\,(1 + r_B) = -100\,(1 + r_S) + 105 = 5 - 100\,r_S$$
$$= \begin{cases} 5 - 100 \cdot (-0.05) = 10 & \text{if } \omega = \omega^-, \\ 5 - 100 \cdot 0.05 = 0 & \text{if } \omega = \omega^+. \end{cases}$$

It is an arbitrage since $\Pi_0 = 0$, $\mathbb{P}(\Pi_T \geqslant 0) = 1$ and $\mathbb{P}(\Pi_T > 0) = \mathbb{P}(\omega^-) > 0$.

(b) The stock outperforms the bond, so we should sell short the bond and buy the stock. Form the portfolio $(x, y) = (2, -2)$. At maturity, we have

$$\Pi_T = x\,S_0\,(1 + r_S) + y\,B_0\,(1 + r_B) = 200\,(1 + r_S) - 210 = 200\,r_S - 10$$
$$= \begin{cases} 200 \cdot (0.05) - 10 = 0 & \text{if } \omega = \omega^-, \\ 200 \cdot 0.1 - 10 = 10 & \text{if } \omega = \omega^+. \end{cases}$$

Clearly, it is an arbitrage portfolio.

$$\square$$

In conclusion, let us demonstrate that if the initial price C_0 of a claim is equal to the initial cost Π_0 of the replicating portfolio in (2.5), then there is no arbitrage. In other words, if the actual initial price is less than or greater than the price in (2.5), then there is an arbitrage portfolio in the base assets B and S and the claim C. One way to argue this statement is to apply the Law of One Price. Indeed, since the payoff C_T is identical to that of the replicating portfolio, Π_T, the present values C_0 and Π_0 have to be the same, or else an arbitrage exists. On the other hand, we can always form an arbitrage portfolio when $C_0 \neq \Pi_0$, as demonstrated in the following example.

Example 2.12. Consider a single-period binomial model with $r_B = 10\%$, $r_S^- = -10\%$, $r_S^+ = 20\%$, $S_0 = \$100$, and $B_0 = \$10$. Contract C has the following payoff: $C^- = 0$ and $C^+ = \$50$. Let C_0 be its initial market price.

(a) Find portfolio (x, y) that replicates the payoff (C^-, C^+) and then calculate its initial cost $\Pi_0 = \Pi_0^{(x,y)}$.

(b) Suppose that $C_0 > \Pi_0$. Construct an arbitrage portfolio.

(c) Suppose that $C_0 < \Pi_0$. Construct an arbitrage portfolio.

Solution. (a) Applying (2.4) gives us the replicating portfolio:

$$x = \frac{50 - 0}{100 \cdot (0.2 - (-0.1))} = \frac{5}{3} \quad \text{and} \quad y = \frac{0 \cdot (1 + 0.2) - 50 \cdot (1 - 0.1)}{10 \cdot (1 + 0.1) \cdot (0.2 - (-0.1))} = -\frac{150}{11}.$$

The initial value of the portfolio is

$$\Pi_0^{(x,y)} = xS_0 + yB_0 = \frac{5}{3} \cdot 100 - \frac{150}{11} \cdot 10 = \frac{1000}{33} \cong \$30.303\,.$$

(b) Suppose that the actual price of the contract C_0 is greater than Π_0. Write and sell the contract for C_0. Use the proceeds to buy $x = \frac{5}{3}$ shares of the stock. If $C_0 < xS_0 = \frac{500}{3}$, then borrow $\frac{50}{3} - \frac{C_0}{10}$ bonds (each bond is worth \$10); otherwise, we invest the balance in bonds. So, we include $\frac{C_0}{10} - \frac{50}{3}$ bonds in the portfolio. As a result, we form the portfolio (x, y, z) that contains $x = \frac{5}{3}$ shares of stock S, $y = \frac{C_0}{10} - \frac{50}{3}$ bonds B, and $z = -1$ contracts C. The initial cost is zero. At the end of the period, we sell stock and pay C_T to the holder of the contract. The balance is positive for every market scenario whenever $C_0 > \frac{1000}{33}$:

$$\Pi_T^{(x,y,z)} = \underbrace{\frac{5}{3}S_T - \frac{150}{11}B_T - C_T}_{=0} + \left(\frac{C_0}{10} - \frac{50}{3} + \frac{150}{11}\right)B_T = \frac{11}{10}\left(C_0 - \frac{1000}{33}\right) > 0.$$

(c) Suppose now that the actual price of the contract C_0 is less than Π_0. Buy the contract for C_0. To get the money, sell short $x = \frac{5}{3}$ stock shares. If $C_0 > xS_0 = \frac{500}{3}$, then borrow $\frac{C_0}{10} - \frac{50}{3}$ bonds; otherwise, we invest the balance in bonds. So, we include $\frac{50}{3} - \frac{C_0}{10}$ bonds in the portfolio. As a result, we form the portfolio (x, y, z) that contains $x = -\frac{5}{3}$ shares of stock S, $y = \frac{50}{3} - \frac{C_0}{10}$ bonds B, and $z = 1$ contracts C. The initial cost is zero. At the end of the period, we receive C_T from the writer of the contract and then buy stock to close the short position in S. The balance is positive for every market scenario whenever $C_0 < \frac{1000}{33}$:

$$\Pi_T^{(x,y,z)} = \underbrace{C_T - \frac{5}{3}S_T + \frac{150}{11}B_T}_{=0} + \left(\frac{50}{3} - \frac{C_0}{10} - \frac{150}{11}\right)B_T = \frac{11}{10}\left(\frac{1000}{33} - C_0\right) > 0. \quad \square$$

2.3.3 Arbitrage in the Binomial Tree Model

In a multi-period model, there is more flexibility for an investment portfolio. The investor may alter the positions in the portfolio at any time by selling some assets and investing the proceeds in others. Therefore, the definition of an arbitrage investment strategy is a bit different compared with the definition of an arbitrage portfolio for the single-period case.

Definition 2.3. An admissible strategy such that $\Pi_0 = 0$ and $\mathbb{P}(\Pi_t > 0) > 0$ for some $t = 1, 2, \ldots$ is called an *arbitrage* (strategy).

The cost of setting up an arbitrage strategy is zero. The self-financing condition means that there are no injections or withdrawals of funds at any intermediate date. The admissibility guarantees that the holder will not face a potential loss. At the maturity date, there is no loss since the terminal value is nonnegative. Moreover, there exists at least one scenario where liquidating the portfolio will result in a positive gain. In summary, an arbitrage strategy is a possibility of having a potential gain at no cost and without potential losses. Since the wealth of an admissible strategy is always nonnegative, the definitions of an arbitrage portfolio and an arbitrage strategy are the same for the single-period binomial model.

Theorem 2.6. *The binomial tree model admits no arbitrage iff* $d < 1 + r < u$.

Proof. First, consider the case of a one-period binomial tree (i.e., $T = 1$). As was justified in Theorem 2.5, the rate r of interest on a risk-free investment has to satisfy $d < 1 + r < u$, or else an arbitrage possibility would arise. Indeed, for the one-step return on the stock we have $1 + r_S^- = d$ and $1 + r_S^+ = u$; the one-step return on the bond is r. There is no arbitrage iff $r_S^- < r_B < r_S^+$, which is equivalent to $d < 1 + r < u$.

Now let us consider a multi-period binomial model. Suppose that $1 + r \leqslant d$ or $u \leqslant 1 + r$ holds. To construct an arbitrage portfolio, we proceed as follows. At time 0, construct a portfolio which is long in stock if $1 + r \leqslant d$ or short in stock if $u \leqslant 1 + r$ with zero initial value. At time 1, close the position in the stock and invest the proceeds in risk-free bonds. As a result of these manipulations, we obtain a positive amount of cash invested in bonds.

Let $d < 1 + r < u$ and suppose that there is an arbitrage strategy, i.e., there is a self-financing strategy with zero initial value such that $\Pi_t \geqslant 0$ for all $t \geqslant 0$ with probability 1 and $\Pi_T > 0$ with nonzero probability at maturity time T. Find the smallest time $t > 0$ for which $\Pi_t(\omega) > 0$ at some state ω. Since each state in the model is a path in the binomial tree, we can find a one-step subtree with two branches, so that $\Pi_{t-1} = 0$ at its root, $\Pi_t \geqslant 0$ at each node growing out of this root with $\Pi_t > 0$ in at least one of these nodes. Note that the path ω is passing through the root and the node where $\Pi_t > 0$. In the one-step case this is impossible if $d < 1 + r < u$, leading to a contradiction. $\square$

2.3.4   Risk-Neutral Probabilities

Although the future stock prices are uncertain, it is natural to compare the expected return on stock and the risk-free rate of return. The expected stock prices under the real-world probability function $\mathbb{P}$ are given by

$$\mathrm{E}[S_t] = S_0(1 + \mathrm{E}[r_S])^t, \quad t = 0, 1, 2, \ldots \tag{2.35}$$

where r_S denotes a one-period rate of return on the stock. Since $r_S = u - 1$ with probability p and $r_S = d - 1$ with probability $1 - p$, we obtain

$$\mathrm{E}[r_S] = p(u - 1) + (1 - p)(d - 1) = pu + (1 - p)d - 1.$$

Suppose that the amount S_0 is invested in a risk-free bank account. It will grow to $S_0(1 + r)^t$ after t steps, where r is the compound risk-free interest rate. Clearly, to compare the expected return on the stock, $\mathrm{E}[S_t/S_0]$, and the risk-free return, $(1 + r)^t$, we only need to compare the average one-step rate $\mathrm{E}[r_S]$ on the stock and the risk-free rate r. There exist three main types of investors.

- A typical *risk-averse* investor requires that $\mathrm{E}[r_S] > r$, arguing that she or he should be rewarded with a higher expected return as compensation for risk.

- A *risk-seeker* investor may be attracted by the reverse situation when $\mathrm{E}[r_S] < r$, if a risky return is very high with small nonzero probability and low with large probability.

- A border situation of a market in which $\mathrm{E}[r_S] = r$ is referred to as *risk-neutral*.

We now introduce a new probability function $\widetilde{\mathbb{P}}$ with the probabilities of one-period upward and downward moves of the stock price $\widetilde{\mathbb{P}}(\text{up}) = \tilde{p}$ and $\widetilde{\mathbb{P}}(\text{down}) = 1 - \tilde{p}$, respectively, such that the risk-neutrality condition

$$\widetilde{\mathrm{E}}[r_S] = \tilde{p}u + (1 - \tilde{p})d - 1 = r \tag{2.36}$$

is satisfied. This implies that

$$\tilde{p} = \frac{1 + r - d}{u - d} \text{ and } 1 - \tilde{p} = \frac{u - r - 1}{u - d}. \tag{2.37}$$

We shall call $\tilde{p}$ and $1 - \tilde{p}$ the *risk-neutral probabilities* of the stock price upward and downward moves, respectively. The corresponding probability function is called the risk-neutral probability function (or measure) and is denoted by $\widetilde{\mathbb{P}}$. Here $\widetilde{\mathrm{E}}$ denotes the mathematical expectation with respect to the probability function $\widetilde{\mathbb{P}}$; it is called the *risk-neutral expectation*.

Theorem 2.7. *The binomial tree model admits no arbitrage iff there exists the risk-neutral probability* $\tilde{p} \in (0, 1)$.

Proof. It is clear from (2.37) that $0 < \tilde{p} < 1$ iff $d < 1 + r < u$. The latter is a necessary and sufficient condition of the absence of arbitrage. $\qquad \square$

To explain why $\tilde{p}$ is called a risk-neutral probability, we are going to compare the real-world expected return $\mathrm{E}[r_S]$ and the risk-neutral expected return $\widetilde{\mathrm{E}}[r_S] = r$. Let us define the risk of the investment in the stock to be the standard deviation of the one-step return r_S:

$$\sigma_S = \sqrt{\mathrm{Var}(r_S)} = \sqrt{\mathrm{E}[r_S^2] - (\mathrm{E}[r_S])^2}.$$

This parameter is often called the *volatility* of stock price return. It follows that

$$\sigma_S^2 = \mathrm{Var}(r_S) = (u - d)^2 p(1 - p). \tag{2.38}$$

Let us compare the expected returns $\mathrm{E}[r_S]$ and $\widetilde{\mathrm{E}}[r_S]$:

$$\mathrm{E}[r_S] - \widetilde{\mathrm{E}}[r_S] = up + d(1 - p) - u\tilde{p} - d(1 - \tilde{p}) = (p - \tilde{p})(u - d). \tag{2.39}$$

Assume that $\mathrm{E}[r_S] \geqslant r$, that is, the expected return is not less than the risk-free return. Combining (2.38) with (2.39) gives

$$\mathrm{E}[r_S] - r = \frac{p - \tilde{p}}{\sqrt{p(1 - p)}} \sigma_S.$$

We say that one asset is riskier than another when it has a higher volatility of return. If the volatility is zero (i.e., we deal with a risk-free asset), then the expected return is just r; if the volatility is nonzero, then we have a higher expected return. This result fits well with reality—if you want a higher expected return you must take on more risk. However, when $p = \tilde{p}$, i.e., we deal with a risk-neutral market, the expected return is always r no matter what value the volatility σ has.

In reality, the risk-neutral probability $\tilde{p}$ has no relation to the real-world probability p. However, the risk-neutral probability function is of great practical importance to us with respect to computing no-arbitrage prices of derivative contracts. Let us consider a single-period binomial model. The fair price C_0 of a derivative contract with payoffs $C^{\pm} = C_T(\omega^{\pm})$

in the two possible outcomes $\omega^{\pm}$ is given by (2.5). By using the notation $r_B = r$, $r_S^+ = u-1$, and $r_S^- = d-1$, we can rewrite (2.5) as follows:

$$C_0 = \frac{1}{1+r}\left[\frac{1+r-d}{u-d}C^+ + \frac{u-r-1}{u-d}C^-\right].$$

Substituting (2.37) in the above equation gives us a simple valuation formula:

$$C_0 = \frac{1}{1+r}\left[\tilde{p}C^+ + (1-\tilde{p})C^-\right] = \frac{1}{1+r}\widetilde{\mathrm{E}}\left[C_T\right] = \frac{B_0}{B_T}\widetilde{\mathrm{E}}\left[C_T\right]. \tag{2.40}$$

In other words, the no-arbitrage price of claim C is given by a risk-neutral expectation of the discounted future payoff function. The discounting factor is $\frac{B_0}{B_T} = \frac{1}{1+r_B}$. The interesting fact is that this formula works for more sophisticated models and general payoff functions.

2.3.5 Martingale Property

Equation (2.40) can be rewritten as follows:

$$\widetilde{\mathrm{E}}\left[\frac{C_T}{B_T}\right] = \frac{C_0}{B_0}. \tag{2.41}$$

In this case, the process $\left\{\frac{C_t}{B_t}\right\}_{t\in\{0,T\}}$ is said to be a *martingale* under the risk-neutral probability function $\widetilde{\mathbb{P}}$. Now we proceed to the multi-period case.

It follows from (2.35) that the expectation of S_t with respect to the risk-neutral probability function $\widetilde{\mathbb{P}}$ is

$$\widetilde{\mathrm{E}}[S_t] = S_0(1+r)^t, \tag{2.42}$$

since $\widetilde{\mathrm{E}}[r_S] = r$. In other words, the expected return on the stock under $\widetilde{\mathbb{P}}$ is equal to the risk-free return over the same time interval from 0 to t.

Equation (2.42) can be extended to any time step in the binomial tree model. Suppose that t time steps have passed and the stock price has changed from S_0 to S_t. Let us find the risk-neutral expectation of the price S_{t+1} given the price S_t. Formally, we need to find the *conditional expectation* of S_{t+1} given S_t. We can write $S_{t+1} = S_t R_{t+1}$. Since in the binomial tree model all single-period returns are i.i.d., the distribution of stock return R_{t+1} does not depend on the time t. The risk-neutral expectation of R_{t+1} is equal to $1+r$. The conditional expectation of S_{t+1} given S_t is

$$\widetilde{\mathrm{E}}[S_{t+1} \mid S_t] = \widetilde{\mathrm{E}}[S_t R_{t+1} \mid S_t] = S_t\widetilde{\mathrm{E}}[R_{t+1} \mid S_t] = S_t\widetilde{\mathrm{E}}[R_{t+1}] = S_t(1+r). \tag{2.43}$$

The above derivation is based on the following two facts. First, S_t is given and hence can be taken out of the expectation. Second, since stock returns are mutually independent, R_{t+1} is independent of S_t, and hence the last expectation becomes an unconditional one. Now we introduce *discounted stock prices*:

$$\overline{S}_t \equiv \frac{S_t}{B_t} = \frac{S_t}{B_0(1+r)^t}, \quad t \geqslant 0.$$

Dividing both sides of (2.43) by the bond price B_{t+1} and using $B_{t+1} = B_t(1+r)$ gives

$$\widetilde{\mathrm{E}}\left[\frac{S_{t+1}}{B_{t+1}} \mid S_t\right] = \frac{S_t(1+r)}{B_{t+1}} = \frac{S_t(1+r)}{B_t(1+r)} = \frac{S_t}{B_t}.$$

Thus, (2.43) takes the form

$$\widetilde{E}\left[\overline{S}_{t+1} \mid \overline{S}_t\right] = \overline{S}_t, \quad t \geqslant 0 \tag{2.44}$$

We say that the discounted stock price process $\{\overline{S}_t\}_{t\geqslant 0}$ is a *martingale* under the risk-neutral probability function $\widetilde{\mathbb{P}}$. We will provide a more complete formalism of martingales in Chapter 6.

2.3.6 Risk-Neutral Log-Normal Model

In Subsection 2.2.6, we derived the log-normal price model as the limiting case of a sequence of binomial tree models as the number of periods N approaches infinity. Let r be the risk-free interest rate under continuous compounding. The equivalent one-period interest rate r_N in the binomial tree model with N periods is $r_N = e^{r\delta_N} - 1$, where $\delta_N = \frac{T}{N}$. As was demonstrated in Subsection 2.2.6, the log-return $L_{[0,N]}^{(N)} = \ln(S_N^{(N)}/S_0)$ is approximately normal, as $N \to \infty$. Let us find the parameters of the limiting normal distribution under the risk-neutral probability measure. In the N-period model, the risk-neutral probability of the upward movement of the stock price is

$$\tilde{p}_N = \frac{r_N + 1 - d_N}{u_N - d_N} = \frac{e^{\delta_N r} - e^{-\sqrt{\delta_N}\sigma}}{e^{\sigma\sqrt{\delta_N}} - e^{-\sigma\sqrt{\delta_N}}}.$$

Introduce the risk-neutral probability function $\widetilde{\mathbb{P}}_N$ with the probability $\tilde{p}_N$ of an upward stock price move. Under this probability function, the normalized log-return, $L_N^* := \frac{L_N - E[L_N]}{\sqrt{\text{Var}(L_N)}}$, where $L_N \equiv L_{[0,N]}^{(N)}$, is expressed in terms of a binomial random variable with probability $\tilde{p}_N$ of success, $Y_N \sim Bin(N, \tilde{p}_N)$, as given in (2.34). It is not difficult to compute the expectation and variance of the log-returns under the risk-neutral probability $\widetilde{\mathbb{P}}_N$:

$$E_{\widetilde{\mathbb{P}}_N}[L_N] = (2\tilde{p}_N - 1)\sigma\sqrt{NT}, \quad \text{Var}_{\widetilde{\mathbb{P}}_N}(L_N) = \tilde{p}_N(1 - \tilde{p}_N)4\sigma^2 T.$$

To find the distribution of $\lim_{N\to\infty} L_N$, we need to know the limiting values of the expectation and variance.

Proposition 2.8. *As $N \to \infty$, we have the following limits:*

$$\tilde{p}_N \to \frac{1}{2}, \quad E_{\widetilde{\mathbb{P}}_N}[L_N] \to rT - \frac{1}{2}\sigma^2 T, \quad \text{Var}_{\widetilde{\mathbb{P}}_N}(L_N) \to \sigma^2 T.$$

Proof. The proof is left as an exercise for the reader. $\square$

By the de Moivre–Laplace theorem, the probability distribution of log-returns converges weakly to the normal distribution, as $N \to \infty$. Under the risk-neutral probability, the asymptotic distribution of L_N, as $N \to \infty$, is $Norm((r - \frac{1}{2}\sigma^2)T, \sigma^2 T)$. Therefore, in the limiting case, the *risk-neutral probability distribution* of the stock price $S(T) = \lim_{N\to\infty} S_N^{(N)}$ is the log-normal distribution:

$$S(T) \stackrel{d}{=} S(0)\, e^{(r-\sigma^2/2)T + \sigma\sqrt{T}Z}, \tag{2.45}$$

where $Z \sim Norm(0,1)$. Recall that the notation $X \stackrel{d}{=} Y$ means that random variables X and Y have the same probability distribution.

The interesting fact is that the limiting distribution does not depend on the real-world expected return μ on the stock. In the risk-neutral binomial tree model, the expected return on the stock is the same as that of the risk-free bond. It is not difficult to check that in the limiting case we observe the same behaviour for the risk-neutral log-normal price model:

$$\widetilde{E}[S(T)] = S(0)\, e^{rT} \quad \text{or, equivalently,} \quad \widetilde{E}[e^{-rT} S(T)] = S(0).$$

2.4 Value at Risk

In the beginning of this chapter, we defined a risky asset as that with uncertain future cash flows. Examples of such assets include stocks, derivative contracts, defaultable bonds, and similar contingent claims subject to default risk. To distinguish risky and risk-free assets, we need to take a look at the distribution of their returns. The return of a risky asset is uncertain. Hence, from the mathematical point of view, it may be viewed as a random variable with nonzero variance. The return on a risk-free asset is certain, so its variance is zero. Therefore, the risk associated with an asset (with return R) can be measured by computing the standard deviation of the return on the asset: $\sigma = \sqrt{\mathrm{Var}(R)}$. A risk-averse investor prefers an asset with lower σ. However, the value of σ may not tell us how large the loss may be. The variance and expectation of the return on a risky asset alone define the shape of the profit and loss distribution only when the asset return has a normal distribution (or Student's t-distribution).

One can use other market *risk metrics* to measure the uncertainty in the portfolio return or loss. Losses are the central object of interest in risk management. The concept of a loss distribution works on all levels of aggregation from a single security to the overall position of a financial institution. Loss distributions can be compared across portfolios. For example, one may be interested in the probability that loss L on a specific financial asset (or a portfolio of assets) over some period of time exceeds a given amount. That is, one may wish to evaluate the probability $\mathbb{P}(L > A)$ for a given loss L and amount A. It is called the probability of under-performing.

Let us reverse the question and find an amount A so that the probability of a loss not exceeding this amount is equal to a given probability, say 95% (although we may consider another confidence level such as 90% or 99%). That is, find A such that $\mathbb{P}(L \leqslant A) = 95\%$. This value is referred to as *Value at Risk* and denoted by VaR.

The VaR is a measure of the risk of loss on a specific portfolio of financial assets. For a given portfolio, probability level, and time horizon, VaR is defined as a *threshold level* such that the probability that the loss on the portfolio over the given time horizon exceeds this level is equal to the given probability.

VaR has two basic parameters: the *significance level* denoted $\alpha \in (0,1)$ (or *confidence level* denoted $1 - \alpha$), and the *risk horizon* denoted h, which is the period of time over which we measure the potential loss. Traditionally, h is measured in trading days. Common parameters for VaR are 1% and 5% significance levels and one-day and 10-day risk horizons, although other combinations are also in use. When VaR is computed, it is assumed that the current position in the portfolio of interest will remain unaltered over the chosen time period. Let Π_t denote the value of the portfolio at the time t. The value of the same portfolio at the future time $t + h$, discounted to time t, is $Z(t, t + h)\Pi_{t+h}$, where $Z(t, t + h)$ is the price of a unit zero-coupon bond that matures at the time $t + h$. For example, for the case with continuously compounded interest, we have that $Z(t, t + h) = \mathrm{e}^{-jh}$ (where j is the daily interest rate). The discounted profit-and-loss (P&L) over a risk horizon of h days is

$$G_h := \mathrm{e}^{-jh}\Pi_{t+h} - \Pi_t.$$

In other words, G_h is the present value of the gain from an investment; hence,

$$L_h := -G_h = \Pi_t - \mathrm{e}^{-jh}\Pi_{t+h}$$

is the present value of the loss. The future value Π_{t+h} is uncertain, and hence the discounted P&L is a random variable. To calculate the VaR of the portfolio, we need to know the distribution of this random variable. Since extreme losses are of special interest, we focus on the lower tail of the P&L probability distribution.

Given the significance level α and risk horizon h, the $100\alpha\%$ h-day VaR is defined as the present value of the minimum possible loss amount so that the probability that it would be exceeded over an h-day time period is at most α. Mathematically, $\text{VaR}_{\alpha,h}$ is the $1 - \alpha$ quantile of the loss L_h. Alternatively, VaR can also be defined as a quantile of the P&L probability distribution. Both approaches are presented below:

$$\text{VaR}_{\alpha,h} = \inf\{x \in \mathbb{R} \, : \, \mathbb{P}(L_h > x) \leqslant \alpha\} = \inf\{x \in \mathbb{R} \, : \, F_{L_h}(x) \geqslant 1 - \alpha\} = \ell_{1-\alpha,h} \quad (2.46)$$
$$= -\inf\{x \in \mathbb{R} \, : \, \mathbb{P}(G_h > x) \leqslant 1 - \alpha\}$$
$$= -\inf\{x \in \mathbb{R} \, : \, F_{G_h}(x) \geqslant \alpha\} = -g_{\alpha,h}. \quad (2.47)$$

Here, we use the following notations:

- $F_{G_h}(x) := \mathbb{P}(G_h \leqslant x)$ is the CDF of the P&L G_h,

- $F_{L_h}(x) := \mathbb{P}(L_h \leqslant x)$ is the CDF of the loss L_h,

- $g_{\alpha,h}$ is the α-quantile of the P&L distribution,

- $\ell_{1-\alpha,h}$ is the $(1 - \alpha)$-quantile of the loss distribution.

For a continuous price model with a continuous and strictly monotonic distribution function F_{G_h}, it is possible to solve the equation $F_{G_h}(x) = \alpha$ for x, and then the VaR is expresses in terms the inverse CDF function:

$$\text{VaR}_{\alpha,h} = -F_{G_h}^{-1}(\alpha).$$

Note that $g_{\alpha,h}$ is negative, and hence the VaR value is positive, i.e., VaR measures losses and is reported as a positive amount that corresponds to the loss.

When VaR is estimated from a P&L distribution, it is given in value terms. However, one may prefer to analyze the return distribution rather than the P&L distribution. In this case, the VaR is expressed as a percentage of the current value of the portfolio. The discounted h-day return on the portfolio is

$$\frac{e^{-jh}\Pi_{t+h} - \Pi_t}{\Pi_t}.$$

To calculate the VaR, we first find $r_{\alpha,h}$, the α-quantile of the return distribution by solving

$$\mathbb{P}\left(\frac{e^{-jh}\Pi_{t+h} - \Pi_t}{\Pi_t} < r_{\alpha,h}\right) = \alpha,$$

and then the VaR is given by

$$\text{VaR}_{\alpha,h} = \begin{cases} -r_{\alpha,h} & \text{as a percentage of the portfolio value } \Pi_t, \\ -r_{\alpha,h}\Pi_t & \text{as a quantity in value terms.} \end{cases}$$

Example 2.13. Assume that the discounted P&L is a normal random variable with mean μ and variance σ^2. Calculate 1% VaR.

Solution. We have that the discounted P&L G is given by $G := \mu + \sigma Z$ with $Z \sim Norm(0, 1)$. We need to find g so that

$$\alpha = 0.01 = \mathbb{P}(G < g) = \mathbb{P}\left(\frac{G - \mu}{\sigma} < \frac{g - \mu}{\sigma}\right) = \mathbb{P}\left(Z < \frac{g - \mu}{\sigma}\right) = \mathcal{N}\left(\frac{g - \mu}{\sigma}\right).$$

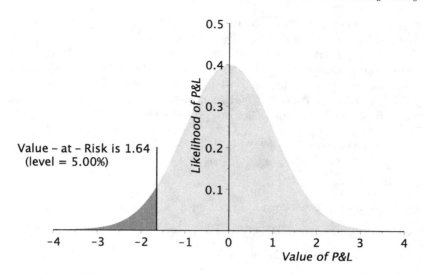

FIGURE 2.5: The Value-at-Risk diagram for a standard normal Profit-and-Loss PDF. The light-grey area to the right of the line represents 95% of the total area under the curve. The dark-grey area to the left of the line represents 5% of the total area under the curve.

Let us find the 0.01-quantile $x_{0.01}$ for the standard normal distribution. We use the table of the standard normal CDF to find x such that $\mathcal{N}(x) = 0.01$. By symmetry, $\mathcal{N}(-x) = 1 - 0.01 = 0.99$. From the table we have $\mathcal{N}(2.33) = 0.9901$ as the closest value. Hence, $x_{0.01} = -2.33$, and the VaR value is given by $-g = -(\mu + \sigma x_{0.01}) = -\mu + 2.33\sigma$. Therefore, among investments whose gains are normally distributed, the VaR criterion would select the one having the largest value of $-\mu + 2.33\sigma$. Figure 2.5 displays the PDF of the P&L and the 5% VaR value for the standard normal distribution.

In general, for an arbitrary value $\alpha \in (0, 1)$, the Value at Risk is

$$\mathrm{VaR}_\alpha = -\mu + \sigma z_\alpha,$$

where z_α is the normal critical value for level $\alpha \in (0, 1)$ defined by $\mathcal{N}(z_\alpha) = 1 - \alpha$. $\qquad\square$

Example 2.14. Let the P&L be $G := s(e^X - 1)$ where $X \sim \mathrm{Norm}(\mu, \sigma^2)$ and $s > 0$. Find $VaR_\alpha(G)$ for $\alpha \in (0, 1)$.

Solution. The P&L G has a log-normal distribution. Let us find its CDF:

$$F_G(x) = \mathbb{P}(s(e^X - 1) \leqslant x) = \mathbb{P}(e^X \leqslant 1 + x/s) = \mathbb{P}(X \leqslant \ln(1 + x/s)$$

$$= \mathbb{P}\left(\frac{X - \mu}{\sigma} \leqslant \frac{\ln(1 + x/s) - \mu}{\sigma}\right) = \mathcal{N}\left(\frac{\ln(1 + x/s) - \mu}{\sigma}\right).$$

The distribution of G is continuous and the CDF $F_G(x)$ is a strictly monotonic function for $x > 0$. To find the VaR, solve $F_L(x) = \alpha$ for x:

$$\mathcal{N}\left(\frac{\ln(1 + x/s) - \mu}{\sigma}\right) = \alpha$$

$$\frac{\ln(1 + x/s) - \mu}{\sigma} = \mathcal{N}^{-1}(\alpha)$$

$$\ln(1 + x/s) = \mu - \sigma z_\alpha$$
$$x = s\left(e^{\mu - \sigma z_\alpha} - 1\right)$$

where we use the property that $\mathcal{N}^{-1}(\alpha) = -\mathcal{N}^{-1}(1-\alpha) = -z_\alpha$. Therefore, the VaR with confidence level α is

$$\text{VaR}_\alpha = s\left(1 - e^{\mu - \sigma z_\alpha}\right).$$

For example, if $s = \$10^6$, $\mu = 0$, $\sigma = 1.5\%$, and $\alpha = 5\%$, then

$$\text{VaR}_{5\%} = 10^6 \cdot \left(1 - e^{-0.015 \cdot 1.645}\right) \cong \$24{,}373.06.$$

That is, the loss will not exceed $\$24{,}373.06$ with confidence 95%. $\qquad\square$

The VaR with significance level α gives us a value that has only a $100\%\alpha$ chance of being exceeded by the loss from an investment. However, this value does not tell us what the actual loss may be. It has been suggested that the conditional expected loss given that it exceeds the VaR is a better metric of the risk. This conditional expected loss is called the *conditional value at risk*, denoted CVaR, or the *expected shortfall*. The CVaR criterion is to choose the investment having the smallest CVaR.

Example 2.15. Assume that the discounted P&L is a normal random variable with mean μ and variance σ^2. Calculate 1% CVaR.

Solution. We have that the discounted P&L is $G = \mu + \sigma Z$ with $Z \sim Norm(0, 1)$. The CVaR is given by

$$\text{CVaR} = \text{E}[-G \mid -G > \text{VaR}] = \text{E}[-G \mid -G > -\mu + 2.33\sigma]$$
$$= \text{E}\left[\sigma\left(\frac{-G + \mu}{\sigma}\right) - \mu \,\middle|\, \frac{-G + \mu}{\sigma} > 2.33\right]$$
$$= \sigma \text{E}\left[\frac{-G + \mu}{\sigma} \,\middle|\, \frac{-G + \mu}{\sigma} > 2.33\right] - \mu = \sigma \text{E}[Z \mid Z > 2.33] - \mu.$$

For a standard normal random variable Z we have

$$\text{E}[Z \mid Z > a] = \frac{\text{E}[Z\,\mathbb{I}_{\{Z>a\}}]}{\mathbb{P}(Z > a)} = \frac{1}{\mathbb{P}(Z > a)} \int_a^\infty z\, n(z)\, \text{d}z$$
$$= \frac{1}{\mathbb{P}(Z > a)} \int_a^\infty \frac{1}{\sqrt{2\pi}} e^{-z^2/2}\, \text{d}(z^2/2) = \frac{1}{\sqrt{2\pi}\,\mathbb{P}(Z > a)} e^{-a^2/2}$$

for any real a. Hence, we obtain that

$$\text{CVaR} = \sigma \frac{1}{\sqrt{2\pi}\,0.01} e^{-2.33^2/2} - \mu \cong -\mu + 2.64\sigma,$$

since $\mathbb{P}(Z > 2.33) \cong 0.01$. $\qquad\square$

2.5   Dividend Paying Stock

Consider a stock (with the price process $\{S(t)\}_{t \geq 0}$) that pays dividends. Every moment a dividend payment is made, the price of one share instantaneously drops down by the amount of the dividend payment or otherwise an arbitrage opportunity would arise. Indeed, one can buy a share of stock right before a dividend payment is made, receive the payment, and then immediately sell the share. There is an arbitrage profit if the stock price is not adjusted when the dividend payment is made. Here, we assume that there is no delay between the ex-dividend date and the date when shareholders receive dividend payments. Note that in the U.S., the Internal Revenue Service (IRS) defines the ex-dividend date as "the first date following the declaration of a dividend on which the buyer of a stock is not entitled to receive the next dividend payment."

Suppose that a dividend payment $\text{div}(t_*)$ is made at time t_*. This payment can be given as a monetary amount or as a percentage of the spot price, i.e., $\text{div}(t_*) = d_* S(t_*)$ with the dividend percentage $0 \leqslant d_* \leqslant 1$. The price of one share immediately after the dividend payment must be $S(t_*) - \text{div}(t_*) = S(t_*) - d_* S(t_*) = (1 - d_*) S(t_*)$ or otherwise an arbitrage opportunity exists. We illustrate this idea with a single-period model. Let S be the stock price at the beginning of the period. At the end of the period, the (pre-dividend) price is Su with probability p or Sd with probability $1 - p$. After the dividend is paid, the price goes down by the dividend amount to become $Su(1 - d_*)$ or $Sd(1 - d_*)$. This situation is illustrated in Figure 2.6.

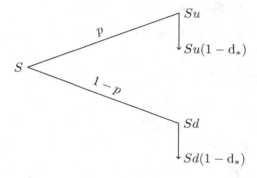

FIGURE 2.6: A single-period binomial model for a stock with dividends.

Suppose that the dividend on a single share is used to purchase

$$\frac{d_* S(t_*)}{(1 - d_*) S(t_*)} = \frac{d_*}{1 - d_*}$$

additional shares at time t_*. So, each share in our portfolio will grow to $1 + \frac{d_*}{1 - d_*} = \frac{1}{1 - d_*}$ shares right after time t_*. The market value of one share is

$$\Pi_t = \begin{cases} S(t) & \text{if } t < t_*, \\ \frac{1}{1 - d_*} S(t) & \text{if } t \geqslant t_*. \end{cases}$$

So, we may use the same asset price model without making dividend adjustments, if the dividends are assumed to be reinvested in the stock.

Let the dividends be paid at times $t_k = k\Delta t$, $k = 1, 2, \ldots, m$, distributed evenly with the step size $\Delta t = \frac{T}{m}$. Let d_m be the dividend percentage. Starting with one share at time 0, the market value of our portfolio at time T after the mth dividend payment is $\Pi_T = \frac{1}{(1 - d_m)^m} S(T)$. Let $m \to \infty$, so that $d_m / \Delta t \to q$ with $q > 0$, then

$$\frac{1}{(1 - d_m)^m} \to e^{qT}.$$

The stock is said to pay dividends continuously at a rate of q. If the dividends are reinvested in the stock, then an investment in one share held at time 0 will increase to become e^{qT} shares at time T. Therefore, we need to start with e^{-qT} shares at time 0 to obtain 1 share at time T.

2.6   Exercises

Exercise 2.1. At time 0, the value of a risk-free bond is $B_0 = 100$, and the stock price is $S_0 = 100$. Suppose that the annual risk-free interest rate is $r = 5\%$, and the one-year return on the stock is

$$r_S = \begin{cases} 10\% & \text{with probability } 60\% \\ -5\% & \text{with probability } 40\% \end{cases}$$

(a) Find positions x and y so that the wealth $\Pi_t = xS_t + yB_t$ of the portfolio (x, y) at time $T = 1$ is

$$\Pi_T = \begin{cases} \$1000 & \text{if the stock price goes up} \\ \$1500 & \text{if the stock price goes down} \end{cases}$$

(b) What is the expected return of the portfolio over the first year?

Exercise 2.2. Consider a binomial model with $r_B = 10\%$, $r_S^- = -10\%$, $r_S^+ = 20\%$, $S_0 = 100$, and $B_0 = 10$. On the (x, y)-plane draw a domain representing the set of all admissible portfolios (i.e., having nonnegative values Π_0 and Π_T) with x stock shares and y bonds.

Exercise 2.3. Consider a one-period binomial model with $r_B = 5\%$, $r_S^- = -10\%$, $r_S^+ = 15\%$, $S_0 = \$100$, and $B_0 = \$100$. Construct a no-short-selling portfolio with an initial value of \$10,000 such that, at all times, the stock investment is at least twice that invested in risk-free bonds.

Exercise 2.4. Consider a portfolio maturing in three months that consists of eight shares of a risky stock and four units of a risk-free bond. Suppose that the stock's price is currently \$150 per share, and, three months from now, its price will be either \$120 or \$180. Also, one bond is currently \$90, and, three months from now, it will be worth \$100.8.

(a) Determine the initial value and the terminal value of the portfolio.

(b) Find the compounded monthly rate of return on the portfolio.

Exercise 2.5. A portfolio maturing in four months consists of nine shares of a risky stock and 10 units of a risk-free bond. Suppose that the stock price is currently \$60 per share, and, four months from now, there is a 25% chance that its price will be 9% higher or a 75% chance that its price will be 9% lower. Also, one bond is currently worth \$90, and, four months from now, it will be 1% higher.

(a) Determine the initial value and the terminal value of the portfolio.

(b) Find the expected portfolio value at maturity.

(c) Find the expected value of the compounded monthly rate of return on the portfolio.

Exercise 2.6. Consider a stock with an initial price of \$150 that will either be 20% higher or 30% lower after each period. Construct a three-period binomial model for the stock price.

Exercise 2.7. Consider a stock with an initial price of \$130 that has a 25% chance it will be 10% higher after each period or 15% lower otherwise. Find the probability distribution of the stock price S_4 and compute $\mathbb{P}(S_4 \geqslant S_0)$.

Exercise 2.8. Find $E[S_T]$ for a binomial tree model with T periods.

Exercise 2.9. Let $P_{n,k} = \binom{n}{k}p^k(1-p)^{n-k}$ for $k = 0, 1, \ldots, n$ and $P_{n,k} = 0$ for all $k < 0$ and all $k > n$. Prove the recurrence:

$$P_{n+1,k} = p \cdot P_{n,k-1} + (1-p) \cdot P_{n,k}.$$

Exercise 2.10. Consider a stock with an initial price of \$70 and suppose that the expected value and volatility of the annual log-return on the stock are 5% and 20%, respectively. In an N-period binomial tree model with $N = 40$:

(a) Calculate the factors u and d as well as the real-world probability p.

(b) Find the expected value of a single-period return on the stock.

(c) Find the expected value of the stock price at the year-end.

Exercise 2.11. To calibrate an N-period binomial tree model, one needs to solve the simultaneous equations (2.24)–(2.26).

(a) Find the exact solution to (2.24)–(2.26).

(b) Let the condition $u_N d_N = 1$ in (2.26) be replaced by $p_N = \frac{1}{2}$. Find u_N and d_N satisfying (2.24)–(2.25) and this new condition.

Exercise 2.12. Let $Z \sim Norm(0, 1)$. Find the mathematical expectation of $X = e^{aZ+b}$ with $a, b \in \mathbb{R}$. Use the result obtained to find the variance $\mathrm{Var}(X) = \mathrm{E}[X^2] - \mathrm{E}[X]^2$.

Exercise 2.13. Consider the log-normal price model $S(T) = S(0)e^{\mu T + \sigma\sqrt{T}Z}$, where $Z \sim Norm(0, 1)$, with drift parameter $\mu = 0.02$ and volatility parameter $\sigma = 0.2$. If $S(0) = 100$, find (a) $\mathrm{E}[S(5)]$, (b) $\mathbb{P}(S(5) > 100)$, (c) $\mathbb{P}(S(5) < 110)$.

Exercise 2.14 (A variant of the one-price theorem). Assume that there are no arbitrage portfolios. Suppose there are two assets X and Y with initial prices X_0 and Y_0. At some time $T > 0$, let $X_T(\omega) \geqslant Y_T(\omega)$ hold for all states of the world and $X_T(\omega') > Y_T(\omega')$ hold for at least one state ω'. Prove that $X_0 > Y_0$. [Hint: Suppose the converse is true and then construct an arbitrage portfolio.]

Exercise 2.15. Consider two assets with respective value functions V_t and W_t, with $t \in [0, T]$. Suppose that $V_0 > W_0$ and $\mathbb{P}(V_T \leqslant W_T) = 1$. Find an arbitrage opportunity.

Exercise 2.16. A market currently consists of a \$120 risky stock and a \$150 risk-free bond. Suppose that, in six months, there is a 20% chance that the stock price will be \$145 or \$110 otherwise, and the bond will be worth \$177.00. Consider a contract with maturity $T = 6$ months that pays \$15 if the stock price decreases and \$85 if the stock price goes up.

(a) Find a portfolio replicating the contract.

(b) Calculate the initial value of the replicating portfolio.

(c) Find the expected terminal value of the replicating portfolio.

Exercise 2.17. Consider a market model with a risky stock and a risk-free bond, which are initially worth \$140 and \$60, respectively. Suppose that, in two months, there's a 75% chance that the stock's price will be \$165 or \$130 otherwise, and the bond will be worth \$72.51.

(a) Find single-period rates of return on the stock and the bond, respectively.

(b) Determine whether an arbitrage opportunity exists. If it does, set up an arbitrage port-folio with the initial value zero.

Exercise 2.18. A market currently consists of a risky $50 stock and $50 risk-free bond. Suppose that, in three months, there is a 20% chance that the stock's price will be $65 or $35 otherwise, and the bond will be worth $64.00. Consider a contract with maturity $T =$ three months and payoff of $60 if the stock goes down and $40 if the stock goes up.

(a) Find a portfolio replicating the contract. Calculate the initial value of the replicating portfolio.

(b) For the portfolio in (a), determine its expected terminal value and the allocation weights.

(c) Find single-period rates of return on the stock and the bond, respectively. Use the result to determine whether an arbitrage opportunity exists.

Exercise 2.19. A market currently consists of a risky $130 stock and a $70 risk-free bond. Suppose that in one year the stock's price will be either $160 or $150, and the bond will be worth $72.10. Show that arbitrage exists and set up an arbitrage portfolio with the initial value zero.

Exercise 2.20. A market currently consists of a risky $150 stock and a $150 risk-free bond. Suppose that in one year the stock's price will be either $155 or $130, and the bond will be worth $156.50. Show that an arbitrage opportunity exists and set up an arbitrage portfolio with the initial value zero.

Exercise 2.21. A market currently consists of a claim, a risky $80 stock, and a $120 risk-free bond. Suppose that in one year's time:

- The claim will have a payoff of either $C^+ = \$95$ or $C^- = \$5$.

- The stock's price will be either $S^+ = \$110$ or $S^- = \$55$.

- The bond will be worth $151.20.

(a) Determine the risk-neutral probability of the stock price increasing in value in one year's time.

(b) Use expectation to find the no-arbitrage initial price (e.g., the fair price) C_0 of the claim.

(c) Suppose the market price of the claim is mispriced at $C_0 = \$48.00$. If you short sell 27 shares of the stock, then you will be able to buy 10 bonds and some claims. Find the number of claims so that the portfolio costs nothing to construct, show it is an arbitrage, and find the profit at maturity.

Exercise 2.22. Assume a single-period model with two base assets. An investor has written a contract with payoff C_T. Additional, she buys a number of shares of the risky stock to *hedge* her short position in the contract and invest in the risk-free bond. The time-t value of the *hedging* portfolio is $V_t = xS_t + yB_t - C_t$ with $t \in \{0, T\}$. Derive explicit formulae for x and y in the hedging portfolio by solving $\mathrm{Var}(V_T) \to \min$ subject to $V_0 = 0$.

Exercise 2.23. Consider a trinomial model with $B_0 = 100$, $B_T = 110$, $S_0 = 100$, $d = 0.9$, $m = 1$, $u = 1.2$.

(a) Is it possible to find a portfolio replicating the payoff with $C_T(\omega^-) = 0$, $C_T(\omega^0) = 0$, $C_T(\omega^+) = 20$?

(b) Is it possible to replicate the payoff with $C_T(\omega^-) = 13$, $C_T(\omega^0) = 12$, $C_T(\omega^+) = 10$?

Exercise 2.24. Consider a single-period binomial model with a risk-free asset B, such that $B_0 = 10$ and $B_T = 11$, and a risky asset with initial price $S_0 = 100$ and terminal price S_T.

(a) Let $S_T(\omega^1) = 90$ and $S_T(\omega^2) = Y$ with $Y > 0$. For what values of Y is the model arbitrage-free?

(b) Let $S_T(\omega^1) = X$ and $S_T(\omega^2) = Y$ with $0 < X < Y$. For what values of X and Y is the model arbitrage-free? Draw a diagram representing all possible solutions.

(c) Suppose that $S_T(\omega^1) = 90$ and $S_T(\omega^2) = 105$. Find an arbitrage opportunity by selling short one share of either X or Y.

(d) Suppose that $S_T(\omega^1) = 110$ and $S_T(\omega^2) = 120$. Find an arbitrage opportunity by selling short one share of either X or Y.

Exercise 2.25. In the setting of Exercise 2.1

(a) find the risk-neutral probabilities $\{\tilde{p}, 1 - \tilde{p}\}$ for this binomial model;

(b) verify that under the risk-neutral probabilities we have

$$\widetilde{E}\left[\frac{S_1}{B_1}\right] = \frac{S_0}{B_0}.$$

In the latter case, the discounted stock price process S_t/B_t is said to be a martingale.

Exercise 2.26. Consider a stock with an initial price of \$150 that has a 70% chance it will be 10% higher after each period or 20% lower otherwise. Let the risk-free rate in the market be 6%.

(a) Determine the risk-neutral probability, $\tilde{p}$, of the stock price increasing in value (in one period).

(b) Find the risk-neutral probability distribution of the stock price S_3 and compute $\mathbb{P}(S_3 \leqslant S_0)$.

Exercise 2.27. Consider a stock with an initial price of \$70 that has a 10% chance it will be 22.5% higher after each period or 27.5% lower otherwise. Let the risk-free rate in the market be 6%.

(a) Determine the risk-neutral probability, $\tilde{p}$, of the stock price increasing in value (in one period).

(b) Compute the risk-neutral expectation of the single-period stock return and then find the risk-neutral expected stock price $\widetilde{E}[S_3]$.

Exercise 2.28. Consider a single-period binomial model with a risk-free asset B, such that $B_0 = 10$ and $B_T = 11$, and a risky asset with $S_0 = 100$, the time-T price of which can follow two possible scenarios: $S_T(\omega^1) = 90$, $S_T(\omega^2) = 120$.

(a) Find the risk-neutral probabilities $\tilde{p}_1 = \widetilde{\mathbb{P}}(\omega^1)$ and $\tilde{p}_2 = \widetilde{\mathbb{P}}(\omega^2)$.

(b) Show that $\widetilde{E}\left[\frac{S_T}{B_T}\right] = \frac{S_0}{B_0}$.

Exercise 2.29. Consider a single-period binomial model with a risk-free asset B, such that $B_0 = 10$ and $B_T = 11$, and a risky asset with $S_0 = 100$, the time-T price of which can follow two possible scenarios: $S_T(\omega^1) = 90$, $S_T(\omega^2) = 120$. Consider a derivative with payoff $C_T = \max\{S_T - 100, 0\}$ (a European option). That is, $C_T(\omega^1) = 0$, $C_T(\omega^2) = 20$. Find the fair initial price C_0 by using:

(a) a replicating portfolio.

(b) the risk-neutral pricing formula

$$C_0 = \frac{B_0}{B_T}\widetilde{\mathbb{E}}[C_T] = \frac{B_0}{B_T}\sum_{k=1}^{2} C_T(\omega^k)\tilde{p}_k$$

where $\tilde{p}_1 = \widetilde{\mathbb{P}}(\omega^1)$ and $\tilde{p}_2 = \widetilde{\mathbb{P}}(\omega^2)$ are risk-neutral probabilities

Exercise 2.30. Consider a four-period binomial tree model with $S_0 = 32$, $u = 2$, $d = 0.5$, $r = 50\%$, $p = 60\%$.

(a) Construct the binomial tree diagram.

(b) Find the real-word and the risk-neutral probability distributions of S_4.

(c) Find $\mathbb{E}[S_4]$ and $\widetilde{\mathbb{E}}[S_4]$.

(d) Find $\mathbb{P}(S_4 > S_0)$ and $\widetilde{\mathbb{P}}(S_4 > S_0)$.

Exercise 2.31. Consider a single-period market model with a risk-free asset B, such that $B_0 = 10$ and $B_T = 11$, and a risky asset with $S_0 = 50$, the time-T price of which can follow three possible scenarios: $S_T(\omega^1) = 70$, $S_T(\omega^2) = 55$, $S_T(\omega^3) = 40$.

(a) Show that the model admits no arbitrage opportunities.

(b) Find risk-neutral probabilities $\tilde{p}_i = \widetilde{\mathbb{P}}(\omega^i)$ of the scenarios, such that

$$\widetilde{\mathbb{E}}[S_T/B_T] = S_0/B_0$$

holds. Find the general solution that will depend on a variable parameter. Find the range for that parameter so that $\tilde{p}_1$, $\tilde{p}_2$, $\tilde{p}_3$ define a probability function.

Exercise 2.32. Consider a single-period market model with three states of the world and two base assets: a risky stock S and risk-free money market account A. Suppose that the possible stock prices at time $t = T$ are as follows:

$$S_T = \begin{cases} S^u & \text{with probability } p_1 > 0, \\ S^m & \text{with probability } p_2 > 0, \\ S^d & \text{with probability } p_3 = 1 - p_1 - p_2 > 0, \end{cases}$$

where $0 < S^d < S^m < S^u$. Let the initial investment in a risk-free bond be equal to the current stock price, i.e., $B_0 = S_0$, Prove that at time $t = T$ we have that $S^d < B_T < S^u$ or else an arbitrage possibility would arise. In the latter case, construct an arbitrage portfolio.

Exercise 2.33. Consider a single-period trinomial model with a risk-free asset B, such that $B_0 = 10$ and $B_T = 11$, and a risky asset with $S_0 = 100$, the time-T price of which can follow three possible scenarios: $S_T(\omega^1) = 90$, $S_T(\omega^2) = 100$, $S_T(\omega^3) = 120$. Consider a derivative with payoff $C_T = \max\{S_T - 100, 0\}$ (a European call option). That is, $C_T(\omega^1) = 0$, $C_T(\omega^2) = 0$, $C_T(\omega^3) = 20$.

(a) Find all super-replicating portfolios (x, y) such that

$$\Pi_T^{(x,y)}(\omega) = xS_T(\omega) + yB_T \geqslant C_T(\omega) \quad \text{for } \omega \in \{\omega^1, \omega^2, \omega^3\}$$

(b) Find $\min \Pi_0^{(x,y)}$ and $\max \Pi_0^{(x,y)}$ on the set of super-replicating portfolios found in part (a).

Exercise 2.34. Consider a market with a risk-free bond B, for which $B_0 = 50$, $B_1 = 55$, and $B_2 = 60$, and a risky stock with the spot price $S_0 = 50$. Suppose that the stock price at times $t = 1$ and $t = 2$ can follow four possible scenarios:

Scenario	S_1	S_2
ω^1	60	70
ω^2	60	55
ω^3	45	45
ω^4	45	40

(a) Find an arbitrage investment strategy if there are no restrictions on short selling.

(b) Is there an arbitrage opportunity if no short selling of the risky asset is allowed?

Exercise 2.35. Given the bond and stock prices in Exercise 2.34, is there an arbitrage strategy if short selling of stock is allowed, but transaction costs of 5% of the transaction volume apply whenever stock is traded (purchased or sold)?

Exercise 2.36. Given the bond and stock prices in Exercise 2.34 except that $S_2(\omega^2) = S_2(\omega^3) = 50$, and the probabilities $\mathbb{P}(\omega^1) = \frac{17}{33}$, $\mathbb{P}(\omega^2) = \frac{5}{33}$, $\mathbb{P}(\omega^3) = \frac{10}{33}$, and $\mathbb{P}(\omega^4) = \frac{1}{33}$, show that the discounted stock price process S_t/B_t, $t = 0, 1, 2$, is a martingale under this probability $\mathbb{P}$, i.e., show that $E\left[\frac{S_{t+1}}{B_{t+1}} \mid S_t\right] = \frac{S_t}{B_t}$ for $t = 0, 1$.

Exercise 2.37. Prove Proposition 2.8.

Exercise 2.38. Consider the log-normal model under the risk-neutral dynamic:

$$S(T) = S(0)e^{(r-\sigma^2/2)T+\sigma\sqrt{T}Z}, \quad \text{where } Z \sim Norm(0, 1).$$

Show that $\widetilde{E}[e^{-rT}S(T)] = S(0)$.

Exercise 2.39. Consider the binomial tree price model under the risk-neutral dynamic. Show that $\widetilde{E}\left[(1+r)^{-N}S_N\right] = S_0$.

3

Portfolio Management

3.1 Expected Utility Functions

3.1.1 Utility Functions

Suppose we have different investment opportunities to choose from. These investments may affect our future wealth. For example, our task is to allocate an initial capital among several risky assets to form an investment portfolio. The future wealth of a risky investment is uncertain and follows some probability distribution. Thus, the investment selection procedure can be reduced to the optimization of such a probability distribution. If the outcomes from all alternatives were certain, then we would select the investment that produces the largest return. In the presence of uncertainty, we may want to minimize the variance of the respective probability distribution, but other criteria can also be applied. So we need a systematic way to rank random wealth levels. In the case of alternatives with uncertain outcomes, we introduce some score function that is calculated as an expected value of a so-called *utility function*.

Suppose we are given a function $u\colon \mathbb{R} \to \mathbb{R}$ so that each possible investment can be assessed by computing the expected utility value $\mathrm{E}[u(V)]$ of the future wealth V. In other words, the value of an investment can be measured by the expected value of the utility of its consequences. To compare possible alternatives, we first compute $\mathrm{E}[u(V)]$ for each possible wealth function V and then choose the option with the highest expected utility value. The specific utility function used depends on personal investment preferences, risk tolerance, and individual financial environment. Formally, the investor with utility function u prefers one investment with the future wealth W to another opportunity with the wealth V iff $\mathrm{E}[u(V)] < \mathrm{E}[u(W)]$;

The simplest example of a utility is the linear function $u(x) = x$. Whoever uses it ranks uncertain wealth functions by their expected values. Indeed, $\mathrm{E}[u(V)] = \mathrm{E}[V]$, and thus $\mathrm{E}[u(V)] < \mathrm{E}[u(W)]$ holds iff $\mathrm{E}[V] < \mathrm{E}[W]$. In general, the linear utility is $u(x) = ax + b$ with $a > 0$. For any random wealth V we have

$$\mathrm{E}[u(V)] = \mathrm{E}[aV + b] = a\mathrm{E}[V] + b = u(\mathrm{E}[V]).$$

Hence the linear utility function has no preference for a deterministic wealth or for a random wealth provided that expected wealth is the same. Therefore, this function reflects expectations of a *risk-indifferent* investor.

Let us consider a less trivial example. Suppose that the current wealth V increases to $V + \Delta V$ and that "the joy of winning" ΔV given by $u(V + \Delta V) - u(V)$ is directly proportional to ΔV and inversely proportional to V:

$$u(V + \Delta V) - u(V) = a\frac{\Delta V}{V} \quad \text{with } a > 0.$$

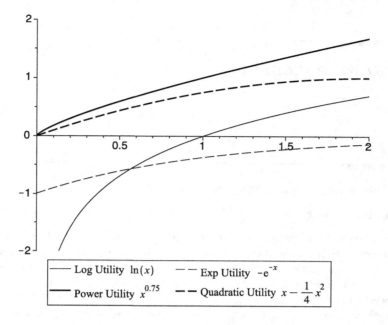

FIGURE 3.1: Sample plots of four commonly used utility functions.

As the change in value ΔV approaches 0, we have

$$\frac{u(V + \Delta V) - u(V)}{\Delta V} \to u'(V).$$

Hence, the utility function satisfies $u'(V) = \frac{a}{V}$. By integrating, we obtain the *logarithmic utility* function.

$$u(v) = a \ln V + b.$$

Here are some of the most commonly used utility functions (see also Figure 3.1):

- the linear utility function $u(x) = ax + b$ with $a > 0$;
- the logarithmic (log) utility function $u(x) = a \ln x + b$ with $a > 0$;
- the exponential utility function $u(x) = b - e^{-ax}$ with $a > 0$;
- the power utility function $u(x) = x^a$ with $0 < a < 1$;
- the quadratic utility function $u(x) = x - ax^2$ with $a > 0$ defined for $x < \frac{1}{2a}$.

As you can see in Figure 3.1, some utility functions can take negative values. This negativity does not matter since an investor ranks investments using relative values. Moreover, the addition of a constant to a utility function and the multiplication of a utility function by a positive constant do not affect the rankings. Indeed, if for some utility function u and two investments V_1 and V_2 we have that $\mathrm{E}[u(V_1)] \leqslant \mathrm{E}[u(V_2)]$, then for any $a > 0$ and $b \in \mathbb{R}$ we obtain

$$\mathrm{E}[au(V_1) + b] = a\mathrm{E}[u(V_1)] + b \leqslant a\mathrm{E}[V_2] + b = \mathrm{E}[au(V_2) + b].$$

So the utility $au(x) + b$ gives the same ranking as the function $u(x)$. In general, given a utility function u, we can define another utility function

$$v(x) = au(x) + b \text{ with } a > 0.$$

This new utility function v is said to be *equivalent* to u. Equivalent utility functions give identical rankings of investment opportunities.

Here are some examples of equivalent utility functions:

- x, $10x$, $10x - 100$;
- $\ln x$, $2\ln x + 10$, $\log_{10} x$ (since $\log_{10} x = \frac{\ln x}{\ln 10}$ and $\ln 10 > 0$), $\ln x^2$ (since $\ln x^2 = 2\ln x$);
- $-e^{-0.01x}$, $1 - e^{-0.01x}$, $1 - e^{1-0.01x}$ (since $e^{1-0.01x} = e \cdot e^{-0.01x}$).

Example 3.1. An investor with total capital W can invest any amount between 0 and W. If an amount is invested, then the same amount is either gained or lost with respective probabilities p and $1 - p$. In other words, with probability p, the investor doubles the initial investment; with probability $1 - p$, the investor loses all the invested money. What amount should be invested if the log utility function $u(V) = \ln V$ is utilized for ranking alternatives?

Solution. Let the amount of xW for some $0 \leqslant x \leqslant 1$ be invested. The investor's final fortune $V(x)$ is either $W + xW$ or $W - xW$ with respective probabilities p and $1 - p$. Hence the expected utility of the final wealth is

$$
\begin{aligned}
E[u(V(x))] &= p\ln(W + xW) + (1 - p)\ln(W - xW) \\
&= p\ln((1 + x)W) + (1 - p)\ln((1 - x)W) \\
&= p\ln(1 + x) + (1 - p)\ln(1 - x) + \ln W.
\end{aligned}
$$

To find the optimal value of x, let us differentiate $E[u(V(x))]$ with respect to x and then find zeros of the derivative obtained:

$$
\begin{aligned}
\frac{\mathrm{d}}{\mathrm{d}x} E[u(V(x))] &= \frac{\mathrm{d}}{\mathrm{d}x} \left(p\ln(1 + x) + (1 - p)\ln(1 - x) \right) \\
&= \frac{p}{1 + x} - \frac{1 - p}{1 - x} = \frac{2p - (1 + x)}{1 - x^2}.
\end{aligned}
$$

If $p \in (0, \frac{1}{2}]$, then the derivative is strictly negative for all $x \in (0, 1)$ and the expected utility attains its maximum value at $x = 0$. In this case, the risk to lose the invested amount is too high, and it is reasonable to invest nothing. If $p \in (\frac{1}{2}, 1)$, then the derivative is zero at $x^* = 2p - 1 \in (0, 1)$. The second derivative of the expected utility function is negative at x^*, hence $x^* = 2p - 1 \in (0, 1)$ is the point of maximum of $V(x)$. Therefore, $100(2p - 1)\%$ of the initial capital is to be invested. For example, for $p = 70\%$, the investor shall invest 40% of the fortune. $\square$

Example 3.2. An investor equipped with the power utility function $u(x) = x - x^2/2$ is selecting between two investments with uncertain returns. If the wealth provided by investment 1 has the continuous uniform distribution on the interval $(0, 1)$ with the PDF $f_1(x) = \mathbb{I}_{(0,1)}(x)$, and the wealth of investment 2 has the PDF $f_2(x) = 6x(1 - x)\mathbb{I}_{(0,1)}(x)$, which investment should the investor choose?

Solution. Calculate the expected utility for each investment:

$$
E[u(V_1)] = \int_0^1 u(x)\, f_1(x)\, \mathrm{d}x = \int_0^1 (x - x^2/2)\, \mathrm{d}x = \frac{1}{3} = \frac{20}{60},
$$

$$
E[u(V_2)] = \int_0^1 u(x)\, f_2(x)\, \mathrm{d}x = \int_0^1 6x(1 - x)(x - x^2/2)\, \mathrm{d}x = \frac{7}{20} = \frac{21}{60}.
$$

Since $E[u(V_1)] < E[u(V_2)]$, the second investment is more preferable. Note that for both investments, the expected wealth is $\frac{1}{2}$. Thus, the investment opportunities are equivalent for a risk-indifferent investor using a linear utility function. $\square$

Example 3.3. The Saint Petersburg Paradox, originally proposed by Nicolaus Bernoulli, is a classic example of how utility functions are used in the decision-making process. Consider a game of chance where a fixed fee is paid to enter, and then a fair coin will be tossed repeatedly until the first head appears ending the game. The payoff starts at $1, and then it is doubled every time a tail appears. See Figure 3.2. As a result, the player wins 2^{k-1} if a head first appears on the kth toss ($k = 1, 2, 3, \ldots$). How much should the player be willing to pay to enter such a game?

Solution. First, let us find the expected value of the payoff. We deal with a sequence of independent trials where the probability of success (i.e., a head occurs) is $\frac{1}{2}$. With probability $p_1 = \frac{1}{2}$, a head first appears on the first toss, and the player wins $1; with probability $p_2 = \frac{1}{4}$, a head first appears on the second toss, and the player wins $2; with probability $p_3 = \frac{1}{8}$, a head first appears on the third toss, and the player wins $4, etc. The probability that a head first appears on the kth toss is $p_k = 2^{-k}$; the payoff is then 2^{k-1}. Let X denote the number of tails before the first head. It follows the geometric probability distribution: $X \sim Geom(1/2)$. The payoff is then equal to 2^X. Therefore, the expected payoff $E = \mathrm{E}[2^X]$ is then

$$E = \sum_{k=1}^{\infty} 2^{k-1} \cdot 2^{-k} = 1 \cdot \frac{1}{2} + 2 \cdot \frac{1}{4} + 4 \cdot \frac{1}{8} + \cdots + 2^{k-1} \cdot \frac{1}{2^k} + \cdots .$$

We can observe that the expected payout for the player of this game is an infinite amount:

$$E = \frac{1}{2} + \frac{1}{2} + \frac{1}{2} + \cdots = \infty .$$

So no matter how large is the fee paid to enter this game, the player will eventually make a profit in the long run repeatedly playing this game. However, it would be unreasonable to pay, for example, one million dollars to play this game of chance. The probability of winning at least one million in a single game is very small:

$$\mathbb{P}(2^X \geqslant 10^6) = \mathbb{P}\left(X \geqslant \log_2(10^6)\right) = \mathbb{P}(X \geqslant 20) = (1 - 0.5)^{20} \cong 9.5367 \cdot 10^{-7}.$$

The classic solution to this "paradox" is to assume that one's valuation of money is different from its face value and depends on his or her wealth. Let us apply the logarithmic utility model to find a reasonable price c charged to enter the game. Let the initial wealth of the player be denoted V_0. After paying the price of c, the wealth becomes $V_0 - c$. If a head

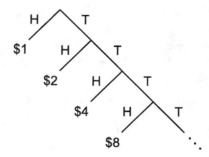

FIGURE 3.2: An outcome tree for the Saint Petersburg game. The game consists of a series of coin tosses offering a 50% chance of winning $1, a 25% chance of $2, a 12.5% chance of $4, and so on. The gamble may continue indefinitely.

first appears in the kth toss (with probability 2^{-k}), then the wealth becomes $V_0 - c + 2^{k-1}$. The expected log utility function of the total wealth $V = V(c)$ after playing the game is

$$E[\ln V] = \sum_{k=1}^{\infty} \ln(V_0 + 2^{k-1} - c)\, \frac{1}{2^k} < \infty.$$

A rational player is willing to play the game only if the game does not decrease the expected utility of his or her wealth:

$$E[\ln V] \geqslant E[\ln V_0].$$

After plotting the expected change in utility,

$$E[\ln V] - E[\ln V_0] = E[\ln V - \ln V_0] = E[\ln(V/V_0)],$$

as a function of the cost c (see Figure 3.3a), we observe that $E[\ln(V(c)/V_0)]$ is a strictly decreasing function of c. There exists a maximum cost c^* so that any price $c < c^*$ gives a positive expected change in utility. The cost c^* depends on the initial capital V_0 and can be found by solving $E[\ln(V(c)/V_0)] = 0$ for c. For example, a person with \$2 in his pocket is willing to pay up to \$2, a person with \$1000 is willing to pay up to \$5.96, and a millionaire is willing to pay up to \$10.93. [Note: To obtain these values, we solve the equation $E[\ln(V(c)/V_0)] = 0$ numerically and then round the values down to the nearest cent.] Figure 3.3b displays the dependence of the cost c^* on the initial wealth V_0 where the horizontal axis has a logarithmic scale. We can observe an almost linear dependence between c^* and $\ln V_0$. □

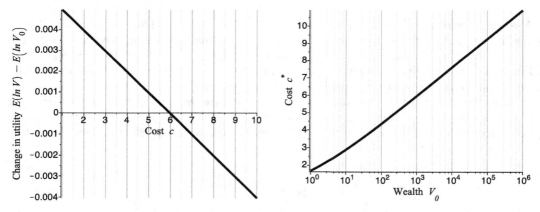

(a) The change in utility vs the cost c for the initial wealth $V_0 = 1000$.

(b) The maximum cost c^* vs the initial wealth V_0.

FIGURE 3.3: The relationship between variables in the Saint Petersburg paradox

3.1.1.1 Risk Aversion

Utility functions are constructed based on the following general principles.

Principle 1. Investors prefer more to less. If there are two certain amounts V_1 and V_2, then an investor prefers the larger one, i.e., $V_1 \leqslant V_2$ implies $u(V_1) \leqslant u(V_2)$. Hence, u is an increasing function.

Principle 2. Investors are averse to risk. A positive deviations ΔV from the wealth V cannot compensate for the equally large and equally probable negative deviations $-\Delta V$ from the wealth, i.e.,

$$u(V) - u(V - \Delta V) \geqslant u(V + \Delta V) - u(V). \qquad (3.1)$$

The left-hand side in (3.1), $u(V) - u(V - \Delta V)$, is "the pain of losing ΔV dollars," and the right-hand side, $u(V + \Delta V) - u(V)$, is "the joy of winning ΔV dollars." The inequality says that the pain of losing outweighs the joy of winning, alternatively that we react more severely to a loss then we do to a gain of the same magnitude. By rearranging terms, we obtain

$$u(V) \geqslant \frac{u(V + \Delta V) + u(V - \Delta V)}{2}. \qquad (3.2)$$

The inequality in (3.2) holds for all V and ΔV if u is a concave function.

In summary, a utility is an increasing, concave function. It is easy to verify that the logarithmic, exponential, power, quadratic and linear utility functions satisfy both principles.

Recall that a function u defined on an interval $[a, b]$ is *concave* if for any $p \in [0, 1]$ and any $x, y \in \mathbb{R}$ there holds

$$u(px + (1 - p)y) \geqslant pu(x) + (1 - p)u(y). \qquad (3.3)$$

A function u is *convex* on $[a, b]$ if the function $-u$ is concave on $[a, b]$. That is, for any p with $0 \leqslant p \leqslant 1$ and any $x, y \in \mathbb{R}$ we have

$$u(px + (1 - p)y) \leqslant pu(x) + (1 - p)u(y).$$

A twice differentiable function u is concave (convex) on an interval $[a, b]$ if its second derivative u'' is non-positive (nonnegative) on $[a, b]$. A utility function is said to be *risk-averse* (on an interval $[a, b]$) if it is concave (on the interval $[a, b]$). A concave function is depicted in Figure 3.4. For a twice-differentiable utility function, the risk-averse condition means that the second derivative of the utility function is non-positive.

Suppose that there are two alternatives for the future wealth: the first provides either x or y each with a probability of $\frac{1}{2}$, whereas the second gives $\frac{1}{2}x + \frac{1}{2}y$ with certainty. Although both alternatives have the same expected value, a risk-averse investor prefers the certain wealth of $\frac{1}{2}x + \frac{1}{2}y$ to a 50-50 chance of x and y:

$$u\left(\frac{x}{2} + \frac{y}{2}\right) \geqslant \frac{1}{2}u(x) + \frac{1}{2}u(y).$$

A generalization of this example provides the following probabilistic interpretation of (3.3). A risk-averse investor prefers the certain wealth $px + (1 - p)y$ to the uncertain wealth equal x or y with respective probabilities p and $1 - p$ for any $p \in (0, 1)$.

Recall *Jensen's inequality*: let u be a concave function, then for any random variable V,

$$E[u(V)] \leqslant u(E[V]). \qquad (3.4)$$

This means that a risk-averse investor prefers a certain wealth of W to an uncertain wealth V with the same expected value $E[V] = W$. This observation relates to the notion of the certainty equivalent.

The *certainty equivalent* of an uncertain wealth V is defined as the amount of a constant wealth C that has the utility level equal to the expected utility of V:

$$u(C) = E[u(V)]. \qquad (3.5)$$

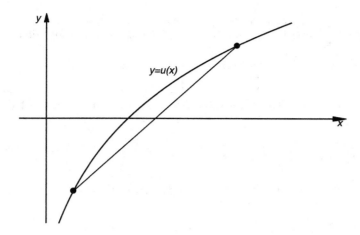

FIGURE 3.4: The concave (or convex-upward) plot of a typical risk-averse utility function. As is seen, every curve segment of a concave plot lies above a chord connecting the endpoints of the segment.

For example, in the Saint Petersburg paradox, if the price c equals c^*, then the certainty equivalent C is V_0. We have $C > V_0$ if $c < c^*$ and $C < V_0$ if $c > c^*$.

The certainty equivalent is the same for all equivalent utility functions. Combining (3.4) and (3.5) gives that the certainty equivalent is always less than the expected value of the wealth for a risk-averse investor with a concave utility function:

$$u(C) \leqslant u(\mathrm{E}[V]) \implies C \leqslant \mathrm{E}[V].$$

The exception is the linear utility $u(x) = a + bx$, for which $C = \mathrm{E}[V]$ since $\mathrm{E}[u(V)] = \mathrm{E}[a + bV] = a + b\mathrm{E}[V] = u(\mathrm{E}[V])$.

Example 3.4. Let $u(x) = 1 - e^{-0.1x}$ and $W \sim \mathit{Unif}(0, 10)$ (in thousands of dollars). Calculate and compare $\mathrm{E}[u(W)]$ and $u(\mathrm{E}[W])$. Find the certainty equivalent C.

Solution. The expected wealth is $\mathrm{E}[W] = \frac{0+10}{2} = \$5\mathrm{K}$. Hence, the utility of the expected wealth is

$$u(\mathrm{E}[W]) = u(5) = 1 - e^{-0.1 \cdot 5} \cong 0.393469.$$

The expected utility is

$$\mathrm{E}[u(W)] = \int_0^{10} u(x) \cdot \frac{1}{10}\, \mathrm{d}x \cong 0.367879.$$

Notice that $\mathrm{E}[u(W)] < u(\mathrm{E}[W])$. To find the certainty equivalent, solve $\mathrm{E}[u(W)] = u(C)$:

$$0.367879 = 1 - e^{-0.1C} \iff C \cong -10\ln(1 - 0.367879) \cong \$4586.74.$$

As we can see, $C < \mathrm{E}[W]$. $\qquad\qquad\square$

The certainty equivalent is a solution to the nonlinear equation (3.5). For some utility functions, we can find a closed-form formula for C, whereas for other functions equation (3.5) can only be solved numerically. Let us find an approximate formula for the certainty equivalent using Taylor's series expansion. Represent the uncertain return V in the form $V = W + \epsilon$, where W is the expected value of the wealth and ϵ is a zero-mean random risk.

A natural way to measure risk aversion is to ask how much an investor is ready to pay to get rid of the zero-mean risk ϵ. This price, called a *risk premium* and denoted π, is defined implicitly by

$$E[u(W + \epsilon)] = u(W - \pi). \tag{3.6}$$

That is, the risk premium is $\pi = E[V] - C$. The risk premium π is nonnegative. Let us consider a small risk ϵ. Expanding the left- and right-hand sides of (3.6) in Taylor's approximations gives

$$E[u(W + \epsilon)] \approx E\left[u(W) + \epsilon u'(W) + \frac{\epsilon^2}{2}u''(W)\right]$$

$$= u(W) + E[\epsilon]\, u'(W) + \frac{E[\epsilon^2]}{2}\, u''(W) = u(W) + \frac{\sigma_\epsilon^2}{2}\, u''(W)$$

and

$$u(W - \pi) \approx u(W) - \pi u'(W),$$

respectively, where $\sigma_\epsilon^2 := E[\epsilon^2]$ is the variance of ϵ. Clearly, $\sigma_\epsilon^2 = \text{Var}(V - W) = \text{Var}(V)$. Substituting these back into (3.6), we obtain

$$\pi \approx \frac{\sigma_\epsilon^2}{2} A_u(W),$$

where

$$A_u(W) := -\frac{u''(W)}{u'(W)}$$

is the *Arrow–Pratt absolute risk aversion coefficient*. We say that Investor 1 (with utility function u_1) is more risk-averse than Investor 2 (with utility function u_2) if, for the same initial wealth W and zero-mean risk ϵ, the risk premium π_1 paid by Investor 1 is larger than the risk premium π_2 of Investor 2, or, equivalently, $A_{u_1}(W) > A_{u_2}(W)$. The certainty equivalent is then given by

$$C = E[V] - \pi \approx E[V] - A_u(E[V])\, \text{Var}(V).$$

The degree of risk aversion can be viewed as a measure of the magnitude of the concavity of the utility function: the stronger the bend in the function, the larger the risk aversion coefficient A. For example, the risk-aversion coefficient for a linear utility function, $u(V) = a + bV$, is zero. The coefficient A is normalized by the derivative u' that appears in the denominator. This makes A independent of linear transformations of the utility function u. Indeed, for any $a \neq 0$ and b we have

$$A_{au+b}(x) = -\frac{(au(x) + b)''}{(au(x) + b)'} = -\frac{au''(x)}{au'(x)} = -\frac{u''(x)}{u'(x)} = A_u(x).$$

The function $A(W)$ shows how risk aversion changes with the wealth level. It is usually argued that absolute risk aversion should be a decreasing function of wealth. That is, many investors are willing to take more risk when they are financially secure. For example, a lottery to gain or lose \$100 is potentially life-threatening for an investor with wealth $W = \$100$, whereas it is negligible for an investor with wealth $W = \$100,000$. The former individual should be willing to pay more than the latter for the elimination of such a risk. Thus, we may require that the risk premium associated with any risk is decreasing in wealth. It can be shown that this holds if and only if the Arrow–Pratt absolute risk aversion coefficient is decreasing in wealth. This requirement means that

$$A'(x) = -\frac{u'''(x)u'(x) - u''(x)^2}{u'(x)^2} < 0.$$

A necessary condition for this to hold is $u'''(x) > 0$.

As a specific example, consider the exponential utility function $u(x) = -e^{-ax}$. Differentiate it to obtain

$$u'(x) = ae^{-ax} \quad \text{and} \quad u''(x) = -a^2 e^{-ax}.$$

Therefore, we have $A(x) = -u''(x)/u'(x) = a$. In this case, the risk aversion remains constant as wealth increases. As another example, consider the power utility function $u(x) = x^a$ with $0 < a < 1$. We have $u'(x) = ax^{a-1}$ and $u''(x) = a(a-1)x^{a-2}$. Thus, $A(x) = (1-a)/x$. So risk aversion decreases as wealth increases. Similarly, for the logarithmic utility function $u(x) = \ln x$, we have $u'(x) = 1/x$, $u''(x) = -1/x^2$, and thus $A(x) = 1/x$.

3.1.2 Mean-Variance Criterion

Suppose that the optimal investment opportunity is chosen by maximizing the expected utility of the wealth. Let us show how the utility maximization method reduces to the mean-variance criterion when an optimal investment is selected by maximizing the expected wealth and minimizing the variance of the wealth. Suppose that the final wealth follows a normal probability distribution and the investor uses an exponential utility function $u(x) = -e^{-ax}$ with $a > 0$. Recall that the mathematical expectation of an exponential function of a normal random variable Z is expressed in terms of the expected value and variance of Z as follows:

$$E[e^Z] = e^{E[Z] + \text{Var}(Z)/2}.$$

If the wealth V is normal, then $-aV$ is also normal with mean $E[-aV] = -aE[V]$ and variance $\text{Var}(-aV) = a^2 \text{Var}(V)$. Therefore, the expected utility of wealth V is

$$E[u(V)] = -\exp\left(-aE[V] + a^2 \text{Var}(V)/2\right) = -\exp\left(-a(E[V] - a\,\text{Var}(V)/2)\right).$$

The exponential function is increasing. Thus, the expected utility is maximized by choosing an investment that maximizes $E[V] - a\,\text{Var}(V)/2$:

$$E[-e^{-aV}] \to \max \iff E[V] - a\,\text{Var}(V)/2 \to \max.$$

This means that alternative investments can be ranked by comparing their means and variances. If there are two investments so that $E[V_1] \geqslant E[V_2]$ and $\text{Var}(V_1) \leqslant \text{Var}(V_2)$, then the first investment results in a larger expected utility than does the second: $E[u(V_1)] \geqslant E[u(V_2)]$.

One can arrive at the same conclusion for the case of a quadratic utility function $u(x) = x - ax^2$ with $a > 0$. Assuming that the wealth V satisfies $V < \frac{1}{2a}$, the expected utility $E[u(V)]$ is maximized by selecting an investment with a larger expected wealth and smaller variance $\text{Var}(V)$. Indeed, the expected utility of the wealth V is

$$E[u(V)] = E[V - aV^2] = E[V] - aE[V^2] = E[V] - a\,\text{Var}(V) + aE[V]^2 = u(E[V]) - a\,\text{Var}(V).$$

Thus, the expected utility $E[u(V)]$ is becoming larger if $E[V]$ is increasing and $\text{Var}(V)$ is decreasing.

To deal with a general utility function u, let us consider the Taylor's expansion of u about the point $E[V]$:

$$u(V) \approx u(E[V]) + u'(E[V])(V - E[V]) + \frac{1}{2}u''(E[V])(V - E[V])^2.$$

Taking the expectation of both parts gives

$$E[u(V)] \approx u(E[V]) + u'(E[V])E[V - E[V]] + \frac{1}{2}u''(E[V])E\left[(V - E[V])^2\right]$$

$$= u(E[V]) + u''(E[V])\,\mathrm{Var}[V]/2. \tag{3.7}$$

Here we use that $E[V - E[V]] = E[V] - E[V] = 0$ and $E\left[(V - E[V])^2\right] = \mathrm{Var}(V)$. Therefore, a reasonable approximation to the optimal investment is given by an investment that maximizes

$$u(E[V]) + u''(E[V])\,\mathrm{Var}[V]/2.$$

Suppose that $u''(x)$ is a nondecreasing function in x. Then, since $u''(x) \leqslant 0$, an optimal investment V can be again selected by both maximizing the expected value $E[V]$ and minimizing the variance $\mathrm{Var}(V)$. Recall that the standard deviation $\sigma_V = \sqrt{\mathrm{Var}(V)}$ characterizes the risk associated with the investment V. Therefore, the mean-variance criterion tells us that the optimal investment is attained by maximizing the expected value of the wealth and minimizing the risk.

3.2   Portfolio Optimization for Two Assets

3.2.1   Portfolio of Two Risky Assets

In a single-period setting, let us consider a model with two risky assets with respective time-t values A_t^1 and A_t^2, where $t \in \{0, T\}$. Each asset, labelled by $i = 1, 2$, is characterized by its initial value A_0^i and the respective single-period return $r_i = \frac{A_T^i - A_0^i}{A_0^i}$. At that we have $A_T^i = A_0^i(1 + r_i)$. The risky returns r_1 and r_2 (as well as the terminal asset prices A_T^1 and A_T^2) are random variables defined on a common probability space with state space Ω and probability function $\mathbb{P}$.

Let us form a portfolio (x_1, x_2) by purchasing x_1 shares of asset 1 and x_2 shares of asset 2. The initial wealth of such a portfolio is $V_0 = x_1 A_0^1 + x_2 A_0^2$. The rate of return r_V is then given by

$$r_V = \frac{V_T - V_0}{V_0} = \frac{x_1(A_T^1 - A_0^1) + x_2(A_T^2 - A_0^2)}{V_0}$$

$$= \frac{x_1(A_T^1 - A_0^1)}{A_0^1}\frac{A_0^1}{V_0} + \frac{x_2(A_T^2 - A_0^2)}{A_0^2}\frac{A_0^2}{V_0}$$

$$= \frac{x_1 A_0^1}{V_0}r_1 + \frac{x_2 A_0^2}{V_0}r_2.$$

Introduce

$$w_1 = \frac{x_1 A_0^1}{V_0} \quad \text{and} \quad w_2 = \frac{x_2 A_0^2}{V_0}, \tag{3.8}$$

which are called the *allocation weights* of funds between the two underlying assets. In other words, $100w_i\%$ of the initial wealth is invested in asset i with $i = 1, 2$. The allocation weights add up to one: $w_1 + w_2 = 1$. If short selling is allowed, then one of the weights may be negative and, hence, the other is greater than one. For a portfolio without short selling, both weights are between zero and one. The total wealth at the end of the period is

$$V_T = (1 + r_V)V_0 = (1 + w_1 r_1 + w_2 r_2)V_0 = (w_1(1 + r_1) + w_2(1 + r_2))V_0.$$

Given the initial wealth V_0, initial prices A_0^1 and A_0^2, and weights w_1 and w_2, we can solve equations in (3.8) for x_1 and x_2 to find the number of shares for each asset:

$$x_1 = \frac{w_1 V_0}{A_0^1} \quad \text{and} \quad x_2 = \frac{w_2 V_0}{A_0^2}.$$

Here, the products $w_1 V_0$ and $w_2 V_0$ represent the amounts initially invested in assets 1 and 2, respectively.

A portfolio with weights (w_1, w_2) can be characterized by the expected return and the variance of the return. Since $r_V = w_1 r_1 + w_2 r_2$, we have that

$$\begin{aligned}
\mathrm{E}[r_V] &= \mathrm{E}[w_1 r_1] + \mathrm{E}[w_2 r_2] \\
&= w_1 \mathrm{E}[r_1] + w_2 \mathrm{E}[r_2], & (3.9) \\
\mathrm{Var}(r_V) &= \mathrm{Var}(w_1 r_1) + \mathrm{Var}(w_2 r_2) + 2\,\mathrm{Cov}(w_1 r_1, w_2 r_2) \\
&= w_1^2\,\mathrm{Var}(r_1) + w_2^2\,\mathrm{Var}(r_2) + 2 w_1 w_2\,\mathrm{Cov}(r_1, r_2) \\
&= w_1^2\,\mathrm{Var}(r_1) + w_2^2\,\mathrm{Var}(r_2) + 2 w_1 w_2\,\mathrm{Corr}(r_1, r_2)\sqrt{\mathrm{Var}(r_1)}\sqrt{\mathrm{Var}(r_2)}. & (3.10)
\end{aligned}$$

Here, we define the *coefficient of correlation* between two returns as follows:

$$\mathrm{Corr}(r_1, r_2) = \frac{\mathrm{Cov}(r_1, r_2)}{\sqrt{\mathrm{Var}(r_1)\,\mathrm{Var}(r_2)}} \in [-1, 1].$$

Note that if the variance of one of the returns is zero (i.e., it is a return on a risk-free asset), then the correlation coefficient is undefined.

Proposition 3.1. *The variance of the return on a portfolio without short selling (i.e., both w_1 and w_2 are nonnegative) cannot exceed the greater of the variances of the underlying asset returns:*

$$0 \leqslant \mathrm{Var}(r_V) \leqslant \max\{\mathrm{Var}(r_1), \mathrm{Var}(r_2)\}.$$

Proof. Since the value of the correlation coefficient is always between -1 and 1, from (3.10) we obtain that

$$\begin{aligned}
\mathrm{Var}(r_V) &\leqslant w_1^2\,\mathrm{Var}(r_1) + w_2^2\,\mathrm{Var}(r_2) + 2 w_1 w_2 \sqrt{\mathrm{Var}(r_1)}\sqrt{\mathrm{Var}(r_2)} \\
&\leqslant \left(w_1\sqrt{\mathrm{Var}(r_1)} + w_2\sqrt{\mathrm{Var}(r_2)}\right)^2 \\
&\leqslant (w_1 + w_2)^2 \max\{\mathrm{Var}(r_1), \mathrm{Var}(r_2)\} = \max\{\mathrm{Var}(r_1), \mathrm{Var}(r_2)\}.
\end{aligned}$$

On the other hand, the variance is always a nonnegative quantity. $\square$

Introduce the following notation for the expected returns, variances of returns, and correlation coefficient:

$$\mu_i := \mathrm{E}[r_i], \quad \sigma_i^2 := \mathrm{Var}(r_i), \quad (i = 1, 2); \quad \rho_{12} := \mathrm{Corr}(r_1, r_2).$$

Moreover, denote $\mu_V := \mathrm{E}[r_V]$ and $\sigma_V^2 := \mathrm{Var}(r_V)$. In this notation, (3.9) and (3.10) take the respective forms:

$$\mu_V = w_1\mu_1 + w_2\mu_2 \quad \text{and} \quad \sigma_V^2 = w_1^2\sigma_1^2 + w_2^2\sigma_2^2 + 2\rho_{12}w_1w_2\sigma_1\sigma_2. \tag{3.11}$$

When the number of base assets is greater than two, it is convenient to use vectors and matrices to describe portfolios and compute their expected returns and risk values. We will be using column vectors for portfolio weights: $\mathbf{w} = \begin{bmatrix} w_1 & w_2 \end{bmatrix}^\top \equiv \begin{bmatrix} w_1 \\ w_2 \end{bmatrix}$. Equation (3.11) can also be written using the matrix-vector notation:

$$\mu_V = \begin{bmatrix} \mu_1 & \mu_2 \end{bmatrix} \begin{bmatrix} w_1 \\ w_2 \end{bmatrix} \quad \text{and} \quad \sigma_V^2 = \begin{bmatrix} w_1 & w_2 \end{bmatrix} \begin{bmatrix} \sigma_1^2 & \sigma_1\sigma_2\rho_{12} \\ \sigma_1\sigma_2\rho_{12} & \sigma_2^2 \end{bmatrix} \begin{bmatrix} w_1 \\ w_2 \end{bmatrix}. \tag{3.12}$$

Example 3.5. Consider two assets with the following probability distributions of their returns:

Scenario ω	Probability $\mathbb{P}(\omega)$	Return r_1	Return r_2
ω^1	0.1	-20%	30%
ω^2	0.6	5%	10%
ω^3	0.3	10%	-20%

Calculate the expected returns μ_1 and μ_2, standard deviations σ_1 and σ_2, and correlation coefficient of returns ρ_{12}.

Solution. To compute the mathematical expectation of a random variable X on a finite sample space Ω, we use the formula

$$\mathrm{E}[X] = \sum_{\omega \in \Omega} X(\omega)\mathbb{P}(\omega).$$

The expected returns are

$$\mu_1 = \mathrm{E}[r_1] = \sum_{i=1}^{3} r_1(\omega^i)\mathbb{P}(\omega^i) = (-0.2) \cdot 0.1 + 0.05 \cdot 0.6 + 0.1 \cdot 0.3 = 0.04 = 4\%,$$

$$\mu_2 = \mathrm{E}[r_2] = \sum_{i=1}^{3} r_2(\omega^i)\mathbb{P}(\omega^i) = 0.3 \cdot 0.1 + 0.1 \cdot 0.6 + (-0.2) \cdot 0.3 = 0.03 = 3\%.$$

Using the fact that $\mathrm{Var}(X) = \mathrm{E}[(X - \mathrm{E}[X])^2]$ and $\mathrm{Cov}(X,Y) = \mathrm{E}[(X - \mathrm{E}[X])(Y - \mathrm{E}[Y])]$, we similarly obtain:

$$\mathrm{Var}(r_1) = (-0.2 - 0.04)^2 \cdot 0.1 + (0.05 - 0.04)^2 \cdot 0.6 + (0.1 - 0.04)^2 \cdot 0.3 = 0.0069,$$
$$\mathrm{Var}(r_2) = (0.3 - 0.03)^2 \cdot 0.1 + (0.1 - 0.03)^2 \cdot 0.6 + (-0.2 - 0.03)^2 \cdot 0.3 = 0.02610,$$
$$\mathrm{Cov}(r_1, r_2) = (-0.2 - 0.04) \cdot (0.3 - 0.03) \cdot 0.1 + (0.05 - 0.04) \cdot (0.1 - 0.03) \cdot 0.6$$
$$+ (0.1 - 0.04) \cdot (-0.2 - 0.03) \cdot 0.3 = -0.0102.$$

The standard deviations are

$$\sigma_1 = \sqrt{\mathrm{Var}(r_1)} = \sqrt{0.0069} \cong 8.307\%, \quad \sigma_2 = \sqrt{\mathrm{Var}(r_2)} = \sqrt{0.02610} \cong 16.156\%.$$

The correlation coefficient ρ_{12} is

$$\rho_{12} = \frac{\mathrm{Cov}(r_1, r_2)}{\sqrt{\mathrm{Var}(r_1)\,\mathrm{Var}(r_2)}} \cong \frac{-0.0102}{0.08307 \cdot 0.16156} \cong -0.76007 = -76.007\%. \qquad \square$$

If the joint probability distribution of asset returns is known, we can find an optimal allocation of a capital by maximizing the expected utility of the future wealth. Let us consider two examples. In the first one, the risky returns follow the joint discrete distribution from Example 3.5. In the second problem, we deal with normally distributed returns.

Example 3.6. Find an optimal allocation of the initial wealth $V_0 = 1000$ between two risky assets from Example 3.5 when attempting to maximize the expected value, $E[u(V_T)]$, of an exponential utility function $u(x) = 1 - e^{-0.01x}$ of the terminal wealth V_T.

Solution. Since the sum of portfolio weights is one, we only need one independent variable to parametrize all portfolios of two assets. Let the portfolio weights be $w_1 = x$ and $w_2 = 1 - x$, respectively, with $x \in (-\infty, \infty)$. The return on such a portfolio is $r_V(x) = xr_1 + (1 - x)r_2$. At the end of the period, the portfolio value is $V_T = V_0(1 + r_V)$. Find the optimal allocation by solving the following maximization problem:

$$E[u(V_T)] = E\left[1 - e^{-0.01V_T}\right] = E\left[1 - e^{-0.01V_0(1+r_V)}\right] = 1 - E\left[e^{-10(1+xr_1+(1-x)r_2)}\right] \to \max_x.$$

It is equivalent to minimizing $E[e^{-10(1+xr_1+(1-x)r_2)}]$ w.r.t. x. Evaluate the mathematical expectation:

$$E[e^{-10(1+xr_1+(1-x)r_2)}] = \sum_{i=1}^{3} p_i\, e^{-10(1+xr_1(\omega^i)+(1-x)r_2(\omega^i))}$$

$$= 0.1e^{-13+5x} + 0.6e^{-11+0.5x} + 0.3e^{-8-3x}.$$

Differentiate the expected value w.r.t. x and equate the obtained derivative to zero:

$$0.5e^{-13+5x} + 0.3e^{-11+0.5x} - 0.9e^{-8-3x} = 0.$$

The resulting equation can be solved numerically to yield the optimal value $x \cong 0.6743109$, where the expected utility attains its maximum value. Therefore, the optimal allocation weights are $w_1 \cong 67.4309\%$ and $w_2 \cong 32.5691\%$. □

Example 3.7. Assume that the returns on two risky assets are normally distributed with parameters as those calculated in Example 3.5. Find an optimal allocation of the initial wealth $V_0 = 1000$ that maximizes the expected utility $E[1 - e^{-0.01V_T}]$.

Solution. As shown in Example 3.6, to find an optimal allocation, we need to maximize the function $1 - E\left[e^{-10(1+xr_1+(1-x)r_2)}\right]$. Since both r_1 and r_2 are normally distributed, the exponent $X = -10(1 + xr_1 + (1 - x)r_2)$ has a normal distribution. Let us find its mean and variance:

$$\begin{aligned}
E[X] &= E[-10(1 + xr_1 + (1 - x)r_2)] = -10(1 + xE[r_1] + (1 - x)E[r_2]) \\
&= -10(1 + x\mu_1 + (1 - x)\mu_2), \\
Var(X) &= Var[-10(1 + xr_1 + (1 - x)r_2)] = (-10)^2\, Var(xr_1 + (1 - x)r_2) \\
&= 100\left(Var(xr_1) + 2\,Cov(xr_1, (1 - x)r_2) + Var((1 - x)r_2)\right) \\
&= 100\left(x^2\, Var(r_1) + 2x(1 - x)\,Cov(r_1, r_2) + (1 - x)^2\, Var(r_2)\right) \\
&= 100\left(x^2\sigma_1^2 + 2x(1 - x)\sigma_1\sigma_2\rho_{12} + (1 - x)^2\sigma_2^2\right).
\end{aligned}$$

Therefore, we have

$$X \sim Norm\left(-10(1 + x\mu_1 + (1 - x)\mu_2),\, 100(x^2\sigma_1^2 + (1 - x)^2\sigma_2^2 + 2x(1 - x)\sigma_1\sigma_2\rho_{12})\right).$$

Since $E[\exp(X)] = \exp(E[X] + \operatorname{Var}(X)/2)$ for a normal variate X, the expected utility $E[u(V_T)]$ is equal to

$$1 - \exp\left(-10(1 + x\mu_1 + (1-x)\mu_2) + 50(x^2\sigma_1^2 + (1-x)^2\sigma_2^2 + 2x(1-x)\sigma_1\sigma_2\rho_{12})\right).$$

Finding a maximum of this function is equivalent to solving the following optimization problem:

$$(1 + x\mu_1 + (1-x)\mu_2) - 5(x^2\sigma_1^2 + (1-x)^2\sigma_2^2 + 2x(1-x)\sigma_1\sigma_2\rho_{12}) \to \max_x \iff$$

$$1 + 0.04x + 0.03(1-x) - (5 \cdot 0.0069)x^2 - (5 \cdot 0.002610)(1-x)^2$$

$$+(10 \cdot 0.08307 \cdot 0.16156 \cdot 0.76007)x(1-x) \to \max_x \iff$$

$$0.8995 + 0.373x - 0.267x^2 \to \max_x.$$

The maximum of the quadratic function $y(x) = 0.8995 + 0.373x - 0.267x^2$ is attained at $x \cong 0.698502$. Thus, the optimal weights are $w_1 \cong 69.8502\%$ and $w_2 \cong 30.1498\%$. $\qquad\square$

3.2.2   Portfolio Lines

On the (σ, μ)-plane, a portfolio V with allocation weights (w_1, w_2) is represented by a point whose coordinates (σ_V, μ_V) are calculated by (3.11). Let us find a set of points on the (σ, μ)-plane that describes all possible portfolios in the two underlying assets. Since $w_1 + w_2 = 1$, all portfolios can be parameterized by a single variable $x \in \mathbb{R}$ so that $w_1 = x$ and $w_2 = 1 - x$. Therefore, the set of all admissible portfolios can be represented by a *portfolio line* (which can shrink to a single point in some extreme cases). Equations (3.11) can be rewritten as follows:

$$\mu_V(x) = x\mu_1 + (1-x)\mu_2, \quad \sigma_V^2(x) = x^2\sigma_1^2 + (1-x)^2\sigma_2^2 + 2x(1-x)\sigma_1\sigma_2\rho_{12} \qquad (3.13)$$

with $x \in (-\infty, \infty)$. For portfolios without short selling (i.e., both weights w_1 and w_2 are nonnegative), we have that $0 \leqslant x \leqslant 1$.

3.2.2.1   Case with $|\rho_{12}| = 1$

First, assume that $\rho_{12} = 1$. From (3.13) we obtain that the variance of return is given by $\sigma_V^2(x) = (x\sigma_1 + (1-x)\sigma_2)^2$, and hence $\sigma_V(x) = |x\sigma_1 + (1-x)\sigma_2|$. The portfolio line is described by two parametric equations

$$\sigma_V(x) = |x(\sigma_1 - \sigma_2) + \sigma_2| \ \text{ and } \ \mu_V(x) = x\mu_1 + (1-x)\mu_2 \ \text{ with } \ x \in \mathbb{R}.$$

Let us assume that $\mu_1 \neq \mu_2$ and $\sigma_1 \neq \sigma_2$ (we leave the other cases as exercises for the reader). We can solve the second equation for x to obtain $x = \frac{\mu_V - \mu_2}{\mu_1 - \mu_2}$. Substituting this expression in the formula for σ_V gives us the following relationship:

$$\sigma_V = \left|\sigma_2 + (\sigma_1 - \sigma_2)\frac{\mu_V - \mu_2}{\mu_1 - \mu_2}\right| \implies \sigma_V = \left|\frac{\sigma_1 - \sigma_2}{\mu_1 - \mu_2}\mu_V + \frac{\sigma_2\mu_1 - \sigma_1\mu_2}{\mu_1 - \mu_2}\right|.$$

As we can see, the standard deviation σ_V is a piecewise-linear function of μ_V:

$$\sigma_V = \begin{cases} a\mu_V + b & \text{if } \mu_V \geqslant -\frac{b}{a}, \\ -(a\mu_V + b) & \text{if } \mu_V < -\frac{b}{a}, \end{cases} \quad \text{where } a = \frac{\sigma_1 - \sigma_2}{\mu_1 - \mu_2} \text{ and } b = \frac{\sigma_2\mu_1 - \sigma_1\mu_2}{\mu_1 - \mu_2}.$$

The plot of σ_V as a function of μ_V is a broken line with two half-lines. It is interesting that the portfolio line contains a portfolio with zero variance (i.e., a risk-free portfolio). Indeed, $\sigma_V = 0$ if $\mu_V = \frac{\sigma_1 \mu_2 - \sigma_2 \mu_1}{\sigma_2 - \sigma_1}$. The weights $w_1 = x$ and $w_2 = 1 - x$ can be obtained by solving the equation $x\sigma_1 + (1 - x)\sigma_2 = 0$ for x. Hence, the weights of a risk-free portfolio are

$$\widehat{w}_1 = \frac{\sigma_2}{\sigma_2 - \sigma_1} \quad \text{and} \quad \widehat{w}_2 = \frac{\sigma_1}{\sigma_1 - \sigma_2} . \tag{3.14}$$

One of the weights is negative. Hence short selling is necessary to construct such a portfolio without risk.

Now let us find what part of the portfolio line corresponds to portfolios without short selling. The portfolios with weights $(0, 1)$ and $(1, 0)$ are the endpoints of such a set. By changing x from 0 to 1, we continuously move the point along the line of portfolios without short selling from one endpoint to the other. Since the portfolio with $\sigma_V = 0$ has a negative weight, the no-short-selling line is a segment on one of the two rays. The final result of our analysis is presented in Figure 3.5a.

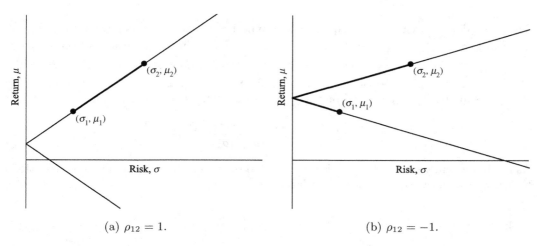

(a) $\rho_{12} = 1$.   (b) $\rho_{12} = -1$.

FIGURE 3.5: A typical portfolio line for the case with $|\rho_{12}| = 1$. The bold part indicates portfolios without short selling.

Similarly, we can construct a portfolio line for the case with $\rho_{12} = -1$. It is described by

$$\sigma_V(x) = |x(\sigma_1 + \sigma_2) - \sigma_2| \quad \text{and} \quad \mu_V(x) = x\mu_1 + (1 - x)\mu_2 \quad \text{with} \quad x \in \mathbb{R} .$$

By excluding x from the above equations, we obtain

$$\sigma_V = \left| \frac{\sigma_1 + \sigma_2}{\mu_1 - \mu_2} \mu_V - \frac{\sigma_2 \mu_1 + \sigma_1 \mu_2}{\mu_1 - \mu_2} \right| .$$

Again, the plot of σ_V as a function of μ_V is a broken line. Now $\sigma_V = 0$ if $\mu_V = \frac{\sigma_1 \mu_2 + \sigma_2 \mu_1}{\sigma_1 + \sigma_2}$. The weights of the risk-free portfolio are

$$\widehat{w}_1 = \frac{\sigma_2}{\sigma_1 + \sigma_2} \quad \text{and} \quad \widehat{w}_2 = \frac{\sigma_1}{\sigma_1 + \sigma_2} . \tag{3.15}$$

Both weights are positive, so no short selling is required to construct a portfolio with a zero variance of return. The no-short-selling line is a broken line segment lying on both half-lines. The result of our analysis is given in Figure 3.5b.

3.2.2.2   Case with $|\rho_{12}| < 1$

Excluding x from (3.13) and expressing σ^2 as a function of μ gives

$$\sigma^2 = \frac{(\mu - \mu_2)^2}{(\mu_1 - \mu_2)^2}\sigma_1^2 + \frac{(\mu - \mu_1)^2}{(\mu_1 - \mu_2)^2}\sigma_2^2 - 2\frac{(\mu - \mu_1)(\mu - \mu_2)}{(\mu_1 - \mu_2)^2}\rho_{12}\sigma_1\sigma_2 \,.$$

After doing some algebra, we can bring this equation to the form:

$$\sigma^2 = A\mu^2 - 2B\mu + C, \text{ where}$$

$$A = \frac{\sigma_1^2 - 2\rho_{12}\sigma_1\sigma_2 + \sigma_2^2}{(\mu_1 - \mu_2)^2},$$

$$B = \frac{\mu_1\sigma_2^2 + \mu_2\sigma_1^2 - 2\rho_{12}\sigma_1\sigma_2(\mu_1 + \mu_2)}{(\mu_1 - \mu_2)^2},$$

$$C = \frac{(\mu_1\sigma_2)^2 + (\mu_2\sigma_1)^2 - 2\rho_{12}\sigma_1\sigma_2\mu_1\mu_2}{(\mu_1 - \mu_2)^2}\,.$$

The curve defined by the above equation is a hyperbola. Indeed, let us rewrite the equation as follows: $\sigma^2 = A(\mu - \frac{B}{A})^2 + D$, where $D = C - \frac{B^2}{A}$. By changing variables from (σ, μ) to $(x = \frac{\sigma}{\sqrt{D}}, y = \frac{\sqrt{A}}{\sqrt{D}}\mu - \frac{B}{\sqrt{AD}})$, one can easily obtain the canonical equation of a hyperbola: $x^2 - y^2 = 1$. A typical portfolio line is given in Figure 3.6. The line has two asymptotes that cross at some point located on the vertical axis. The μ coordinate of this point is

$$\mu = \frac{B}{A} = \frac{\mu_1\sigma_2^2 + \mu_2\sigma_1^2 - 2\rho_{12}\sigma_1\sigma_2(\mu_1 + \mu_2)}{\sigma_1^2 - 2\rho_{12}\sigma_1\sigma_2 + \sigma_2^2}\,.$$

As it will be demonstrated later, it is the expected return on the minimum variance portfolio that is obtained by minimizing the variance σ_V^2.

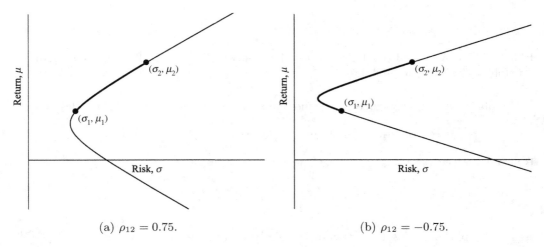

(a) $\rho_{12} = 0.75$.   (b) $\rho_{12} = -0.75$.

FIGURE 3.6: A typical portfolio line for the case with $-1 < \rho_{12} < 1$. The bold part indicates portfolios without short selling.

As is seen from Figures 3.5 and 3.6, the plot of a portfolio line is a hyperbola for the case with $\rho_{12} \in (-1, 1)$ and a broken line for the extreme case with $|\rho_{12}| = 1$. The evolution of a portfolio line when $\mu_{1,2}$ and $\sigma_{1,2}$ are fixed and ρ_{12} is changing from -1 to 1 is represented in Figure 3.7.

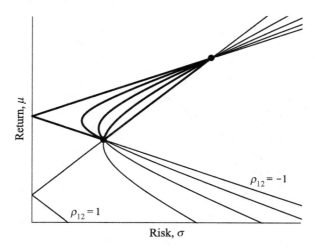

FIGURE 3.7: Portfolio lines for varying ρ_{12} and fixed μ's and σ's. The parameter ρ_{12} is changing from -1 to 1 with the step size of 0.5. The bold parts indicate portfolios without short selling.

3.2.2.3 Case with a Risk-Free Asset

Suppose that one of two assets (say, asset 2) in our portfolio is risk-free, that is, the variance of its return is zero: $\sigma_2^2 = \mathrm{Var}(r_2) = 0$. Hence, the return r_2 is constant: $r_2 \equiv r$. The formula of the variance in (3.13) reduces to $\sigma_V^2(x) = x^2 \sigma_1^2$ or just $\sigma_V(x) = |x|\sigma_1$. So the standard deviation σ_V of such a portfolio depends on the weight w_1 of the risky asset as follows: $\sigma_V = |w_1|\sigma_1$. The portfolio line is described by a piecewise linear function: $\sigma_V = \sigma_1 \left| \frac{\mu_V - r}{\mu_1 - r} \right|$. Thus, the portfolio plot is a broken line with its vertex at the point that corresponds to the risk-free asset (see Figure 3.8).

3.2.3 The Minimum Variance Portfolio

As is seen in Figures 3.5 and 3.7, there is always a portfolio with the smallest possible variance σ_V^2. We already found risk-free portfolios with zero variance for the case with $|\rho_{12}| = 1$. Let us now find the general solution to this problem.

Theorem 3.2. *Suppose that $|\rho_{12}| < 1$ or $\sigma_1 \neq \sigma_2$ holds. The portfolio with the minimum variance is attained at*

$$\widehat{w}_1 = \frac{\sigma_2^2 - \rho_{12}\sigma_1\sigma_2}{\sigma_1^2 + \sigma_2^2 - 2\rho_{12}\sigma_1\sigma_2} \quad and \quad \widehat{w}_2 = \frac{\sigma_1^2 - \rho_{12}\sigma_1\sigma_2}{\sigma_1^2 + \sigma_2^2 - 2\rho_{12}\sigma_1\sigma_2}. \quad (3.16)$$

The variance of the portfolio is

$$\sigma_{\mathrm{mv}}^2 = \frac{(1 - \rho_{12}^2)\sigma_1^2\sigma_2^2}{\sigma_1^2 + \sigma_2^2 - 2\rho_{12}\sigma_1\sigma_2}. \quad (3.17)$$

The expected return of the portfolio is

$$\mu_{\mathrm{mv}} = \frac{\mu_1\sigma_2^2 + \mu_2\sigma_1^2 - 2\rho_{12}\sigma_1\sigma_2(\mu_1 + \mu_2)}{\sigma_1^2 - 2\rho_{12}\sigma_1\sigma_2 + \sigma_2^2}. \quad (3.18)$$

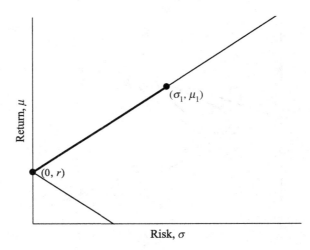

FIGURE 3.8: Portfolio line for one risky and one risk-free asset. The risk-free rate of return is r. The bold part indicates portfolios without short selling.

Proof. By differentiating the variance σ_V^2 given by (3.13) w.r.t. x and equating the derivative to zero, we obtain the following linear equation for x:

$$\frac{\mathrm{d}\sigma_V^2}{\mathrm{d}x} = 2(\sigma_1^2 + \sigma_2^2 - 2\rho_{12}\sigma_1\sigma_2)x - 2(\sigma_2^2 - \rho_{12}\sigma_1\sigma_2) = 0\,.$$

The solution is

$$x_0 = \frac{\sigma_2^2 - \rho_{12}\sigma_1\sigma_2}{\sigma_1^2 + \sigma_2^2 - 2\rho_{12}\sigma_1\sigma_2}\,. \tag{3.19}$$

Thus, for the weights $\widehat{w}_1 = x_0$ and $\widehat{w}_2 = 1 - x_0$, we immediately obtain (3.16). Since the second derivative of σ_V^2 w.r.t. x is positive everywhere:

$$\frac{\mathrm{d}^2\sigma_V^2}{\mathrm{d}x^2} = 2(\sigma_1^2 + \sigma_2^2 - 2\rho_{12}\sigma_1\sigma_2) > 0\,,$$

the variance σ_V^2 attains its smallest value at x_0. Substituting x_0 in (3.11) gives us equation (3.17) for the minimum variance.

The formulae in (3.16) and (3.17) work for both cases when $|\rho_{12}| = 1$ or $|\rho_{12}| < 1$. If $|\rho_{12}| = 1$, then $\sigma_{\mathrm{mv}}^2 = 0$ in (3.17) and the formulae in (3.16) reduce to (3.14) or (3.15) depending on the sign of ρ_{12}. □

3.2.3.1 Case without Short Selling

While proving Theorem 3.2, we did not take into account whether short sells are allowed. Let us find the minimum variance portfolio without short sells, i.e., with nonnegative weights. The function σ_V^2 in (3.13) attains its minimum value on $[0, 1]$ either at one of the boundary points $x \in \{0, 1\}$ or at the point x_0 given by (3.19) provided $0 < x_0 < 1$. Both weights $\widehat{w}_1$ and $\widehat{w}_2$ in (3.16) are positive iff $\rho_{12}\sigma_2 < \sigma_1$ and $\rho_{12}\sigma_1 < \sigma_2$, or, equivalently, if $\rho_{12} < \min\{\frac{\sigma_1}{\sigma_2}, \frac{\sigma_2}{\sigma_1}\}$. If that is the case, then it is possible to construct a portfolio without short selling with risk lower than that of any of the individual assets.

Otherwise, when $\rho_{12} \geqslant \min\{\frac{\sigma_1}{\sigma_2}, \frac{\sigma_2}{\sigma_1}\}$, the minimum variance portfolio without short selling is composed of shares of only one of the base assets. If $\sigma_1 < \sigma_2$ (hence $x_0 > 1$), then the portfolio has only shares of asset 1 and its variance is σ_1^2. If $\sigma_2 < \sigma_1$ (hence $x_0 < 0$), then the portfolio has only shares of asset 2 and its variance is σ_2^2. For the special case with $\sigma_1 = \sigma_2$ and $\rho_{12} = 1$, the variance is the same for any portfolio: $\sigma_V^2 = \sigma_1^2 = \sigma_2^2$.

3.2.4 Selection of Optimal Portfolios

A typical problem of a risk manager is the selection of optimal investment portfolios. First, let us consider a situation with one risky asset (e.g., a stock) and one risk-free asset (e.g., a bond). Let the expected return and volatility of the risky asset be respectively $\mu_1 = 10\%$ and $\sigma_1 = 20\%$. Additionally, where necessary, we assume that the return on the risky asset, denoted r_1, has a normal distribution. That is, $r_1 \sim Norm(0.1, 0.2^2)$. The return on the risk-free asset is $r_2 = r = 5\%$. Suppose that $100x\%$ of the initial capital V_0 is invested in the risky asset. The terminal wealth is

$$V_T = V_T(x) = V_0(1 + r_V(x)) = V_0\left(1 + xr_1 + (1-x)r\right)$$

where r_V is the portfolio return. Let us find the optimal portfolio that maximizes the expected utility $E[1 - \exp(-\alpha V_T)]$ of the terminal wealth. Here, $\alpha > 0$ is the risk-averse parameter. Since r_1 is a normal random variable and r is a constant, the portfolio return r_V has a normal distribution with mean $\mu_V(x) = \mu_1 x + r(1-x) = 0.05 + 0.05x$ and variance $\sigma_V^2(x) = \sigma^2 x^2 = 0.04x^2$.

Using the normal moment generating function (MGF), we calculate the expected utility as follows:

$$E\left[1 - e^{-\alpha V_T(x)}\right] = 1 - E\left[e^{-\alpha V_T(x)}\right] = 1 - e^{-\alpha V_0}E\left[e^{-\alpha V_0 r_V(x)}\right]$$

$$= 1 - e^{-\alpha V_0}e^{-\alpha V_0 \mu_V(x) + (-\alpha V_0)^2 \sigma_V^2(x)/2}$$

$$= 1 - e^{-\alpha V_0}e^{-\alpha V_0(\mu_V(x) - \alpha V_0 \sigma_V^2(x)/2)}.$$

Since $\alpha V_0 > 0$ holds, $1 - e^{-\alpha V_0}e^{-\alpha V_0 y}$ is an increasing function of y. Thus, to maximize the expected utility, we need to find a maximum of the function

$$y(x) := \mu_V(x) - \frac{\alpha V_0 \sigma_V^2(x)}{2} = 0.05 + 0.05x - \frac{\alpha V_0 0.04x^2}{2}.$$

Suppose that $\alpha V_0 = 1$ holds. The solution to the optimization problem

$$0.05 + 0.05x - 0.04x^2/2 \to \max_x$$

is $x^* = 0.05/0.04 = 1.25$. The optimal weights are

$$\widehat{w}_1 = x^* \cong 1.25 \quad \text{and} \quad \widehat{w}_2 = 1 - x^* \cong -0.25.$$

If we let the risk-free rate of return r vary, then the optimization problem takes the form

$$(0.1 - r)x + r - 0.04x^2/2 \to \max_x.$$

The optimal weights are $\widehat{w}_1 = 2.5 - 25r$ and $\widehat{w}_2 = 25r - 1.5$. They are nonnegative (and hence it is a portfolio without short selling) iff $6\% \leqslant r \leqslant 10\%$.

Now, we consider the case with two risky assets. Suppose that the returns r_1 and r_2 are characterized by five parameters, namely, the expected returns μ_1 and μ_2, the variances of returns σ_1^2 and σ_2^2, and the correlation coefficient $\rho_{12} = \text{Corr}(r_1, r_2)$.

There are many criteria that can be used to select an optimal portfolio. We consider four examples: minimization of the risk, minimization of the Value at Risk, maximization of the Share ratio (which is equivalent to minimization of a loss probability), and maximization of an expected utility of the terminal wealth. All examples will be illustrated with the following data:

$$\mu_1 = 10\%, \ \sigma_1 = 20\%, \ \mu_2 = 15\%, \ \sigma_2 = 40\%, \ \text{and} \ \rho_{12} = -20\%. \qquad (3.20)$$

Here, we make an additional assumption that the returns r_1 and r_2 are normally distributed. Since $\rho_{12} \neq 0$, they are dependent random variables. Figure 3.9 demonstrates the location of four optimal portfolios on the portfolio line.

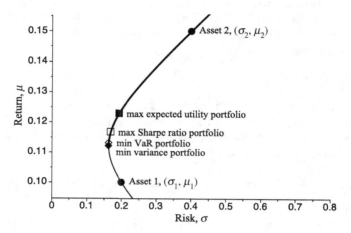

FIGURE 3.9: Selection of optimal portfolios with two risky assets: the minimum variance portfolio (solid diamond symbol), the minimum VaR portfolio (diamond symbol), the minimum loss probability portfolio (box symbol) the maximum expected utility portfolio (solid box symbol).

3.2.4.1   Minimum Variance Portfolio

The variance of the terminal portfolio value is

$$\text{Var}(V_T) = \text{Var}\left(V_0\left(1 + r_V\right)\right) = V_0^2 \, \text{Var}(r_V).$$

So, minimization of $\text{Var}(V_T)$ is equivalent to minimization of $\sigma_V^2 = \text{Var}(r_V)$. Let us find the weights that minimize the variance of the portfolio return. Solve $\frac{d\sigma_V^2(x)}{dx} = 0$ where $\sigma_V^2(x)$ is given by (3.13) to obtain:

$$x_0 = \frac{\sigma_2^2 - \rho_{12}\sigma_1\sigma_2}{\sigma_1^2 + \sigma_2^2 - 2\rho_{12}\sigma_1\sigma_2} = \frac{0.4^2 - (-0.2) \cdot 0.2 \cdot 0.4}{0.2^2 + 0.4^2 - 2 \cdot (-0.2) \cdot 0.2 \cdot 0.4} \cong 0.7586 \, .$$

Thus, $\widehat{w}_1 = x_0 \cong 75.86\%$ and $\widehat{w}_2 = 1 - x_0 \cong 24.14\%$. The expected return and the variance of return are, respectively,

$$\mu_V = 0.1 \cdot 0.7586 + 0.15 \cdot 0.2414 \cong 0.1121 = 11.21\% \, ,$$

$$\sigma_V^2 = 0.2^2 \cdot 0.7586^2 + 0.4^2 \cdot 0.2414^2 + 2 \cdot (-0.2) \cdot 0.2 \cdot 0.4 \cdot 0.7586 \cdot 0.2414 \cong 0.02648 \, .$$

Thus, $\sigma_V \cong 0.1627 = 16.27\%$. Notice that $\mu_1 < \mu_V < \mu_2$ and $\sigma_V < \min\{\sigma_1, \sigma_2\} = \min\{0.2, 0.4\} = 0.2$. That is, the volatility of the constructed portfolio is less than volatilities of the base assets. We managed to lower the risk by diversifying the portfolio.

3.2.4.2 Minimum VaR Portfolio

The Value at Risk VaR_α with significance level α for the portfolio return is implicitly defined by

$$\mathbb{P}(r_V < -\text{VaR}_\alpha) = \alpha.$$

Assuming that the portfolio return is normally distributed, the VaR is given by $\text{VaR}_\alpha = -\mu_V + z_\alpha \sigma_V$ where z_α is the standard normal critical value, μ_V and σ_V are, respectively, the expected return and volatility of the return r_V (see Example 2.13 in Chapter 2). Let us find the allocation weights that minimize VaR_α:

$$\text{VaR}_\alpha(x) = -\mu_V(x) + z_\alpha \sigma_V(x) \to \min_x.$$

Let $\alpha = 5\%$ and hence $z_\alpha = 1.645$. Differentiate the function

$$f(x) := -\mu_1 x - \mu_2(1-x) + 1.645\sqrt{\sigma_1^2 x^2 + \sigma_2^2(1-x)^2 + 2\sigma_1\sigma_2\rho_{12}x(1-x)},$$

plug in the data from (3.20), and then numerically solve $f'(x) = 0$ to find the optimal weights:

$$\widehat{w}_1 \cong 73.73\% \quad \text{and} \quad \widehat{w}_2 \cong 26.27\%.$$

For this portfolio, we have

$$\text{VaR}_{5\%} \cong 15.51\%, \quad \mu_V \cong 11.31\%, \quad \sigma_V \cong 16.31\%.$$

3.2.4.3 Maximum Sharpe Ratio Portfolio. Market Portfolio and Market Line

Let us find the allocation weights when attempting to minimize the probability that the return on the portfolio is less than a certain threshold r:

$$\mathbb{P}(r_V \leqslant r) \to \min.$$

Given that r_1 and r_2 follow a bivariate normal distribution, the probability distribution of r_V is $Norm(\mu_V(x), \sigma_V^2(x))$. Therefore,

$$\mathbb{P}(r_V(x) \leqslant r) = \mathbb{P}\left(\frac{r_V(x) - \mu_V(x)}{\sigma_V(x)} \leqslant \frac{r - \mu_V(x)}{\sigma_V(x)}\right)$$

$$= \mathbb{P}\left(Z \leqslant \frac{r - \mu_V(x)}{\sigma_V(x)}\right) = \mathcal{N}\left(\frac{r - \mu_V(x)}{\sigma_V(x)}\right),$$

where Z denotes a standard normal random variable and $\mathcal{N}$ is a standard normal CDF. Since a normal CDF is a strictly increasing function of its argument, it is sufficient to solve

$$\frac{\mu_V(x) - r}{\sigma_V(x)} \to \max_x. \tag{3.21}$$

Equation (3.21) relates to the so-called *Sharpe ratio*. The Sharpe ratio is a measure of the excess return (or risk premium) per unit of risk in an investment portfolio. It is named after William Forsyth Sharpe. The Sharpe ratio is defined as

$$\frac{\mathrm{E}[r_V - r]}{\sqrt{\mathrm{Var}(r_V - r)}}, \tag{3.22}$$

where r is the return on a benchmark asset, such as a risk-free security, $\mathrm{E}[r_V - r]$ is the

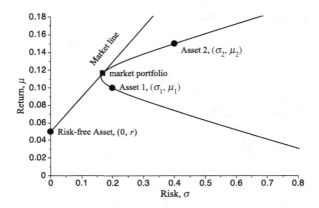

FIGURE 3.10: The market line and the market portfolio for the case with two risky assets.

expected value of the excess of the portfolio return r_V over the benchmark return, and $\mathrm{Var}(r_V - r)$ is the variance of the excess return. Since r is constant, we have

$$\mathrm{E}[r_V - r] = \mathrm{E}[r_V] - r = \mu_V - r \quad \text{and} \quad \mathrm{Var}(r_V - r) = \mathrm{Var}(r_V) = \sigma_V^2.$$

The Sharpe ratio is used to characterize how well the return of a portfolio compensates the investor for the risk taken. When comparing two portfolios against the same benchmark asset, the portfolio with the higher Sharpe ratio gives more return for the same level of risk. Investors are often advised to pick investments with high Sharpe ratios.

The solution to (3.21) can be obtained by using standard methods of calculus: differentiate the left-hand side of (3.21) w.r.t. x, equate the derivative to zero and then solve the obtained equation for x. As a result, we obtain the following allocation weights:

$$
\begin{aligned}
\widehat{w}_1 &= \frac{(\mu_1 - r)\sigma_2^2 - (\mu_2 - r)\rho_{12}\sigma_1\sigma_2}{(\mu_1 - r)\sigma_2^2 + (\mu_2 - r)\sigma_1^2 - (\mu_1 + \mu_2 - 2r)\rho_{12}\sigma_1\sigma_2}, \\
\widehat{w}_2 &= \frac{(\mu_2 - r)\sigma_1^2 - (\mu_1 - r)\sigma_1\sigma_2\rho_{12}}{(\mu_1 - r)\sigma_2^2 + (\mu_2 - r)\sigma_1^2 - (\mu_1 + \mu_2 - 2r)\sigma_1\sigma_2\rho_{12}}.
\end{aligned}
\tag{3.23}
$$

Let the risk-free rate be $r = 5\%$. Substituting (3.20) into (3.23) gives us the following solution: $\widehat{w}_1 = \frac{2}{3}$ and $\widehat{w}_2 = \frac{1}{3}$. The expected return and risk are $\mu_V \cong 11.67\%$ and $\sigma_V \cong 16.87\%$, respectively.

The portfolio with weights in (3.23) provides the maximum value of the Sharpe ratio among all portfolios in the two risky assets. On the (σ, μ)-plot, the Sharpe ratio $\frac{\mu_V - r}{\sigma_V}$ represent the slope of a straight half-line starting from $(0, r)$ and passing through the point (σ_V, μ_V). By maximizing the Sharpe ration, we can find the line with the highest slope. If the point (σ_V, μ_V) corresponds to a portfolio in two risky assets that lies on the portfolio line, then, by maximizing the slope, we obtain a half-line which is tangent to the portfolio line. Therefore, the maximum Sharpe ratio portfolio with weights in (3.23) corresponds to the point of tangency (see Figure 3.10). This unique portfolio is called the *market portfolio*. We will use the notation $(\sigma_\mathcal{M}, \mu_\mathcal{M})$ for its position on the risk-reward plane.

The straight half-line starting from $(0, r)$ and passing through $(\sigma_\mathcal{M}, \mu_\mathcal{M})$ is called the *market line*. Its equation is

$$\mu = r + \frac{\mu_\mathcal{M} - r}{\sigma_\mathcal{M}}\sigma. \tag{3.24}$$

This half-line is a feasible region for all optimal portfolios that include a risk-free component and a risky investment in the market portfolio. For all portfolios on this line, the Sharpe ratio attains its maximum value. To find an optimal allocation, the investor first sets the target rate of return $\mu \geqslant r$ and then finds an optimal combination of the risk-free asset and the market portfolio. Suppose that $100w_0\%$ of the initial wealth is invested in the risk-free asset with return r and $100(1 - w_0)\%$ of the capital is invested in the market portfolio with return $r_\mathcal{M}$. The desired w_0 is computed from the expected return:

$$\mu = \mathrm{E}[w_0 r + (1 - w_0)r_\mathcal{M}] = w_0 r + (1 - w_0)\mu_\mathcal{M} \implies w_0 = \frac{\mu_\mathcal{M} - \mu}{\mu_\mathcal{M} - r}.$$

Thus, the volatility (risk) of the resulting portfolio is

$$\sigma = \sqrt{\mathrm{Var}(w_0 r + (1 - w_0)r_\mathcal{M})} = |1 - w_0|\sigma_\mathcal{M} = \left(\frac{\mu - r}{\mu_\mathcal{M} - r}\right)\sigma_\mathcal{M}.$$

The values μ and σ found above satisfy equation (3.24), so this portfolio lies on the market line. For $\mu \in [r, \mu_\mathcal{M}]$, the weight w_0 is nonnegative and hence we are investing in the risk-free security. For $\mu > \mu_\mathcal{M}$, the weight w_0 is negative, which means that we are borrowing money using the risk-free security

3.2.4.4   Maximum Expected Utility Portfolio

Let us find a portfolio that maximizes the mathematical expectation of the exponential utility function, $u(V) = 1 - e^{-\alpha V}$ with $\alpha > 0$, of the wealth $V_T = (1 + r_V)V_0$ with some initial capital $V_0 > 0$:

$$\mathrm{E}[u(V_T)] = \mathrm{E}\left[1 - e^{-\alpha V_0 (1 + r_V)}\right] \to \max. \tag{3.25}$$

Let the portfolio weights be $w_1 = x$ and $w_2 = 1 - x$ with $x \in \mathbb{R}$. The terminal wealth is then a function of x: $V_T(x) = (1 + r_V(x))V_0$. Assume that the rate of return $r_V(x)$ is normally distributed with mean $\mu_V(x)$ and variance $\sigma_V^2(x)$. Using (3.13) gives $\mu_V(x) = \mu_2 + x(\mu_1 - \mu_2)$ and $\sigma_V^2(x) = Ax^2 + 2Bx + C$, where

$$A = \sigma_1^2 + \sigma_2^2 - 2\rho_{12}\sigma_1\sigma_2, \quad B = (\rho_{12}\sigma_1\sigma_2 - \sigma_2^2), \quad C = \sigma_2^2. \tag{3.26}$$

Using the normal MGF, we compute the expected utility as follows:

$$\mathrm{E}\left[1 - e^{-\alpha V_T(x)}\right] = 1 - \mathrm{E}\left[e^{-\alpha V_T(x)}\right] = 1 - e^{-\alpha V_0}\mathrm{E}\left[e^{-\alpha V_0 r_V(x)}\right]$$

$$= 1 - e^{-\alpha V_0}e^{-\alpha V_0 \mu_V(x) + \alpha^2 V_0^2 \sigma_V^2(x)/2}$$

$$= 1 - e^{-\alpha V_0}e^{-\alpha V_0(\mu_V(x) - \alpha V_0 \sigma_V^2(x)/2)}.$$

To solve (3.25), it suffices to maximize

$$\mu_V(x) - \frac{\alpha V_0 \sigma_V^2(x)}{2}. \tag{3.27}$$

Substituting the above expressions for μ_V and σ_V gives us the following target function to be maximized:

$$f(x) := \mu_2 + x(\mu_1 - \mu_2) - \frac{\alpha V_0}{2}(Ax^2 + 2Bx + C).$$

Differentiate f with respect to x and equate the derivative to zero to obtain the following linear equation:

$$f'(x) = 0 \iff (\mu_1 - \mu_2) - \alpha V_0 (Ax + B) = 0.$$

The solution is

$$x^* = -\frac{B}{A} + \frac{1}{A}\frac{\mu_1 - \mu_2}{\alpha V_0} = \frac{(\mu_1 - \mu_2)/(\alpha V_0) + \sigma_2^2 - \rho_{12}\sigma_1\sigma_2}{\sigma_1^2 + \sigma_2^2 - 2\rho_{12}\sigma_1\sigma_2}.$$

Since $f''(x) = -\alpha V_0 A < 0$, the function f is a concave function. Therefore, f attains its maximum at x^*.

Suppose that $\alpha V_0 = 1$ and do computations using the data in (3.20). The optimal weights are

$$\widehat{w}_1 = x^* \cong 0.5431 \quad \text{and} \quad \widehat{w}_2 = 1 - x^* \cong 0.4569.$$

The expected return is $\mu_V(x_0) \cong 12.28\%$; the risk of return is $\sigma_V(x_0) \cong 19.30\%$.

We can also consider other utility functions. However, in most cases we need to use a computational method to find the maximum of the expected utility function. The Taylor series approximation (3.7) can also be applied, as is demonstrated in the next example. Let us find an optimal portfolio when attempting to maximize the expected value of the square-root utility function of $V_T = (1 + r_V)V_0$:

$$\mathrm{E}[\sqrt{V_T}] = \sqrt{V_0}\,\mathrm{E}\left[\sqrt{1 + r_V}\right] \to \max. \tag{3.28}$$

Expand $\sqrt{1 + r_V}$ in a Taylor series about the point $\mu_V(x)$ to obtain

$$\mathrm{E}\left[\sqrt{1 + r_V(x)}\right] \approx \sqrt{1 + \mu_V(x)} - \frac{1}{8}(1 + \mu_V(x))^{-3/2}\sigma_V^2(x),$$

where $r_V(x) = xr_1 + (1 - x)r_2$, and $\mu_V(x)$ and $\sigma_V^2(x)$ are given by (3.13). Now the maximization problem (3.28) reduces to

$$f(x) := \sqrt{1 + x\mu_1 + (1 - x)\mu_2} - \frac{x^2\sigma_1^2 + (1 - x)^2\sigma_2^2 + 2x(1 - x)\sigma_1\sigma_2\rho_{12}}{8(1 + x\mu_1 + (1 - x)\mu_2)^{3/2}} \to \max_x.$$

Equating the derivative of the function f to zero and solving numerically the equation obtained with parameters in (3.20) give the following optimal weights: $\widehat{w}_1 \cong 25.64\%$ and $\widehat{w}_2 \cong 74.36\%$. The resulting portfolio has the following expected return and standard deviation of return: $\mu_V \cong 13.72\%$ and $\sigma_V \cong 29.16\%$.

Concluding remarks. Let us assume without loss of generality that $\sigma_1 < \sigma_2$. We would then expect that $\mu_1 < \mu_2$ to be compensated for a higher level of risk. To minimize the risk σ_V we should invest $x_0 = -\frac{B}{A}$ in the low-risk asset (where A and B are given in (3.26)). To maximize the expected exponential utility, we invest $x^* = -\frac{B}{A} + \frac{1}{A}\frac{\mu_1 - \mu_2}{\alpha V_0}$. We have

$$x^* = x_0 - \frac{1}{A}\frac{\mu_2 - \mu_1}{\alpha V_0} < x_0,$$

since $\mu_1 < \mu_2$ holds. There are several important insights here. First, we see that maximizing the expected utility is *not* the same as minimizing the variance, even for a risk-averse investor. Risk-averse investors are willing to take on a certain amount of risk *provided they are adequately compensated*. To see this, note that the difference $x_0 - x^*$ is increasing in $\mu_2 - \mu_1$: the higher the compensation being offered, the more the risk-averse investor will allocate to the riskier asset. Risk aversion is therefore not the same as complete risk avoidance. If V_0 is large, then $x_0 - x^*$ is small; the increase in wealth is not worth the possibility of losing large sums when marginal returns to wealth are small (as they are for wealthy individuals). The parameter α measures the risk aversion. As α increases, the difference $x_0 - x^*$ decreases. Thus, for an extremely risk-averse investor, the value of x^* is close to x_0. Lastly, we can observe that if σ_2 is large, then A is large as well, and hence the difference $x_0 - x^*$ is small. In other words, if the second asset is too risky, we prefer to invest less in it.

3.3 Portfolio Optimization for N Assets

3.3.1 Portfolios of Several Assets

Consider a market model with N different assets whose respective time-t values are $A_t^1, A_t^2, \ldots, A_t^N$ for $t \in \{0, T\}$. The return on the ith asset is $r_i = \frac{A_T^i - A_0^i}{A_0^i}$. Suppose a portfolio is constructed from these N base assets. Let x_i be the number of shares of asset i with $i = 1, 2, \ldots, N$. The time-t portfolio value is $V_t = \sum_{i=1}^{N} x_i A_t^i$ for $t \in \{0, T\}$. The return on the portfolio is a linear combination of the returns on the assets:

$$
r_V = \frac{V_T - V_0}{V_0} = \sum_{i=1}^{N} \frac{x_i(A_T^i - A_0^i)}{V_0}
$$

$$
= \sum_{i=1}^{N} \frac{x_i A_0^i}{V_0} \frac{A_T^i - A_0^i}{A_0^i} = \sum_{i=1}^{N} \frac{x_i A_0^i}{V_0} r_i .
$$

Define the allocation weights $w_i = \frac{x_i A_0^i}{V_0}$ with $i = 1, 2, \ldots, N$ of funds between the N base assets. The formula for the return r_V takes the following compact form:

$$
r_V = w_1 r_1 + w_2 r_2 + \cdots + w_N r_N = \sum_{i=1}^{N} w_i r_i .
$$

Let us denote

$$
\mathbf{w} := \begin{bmatrix} w_1 & w_2 & \cdots & w_N \end{bmatrix}^\top \in \mathbb{R}^N ,
$$

where $\mathbf{x}^\top$ denotes the transpose of a vector $\mathbf{x}$. In this section, we operate with column vectors.

The sum of the weights is one. This fact can be written in vector form:

$$
\mathbf{u}^\top \mathbf{w} = 1, \text{ where } \mathbf{u} := \begin{bmatrix} 1 & 1 & \cdots & 1 \end{bmatrix}^\top \in \mathbb{R}^N . \tag{3.29}
$$

We denote $\mu_i = \mathrm{E}[r_i]$—the expected return on asset i, $\sigma_i^2 = \mathrm{Var}(r_i)$—the variance of the return on asset i, and $c_{ij} = \mathrm{Cov}(r_i, r_j)$—the covariance between returns r_i and r_j for $i, j = 1, 2, \ldots, N$. The expected returns and covariances between returns can be respectively arranged into an $N \times 1$ column vector and an $N \times N$ matrix:

$$
\mathbf{m} := \begin{bmatrix} \mu_1 \\ \mu_2 \\ \vdots \\ \mu_N \end{bmatrix} \quad \text{and} \quad \mathbf{C} := \begin{bmatrix} c_{11} & c_{12} & \cdots & c_{1N} \\ c_{21} & c_{22} & \cdots & c_{2N} \\ \vdots & \vdots & \ddots & \vdots \\ c_{N1} & c_{N2} & \cdots & c_{NN} \end{bmatrix} .
$$

Assume that there are at least two assets with different expected returns. That is, there exists no μ such that $\mathbf{m} = \mu\mathbf{u}$.

The matrix $\mathbf{C}$ is called a *covariance matrix*. Recall that the covariance $\sigma_{XY} \equiv \mathrm{Cov}(X, Y)$ of two random variables X and Y can be factorized into a product of the standard deviations, σ_X and σ_Y, and the coefficient of correlation between X and Y denoted by $\mathrm{Corr}(X, Y) \equiv \rho_{XY}$ as follows: $\sigma_{XY} = \sigma_X \rho_{XY} \sigma_Y$. Therefore, for any $i, j = 1, 2, \ldots, N$, the covariance c_{ij} is a product of the standard deviations σ_i and σ_j and the correlation coefficient $\rho_{ij} :=$ $\mathrm{Corr}(r_i, r_j)$:

$$
c_{ij} = \sigma_i \sigma_j \rho_{ij} .
$$

As a result, the covariance matrix $\mathbf{C}$ can be represented as a product of two diagonal matrices filled with standard deviations of returns, σ_i, and a correlation matrix whose entries are coefficients of correlation between returns, ρ_{ij}:

$$
\mathbf{C} = \begin{bmatrix} \sigma_1 & 0 & \cdots & 0 \\ 0 & \sigma_2 & \cdots & 0 \\ \vdots & \vdots & \ddots & \vdots \\ 0 & 0 & \cdots & \sigma_N \end{bmatrix} \begin{bmatrix} 1 & \rho_{12} & \cdots & \rho_{1N} \\ \rho_{21} & 1 & \cdots & \rho_{2N} \\ \vdots & \vdots & \ddots & \vdots \\ \rho_{N1} & \rho_{N2} & \cdots & 1 \end{bmatrix} \begin{bmatrix} \sigma_1 & 0 & \cdots & 0 \\ 0 & \sigma_2 & \cdots & 0 \\ \vdots & \vdots & \ddots & \vdots \\ 0 & 0 & \cdots & \sigma_N \end{bmatrix}.
$$

Here, we use the fact that $\mathrm{Corr}(X, X) = 1$ for every random variable X, hence $\rho_{ii} = 1$ for all $i = 1, 2, \ldots, N$.

The covariance matrix is symmetric (i.e., $\mathbf{C} = \mathbf{C}^\top$) and positive definite, i.e., $\mathbf{w}^\top \mathbf{C} \mathbf{w} > 0$ for every nonzero vector $\mathbf{w} \in \mathbb{R}^N$. Since $\mathbf{C}$ is positive definite, it is a nonsingular matrix, and hence its inverse matrix $\mathbf{C}^{-1}$ exists. There exist several necessary and sufficient criteria to determine if a real symmetric matrix $\mathbf{C}$ is positive definite, including the following.

- All eigenvalues of $\mathbf{C}$ are positive.

- All leading principal minors are positive. The kth leading principal minor of $\mathbf{C}$ is the determinant of the upper-left k-by-k corner of $\mathbf{C}$, where $k = 1, 2, \ldots, N$. This method is known as *Sylvester's criterion*.

- There exists a unique lower triangular matrix $\mathbf{L}$, with strictly positive diagonal elements, that allows the factorization of $\mathbf{C}$ into $\mathbf{C} = \mathbf{L}\mathbf{L}^\top$. It is called the *Cholesky factorization*.

Note that in general $\mathbf{C}$ can be a semi-positive definite matrix, meaning that $\mathbf{w}^\top \mathbf{C} \mathbf{w} \geqslant 0$ for all $\mathbf{w} \in \mathbb{R}^N$.

Let us find the mathematical expectation and variance of r_V by applying well-known equations for the mathematical expectation and variance of a sum of (dependent) random variables. The expected return on the portfolio V with weight vector $\mathbf{w}$ is

$$
\mu_V = \mathrm{E}[r_V] = \mathrm{E}\left[\sum_{i=1}^N w_i r_i\right] = \sum_{i=1}^N \mathrm{E}[w_i r_i] = \sum_{i=1}^N w_i \mu_i. \tag{3.30}
$$

The variance of r_V is

$$
\sigma_V^2 = \mathrm{Var}(r_V) = \mathrm{Var}\left(\sum_{i=1}^N w_i r_i\right)
$$

$$
= \mathrm{Cov}\left(\sum_{i=1}^N w_i r_i, \sum_{j=1}^N w_j r_j\right) = \sum_{i=1}^N \sum_{j=1}^N \mathrm{Cov}(w_i r_i, w_j r_j)
$$

$$
= \sum_{i=1}^N \sum_{j=1}^N w_i w_j \, \mathrm{Cov}(r_i, r_j) = \sum_{i=1}^N \sum_{j=1}^N w_i w_j c_{ij}. \tag{3.31}
$$

The above equations can also be written in the matrix-vector form:

$$
\mu_V = \mathbf{m}^\top \mathbf{w}, \tag{3.32}
$$

$$
\sigma_V^2 = \mathbf{w}^\top \mathbf{C} \mathbf{w}. \tag{3.33}
$$

Note that we do not assume any probability distribution for the vector of returns. Our analysis of portfolios is entirely based on the knowledge of the vector of expected returns $\mathbf{m}$ and covariance matrix $\mathbf{C}$.

Example 3.8. Show that $\mathbf{R} = \begin{bmatrix} 1 & 0.5 & -0.5 \\ 0.5 & 1 & 0.25 \\ -0.5 & 0.25 & 1 \end{bmatrix}$ is a positive-definite correlation matrix.

Solution. The matrix $\mathbf{R}$ is symmetric, since for any $i, j = 1, 2, 3$ the (i, j)th entry $R_{i,j}$ equals $R_{j,i}$. Let us verify that all its leading principal minors are positive:

$$|1| = 1 > 0,$$

$$\begin{vmatrix} 1 & 0.5 \\ 0.5 & 1 \end{vmatrix} = 0.75 > 0,$$

$$\begin{vmatrix} 1 & 0.5 & -0.5 \\ 0.5 & 1 & 0.25 \\ -0.5 & 0.25 & 1 \end{vmatrix} = \frac{5}{16} > 0.$$

Thus, according to Sylvester's criterion, $\mathbf{R}$ is a positive definite matrix. Alternatively, we can find the eigenvalues of $\mathbf{R}$: $\lambda_1 = \frac{5}{4}$, $\lambda_{2,3} = \frac{7}{8} \pm \frac{\sqrt{33}}{8}$. They are positive. Thus, $\mathbf{R} > 0$. Lastly, since all elements of $\mathbf{R}$ on the main diagonal are equal to 1, and all its off-diagonal entries are between -1 and 1, the matrix $\mathbf{R}$ is a correlation matrix. $\qquad\square$

Example 3.9. Construct a covariance matrix using the correlation matrix $\mathbf{R}$ from Example 3.8 and vector of standard deviations $\boldsymbol{\sigma} = \begin{bmatrix} 0.1 & 0.15 & 0.2 \end{bmatrix}^{\top}$.

Solution. The correlation matrix $\mathbf{C}$ is equal to the product $\mathbf{D}\mathbf{R}\mathbf{D}$ where

$$\mathbf{D} = \begin{bmatrix} \sigma_1 & 0 & 0 \\ 0 & \sigma_2 & 0 \\ 0 & 0 & \sigma_3 \end{bmatrix} = \begin{bmatrix} 0.1 & 0 & 0 \\ 0 & 0.15 & 0 \\ 0 & 0 & 0.2 \end{bmatrix}$$

is a diagonal matrix with standard deviations on the main diagonal. Thus, we compute

$$\mathbf{C} = \begin{bmatrix} 0.1 & 0 & 0 \\ 0 & 0.15 & 0 \\ 0 & 0 & 0.2 \end{bmatrix} \cdot \begin{bmatrix} 1 & 0.5 & -0.5 \\ 0.5 & 1 & 0.25 \\ -0.5 & 0.25 & 1 \end{bmatrix} \cdot \begin{bmatrix} 0.1 & 0 & 0 \\ 0 & 0.15 & 0 \\ 0 & 0 & 0.2 \end{bmatrix}$$

$$= \begin{bmatrix} 0.01 & 0.0075 & -0.01 \\ 0.0075 & 0.0225 & 0.0075 \\ -0.01 & 0.0075 & 0.04 \end{bmatrix}.$$

$\qquad\square$

In the next subsections, we shall solve the following four multi-asset optimization problems.

1. Find a portfolio with the minimum variance. It will be called the *minimum variance portfolio.*

2. Find a portfolio with the minimum variance among all portfolios whose expected return is fixed and equal to a given number. We may obtain different solutions for portfolios with or without short sells. The set of such portfolios parameterized by the expected return is called the *minimum variance (portfolio) line.*

3. Find a portfolio that maximizes the expected exponential utility of return. We will show that this portfolio converges to the minimum variance portfolio as the risk aversion increases.

4. Lastly, we derive weights of a market portfolio that maximizes the Share ratio.

3.3.2   The Minimum Variance Portfolio

To find the minimum variance portfolio, we need to solve

$$\sigma_V^2(\mathbf{w}) = \mathbf{w}^\top \mathbf{C} \mathbf{w} \to \min_{\mathbf{w}} \tag{3.34}$$

subject to the constraint

$$\mathbf{u}^\top \mathbf{w} = 1. \tag{3.35}$$

Let us use the method of Lagrange multipliers. First, we find the critical points of the function

$$F(\mathbf{w}, \lambda) := \mathbf{w}^\top \mathbf{C} \mathbf{w} - \lambda(\mathbf{u}^\top \mathbf{w} - 1).$$

The partial derivatives of F with respect to w_i for $i = 1, 2, \ldots, N$ are

$$\frac{\partial F}{\partial w_i}(\mathbf{w}, \lambda) = \frac{\partial}{\partial w_i} \left(\sum_{i=1}^{N} \sum_{j=1}^{N} w_i w_j c_{ij} - \lambda \sum_{i=1}^{N} w_i + \lambda \right)$$

$$= 2 \sum_{j=1}^{N} w_j c_{ij} - \lambda,$$

where λ is a Lagrange multiplier. Equating them to zero gives us the following linear equations:

$$2 \sum_{j=1}^{N} w_j c_{ij} - \lambda = 0 \quad \text{for all} \quad i = 1, 2, \ldots, N.$$

Let $\mathbf{c}_i^\top$ denote the ith row of matrix $\mathbf{C}$. Then the above equations can be rewritten in vector form:

$$2\mathbf{c}_i^\top \mathbf{w} - \lambda = 0 \quad \text{for} \quad i = 1, 2, \ldots, N.$$

Combine these N equations to obtain the final equation in matrix-vector form:

$$2\mathbf{C}\mathbf{w} - \lambda\mathbf{u} = \mathbf{0}.$$

Multiplying both parts by the inverse matrix $\mathbf{C}^{-1}$ on the left gives:

$$2\mathbf{C}^{-1}\mathbf{C}\mathbf{w} - \lambda\mathbf{C}^{-1}\mathbf{u} = 2\mathbf{w} - \lambda\mathbf{C}^{-1}\mathbf{u} \quad \text{and} \quad \mathbf{C}^{-1}\mathbf{0} = \mathbf{0} \implies 2\mathbf{w} - \lambda\mathbf{C}^{-1}\mathbf{u} = \mathbf{0}.$$

Solve this equation for $\mathbf{w}$ to obtain that $\mathbf{w} = \frac{\lambda}{2}\mathbf{C}^{-1}\mathbf{u}$. The only missing variable is λ. Substitute the expression for $\mathbf{w}$ in the constraint (3.35) to obtain

$$1 = \frac{\lambda}{2}\mathbf{u}^\top \mathbf{C}^{-1}\mathbf{u} \implies \lambda = \frac{2}{\mathbf{u}^\top \mathbf{C}^{-1}\mathbf{u}}.$$

Finally, we obtain weights of the minimum variance (mv) portfolio:

$$\mathbf{w}_{\text{mv}} = \frac{\mathbf{C}^{-1}\mathbf{u}}{\mathbf{u}^\top \mathbf{C}^{-1}\mathbf{u}}. \tag{3.36}$$

Since the matrix of second derivatives of the function $\sigma_V^2(\mathbf{w})$ equals $2\mathbf{C}$ and $\mathbf{C}$ is positive definite, the function $F(\mathbf{w}, \lambda)$ is a convex function of $\mathbf{w}$ for every value of λ. Therefore, the function $\sigma_V^2(\mathbf{w})$ has a minimum at $\mathbf{w}_{\text{mv}}$. The minimum variance can be computed by putting the weights $\mathbf{w}_{\text{mv}}$ in (3.31):

$$\sigma_{\text{mv}}^2 = \mathbf{w}_{\text{mv}}^\top \mathbf{C} \mathbf{w}_{\text{mv}} = \frac{1}{\mathbf{u}^\top \mathbf{C}^{-1}\mathbf{u}}.$$

As an example, let us consider the case of two assets. The covariance matrix $\mathbf{C}$ and its inverse $\mathbf{C}^{-1}$ are

$$\mathbf{C} = \begin{bmatrix} \sigma_1^2 & \rho_{12}\sigma_1\sigma_2 \\ \rho_{12}\sigma_1\sigma_2 & \sigma_2^2 \end{bmatrix} \text{ and } \mathbf{C}^{-1} = \frac{1}{1-\rho_{12}^2} \begin{bmatrix} \frac{1}{\sigma_1^2} & -\frac{\rho_{12}}{\sigma_1\sigma_2} \\ -\frac{\rho_{12}}{\sigma_1\sigma_2} & \frac{1}{\sigma_2^2} \end{bmatrix}.$$

The weight vector is then

$$\begin{aligned} \mathbf{w}_{\mathrm{mv}}^\top = [w_1, \ w_2] &= \frac{1}{\frac{1}{\sigma_1^2} - \frac{2\rho_{12}}{\sigma_1\sigma_2} + \frac{1}{\sigma_2^2}} \left[\frac{1}{\sigma_1^2} - \frac{\rho_{12}}{\sigma_1\sigma_2}, \ \frac{1}{\sigma_2^2} - \frac{\rho_{12}}{\sigma_1\sigma_2} \right] \\ &= \left[\frac{\sigma_2^2 - \rho_{12}\sigma_1\sigma_2}{\sigma_1^2 - 2\rho_{12}\sigma_1\sigma_2 + \sigma_2^2}, \ \frac{\sigma_1^2 - \rho_{12}\sigma_1\sigma_2}{\sigma_1^2 - 2\rho_{12}\sigma_1\sigma_2 + \sigma_2^2} \right]. \end{aligned}$$

The resulting expression is identical to that of (3.16).

3.3.3 Minimum Variance Portfolio Line

Now, we consider a set of portfolios with fixed expected return μ, i.e., we have the additional constrain $\mathbf{m}^\top\mathbf{w} = \mu$. To find the minimum variance portfolio in such a set, we need to minimize $\sigma_V^2(\mathbf{w}) = \mathbf{w}^\top\mathbf{C}\mathbf{w}$ subject to the constraints

$$\mathbf{u}^\top\mathbf{w} = 1 \text{ and } \mathbf{m}^\top\mathbf{w} = \mu. \tag{3.37}$$

As a result, we obtain a family of minimum variance portfolios $\mathbf{w} = \widehat{\mathbf{w}}(\mu)$ parameterized by μ. On the risk-return plot, such a family is represented by a continuous line called the *minimum variance line*.

Again, to find the equation of the minimum variance line we apply the method of Lagrange multipliers. Introduce the function

$$G(\mathbf{w}, \lambda_1, \lambda_2) := \mathbf{w}^\top\mathbf{C}\mathbf{w} - \lambda_1(\mathbf{u}^\top\mathbf{w} - 1) - \lambda_2(\mathbf{m}^\top\mathbf{w} - \mu),$$

where λ_1 and λ_2 are Lagrange multipliers. Differentiate G w.r.t. weight w_i and equate the derivative to zero:

$$\frac{\partial G}{\partial w_i} = 2\sum_{j=1}^N w_j c_{ij} - \lambda_1 - \lambda_2\mu_i = 0 \text{ for } i = 1, 2, \dots, N.$$

The equations can be written as

$$2\mathbf{c}_i^\top\mathbf{w} - \lambda_1 - \lambda_2\mu = 0 \text{ for } i = 1, 2, \dots, N.$$

Combine them to obtain the following matrix-vector equation:

$$2\mathbf{C}\mathbf{w} - \lambda_1\mathbf{u} - \lambda_2\mathbf{m} = 0.$$

By solving for the weights $\mathbf{w}$, we obtain:

$$\mathbf{w} = \mathbf{C}^{-1}\left(\frac{\lambda_1}{2}\mathbf{u} + \frac{\lambda_2}{2}\mathbf{m}\right) = \frac{\lambda_1}{2}\mathbf{C}^{-1}\mathbf{u} + \frac{\lambda_2}{2}\mathbf{C}^{-1}\mathbf{m}. \tag{3.38}$$

The constraints (3.37) are revealed from equations $\frac{\partial G}{\partial \lambda_i} = 0$ for $i = 1, 2$. Now substitute the expression (3.38) for $\mathbf{w}$ into the constraints (3.37) to obtain the following system of equations:

$$\begin{cases} \frac{1}{2}\mathbf{u}^\top\mathbf{C}^{-1}\mathbf{u}\lambda_1 + \frac{1}{2}\mathbf{u}^\top\mathbf{C}^{-1}\mathbf{m}\lambda_2 = 1, \\ \frac{1}{2}\mathbf{m}^\top\mathbf{C}^{-1}\mathbf{u}\lambda_1 + \frac{1}{2}\mathbf{m}^\top\mathbf{C}^{-1}\mathbf{m}\lambda_2 = \mu. \end{cases} \tag{3.39}$$

Recall that a 2-by-2 system of linear equations

$$\begin{cases} a_{11}x_1 + a_{12}x_2 = b_1, \\ a_{21}x_1 + a_{22}x_2 = b_2 \end{cases}$$

admits a unique solution $x_1 = \frac{1}{D}\begin{vmatrix} b_1 & a_{12} \\ b_2 & a_{22} \end{vmatrix}$ and $x_2 = \frac{1}{D}\begin{vmatrix} a_{11} & b_1 \\ a_{21} & b_2 \end{vmatrix}$ with $D := \begin{vmatrix} a_{11} & a_{12} \\ a_{21} & a_{22} \end{vmatrix}$

provided $D \neq 0$. Here, $\begin{vmatrix} a & b \\ c & d \end{vmatrix} = ad - bc$ denotes the determinant of a 2-by-2 matrix. Solve the system (3.39) for λ_1 and λ_2 and then plug the solution into (3.38) to obtain the final formula for the portfolio weights:

$$\widehat{\mathbf{w}} = \frac{1}{D}\begin{vmatrix} 1 & \mathbf{u}^\top\mathbf{C}^{-1}\mathbf{m} \\ \mu & \mathbf{m}^\top\mathbf{C}^{-1}\mathbf{m} \end{vmatrix}\mathbf{C}^{-1}\mathbf{u} + \frac{1}{D}\begin{vmatrix} \mathbf{u}^\top\mathbf{C}^{-1}\mathbf{u} & 1 \\ \mathbf{m}^\top\mathbf{C}^{-1}\mathbf{u} & \mu \end{vmatrix}\mathbf{C}^{-1}\mathbf{m} \tag{3.40}$$

with $D := \begin{vmatrix} \mathbf{u}^\top\mathbf{C}^{-1}\mathbf{u} & \mathbf{u}^\top\mathbf{C}^{-1}\mathbf{m} \\ \mathbf{m}^\top\mathbf{C}^{-1}\mathbf{u} & \mathbf{m}^\top\mathbf{C}^{-1}\mathbf{m} \end{vmatrix} \neq 0$. Note that the solution does not exist if $\mathbf{m} = \nu\mathbf{u}$ for some ν.

The determinants in (3.40) are linear functions of μ. Therefore, the weights of portfolios on the minimum variance line depend on μ linearly as well: $\widehat{\mathbf{w}} = \mu\mathbf{a} + \mathbf{b}$ with

$$\mathbf{a} := \frac{\mathbf{u}^\top\mathbf{C}^{-1}\mathbf{u}\mathbf{C}^{-1}\mathbf{m} - \mathbf{u}^\top\mathbf{C}^{-1}\mathbf{m}\mathbf{C}^{-1}\mathbf{u}}{D},$$

$$\mathbf{b} := \frac{\mathbf{m}^\top\mathbf{C}^{-1}\mathbf{m}\mathbf{C}^{-1}\mathbf{u} - \mathbf{m}^\top\mathbf{C}^{-1}\mathbf{u}\mathbf{C}^{-1}\mathbf{m}}{D}.$$

This observation allows us to describe the shape of the minimum variance line. Let us select any two different portfolios with respective weights $\mathbf{w}'$ and $\mathbf{w}''$ on the line. Then the minimum variance line consists of portfolios with weights $\mathbf{w} = x\mathbf{w}' + (1 - x)\mathbf{w}''$ for $x \in \mathbb{R}$. Indeed, the weights of the two chosen portfolios satisfy $\mathbf{w}' = \mu'\mathbf{a} + \mathbf{b}$ and $\mathbf{w}'' = \mu''\mathbf{a} + \mathbf{b}$ for some $\mu' \neq \mu''$. Every linear combination of the weights $\mathbf{w}'$ and $\mathbf{w}''$ satisfies the same equation:

$$\mathbf{w} = x\mathbf{w}' + (1 - x)\mathbf{w}'' = (x\mu' + (1 - x)\mu'')\mathbf{a} + (x + (1 - x))\mathbf{b} = \mu_x\mathbf{a} + \mathbf{b}$$

with $\mu_x = x\mu' + (1 - x)\mu''$. Conversely, for every $\mu \in \mathbb{R}$ there exist $x \in \mathbb{R}$ so that $\mu = x\mu' + (1 - x)\mu''$. Therefore, the portfolios with weights $x\mathbf{w}' + (1 - x)\mathbf{w}''$, $x \in \mathbb{R}$, exhaust the minimum variance line. This result means that the minimum variance line has the same shape as that describing a set of portfolios constructed from two assets. The shape of the line is a hyperbola, and it does not depend on the number of assets. The set of admissible portfolios in the $N \geqslant 3$ bases assets is represented by a planar domain bounded by the minimum variance line. The shape of this domain is known as the *Markowitz bullet*. All elementary portfolios consisting of individual assets lie inside the bullet, as shown in Figure 3.11.

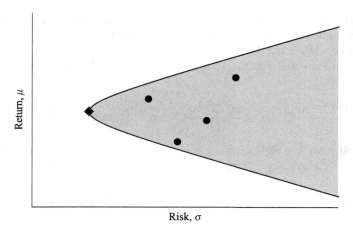

FIGURE 3.11: The set of admissible portfolios (the Markowitz bullet) in four underlying assets (which are marked by solid circles) bounded by the minimum variance line. The minimum variance portfolio is marked by a diamond.

Example 3.10. Let us consider a portfolio in three underlying assets whose expected returns, standard deviations of returns, and correlations between returns are as follows:

$$\mu_1 = 0.1, \qquad \sigma_1 = 0.2, \qquad \rho_{12} = \rho_{21} = -0.2,$$
$$\mu_2 = 0.15, \qquad \sigma_2 = 0.3, \qquad \rho_{23} = \rho_{32} = 0.2,$$
$$\mu_3 = 0.3, \qquad \sigma_3 = 0.4, \qquad \rho_{31} = \rho_{13} = -0.4.$$

(a) Find the minimum variance portfolio.

(b) Find the minimum variance portfolio line.

Solution. First, to apply the formulae in (3.36) and (3.40), we arrange the expected returns μ_i in a vector $\mathbf{m}$ and construct the covariance matrix $\mathbf{C}$ with entries $C_{ij} = \sigma_i \sigma_j \rho_{ij}$:

$$\mathbf{m} = \begin{bmatrix} 0.10 \\ 0.15 \\ 0.30 \end{bmatrix}, \quad \mathbf{C} = \begin{bmatrix} 0.040 & -0.012 & -0.032 \\ -0.012 & 0.090 & 0.024 \\ -0.032 & 0.024 & 0.160 \end{bmatrix}.$$

The matrix $\mathbf{C}$ is positive definite, hence the inverse matrix $\mathbf{C}^{-1}$ exists:

$$\mathbf{C}^{-1} \cong \begin{bmatrix} 30.3030 & 2.5252 & 5.6818 \\ 2.5252 & 11.7845 & -1.2626 \\ 5.6818 & -1.2626 & 7.5758 \end{bmatrix}.$$

The weights of the minimum variance portfolio are

$$\mathbf{w}_{\mathrm{mv}} = \frac{\mathbf{u}^\top \mathbf{C}^{-1}}{\mathbf{u}^\top \mathbf{C}^{-1} \mathbf{u}} \cong \begin{bmatrix} 0.6060 & 0.2053 & 0.1887 \end{bmatrix}^\top.$$

The expected return μ_{mv} and standard deviation (the risk) of the return σ_{mv} of the minimum variance portfolio are, respectively,

$$\mu_{\mathrm{mv}} = \mathbf{m}^\top \mathbf{w}_{\mathrm{mv}} \cong 14.80\% \quad \text{and} \quad \sigma_{\mathrm{mv}} = \sqrt{\mathbf{w}_{\mathrm{mv}}^\top \mathbf{C} \mathbf{w}_{\mathrm{mv}}} \cong 12.54\%.$$

To describe the minimum variance line, we need to find the weight vectors of any two portfolios on the line. We have already found one of them—the minimum variance portfolio. Since $\mu_{\mathrm{mv}} \neq 0$, the other portfolio on the minimum variance line to be selected can be a portfolio with zero expected return. To find its weights, apply equation (3.40) where we put $\mu = 0$ to obtain

$$\mathbf{w}_0 = \frac{\mathbf{m}^\top \mathbf{C}^{-1} \mathbf{m}}{D} \mathbf{C}^{-1} \mathbf{u} - \frac{\mathbf{m}^\top \mathbf{C}^{-1} \mathbf{u}}{D} \mathbf{C}^{-1} \mathbf{m} \cong \begin{bmatrix} 1.1459 & 0.4721 & -0.6180 \end{bmatrix}^\top .$$

Now the portfolios with weights $x\mathbf{w}_{\mathrm{mv}} + (1 - x)\mathbf{w}_0$ where $x \in \mathbb{R}$ exhaust the minimum variance line. For example, by setting $x = 1$, we can obtain the minimum variance portfolio.

$\square$

Since $w_3 = 1 - w_1 - w_2$, all portfolios in three basis assets from the above example can be described by the two weights w_1 and w_2. On the (w_1, w_2)-plane, every portfolio line is represented by a straight line. Figure 3.12 visualizes the set of admissible portfolios from Example 3.10 on the (w_1, w_2)-plane (the left plot) and the (σ, μ)-plane (the right plot). The three dashed lines represents portfolio in two underlying assets only. For example, the line described by the equation $w_1 = 0$ represents all portfolios in the basis assets 2 and 3 only; the line $w_1 + w_2 = 1$ represents all portfolios in the basis assets 1 and 2 only, etc. The triangle bounded by the three dashed lines on the left plot contains all portfolios without short selling (with $\mathbf{w} \geqslant 0$). The bold line represents the minimum variance line, and the minimum variance portfolio is marked by a diamond. As we can see in the left plot of Figure 3.12, the minimum variance portfolio lies inside the triangle bounded by dashed lines, since it is without short selling, and all of its weights are positive.

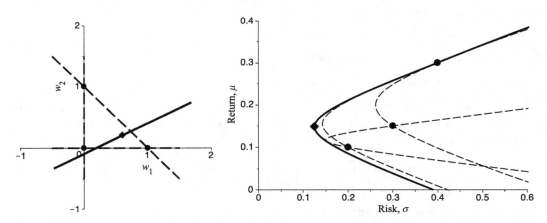

FIGURE 3.12: The minimum variance line from Example 3.10 is plotted as a bold line. The minimum variance portfolio is marked by a diamond. The solid circles represent the basis assets. The dashed lines represent portfolio lines.

3.3.4   Case without Short Selling

The case without short selling is very similar to that considered in the previous section. No short selling means that all positions in an investment portfolio have to be nonnegative. We can find the minimum variance portfolio line and the minimum variance portfolio by solving respective quadratic programming problems that have one additional condition: the weights w_i are now nonnegative. To find the minimum variance portfolio, we need to solve

(3.34):

$$\sigma_V^2(\mathbf{w}) := \mathbf{w}^\top \mathbf{C} \mathbf{w} \to \min_w$$

subject to the constraints

$$\mathbf{u}^\top \mathbf{w} = 1, \ \mathbf{w} \geqslant \mathbf{0}. \tag{3.41}$$

Here, the meaning of $\mathbf{w} \geqslant \mathbf{0}$ is that all $w_i \geqslant 0$. To obtain a family of minimum variance portfolios parameterized by the expected return μ, we need to minimize $f(\mathbf{w})$ subject to the constraints

$$\mathbf{u}^\top \mathbf{w} = 1, \ \mathbf{m}^\top \mathbf{w} = \mu, \ \text{and} \ \mathbf{w} \geqslant \mathbf{0}. \tag{3.42}$$

The constraints in (3.41) and (3.42) are almost the same as are in (3.35) and (3.37), respectively. These quadratic optimization problems can be solved numerically. Computer systems such as MAPLE^{TM}, MATHEMATICA^{TM}, and MATLAB^{TM} can be applied to solve the minimization problems (3.34)–(3.41) and (3.34)–(3.42).

Let us consider the case with three assets from Example 3.10. The weights can be parameterized by two real variables $w_1, w_2 \in [0, 1]$ with $w_1 + w_2 \leqslant 1$ and hence $w_3 = 1 - w_1 - w_2 \geqslant 0$. On the (w_1, w_2)-plane, the set of admissible portfolios without short selling is represented by a triangle with vertices $(0, 0)$, $(0, 1)$, and $(1, 0)$.

The expected return on a portfolio with nonnegative weights $\mathbf{w}$ is bounded above and below by $\max \mu_i$ and $\min \mu_i$, respectively. Therefore, the minimum variance line is a bounded curve on the (σ, μ)-plane. It connects two points corresponding to the assets with the lowest μ and highest μ, respectively. The set of admissible portfolios (with nonnegative weights) is represented by a planar domain bounded by the minimum variance line and no-short-selling parts of the portfolio lines corresponding to different pairs of the underlying assets (see Figure 3.13).

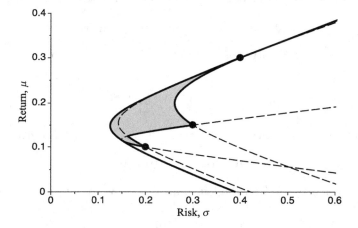

FIGURE 3.13: The set of admissible portfolios in three underlying assets without short selling.

3.3.5 Maximum Expected Utility Portfolio

Suppose that the asset returns r_j, $j = 1, 2, \ldots, N$ follow a multivariate normal probability distribution. The portfolio return r_V is a linear combination of single-asset returns, and hence, as a sum on normal variates, it also has a normal distribution:

$$r_V \sim Norm(\mu_V, \sigma_V^2) \implies r_V \overset{d}{=} \mu_V + \sigma_V Z, \ \text{with} \ Z \sim Norm(0, 1).$$

Consider an investor who uses an exponential utility function $u(x) = -e^{-\alpha x}$ with $\alpha > 0$ to find an optimal portfolio. The allocation weights can be found as a solution to the optimization problem where we are attempting to maximize the expected utility of the terminal wealth $V_T = V_0(1 + r_V)$:

$$\mathrm{E}\left[-e^{-\alpha V_0(1+r_V)}\right] \to \max_{\mathbf{w}} \quad \text{subject to } \mathbf{u}^\top \mathbf{w} = 1. \tag{3.43}$$

Using the normal moment generating function, we can compute the mathematical expectation of the exponential function:

$$\mathrm{E}\left[-e^{-\alpha V_0 - \alpha V_0 r_V}\right] = -e^{-\alpha V_0}\mathrm{E}\left[e^{-\alpha V_0(\mu_V + \sigma_V Z)}\right] = -e^{-\alpha V_0 - \alpha V_0(\mu_V - \alpha V_0 \sigma_V^2/2)}.$$

Therefore, the optimization problem (3.43) is equivalent to

$$\mu_V - \frac{\alpha_0}{2}\sigma_V^2 \to \max_{\mathbf{w}} \quad \text{subject to } \mathbf{u}^\top \mathbf{w} = 1,$$

where $\alpha_0 = \alpha V_0$. It is the same optimization problem as that for the two-asset case (see equation (3.27)).

Theorem 3.3. *The vector of weights for the maximum expected utility (meu) portfolio is*

$$\mathbf{w}_{\text{meu}} = \frac{1}{\alpha_0}\mathbf{C}^{-1}\mathbf{m} + \left(\frac{1 - \frac{1}{\alpha_0}\mathbf{u}^\top \mathbf{C}^{-1}\mathbf{m}}{\mathbf{u}^\top \mathbf{C}^{-1}\mathbf{u}}\right)\mathbf{C}^{-1}\mathbf{u}. \tag{3.44}$$

Proof. Let us use the method of Lagrange multipliers. First, we find the critical points of the function

$$F(\mathbf{w}, \lambda) := \mathbf{m}^\top \mathbf{w} - \frac{\alpha_0}{2}\mathbf{w}^\top \mathbf{C}\mathbf{w} - \lambda(\mathbf{u}^\top \mathbf{w} - 1).$$

The partial derivatives of F with respect to w_i for $i = 1, 2, \ldots, N$ are

$$\frac{\partial F}{\partial w_i}(\mathbf{w}, \lambda) = m_i - \alpha_0 \sum_{j=1}^{N} w_j c_{ij} - \lambda.$$

Equating them to zero and writing the system in matrix-vector form give us

$$\mathbf{m} - \alpha_0 \mathbf{C}\mathbf{w} - \lambda\mathbf{u} = \mathbf{0}.$$

Multiply this equation by $\mathbf{C}^{-1}$ on the left and solve for $\mathbf{w}$ to obtain

$$\mathbf{w} = \frac{1}{\alpha_0}\mathbf{C}^{-1}(\mathbf{m} - \lambda\mathbf{u}).$$

Substitute the expression for $\mathbf{w}$ in the constraint $\mathbf{u}^\top \mathbf{w} = 1$ to obtain

$$1 = \frac{1}{\alpha_0}\mathbf{u}^\top \mathbf{C}^{-1}(\mathbf{m} - \lambda\mathbf{u}) \implies \lambda = \frac{\mathbf{u}^\top \mathbf{C}^{-1}\mathbf{m} - \alpha_0}{\mathbf{u}^\top \mathbf{C}^{-1}\mathbf{u}}.$$

Finally, combining the formulae for $\mathbf{w}$ and λ we arrive at equation (3.44). $\qquad \square$

Let us find out what happens to the weights of the maximum utility portfolio in the limiting case as $\alpha \to \infty$. Recall that α measures the risk-aversion. So, in the limiting case, we deal with a complete aversion to risk. The term $\frac{1}{\alpha_0}\mathbf{C}^{-1}\mathbf{m}$ converges to a zero vector, as well as $\frac{1}{\alpha_0}\mathbf{u}^\top \mathbf{C}^{-1}\mathbf{m} \to 0$, when $\alpha \to \infty$ and V_0 is fixed. So, the solution in (3.44) converges to $\frac{\mathbf{C}^{-1}\mathbf{u}}{\mathbf{u}^\top \mathbf{C}^{-1}\mathbf{u}}$. Thus, as $\alpha \to \infty$, the maximum expected utility portfolio converges to the minimum variance portfolio.

We can also find a portfolio that maximizes the expected utility of the return r_V:

$$\mathrm{E}\left[-e^{-\alpha r_V}\right] \to \max.$$

The solution to this optimization problem has the same form as in (3.44) with α in place of α_0 (i.e., set $V_0 = 1$).

Example 3.11. Let us consider a portfolio in three underlying assets from Example 3.10. Find the maximum expected utility portfolio for $\alpha = 0.001$ and $V_0 = 1000$.

Solution. Apply (3.44) with $\alpha_0 = \alpha V_0 = 0.001 \cdot 1000 = 1$ to find the weights of the maximum expected utility portfolio:

$$\mathbf{w}_{\mathrm{meu}} = \mathbf{C}^{-1}\mathbf{m} + \left(\frac{1 - \mathbf{u}^\top \mathbf{C}^{-1}\mathbf{m}}{\mathbf{u}^\top \mathbf{C}^{-1}\mathbf{u}}\right)\mathbf{C}^{-1}\mathbf{u} \cong \begin{bmatrix} 0.0196 & -0.08444 & 1.06485 \end{bmatrix}^\top.$$

The expected return μ_{meu} and standard deviation (the risk) of the return σ_{meu} are, respectively,

$$\mu_{\mathrm{meu}} = \mathbf{m}^\top \mathbf{w}_{\mathrm{meu}} \cong 30.8747\% \quad \text{and} \quad \sigma_{\mathrm{meu}} = \sqrt{\mathbf{w}_{\mathrm{meu}}^\top \mathbf{C} \mathbf{w}_{\mathrm{meu}}} \cong 42.0082\%.$$

When we plug $\mu = \mu_{\mathrm{meu}}$ in the minimum variance line equation (3.40) with parameters as in Example 3.10, we obtain the weights $\mathbf{w}_{\mathrm{meu}}$. It means that the maximum expected utility portfolio lies on the minimum variance line. Indeed, if there existed another portfolio with the risk $\sigma = \sigma_{\mathrm{meu}}$ and expected return $\mu < \mu_{\mathrm{meu}}$, it would have a larger value of the expected utility $\mathrm{E}[-e^{\alpha V_T}]$, which contradicts to the fact that the portfolio with weights $\mathbf{w}_{\mathrm{meu}}$ has the maximum expected utility among all admissible portfolios. □

3.3.6 Efficient Frontier and Capital Market Line

Given a choice between two risky assets, a rational investor will choose an asset with higher expected return μ and lower risk σ.

An asset with (μ_1, σ_1) is said to *dominate* another asset with (μ_2, σ_2) whenever $\mu_1 \geqslant \mu_2$ and $\sigma_1 \leqslant \sigma_2$. A portfolio in risky assets is called *efficient* if there is no other portfolio, except itself, that dominates it. The set of efficient portfolios among all attainable portfolios is called the *efficient frontier*. In particular, an efficient portfolio has the highest expected return μ among all attainable portfolios with the same level of risk σ and has the lowest σ among all attainable portfolios with the same μ.

Let us consider the case with two risky assets. The set of admissible portfolios is represented by a portfolio line on the (σ, μ)-plane. The line is passing through the two base assets (σ_1, μ_1) and (σ_2, μ_2). As was proved in the previous section, there is a portfolio with the minimum possible variance σ_{mv}^2 given in (3.17). For every $\sigma > \sigma_{\mathrm{mv}}$, there are two portfolios on the portfolio line, (σ, μ_1) and (σ, μ_2), with $\mu_1 < \mu_2$. A rational investor would choose the portfolio (σ, μ_2) with a higher expected return. Therefore, in the case of two risky assets, the efficient frontier is the upper half of the portfolio line with the minimum variance portfolio $(\sigma_{\mathrm{mv}}, \mu_{\mathrm{mv}})$ as an endpoint. If one of the two assets is risk-free, then the portfolio line is a broken line with its vertex at the risk-free asset. The efficient frontier is the upper half-line. Both cases are represented in Figure 3.14.

In the situation with multiple risky assets ($N > 2$), the set of admissible portfolios (the Markowitz bullet) is a planar domain on the (σ, μ)-plane bounded by the minimum

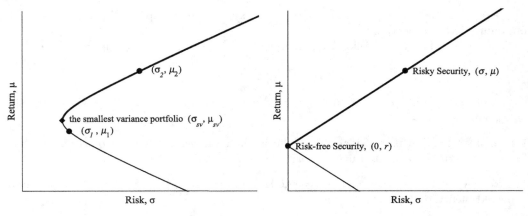

(a) The case with two risky assets.   (b) The case with one risky and one risk-free asset

FIGURE 3.14: The efficient frontier (the bold line) for two assets.

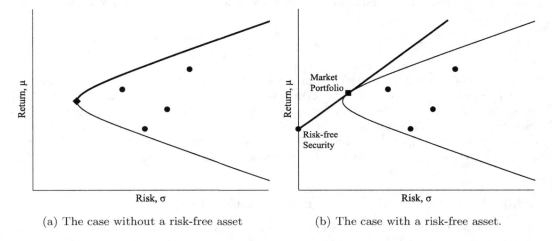

(a) The case without a risk-free asset   (b) The case with a risk-free asset.

FIGURE 3.15: The efficient frontier (the bold line) constructed from multiple risky assets.

variance line. Fix the value of $\sigma \geq \sigma_{\mathrm{mv}}$ and consider all admissible portfolios V with the standard deviation $\sigma_V = \sigma$. On the (σ, μ)-plane, this set is a straight line segment enclosed by the minimum variance line. By maximizing the expected return, we find that the efficient portfolio with risk σ is located on the upper half of the minimum variance line. We can repeat if for all admissible portfolios to show that efficient portfolios are all lying on the upper half of the minimum variance line (see Figure 3.15a).

Lastly, let us assume that a risk-free asset labelled B with the rate of return r is available in addition to N risky assets. Let $100w_0\%$ of the capital be allocated in the risk-free security. Then, $100(1 - w_0)\%$ is the risky portion formed from the N risky assets with the return $r_{\mathcal{M}}$. The total return on this portfolio with $N + 1$ assets is

$$r_V = w_0 r + (1 - w_0) \underbrace{\sum_{i=1}^{N} w_i r_i}_{=:r_{\mathcal{M}}} = w_0 r + (1 - w_0) r_{\mathcal{M}},$$

where $w_1, w_2, \ldots, w_N$ are the allocation weights for the risky assets so that $w_1 + \cdots + w_N = 1$ holds.

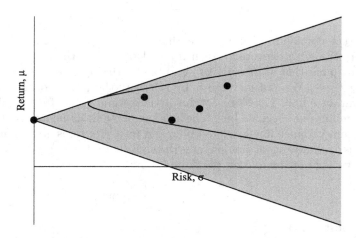

FIGURE 3.16: The set of admissible portfolios constructed from four risky assets and one risk-free asset.

As is shown in Subsection 3.2.2, all portfolios with $r_V = w_0 r + (1 - w_0) r_{\mathcal{M}}$ consisting of one risk-free and one risky asset (the risky portfolio $V_{\mathcal{M}}$ of an investment portfolio can be considered as a new asset) form a broken line having upper and lower half-lines with the common vertex at the point with coordinates $(0, r)$. The efficient frontier constructed from such portfolios is the upper half-line like that in Figure 3.15b. By taking a risky portfolio $V_{\mathcal{M}}$ anywhere in the Markowitz bullet, we can construct the set of admissible portfolios that is represented on the (σ, μ)-plane by a cone bounded by two half-lines, as is shown in Figure 3.16.

The efficient frontier of the portfolios containing a risk-free asset in addition to N risky ones is the upper half-line which is passing through the point representing the risk-free asset and tangent to the minimum variance line. Indeed, to minimize the risk, the portfolio $V_{\mathcal{M}}$ in risky assets has to be selected on the minimum variance line. To maximize the return, the portfolio $V_{\mathcal{M}}$ has to be selected so that the upper half-line has the largest possible slope. If the risk-free return r is not too high, the largest possible slope is achieved when the upper half-line is tangent to the Markowitz bullet. If r is too high, then the efficient frontier is obtained in the limiting case as the portfolio $V_{\mathcal{M}}$ selected on the upper half of the minimum variance line goes to infinity. The efficient frontier is no longer tangent to the Markowitz bullet but is parallel to its asymptote (recall that the shape of the bullet is a hyperbola). The tangency point with coordinates $(\sigma_{\mathcal{M}}, \mu_{\mathcal{M}})$ is the so-called *market portfolio*. The efficient frontier is called the *capital market line*. A rational investor forming her portfolio with a risk-free asset with return r and risky assets available on the market selects the portfolio on this line. Figure 3.15b shows the market line for portfolios with one risk-free and several risky assets.

Theorem 3.4. *Let the risk-free rate of return r be less than the return on the minimum variance portfolio μ_{mv}:*

$$r < \frac{\mathbf{m}^\top \mathbf{C}^{-1} \mathbf{u}}{\mathbf{u}^\top \mathbf{C}^{-1} \mathbf{u}}.$$

Then the market portfolio exists, and its vector of weights is

$$\mathbf{w}_{\mathcal{M}} = \frac{\mathbf{C}^{-1}(\mathbf{m} - r\mathbf{u})}{\mathbf{u}^{\top}\mathbf{C}^{-1}(\mathbf{m} - r\mathbf{u})}. \tag{3.45}$$

Proof. The market portfolio maximizes the Sharpe ratio $\frac{\mu_V - r}{\sigma_V}$. Geometrically, the Sharpe ratio of the portfolio with expected return μ_V and risk σ_V is the slope of a straight half-line on the (σ, μ)-plain that starts at $(0, r)$ and passes through (σ_V, μ_V). The slope of this line is given by $\frac{\mu_V - r}{\sigma_V}$. We can increase it by selecting the portfolio on the upper half of the minimum variance line. The slope attains its maximum value if the straight half-line is tangent to the minimum variance line. The market portfolio is hence the tangency point. Since the minimum variance line is a hyperbola and its asymptotes are crossing at the point $(0, \mu_{\mathrm{mv}})$, there is such a tangent line passing through $(0, r)$ iff $r < \mu_{\mathrm{mv}}$.

To find the weights of the market portfolio, we need to solve the optimization problem

$$\frac{\mu_V - r}{\sigma_V} = \frac{\mathbf{m}^{\top}\mathbf{w} - r}{\sqrt{\mathbf{w}^{\top}\mathbf{C}\mathbf{w}}} \to \max_{\mathbf{w}} \tag{3.46}$$

subject to the constraint $\mathbf{u}^{\top}\mathbf{w} = 1$.

Let us use the method of Lagrange multipliers. First, we find the critical points of

$$F(\mathbf{w}, \lambda) := \frac{\mathbf{m}^{\top}\mathbf{w} - r}{\sqrt{\mathbf{w}^{\top}\mathbf{C}\mathbf{w}}} - \lambda(\mathbf{u}^{\top}\mathbf{w} - 1).$$

The partial derivatives of F with respect to w_i for $i = 1, 2, \ldots, N$ are

$$\frac{\partial F}{\partial w_i}(\mathbf{w}, \lambda) = \frac{\mu_i\sqrt{\mathbf{w}^{\top}\mathbf{C}\mathbf{w}} - (\mathbf{m}^{\top}\mathbf{w} - r)\frac{\mathbf{c}_i^{\top}\mathbf{w}}{\sqrt{\mathbf{w}^{\top}\mathbf{C}\mathbf{w}}}}{\mathbf{w}^{\top}\mathbf{C}\mathbf{w}} - \lambda,$$

where $\mathbf{c}_i^{\top}$ denotes the ith row of matrix $\mathbf{C}$. Equate the derivatives to zero and write down the system in matrix-vector form:

$$\frac{\sigma_V\mathbf{m} - \left(\frac{\mu_V - r}{\sigma_V}\right)\mathbf{C}\mathbf{w}}{\sigma_V^2} - \lambda\mathbf{u} = 0. \tag{3.47}$$

Multiply equation (3.47) by $\mathbf{w}^{\top}$ on the left:

$$\frac{\sigma_V\mathbf{w}^{\top}\mathbf{m} - \left(\frac{\mu_V - r}{\sigma_V}\right)\mathbf{w}^{\top}\mathbf{C}\mathbf{w}}{\sigma_V^2} - \lambda\mathbf{w}^{\top}\mathbf{u} = 0$$

$$\frac{\sigma_V\mathbf{w}^{\top}\mathbf{m} - \left(\frac{\mu_V - r}{\sigma_V}\right)\sigma_V^2}{\sigma_V^2} - \lambda = 0$$

$$\sigma_V\mu_V - (\mu_V - r)\sigma_V = \lambda\sigma_V^2$$

$$r\sigma_V = \lambda\sigma_V^2 \implies \lambda = \frac{r}{\sigma_V}.$$

Plug this result back in (3.47) to obtain:

$$\frac{\sigma_V\mathbf{m} - \left(\frac{\mu_V - r}{\sigma_V}\right)\mathbf{C}\mathbf{w}}{\sigma_V^2} - \frac{r}{\sigma_V}\mathbf{u} = 0$$

$$(\mathbf{m} - r\mathbf{u}) - \left(\frac{\mu_V - r}{\sigma_V^2}\right)\mathbf{C}\mathbf{w} = 0$$

$$\left(\frac{\mu_V - r}{\sigma_V^2}\right)\mathbf{w} = \mathbf{C}^{-1}(\mathbf{m} - r\mathbf{u}).$$

Multiply the last equation by $\mathbf{u}^\top$ on the left to obtain

$$\left(\frac{\mu_V - r}{\sigma_V^2}\right)\mathbf{u}^\top\mathbf{w} = \mathbf{u}^\top\mathbf{C}^{-1}(\mathbf{m} - r\mathbf{u}) \implies \frac{\mu_V - r}{\sigma_V^2} = \mathbf{u}^\top\mathbf{C}^{-1}(\mathbf{m} - r\mathbf{u}).$$

Therefore, we have

$$\left(\mathbf{u}^\top\mathbf{C}^{-1}(\mathbf{m} - r\mathbf{u})\right)\mathbf{w} = \mathbf{C}^{-1}(\mathbf{m} - r\mathbf{u}).$$

Solving it for $\mathbf{w}$ gives us equation (3.45). $\qquad\square$

The expected return, $\mu_{\mathcal{M}}$, and variance of return, $\sigma_{\mathcal{M}}^2$, of the market portfolio can be found by using (3.32) and (3.33). The capital market line that starts at the risk-free asset (represented by the point $(0, r)$ on the (σ, μ)-plane) and passes through the market portfolio with expected return $\mu_{\mathcal{M}}$ and standard deviation of return $\sigma_{\mathcal{M}}$ satisfies the equation

$$\frac{\mu - r}{\mu_{\mathcal{M}} - r} = \frac{\sigma - 0}{\sigma_{\mathcal{M}} - 0} \iff \mu = r + \frac{\mu_{\mathcal{M}} - r}{\sigma_{\mathcal{M}}}\sigma.$$

Example 3.12. Let us consider a portfolio in three underlying assets from Example 3.10.

(a) Construct the efficient frontier.

(b) Find the market portfolio if the risk-free rate of return is $r = 5\%$.

(c) Construct an efficient portfolio with $\mu = 20\%$ composed of:

(i) the three risky underlying assets;

(ii) the three risky assets and a risk-free asset with the rate of return of 5%.

Solution. The efficient frontier is the upper half of the minimum variance line. As was discussed above, a portfolio line can be constructed from any two portfolios by taking a linear combination $x\mathbf{w}_1 + (1 - x)\mathbf{w}_2$ with $x \in \mathbb{R}$ of their weights. We already found two efficient portfolios, namely, the minimum variance portfolio $\mathbf{w}_{\mathrm{mv}}$ and the maximum expected utility portfolio $\mathbf{w}_{\mathrm{meu}}$ in Examples 3.10 and 3.11, respectively. Thus, all efficient portfolios can be parametrized as follows:

$$\mathbf{w}_{\mathrm{eff}}(x) = x\mathbf{w}_{\mathrm{meu}} + (1 - x)\mathbf{w}_{\mathrm{mv}}$$

$$\cong \begin{bmatrix} 0.605960 - 0.586369x & 0.205298 - 0.289735x & 0.188742 + 0.876104x \end{bmatrix}^\top \quad (3.48)$$

with $x \geqslant 0$.

The market portfolio can be found in two ways. First, we can apply (3.45) to obtain

$$\mathbf{w}_{\mathcal{M}} = \begin{bmatrix} 0.5118 & 0.1588 & 0.3294 \end{bmatrix}^\top.$$

Alternatively, we can use the fact that the market portfolio is efficient, and hence it lies on the efficient frontier which is described by the weight function $\mathbf{w}_{\mathrm{eff}}(x)$, $x \geqslant 0$ in (3.48). Let us find the expected return and variance of the portfolio $\mathbf{w}_{\mathrm{eff}}(x)$:

$$\mu_{\mathrm{eff}}(x) = \mathbf{m}^\top\mathbf{w}_{\mathrm{eff}}(x) \cong 0.148013 + 0.160734x,$$

$$\sigma_{\mathrm{eff}}^2(x) = \mathbf{w}_{\mathrm{eff}}^\top(x)\mathbf{C}\mathbf{w}_{\mathrm{eff}}(x) \cong 0.015735 + 0.160734x^2.$$

The Sharpe ratio $\dfrac{\mu_{\mathrm{eff}}(x) - r}{\sigma_{\mathrm{eff}}(x)}$ attains its maximum value at $x^* \cong 0.160541$. Thus, the market portfolio has the weight vector $\mathbf{w}_{\mathcal{M}} = \mathbf{w}_{\mathrm{eff}}(x^*)$. Its expected return and risk are, respectively, $\mu_{\mathcal{M}} = \mu_{\mathrm{eff}}(x^*) = 17.3818\%$ and $\sigma_{\mathcal{M}} = \sigma_{\mathrm{eff}}(x^*) = 14.0988\%$. The second approach can be used to find other efficient portfolios such as the minimum VaR portfolio. As we can see, a multi-asset optimization problem can be reduced to finding a maximum or a minimum of a single-variable function.

Now, let us find two efficient portfolios $\mathbf{w}_{\text{eff},1}$ and $\mathbf{w}_{\text{eff},2}$ with the expected return of 20%. The first portfolio does not contain a risk-free asset, whereas the second one includes it. The weight vector $\mathbf{w}_{\text{eff},1}$ is given by (3.48) with some x. Solve $\mu_{\text{eff}}(x) = 0.1$ to obtain:

$$0.148013 + 0.160734x = 0.2 \implies x_1 \cong 0.323433.$$

Thus, $\mathbf{w}_{\text{eff},1} = \mathbf{w}_{\text{eff}}(x_1) = \begin{bmatrix} 0.416309 & 0.111588 & 0.472103 \end{bmatrix}^\top$. Its risk is $\sigma_{\text{eff},1} = 18.0414\%$.

The second portfolio lies on the market line. Its expected return is given by $\mu_{\text{eff},2} = w_0 r + (1 - w_0)\mu_{\mathcal{M}}$, where $100w_0\%$ of the wealth is invested in the risk-free asset and $100(1 - w_0)\%$ is the portion of the market portfolio $\mathbf{w}_{\mathcal{M}}$. Solve $w_0 r + (1 + w_0)\mu_{\mathcal{M}} = 0.2$ to obtain $w_0 \cong -0.211460$. Since $w_0 < 0$, we need to borrow using the risk-free asset. The weights of the risky assets are given by $(1 - w_0)\,\mathbf{w}_{\mathcal{M}}$. The resulting weights for all four assets are

$$\mathbf{w}_{\text{eff},2} = \begin{bmatrix} w_0 & w_1 & w_2 & w_3 \end{bmatrix}^\top = \begin{bmatrix} -0.211460 & 0.620055 & 0.192360 & 0.399045 \end{bmatrix}^\top.$$

The risk is $\sigma_{\text{eff},2} = |1 - w_0| \cdot \sigma_{\mathcal{M}} \cong 17.0802\%$. As we can see, it is less risky than the portfolio $\mathbf{w}_{\text{eff},1}$ thanks to the addition of a risk-free asset. $\qquad \square$

3.4 The Capital Asset Pricing Model

The Capital Asset Pricing Model (CAPM) attempts to relate r_i, the return on asset i, to $r_{\mathcal{M}}$, the return of the entire market, which can be measured by some index such as Standard and Poor's index of 500 stocks (S&P 500). In the Markowitz portfolio model, the market portfolio can be used as a good approximation to such a market index. Indeed, every rational investor will select a portfolio on the capital market line since it is the efficient frontier constructed from a risk-free asset and several risky assets. Therefore, every investor will be holding a portfolio with the same relative proportions of risky assets. This means that for each risky asset its weight in the market portfolio is equal to the relative share of the asset in the whole market.

The CAPM assumes that the dependence between r_i and $r_{\mathcal{M}}$ takes the following form:

$$r_i = r + \beta_i(r_{\mathcal{M}} - r) + \epsilon_i, \tag{3.49}$$

where β_i is a constant called the *beta factor* for asset i, r is a risk-free rate of return, and ϵ_i is a residual random variable having a normal distribution with mean zero. The residual ϵ_i is assumed to be independent of $r_{\mathcal{M}}$.

There are several ways to compute beta factors.

(1) Suppose that the joint probability distribution of r_i and $r_{\mathcal{M}}$ is given. Compute the covariance of r_i and $r_{\mathcal{M}}$ by employing (3.49) and using the independence of $r_{\mathcal{M}}$ and ϵ_i:

$$\text{Cov}(r_i, r_{\mathcal{M}}) = \underbrace{\text{Cov}(r, r_{\mathcal{M}})}_{=0} + \beta_i \underbrace{\text{Cov}(r_{\mathcal{M}}, r_{\mathcal{M}})}_{=\text{Var}(r_{\mathcal{M}})} - \beta_i \underbrace{\text{Cov}(r, r_{\mathcal{M}})}_{=0} + \underbrace{\text{Cov}(\epsilon_i, r_{\mathcal{M}})}_{=0} = \beta_i \text{Var}(r_{\mathcal{M}}).$$

Therefore, the beta factor of asset i is given by

$$\beta_i = \frac{\text{Cov}(r_i, r_{\mathcal{M}})}{\text{Var}(r_{\mathcal{M}})}. \tag{3.50}$$

(2) Consider a market model with a set of market scenarios Ω. Suppose that for each market scenario $\omega \in \Omega$, the values of returns on asset i and the market portfolio $\mathcal{M}$ are given. We can plot the value of $r_i(\omega^j)$ against $r_{\mathcal{M}}(\omega)$ for each $\omega \in \Omega$ and then find the line of best fit, also known as the regression line. Employ the model $r_i = \alpha + \beta r_{\mathcal{M}} + \epsilon_i$. So the residual random variable $\epsilon_i \colon \Omega \to \mathbb{R}$ is the difference between the actual return r_i and the predicted return $\alpha + \beta r_{\mathcal{M}}$. The line of best fit is defined by

$$\mathrm{E}[\epsilon_i^2] \to \min_{\alpha, \beta} .$$

The expected value of ϵ_i^2 is given by

$$\mathrm{E}[\epsilon_i^2] = \mathrm{E}[r_i^2] - 2\beta \mathrm{E}[r_i r_{\mathcal{M}}] + \beta^2 \mathrm{E}[r_{\mathcal{M}}^2] + \alpha^2 - 2\alpha \mathrm{E}[r_i] + 2\alpha\beta \mathrm{E}[r_{\mathcal{M}}] .$$

A necessary condition for a minimum of $\mathrm{E}[\epsilon_i^2]$ as a function of α and β is that the partial derivatives w.r.t. α and β should be zero at the point of minimum, (α_i, β_i):

$$\frac{\partial}{\partial \alpha} \mathrm{E}[\epsilon_i^2] = 0 \iff \alpha + \beta \mathrm{E}[r_{\mathcal{M}}] = \mathrm{E}[r_i],$$

$$\frac{\partial}{\partial \beta} \mathrm{E}[\epsilon_i^2] = 0 \iff \alpha \mathrm{E}[r_{\mathcal{M}}] + \beta \mathrm{E}[r_{\mathcal{M}}^2] = \mathrm{E}[r_i r_{\mathcal{M}}] .$$

As a result, we obtain a system of linear equations that can be solved to find

$$\beta_i = \frac{\mathrm{Cov}(r_i, r_{\mathcal{M}})}{\mathrm{Var}(r_{\mathcal{M}})}, \quad \alpha_i = \mathrm{E}[r_i] - \beta_i \mathrm{E}[r_{\mathcal{M}}] .$$

Note that for the beta factor we obtained the same expression as that in (3.50).

(3) Suppose that historical data of returns on some portfolio V and the market portfolio M, $\{r_V^{(j)}, r_{\mathcal{M}}^{(j)}\}_{j=1,2,\ldots,N}$, are available. Let us find the line of best fit by minimizing the sum of squared residuals:

$$\sum_{j=1}^{N} \left(r_V^{(j)} - (\alpha + \beta r_{\mathcal{M}}^{(j)}) \right)^2 \to \min_{\alpha, \beta} .$$

The solution to this minimization problem is

$$\beta_i = \frac{N \sum_j r_V^{(j)} r_{\mathcal{M}}^{(j)} - \left(\sum_j r_V^{(j)} \right) \left(\sum_j r_{\mathcal{M}}^{(j)} \right)}{N \sum_j (r_{\mathcal{M}}^{(j)})^2 - \left(\sum_j r_{\mathcal{M}}^{(j)} \right)^2}, \quad \alpha_i = \frac{\sum_j r_V^{(j)} - \beta_i \sum_j r_{\mathcal{M}}^{(j)}}{N} .$$

The beta factors for individual assets can be computed by (3.50) or from historical data. The beta factor of a portfolio V in N assets with weights $w_1, \ldots, w_N$ is given by

$$\beta_V = w_1 \beta_1 + \cdots + w_N \beta_N .$$

Indeed, the covariance function is bilinear; therefore

$$\beta_V = \frac{\mathrm{Cov}(r_V, r_{\mathcal{M}})}{\mathrm{Var}(r_V)} = \frac{\mathrm{Cov}(w_1 r_1 + \cdots + w_n r_N, r_{\mathcal{M}})}{\mathrm{Var}(r_V)}$$

$$= \frac{w_1 \mathrm{Cov}(r_1, r_{\mathcal{M}}) + \cdots + w_N \mathrm{Cov}(r_N, r_{\mathcal{M}})}{\mathrm{Var}(r_V)} = w_1 \beta_1 + \cdots + w_N \beta_N .$$

The beta factor of the market portfolio is equal to one.

By taking the mathematical expectation of both parts of (3.49), we obtain

$$\mu_i = r + \beta_i (\mu_{\mathcal{M}} - r) ,$$

where $\mu_i = \mathrm{E}[r_i]$ and $\mu_{\mathcal{M}} = \mathrm{E}[r_{\mathcal{M}}]$. The expected return plotted against the beta factor of any portfolio will form a straight line on the (β, μ)-plane, called the *asset market line*.

3.5   Exercises

Exercise 3.1. Show that the functions $u_1(x) = a \ln x$ and $u_2(x) = 1 - e^{-ax}$ with $a > 0$ both satisfy the definition of a utility function, i.e., each of them is an increasing, concave function.

Exercise 3.2. Show that the functions $u_3(x) = x^a$ with $0 < a < 1$ and $u_4(x) = x - bx^2$ with $b > 0$ and $x < \frac{1}{2b}$ both satisfy the definition of a utility function, i.e., each of them is an increasing, concave function.

Exercise 3.3. Suppose that $u(w) = \sqrt{w}$ and W is uniformly distributed on the set $\{4, 25, 64, 100\}$. Calculate $E[u(W)]$ and $u(E[W])$.

Exercise 3.4. The investor must choose one of three investments. Her fortune after investment k is a random variable V_k with density function f_k with $k = 1, 2, 3$, where $f_1(x) = e^{-x} \mathbb{I}_{\{x \geq 0\}}$, $f_2(x) = \frac{1}{2} \mathbb{I}_{\{0 \leq x \leq 2\}}$, and $f_3(x) = x \mathbb{I}_{\{0 \leq x < 1\}} + (2 - x) \mathbb{I}_{\{1 \leq x \leq 2\}}$.

(a) Which investment should she choose if her utility function is $u(x) = 1 - e^{-x}$?

(b) Which investment is more preferable for a risk-indifferent investor with a linear utility function?

Exercise 3.5. An investor with capital W can invest an amount $V = aW$ for some $0 \leq a \leq 1$. If V is invested, then after one year the invested amount is doubled with probability p or lost with probability $1 - p$. Suppose that the remaining capital $W - aW$ can be put in a risk-free bank account to earn interest at an annual rate of interest r. How much should be invested by an investor using:

(a) a log utility function $u(V) = \ln V$,

(b) an exponential utility function $u(V) = 1 - e^{-0.1V}$?

Exercise 3.6. Consider an investment whose future wealth V, in dollars, follows the continuous probability distribution with the triangular density function

$$f(x) = \begin{cases} \dfrac{2(x - 2400)}{180{,}000} & \text{if } 2400 < x < 2600, \\[2mm] \dfrac{2(3300 - x)}{630{,}000} & \text{if } 2600 \leq x < 3300. \end{cases}$$

Let the utility function be $u(x) = \left(\dfrac{x}{2600}\right)^{0.3}$.

(a) Find the expected future wealth $E[V]$.

(b) Find the expected utility of the future wealth $E[u(V)]$.

(c) Find the risk premium. That is, the amount a risk-averse investor would be willing to pay to rid their investment of its uncertainty.

Exercise 3.7. Consider an investment whose future wealth V takes one of the values $(V_1, V_2, V_3, V_4) = (\$2200, \$2400, \$2600, \$3000)$ with probabilities $(p_1, p_2, p_3, p_4) = \left(\frac{4}{14}, \frac{4}{14}, \frac{1}{14}, \frac{5}{14}\right)$, respectively. Let the utility function be $u(x) = \ln x$.

(a) Find the expected utility of the future wealth $E[u(V)]$.

(b) Find the utility of the expected future wealth $u(E[V])$. Explain why this value is greater than the result in part (a).

(c) Find the certainty equivalent C such that $u(C) = E[u(V)]$. That is, the amount a rational investor would be willing to accept risk-free instead of entering into the investment.

Exercise 3.8. Consider an investment whose future wealth V, in dollars, follows the continuous uniform distribution $Unif(2000, 2500)$. Let the utility function be $u(x) = x^{0.2}$.

(a) Find the expected utility of the future wealth $E[u(V)]$.

(b) Find the certainty equivalent C such $u(C) = E[u(V)]$. That is, the amount a rational investor would be willing to accept risk-free instead of entering into the investment.

Exercise 3.9. Paul's current wealth is \$290 and he is considering the purchase of a lottery ticket that costs \$4. There is a 40% chance that the ticket will have a \$50 payoff and \$0 otherwise. Paul's utility is $u(x) = 1 - e^{-0.004x}$.

(a) If Paul was to buy the ticket, what would the relative change in expected utility be? That is, find $\dfrac{E[u(V)] - u(V_0)}{u(V_0)}$. Should Paul by the ticket or not?

(b) What maximum price should Paul be willing to pay for the lottery ticket?

Exercise 3.10. The exponential utility function $u(x) = -e^{-ax}$ is used to analyze an uncertain investment V with the rate of return $r = \frac{V - V_0}{V_0}$ that has mean μ and variance σ^2.

(a) Find the ARA coefficient A_u and then use it to estimate the risk premium π and the certainty equivalent C.

(b) Calculate π and C assuming that V follows a normal probability distribution.

Exercise 3.11. Let for $w > 0$,

$$u(w) = \begin{cases} \frac{w^{1-\alpha} - 1}{1-\alpha} & \text{if } \alpha \neq 1, \\ \ln(w) & \text{if } \alpha = 1. \end{cases}$$

(a) For what values of α is u a reasonable utility function?

(b) Show that the coefficient of risk aversion for this utility function is $\dfrac{\alpha}{w}$.

(c) Show that $\lim_{\alpha \to 1} \frac{w^{1-\alpha} - 1}{1-\alpha} = \ln(w)$, for any $w > 0$.

Exercise 3.12. I have power utility with $\alpha = 0.5$ and my current wealth is $W = 1,000$.

(a) Find that value of X for which the joy of winning \$$X$ is equal to the pain of loosing \$100.

(b) Repeat (a) in the case that my initial wealth is $W = 500$ and comment on any difference.

(c) Repeat (a) using the logarithmic utility function and comment on any difference.

Exercise 3.13. Paul has exponential utility $u(x) = 1 - e^{-0.01x}$ and his current wealth is $w = 100$. He is considering the purchase of a lottery ticket that costs \$20 and either pays off \$40 or \$0 with equal probability.

(a) Find Paul's expected utility from buying the ticket and answer whether Paul should buy the ticket.

(b) What is the maximum price he would be willing to pay?

(c) Paul is offered the choice between (i) a lottery ticket that pays $100 or $0, with respective probabilities 10% and 90%, or (ii) $X in cash. For what values of X would he choose (ii)?

Exercise 3.14. I have power utility $u(x) = x^{0.25}$ and my current wealth is $\$1,000$. I am offered an even-money bet at $\$600$, which means that if I place the bet then one of two things will happen—I will either win $\$600$ with probability p or lose $\$600$ with probability $1 - p$. If I do not place the bet then I just sit here with my current wealth. Determine the minimum value of p that would induce me to take the bet.

Exercise 3.15. Consider an even-money bet at $\$h$ (see previous problem for definition of even-money bet). Let p be the probability that I win the bet and let p^* be the minimum value of p that would induce me to place the bet.

(a) Show that

$$p^* = \frac{u(w) - u(w - h)}{u(w + h) - u(w - h)}.$$

(b) Prove that $p^* > \frac{1}{2}$, and explain why this makes sense.
    Hint: $u(w + h) - u(w - h) = [u(w + h) - u(w)] + [u(w) - u(w - h)]$

Exercise 3.16. I have exponential utility and am offered an even-money bet at $\$h$.

(a) Show that the minimum probability that would induce me to take the bet is

$$p^* = \frac{e^{\alpha h} - 1}{e^{\alpha h} - e^{-\alpha h}}.$$

(b) Evaluate the limit of p^* as $\alpha \to \infty$, and explain why your answer makes sense.

(c) Evaluate the limit of p^* as $\alpha \to 0$, and explain why your answer makes sense.

(d) Evaluate the limit of p^* as $h \to 0$, and explain why your answer makes sense.

Exercise 3.17. I have exponential utility and my current wealth is w. I am considering the purchase of a lottery ticket that costs $\$c$ and will either be worth $\$40$ with probability p, or $\$0$ with probability $1 - p$. Let c^* the maximum price that I am willing to pay.

(a) Show that $c^* = -\frac{\ln(pe^{-40\alpha} + (1-p))}{\alpha}$.

(b) Show that $\lim_{p \to 0} c^* = 0$ and $\lim_{p \to 1} c^* = 40$. Explain these results.

Exercise 3.18. I have exponential utility and my current wealth is w. I am considering the purchase of a lottery ticket that costs $\$c$ and will either be worth $\$40$ with 60% probability, or $\$0$ with 40% probability. Let c^* the maximum price that I am willing to pay.

(a) Show that $c^* = -\frac{\ln(0.6e^{-40\alpha} + 0.4)}{\alpha}$.

(b) Show that $\lim_{\alpha \to \infty} c^* = 0$ and $\lim_{\alpha \to 0} c^* = 24$. Explain these results.

Exercise 3.19. Suppose that W is a random variable taking on two possible values, w_1 and w_2 s.t. $0 < w_1 < w_2$, with respective probabilities p and $1 - p$. Sketch a graph that clearly illustrates (i) the certainty equivalent of W, (ii) the point $(E[W], E[u(W)])$ and (iii) the point $(E[W], u(E[W]))$.

Exercise 3.20. Consider a version of the Saint Petersburg game where the cost c equals the initial wealth V_0. Determine the certainty equivalent of the lottery for logarithmic utility. [*Hint:* $\sum_{k=1}^{\infty} \frac{k}{2^k} = 2$.]

Exercise 3.21. Consider a single-period market model with risky stock S and risk-free bond B. Let $S_0 = B_0 = \$100$, the bond rate be $r_B = 10\%$, and the rate r_S have a normal distribution with mean μ and standard deviation $\sigma = 15\%$. Find the optimal portfolio $\varphi = (x, y)$ with initial value $V_0 = \$1000$ that maximizes $E[-\exp(-0.1\, V_T)]$.

Exercise 3.22. Consider a market model with three scenarios $\{\omega^1, \omega^2, \omega^3\}$ and two risky assets with returns r_1 and r_2. Let the probabilities of the scenarios and values of the returns be as follows:

ω	$\mathbb{P}(\omega)$	$r_1(\omega)$	$r_2(\omega)$
ω^1	0.5	10%	5%
ω^2	0.3	5%	10%
ω^3	0.2	15%	−5%

(a) Find the expected values and standard deviations of the returns. Find the coefficient of correlation between r_1 and r_2.

(b) Find the minimum variance portfolio, its expected return and volatility.

Exercise 3.23. Consider a single-period market that contains a \$250 risk-free bond with a bond rate of 10% and risky \$200 stock. Suppose that the rate of return on the stock is normally distributed with mean $\mu_S = 5\%$ and standard deviation $\sigma_S = 15\%$.

(a) State the expected return, μ_{rv} and associated risk (as measured by the standard deviation), σ_{rv}^2, in terms of w if w is the stock's weight allocation in the portfolio.

(b) Suppose that the utility function used was $u(x) = -e^{-0.005x}$ and the portfolio was constructed so that it's initial value was \$8000 and the expected utility of future wealth was maximized. Determine the following:

   (i) The percentage of the portfolio's initial value invested in the stock and in the bond.

   (ii) The number of stock shares and bonds in the portfolio.

   (iii) The expected rate of return on the portfolio.

   (iv) The risk in the portfolio.

Exercise 3.24. Consider an investment of \$1000 in two risky assets whose returns follow a bivariate normal distribution with the following expected values and standard deviations:

$$\mu_1 = 0.1, \ \sigma_1 = 0.2, \ \mu_2 = 0.15, \ \sigma_1 = 0.3.$$

The correlation coefficient between the returns is $\rho = -0.5$.

(a) Suppose that the allocation weights of an investment portfolio for assets 1 and 2 are, respectively, $w_1 = x$ and $w_2 = 1 - x$ for some $x \in \mathbb{R}$. Show that the terminal value V_T of a portfolio is normal. Find the expected value and variance of V_T.

(b) Find the optimal portfolio when employing the utility function

$$u(V) = 1 - e^{-0.01V}.$$

Exercise 3.25. Consider a market model with four scenarios and two risky assets with rates of return r_1 and r_2, respectively. Let the distribution of the rates is as follows:

Probability	r_1	r_2
25%	6%	6%
25%	1%	5%
20%	7%	1%
30%	2%	3%

For the portfolio of these two assets with the allocation weights $w_1 = 60\%$ and $w_2 = 40\%$, find:

(a) the expected return, $\mu_V = \mathrm{E}[r_V]$;

(b) the risk, $\sigma_V = \sqrt{\mathrm{Var}(r_V)}$.

Exercise 3.26. Consider a portfolio consisting of the following two risky assets.

Asset i	μ_i (Return on Asset i)	σ_i (Risk in Asset i)
1	16%	18%
2	12%	10%

(a) If the coefficient of correlation between the returns is $\rho = 100\%$, what are the allocation weights in a minimum variance portfolio? Find the expected rate of return on the minimum variance portfolio.

(b) If the coefficient of correlation between the returns is $\rho = -100\%$, what are the allocation weights in a minimum variance portfolio? Find the expected rate of return on the minimum variance portfolio.

Exercise 3.27. Consider a portfolio consisting of two risky assets with $\mu_1 = 10\%$, $\mu_2 = 15\%$, $\sigma_1 = 20\%$, $\sigma_2 = 25\%$, $\rho = -50\%$.

(a) Find an optimal allocation of the initial wealth $V_0 = \$1000$ between two risky assets, when attempting to maximize $\mathrm{E}[-e^{-0.1\,V_T}]$. Assume that the asset returns are normally distributed.

(b) Find the allocation weights for the minimum variance portfolio.

(c) Compute and compare μ_V and σ_V for portfolios obtained in (a) and (b). What happens with the maximum expected utility portfolio (a) when $V_0 \to \infty$?

Exercise 3.28. Consider a portfolio consisting of the following two risky assets.

Asset i	μ_i	σ_i
1	13%	16%
2	17%	24%

Let the coefficient of correlation between asset returns be $\rho = 80\%$.

(a) If the allocation weights for Assets 1 and 2 are 25% and 75%, respectively, find the expected return and associated risk (as measured by the standard deviation) in the portfolio.

(b) Suppose that the expected rate of return on the portfolio is 20%. Find the allocation weights and the risk in the portfolio.

(c) Find the allocation weights for the minimum variance portfolio. Find μ_V and σ_V for the portfolio.

Exercise 3.29. Find an optimal allocation of the initial wealth $V_0 = \$1000$ between two risky assets with $\mu_1 = 10\%$, $\mu_2 = 15\%$, $\sigma_1 = 20\%$, $\sigma_2 = 25\%$, $\rho = -50\%$, when attempting to minimize the probability $\mathbb{P}(r_V < r_0)$ with $r_0 = 5\%$. Compute μ_V and σ_V for the portfolio obtained. Assume that the asset returns are normally distributed.

Exercise 3.30. Plot portfolio lines with and without short selling for the case with two assets if

(a) $|\rho_{12}| = 1$, $\mu_1 = \mu_2$, and $\sigma_1 \neq \sigma_2$,

(b) $|\rho_{12}| = 1$, $\mu_1 \neq \mu_2$, and $\sigma_1 = \sigma_2$,

(c) $\mu_1 = \mu_2$, and $\sigma_1 = \sigma_2$.

Exercise 3.31. Consider a portfolio consisting of two risky assets with initial values of $\$300$ and $\$400$, whose rates of returns follow a bivariate normal distribution so that

μ_1	σ_1	μ_2	σ_2	ρ
20%	14%	8%	10%	-35%

Suppose that the initial value of the portfolio was $\$7500$.

(a) Determine the allocation weights w_1 and w_2 in a maximum expected utility portfolio if the utility used is $u(v) = -e^{-0.005v}$.

(b) Using the results in part (a), determine the number of shares x_1 and x_2 of Assets 1 and 2, respectively, in the portfolio.

(c) Determine the expected rate of return and the risk for the maximum expected utility portfolio constructed in part (a).

Exercise 3.32. Consider a portfolio consisting of the following two risky assets with the following parameters:

μ_1	σ_1	μ_2	σ_2	ρ
12%	20%	11%	22%	-40%

Let $100x\%$ of the initial capital is invested in the first risky asset and $100(1-x)\%$ in the second one, for some $x \in (-\infty, \infty)$.

(a) Find the expected value, $\mu(x)$, and variance, $\sigma^2(x)$, of the rate of return of a portfolio as functions of x:

(b) Solve $\dfrac{d\sigma^2(x)}{dx} = 0$ for x to find the weights of the smallest variance portfolio.

(c) Find the expected return and associated risk (as measured by the standard deviation) for the portfolio found in part (b)

(d) Find a portfolio that minimizes the probability $\mathbb{P}(r_V < r_f)$ where r_V is the rate of return on a portfolio in the two risky assets and $r_f = 2\%$ is the risk-free rate of return. Assume that r_V has a normal probability distribution. Recall that in this case, you need to maximize $\dfrac{\mu(x) - r_f}{\sqrt{\sigma^2(x)}}$ to find the portfolio weights.

(e) Find the expected return and associated risk (as measured by the standard deviation) for the portfolio found in part (d)

Exercise 3.33. The data below corresponds to the historical performance of the Canadian bond market (μ_1, σ_1) and Canadian stock market (μ_2, σ_2) .

μ_1	σ_1	μ_2	σ_2	ρ
4.2%	4.4%	7.7%	13.9%	−8%

Let $\mu(w)$ and $\sigma^2(w)$ denote the mean and variance of my portfolio, assuming I invest $100w\%$ of my wealth in the bond market (I am not allowed to take short positions).

(a) Sketch $\mu(w)$ and $\sigma^2(w)$.

(b) Determine the volatility of the portfolio whose expected return is 6%.

(c) Plot the points $(\sigma(w), \mu(w))$ for $w = 0$, $w = w^*$ and $w = 1$, where w^* is the variance-minimizing allocation to the bond market.

(d) Use your points from (c) to construct a rough sketch of the set of feasible risk-return pairs (i.e., the parametric curve $\{(\sigma(w), \mu(w)) : 0 \leqslant w \leqslant 1\}$).

Exercise 3.34. The data below corresponds to the historical performance of some stock and at-the-money put options on the stock (assuming the puts are held until maturity).

μ_1	σ_1	μ_2	σ_2	ρ
4.4%	21.3%	−38%	99.6%	−87%

Let $\mu(w)$ and $\sigma^2(w)$ denote the mean and variance of my portfolio, assuming I invest $100w\%$ of my wealth in the stock.

(a) What is the expected return on the minimum-variance portfolio?

(b) Plot the points $(\sigma(w), \mu(w))$ for $w = 0$, $w = w^*$ and $w = 1$, where w^* is the variance-minimizing allocation to the stock.

(c) Use your points from (b) to construct a rough sketch of the set of feasible risk-return pairs (i.e., the parametric curve $\{(\sigma(w), \mu(w)) : 0 \leqslant w \leqslant 1\}$).

(d) Is there any reason for a risk-averse investor to consider investing in put options here?

Exercise 3.35. Suppose that the returns on the Canadian bond and stock markets are normally distributed (with parameters as in Exercise 3.33), that I have exponential utility $u(v) = 1 - e^{-\alpha v}$. My current wealth is $V_0 = \$1000$. Define the risk-adjusted return on a investment with the rate $r \sim Norm(\mu, \sigma^2)$ by $\mu - \alpha V_0 \frac{\sigma^2}{2}$.

(a) Determine the risk-adjusted returns on the bond and stock markets assuming $\alpha = 0.001$. If I am not able to diversify, should I choose bonds or equities?

(b) Determine the risk-adjusted returns on the bond and stock markets assuming $\alpha = 0.01$. If I am not able to diversify, should I choose bonds or equities?

(c) Compare your answers from (a) and (b).

(d) For what values of α does the risk-adjusted return on the bond market exceed that on the stock market?

Exercise 3.36. Suppose that the returns on the Canadian bond and stock markets are normally distributed (with parameters as in Exercise 3.33), that I have exponential utility $u(v) = 1 - e^{-\alpha v}$, and that my current wealth is $V_0 = \$1000$. Let $\mu(w)$ and $\sigma^2(w)$ denote the mean and variance of my portfolio, assuming I invest $100w\%$ of my wealth in the bond (I am not allowed to take short positions).

(a) Sketch the objective function, $\mu(w) - \frac{\alpha V_0}{2}\sigma^2(w)$, for utility maximization in the case $\alpha = 0.01$. What is the utility-maximizing allocation?

(b) Sketch the objective function for utility maximization in the case $\alpha = 0.001$. What is the utility-maximizing allocation?

(c) Compare your answers from (a) and (b).

(d) Show that the maximum of $\mu(w) - 500\alpha\sigma^2(w)$ occurs at $\tilde{w} = 0.8909 - \frac{0.001574}{\alpha}$.

(e) For what values of α is the utility-maximizing allocation 100% equities?

(f) Evaluate $\lim_{\alpha \to \infty} \tilde{w}$, where $\tilde{w}$ is given in (d). Does your answer make intuitive sense?

(g) Sketch the utility-maximizing allocation to bonds, as a function of α. How does the utility-maximizing allocation compare to the variance-minimizing allocation?

Exercise 3.37. Suppose that there are two assets in which I can invest. The first is a risk-free asset whose expected return is $r > 0$ and whose volatility is zero. The second is a risky asset whose return is normally distributed with mean $\mu > r$ and volatility $\sigma > 0$. Let $\mu(w)$ and $\sigma^2(w)$ denote the mean and variance of my portfolio's return, assuming I invest $100w\%$ of my wealth in the risky asset.

(a) Sketch $\sigma^2(w)$ and determine the variance-minimizing allocation.

(b) Show that if $\lambda > 0$ is a constant, then $\mu(w) - \lambda\sigma^2(w)$ attains its minimum at $\tilde{w} = \frac{\mu - r}{\lambda\sigma^2}$.

(c) Use your answer from (b) to explain why it is always optimal for a utility maximizer (with the exponential utility) to invest a positive amount in the risky asset, regardless of their degree of risk aversion or initial wealth

Exercise 3.38. Suppose I can invest in two assets: a risk-free asset that has a guaranteed return of 2% and a risky asset whose return is normally distributed with mean $\mu > r$ and volatility $\sigma > 0$. Further suppose that I have exponential utility with $\alpha = 0.001$ and my current level of wealth is $1000.

(a) Assume that the volatility of the risky asset is 15% and that I am not able to diversify. For what values of μ would I choose the risky asset over the risk-free asset.

(b) Repeat your calculation from (a) assuming volatility is 30%, and comment on any difference.

(c) Assume that the expected return on the risky asset is 7% and that I am not able to diversify. For what values of σ would I choose the risky asset over the risk-free asset?

(d) Repeat your calculation from (c) assuming the expected return is 12%, and comment on any difference.

Exercise 3.39. Show that the optimal allocation of one's investment portfolio $V_T = V_0(1 + r_V)$ does not depend on the initial capital V_0 when attempting to maximize the mathematical expectation of:

(a) a log utility function $u(V_T) = \ln V_T$,

(b) a power utility function $u(V_T) = (V_T)^a$ with $0 < a < 1$.

In other words, show that the maximization of $E[u(V_T)]$ reduces to the maximization of $E[u(1 + r_V)]$.

Exercise 3.40. Consider the matrix

$$
\mathbf{C} = \begin{bmatrix} 1 & x & -0.3 \\ x & 1 & 0.5 \\ -0.3 & 0.5 & 1 \end{bmatrix}
$$

(a) Show that $\mathbf{C}$ cannot be a covariance matrix if $x = 0.75$.

(b) For what values of x is $\mathbf{C}$ positive-definite?

Exercise 3.41. Consider three assets whose returns have the following standard deviation and correlation coefficients:

$$
\sigma_1 = 0.2, \ \sigma_2 = 0.25, \ \sigma_3 = 0.15, \ \rho_{12} = -0.4, \ \rho_{13} = 0.3, \ \rho_{23} = 0.7.
$$

(a) Find the covariance matrix $\mathbf{C}$.

(b) For what values of σ_3 is the matrix $\mathbf{C}$ positive definite?

Exercise 3.42. Consider three assets with the covariance matrix defined in Exercise 3.41. Let their expected returns be $\mu_1 = 4\%$, $\mu_2 = 5\%$, and $\mu_3 = 3\%$, respectively.

(a) Find the weights in the minimum variance portfolio constructed using the assets.

(b) Suppose that an exponential utility function with $\alpha = 0.01$ is used to find an optimal allocation of the capital $V_0 = \$1000$. Find the weights in the maximum expected utility portfolio.

(c) Compute and compare the expected return and volatility of the portfolios constructed in parts (a) and (b).

Exercise 3.43. Consider three risky assets from Exercises 3.41 and 3.42. Suppose that there exists a risk-free asset with a guaranteed return of 2%.

(a) Find the market portfolio and compute its expected return and volatility.

(b) If I invest 40% of my capital in the risk-free asset and 60% in the market portfolio, what is the expected return and volatility of my portfolio?

(c) What percentage of capital must be invested in the risk-free asset if the risk in the four-asset portfolio is to be equivalent to the risk in the three-asset minimum variance portfolio from part (a) of Exercises 3.42?

4

Primer on Derivative Securities

A *derivative* is a financial contract whose value depends on (or derives from) the values of other basis assets and market variables such as stocks, bonds, commodities, interest rates, exchange rates, market indices, etc. Such variables and assets are called *underlyings*. There are three broad categories of traders interested in derivatives:

hedgers who use derivatives to reduce the risk that they face from potential future movements in prices and rates;

speculators who use derivatives to bet on the future directions of asset values and market variables;

arbitrageurs who take offsetting positions in two or more instruments to lock in a risk-free profit.

In this chapter we consider two main types of derivatives, namely, *forwards* and *options*.

4.1 Forward Contracts

A *forward contract* is a commitment to buy or to sell a given amount of a commodity or an asset for a fixed price on a fixed future date, all specified in advance. The price fixed now for future exchange is called the *forward price*; the fixed date is called the *delivery date* or the *expiration date*. A forward contract is a direct agreement between two parties. Both parties are *obliged* to fulfill the contract. The party to the contract who agrees to sell the asset is said to enter into a *short forward position*. The other party who has to buy the asset is said to take a *long forward position*. The exchange flows between the two parties are illustrated in Figure 4.1.

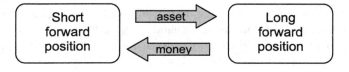

FIGURE 4.1: A forward contract diagram.

Key features of forward contracts are as follows:

- they can be customized to fit specific needs of the parties;

- there is a risk that the other party may not make the required delivery or payment (it is called the counterparty default risk);

- they are traded over the counter rather than on exchanges, i.e., transactions take place in unregulated markets consisting of banks, government, and corporations;

- no money changes hands until maturity, and the forwards tend to be held to maturity.

The asset on which a forward contract is written may be physically delivered to the buyer, or the contract may be settled by paying the difference between the forward price and the actual asset price in cash. In the former case, the forward contract needs to specify delivery logistics, such as the exact time, date, and place for delivery.

The main motivation for entering into a forward contract is to become independent of the uncertain future price of a risky asset. Typically, there is no premium paid for entering into a forward contract, and hence the initial value of a contract is zero.

Let $F(t,T)$ with $t < T$ denote the delivery price for a forward contract, which is agreed upon at time t, and whose delivery time is T. Let S denote the price of the underlying asset on which the forward contract is written. In particular, $S(t)$ is the asset price at time t when the contract is initiated, and $S(T)$ is the price at the delivery time T. Since the party who is being long can buy the asset for the forward price $F(t,T)$ to sell it immediately for the spot price $S(T)$ at the market, the payoff of the long forward contract is $S(T) - F(t,T)$. Therefore, the payoff of the short forward contract is $F(t,T) - S(T)$.

At the delivery time T there are two possibilities:

1. If $F(t,T) < S(T)$, then the party taking a long forward position benefits from a positive payoff. The instant profit is $S(T) - F(t,T)$ since the asset bought (at time T) at the forward price $F(t,T)$ can be immediately sold for $S(T)$. The party with a short position has negative payoff $F(t,T) - S(T)$ by selling the asset below the market price $S(T)$; this represents a loss in the amount of $S(T) - F(t,T)$.

2. If $F(t,T) > S(T)$, the situation is exactly reversed. The party taking a short forward position benefits from a positive payoff since the asset is sold at price $F(t,T)$ which is above its market price $S(T)$. The payoff to the party taking a long position is negative, representing a loss in the amount of $F(t,T) - S(T)$.

The payoff diagrams for the two parties respectively taking short and long forward positions are illustrated in Figure 4.2.

Example 4.1. Find the payoffs for a long forward contract and a short forward contract on the Dow Jones Industrial Average (DJI) at a forward price of \$26,000 after six months if the index in six months takes on one of the following values: \$25,500, \$25,800, \$26,100, and \$26,400.

Solution. Clearly, if the index value is less than \$26,000, the long forward loses money and the short forward makes profit. On the other hand, if the index value is greater than \$26,000, the long forward makes profit and the short forward loses money. The future payoff values for both contracts are given in the table below.

Index value in six months	Long forward payoff	Short forward payoff
\$25,500	−\$500	\$500
\$25,800	−\$200	\$200
\$26,100	\$100	−\$100
\$26,400	\$400	−\$400

□

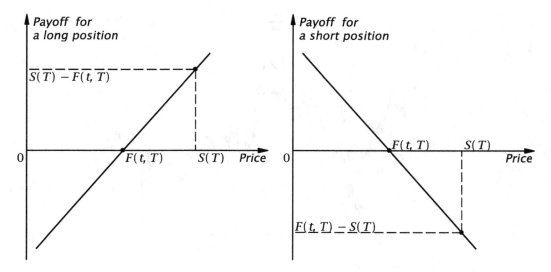

FIGURE 4.2: Payoff diagrams for the long and short positions in a forward contract.

4.1.1   No-Arbitrage Evaluation of Forward Contracts

Consider a market with a risk-free bank account B (with a constant interest rate r) and a risky stock S. Suppose that the discounting function $d(t) = \frac{B(0)}{B(t)}$ and the accumulation function $a(t) = \frac{1}{d(t)} = \frac{B(t)}{B(0)}$ for the bank account are given by (1.20) and (1.19), respectively. Thus, for any time period $[t, T]$ with $0 \leqslant t \leqslant T$ we have

- $B(T) = B(t)(1 + r)^{T-t}$ and $d(T - t) = (1 + r)^{-(T-t)}$ if the interest is compounded annually;

- $B(T) = B(t)(1 + r/m)^{(T-t)m}$ and $d(T - t) = (1 + r/m)^{-(T-t)m}$ if the interest is compounded m times per year;

- $B(T) = B(t)e^{r(T-t)}$ and $d(T - t) = e^{-r(T-t)}$ if the interest is compounded continuously.

In what follows, we assume that the bank account allows for borrowing or investing any amount of money at the same interest rate. We can also make or withdraw an investment at any moment without losing interest or paying an additional fee.

Suppose that a forward contract on the stock is written at time t and has delivery time T with $0 \leqslant t < T$. Let us find the forward price of such a contract which guarantees the absence of arbitrage and has no initial premium.

4.1.1.1   Forward Price for an Asset without Carrying Costs and Dividends

Theorem 4.1. *For an asset having no carrying costs and paying no dividends, the no-arbitrage time-t forward price for delivery time T (with $t \leqslant T$) is equal to the time-T value of the current spot price:*

$$F(t, T) = \frac{S(t)}{d(T - t)}. \tag{4.1}$$

In particular, if the interest is compounded continuously, then the no-arbitrage forward price is $F(t, T) = S(t)e^{r(T-t)}$.

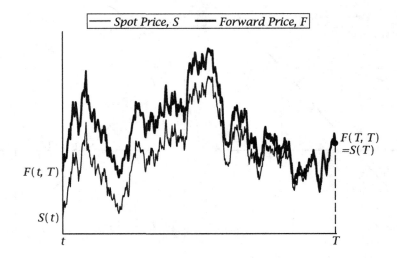

FIGURE 4.3: The dynamics of spot and forward prices.

Proof. We prove (4.1) by contradiction. First, suppose that $F(t,T) > \frac{S(t)}{d(T-t)}$. Let us form the following portfolio at time t: borrow $S(t)$ dollars from the risk-free bank account, buy one share of stock for $S(t)$, and enter into a short forward contract (with the agreement to sell one share of stock for $F(t,T)$). The initial value of such a portfolio is zero. At the delivery time T, we sell the stock for $F(t,T)$ and pay $\frac{S(t)}{d(T-t)}$ to return the amount borrowed with interest. As a result, we end up with a positive risk-free profit of $F(t,T) - \frac{S(t)}{d(T-t)}$. This contradicts the no-arbitrage principle.

Let $F(t,T) < \frac{S(t)}{d(T-t)}$ hold. Then we take a short position in the stock (borrow one share), sell the share for $S(t)$ and invest the proceeds in the bank account, and enter into a long forward contract (with the agreement to buy one share of stock for $F(t,T)$). Again the initial value of the portfolio is zero. At the delivery time T, we buy one stock share for $F(t,T)$ and close out the short position in the stock; we cash out the risk-free investment with interest to get $\frac{S(t)}{d(T-t)}$. As a result, we end up with a positive risk-free profit of $\frac{S(t)}{d(T-t)} - F(t,T)$. Again, this contradicts the no-arbitrage principle. Therefore, the only possibility is that the forward price $F(t,T)$ is equal to $\frac{S(t)}{d(T-t)}$. $\qquad\square$

Since for any $t < T$ the discount factor $d(T-t)$ is less than 1, the forward price is always larger than the spot price: $F(t,T) = \frac{S(t)}{d(T-t)} > S(t)$. As $t \to T$, the difference $F(t,T) - S(t)$ converges to zero, i.e., $F(T,T) = S(T)$. A visualization of this fact is given in Figure 4.3, which depicts a sample path simulation of a stock price process and its corresponding forward price up to maturity T. Note that $S(t)$ is always below $F(t,T)$ until T, at which point they coincide.

Example 4.2. Consider a 10-month forward contract on stock with a current price of \$500. Assume that the risk-free rate of interest is $r^{(12)} = 6\%$.

(a) Show that the no-arbitrage forward price is $F(0, \frac{10}{12}) = \$525.57$.

(b) Find an arbitrage opportunity if the delivery price is \$500 and it costs nothing to enter into the forward contract.

Solution.

(a) Applying (4.1) with monthly compounding gives the no-arbitrage forward price:

$$F\left(0, \frac{10}{12}\right) = 500(1 + 0.06/12)^{10} \cong \$525.57.$$

(b) Take a short position in the stock by borrowing one share; sell the share for $S(0) = \$500$ and invest the proceeds in the bank account; enter into a long forward contract with the agreement to sell one share of stock for \$500 in 10 months. At the delivery time $T = \frac{10}{12}$, we cash out the risk-free investment with interest to obtain $500(1+0.06/12)^{10} \cong \$525.57$, buy one stock share for \$500, and close out the short position in the stock. As a result, we end up with a positive arbitrage profit of $\$525.57 - \$500 = \$25.57$. $\qquad\square$

4.1.1.2   Forward Price for an Asset with Carrying Costs or Dividends

Holding physical assets (e.g., gold or oil) entail storage costs such as facility rental and insurance fees. Holding securities (e.g., stock or bond) may, alternatively, entail negative costs representing dividend or coupon payments. These costs (or incomes) affect the no-arbitrage forward price.

First, consider a stock that pays dividends. Let the dividend payments

$$\operatorname{div}(t_1), \operatorname{div}(t_2), \ldots, \operatorname{div}(t_m)$$

be made on each share of stock at times $t_1, t_2, \ldots, t_m \in [t, T]$, respectively. The present (time-t) value of the dividends paid over $[t, T]$, denoted $\operatorname{div}(t, T)$, is given by

$$\operatorname{div}(t, T) = \operatorname{div}(t_1)d(t_1 - t) + \operatorname{div}(t_2)d(t_2 - t) + \cdots + \operatorname{div}(t_m)d(t_m - t),$$

where $d(t)$ is a discounting function. We can modify the formula (4.1) for the forward price by subtracting the present value of the dividend payments from the spot price $S(t)$. Indeed, let us consider the following strategy: buy one share of stock and enter into a short forward contract to sell the asset for $F(t, T)$ at time T. The net present value of such a strategy (at time t) is $\operatorname{NPV}(t) = -S(t) + \operatorname{div}(t, T) + F(t, T)d(T - t)$. There is no arbitrage iff the NPV is zero. Therefore, $F(t, T) = (S(t) - \operatorname{div}(t, T))/d(T - t)$. The no-arbitrage forward price formula can also be justified by constructing arbitrage strategies when $F(t, T) \neq (S(t) - \operatorname{div}(t, T))/d(T - t)$, similar to those in the proof of Theorem 4.1.

Theorem 4.2. *For a stock paying dividends, the no-arbitrage forward price at time $t < T$ is*

$$F(t, T) = \frac{S(t) - \operatorname{div}(t, T)}{d(T - t)}. \tag{4.2}$$

Proof. First, suppose that $F(t, T) > \frac{S(t) - \operatorname{div}(t,T)}{d(T-t)}$. Then at time t we form the following portfolio. Borrow $S(t)$ dollars from the bank account, buy one share of stock for $S(t)$, and enter into a short forward contract. The initial value of such a portfolio is zero. Cash all dividend payments and invest them in the risk-free bank account. At the delivery time T, we sell the stock for $F(t, T)$, collect $\frac{\operatorname{div}(t,T)}{d(T-t)}$, and pay $\frac{S(t)}{d(T-t)}$ to return the amount borrowed with interest. The risk-free profit is $F(t, T) + \frac{\operatorname{div}(t,T)}{d(T-t)} - \frac{S(t)}{d(T-t)} > 0$. This contradicts the no-arbitrage principle.

Let $F(t, T) < \frac{S(t) - \operatorname{div}(t,T)}{d(T-t)}$. We take a short position in the stock (borrow one share), sell the share for $S(t)$, and invest the proceeds in the bank account. At the same time we enter into a long forward contract. Again the initial value of the portfolio is zero. Every time a dividend payment is due, we borrow the necessary amount and pay a dividend to the stockholder. As a result, by time T we owe $\frac{\text{div}(t,T)}{d(T-t)}$. At the delivery time T, we buy one share of S for $F(t,T)$ and close out the short position in the stock; we cash out the risk-free investment with interest $\frac{S(t)}{d(T-t)}$ and clear the loan. As a result, we end up with a positive risk-free profit of $\frac{S(t)}{d(T-t)} - \frac{\text{div}(t,T)}{d(T-t)} - F(t,T) > 0$.   □

Example 4.3. Consider a 10-month forward contract on stock with a price of \$500. Assume that the risk-free rate of interest is $r^{(12)} = 6\%$. Let a dividend payment of \$50 be paid five months after the initialization of the forward contract.

(a) Find the no-arbitrage forward price $F(0, \frac{10}{12})$.

(b) Find an arbitrage opportunity if the delivery price is \$490 and the forward contract has no initial premium.

Solution.

(a) The present value of the dividend payment is

$$\text{div}(0, 10/12) = 50 \cdot (1 + 0.06/12)^{-5} = 50 \cdot (1 + 0.005)^{-5} = 48.768533,$$

since the one-month interest rate is $r^{(12)}/12 = 0.06/12 = 0.005$. Therefore, the no-arbitrage forward price is

$$F(0, 10/12) = (500 - 48.768533) \cdot (1 + 0.005)^{10} = 474.3075034 \cong \$474.31.$$

(b) Borrow \$500 by making two loans: $50 \cdot 1.005^{-5} \cong \$48.77$ for five months and \$500 − \$48.77 = \$451.23 for 10 months. Buy one stock share for \$500. Enter into a short forward contract to sell the stock in 10 months for \$490. In five months, collect the dividend payment of \$50 and use it to repay the first loan. In 10 months, sell the stock share for \$490. Repay the second loan: $451.23 \cdot 1.005^{10} \cong \$474.31$. The risk-free profit realized is \$490 − \$474.31 = \$15.69.   □

Example 4.4. Consider a coupon bond with a face value of \$1,000, a quarterly coupon of $c^{(4)} = 6\%$, and several years to maturity. Currently, this bond is selling for \$845, and the previous coupon has just been paid. What is the forward price for delivery of this bond in two years? Assume that the rate of interest compounded continuously is constant at 3%.

Solution. There will be $m =$ eight quarterly coupon payments in the next two years. Each coupon is equal to $C = 1000 \times \frac{0.06}{4} = \$15$. The interest is compounded continuously at the rate of 3%, and hence the equivalent quarterly rate is

$$j = e^{0.03/4} - 1 = 0.007528195.$$

Therefore, the present value of the eight quarterly coupons is

$$C \cdot a_{\overline{m}|j} = 15 \cdot \frac{1 - 1.007528195^{-8}}{0.007528195} = 116.0347130.$$

The no-arbitrage forward price is then

$$F(0,2) = (845 - 116.0347130)e^{0.03 \cdot 2} = 774.0419832 \cong \$774.04.$$

□

Example 4.5. The price of a stock is \$200 on January 1, and \$5 dividend is to be paid on each of March 1, June 1, and October 1. Let the risk-free continuously compounded interest rate be 4%. Joe believes the stock price will go up and takes a long position in a one-year forward contract on the stock.

(a) What is the fair delivery price?

(b) On July 1, the stock price has risen to \$300. Joe feels that now is the time to cash out. Explain how he can use a second forward contract (issued on July 1) to lock in a risk-free profit. Find his risk-free profit realized on January 1 next year.

Solution.

(a) The discounted value of the dividend payments is

$$\text{div}(0,1) = 5 \cdot \left(e^{-0.04 \cdot \frac{2}{12}} + e^{-0.04 \cdot \frac{5}{12}} + e^{-0.04 \cdot \frac{9}{12}} \right)$$

$$= 14.736362 \,.$$

The forward price is

$$F(0,1) = (200 - \text{div}(0,1))e^{0.04} = 192.824390 \cong \$192.82 \,.$$

(b) Joe should enter into a short forward contract issued in July 1 with the delivery time on January 1 of the next year at no cost. The fair delivery price is

$$F(1/2,1) = \frac{300 - \text{div}(1/2,1)}{\exp(-0.04 \cdot \frac{1}{2})} = 301.010151 \cong \$301.01 \,.$$

On January 1, Joe will buy the stock for $F(0,1) = \$192.82$ (thanks to the long forward contract) and sell it immediately for $F(1/2,1) = \$301.01$ (thanks to the short forward contract). The risk-free profit is \$301.01 − \$192.82 = \$108.19.

$\square$

Consider an asset with carrying costs, and let the cost payments $c(t_1), c(t_2), \ldots, c(t_m)$ per unit for holding the asset be made at times $t_1, t_2, \ldots, t_m \in [t, T]$, respectively. These instalments can be viewed as negative dividend payments: $\text{div}(t_k) = -c(t_k)$ for $k = 1, 2, \ldots, m$. Thus, we can obtain the forward price formula by replacing $\text{div}(t, T)$ in (4.2) by $-c(t, T)$ where

$$c(t,T) := c(t_1)d(t_1 - t) + c(t_2)d(t_2 - t) + \cdots + c(t_m)d(t_m - t)$$

denotes the present (time-t) value of all carrying cost payments made over the time interval $[t, T]$. The forward price takes the following form:

$$F(t,T) = \frac{S(t)}{d(T-t)} + \sum_{k=1}^{m} \frac{c(t_k)}{d(T-t_k)} = \frac{S(t) + c(t,T)}{d(T-t)} \,. \tag{4.3}$$

Example 4.6. Oil currently trades at \$105 per barrel and the continuously compounded interest rate is 2.5% per annum.

(a) Assuming zero storage costs, what is the fair delivery price for one-year oil?

(b) Assuming it costs \$1 per month (payable at the beginning of each month) to store one barrel of oil, what is the fair delivery price for one-year oil?

Solution.

(a) The no-arbitrage forward price is

$$F(0,1) = 105 \cdot e^{0.025} = 107.6580877 \cong \$107.66.$$

(b) The equivalent monthly rate is $j = e^{0.025/12} - 1 = 0.002085505$. The present value of storage cost payments to be made during the next year is

$$c(0,1) = 1 + a_{\overline{11}|j} = 1 + \frac{1 - 1.002085505^{-11}}{0.002085505} = 11.86359155.$$

Thus, the no-arbitrage forward price is

$$F(0,1) = (S(0) + c(0,1))e^{0.025} = (105 + 11.86359155)e^{0.025} = 119.8220076 \cong \$119.82.$$

$\square$

4.1.2   Value of a Forward Contract

Consider a long forward contract with delivery time T and forward price $F(0,T)$ that is initiated at time 0. Initially, the value of such a forward contract is zero: $V(0) = 0$. At the delivery date, the value is equal to the payoff: $V(T) = S(T) - F(0,T)$. As time goes by, the spot price of the underlying asset may change; hence the value of the forward contract, $V(t)$, may vary as well.

Theorem 4.3. *The no-arbitrage value $V(t)$, $0 \leqslant t \leqslant T$, of a long forward contract that is initiated at time $t = 0$ and has delivery time T and delivery price $F(0,T)$ is given by*

$$V(t) = (F(t,T) - F(0,T))\, d(T - t). \tag{4.4}$$

Proof. At time t, consider two portfolios: one only has a long forward contract initiated at time 0 with value $V(t)$; the other consists of the long forward contract initiated at time t (whose value is zero at time t) and the risk-free investment of $(F(t,T) - F(0,T))d(T - t)$. At the delivery time T, both portfolios have the same value:

$$S(T) - F(0,T) = S(T) - F(t,T) + \frac{(F(t,T) - F(0,T))d(T - t)}{d(T - t)}.$$

By the Law of One Price, the time-t values of these portfolios have to be the same or else an arbitrage opportunity exists. $\square$

The value $V(t)$ of the long forward contract can be expressed in terms of stock prices by combining equation (4.4) with (4.1) or (4.2). Consider several important cases.

- For an asset paying no dividends/coupons and having no carrying costs, Equations (4.1) and (4.4) give

$$V(t) = S(t) - S(0)\frac{d(T - t)}{d(T)} = S(t) - \frac{S(0)}{d(t)}.$$

In particular, if interest is compounded continuously, we have $V(t) = S(t) - S(0)\, e^{rt}$.

- For a stock paying dividends (or a bond with coupon payments), equations (4.2) and (4.4) give

$$V(t) = S(t) - \frac{S(0) - \mathrm{div}(0,t)}{d(t)},$$

  where $\mathrm{div}(0,t)$ is the time-0 value of dividend payments paid over $[0,t]$.

- For an asset having carrying costs, (4.3) and (4.4) give

$$V(t) = S(t) - \frac{S(0) + c(0,t)}{d(t)},$$

  where $c(0,t)$ is the time-0 value of carrying cost payments mode over $[0,t]$.

Example 4.7. Let $S(0) = \$100$ and $r = 5\%$ (compounded continuously). A long forward contract with the delivery time $T =$ one year is initiated at time 0. If $S(0.5) = \$110$, what is the value $V(0.5)$ of the long forward contract?

Solution. First, apply (4.1) to find $F(0,1) = 100\,e^{0.05 \cdot 1} \cong \$105.13$. The forward price at time $t = 0.5$ is $F(0.5,1) = 110\,e^{0.05 \cdot 0.5} \cong \$112.79$. Therefore, the value $V(0.5)$ of the long forward contract is

$$V(0.5) = (F(0.5,1) - F(0,1))\,e^{-0.05 \cdot (1-0.5)} \cong \$7.47. \qquad \square$$

One can also consider a forward contract with a fixed delivery price K that may differ from the forward price $F(t,T)$ given by (4.1).

Theorem 4.4. *The no-arbitrage value $V_K(t)$, $0 \leqslant t \leqslant T$, of a long forward contract initiated at time $t = 0$ and having the delivery time T and delivery price K is given by*

$$V_K(t) = (F(t,T) - K)\,d(T - t).$$

The proof is very similar to that of Theorem 4.3 and is left as an exercise for the reader (see Exercise 4.12). As we can see, the initial price of such a forward contract is nonzero iff the delivery price $K \neq F(0,T)$.

4.2   Basic Options Theory

The holder of a forward contract is obliged to trade at the maturity of the contract. Unless the position is closed before the delivery date (i.e., the contract is sold to another party), the holder of a long forward must take possession of the asset regardless of whether the asset has risen or fallen (or pay the difference in prices).

An *option* is a contract that gives its buyer the right, not the obligation, to trade in the future at a previously agreed price. Options can be written on numerous products, such as stocks, funds, indices, gold, wheat, foreign currency, tulip bulbs, movie scripts, etc. In this chapter we deal with stock options (also called *equity options*), which give the holder the right to buy (or to sell) a stock for a specified price during a specified period.

A *call* option gives its holder the right to buy an asset (called the *underlying*) at a specified price (called the *exercise price* or the *strike price*) before a specified future date (called the *expiration/expiry date* or the *maturity date*). A *put* option gives its holder the right to sell the underlying asset at the strike price before the expiry date. The buyer of an option has the right, not the obligation to exercise the option. However, the seller (also called the writer) of the option must sell or buy the asset at the strike price once the option is exercised. Typically, the strike price, denoted K, is fixed in advance. The expiry date will be denote by T.

In this chapter we consider two most important style of options, namely, *European* and *American* (style) options. The terms American and European refer to the type of option, not the geographical region where the options are bought or sold. A *European/American call option (put option)* is a contract giving the holder the right to buy the underlying asset (to sell the underlying asset) at the fixed strike price K. The main difference between American- and European-style options relates to when the options may be exercised. Both European call and put options may only be exercised at the expiration date of the option. That is, a *European call* or *put option* gives the right to buy or, respectively, to sell the underlying asset for the strike price K at the exercise time T. American options may be exercised at any time before or at the expiration date. That is, an a *American call* or *put option* gives the right to buy or, respectively, to sell the underlying asset for the strike price K at any time between now and the expiration time T.

Since the holder of a European call (or put) option may only exercise his or her right and sell (or buy) the asset at the time T when the option is expiring, there is no difference between the exercise date and the expiration date for a European-style derivative. There is a difference between these two time moments for *American options*, which can be exercised at any time prior to the expiry date.

Note that these types of call and put options are commonly referred as *standard options* or *"vanilla options."* Options with a more complex payoff structure are categorized as "exotic options." Spreads, binary (digital) options, and other examples of European-style exotic derivatives are introduced in the second half of this chapter. More sophisticated path-dependent derivatives such as barrier and lookback options are discussed in Part II.

4.2.1   Payoffs of Standard Options

The holder of a standard European call or put option has two possible scenarios at the expiry date T.

$S(T) \geqslant K$. That is, the spot price $S(T)$ exceeds the strike price K. The holder of a call option should exercise the option to receive $S(T) - K$ (since the asset bought for K dollars can be immediately sold for $S(T)$ dollars). The holder of a put option should not exercise the option. The put payoff is zero.

$S(T) \leqslant K$. That is, the strike price K exceeds the spot price $S(T)$. The holder of a put option should exercise the option to receive $K - S(T)$ (since the asset can be bought for $S(T)$ dollars to be immediately sold for K dollars). The holder of a call option should not exercise the option. The call payoff is zero.

Therefore, the payoff function, denoted $\Lambda(S)$, of the holder of a European option is a nonnegative and piecewise-linear function:

$$\Lambda_{Call}(S) = \begin{cases} S - K & \text{if } S > K, \\ 0 & \text{if } S \leqslant K, \end{cases} \qquad \Lambda_{Put}(S) = \begin{cases} K - S & \text{if } S < K, \\ 0 & \text{if } S \geqslant K. \end{cases}$$

We can write the payoff functions in compact form:

$$\Lambda_{Call}(S) = (S - K)^+ \quad \text{and} \quad \Lambda_{Put}(S) = (K - S)^+,$$

where $x^+ := \max\{x, 0\}$ is the positive part function. The call and put payoffs as functions of the spot price S are plotted in Figure 4.4.

Additionally, the diagrams in Figure 4.4 illustrate the often used terminology related to moneyness. *Moneyness* refers to the relative position of the current price (or future price) of an underlying asset with respect to the strike price of the option. An option is said to be *in the money* if exercising the option yields a profit (i.e., its payoff is positive). An option is said to be *out of the money* if exercising the option is unprofitable (i.e., the payoff function is zero). An option is said to be *at the money* when the strike price is equal to the spot price of the underlying asset. Note that when an option is in the money, it does not mean that the option is overall profitable, since the positive payoff may still less than the premium paid for the option contract.

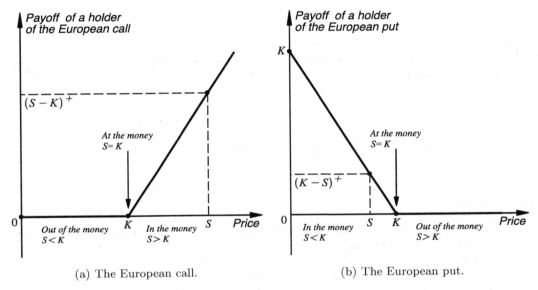

(a) The European call.   (b) The European put.

FIGURE 4.4: Payoff diagrams for European options.

It costs a trader nothing to enter into a forward contract (with the no-arbitrage delivery price), whereas the purchase of an option requires an up-front payment which is the option value at inception. Indeed, if no premium were paid, the option holder would lose no money and make a positive profit whenever the option is in the money. According to the no-arbitrage principle, the price paid for an option has to be nonnegative, although the option price should be strictly positive if the option is exercised (in the money) with a nonzero probability. Let $C^E \geq 0$ and $P^E \geq 0$ respectively denote the prices paid by a buyer of the European call and put options. For convenience, suppose that the option is written at time 0. By taking into account the time value of the price, we have the following formulae for the overall profit of an option buyer at time T:

$$\text{Profit}_{Call} = (S(T) - K)^+ - C^E e^{rT} \text{ and } \text{Profit}_{Put} = (K - S(T))^+ - P^E e^{rT}.$$

The gain of a holder of the European call or put option is plotted in Figure 4.5. Here and later in this section, we assume that the interest is compounded continuously at a risk-free rate r, i.e., $d(t) = e^{-rt}$.

Similarly to a forward contract, every option contract has two parties: the buyer of an option who has the right to exercise the option, and the writer who is obliged to deliver the underlying asset if the option is exercised at maturity. We say that the buyer takes the long position, and the writer takes the short position. This situation is illustrated in Figure 4.6 for both call and put options. The writer receives a cash premium up front but has potential

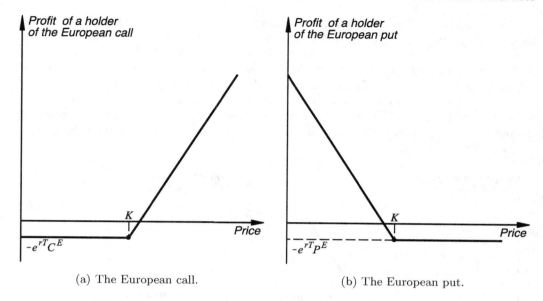

(a) The European call.   (b) The European put.

FIGURE 4.5: Profit diagrams for standard European options.

liabilities later. The writer's profit or loss is the reverse of that for the purchaser (the holder) of the option.

Example 4.8. Find the expected value of the gain for a holder of a European put option with strike price $K = \$100$ and time to expiry $T = 0.5$ years, if the risk-free continuously compounded rate is $r = 10\%$, the current price of the underlying security is $S(0) = \$95$, the option is bought for \$7, and the price $S(0.5)$ takes on one of four values: \$80, \$90, \$100, \$110, with equal probabilities.

Solution. The expected value of the gain (or loss) to the option holder is the difference between the expected payoff and the option premium paid. The option is in the money only when $S(0.5) = \$80$ or $S(0.5) = \$90$. The respective payoffs are $(100 - 80)^+ = \$20$ and $(100 - 90)^+ = \$10$. The expected gain is

$$\mathrm{E[Gain]} = -P^E\,\mathrm{e}^{rT} + \mathrm{E[Payoff]}$$
$$= -7\,\mathrm{e}^{0.1 \cdot 0.5} + 20 \cdot 0.25 + 10 \cdot 0.25 \cong -7.36 + 5 + 2.50 = \$0.14. \qquad \square$$

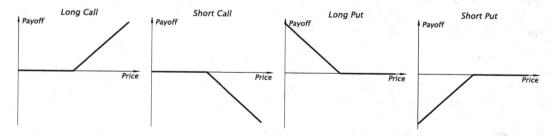

FIGURE 4.6: Payoff diagrams for long and short European call and put options.

4.2.2 Put-Call Parities

We now derive an important relation between the no-arbitrage prices of European put and call options called the *put-call parity*. A similar relation can be found for other pairs of put and call options. Such a parity can be used to find the price of one option (call or put) given the price of the other option. Also, the parity can be used to find out if an arbitrage opportunity exists.

Suppose we have European call and put options issued at time $t = 0$. Assume that both options have the same expiry date T and exercise price K. Consider the following two portfolios:

Portfolio 1 consists of one European call and the risk-free investment whose time-0 value is Ke^{-rT}.

Portfolio 2 consists of one European put and one share of stock.

At the expiry time T, both portfolios are worth $\max(S(T), K)$. Indeed, if $S(T) \geqslant K$, then the two portfolios have the following values at maturity:

$$\Pi_1(T) = S(T) - K + K = S(T) \text{ and } \Pi_2(T) = 0 + S(T) = S(T).$$

If $S(T) \leqslant K$, then

$$\Pi_1(T) = 0 + K = K \text{ and } \Pi_2(T) = K - S(T) + S(T) = K.$$

By the Law of One Price, if $\Pi_1(T, \omega) = \Pi_2(T, \omega)$ for any outcome $\omega \in \Omega$, then $\Pi_1(0) = \Pi_2(0)$. The portfolios must therefore have identical values today:

$$\underline{C^E + Ke^{-rT}} = \Pi_1(0) = \Pi_2(0) = \underline{P^E + S(0)}.$$

As a result, we obtain the *put-call parity*

$$C^E - P^E = S(0) - K e^{-rT}. \tag{4.5}$$

Assuming that interest is compounded annually, the parity has the following form:

$$C^E - P^E = S(0) - K (1 + r)^{-T}.$$

The put-call parity can also be written in terms of a risk-free ZCB price:

$$C^E - P^E = S(0) - K Z(0, T).$$

Alternatively, the call-put parity can be derived using the representation of a long forward payoff as a sum of payoffs of a long call option and a short put option (see Figure 4.7). Recall that the time value of a long forward contract with delivery price K and delivery time T is given by

$$V_K(t) = (F(t, T) - K)e^{-r(T-t)}, \ 0 \leqslant t \leqslant T.$$

In particular, if $K \neq F(0, T)$, then such a contact has a nonzero initial value

$$V_K(0) = (F(0, T) - K) e^{-rT} = (e^{rT}S(0) - K) e^{-rT} = S(0) - Ke^{-rT}.$$

At the expiry date, the payoff of such a long forward contract can be represented as a sum of a long call payoff and a short put payoff with strike price K. At time 0, the portfolio with one long call and one short put should have the same value as the long forward contract. Therefore,

$$C^E - P^E = S(0) - Ke^{-rT}.$$

In the case of a stock paying dividends, we have $V_K(0) = S(0) - \text{div}(0, T) - Ke^{-rT}$. Thus,

$$C^E - P^E = S(0) - \text{div}(0, T) - Ke^{-rT}. \tag{4.6}$$

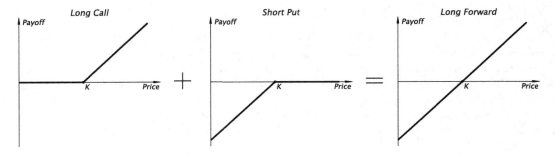

FIGURE 4.7: The representation of a long forward payoff as a sum of a long European call and a short European put. The strike price equals the delivery price.

4.2.3 Properties of European Options

The option price depends on some variables including the following:

Variables describing the contract: the strike price K and expiry time T.

Variables describing the underlying: the initial price $S(0)$ and dividend yield q.

Market variables: the interest rate r.

One can derive various properties of the no-arbitrage prices C^E and P^E solely using the no-arbitrage principle. Such properties are general and independent of the pricing model. For example, one can obtain bounds on the option price (as a function of $S(0)$ and K), prove monotonicity of the option price (as a function of $S(0)$, K, and T), and prove convexity of the option price (as a function of $S(0)$ and K).

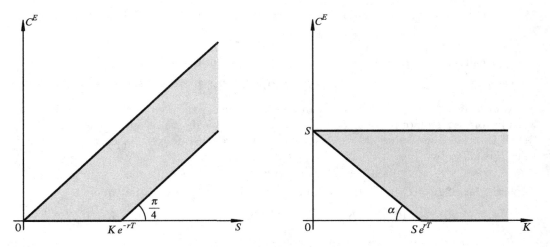

FIGURE 4.8: Lower and upper bounds on the European call price. The shaded areas represent regions of admissible option prices. Here, S denotes the spot price, and $\alpha = \arctan e^{-rT}$.

Theorem 4.5 (Upper and lower bounds on European options prices). *The no-arbitrage initial prices of the European call, C^E, and European put, P^E, satisfy*

$$(S(0) - Ke^{-rT})^+ \leqslant C^E < S(0), \tag{4.7}$$

$$(Ke^{-rT} - S(0))^+ \leqslant P^E < Ke^{-rT}. \tag{4.8}$$

Proof. Suppose that $C^E \geqslant S(0)$. Let us sell a call option and buy one stock share. The balance $C^E - S(0) \geqslant 0$ is invested in the risk-free bank account. At the expiry time T, we sell the stock for $\min\{S(T), K\}$. Our arbitrage profit is $\min\{S(T), K\} + (C^E - S(0))\, e^{rT} > 0$. Thus, the no-arbitrage price of the European call satisfies the upper bound: $C^E < S(0)$.

Using the put-call parity and nonnegativity of option prices, we obtain the lower bounds on the option prices:

$$\begin{cases} C^E \geqslant 0, \\ C^E = S(0) - Ke^{-rT} + P^E \geqslant S(0) - Ke^{-rT} \end{cases} \implies C^E \geqslant \max\{0, S(0) - Ke^{-rT}\},$$

$$\begin{cases} P^E \geqslant 0, \\ P^E = Ke^{-rT} - S(0) + C^E \geqslant Ke^{-rT} - S(0) \end{cases} \implies P^E \geqslant \max\{0, Ke^{-rT} - S(0)\}.$$

The upper bound on C^E and the put-call parity give us the upper bound on P^E. $\qquad\square$

The results of Theorem 4.5 are illustrated in Figures 4.8 and 4.9.

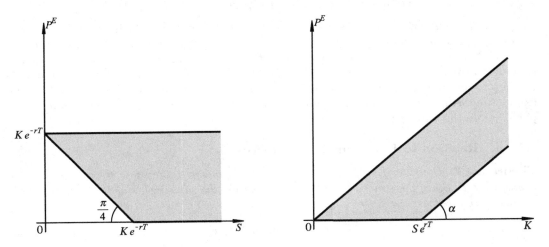

FIGURE 4.9: Lower and upper bounds on the European put price. The shaded areas represent regions of admissible option prices. Here, S denotes the spot price, and $\alpha = \arctan e^{-rT}$.

Theorem 4.6 (Monotonicity of European option prices). *The no-arbitrage price C^E (P^E) of the European call (put) is a nondecreasing (nonincreasing) function of the spot price $S = S(0)$ and a nonincreasing (nondecreasing) function of the strike price K:*

$$K' < K'' \implies C^E(K') \geqslant C^E(K'') \text{ and } P^E(K') \leqslant P^E(K''),$$

$$S' < S'' \implies C^E(S') \leqslant C^E(S'') \text{ and } P^E(S') \geqslant P^E(S'').$$

Proof. Suppose that $C^E(K') < C^E(K'')$ for some $K' < K''$. Write and sell a call with strike K''; buy a call with strike K'; invest the difference $C^E(K'') - C^E(K')$ without risk. The payoff of the combination of two options is nonnegative: $(S - K'')^+ \leqslant (S - K')^+$ since $K' < K''$. At the expiry time T, the total arbitrage profit is

$$(C^E(K'') - C^E(K'))\, e^{rT} + (S(T) - K')^+ - (S(T) - K'')^+ > 0.$$

Similarly, we prove that $P^E(K') \leqslant P^E(K'')$: suppose the converse and then form an arbitrage portfolio with a long put struck at K'', a short put struck at K', and a risk-free investment of $P^E(K') - P^E(K'')$.

Suppose that $C^E(S') > C^E(S'')$ for some $S' < S''$. Let $S(0)$ be the initial stock price. Consider two portfolios with $x' = \frac{S'}{S(0)}$ and $x'' = \frac{S''}{S(0)}$ shares of stock, respectively. The respective costs of the portfolios are $S' = x'S(0)$ and $S'' = x''S(0)$. Proceed as follows: write and sell a call on a portfolio with x' shares of stock; buy a call on a portfolio with x'' shares; invest the balance $C^E(S') - C^E(S'')$ without risk. Both options have the same strike price and expiry time. Again, the payoff of the combination of two options is nonnegative. Indeed, if $S(T) < \frac{K}{x''}$, then both options will not be exercised; if $\frac{K}{x''} \leqslant S(T) < \frac{K}{x'}$, then exercise the long call, the short call will not be exercised; if $\frac{K}{x'} \leqslant S(T)$, then both options are exercised, we cover the liability of the short call by the proceeds of the long call. At the maturity time T, the cumulative arbitrage profit is

$$(C^E(S') - C^E(S''))\, e^{rT} + (x''S(T) - K)^+ - (x'S(T) - K)^+ > 0.$$

Similarly, we prove that $P^E(S') \geqslant P^E(S'')$ if $S' < S''$. $\qquad\square$

4.2.4 Early Exercise and American Options

An *American call option (American put option)* is a contract giving the holder the right not the obligation to *buy* (to *sell*) the underlying asset at the strike price K fixed in advance at *any time* between now and the expiration time T. The main difference between a European-style derivative and an American-style derivative is that the latter can be exercised at any time up to and including the expiry time, whereas the former can only be exercised at the expiry time. American options are similar to callable bonds that can be redeemed at some point before the date of maturity.

4.2.4.1 Relation to European Option Prices

Theorem 4.7. *Consider European and American options with the same strike price K and expiry time T. Since the American option gives at least the same rights as the corresponding European counterpart, we have*

$$C^E \leqslant C^A \quad and \quad P^E \leqslant P^A.$$

Proof. Suppose that $C^E > C^A$. Let us form the following portfolio: buy one American call (worth C^A), and write and sell one European call (worth C^E). The difference $C^E - C^A > 0$ is invested in a risk-free bank account. Now we wait until the expiry date. If the stock price at the expiry date is larger than the strike price, then exercise the American option and sell the stock share to the holder of the European call for the same price K. If $S(T) \leqslant K$, then both options are not exercised. At time T withdraw the investment. The final balance of $(C^E - C^A)\, e^{rT} > 0$ is a risk-free profit. Similarly, we prove that $P^E \leqslant P^A$. $\qquad\square$

Consider European and American call options with the same strike price K and expiry time T on a *nondividend* paying stock. Theorem 4.7 tells us that $C^A \geqslant C^E$. In fact, an American call is worth the same as a European call with the same term. Indeed, suppose that $C^A > C^E$. Form the following portfolio: write and sell one American call (for C^A) and buy one European call (for C^E), investing the balance $C^A - C^E$ without risk.

Scenario 1: The American call is exercised at time $t \leqslant T$.

- Borrow one stock share and sell it for K to settle your obligation as a writer of the call.

- Invest K at rate r. ıthe stock.
- Your arbitrage profit is $(C^A - C^E)\, \mathrm{e}^{rT} + K\,(\mathrm{e}^{r(T-t)} - 1) > 0$.

Scenario 2: The American call is not exercised at all. You will end up with the European call, and your arbitrage profit is $(C^A - C^E)\mathrm{e}^{rT} + (S(T) - K)^+ > 0$.

Therefore, the American call price should be equal to the value of its European-style counterpart: $C^E = C^A$. So it is not wise to exercise an American call early. Note that the situation is different when options are written on a dividend-paying stock.

4.2.4.2 Put-Call Parity Estimate

Consider American put and call options with the same strike K and expiry T written on a stock paying no dividends (hence, $C^E = C^A$). Let us first obtain the upper bound:

$$C^A - P^A = C^E - P^A \leqslant C^E - P^E = S(0) - K\mathrm{e}^{-rT}.$$

Now let us show that $C^A - P^A \geqslant S(0) - K$. Suppose that it fails. That is,

$$C^A - P^A < S(0) - K \quad \Longleftrightarrow \quad P^A - C^A + S(0) > K.$$

Form the following portfolio:

- write and sell a put (for P^A);
- buy a call (for C^A);
- sell short one share (for $S(0)$);
- invest the balance $P^A - C^A + S(0) > K$ without risk.

There are two scenarios.

Scenario 1: The put is exercised at time $t \leqslant T$.

- Withdraw K from the risk-free investment to buy a share of stock (from the holder of the put).
- Use the share to close out the short position in the stock.
- We still have the call and a positive cash balance:

$$(P^A - C^A + S(0))\mathrm{e}^{rt} - K > K\mathrm{e}^{rt} - K \geqslant 0.$$

Scenario 2: The put has not been exercised at all.

- At time T we exercise the call option, buy one share of stock for K, and close out the short position in the stock.
- The balance at time T is

$$(P^A - C^A + S(0))\mathrm{e}^{rT} - K > K\mathrm{e}^{rT} - K = K(\mathrm{e}^{rT} - 1) > 0.$$

As a result, we proved the put-call parity estimate for American call and put options:

$$S(0) - K \leqslant C^A - P^A \leqslant S(0) - K\mathrm{e}^{-rT}. \tag{4.9}$$

With the help of the put-call parity estimate, we can obtain bounds on the American option prices. Let us consider the case of a stock paying no dividends. As we proved before, the European and American calls with the same strike and expiry have the same price, i.e., $C^A = C^E$. Therefore, the bounds on the European call price are valid for the American call price C^A as well:

$$(S(0) - Ke^{-rT})^+ \leqslant C^A < S(0).$$

Consider the American put. Since one can purchase and then immediately exercise the American option, the price cannot be less than the immediate payoff (otherwise the risk-free profit is $(K - S(0))^+ - P^A$). Thus, $P^A \geqslant (K - S(0))^+$. By using (4.9), we obtain that

$$P^A \leqslant C^A + K - S(0) = C^E + K - S(0) \leqslant S(0) + K - S(0) = K.$$

Therefore, $P^A \leqslant K$. By combining both of the bounds, we obtain

$$(K - S(0))^+ \leqslant P^A \leqslant K.$$

4.2.4.3   Monotonicity Properties of American Option Prices

Similar to the European case, one can derive various properties of C^A and P^A based on the no-arbitrage principle. These properties are general and independent of the asset price model used. For example,

- the option price is a monotonic function of spot price $S = S(0)$, strike price K, and expiry time T:

$$\left.\begin{array}{ccc} C^A(S) \nearrow & C^A(K) \searrow & C^A(T) \nearrow \\ P^A(S) \searrow & P^A(K) \nearrow & P^A(T) \nearrow \end{array}\right\} \quad \text{as } S \nearrow, K \nearrow, T \nearrow;$$

- the option price is a convex function of spot price S and strike price K.

4.2.5   Nonstandard European-Style Options

Standard European options can be used as building blocks to create more complex financial instruments. An investor with specific views on the future behaviour of stock prices may be interested in derivatives with payoff profiles that are different from those of the standard European call and put options. In general, any continuous piecewise-linear function can be manufactured from European call and put payoffs with different strikes. So being given a specific payoff profile, we can design a portfolio of standard options with the same payoff function.

A *spread strategy* involves taking a position in two or more options of the same type. An investor who expects the stock price to rise may form the following portfolio: Buy a call option with strike price K_1 and then, to reduce the premium paid for the call option, write and sell another call option with the same exercise date but with the strike price $K_2 > K_1$. The resulting portfolio is called a *bull spread* (see Figure 4.10). Payoffs of the two European calls and the resulting portfolio are described in the table below.

Stock Price S:	Long Call Payoff:	Short Call Payoff:	Total Payoff:
$S \leqslant K_1$	0	0	0
$K_1 < S < K_2$	$S - K_1$	0	$S - K_1$
$K_2 \leqslant S$	$S - K_1$	$K_2 - S$	$K_2 - K_1$

The payoff is positive for high future stock prices. Clearly, according to the Law of One Price, the no-arbitrage price of the bull spread is equal to $C^E(K_1) - C^E(K_2)$.

FIGURE 4.10: The bull spread.

A *bear spread* would satisfy an investor who believes that the stock price will decline. The bear spread is equivalent to a portfolio that has one long put option with strike price K_2 and one short put option with strike $K_1 < K_2$ (see Figure 4.11). Both options have the same exercise date. The payoffs are described in the table below.

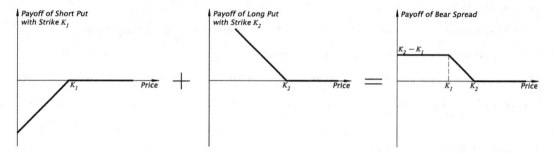

FIGURE 4.11: The bear spread.

Stock Price S:	Short Put Payoff:	Long Put Payoff:	Total Payoff:
$S \leqslant K_1$	$S - K_1$	$K_2 - S$	$K_2 - K_1$
$K_1 < S < K_2$	$K_2 - S$	0	$K_2 - S$
$K_2 \leqslant S$	0	0	0

The payoff is positive for low future stock prices. The no-arbitrage price of the bear spread is equal to $P^E(K_2) - P^E(K_1)$.

An investor who expects that the future stock price will change insignificantly and stay in an interval $[K_1, K_3]$ may choose a *butterfly spread* that is constructed from three options of the same kind expiring on the same date with strike prices K_i, $i = 1, 2, 3$, so that $K_1 < K_2 < K_3$, where $K_2 = \frac{K_1 + K_3}{2}$. One way to construct a butterfly spread is to combine one long call with strike K_1, two short calls with strike K_2, and one long call with strike K_3 (see Figure 4.12).

The payoff to the holder of the butterfly spread is

$$
\Lambda_{BS}(S) = \begin{cases} 0 & \text{if } S \leqslant K_1, \\ S - K_1 & \text{if } K_1 < S \leqslant K_2, \\ K_3 - S & \text{if } K_2 < S \leqslant K_3, \\ 0 & \text{if } S > K_3. \end{cases}
$$

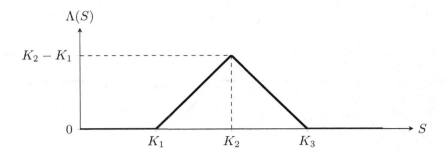

FIGURE 4.12: The butterfly spread.

The no-arbitrage price is $C^E(K_1) + C^E(K_3) - 2C^E(K_2)$. Note that a portfolio with one long stock, one short put with strike K_1, two long puts with strike K_2 and one short put with strike K_3 has the same payoff profile as that constructed from call options.

A *combination* is an option portfolio that involves taking a position in both calls and puts on the same underlying security. Some examples are as follows.

Straddle combines one put and one call with the same strike and expiry date.

Strip combines one long call and two long puts with the same strike and expiry date.

Strap combines two long calls and one long put with the same strike and expiry date.

Strangle is a combination of a put and a call with the same expiry date but different strike prices.

Example 4.9. Consider a European derivative with payoff

$$\Lambda(S) = \begin{cases} 8 - S & \text{if } S \leqslant 8, \\ 2S - 24 & \text{if } S \geqslant 12, \end{cases} \quad \text{and} \quad \Lambda(S) = 0 \text{ if } 8 < S < 12.$$

(a) Sketch the payoff function as a function of S.

(b) What portfolio of standard European options with the common expiry $T = 2$ is equivalent to this derivative?

Solution. The plot of the function $\Lambda(S)$ is provided below.

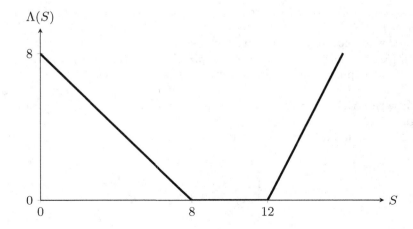

As we can see, a portfolio combining one long put with strike $K_1 = 8$ and two long calls with strike $K = 12$ has the same payoff structure: $\Lambda(S) = (8 - S)^+ + 2(S - 12)^+$. So, the derivative with payoff $\Lambda(S)$ is a straddle-type combination.

$\square$

4.3 Fundamentals of Option Pricing

4.3.1 Pricing of European-Style Derivatives in the Binomial Tree Model

For a European-style derivative, the payoff Λ is a function of the underlying price S at maturity time T. Examples of such derivatives include:

- a European call with $\Lambda(S) = (S - K)^+$ and a European put with $\Lambda(S) = (K - S)^+$, where K is the strike price;

- a long forward with $\Lambda(S) = S - K$ and a short forward with $\Lambda(S) = K - S$, where K is the delivery price;

- spreads and combinations such as a straddle, strip, strap, and strangle, whose payoffs can be expressed in terms of standard call and put payoffs;

- binary call and put options (cash-or-nothing options) with respective payoffs

$$\Lambda(S) = \mathbb{I}_{\{S \geqslant K\}} = \begin{cases} 1 & \text{if } S \geqslant K, \\ 0 & \text{otherwise,} \end{cases} \quad \text{and} \quad \Lambda(S) = \mathbb{I}_{\{S \leqslant K\}} = \begin{cases} 1 & \text{if } S \leqslant K, \\ 0 & \text{otherwise.} \end{cases}$$

Suppose that the derivative is issued at time 0. Our objective is to calculate no-arbitrage (fair) prices of such derivatives. Recall that in the binomial tree model, the time-t asset price is denoted by S_t. Similarly, we use the notation V_t for the t-time derivative price.

Clearly, at the maturity date T, the no-arbitrage value of a European-style derivative is equal to the payoff: $V_T(\omega) = \Lambda(S_T(\omega))$ for all $\omega \in \Omega$. Therefore, the time-T value of the derivative is a function of the price S_T, i.e., $V_T = \Lambda(S_T)$ holds. In fact, one can prove that at every time $t \in [0, T]$, the value of a European-style derivative is a function of the current asset price S_t and the calendar time t, i.e., $V_t = V_t(S_t)$. A rigorous proof of this property for the binomial tree model is given in Chapter 7. Since future prices of a risky asset are random variables, the derivative price is also random at any time $t > 0$. However, the initial price $V_0 = V_0(S_0)$ is a certain quantity.

4.3.1.1 Replication and Pricing of Options in the Single-Period Binomial Model

Consider a single-period binomial model with two states of the world: $\Omega = \{\omega^+, \omega^-\}$. At time $t = 0$ the stock price is S_0; at time $t = 1$ the stock price equals $S^+ := S_1(\omega^+) = S_0 u$ with the probability p and $S^- := S_1(\omega^-) = S_0 d$ with the probability $1 - p$. Recall that p and $1 - p$ are called the physical or real-world probabilities. The time values of a risk-free bond are B_0 and $B_1 = B_0(1 + r)$. Here, r is a risk-free one-period interest rate; d and u are, respectively, downward and upward one-period returns on the stock S.

Suppose we wish to find an initial no-arbitrage price of a derivative with payoff $\Lambda(S)$ at maturity time $T = 1$. In Section 2.2, we constructed a portfolio (x, y) in the stock and bond that replicates the payoff function by solving the following system of two linear equations:

$$\begin{cases} xS^+ + yB_1 = \Lambda(S^+) \\ xS^- + yB_1 = \Lambda(S^-). \end{cases}$$

The solution gives the replicating portfolio with the positions

$$\begin{cases} x = \dfrac{\Lambda(S^+) - \Lambda(S^-)}{S^+ - S^-} = \dfrac{\Lambda(S^+) - \Lambda(S^-)}{S_0(u - d)} \\ y = \dfrac{\Lambda(S^-)S^+ - \Lambda(S^+)S^-}{B_1(S^+ - S^-)} = \dfrac{\Lambda(S^-)u - \Lambda(S^+)d}{B_0(1 + r)(u - d)}. \end{cases} \tag{4.10}$$

By construction, the portfolio (x, y) and the derivative with payoff Λ have the same value at time 1. According to the law of one price they should have the same value at time 0 or else an arbitrage opportunity will arise, i.e.,

$$\forall \omega \in \Omega \ \Pi_1^{(x,y)}(\omega) = V_1(\omega) = \Lambda(S_1(\omega)) \implies V_0(S_0) = \Pi_0^{(x,y)}.$$

Thus, the initial no-arbitrage value of the derivative is

$$\begin{aligned} V_0(S_0) = \Pi_0^{(x,y)} &= xS_0 + yB_0 = \frac{\Lambda(S^+) - \Lambda(S^-)}{S_0(u - d)} S_0 + \frac{\Lambda(S^-)u - \Lambda(S^+)d}{B_0(1 + r)(u - d)} B_0 \\ &= \frac{\Lambda(S^+) - \Lambda(S^-)}{u - d} + \frac{\Lambda(S^-)u - \Lambda(S^+)d}{(1 + r)(u - d)} \\ &= \frac{1}{1 + r}\left(\Lambda(S^+)\frac{(1 + r) - d}{u - d} + \Lambda(S^-)\frac{u - (1 + r)}{u - d} \right) \\ &= \frac{1}{1 + r}\left(\Lambda(S_0 u)\frac{1 + r - d}{u - d} + \Lambda(S_0 d)\frac{u - r - 1}{u - d} \right). \end{aligned}$$

Notice the above expression can be written as a mathematical expectation:

$$V_0(S_0) = \frac{1}{1 + r}\left(\Lambda(S_0 u)\tilde{p} + \Lambda(S_0 d)(1 - \tilde{p}) \right) = \frac{1}{1 + r}\tilde{\mathbb{E}}[\Lambda(S_1)] \tag{4.11}$$

$$= \frac{1}{1 + r}\left(V_1(S_0 u)\tilde{p} + V_1(S_0 d)(1 - \tilde{p}) \right) = \frac{1}{1 + r}\tilde{\mathbb{E}}[V_1(S_1)] \tag{4.12}$$

where $\tilde{p} := \frac{1 + r - d}{u - d}$ and $1 - \tilde{p} = \frac{u - r - 1}{u - d}$. This formula is called the *risk-neutral pricing formula*. The numbers $\tilde{p}$ and $1 - \tilde{p}$ are called the *risk-neutral probabilities*. Recall that $0 < \tilde{p} < 1$ iff $d < 1 + r < u$.

Example 4.10 (Replication and pricing in one period). Consider a single-period binomial model with the following parameters: $S_0 = \$100$, $B_0 = \$1$, $d = 0.9$, $u = 1.15$, $r = 0.05$. Replicate a call option with strike price $K = \$95$ and expiry time $T = 1$. Find the option price.

Solution. Application of (4.10) gives positions x and y of the replicating portfolio for the European call option:

$$x = \frac{(100 \cdot 1.15 - 95)^+ - (100 \cdot 0.9 - 95)^+}{100 \cdot (1.15 - 0.9)} = \frac{20 - 0}{100 \cdot 0.25} = \frac{4}{5} = 0.8,$$

$$y = \frac{0 \cdot 1.15 - 20 \cdot 0.9}{1 \cdot 1.05 \cdot (1.15 - 0.9)} = \frac{-18}{1.05 \cdot 0.25} \cong -68.5714.$$

So to replicate the European call, we need $\frac{4}{5}$ shares of stock and a loan in the amount of \$68.57. The current price of the call option is equal to the portfolio value:

$$C_0^E = xS_0 + yB_0 \cong 0.8 \cdot 100 - 68.5714 \cdot 1 \cong \$11.43.$$

On the other hand, we have $\tilde{p} = (1.05 - 0.9)/(1.15 - 0.9) = 0.6$ and $1 - \tilde{p} = 0.4$. So, the risk-neutral pricing formula in (4.11) gives

$$C_0^E = \frac{1}{1.05} \cdot (20 \cdot 0.6 + 0 \cdot 0.4) = \frac{20 \cdot 0.6}{1.05} \cong \$11.43. \qquad \square$$

It is easy to see that a mispriced derivative security leads to arbitrage. Consider a derivative contract with payoff Λ in a single-period binomial model. If the trading price of the derivative, say $\widehat{V}_0$, is not equal to the risk-neutral price V_0 then one can construct an arbitrage portfolio (x, y, z) with x stock shares, y dollars in the interest-bearing risk-free account, and z derivative contracts as follows. Suppose $\widehat{V}_0 > V_0$. Form a portfolio with $x = \Delta$, $y = \beta + \widehat{V}_0 - V_0$, and $z = -1$, where

$$\Delta = \frac{\Lambda(S^+) - \Lambda(S^-)}{S_0(u - d)} \quad \text{and} \quad \beta = \frac{\Lambda(S^-)u - \Lambda(S^+)d}{(1 + r)(u - d)}.$$

The initial wealth is zero. At expiry, $\Pi_1 = (\widehat{V}_0 - V_0)(1 + r) > 0$. Suppose $\widehat{V}_0 < V_0$. Form a portfolio with x, y, z that are negative for the case with $\widehat{V}_0 > V_0$, i.e., the following portfolio: $(-\Delta, -\beta + (V_0 - \widehat{V}_0), 1)$. Again, the initial wealth is zero. At expiry, $\Pi_1 = (V_0 - \widehat{V}_0)(1 + r) > 0$.

4.3.1.2   Pricing in the Binomial Tree Model

Now we turn our attention to a multi-period binomial tree model. Recall that at time $t \in \{0, 1, 2, \dots\}$, the stock price has the probability distribution given in (2.15):

$$S_t = S_0 u^n d^{t-n} \text{ with probability } \binom{t}{n} p^t (1 - p)^{t-n} \text{ for } n = 0, 1, 2, \dots, t,$$

where S_0 is the initial stock price. The time-t bond value is $B_t = B_0(1 + r)^t$.

Our main objective is the pricing of a derivative contract with maturity time $T \in \{1, 2, \dots\}$. Let V_t denote the no-arbitrage derivative price at time $t \in \mathbf{T}$ where $\mathbf{T} = \{0, 1, \dots, T\}$. These prices form a derivative price process $\{V_t\}_{t \in \mathbf{T}}$, which is a *stochastic process* meaning that for any time t, the value V_t is a random variable defined on the state space Ω. So we write $V_t = V_t(\omega)$ with $\omega \in \Omega$.

To find the initial no-arbitrage price of the derivative, V_0, one can again use replication. However, a static replicating portfolio is *not sufficient* for a multi-period model. To replicate the whole derivative price process, we need to apply a self-financing dynamic portfolio strategy in the stock and the bond. An admissible self-financing strategy $\{(x_t, y_t)\}_{t \in \{0, 1, \dots, T-1\}}$ is said to *replicate the derivative* if the terminal portfolio value, Π_T, and the terminal derivative value, V_T, coincide for every market scenario $\omega \in \Omega$:

$$x_{T-1}(\omega)S_T(\omega) + y_{T-1}(\omega)B_T = V_T(\omega).$$

By the law of one price, the initial derivative price has to be equal to the initial value of the replicating strategy to avoid arbitrage:

$$V_0 = x_0 S_0 + y_0 B_0.$$

Additionally, for every time $t \in \mathbf{T}$, the replicating strategy and the derivative must have the same value for all market scenarios: $V_t = x_t S_t + y_t B_t$. This problem will be investigated in full detail in later chapters. At this point, our goal is to obtain a closed-form formula for the initial derivative price. So we will use a more straightforward approach.

Consider a European-style derivative contract with payoff function Λ, which is contingent on the stock price process. At the maturity time T, $V_T = \Lambda(S_T)$ holds. As will be demonstrated in later chapters, the time-t price V_t of any European-style derivative is a function of the time-t asset price: $V_t = V_t(S_t)$.

In the binomial tree model, the evolution of a stock price process is described by a recombining binomial tree. Such a tree can be viewed as a combination of single-period subtrees with two branches, where $S_{t+1} = S_t u$ for upward and $S_{t+1} = S_t d$ for downward branches starting at the node (t, S_t). For each subtree, we can use the one-period pricing formula (4.11). Therefore, in the binomial tree model the derivative prices can be computed using backward-in-time recurrence:

$$V_t(S_t) = \frac{1}{1+r}\left(\tilde{p}V_{t+1}(S_t u) + (1 - \tilde{p})V_{t+1}(S_t d)\right) \tag{4.13}$$

for $t = T - 1, T - 2, \ldots, 0$. Before beginning to apply the above recurrence, the derivative prices $V_T(S_T)$ at maturity for each of $T + 1$ nodes $S_{T,n} \equiv S_0 u^n d^{T-n}$ with $n = 0, 1, \ldots, T$ are simply given by the payoff function:

$$V_T(S_{T,n}) = \Lambda(S_{T,n}).$$

For the first backward time step, we set $t = T - 1$ in (4.13) and obtain T derivative prices $V_{T-1}(S_{T-1})$ for $S_{T-1} = S_{T-1,n} \equiv S_0 u^n d^{T-1-n}$ with $n = 0, 1, \ldots, T - 1$. In similar fashion, we apply (4.13) to obtain the derivative prices at times $t = T - 2, \ldots, 1, 0$. For example, by applying (4.13) $T - t$ times (for $0 \leqslant t \leqslant T$) we arrive at $t + 1$ derivative prices $V_t(S_{t,n})$ at time t for prices $S_{t,n} := S_0 u^n d^{t-n}$ with $n = 0, 1, \ldots, t$. After applying the recurrence relation T times we arrive at a single price $V_0 = V_0(S_0)$. As a result, we have the following recursive algorithm for calculating prices V_t^E of a European-style derivative:

$$V_T^E(S_T) = \Lambda(S_T) \quad \text{for } S_T \in \{S_0 u^n d^{T-n}, \ n = 0, 1, \ldots, T\}, \tag{4.14}$$

$$V_t^E(S_t) = \frac{1}{1+r}\left[\tilde{p}V_{t+1}^E(S_t u) + (1 - \tilde{p})V_{t+1}^E(S_t d)\right] \tag{4.15}$$

$$\text{for } t = T - 1, T - 2, \ldots, 1, 0 \text{ and } S_t \in \{S_0 u^n d^{t-n}, \ n = 0, 1, \ldots, t\}.$$

The process is illustrated in Figure 4.13.

To better clarify how one applies the recursive algorithm (4.14)–(4.15), let us consider a two-period binomial tree model. To compute the derivative prices $V_t \equiv V_t(S_t)$, $t = 0, 1, 2$, we proceed backward in time starting at the maturity $T = 2$. For every price S_2, derivative prices are equal to payoff values: $V_2(S_2) = \Lambda(S_2)$. At time $t = 1$, the stock price takes on one of two possible values: $S_0 u$ or $S_0 d$. The derivative prices can be computed using the one-period pricing formula as follows:

$$\begin{aligned}
V_1(S_0 u) &= \frac{1}{1+r}\left(\tilde{p}V_2(S_0 uu) + (1 - \tilde{p})V_2(S_0 ud)\right) \\
&= \frac{1}{1+r}\left(\tilde{p}\Lambda(S_0 u^2) + (1 - \tilde{p})\Lambda(S_0 ud)\right), \\
V_1(S_0 d) &= \frac{1}{1+r}\left(\tilde{p}V_2(S_0 du) + (1 - \tilde{p})V_2(S_0 dd)\right) \\
&= \frac{1}{1+r}\left(\tilde{p}\Lambda(S_0 du) + (1 - \tilde{p})\Lambda(S_0 d^2)\right).
\end{aligned}$$

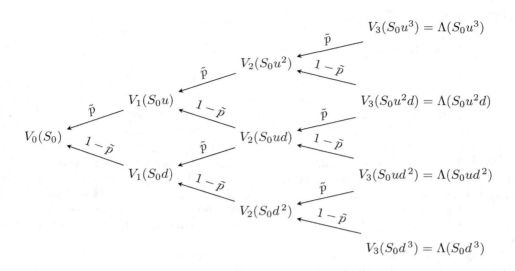

FIGURE 4.13: Valuation of a European derivative on a binomial tree model with three periods.

Here, we used $V_2(S) = \Lambda(S)$. Finally, at time $t = 0$, we again apply the one-period pricing formula to obtain the final expression for the no-arbitrage pricing formula:

$$V_0(S_0) = \frac{1}{1+r}\big(\tilde{p}V_1(S_0u) + (1 - \tilde{p})V_1(S_0d)\big)$$
$$= \frac{1}{1+r}\Big(\frac{\tilde{p}}{1+r}\big(\tilde{p}\Lambda(S_0u^2) + (1 - \tilde{p})\Lambda(S_0ud)\big)$$
$$+ \frac{1 - \tilde{p}}{1+r}\big(\tilde{p}\Lambda(S_0du) + (1 - \tilde{p})\Lambda(S_0d^2)\big)\Big). \tag{4.16}$$

Example 4.11 (Pricing in two periods). Consider the two-period binomial model from Example 4.10. Find the no-arbitrage prices C_0^E and C_1^E of the call price with the strike price $K = \$95$ and expiry time $T = 2$.

Solution. First, compute the risk-neutral probabilities:

$$\tilde{p} = \frac{1.05 - 0.9}{1.15 - 0.9} = 0.6, \quad 1 - \tilde{p} = 0.4.$$

Second, we calculate the prices $C_1^E(S_1)$ when $S_1 = S_0u = \$115$ and $S_1 = S_0d = \$90$:

$$C_1^E(90) = \frac{1}{1+r}\big(\Lambda(90\,u)\,\tilde{p} + \Lambda(90\,d)\,(1 - \tilde{p})\big)$$
$$= \frac{1}{1.05} \cdot \big((90 \cdot 1.15 - 95)^+ \cdot 0.6 + (90 \cdot 0.9 - 95)^+ \cdot 0.4\big)$$
$$= \frac{8.5 \cdot 0.6 + 0 \cdot 0.4}{1.05} \cong \$4.85714,$$
$$C_1^E(115) = \frac{1}{1.05} \cdot \big((115 \cdot 1.15 - 95)^+ \cdot 0.6 + (115 \cdot 0.9 - 95)^+ \cdot 0.4\big)$$
$$= \frac{37.25 \cdot 0.6 + 8.5 \cdot 0.4}{1.05} \cong \$24.52381.$$

Third, calculate $C_0^E(100)$:

$$C_0^E(100) = \frac{1}{1+r}\left(C_1^E(100\,u)\,\tilde{p} + C_1^E(100\,d)\,(1-\tilde{p})\right)$$

$$\cong \frac{1}{1.05} \cdot (24.52381 \cdot 0.6 + 4.85714 \cdot 0.4) \cong \$15.86394. \qquad \square$$

A T-period recombining binomial tree has $t+1$ nodes at any time $t \in \{0, 1, \ldots, T\}$, and hence there are $\sum_{t=0}^{T}(t+1) = \frac{(T+1)(T+2)}{2}$ nodes in total. The payoff function Λ is only evaluated at the $T+1$ terminal nodes, and then we apply the backward-in-time recursion at $\frac{T(T+1)}{2}$ internal nodes to calculate the no-arbitrage derivative value for every possible asset price. Thus, even if we only need the initial value V_0, the derivative price has to be found at every tree node using in (4.13). The total computational cost is of order $\mathcal{O}(T^2)$, but it can be substantially reduced if we need to compute the initial value V_0 only. In what follows, we derive a direct binomial derivative price formula that only requires the payoff values.

To explain this idea, let us consider a two-period binomial tree model and return to equation (4.16). Combine the payoff values as follows:

$$V_0(S_0) = \frac{1}{(1+r)^2}\left(\tilde{p}^2\Lambda(S_0 u^2) + \tilde{p}(1-\tilde{p})\Lambda(S_0 ud) + (1-\tilde{p})\tilde{p}\Lambda(S_0 du) + (1-\tilde{p})^2\Lambda(S_0 d^2)\right)$$

$$= \frac{1}{(1+r)^2}\left(\tilde{p}^2\Lambda(S_0 u^2) + 2\tilde{p}(1-\tilde{p})\Lambda(S_0 ud) + (1-\tilde{p})^2\Lambda(S_0 d^2)\right). \tag{4.17}$$

As we can see, the initial value $V_0(S_0)$ is given as a weighted sum of payoff values .

Theorem 4.8 (The Binomial Derivative Price Formula). *In the binomial tree model, the no-arbitrage initial price of a European-style derivative with payoff $\Lambda(S_T)$, at (discrete) maturity time $T \geqslant 1$, is given by*

$$V_0(S_0) = \frac{1}{(1+r)^T}\sum_{n=0}^{T}\binom{T}{n}\tilde{p}^n(1-\tilde{p})^{T-n}\Lambda\left(S_0 u^n d^{T-n}\right). \tag{4.18}$$

Proof. We prove the assertion by induction. As demonstrated in (4.11) and (4.17), the formula (4.18) is valid for $T = 1$ and $T = 2$. Suppose it is valid for $T \geqslant 1$; let us prove that it holds for $T + 1$. Applying (4.12) gives

$$V_0(S_0) = \frac{1}{1+r}\left[V_1(S_0 u)\,\tilde{p} + V_1(S_0 d)\,(1-\tilde{p})\right]. \tag{4.19}$$

That is, the price V_0 is a weighted sum of $V_1(S_0 u)$ and $V_1(S_0 d)$, which are the prices of the derivative issued at time $t = 1$ and expiring at time $t = T + 1$ provided that the current stock price is, respectively, $S_0 u$ and $S_0 d$. Since there are T periods from the issue time $t = 1$ to the expiry time $t = T + 1$, we can use (4.18) to evaluate the derivatives:

$$V_1(S_0 d) = \frac{1}{(1+r)^T}\sum_{n=0}^{T}\binom{T}{n}\tilde{p}^n(1-\tilde{p})^{T-n}\Lambda\left((S_0 d)\,u^n d^{T-n}\right),$$

$$V_1(S_0 u) = \frac{1}{(1+r)^T}\sum_{n=0}^{T}\binom{T}{n}\tilde{p}^n(1-\tilde{p})^{T-n}\Lambda\left((S_0 u)\,u^n d^{T-n}\right). \tag{4.20}$$

By combining (4.19) and (4.20), we obtain

$$
\begin{aligned}
V_0(S_0) = \frac{1}{(1+r)^{T+1}} \Bigg[& \tilde{p}^0 (1-\tilde{p})^{T+1} \binom{T}{0} \Lambda \left(S_0 u^0 d^{T+1} \right) \\
& + \tilde{p}^{T+1} (1-\tilde{p})^0 \binom{T}{T} \Lambda \left(S_0 u^{T+1} d^0 \right) \\
& + \sum_{n=1}^{T} \left(\binom{T}{n-1} + \binom{T}{n} \right) \tilde{p}^n (1-\tilde{p})^{T+1-n} \Lambda \left(S_0 u^n d^{T+1-n} \right) \Bigg].
\end{aligned}
$$

Using the identities $\binom{T}{0} = \binom{T}{T} = 1$ and $\binom{T}{n-1} + \binom{T}{n} = \binom{T+1}{n}$ gives the derivative price formula for $T+1$ time periods:

$$
V_0(S_0) = \frac{1}{(1+r)^{T+1}} \sum_{n=0}^{T+1} \binom{T+1}{n} \tilde{p}^n (1-\tilde{p})^{(T+1)-n} \Lambda \left(S_0 u^n d^{(T+1)-n} \right). \qquad \square
$$

As follows from (2.15), the asset price S_T can be expressed in terms of a binomial random variable $S_T = S_0 u^{Y_T} d^{T-Y_T}$, where $Y_T \sim Bin(T, \tilde{p})$, so that

$$
Y_T = n \text{ with } \tilde{\mathbb{P}}\text{-probability } \binom{T}{n} \tilde{p}^n (1-\tilde{p})^{T-n} \text{ for all } n = 0, 1, 2, \ldots, T.
$$

Therefore, the option price formula in (4.18) admits a very compact form:

$$
V_0(S_0) = \frac{1}{(1+r)^T} \tilde{\mathbb{E}} \left[\Lambda \left(S_0 u^{Y_T} d^{T-Y_T} \right) \right] = \frac{1}{(1+r)^T} \tilde{\mathbb{E}} \left[\Lambda(S_T) \right]. \qquad (4.21)
$$

In other words, the initial no-arbitrage price of a European-style derivative is equal to the discounted risk-neutral expectation of the payoff at the maturity (exercise) date.

Example 4.12 (Pricing two nonstandard options). Consider a binomial tree model with $S_0 = \$100$, $u = 1.2$, $d = 0.9$, and $r = 0.1$. Find no-arbitrage present values of the following European derivatives:

(a) the butterfly spread with the payoff

$$
\Lambda_B(S) = \begin{cases} 0, & \text{if } S \notin [100, 140], \\ 140 - S, & \text{if } S \in [120, 140], \\ S - 100, & \text{if } S \in [100, 120); \end{cases}
$$

(b) the binary (cash-or-nothing) call option with the payoff

$$
\Lambda_D(S) = \begin{cases} 0, & \text{if } S < 120, \\ 20, & \text{if } S \geqslant 120. \end{cases}
$$

Both derivatives are exercised at time $T = 3$.

Solution. The risk-neutral probabilities are

$$
\tilde{p} = \frac{1.1 - 0.9}{1.2 - 0.9} = \frac{0.2}{0.3} = \frac{2}{3} \text{ and } 1 - \tilde{p} = \frac{1}{3}.
$$

Application of Equations (4.18) and (4.21) gives us the price of the butterfly spread:

$$\frac{1}{(1+r)^3}\widetilde{E}[\Lambda_B(S_3)] = \frac{1}{1.1^3}\sum_{n=0}^{3}\binom{3}{n}\left(\frac{2}{3}\right)^n\left(\frac{1}{3}\right)^{3-n}\Lambda_B\left(100\cdot 1.2^n\cdot 0.9^{3-n}\right)$$

$$= \frac{1}{1.1^3}\cdot\left(\frac{1}{27}\cdot\Lambda_B(72.9) + 3\cdot\frac{2}{27}\cdot\Lambda_B(97.2) + 3\cdot\frac{4}{27}\cdot\Lambda_B(129.6) + \frac{8}{27}\cdot\Lambda_B(172.8)\right)$$

$$= \frac{1}{1.331}\cdot\left(\frac{1}{27}\cdot 0 + 3\cdot\frac{2}{27}\cdot 0 + 3\cdot\frac{4}{27}\cdot 10.4 + \frac{8}{27}\cdot 0\right)$$

$$= \frac{1}{1.331}\cdot\frac{4}{9}\cdot 10.4 \cong \$3.4727.$$

The pricing formula for the cash-or-nothing call option is simplified as follows:

$$\frac{1}{(1+r)^3}\widetilde{E}[\Lambda_D(S_3)] = \frac{1}{1.1^3}\widetilde{E}[20\cdot\mathbb{I}_{\{S_3\geqslant 120\}}]$$

$$= \frac{1}{1.1^3}\cdot 20\cdot\widetilde{\mathbb{P}}(S_3\geqslant 120) = \frac{20}{1.1^3}\cdot\left(\widetilde{\mathbb{P}}(S_3 = 172.8) + \widetilde{\mathbb{P}}(S_3 = 129.6)\right)$$

$$= \frac{20}{1.1^3}\cdot\left(\frac{8}{27} + \frac{12}{27}\right) = \frac{20\cdot 20}{1.1^3\cdot 27} \cong \$11.1306. \qquad\square$$

For standard European call and put options, the option price formula in (4.18) can be written in a simpler form. Introduce the cumulative distribution function (CDF for short) $\mathcal{B}$ of the binomial probability distribution $Bin(n,p)$:

$$\mathcal{B}(m;n,p) = \sum_{k=0}^{m}\mathcal{b}(k;n,p) = \sum_{k=0}^{m}\binom{n}{k}p^k(1-p)^{n-k}, \; m = 0, 1, \ldots, n.$$

Here $\mathcal{b}$ denotes the probability mass function (PMF) of the binomial distribution; it is given by $\mathcal{b}(k;n,p) = \binom{n}{k}p^k(1-p)^{n-k}$ for $k = 0, 1, \ldots, n$. Recall that the CDF F of a random variable X is defined by $F(x) = \mathbb{P}(X \leqslant x)$; the PMF of a discrete random variable X is $p(x) = \mathbb{P}(X = x)$.

Theorem 4.9 (The Cox–Ross–Rubinstein (CRR) Option Price Formula). *In the binomial tree model, the no-arbitrage initial price of standard European call and put options with strike price K and expiry time T are, respectively, given by*

$$C_0^E(S_0) = S_0(1 - \mathcal{B}(m_T;T,\tilde{q})) - \frac{K}{(1+r)^T}(1 - \mathcal{B}(m_T;T,\tilde{p})), \qquad (4.22)$$

$$P_0^E(S_0) = \frac{K}{(1+r)^T}\mathcal{B}(m_T;T,\tilde{p}) - S_0\mathcal{B}(m_T;T,\tilde{q}), \qquad (4.23)$$

where $\tilde{p} = \frac{1+r-d}{u-d}$, $\tilde{q} = \frac{u}{1+r}\tilde{p}$, *and*

$$m_T = \max\left\{m \; : \; 0\leqslant m\leqslant T; \; S_0\,u^m\,d^{T-m}\leqslant K\right\} = \left\lfloor\frac{\ln(K/S_0) - T\ln d}{\ln(u/d)}\right\rfloor. \qquad (4.24)$$

Proof. The proof is left as an exercise for the reader. A more general result is proved in later chapters. $\square$

4.3.2 Pricing of American Options in the Binomial Tree Model

Consider a T-period binomial tree model. Let $\Lambda(S)$ be the payoff function of an American derivative security, which is expiring at time T. The option can be exercised at any time $t \in \{1, 2, \ldots, T\}$ with the immediate payoff $\Lambda(S_t)$. Denote by $V_t^A(S_t)$ the value of the American derivative at time t given that the option has not been exercised previously.

At the expiry time $t = T$ the option is worth $V_T^A(S_T) = \Lambda(S_T)$ given that it has not been exercised until maturity. At any time $t = \{T-1, T-2, \ldots, 1\}$ before expiry, the option holder has the choice to exercise the option immediately with the **intrinsic value** $V_t^{\text{intr}}(S_t) = \Lambda(S_t)$ or to postpone until time $t+1$, when the value of the option will become $V_{t+1}^A(S_{t+1})$. If the holder decides to postpone exercising the option for one period, then the option value is given by the **continuation value** at time t that can be computed by considering a one-step European contingent claim to be priced at time t:

$$V_t^{\text{cont}}(S_t) = \frac{1}{1+r}\left[\tilde{p}V_{t+1}^A(S_t u) + (1-\tilde{p})V_{t+1}^A(S_t d)\right].$$

Since, the option holder has always the choice between immediately exercising and postponing, the American option should be worth the higher of the intrinsic value and the continuation value:

$$V_t^A(S_t) = \max\{V_t^{\text{intr}}(S), V_t^{\text{cont}}(S_t)\}$$
$$= \max\left\{\Lambda(S_t), \frac{1}{1+r}\left[\tilde{p}V_{t+1}^A(S_t u) + (1-\tilde{p})V_{t+1}^A(S_t d)\right]\right\}.$$

Thus, we have the following recursive pricing formulae:

$$V_T^A(S_T) = \Lambda(S_T), \quad \text{for } S_T \in \{S_0 u^n d^{T-n}, \ n = 0, 1, \ldots, T\} \tag{4.25}$$

$$V_t^A(S_t) = \max\{V_t^{\text{intr}}(S_t), V_t^{\text{cont}}(S_t)\}$$
$$= \max\left\{\Lambda(S_t), \frac{1}{1+r}\left[\tilde{p}V_{t+1}^A(S_t u) + (1-\tilde{p})V_{t+1}^A(S_t d)\right]\right\}, \tag{4.26}$$

for $t = T-1, T-2, \ldots, 1$ and $S_t \in \{S_0 u^n d^{t-n}, \ n = 0, 1, \ldots, t\}$.

Assuming that the option cannot be exercised at time 0, we have the following initial no-arbitrage price:

$$V_0^A(S_0) = \frac{1}{1+r}\tilde{\mathbb{E}}[V_1^A(S_1)] = \frac{1}{1+r}\left[\tilde{p}V_1^A(S_0 u) + (1-\tilde{p})V_1^A(S_0 d)\right]. \tag{4.27}$$

Example 4.13 (The case without dividends). Find no-arbitrage prices of the American put and European put options with the common strike price $K = \$100$ both expiring at time $T = 2$ on a stock with the initial price $S_0 = \$100$ in a two-period binomial tree model with $u = 1.2$, $d = 0.9$, and $r = 0.1$.

Solution. First, construct the binomial tree with two periods (see Figure 4.14) and find the risk-neutral probabilities: $\tilde{p} = \frac{2}{3}$ and $1 - \tilde{p} = \frac{1}{3}$ (see Figure 4.14). Second, calculate prices of the put options starting from the expiry date $T = 2$ and going backward in time:

$$t = 2: \quad P_2^E(81) = P_2^A(81) = (100 - 81)^+ = \$19$$
$$P_2^E(108) = P_2^A(108) = (100 - 108)^+ = \$0$$
$$P_2^E(144) = P_2^A(144) = (100 - 144)^+ = \$0,$$

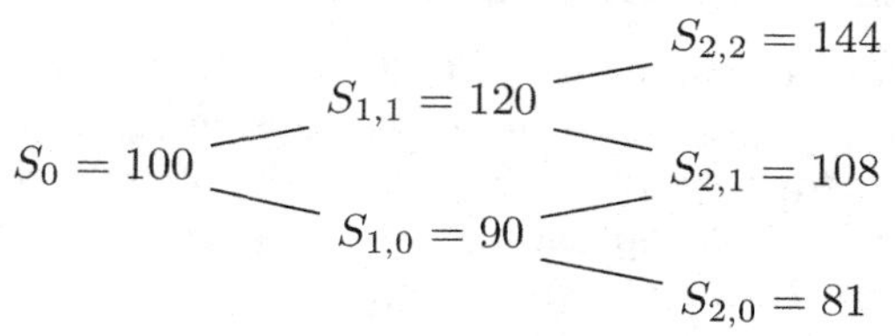

$$S_{2,2} = 144$$
$$S_{1,1} = 120$$
$$S_0 = 100$$
$$S_{2,1} = 108$$
$$S_{1,0} = 90$$
$$S_{2,0} = 81$$

FIGURE 4.14: Stock prices.

$$t = 1 : \quad P_1^E(90) = \frac{1}{1.1}\left(\frac{2}{3}P_2^E(108) + \frac{1}{3}P_2^E(81)\right) \cong \$5.75758$$

$$P_1^E(120) = \frac{1}{1.1}\left(\frac{2}{3}P_2^E(144) + \frac{1}{3}P_2^E(108)\right) = \$0$$

$$P_1^A(90) = \max\left\{(100 - 90)^+, \frac{1}{1.1}\left(\frac{2}{3}P_2^A(108) + \frac{1}{3}P_2^A(81)\right)\right\}$$
$$= \max\{10, 5.75758\} = \$10$$

$$P_1^A(120) = \max\left\{(100 - 120)^+, \frac{1}{1.1}\left(\frac{2}{3}P_2^A(144) + \frac{1}{3}P_2^A(108)\right)\right\}$$
$$= \max\{0, 0\} = \$0.$$

At time $t = 0$ we have:

$$P_0^E(100) = \frac{1}{1.1}\left(\frac{2}{3}P_1^E(120) + \frac{1}{3}P_1^E(90)\right) = \$1.74472$$

$$P_0^A(100) = \frac{1}{1.1}\left(\frac{2}{3}P_1^A(120) + \frac{1}{3}P_1^A(90)\right) = \frac{10}{3 \cdot 1.1} \cong \$3.03030.$$

The solution can be represented in the form of a binomial tree (see Figure 4.15). The American put option should be exercised whenever the continuation value $V_{\text{cont}}(t, S)$ falls below the intrinsic value $(K - S)^+$. In this case, the American option should be exercised early at time $t = 1$ if $S_1 = 90$. If $S_1 = 120$, then the American put option should not be exercised at all, since the spot price S_t exceeds the strike price K at both times $t = 1$ and $t = 2$. $\qquad\square$

4.3.3 Option Pricing in the Log-Normal Model: The Black–Scholes–Merton Formula

In this section, we present the European option pricing formula under the log-normal model. The culminating point will be the derivation of the famous Black–Scholes–Merton formulae for the no-arbitrage prices of standard European call and put options.

The log-normal pricing model can be derived as the limiting case of a sequence of suitably scaled binomial tree models. Consider a sequence of binomial prices $S_N^{(N)}$ first presented in Section 2.2.6. In the risk-neutral setting, the prices $S_N^{(N)}$ converge in distribution (as $N \to \infty$) to the log-normal price $S(T)$ defined by

$$S(T) = S_0 e^{(r - \sigma^2/2)T + \sigma\sqrt{T}Z} \text{ with } Z \sim \text{Norm}(0, 1).$$

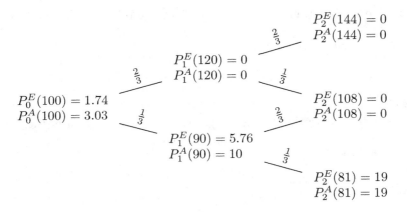

FIGURE 4.15: Pricing in a two-period binomial tree model.

Recall that S_0 is the initial asset price, r is the risk-free interest rate compounded continuously, T is the maturity time (in years), and σ is the (annual) volatility of the asset price. Let us investigate if it is possible to find the no-arbitrage option price formula for the log-normal model by proceeding to the continuous-time limit in (4.21), (4.22), and (4.23).

The first approach to be considered is based on a property of the convergence in distribution and is stated below without a proof.

Theorem 4.10. *Suppose* $X_n \xrightarrow{d} X$, *as* $n \to \infty$. *Let* $f : \mathbb{R} \to \mathbb{R}$ *be a bounded and continuous function. Then*

$$E[f(X_n)] \to E[f(X)], \quad as \ n \to \infty.$$

For the binomial tree model, we derived a general no-arbitrage price formula (4.21) of a European-type derivative with payoff Λ. By applying (4.21) to the sequence of scaled binomial tree models with terminal prices $S_N^{(N)}$ we obtain a sequence of no-arbitrage derivative prices:

$$V_0^{(N)} = (1 + r_N)^{-N} \widetilde{E}\left[\Lambda(S_N^{(N)})\right], \quad N \geqslant 1.$$

Suppose that Λ is a continuous and bounded function such that the European put payoff $\Lambda(S) = (K - S)^+$. Since the sequence of binomial prices $S_N^{(N)}$ converges in distribution to the log-normal price $S(T)$, we immediately obtain that

$$\widetilde{E}\left[\Lambda(S_N^{(N)})\right] \to \widetilde{E}[\Lambda(S(T))], \quad as \ N \to \infty.$$

By the definition of $r_N = \mathrm{e}^{r\,T/N} - 1$, the discount factor is $(1 + r_N)^{-N} = \mathrm{e}^{-rT}$ for every $N \geqslant 1$. Therefore, we have that

$$V_0^{(N)} \to V(0) := \mathrm{e}^{-rT}\widetilde{E}[\Lambda(S(T))] = \mathrm{e}^{-rT}\widetilde{E}\left[\Lambda\big(S_0\mathrm{e}^{(r-\sigma^2/2)T+\sigma\sqrt{T}Z}\big)\right],$$

as N goes to ∞. Since for each N the price $V_0^{(N)}$ admits no-arbitrage, we may expect that in the limiting case the derivative price $V(0)$ admits no-arbitrage as well. A rigorous proof of this fact will be provided in Volume II. The general formula for the no-arbitrage price of a European-style derivative under the log-normal model becomes

$$V(0, S_0) = \mathrm{e}^{-rT}\widetilde{E}[\Lambda(S(T))].$$

Let us apply the above formula to find the no-arbitrage price of a European put:

$$P^E(0, S_0) = \widetilde{E}\left[e^{-rT}\left(K - S_0 e^{(r-\frac{1}{2}\sigma^2)T + \sigma\sqrt{T}Z} \right)^+ \right]$$

$$= \widetilde{E}\left[\left(e^{-rT}K - S_0 e^{-\frac{1}{2}\sigma^2 T + \sigma\sqrt{T}Z} \right)^+ \right].$$

Now the trick is to remove the function $(x)^+$ inside the expectation in the above formula by using the representation $(x)^+ = x\,\mathbb{I}_{\{x \geqslant 0\}}$. The condition $e^{-rT}K - S_0 e^{-\frac{1}{2}\sigma^2 T + \sigma\sqrt{T}Z} \geqslant 0$ is simplified as follows:

$$S_0 e^{-\frac{1}{2}\sigma^2 T + \sigma\sqrt{T}Z} \leqslant e^{-rT}K \tag{4.28}$$

$$\Longleftrightarrow \quad -\frac{1}{2}\sigma^2 T + \sigma\sqrt{T}Z \leqslant -rT + \ln\left(\frac{K}{S_0}\right)$$

$$\Longleftrightarrow \quad Z \leqslant -\frac{\ln\left(\frac{S_0}{K}\right) + rT - \frac{1}{2}\sigma^2 T}{\sigma\sqrt{T}}. \tag{4.29}$$

Introduce the following notation:

$$d_{\pm} = \frac{\ln\left(\frac{S_0}{K}\right) + (r \pm \frac{1}{2}\sigma^2)T}{\sigma\sqrt{T}}. \tag{4.30}$$

Note that $d_+ - d_- = \sigma\sqrt{T}$ holds. Now, the European put price formula takes the form

$$P^E(0, S_0) = \widetilde{E}\left[\left(e^{-rT}K - S_0 e^{-\frac{1}{2}\sigma^2 T + \sigma\sqrt{T}Z} \right) \mathbb{I}_{\{Z \leqslant -d_-\}} \right]$$

$$= \widetilde{E}\left[e^{-rT}K \mathbb{I}_{\{Z \leqslant -d_-\}} \right] - \widetilde{E}\left[S_0 e^{-\frac{1}{2}\sigma^2 T + \sigma\sqrt{T}Z} \mathbb{I}_{\{Z \leqslant -d_-\}} \right]$$

$$= e^{-rT}K \widetilde{E}\left[\mathbb{I}_{\{Z \leqslant -d_-\}} \right] - S_0 \widetilde{E}\left[e^{-\frac{1}{2}\sigma^2 T + \sigma\sqrt{T}Z} \mathbb{I}_{\{Z \leqslant -d_-\}} \right]. \tag{4.31}$$

There are two expectations in (4.31) to take care of. The first one is easy to compute:

$$\widetilde{E}\left[\mathbb{I}_{\{Z \leqslant -d_-\}} \right] = \widetilde{\mathbb{P}}(Z \leqslant -d_-) = \mathcal{N}(-d_-).$$

Recall that $\mathcal{N}(x)$ denotes the cumulative distribution function (CDF) of the standard normal distribution $Norm(0,1)$ given by $\mathcal{N}(z) = \frac{1}{\sqrt{2\pi}}\int_{-\infty}^z e^{-\frac{x^2}{2}}\,dx$. The other expectation in (4.31) can be calculated by using the brute force of calculus:

$$\widetilde{E}\left[e^{-\frac{1}{2}\sigma^2 T + \sigma\sqrt{T}Z} \mathbb{I}_{\{Z \leqslant -d_-\}} \right] = \int_{-\infty}^{-d_-} e^{-\frac{1}{2}\sigma^2 T + \sigma\sqrt{T}z} \mathcal{N}'(z)\,dz$$

$$= \int_{-\infty}^{-d_-} \frac{1}{\sqrt{2\pi}} e^{-\frac{1}{2}\sigma^2 T + \sigma\sqrt{T}z - \frac{z^2}{2}}\,dz = \int_{-\infty}^{-d_-} \frac{1}{\sqrt{2\pi}} e^{-\frac{1}{2}(z - \sigma\sqrt{T})^2}\,dz$$

$$= \int_{-\infty}^{-(d_- + \sigma\sqrt{T})} \frac{1}{\sqrt{2\pi}} e^{-\frac{z^2}{2}}\,dz = \mathcal{N}(-(d_- + \sigma\sqrt{T})) = \mathcal{N}(-d_+).$$

The reader will note that this also follows directly by simply using the following identity:

$$E\left[e^{BX} \mathbb{I}_{\{X < A\}} \right] \equiv \int_{-\infty}^A e^{Bx}\varphi_{\mu,\sigma}(x)\,dx = e^{\mu B + \frac{1}{2}\sigma^2 B^2}\mathcal{N}\left(-\sigma B + \frac{A - \mu}{\sigma} \right),$$

where $X \sim Norm(\mu, \sigma^2)$. As a result, we obtain the Black–Scholes–Merton formula for the no-arbitrage price of a European put option:

$$P^E(0, S_0) = \mathrm{e}^{-rT} K \mathcal{N}(-d_-) - S_0 \mathcal{N}(-d_+). \tag{4.32}$$

The formula for a call option price can be obtained by using the put-call parity (4.5) as follows:

$$C^E(0, S_0) = P^E(0, S_0) + S_0 - \mathrm{e}^{-rT}K = S_0(1 - \mathcal{N}(-d_+)) - \mathrm{e}^{-rT}K(1 - \mathcal{N}(-d_-)).$$

By using the identity $\mathcal{N}(x) = 1 - \mathcal{N}(-x)$, we finally obtain the pricing formula for a European call option:

$$C^E(0, S_0) = S_0 \mathcal{N}(d_+) - \mathrm{e}^{-rT} K \mathcal{N}(d_-). \tag{4.33}$$

In the case of standard European call and put options under the binomial tree model, the general option price formula in (4.21) reduces to the CRR option price formula (4.22) or (4.23). Thus, another way to derive the Black–Scholes–Merton formula is to directly proceed to the limiting case in the CRR option price formula as the number of periods N approaches ∞. To illustrate this idea, let us consider the case of the European put option. For the sequence of scaled binomial tree models introduced in Section 2.2.6, the CRR option price formula in (4.23) takes the form

$$P_0^{(N)}(S_0) = \frac{K}{(1 + r_N)^N} \mathcal{B}(m_N(K); N, \tilde{p}_N) - S_0 \mathcal{B}(m_N(K); N, \tilde{q}_N),$$

where

$$\tilde{p}_N = \frac{\mathrm{e}^{\delta_N r} - \mathrm{e}^{-\sqrt{\delta_N}\sigma}}{\mathrm{e}^{\sigma\sqrt{\delta_N}} - \mathrm{e}^{-\sigma\sqrt{\delta_N}}}, \quad \tilde{q}_N = \frac{\mathrm{e}^{\sqrt{\delta_N}\sigma}}{\mathrm{e}^{\delta_N r}} \tilde{p}_N,$$

and

$$m_N(K) = \left\lfloor \frac{\ln(K/S_0) - N \ln d_N}{\ln(u_N/d_N)} \right\rfloor = \left\lfloor \frac{\ln(K/S_0) + N\sigma\sqrt{\delta_N}}{2\sigma\sqrt{\delta_N}} \right\rfloor.$$

As $N \to \infty$, the probabilities converge to one half: $\tilde{p}_N \to \frac{1}{2}$ and $\tilde{q}_N \to \frac{1}{2}$. According to the de Moivre–Laplace Theorem 2.2, the sequence of standardized binomial random variables $Y_n^* := \frac{Y_n - \mathrm{E}[Y_n]}{\sqrt{\mathrm{Var}(Y_n)}}$ (where expectations are calculated under either $\widetilde{\mathbb{P}}$ or $\widehat{\mathbb{P}}$) converges in distribution to a standard normal random variable. That is, for each $x \in \mathbb{R}$, $F_{Y_n^*}(x) \to \mathcal{N}(x)$. When the limiting distribution is the normal one, we can use the following property stated without a proof.

Theorem 4.11. *Suppose* $X_n \xrightarrow{d} Z$, *as* $n \to \infty$, *with* $Z \sim Norm(0, 1)$. *Let a real sequence* $\{a_n\}_{n \geqslant 1}$ *converge to* $a \in \mathbb{R}$. *Then* $F_{X_n}(a_n) \to \mathcal{N}(a)$, *as* $n \to \infty$.

To apply the above theorem, we need to find the limit $\lim_{N \to \infty} \frac{m_N(K) - \mathrm{E}[Y_N]}{\sqrt{\mathrm{Var}(Y_N)}}$ under the probability functions $\widetilde{\mathbb{P}}$ and $\widehat{\mathbb{P}}$.

Proposition 4.12.

$$\frac{m_N(K) - N\tilde{p}_N}{\sqrt{N\tilde{p}_N(1 - \tilde{p}_N)}} \to -d_- \quad and \quad \frac{m_N(K) - N\tilde{q}_N}{\sqrt{N\tilde{q}_N(1 - \tilde{q}_N)}} \to -d_+, \quad as \ N \to \infty.$$

Proof. The proof is left as an exercise for the reader. $\square$

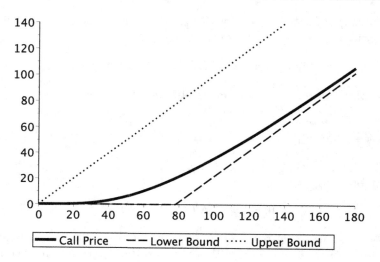

FIGURE 4.16: The Black–Scholes price of the European call option as a function of the initial stock price S. The parameters used are: $K = \$100$, $T = 5$, $\sigma = 30\%$, $r = 5\%$. The upper bound is S; the lower bound is $(S - Ke^{-rT})^+$.

By combining all the above results, we have:

$$\mathcal{B}(m_N(K); N, \tilde{p}_N) = \widetilde{\mathbb{P}}(Y_N \leqslant m_N(K)) = \widetilde{\mathbb{P}}\left(Y_N^* \leqslant \frac{m_N(K) - N\tilde{p}_N}{\sqrt{N\tilde{p}_N(1 - \tilde{p}_N)}} \right) \to \mathcal{N}(-d_-),$$

$$\mathcal{B}(m_N(K); N, \tilde{q}_N) = \widehat{\mathbb{P}}(Y_N \leqslant m_N(K)) = \widehat{\mathbb{P}}\left(Y_N^* \leqslant \frac{m_N(K) - N\tilde{q}_N}{\sqrt{N\tilde{q}_N(1 - \tilde{q}_N)}} \right) \to \mathcal{N}(-d_+).$$

This proves that the CRR put option prices converge to the Black–Scholes price of the put option:

$$P_0^{(N)} \to Ke^{-rT}\mathcal{N}(-d_-) - S_0\mathcal{N}(-d_+), \quad \text{as } N \to \infty.$$

Formulae for the time-t value of the European call and put options can be obtained from (4.32) and (4.33) by making changes: $T \to T - t$ and $S_0 \to S_t$. That is, the maturity time is replaced by the *time to maturity,* and the initial stock value is replaced by the time-t stock price denoted S_t. The resulting formulae are

$$C^E(t, S_t) = S_t\,\mathcal{N}(d_+) - e^{-r(T-t)}K\mathcal{N}(d_-), \tag{4.34}$$

$$P^E(t, S_t) = e^{-r(T-t)}K\mathcal{N}(-d_-) - S_t\,\mathcal{N}(-d_+), \tag{4.35}$$

where

$$d_\pm = \frac{\ln(S_t/K) + (r \pm \sigma^2/2)(T - t)}{\sigma\sqrt{T - t}}. \tag{4.36}$$

Typical plots of Black–Scholes prices of European call and put options along with upper and lower bounds are provided in Figures 4.16 and 4.17.

4.3.4 Greeks and Hedging of Options

The writer of a European option receives a cash premium upfront but has potential liabilities later on the option exercise date. The writer's profit is the reverse of that for the purchaser

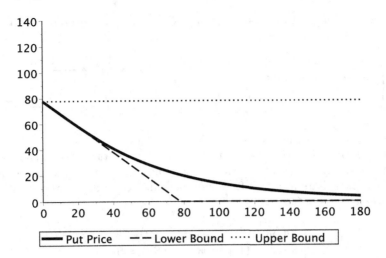

FIGURE 4.17: The Black–Scholes price of the European put option as a function of the initial stock price S. The parameters used are $K = \$100$, $T = 5$, $\sigma = 30\%$, $r = 5\%$. The upper bound is Ke^{-rT}; the lower bound is $(Ke^{-rT} - S)^+$.

of the option and is given by $C^E e^{rT} - (S(T) - K)^+$, where C^E is the premium received from the purchaser of the option. The writer can invest the premium in a risk-free bank account. If the option will end up deep in the money when $S(T) > K$, the writer of the option will be exposed to the risk of a large loss. Theoretically, the loss of the writer may be unlimited. For a European put option, the writer's loss is limited:

$$\text{loss} = -\left(P^E e^{rT} - (K - S(T))^+\right) \leqslant K - P^E e^{rT},$$

but it still may be very large compared to the premium P^E. The writer of an option may reduce this risk over a small time interval by forming a suitable portfolio in the underlying security called a hedge or a hedging portfolio. For a binomial model, such a portfolio is constructed by replicating the option payoff. In reality, it is impossible to hedge perfectly by designing a single (static) portfolio to be held for the whole period. The hedge has to be rebalanced dynamically to adapt it to changes in risk factors that affect the option value. This leads to a hedging strategy that will be studied in more detail in Part II and Volume II. In this section, we discuss static hedging over a single short time interval.

4.3.4.1 Delta of a Derivative in the Binomial Model

Consider a binomial tree model with a risk-free bank account and a risky stock. The current price of the stock is denoted by S. The no-arbitrage time-0 price, $V_0(S)$, of a derivative contract is given by the following risk-neutral one-step pricing formula:

$$V_0(S) = \frac{1}{1+r}\widetilde{\mathrm{E}}[V_1(S_1)] = \frac{1}{1+r}\left(\tilde{p}V_1(Su) + (1-\tilde{p})V_1(Sd)\right),$$

where $\tilde{p} := \frac{r+1-d}{u-d}$ is the risk-neutral probability, u and d are binomial price factors, and $V_1(S_1)$ is the value of the derivative at the end of the first period. Alternatively, we can construct a replicating portfolio with Δ_0 shares of stock and β_0 dollars in a risk-free bank account:

$$
\begin{cases} Su\Delta_0 + (1+r)\beta_0 = V_1(Su) \\ Sd\,\Delta_0 + (1+r)\,\beta_0 = V_1(Sd) \end{cases} \implies \begin{cases} \Delta_0 = \dfrac{V_1(Su) - V_1(Sd)}{Su - Sd} \\ \beta_0 = \dfrac{V_1(Sd)u - V_1(Su)d}{(1+r)(u-d)} \end{cases}
$$

The no-arbitrage price of the derivative is $V_0(S) = \Delta_0 S + \beta_0$.

A writer sells the derivative contract. To hedge her short position, she buys Δ_0 shares of stock and invests the remainder without risk. The initial wealth of the portfolio with Δ_0 shares of stock, β_0 dollars in a risk-free bank account, and one short derivative is zero. At time 1, the wealth becomes

$$
\Pi_1 = \begin{cases} Su\Delta_0 + (1+r)\beta_0 - V_1(Su) = 0 & \text{if } S_1 = Su, \\ Sd\Delta_0 + (1+r)\beta_0 - V_1(Sd) = 0 & \text{if } S_1 = Sd. \end{cases}
$$

As is seen, in either case, the time-1 value of the portfolio is zero. So the portfolio hedges all future risks over the first period. At the end of the first period, the portfolio has to be re-balanced to hedge over the period that follows and so on. As a result, the writer constructs a discrete-time delta hedging strategy, $\{\Delta_n\}_{n \geqslant 0}$. This topic will be addressed in more detail in later chapters.

4.3.4.2   Delta of a Derivative in the Log-Normal Model

Recall that the log-normal model can be obtained as a limiting case of the binomial tree model when the number of periods N goes to infinity. Consider the parametrization of a binomial tree model from Section 2.2.6:

$$
u = e^{\sigma\sqrt{\delta}},
$$

$$
d = e^{-\sigma\sqrt{\delta}},
$$

$$
\tilde{p} = \frac{e^{\delta r} - e^{-\sqrt{\delta}\sigma}}{e^{\sigma\sqrt{\delta}} - e^{-\sigma\sqrt{\delta}}},
$$

where $\delta > 0$ is the length of one period of time. Suppose that the binomial model is applied on the single period $[t, t+\delta]$ with $t \geqslant 0$. The number of stock shares in the hedging portfolio at time t is

$$
\Delta_t \approx \frac{V(t+\delta, Se^{\sigma\sqrt{\delta}}) - V(t+\delta, Se^{-\sigma\sqrt{\delta}})}{Se^{\sigma\sqrt{\delta}} - Se^{-\sigma\sqrt{\delta}}},
$$

where $V(t, s)$ is the value of the derivative at time t when the time-t price of the stock is s. To determine, under the log-normal price model, the number of stock shares in the hedging strategy, we need to let δ go to zero. L'Hôpital's rule and the chain rule for differentiating a two-variable function give

$$
\lim_{\delta \to 0} \frac{V(t+\delta, Se^{\sigma\sqrt{\delta}}) - V(t+\delta, Se^{-\sigma\sqrt{\delta}})}{Se^{\sigma\sqrt{\delta}} - Se^{-\sigma\sqrt{\delta}}}
$$

$$
= \lim_{h \to 0} \frac{V(t+h^2, Se^{\sigma h}) - V(t+h^2, Se^{-\sigma h})}{Se^{\sigma h} - Se^{-\sigma h}}
$$

$$
= \lim_{h \to 0} \frac{S\sigma e^{\sigma h}\frac{\partial}{\partial s}V(t+h^2, s)|_{s=Se^{\sigma h}} + S\sigma e^{-\sigma h}\frac{\partial}{\partial s}V(t+h^2, s)|_{s=Se^{-\sigma h}}}{S\sigma e^{\sigma h} + S\sigma e^{-\sigma h}}
$$

$$= \frac{\lim_{h\to 0} e^{\sigma h} \frac{\partial}{\partial s} V(t + h^2, Se^{\sigma h}) + \lim_{h\to 0} e^{-\sigma h} \frac{\partial}{\partial s} V(t + h^2, Se^{\sigma h})}{\lim_{h\to 0} e^{\sigma h} + \lim_{h\to 0} e^{-\sigma h}}$$

$$= \frac{\partial}{\partial s} V(t, s)|_{s=S} = \frac{\partial}{\partial S} V(t, S).$$

Therefore, a portfolio with $\Delta_t = \frac{\partial}{\partial S} V(t, S)$ stock shares allows the writer to hedge the risk associated with a short position in the derivative. Such a hedge works only over a short time interval while the stock price does not deviate too much from its initial value S. The partial derivative of the price $V(t, S)$ w.r.t. the current stock price S is called the *delta* of the derivative security.

Consider a European call option in the log-normal model. The no-arbitrage price of a European call at time 0 is given by

$$C^E(S) \equiv C^E(0, S) = S\mathcal{N}(d_+) - Ke^{-rT}\mathcal{N}(d_-),$$

where $d_\pm = \dfrac{\ln(S/K) + (r \pm \sigma^2/2)T}{\sigma\sqrt{T}}$ and S is the current stock price. The delta of the call option is given by the derivative of the price $C^E(S)$ w.r.t. S:

$$\frac{\partial}{\partial S} C^E(S) = \mathcal{N}(d_+) + S\frac{\partial}{\partial S}\mathcal{N}(d_+) - Ke^{-rT}\frac{\partial}{\partial S}\mathcal{N}(d_-).$$

The partial derivatives of $\mathcal{N}(d_\pm)$ w.r.t. S are given by

$$\frac{\partial}{\partial S}\mathcal{N}(d_+) = \mathcal{N}'(d_+)\frac{\partial}{\partial S}d_+(S) = \mathcal{N}'(d_+)\frac{1}{S\sigma\sqrt{T}},$$

$$\frac{\partial}{\partial S}\mathcal{N}(d_-) = \mathcal{N}'(d_-)\frac{\partial}{\partial S}d_-(S) = \mathcal{N}'(d_-)\frac{1}{S\sigma\sqrt{T}},$$

where we use the identity $\dfrac{\partial d_\pm}{\partial S} = \dfrac{\partial}{\partial S}\left(\dfrac{\ln(S)}{\sigma\sqrt{T}}\right) = \dfrac{1}{S\sigma\sqrt{T}}$. As a result, we have

$$\frac{\partial}{\partial S} C^E(S) = \mathcal{N}(d_+) + \frac{1}{\sigma\sqrt{T}}\left(\mathcal{N}'(d_+) - \frac{K}{S}e^{-rT}\mathcal{N}'(d_-)\right). \tag{4.37}$$

The derivative of the normal CDF gives the normal PDF: $\mathcal{N}'(x) = \frac{1}{\sqrt{2\pi}}e^{-x^2/2}$. Applying this formula in (4.37) gives

$$\frac{K}{S}e^{-rT}\mathcal{N}'(d_-) = \frac{1}{\sqrt{2\pi}}e^{-\ln(S/K)-rT-d_-^2/2} = \frac{1}{\sqrt{2\pi}}e^{-d_+^2/2} = \mathcal{N}'(d_+).$$

This simplifies the expression in (4.37) to finally give us a formula for the delta of a European call option in the log-normal model as:

$$\text{delta}_{C^E} = \frac{\partial C^E}{\partial S} = \mathcal{N}(d_+). \tag{4.38}$$

The delta of a European put can be easily found by differentiating the put-call parity (4.5) and using the delta of a standard call given in (4.38).

4.3.4.3   Delta Hedging

Consider a portfolio whose value $\Pi = \Pi(S)$ is a function of the current stock price S. Suppose that the price changes from S to $S + \delta S$ over a short time interval. According to Taylor's Theorem, the value of the portfolio is approximately

$$\Pi(S + \delta S) \cong \Pi(S) + \frac{d\Pi(S)}{dS}\delta S.$$

Hence, the change in value of the portfolio is

$$\delta\Pi(S) := \Pi(S + \delta S) - \Pi(S) \cong \frac{\mathrm{d}\Pi(S)}{\mathrm{d}S}\,\delta S.$$

The risk associated with this change in value is

$$\sqrt{\mathrm{Var}(\delta\Pi(S))} \cong \left|\frac{\mathrm{d}\Pi(S)}{\mathrm{d}S}\right| \sqrt{\mathrm{Var}(\delta S)}.$$

The risk will be eliminated if $\frac{\mathrm{d}}{\mathrm{d}S}\Pi(S) = 0$. Note that the derivative $\frac{\mathrm{d}}{\mathrm{d}S}\Pi(S)$ is called the *delta of the portfolio.*

Delta hedging is the process of reducing the risk associated with price changes in the underlying security by keeping the delta of the portfolio as close to zero as possible. This can be achieved by offsetting long and short positions. For example, a short call position may be delta-hedged by a long position in the underlying security. Such a portfolio is called *delta-neutral.*

Consider a portfolio (x, y, z) composed of x shares of stock with current price S, a cash amount y dollars invested without risk, and a short derivative security (i.e., $z = -1$) to be hedged with the initial wealth

$$\Pi(S) \equiv \Pi_0^{(x,y,z)} = xS + y - V(S),$$

where $V(S)$ is the current price of the derivative. Then, the delta of the portfolio is $\frac{\mathrm{d}}{\mathrm{d}S}\Pi(S) = x - \frac{\mathrm{d}}{\mathrm{d}S}V(S)$. If $\frac{\mathrm{d}}{\mathrm{d}S}\Pi(S) = 0$, then $x = \frac{\mathrm{d}}{\mathrm{d}S}V(S)$, where $\frac{\mathrm{d}}{\mathrm{d}S}V(S)$ is the delta of the derivative security. For example, construct a portfolio that hedges the short position in a European call option:

$$(x, y, z) = (\mathcal{N}(d_+), y, -1), \text{ where } d_+ = d_+(S).$$

For any cash amount y, the delta of this portfolio is zero. Consequently, its value is not varying much under small changes about the initial value S of the stock share. It is convenient to choose y so that the initial wealth is zero: $S\mathcal{N}(d_+) + y - C^E(S) = 0$. Application of the Black–Scholes formula for $C^E(S)$ gives

$$y = C^E(S) - \mathcal{N}(d_+)S = -Ke^{-rT}\mathcal{N}(d_-).$$

Example 4.14 (Delta-neutral portfolio). An investor writes and sells 1000 one-month call options with strike price $K = \$50$ on stock with an initial price $S_0 = \$50$. Assuming a log-normal price model with $r = 5\%$ and $\sigma = 20\%$, construct a delta-neutral portfolio with initial wealth zero.

Solution. Application of (4.33) with $S_0 = \$50$, $K = \$50$, $T = \frac{1}{12}$, $r = 0.05$, and $\sigma = 0.2$ gives the price of one call option, C^E:

$$d_+ = \frac{\ln\left(\frac{50}{50}\right) + (0.05 + 0.2^2/2)\cdot\frac{1}{12}}{0.2\sqrt{1/12}} \cong 0.1010363,$$

$$d_- = d_+ - 0.2\sqrt{1/12} \cong 0.0433013,$$

$$\mathcal{N}(d_+) \cong 0.54024,$$

$$\mathcal{N}(d_-) \cong 0.51727,$$

$$C^E = 50\mathcal{N}(d_+) - e^{-0.05/12}50\mathcal{N}(d_-) \cong \$1.25603.$$

The price of 1000 call options is $1000\,C^E \cong 1000 \cdot 1.25603 = \$1,256.03$. The delta of one

European call is $\Delta = \mathcal{N}(d_+) \cong 0.54024$. Therefore, to construct the hedge, the investor buys $1000\,\Delta = 540.24$ stock shares for $50 \cdot 540.24 = \$27,012$ dollars. To cover expenses, the investor borrows $\$27,012 - \$1,256.03 = \$25,755.97$. So the hedging portfolio consists of $x = 540.24$ stock shares, $y = -25,755.97$ dollars in the form of a risk-free loan, and $z = -1000$ short call options. The time-0 value of the portfolio is zero. $\qquad\square$

4.3.4.4 Greeks

As was demonstrated in the previous section, the delta of an option is given by the derivative of the option price w.r.t. the current price S of the underlying. The delta of an option is the sensitivity of the option price to a change in the price of the underlying security. Similarly, we can introduce other so-called *Greek parameters* that measure the sensitivity of a derivative such as an option (or a portfolio of securities) w.r.t. a small change in a given underlying parameter. In the log-normal model, European put and call options are characterized by four parameters: the current underlying price S, current time t, interest rate r, and volatility σ (note that the strike price K and exercise time T are fixed once the option is written). Consider a portfolio consisting of the underlying security and some European-type derivatives based on this security. The most commonly used Greeks for such a portfolio with the value function $V = V(t, S; \sigma, r)$ are described just below.

delta$_V = \frac{\partial V}{\partial S}$ is the sensitivity of the portfolio value to a change in the price of the underlying security.

gamma$_V = \frac{\partial^2 V}{\partial S^2}$ is the sensitivity of the portfolio's delta to a change in the price of the underlying security.

theta$_V = \frac{\partial V}{\partial t}$ is the sensitivity of the portfolio value to a *negative* change in time to maturity, $T - t$.

vega$_V = \frac{\partial V}{\partial \sigma}$ is the sensitivity of the portfolio value to a change in volatility.

rho$_V = \frac{\partial V}{\partial r}$ is the sensitivity of the portfolio value to the interest rate.

For small changes δS, δt, $\delta\sigma$, δr of the respective variables, the change in value of the portfolio value is approximated by using Taylor's Theorem:

$$\delta V = V(t + \delta t, S + \delta S; \sigma + \delta\sigma, r + \delta r) - V(t, S; \sigma, r)$$
$$\cong \frac{\partial V}{\partial S}\delta S + \frac{1}{2}\frac{\partial^2 V}{\partial S^2}(\delta S)^2 + \frac{\partial V}{\partial t}\delta t + \frac{\partial V}{\partial \sigma}\delta\sigma + \frac{\partial V}{\partial r}\delta r$$
$$= \text{delta}_V\,\delta S + \frac{1}{2}\,\text{gamma}_V\,(\delta S)^2 + \text{theta}_V\,\delta t + \text{vega}_V\,\delta\sigma + \text{rho}_V\,\delta r.$$

To immunize the portfolio against small changes of a selected variable, we let the corresponding Greek be equal to zero. The resulting portfolio is said to be *neutral* relative to the Greek selected. In particular, a *delta-neutral* portfolio is protected against small changes of the underlying security price; a *vega-neutral* portfolio is not sensitive to volatility movements.

The delta-gamma approximation

$$\delta V \cong \text{delta}_V\,\delta S + \frac{1}{2}\,\text{gamma}_V\,(\delta S)^2$$

is often used in historical Value-at-Risk (VaR) calculations for portfolios that include options.

The Black–Scholes formula (4.34) allows us to evaluate the Greeks explicitly for a single European call option in the same manner as it was done for the delta:

$$\text{delta}_{C^E} = \mathcal{N}(d_+)$$

$$\text{gamma}_{C^E} = \frac{1}{S\sigma\sqrt{T-t}}\mathcal{N}'(d_+)$$

$$\text{theta}_{C^E} = -\frac{S\sigma}{2\sqrt{T-t}}\mathcal{N}'(d_+) - rKe^{-r(T-t)}\mathcal{N}(d_-)$$

$$\text{vega}_{C^E} = S\sqrt{T-t}\mathcal{N}'(d_+)$$

$$\text{rho}_{C^E} = (T-t)Ke^{-r(T-t)}\mathcal{N}(d_-)$$

where $d_\pm$ is given in (4.36). From the above formulae, one can obtain the following relation:

$$\text{theta}_{C^E} + rS\,\text{delta}_{C^E} + \frac{\sigma^2 S^2}{2}\,\text{gamma}_{C^E} - rC^E = 0.$$

Therefore, the price $C \equiv C^E$ of a European call (as well as the price of any European-style derivative security) satisfies the following partial differential equation (PDE) known as the *Black–Scholes PDE*:

$$\frac{\partial C}{\partial t} + \frac{1}{2}\sigma^2 S^2 \frac{\partial^2 C}{\partial S^2} + rS\frac{\partial C}{\partial S} - rC = 0. \tag{4.39}$$

In Volume II, we will see that equation (4.39) is derived for a continuous-time asset price model by using a self-financing portfolio replication strategy and a no-arbitrage argument.

The Greeks for put options can be calculated in the same manner by differentiating the expression in (4.35) or via the put-call parity. Given European call and put options for the same underlying, strike price, and time to maturity, and with no dividend yield, the sum of the absolute values of the call and put deltas is equal to 1. This is due to the put-call parity: a long call plus a short put (a call minus a put) replicates a forward whose delta is equal to 1. Indeed, let us differentiate the put-call parity (4.5) w.r.t. S:

$$C^E(S) - P^E(S) = S - Ke^{-rT},$$

$$\frac{\partial}{\partial S}C^E(S) - \frac{\partial}{\partial S}P^E(S) = \frac{\partial}{\partial S}S - \frac{\partial}{\partial S}Ke^{-rT},$$

$$\text{delta}_{C^E} - \text{delta}_{P^E} = 1.$$

Therefore, the delta of a European put is

$$\text{delta}_{P^E} = \text{delta}_{C^E} - 1 = -(1 - \mathcal{N}(d_+)) = -\mathcal{N}(-d_+).$$

4.3.5 Black–Scholes Equation

The log-normal pricing model is obtained as a limiting case of binomial tree models when the number of periods approaches infinity. In previous sections, we derived the Black–Scholes pricing formulae for European call and put options as the limit of the respective binomial pricing formulae. Let us now demonstrate how the Black–Scholes PDE (4.39) can be deduced from a single period recurrence relationship for the option value when the length of a period converges to zero. Note that the Black–Scholes equation will also be derived based on the replication argument in Volume II, where all required tools from stochastic calculus will be introduced as well.

Consider a European option with expiration time T and payoff function $\Lambda(S)$. Let $V(t, S)$ denote the option value at time $t \in (0, T]$ for the current stock price S. Assume that the stock price process follows the log-normal model. Let σ be the annual volatility of the underlying asset, and r be the continuously compounded risk-free rate of interest. On a small time interval $[t - \delta, t]$ with $\delta = \frac{1}{n}$, where n is the number of periods per year, the continuous-time model can be approximated by a single-period binomial model with factors $u_n = e^{\sigma\sqrt{\delta}}$ and $d_n = e^{-\sigma\sqrt{\delta}}$, and rate $r_n = e^{r\delta} - 1$. Additionally, suppose the stock pays dividends continuously at a rate (yield) of q. The dividend payment accumulated during the time period of length δ is approximately equal to $100(1 - e^{-q\delta})\%$ of the current stock price. Let S be the stock price just before the dividend payment. Then, according to the no-arbitrage principle, the asset price just after the dividend payment is equal to $Se^{-q\delta}$. The final risk-neutral dynamics of the stock price in a single-period binomial period is illustrated in Figure 4.18

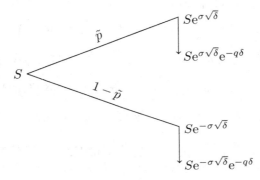

FIGURE 4.18: A single-period binomial model for a stock with continuous dividends.

The single-period binomial approximation of the option price (on the interval $[t - \delta, t]$) can be written as follows:

$$e^{r\delta}V(t - \delta, S) = \tilde{p}_n V\left(t, Se^{\sigma\sqrt{\delta}-q\delta}\right) + (1 - \tilde{p}_n) V\left(t, Se^{-\sigma\sqrt{\delta}-q\delta}\right), \qquad (4.40)$$

where $\tilde{p}_n = \frac{1+r_n-d_n}{u_n-d_n}$. Let us find expansions of the risk-neutral probabilities $\tilde{p}_n$ and $1 - \tilde{p}_n$ for small δ. Using the truncated Maclaurin expansion of the exponential function,

$$e^x = 1 + x + \frac{x^2}{2} + \frac{x^3}{6} + \frac{x^4}{24} + \mathcal{O}(x^5),$$

we obtain

$$u_n = e^{\sigma\sqrt{\delta}} = 1 + \sigma\delta^{1/2} + \frac{\sigma^2}{2}\delta + \frac{\sigma^3}{6}\delta^{3/2} + \frac{\sigma^4}{24}\delta^2 + \mathcal{O}(\delta^{5/2}),$$

$$d_n = e^{-\sigma\sqrt{\delta}} = 1 - \sigma\delta^{1/2} + \frac{\sigma^2}{2}\delta - \frac{\sigma^3}{6}\delta^{3/2} + \frac{\sigma^4}{24}\delta^2 + \mathcal{O}(\delta^{5/2}),$$

and hence

$$u_n - d_n = 2\sigma\delta^{1/2}\left(1 + \frac{\sigma^2}{6}\delta + \mathcal{O}(\delta^2)\right) \implies \frac{1}{u_n - d_n} = \frac{\delta^{-1/2}}{2\sigma}\left(1 - \frac{\sigma^2}{6}\delta + \mathcal{O}(\delta^2)\right),$$

where we also use the geometric series $\frac{1}{1+a\delta+\mathcal{O}(\delta^2)} = 1 - a\delta + \mathcal{O}(\delta^2)$ with $0 < a\delta < 1$.

Therefore, the risk-neutral probabilities are given by

$$\tilde{p}_n = \frac{\delta^{-1/2}}{2\sigma} \left((1 + r\delta) - (1 - \sigma\delta^{1/2} + \mathcal{O}(\delta)) \right) \left(1 - \frac{\sigma^2}{6}\delta + \mathcal{O}(\delta^2) \right)$$

$$= \frac{1}{2\sigma} \left(\sigma + (r - \sigma^2/2)\delta^{1/2} + \frac{\sigma^3}{6}\delta + \mathcal{O}(\delta^{3/2}) \right) \left(1 - \frac{\sigma^2}{6}\delta + \mathcal{O}(\delta^2) \right)$$

$$= \frac{1}{2} + \left(r - \frac{\sigma^2}{2} \right) \frac{\delta^{1/2}}{2\sigma} + \mathcal{O}(\delta^{3/2}), \tag{4.41}$$

$$1 - \tilde{p}_n = \frac{1}{2} - \left(r - \frac{\sigma^2}{2} \right) \frac{\delta^{1/2}}{2\sigma} + \mathcal{O}(\delta^{3/2}). \tag{4.42}$$

Now, apply Taylor's formula to the option value function to obtain

$$V(t - \delta, S) = V - V_t\delta + \mathcal{O}(\delta^2), \tag{4.43}$$

$$V(t, Se^{\sigma\sqrt{\delta} - q\delta}) = V\left(t, S + (\sigma\sqrt{\delta} - q\delta + \sigma^2\delta/2 + \mathcal{O}(\delta^{3/2}))S \right)$$

$$= V + \left(\sigma\sqrt{\delta} - q\delta + \sigma^2\delta/2 \right) SV_S + (\sigma^2\delta/2)S^2V_{SS} + \mathcal{O}(\delta^{3/2}), \tag{4.44}$$

$$V(t, Se^{-\sigma\sqrt{\delta} - q\delta}) = V\left(t, S + (-\sigma\sqrt{\delta} - q\delta + \sigma^2\delta/2 + \mathcal{O}(\delta^{3/2}))S \right)$$

$$= V + \left(-\sigma\sqrt{\delta} - q\delta + \sigma^2\delta/2 \right) SV_S + (\sigma^2\delta/2)S^2V_{SS} + \mathcal{O}(\delta^{3/2}), \tag{4.45}$$

where $V \equiv V(t,S)$, $V_t \equiv \frac{\partial V(t,S)}{\partial t}$, $V_S \equiv \frac{\partial V(t,S)}{\partial S}$, and $V_{SS} \equiv \frac{\partial^2 V(t,S)}{\partial S^2}$. Substituting (4.41)–(4.42) and (4.43)–(4.45) in (4.40) gives

$$(1 + r\delta + \mathcal{O}(\delta^2)) \ (V - \delta V_t + \mathcal{O}(\delta^2))$$

$$= \Bigg[\left(\frac{1}{2} + \left(r - \frac{\sigma^2}{2} \right) \frac{\delta^{1/2}}{2\sigma} + \mathcal{O}(\delta^{3/2}) \right)$$

$$\times \left(V + \left(\sigma\sqrt{\delta} - q\delta + \frac{\sigma^2\delta}{2} \right) SV_S + \frac{\sigma^2\delta}{2}S^2V_{SS} + \mathcal{O}(\delta^{3/2}) \right)$$

$$+ \left(\frac{1}{2} - \left(r - \frac{\sigma^2}{2} \right) \frac{\delta^{1/2}}{2\sigma} + \mathcal{O}(\delta^{3/2}) \right)$$

$$\times \left(V + \left(-\sigma\sqrt{\delta} - q\delta + \frac{\sigma^2\delta}{2} \right) SV_S + \frac{\sigma^2\delta}{2}S^2V_{SS} + \mathcal{O}(\delta^{3/2}) \right) \Bigg].$$

Simplify the above expression and collect terms of the same magnitude (w.r.t. δ) to obtain

$$V - V_t\delta + rV\delta + \mathcal{O}(\delta^2)$$

$$= V + \underbrace{\left(\frac{\sigma}{2}SV_S - \frac{\sigma}{2}SV_S + \left(r - \frac{\sigma^2}{2} \right) \frac{V}{2\sigma} - \left(r - \frac{\sigma^2}{2} \right) \frac{V}{2\sigma} \right)}_{=0} \sqrt{\delta}$$

$$+ \underbrace{\left(\left(r - \frac{\sigma^2}{2} \right) SV_S + \left(\frac{\sigma^2}{2} - q \right) SV_S + \frac{\sigma^2}{2}\delta S^2V_{SS} \right)}_{=(r-q)SV_S + \sigma^2 S^2/2 V_{SS}} \delta + \mathcal{O}(\delta^{3/2}).$$

Cancel V on both sides and divide both sides by δ to obtain

$$V_t + \frac{\sigma^2}{2}S^2V_{SS} + (r - q)SV_S - rV + \mathcal{O}(\sqrt{\delta}) = 0.$$

Now, taking a limit as $\delta \to 0$ gives the Black–Scholes equation

$$\frac{\partial V}{\partial t} + \frac{\sigma^2}{2} S^2 \frac{\partial^2 V}{\partial S^2} + (r - q)S\frac{\partial V}{\partial S} - rV = 0 \tag{4.46}$$

where $V = V(t, S)$ for $0 \leqslant t \leqslant T$ and $S > 0$. The equation is accompanied by the terminal condition at maturity:

$$V(T, S) = \Lambda(S) \text{ for } S > 0.$$

4.4   Exercises

Exercise 4.1. Consider a 12-month long forward contract on a stock currently priced at $90. The risk-free interest rate is 7% per annum, compounded continuously. Suppose that dividend payments of $8 per share are expected after four months and $7 after eight months.

(a) Determine the forward price at time 0.

(b) What is the value of this forward contract six months from now if the stock price at that time is $95?

(c) Suppose that the value of this forward contract six months from now is $5. Determine whether there is an arbitrage opportunity. If one exists in this situation, construct an arbitrage portfolio and find the profit realized.

Exercise 4.2. A manufacture of heavy electrical equipment wishes to take the long side of a forward contract for delivery of copper in six months. The current price of copper is $269.53 per 100 pounds, and six Month Treasury Bill Rate is at 0.65%. What is the appropriate forward price of the copper contract?

Exercise 4.3. The current price of sugar is 12 cents per pound. We wish to find the forward price of sugar to be delivered in five months. The carrying cost of sugar is 1 cent per pound per month, to be paid at the beginning of the month, and the rate of interest compounded monthly is constant at 9% per annum.

Exercise 4.4. Consider a coupon bond with a face of $10,000, a semi-annual coupon of $c^{(2)} = 8\%$, and several years to maturity. Currently, this bond is selling for $9,620, and the previous coupon has just been paid. What is the forward price for delivery of this bond in one year? Assume that the rate of interest compounded semi-annually is constant at 9%.

Exercise 4.5. Corn currently costs $200 per ton, and it costs $30 per month (paid at the end of month) to store one ton of corn. The risk-free rate of interest is 1% per annum, compounded continuously. You enter a one-year long forward contract at the fair delivery price of $563.67.

(a) Verify that the fair delivery price is indeed $563.67.

(b) How high must the spot price of corn be in six months, in order for the contract to have a positive value at that time?

Exercise 4.6. Microsoft stock is currently trading at $41.54 and the risk-free rate of interest is 3% per annum, compounded continuously.

(a) Assuming Microsoft is not scheduled to pay any dividends over the next six months, what is the fair delivery price on a six-month forward contract for Microsoft stock?

(b) Assuming Microsoft is due to pay a dividend of $3 two months from now, what is the fair delivery price for Microsoft stock with delivery in six months?

(c) What is the value of the long contract from (b) one month from now, assuming Microsoft is trading at $42 at that time?

(d) What is the value of the long contract from (b) three months from now, assuming Microsoft is trading at $38 at that time?

Exercise 4.7. A commodity is currently trading at $4600. Assume that the risk-free rate of interest is 4.5% compounded monthly and there is no carrying cost.

(a) Find the forward price of the commodity for delivery in 12 months from now.

(b) Suppose that in two months from now the interest rate increased by 0.5% and the commodity price decreased by $100. Find the no-arbitrage value of the long forward contract from (a).

Exercise 4.8. A treasury bond has a face value of $14000, semi-annual coupons paid at the annual rate of 3.5%, and several years to maturity. Currently this bond is selling for $8350, and the previous coupon has just been paid. Assuming that the risk-free interest rate of 4% is compounded continuously, find the forward price for delivery of this bond in six years.

Exercise 4.9. A commodity is currently trading at $400. The carrying cost is $10 per month paid at the beginning of each month. Assume that the risk-free rate of interest is 6% compounded monthly.

(a) Find the forward price of the commodity for delivery in nine months assuming that today is the beginning of a month, and the carrying cost payment has not been made yet.

(b) Find the forward price of the commodity one month from now if the asset is traded at $450 at that time. What is the value of the long contract from (a)?

Exercise 4.10. As of January 1, the price of a stock is $190. A dividend payment of $2.5 is made on each of April 1, July 1, and October 1. Let the risk-free continuously compounded interest rate be 8%. Kate believes the price of the stock is going to increase, and, therefore, she takes a long position in a one-year forward contract on the stock.

(a) Find the forward price of the stock for delivery in one year.

(b) On June 1, the stock price has risen to $255. What is the current fair value of the forward contract initiated on January 1?

(c) On June 1, Kate feels that now is the time to cash out. Explain how she can use a second forward contract (issued on June 1) to lock in a risk-free profit. Find the risk-free profit realized on January 1 next year.

(d) In fact, Kate did not enter a second forward on June 1. On September 1, the stock price has fallen to $155. She is now concerned that the stock price would keep falling. Explain how she can use a second forward contract (issued on September 1) to lock in her loss. Find the loss realized on January 1 next year.

Exercise 4.11. Suppose that the stock pays dividends continuously at a rate of $q > 0$ and the dividends are reinvested in the stock. An investment in one share held at time 0 will increase to become e^{qT} shares at time $T > 0$. Therefore, we need to start with e^{-qT} shares at time 0 to obtain 1 share at time T. Assuming that interest is compounded continuously at rate r, show that the no-arbitrage T-year forward price at time 0 is

$$F(0,T) = S(0)e^{(r-q)T}. \tag{4.47}$$

Exercise 4.12. Prove the formula from Theorem 4.4 for the no-arbitrage value of a long forward contract initiated at time $t = 0$ and having delivery time T and delivery price K:

$$V_K(t) = (F(t,T) - K) D(T - t).$$

Exercise 4.13. If the interest rate r compounded continuously increases, then how will the present value of a long forward contract on a stock with spot value S_0 and delivery price K change? Estimate the change in value of the contract.

Exercise 4.14. Suppose that the lending rate r_ℓ and the borrowing rate r_b are different for investors so that $r_\ell < r_b$.

(a) Explain why it is mutually beneficial to both parties to enter the forward contract on some asset with spot price S_0 if and only if the fair deliver price $F \in [S_0 e^{r_\ell T}, S_0 e^{r_b T}]$. In other words, show that any no-arbitrage forward price $F(0, T)$ satisfies

$$S_0 e^{r_\ell T} \leqslant F(0, T) \leqslant S_0 e^{r_b T}.$$

Assume that there are no recurring payments associated with the asset.

(b) Explain how you could exploit a forward price F such that $F > S_0 e^{r_b T}$.

(c) Explain how you could exploit a forward price F such that $F < S_0 e^{r_\ell T}$.

Exercise 4.15. In this problem we consider the more realistic setting where people are able to borrow at rate r_b and lend at rate r_ℓ, where $r_b > r_\ell$. In Exercise 4.14, we show that when borrowing and lending rates differ there is no longer a unique fair delivery price, rather there is an interval of fair delivery prices. Throughout the problem we consider an asset whose current spot price is S_0 and for which storage is free. In addition, let F denote the delivery price on a forward with delivery in T years.

(a) Paul owns one unit of the asset and has no use for it. He is considering either (i) selling it now on the spot market or (ii) selling it in T years via a forward contract. By comparing the proceeds associated with each option in T-year dollars, determine those values of F that would lead him to choose option (ii).

(b) Suppose that Jillian does not currently own any units of the asset and knows that she will need one unit in T years. She is considering either (i) buying it now in the spot market or (ii) buying it in T years via a forward contract. By comparing the cost of each option in today's dollars, determine those values of F that would lead her to choose option (ii)?

Exercise 4.16. Oil is currently trading at $100 per barrel and storage is free. You are able to borrow and lend at 5% and 3%, respectively (both rate are annual rates with continuous compounding).

(a) How would you exploit a six-month forward price for oil of $103 per barrel?

(b) How would you exploit a six-month forward price for oil of $101 per barrel?

Exercise 4.17. Suppose that the forward price of the Treasury bond from Exercise 4.4 is equal to $9,200. Find an arbitrage opportunity.

Exercise 4.18. Let the initial asset price be $100 and the risk-free rate be $r = 5\%$ compounded continuously. A long forward contract with delivery time $T = 1$ is exchanged at time 0.

(a) Find the forward price $F(0, 1)$.

(b) What is the value $V(0.5)$ of this long forward contract at time 0.5, if the current spot price is $110?

Exercise 4.19. One ounce of gold currently costs $1,200, and storing gold costs nothing. The interest rate is 2.5% per annum, compounded continuously. Blake believes the price of gold is going to go up, and takes a long position in a one-year forward contract on gold in order to (hopefully) profit from this view. Assume the contracted delivery price is the fair price.

(a) Verify that the fair delivery price is $1,230.38.

(b) Suppose that eight months before the delivery time the price of gold has risen to $1,500 per ounce, and Blake feels that now is the time to cash out. Explain how he can use a second forward contract to lock in a risk-free profit of $294.83 (the profit being realized on the delivery date).

(c) Suppose that eight months before the delivery time the price of gold has fallen to $900, and Blake is beginning to lose sleep. Explain how he can use a second forward contract to limit his loss to $315.25 (the loss being realized eight months on the delivery date).

Exercise 4.20. Consider a 10-month forward contract on stock with a price of $500. Assume that the risk-free rate of interest is $r^{(12)} = 6\%$.

(a) Find the forward price $F = F\left(0, \frac{10}{12}\right)$.

(b) Find an arbitrage opportunity if $F = \$510$.

(c) Find an arbitrage opportunity if $F = \$560$.

Exercise 4.21. Prove by an arbitrage argument that $P^E(K)$ is a nondecreasing function of strike K. That is, if $K' < K''$, then $P^E(K') \leqslant P^E(K'')$ holds.

Exercise 4.22. Prove by an arbitrage argument that if the stock pays dividends whose discounted value is $\mathrm{div}(0, T)$, then

$$\max\{S(0) - Ke^{-rT} - \mathrm{div}(0,T), S(0) - K\} \leqslant C^A,$$
$$\max\{Ke^{-rT} + \mathrm{div}(0,T) - S(0), K - S(0)\} \leqslant P^A.$$

Exercise 4.23. Suppose that the stock pays dividends between time 0 and the expiry time T, whose discounted value (at time 0) is $\mathrm{div}(0, T)$. Show that

$$S(0) - \mathrm{div}(0,T) - K \leqslant C^A - P^A \leqslant S(0) - Ke^{-rT}.$$

Exercise 4.24. Show that if the stock pays dividends continuously at a rate q, then

$$S(0)e^{-qT} - K \leqslant C^A - P^A \leqslant S(0) - Ke^{-rT}.$$

Exercise 4.25. Prove by an arbitrage argument that $P^A(S)$ is a nonincreasing function of the current price $S = S(0)$. That is, if $S' < S''$, then $P^A(S') \geqslant P^A(S'')$ holds.

Exercise 4.26. Assume the spot is S_0, and the simple interest rate is r. Show that the call and put options both struck at $(1 + rT)S_0$ and both expiring at time $t = T$ are of equal value at present time $t = 0$.

Exercise 4.27. Investor A sells a put option for $9.00, and investor B sells a call option for $12.07. Both options have the same strike price of $42 and can be exercised in 13 months. Suppose the stock price on the exercise date is $38, and the continuously compounded interest rate is 7%.

(a) What is the total profit of investor A on the exercise date?

(b) What is the total profit of investor B on the exercise date?

Exercise 4.28. Investor A buys a put option for $2.80, and investor B buys a call option for $7.04. Both options have the same strike price of $58 and can be exercised in 13 months. Suppose the stock price on the exercise date is $62, and the continuously compounded interest rate is 7%.

(a) What is the total profit of investor A on the exercise date?

(b) What is the total profit of investor B on the exercise date?

Exercise 4.29. Suppose that you sell for $6 each two call options with a strike price of $64, and buy for $11 each six put options with a strike price of $72. Suppose all options have the same expiration date of six months, and the continuously compounded interest rate is 3%. What is your profit on the exercise date if the stock price on the exercise date is $51?

Exercise 4.30. Consider a European put option with strike $K = \$47$ and expiry $T =$ one month on a stock that follows a binomial model. The current spot price is $44. On the exercise date, the stock price is $49 with probability 30%, or $34 with probability 70%. Let the risk-free rate of interest compounded continuously be 2%.

(a) Find the no-arbitrage value of this put option.

(b) Find the expected gain to a buyer of this put option on the exercise date.

Exercise 4.31. Consider a European call option with strike $K = \$42$ and expiry $T =$ nine months on a stock that follows a trinomial model. On the exercise date, the stock price is: $53 with probability 0.1, $48 with probability 0.4, or $38 with probability 0.5. If the call has a premium of $3.1, and the risk-free rate of interest compounded continuously is 5%, what is the expected gain to a buyer of this call option on the exercise date?

Exercise 4.32. (a) A stock currently sells for $33.15. A six-month call option with a strike price of $35 has a premium of $5.8. Let the continuously compounded risk-free rate be 3%. What is the price of an associated six-month put option with the same strike?

(b) A stock currently sells for $32.1. A six-month call option with a strike price of $31.65 has a premium of $2.24, and a six-month put with the same strike has a premium of $0.79. Let the continuously compounded risk-free rate be 4%. What is the present value of dividends payable over the next six months?

Exercise 4.33. Suppose that you sell for $10 a call option with a strike price of $73, and you purchase for $52 a share of the stock. Assume a zero rate of interest.

(a) What is the maximum profit possible on the exercise date?

(b) What is the stock price at the exercise date that will result in you breaking even?

Exercise 4.34. Suppose that you sell for $15 a call option with a strike price of $49, sell for $7 a call option with a strike price of $59, and buy for $10 each two call options with a strike price of $54. Assume a zero rate of interest.

(a) What is the maximum profit possible on the exercise date?

(b) What is the maximum loss possible on the exercise date?

(c) What is the maximum stock price at the exercise date that will result in you breaking even?

Exercise 4.35. Consider a stock with an initial price of $75. Suppose that the risk-free rate of interest compounded continuously is 4%. The table below contains no-arbitrage prices of three European call options that will expire in 11 months.

Strike price	$69	$75	$81
Call Value	$15.25	$12.31	$9.86

Find no-arbitrage values of the following European-style derivatives on the same asset with the same expiry date of 11 months

(a) A standard European put option with strike $K = \$75$.

(b) A bull spread with payoff $\Lambda(S) = \begin{cases} 0 & S < 75\,, \\ \frac{3.5}{6} \cdot (S - 75) & 75 \leqslant S \leqslant 81\,, \\ 3.5 & S > 81\,. \end{cases}$

(c) A bear spread with payoff $\Lambda(S) = \begin{cases} 2.5 & S < 69\,, \\ \frac{2.5}{12} \cdot (81 - S) & 69 \leqslant S \leqslant 81\,, \\ 0 & S > 81\,. \end{cases}$

(d) A butterfly spread with payoff $\Lambda(S) = \begin{cases} 0 & S < 69\,, \\ 3.5 \cdot (75 - S) & 69 \leqslant S \leqslant 75\,, \\ 3.5 \cdot (S - 81) & 75 \leqslant S \leqslant 81\,, \\ 0 & S > 81\,. \end{cases}$

Exercise 4.36. I currently own one share of XYZ stock that is currently trading at $192.96. I do not expect the price to fall substantially over the near future, but I do realize that it is a possibility. As such I am considering selling a bit of my stock and using the proceeds to purchase a three-month put option struck at $175 (currently trading at $2.49). Assume that I can buy/sell fractions of stocks and puts.

(a) How many shares do I need to sell in order to purchase exactly one put?

(b) Suppose that I sell enough shares to purchase exactly one put. Show that if V_T is the value of my portfolio in three months, then

$$V_T = \begin{cases} 175 - 0.0129 S_T & S_T \leqslant 175\,, \\ 0.9871 S_T & S_T > 175\,. \end{cases}$$

(c) Sketch V_T as a function of S_T.

(d) For what values of S_T is $V_T > S_T$?

(e) When a person who owns a stock purchases a put on that stock, we say they have purchased a *protective put*. Explain this terminology.

Exercise 4.37. Sketch the payoff to each of the following portfolios, as a function of the stock price at expiry.

(a) Short position in one call option struck at K_1, long position in two call options struck at $K_2 > K_1$. This is called a *backspread*.

(b) Long position in one stock, long position in one put option struck at K_1, short position in one call option struck at $K_2 > K_1$. This is called a *collar*.

Exercise 4.38. Three-month call options on XYZ stock, struck at $195, are currently trading at $6.15. XYZ stock is currently trading at $192.96 and the three-month interest rate is 1.2% compounded continuously.

(a) How much should a three-month put, struck at $195, cost?

(b) How could you exploit a three-month put at $195 that was trading at $12? Be specific.

(c) How could you exploit a three-month put at $195 that was trading at $5? Be specific.

Exercise 4.39. An option is said to be at-the-money if the strike price is equal to the current stock price. Prove that an at-the-money call should always be more valuable than an at-the-money put with the same maturity. [*Hint*: Use put-call parity.]

Exercise 4.40. Find and plot payoff functions for a straddle, strip, strap, and strangle. Assume a long position is taken.

Exercise 4.41. Consider a single-period binomial model from Example 4.10. That is, $S_0 = \$100$, $B_0 = \$1$, $d = 0.9$, $u = 1.15$, and $r = 5\%$. Replicate the put option with strike price $K = \$95$ and expiry time $T = 1$. Find the option price.

Exercise 4.42. Assume a single-period economy with two states. The price of the forward contract struck at K can be shown to be given by $V_t = V_t(S_t, K) = S_t - (1 + r(T-t))^{-1}K$, where S_t is the stock price at time $t \in \{0, T\}$. Find a replicating portfolio in the stock and the zero-coupon bond, with nominal value assumed as $B_T = 1$ for the bond.

Exercise 4.43. A stock is worth $200 today and will be worth either $190 or $220 tomorrow with equal probability. Assuming a zero rate of interest, find the prices of one-day call options struck at $190, $200, and $220, respectively.

Exercise 4.44. A market model consists of a risky stock with the initial price of $140 and a risk-free bond with the initial value of $150. Suppose that, in four months, the stock price will be $S^+ = \$155$ or $S^- = \$115$, and the bond will be worth $163.50. Consider a put option with expiry $T = $ four months and strike $K = \$123$.

(a) Find a portfolio replicating the contract and state.

(b) Find the fair value of the put option by calculating the initial value of the replication portfolio.

Exercise 4.45. Let the asset price process $\{S_t\}_{t=0,1,\ldots,T}$ follow the binomial tree model with the parameters $S_0 = \$50$, $u = 1.2$, $d = 0.9$. Suppose that the risk-free interest rate is $r = 0.1$.

(a) Find the values of the replicating portfolios for a put and a call with strike price $K = \$55$ and exercise time $T = 1$.

(b) Compute the values of the put and call options from (a) using with the replicating portfolios constructed in (a).

(c) Compute the value of a European derivative security with the payoff function

$$\Lambda(S) = \begin{cases} 0 & \text{if } S \leqslant 40 \text{ or } S \geqslant 60, \\ S - 40 & \text{if } 40 < S < 50, \\ 60 - S & \text{if } 50 \leqslant S < 60. \end{cases}$$

and exercise time $T = 2$ using the risk-neutral pricing formula.

Exercise 4.46. Compute the price of an American put with strike price $K = \$13$ expiring at time $T = 2$ on a stock with $S(0) = \$12$ in a binomial tree model with $u = 1.1$, $d = 0.95$, and $r = 0.03$.

Exercise 4.47. Derive the Black–Scholes formula (4.33) for the price of a European call option by calculating the risk-neutral expectation:

$$e^{-rT}\widetilde{\mathrm{E}}[(S(T) - K)^+] = e^{-rT}\widetilde{\mathrm{E}}\left[\left(S(0)e^{(r-\frac{1}{2}\sigma^2)T + \sigma\sqrt{T}Z} - K\right)^+\right].$$

Exercise 4.48. Prove the formulae for the Greeks of a European call option.

Exercise 4.49. Derive the formulae for the Greeks of a European put option.

Exercise 4.50. The share-or-nothing call and put options with strike K at time T have the respective payoff functions $f(S) = S\mathbb{I}_{\{S>K\}}$ and $f(S) = S\mathbb{I}_{\{S<K\}}$.

(a) Assuming the Black–Scholes model, find the non-arbitrage prices of the share-or-nothing put and call options.

(b) Find the put-call parity for the share-or-nothing options.

Exercise 4.51. Consider the Black–Scholes price formula for the price $C^E = C^E(t, S; \sigma)$ of a vanilla European call on a stock paying dividends. Find the following limits:

(a) $\lim_{S\to 0+} C^E$, (b) $\lim_{S\to\infty} C^E$, (c) $\lim_{t\to T-} C^E$, (d) $\lim_{\sigma\to 0+} C^E$, (e) $\lim_{\sigma\to\infty} C^E$.

Exercise 4.52. Consider the log-normal asset price model with $S_0 = \$80$, $\sigma = 30\%$, and $r = 6\%$. Construct (a) a delta-gamma neutral portfolio and (b) a delta-rho neutral portfolio to hedge a short position on 500 calls expiring after 90 days with strike price \$80, taking as an additional component a call option expiring after 120 days with strike price \$85. Find the changes in value of these portfolios after five days if one of the following three scenarios will happen: (i) the stock drops to \$75; (ii) the interest rate jumps to 9%; (iii) the stock drops to \$75 and the interest rate jumps to 9%.

Exercise 4.53. In the framework of the log-normal asset price model consider a strangle option with the payoff function

$$\Lambda(S) = \begin{cases} K_1 - S & \text{if } S \leqslant K_1, \\ 0 & \text{if } K_1 < S < K_2, \\ S - K_2 & \text{if } S \geqslant K_2. \end{cases}$$

(a) Plot the payoff function.

(b) Find a portfolio consisting of standard European calls and puts that replicates the payoff.

(c) Find the arbitrage-free price by using the Black–Scholes price formulae for vanilla options.

(d) Find the delta.

Exercise 4.54. Assume the interest rate is zero and also assume that both call and put options with respective values $C_t(S_t, K)$ and $P_t(S_t, K)$ are differentiable functions of the strike K.

(a) Derive a relation between $\partial C_t(S_t, K)/\partial K$ and $\partial P_t(S_t, K)/\partial K$ for $t = 0$ and $t = T$.

(b) Find the values of these partial derivatives at maturity $t = T$.

[Hint: Use the put-call parity.]

Exercise 4.55. Using a similar integration procedure as was shown for the call option price, provide a complete derivation of the Black–Scholes pricing formula of a European put under the log-normal model. Assume strike K, spot S, time to maturity T, interest rate r, and volatility σ.

Exercise 4.56. A European *binary* option is a so-called "all-or-nothing" claim on an underlying asset. For example, one share of a *cash-or-nothing binary call* has a payoff of exactly one dollar if the asset price ends up above the strike and zero otherwise, i.e., the payoff function is

$$\Lambda(S) = \mathbb{I}_{\{S \geqslant K\}} \equiv \begin{cases} 1 & \text{if } S \geqslant K, \\ 0 & \text{if } S < K. \end{cases}$$

Similarly, a *cash-or-nothing binary put* has payoff $\mathbb{I}_{\{S < K\}}$. Assume the log-normal model for asset prices.

(a) Derive the (Black–Scholes) exact pricing formulas for both the binary call $C(S, T)$ and the put $P(S, T)$ in terms of the spot S and time to maturity T.

(b) Give the relationship between the binary put and call price where both options have the same strike K and maturity T.

(c) Derive the exact formula for the Greek delta of the binary put and call: $\Delta_c = \partial C/\partial S$ and $\Delta_p = \partial P/\partial S$.

Exercise 4.57. Consider the European-style option with payoff

$$\Lambda(S) = \mathbb{I}_{\{K_1 \leqslant S \leqslant K_2\}} \equiv \begin{cases} 1 & \text{if } K_1 \leqslant S \leqslant K_2, \\ 0 & \text{otherwise,} \end{cases}$$

where $0 < K_1 < K_2$. Assume the log-normal model for stock prices.

(a) Find the option value as a function of spot S, strikes K_1 and K_2, expiration time T, rate r and volatility σ. [Hint: You may decompose the payoff in terms of binary options with appropriate indicator functions.]

(b) Derive formulas for the following sensitivities of the option value in (a): $\Delta = \partial V/\partial S$, $\Gamma = \partial^2 V/\partial S^2$, and $\Theta = \partial V/\partial T$.

Exercise 4.58. ABC stock is currently trading at \$880.30, and historically its volatility has been 15%. The risk-free rate of interest for any maturity is 1.5% compounded continuously.

(a) According to the Black–Scholes model, what is the value of a six-month call option on ABC struck at \$200? You may use the fact that $\mathcal{N}(x) \approx 1$ for $x \geqslant 4$.

(b) What is the value of an at-the-money option on ABC expiring in six months?

(c) Which is more sensitive (in relative terms) to changes in the ABC price—an at-the-money call or an out-of-the-money call struck at \$900?

(d) Which is more sensitive (in relative terms) to changes in the ABC price—an at-the-money call expiring in one month or an at-the-money call expiring in six months?

(e) XYZ stock is also trading at \$880.30 but its historical volatility is three times higher than ABC's. Which is more valuable — an at-the-money call option on ABC or an at-the-money call option on XYZ? Is the same true for at-the-money put options?

Exercise 4.59. Sketch the value of each of the following portfolios, as a function of the stock price at expiry.

(a) Short position in one call option struck at K_1, long position in two call options struck at $K_2 > K_1$. This is called a *backspread*.

(b) Long position in one stock, long position in one put option struck at K_1, short position in one call option struck at $K_2 > K_1$. This is called a *collar*.

Part II

Discrete-Time Modelling

5

Single-Period Arrow–Debreu Models

5.1 Specification of the Model

5.1.1 Finite-State Economy. Vector Space of Payoffs. Securities

We now turn to a multinomial generalization of the one-period binomial market model. Although single-period models cannot give a realistic representation of a complex, dynamically changing stock market, we use such models to illustrate many important economic principles. Let us begin with the main assumptions.

- Any economic activity such as trading and consumption takes place only at two times: the initial time $t = 0$ and the terminal time $t = T$.

- The economic environment at time $t = 0$ is completely known.

- At time $t = T$, the economy can be in one of M different states of the world, denoted by ω^j, $j = 1, 2 \ldots, M$, which constitute the state space

$$\Omega = \{\omega^1, \omega^2, \ldots, \omega^M\}.$$

- For each state, or outcome, $\omega^j \in \Omega$, there exists an occurrence probability $p_j > 0$ (called a state probability) such that

$$p_1 + p_2 + \cdots + p_M = 1$$

holds. These probabilities constitute the *real-world or physical probability measure* $\mathbb{P} \colon 2^\Omega \to [0, 1]$, where $\mathbb{P}(\omega^j) \equiv \mathbb{P}(\{\omega^j\}) = p_j$, $j = 1, 2, \ldots, M$, and

$$\mathbb{P}(E) = \sum_{\omega \in E} \mathbb{P}(\omega) \text{ for any event } E \subset \Omega.$$

The probability function is known in advance, but the future state of the world is unknown at time $t = 0$ and is only revealed at time $t = T$.

The above model, specified by the set of market scenarios Ω and the probability distribution function $\mathbb{P}$, is called a *multinomial* one-period model. In fact, the pair $(\Omega, \mathbb{P})$ defines a *finite probability space*. At first glance, a one-period model seems to be unrealistic. However, one-period models are useful for modelling the case with an investor pursuing a *buy-and-hold* strategy. The investor sets up an investment portfolio at time $t = 0$ and holds it to liquidate at time $t = T$. More importantly for this text, such models offer an ideal setting for introducing some of the important concepts underlying asset pricing theory in mathematical finance.

Any financial contract in a single-period model is defined by its initial price and the payoff function at terminal time T. Let us begin with the definition of a payoff.

Definition 5.1. The *payoff function* X of a financial contract is a function $X : \Omega \to \mathbb{R}$. The value $X(\omega)$ is the payment due at time T when the state of the world $\omega \in \Omega$ is reached. Since the state ω is uncertain, X is a random variable defined on the finite probability space $(\Omega, \mathbb{P})$. The M-dimensional vector

$$\mathbf{X} \equiv X(\Omega) := \left[X(\omega^1), \quad X(\omega^2), \quad \ldots \quad , X(\omega^M) \right]$$

is called a *payoff vector* or a *cash-flow vector*.

The simplest example of a financial contract is a so-called Arrow–Debreu security (named after Kenneth Arrow and Gérard Debreu). There are M such securities, each corresponding to a different market scenario. The Arrow–Debreu (AD for short) payoff $\mathcal{E}^j$, $j = 1, 2, \ldots, M$, pays one unit of currency (or one unit of another *numéraire*[1]) if the state of the world ω^j is reached and zero otherwise. As a random variable, the jth AD security corresponds to the indicator random variable $\mathcal{E}^j = \mathbb{I}_{\{\omega^j\}}$, i.e.,

$$\mathcal{E}^j(\omega) := \mathbb{I}_{\{\omega^j\}}(\omega) = \begin{cases} 0, & \text{if } \omega \neq \omega^j, \\ 1, & \text{if } \omega = \omega^j. \end{cases}$$

Recall that $\mathbb{I}_A$ is the indicator random variable for event A; that is, $\mathbb{I}_A(\omega) = 1$ if $\omega \in A$ and $\mathbb{I}_A(\omega) = 0$ if $\omega \notin A$.

Proposition 5.1. *The set of payoffs denoted by $\mathcal{L}(\Omega)$ is a vector space. The AD securities $\{\mathcal{E}^1, \mathcal{E}^2, \ldots, \mathcal{E}^M\}$ form a vector-space basis of $\mathcal{L}(\Omega)$.*

Proof. Clearly, any linear combination of two payoffs is again a payoff. Indeed, for any $a, b \in \mathbb{R}$ and $X, Y \in \mathcal{L}(\Omega)$, the function $aX + bY$ defined by $(aX + bY)(\omega) := aX(\omega) + bY(\omega)$ is a payoff from $\mathcal{L}(\Omega)$. The set $\mathcal{L}(\Omega)$ contains a zero payoff $X \equiv 0$. Therefore, $\mathcal{L}(\Omega)$ satisfies the definition of a vector space. Now, take any $X \in \mathcal{L}(\Omega)$. We have the following representation:

$$X(\omega) = \sum_{j=1}^{M} X(\omega^j) \mathbb{I}_{\{\omega^j\}}(\omega) \text{ for } \omega \in \Omega,$$

where the sum on the right-hand side may only contain one nonzero term. Therefore, the payoff X can be expressed as a linear combination of AD securities:

$$X = x_1 \mathcal{E}^1 + x_2 \mathcal{E}^2 + \cdots + x_M \mathcal{E}^M, \text{ where } x_j := X(\omega^j).$$

Finally, let us prove that the functions $\mathcal{E}^1, \mathcal{E}^2, \ldots, \mathcal{E}^M$ are linearly independent. Suppose that $a_1 \mathcal{E}^1 + a_2 \mathcal{E}^2 + \cdots + a_M \mathcal{E}^M \equiv 0$ holds for some real numbers $a_1, a_2, \ldots, a_M$. In this case, for every $j = 1, 2, \ldots, M$, we have $a_1 \mathcal{E}^1(\omega^j) + \cdots + a_M \mathcal{E}^M(\omega^j) = a_j = 0$, since $\mathcal{E}^i(\omega^j) = 1$ iff $i = j$. Therefore, the linear combination $a_1 \mathcal{E}^1 + a_2 \mathcal{E}^2 + \cdots + a_M \mathcal{E}^M$ is zero iff all a_j are zero. $\qquad \square$

In what follows, we shall use the following terminology. A vector $\mathbf{v} \in \mathbb{R}^n$ is said to be:

strictly positive (denoted $\mathbf{v} \gg 0$) if all its entries are positive ($v_i > 0$ for all $i = 1, 2, \ldots, n$);

positive (denoted $\mathbf{v} > 0$) if all its entries are nonnegative and at least one entry is strictly positive ($v_i \geqslant 0$ for all $i = 1, 2, \ldots, n$ and $v_j > 0$ for at least one j);

---

[1] A *numéraire* is a basic standard by which values are measured. Normally, we use dollars or another currency to compare values of various objects. However, other numéraires can be used, such as a bushel of grain, an ounce of gold, a barrel of oil, or a cowrie shell.

nonnegative (denoted $\mathbf{v} \geqslant 0$) if all its entries are nonnegative ($v_i \geqslant 0$ for all $i = 1, 2, \ldots, n$).

So, we have the following inclusions:

$$\{\text{strictly positive vectors}\} \subseteq \{\text{positive vectors}\} \subseteq \{\text{nonnegative vectors}\}.$$

The same terminology will be used for discrete random variables defined on a finite probability space. For example, a random variable $X \colon \Omega \to \mathbb{R}$ is said to be positive if the vector of all its values $X(\omega)$, $\omega \in \Omega$ is positive. That is, $X(\omega) \geqslant 0$ for all $\omega \in \Omega$ and $X(\omega^*) > 0$ for at least one ω^*.

Definition 5.2. A *contingent claim*, or a claim for short, is any nonnegative payoff X, i.e., $X(\omega) \geqslant 0$ holds for all $\omega \in \Omega$. We will denote the set of all claims by $\mathcal{L}^+(\Omega)$.

For example, in the binomial model with $M = 2$ states, every payoff X is represented by a vector $[x_1, x_2] \in \mathbb{R}^2$, where $x_1 = X(\omega^1)$ and $x_2 = X(\omega^2)$. Every claim is given by a positive vector in $\mathbb{R}^2$. That is, $\mathcal{L}(\Omega) = \mathbb{R}^2$ and $\mathcal{L}^+(\Omega) = \mathbb{R}_+^2$, where $\mathbb{R}_+ \equiv [0, \infty)$. Note that the set of claims $\mathcal{L}^+(\Omega)$ is convex since a linear combination of claims with nonnegative weights a, b is also a claim:

$$a, b \in [0, \infty), \ \ X, Y \in \mathcal{L}^+(\Omega) \implies aX + bY \in \mathcal{L}^+(\Omega).$$

Definition 5.3. An *asset* or a (financial) security, denoted by S, is described by

- its initial price $S_0 > 0$ at time $t = 0$, which is constant for all $\omega \in \Omega$,

- its payoff $S_T \in \mathcal{L}^+(\Omega)$ at time $t = T$ (that is, the security pays $S_T(\omega)$ at time T if the market is in a given state ω).

We also say that asset S has the *price process* $\{S_t\}_{t \in \{0, T\}}$, which is simply a function (random variable)

$$S \colon \{0, T\} \times \Omega \to [0, \infty).$$

An asset is said to be *risky* if there exists at least two states of the world, say ω^j and ω^k, such that $S_T(\omega^j) \neq S_T(\omega^k)$. So, the payoff of a risky asset, S_T, is a nonconstant random variable and hence its value $S_T(\omega)$ depends on the outcome ω realized at time T, i.e., its value is uncertain at time 0. Otherwise, if the asset has a given value $S_T(\omega)$ that is the same for all $\omega \in \Omega$, it is called a *risk-free* asset, i.e., it is constant on all outcomes and hence has a certain time-T value. Such an asset represents a bank account, a money market account, or a risk-free zero-coupon bond. We will denote a risk-free asset by B. Since the payoff of a risk-free asset is the same for all scenarios, its return is constant. Let $r := \frac{B_T - B_0}{B_0} = \frac{B_T}{B_0} - 1 > -1$ denote the one-period risk-free return. Then the time-T value of the risk-free asset is $B_T = B_0(1 + r)$, with $B_0 > 0$ being the initial (time-0) value of the asset.

The immediate result of Proposition 5.1 is that the payoff function of security S can be *replicated* by Arrow–Debreu payoffs. That is, for every financial contract there exists a portfolio of AD securities such that the payoff function of the portfolio is the same as that of the contract:

$$S_T(\omega) = s_1 \mathcal{E}^1(\omega) + s_2 \mathcal{E}^2(\omega) + \cdots + s_M \mathcal{E}^M(\omega) \text{ for all } \omega \in \Omega,$$

where $S_T(\omega^j) = s_j$, $j = 1, \ldots, M$, are the portfolio positions in the respective AD securities.

5.1.2   Initial Price Vector and Payoff Matrix

Let there be N base assets $S^1, S^2, \ldots, S^N$ traded on a market. For many applications risky assets are stocks and risk-free ones are bonds or bank accounts. The initial prices $S_0^i > 0$, $i = 1, 2, \ldots, N$, are known; the terminal values of the respective N assets, $S_T^1, \ldots, S_T^N$, are random variables whose values are generally uncertain at time 0. Typically, one of these N assets can be risk-free, i.e., its terminal payoff is a constant random variable with future value at time T being independent of $\omega \in \Omega$. A one-period N-by-M model with N base assets and M states of the world is described by:

- the N-by-1 *initial price vector* $\mathbf{S}_0 = [S_0^1, S_0^2, \cdots, S_0^N]^\top$;

- the N-by-M *payoff matrix* $\mathbf{S}_T(\Omega)$ (also called the *cash-flow matrix*, or the *dividend matrix*) given by

$$\left[\mathbf{S}_T(\omega^1) \,|\, \mathbf{S}_T(\omega^2) \,|\, \cdots \,|\, \mathbf{S}_T(\omega^M)\right] = \begin{bmatrix} S_T^1(\omega^1) & S_T^1(\omega^2) & \cdots & S_T^1(\omega^M) \\ S_T^2(\omega^1) & S_T^2(\omega^2) & \cdots & S_T^2(\omega^M) \\ \vdots & \vdots & \ddots & \vdots \\ S_T^N(\omega^1) & S_T^N(\omega^2) & \cdots & S_T^N(\omega^M) \end{bmatrix}.$$

That is, the (i, j)-entry of $\mathbf{S}_T(\Omega)$ is $S_T^i(\omega^j)$—the amount paid by the ith asset in state w^j at time T; the jth column is denoted by $\mathbf{S}_T(\omega^j)$, i.e., the column vector whose components correspond to the time-T value of the assets in the given state ω^j. The ith row of the payoff matrix is the *payoff vector* $S_T^i(\Omega)$ of the ith asset. In what follows, we also denote the payoff matrix by $\mathbf{D}$, where $D_{i,j} = S_T^i(\omega^j)$; $i = 1, \ldots, N$, $j = 1, \ldots, M$.

Both $\mathbf{S}_0$ and $\mathbf{S}_T(\Omega)$ are known to all investors. However, the state of the world, ω, is not known in advance at time $t = 0$. Thus, $\mathbf{S}_T = [S_T^1, S_T^2, \cdots, S_T^N]^\top$ is a random N-by-1 vector defined on Ω. The model is depicted in Figure 5.1.

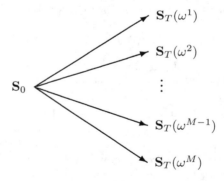

FIGURE 5.1: A scenario tree for the Arrow–Debreu model with M states of the world.

Example 5.1 (Single-period binomial model). Consider a 2-by-2 economy with two states (one up and one down state), $\Omega = \{\omega^1, \omega^2\} \equiv \{\omega^+, \omega^-\}$, and two base assets. The first asset $S^1 \equiv B$ is a risk-free zero-coupon bond paying 1 unit of currency at time T; the second asset $S^2 \equiv S$ is a risky stock. Hence, the terminal time-T price vector $\mathbf{S}_T = [B_T, S_T]^\top$ has first component $S_T^1(\omega^+) = S_T^1(\omega^-) \equiv B_T = 1$. The second component is the time-T stock price, $S_T^2 \equiv S_T$, which is assumed to be given by $S_T(\omega^+) = S_0 u$ and $S_T(\omega^-) = S_0 d$ in the respective states, where $0 < d < u$ are, respectively, down- and up-movement stock price factors and $r > 0$ is the risk-free interest rate. The price dynamics of the two assets is depicted by the following diagram:

$$\mathbf{S}_T(\omega^+) = \begin{bmatrix} 1 \\ S_0 u \end{bmatrix}$$

$$\mathbf{S}_0 = \begin{bmatrix} (1+r)^{-1} \\ S_0 \end{bmatrix}$$

$$\mathbf{S}_T(\omega^-) = \begin{bmatrix} 1 \\ S_0 d \end{bmatrix}$$

Thus, the initial price vector and the payoff matrix are, respectively,

$$\mathbf{S}_0 = \begin{bmatrix} B_0 \\ S_0 \end{bmatrix} = \begin{bmatrix} (1+r)^{-1} \\ S_0 \end{bmatrix},$$

$$\mathbf{D} = \begin{bmatrix} \mathbf{S}_T(\omega^+) \mid \mathbf{S}_T(\omega^-) \end{bmatrix} = \begin{bmatrix} B_T(\omega^+) & B_T(\omega^-) \\ S_T(\omega^+) & S_T(\omega^-) \end{bmatrix} = \begin{bmatrix} 1 & 1 \\ S_0 u & S_0 d \end{bmatrix}.$$

5.1.3 Portfolios of Base Securities

A market portfolio is simply a collection of assets traded on the market. To describe a portfolio in our model, the number of units held in the portfolio needs to be calculated for every base asset. Since there are N base assets, any portfolio can be represented by a vector with N entries.

Definition 5.4. A *portfolio* of N base assets is a 1-by-N vector $\varphi = [\varphi_1, \varphi_2, \cdots, \varphi_N] \in \mathbb{R}^N$, where φ_i, called the position in the ith asset, is the number of units of the ith base asset held in the portfolio. If $\varphi_i > 0$, then the position is said to be *long*; if $\varphi_i < 0$, then the position is said to be *short*. The *portfolio value* denoted Π_t^φ or $\Pi_t[\varphi]$ gives the total value of the portfolio φ at time $t \in \{0, T\}$. The initial (time-0) value is just the inner product[2] of the vectors φ and $\mathbf{S}_0$:

$$\Pi_0^\varphi = \varphi \, \mathbf{S}_0 = \varphi_1 S_0^1 + \varphi_2 S_0^2 + \cdots + \varphi_N S_0^N. \tag{5.1}$$

The terminal (time-T) value is a function of $\omega \in \Omega$ given by

$$\Pi_T^\varphi(\omega) = \varphi \, \mathbf{S}_T(\omega) = \varphi_1 S_T^1(\omega) + \varphi_2 S_T^2(\omega) + \cdots + \varphi_N S_T^N(\omega). \tag{5.2}$$

The initial and terminal portfolio values constitute a portfolio value process $\{\Pi_t^\varphi\}_{t \in \{0,T\}}$. The terminal value Π_T^φ of a portfolio is a random variable defined on Ω. In the state ω^j, the terminal portfolio value is given by the inner product of the portfolio vector φ and the jth column of the cash-flow matrix: $\Pi_T^\varphi(\omega^j) = \varphi \, \mathbf{S}_T(\omega^j)$. Thus, the 1-by-$M$ payoff vector of portfolio terminal values is given by a product of the 1-by-N portfolio position vector and the N-by-M payoff matrix:

$$\Pi_T^\varphi(\Omega) := \begin{bmatrix} \Pi_T^\varphi(\omega^1), \Pi_T^\varphi(\omega^2), \ldots, \Pi_T^\varphi(\omega^M) \end{bmatrix} = \varphi \, \mathbf{D}.$$

As noted above, any real random variable (or payoff) $X : \Omega \to \mathbb{R}$ has the representation $X = \sum_{j=1}^M X(\omega^j) \mathcal{E}^j$, i.e., $X(\omega) = \sum_{j=1}^M X(\omega^j) \mathcal{E}^j(\omega)$, for any $\omega \in \Omega$. Hence, the random variable corresponding to the terminal portfolio value is also expressible in terms of Arrow–Debreu securities as follows:

$$\Pi_T^\varphi(\omega) = \sum_{j=1}^M \Pi_T^\varphi(\omega^j) \, \mathcal{E}^j(\omega) = \sum_{j=1}^M \left(\varphi \, \mathbf{S}_T(\omega^j) \right) \mathcal{E}^j(\omega), \quad \omega \in \Omega.$$

---

[2]Note that the matrix product $\boldsymbol{a} \, \boldsymbol{b}$ is a scalar and is equivalent to the usual dot (inner) product of two vectors where $\boldsymbol{a}$ is 1-by-N and $\boldsymbol{b}$ is N-by-1.

Since the right-hand sides of (5.1) and (5.2) are linear functions of the portfolio vector, Π_t^φ is a linear function of φ, as is stated in the following proposition.

Proposition 5.2. *The portfolio value Π_t^φ with $t \in \{0, T\}$ is a linear function of $\varphi \in \mathbb{R}^N$.*

Proof. Consider two portfolios $\varphi, \psi \in \mathbb{R}^N$. Then, for any $a, b, \in \mathbb{R}$, the value of the combined portfolio $a\varphi + b\psi$ has time-T value

$$\Pi_T^{a\varphi+b\psi}(\omega) = \sum_{j=1}^{M} \left((a\varphi + b\psi)\mathbf{S}_T(\omega^j)\right) \mathcal{E}^j(\omega)$$

$$= \sum_{j=1}^{M} \left(a\varphi\mathbf{S}_T(\omega^j) + b\psi\mathbf{S}_T(\omega^j)\right) \mathcal{E}^j(\omega)$$

$$= a \sum_{j=1}^{M} \left(\varphi\mathbf{S}_T(\omega^j)\right) \mathcal{E}^j(\omega) + b \sum_{j=1}^{M} \left(\psi\mathbf{S}_T(\omega^j)\right) \mathcal{E}^j(\omega)$$

$$= a\Pi_T^\varphi(\omega) + b\Pi_T^\psi(\omega),$$

for all $\omega \in \Omega$. The initial value of the combined portfolio is

$$\Pi_0^{a\varphi+b\psi} = (a\varphi + b\psi)\mathbf{S}_0$$

$$= a\varphi\mathbf{S}_0 + b\psi\mathbf{S}_0 = a\Pi_0^\varphi + b\Pi_0^\psi.$$

That is, $\Pi_t^{a\varphi+b\psi} = a\Pi_t^\varphi + b\Pi_t^\psi$, for each $t \in \{0, T\}$.   □

5.2   Analysis of the Arrow–Debreu Model

5.2.1   Redundant Assets and Attainable Securities

Suppose that there exists a nonzero portfolio φ such that $\Pi_T^\varphi(\omega) = 0$ for all $\omega \in \Omega$. Let $\varphi_j \neq 0$ for some j. Then, the claim of the jth asset, S_T^j, can be expressed as a linear combination of the other asset claims:

$$\sum_{i=1}^{N} \varphi_i S_T^i = 0 \implies \varphi_j S_T^j = -\sum_{i \neq j} \varphi_i S_T^i \implies S_T^j = \sum_{i \neq j} \left(-\frac{\varphi_i}{\varphi_j}\right) S_T^i.$$

If this is the case, the asset S^j is said to be *redundant*. There exists redundancy iff the payoff vector of one base asset is a linear combination of payoffs of other base assets. That is, there exists a redundant asset iff one row of the dividend matrix $\mathbf{D}$ is a linear combination of other rows of $\mathbf{D}$. Therefore, there exists a redundant base asset iff $\text{rank}(\mathbf{D}) < N$. A redundant asset can be found by finding a nontrivial solution φ to the system of M linear equations in N unknowns $\varphi_1, \ldots, \varphi_N$:

$$\varphi \mathbf{D} = 0 \iff \begin{cases} \varphi_1 S_T^1(\omega_1) + \varphi_2 S_T^2(\omega_1) + \cdots + \varphi_N S_T^N(\omega_1) = 0, \\ \varphi_1 S_T^1(\omega_2) + \varphi_2 S_T^2(\omega_2) + \cdots + \varphi_N S_T^N(\omega_2) = 0, \\ \quad\vdots \\ \varphi_1 S_T^1(\omega_M) + \varphi_2 S_T^2(\omega_M) + \cdots + \varphi_N S_T^N(\omega_M) = 0. \end{cases}$$

That is, there is a redundant asset in the market model iff there exists a nonzero portfolio $\varphi \neq \mathbf{0}$ that replicates the zero claim.

Let us consider the case where $N = M$. The payoff matrix $\mathbf{D}$ is then a square one. Recall that the rank of a square matrix is less than its size iff the matrix is singular. Therefore, it is sufficient to calculate the determinant of the payoff matrix and apply the following criterion:

$$\det(\mathbf{D}) = 0 \iff \text{there is a redundant asset.}$$

Example 5.2. Consider a 3-by-3 economy (with three assets and three states) with given payoff matrix

$$\mathbf{D} = \begin{bmatrix} 10 & 8 & 6 \\ 13 & 7 & 0 \\ 7 & 9 & 12 \end{bmatrix}. \tag{5.3}$$

Find a redundant base asset, if any.

Solution. Let us calculate the determinant of the payoff matrix in (5.3):

$$\det(\mathbf{D}) = 10 \cdot 7 \cdot 12 + 13 \cdot 6 \cdot 9 + 7 \cdot 8 \cdot 0 - 7 \cdot 7 \cdot 6 - 9 \cdot 10 \cdot 0 - 13 \cdot 8 \cdot 12 = 0.$$

Hence there is redundancy. Indeed, the sum of the second and third rows equals twice the first row. Thus, $S_T^1 = \frac{1}{2}S_T^2 + \frac{1}{2}S_T^3$, i.e., asset S^1 is redundant (or, equivalently, S^2 is redundant since $S_T^2 = 2S_T^1 - S_T^3$). $\square$

The terminal value of a portfolio in base assets is a payoff function. In other words, each portfolio $\varphi \in \mathbb{R}^N$ generates a payoff $X = \Pi_T^\varphi$ from the vector space $\mathcal{L}(\Omega)$. Let us reverse this situation and find out if it is possible to represent a given payoff $X \in \mathcal{L}(\Omega)$ as the terminal value Π_T^φ of some portfolio $\varphi \in \mathbb{R}^N$.

Definition 5.5. A payoff $X \in \mathcal{L}(\Omega)$ is said to be *attainable* if there exists a portfolio $\varphi \in \mathbb{R}^N$ such that $\Pi_T^\varphi(\omega) = X(\omega)$ for all $\omega \in \Omega$. Any such portfolio φ is called a *hedge* or a *replicating portfolio* for the payoff X. The set of attainable payoffs in a given N-by-M economy will be denoted by $\mathcal{A}(\Omega)$.

The set of attainable payoffs $\mathcal{A}(\Omega) = \{\varphi \mathbf{D} \,|\, \varphi \in \mathbb{R}^N\}$ is a subset of $\mathcal{L}(\Omega)$. Moreover, $\mathcal{A}(\Omega)$ is a vector subspace of $\mathcal{L}(\Omega)$. Indeed, thanks to Proposition 5.2, if X and Y are elements of $\mathcal{A}(\Omega)$, then any linear combination of X and Y is an element of $\mathcal{A}(\Omega)$. Indeed, let the portfolio φ_X and φ_Y replicate the payoffs X and Y, respectively. Then, the portfolio $a\varphi_X + b\varphi_Y$ replicates the payoff $aX + bY$:

$$\Pi_T[a\varphi_X + b\varphi_Y] = a\Pi_T[\varphi_X] + b\Pi_T[\varphi_Y] = aX + bY.$$

Finally, the zero payoff $X \equiv 0$ is attainable since the terminal value of a zero portfolio $\mathbf{0} = [0, 0, \ldots, 0] \in \mathbb{R}^N$ is zero:

$$\Pi_T^0 = \sum_{i=1}^N 0 \cdot S_T^i \equiv 0.$$

To find a portfolio vector $\varphi = [\varphi_1, \varphi_2, \ldots, \varphi_N] \in \mathbb{R}^N$ that replicates any given payoff vector $X(\Omega) \equiv \mathbf{x} = [x_1, x_2, \ldots, x_M] \in \mathbb{R}^M$, we need to solve the following system of linear equations:

$$\begin{cases} \varphi_1 S_T^1(\omega^1) + \varphi_2 S_T^2(\omega^1) + \cdots + \varphi_N S_T^N(\omega^1) = x_1 \\ \varphi_1 S_T^1(\omega^2) + \varphi_2 S_T^2(\omega^2) + \cdots + \varphi_N S_T^N(\omega^2) = x_2 \\ \vdots \\ \varphi_1 S_T^1(\omega^M) + \varphi_2 S_T^2(\omega^M) + \cdots + \varphi_N S_T^N(\omega^M) = x_M \end{cases} \tag{5.4}$$

This system can be written in a compact (row) vector-matrix form:

$$\boldsymbol{\varphi}\,\mathbf{D} = \mathbf{x}.$$

Note that this is written in terms of row vectors. It may also be convenient to work with the equivalent transposed equation in terms of column vectors, $\mathbf{D}^\top \boldsymbol{\varphi}^\top = \mathbf{x}^\top$. If the solution to the system exists, then the payoff $X(\Omega)$ is attainable; otherwise it is said to be *unattainable*. Note that the system (5.4) may have multiple solutions. If that is the case, there are infinitely many replicating portfolios for the claim X.

Replication is a key component to pricing financial securities. The initial price of any asset (or derivative claim) with an attainable payoff $X \in \mathcal{A}(\Omega)$ can be evaluated by replicating X in the base assets as follows:

(a) Find a portfolio $\boldsymbol{\varphi}_X$ in the base assets that replicates X, i.e., $\Pi_T[\boldsymbol{\varphi}_X] \equiv \boldsymbol{\varphi}_X\,\mathbf{S}_T = X$.

(b) Set the initial price $\pi_0(X)$ of the security with payoff $X(\Omega)$ equal to the initial cost $\Pi_0[\boldsymbol{\varphi}_X]$ of setting up the portfolio $\boldsymbol{\varphi}_X$:

$$\pi_0(X) = \Pi_0[\boldsymbol{\varphi}_X] = \boldsymbol{\varphi}_X\,\mathbf{S}_0.$$

To proceed with this approach, we need to answer the following questions regarding uniqueness of replication, market completeness and the law of one price.

1. Under what condition does there exist a unique portfolio that replicates a given payoff and hence gives the initial security price uniquely?

2. Can every payoff be replicated and hence every claim's initial price be evaluated based on replication?

3. Suppose that a given payoff is replicated nonuniquely. Under what condition do all portfolios replicating the same payoff have the same initial value?

The following lemma answers the first question. The other questions are to be investigated in the next sections.

Lemma 5.3. *Every attainable payoff $X \in \mathcal{A}(\Omega)$ has a unique replicating portfolio $\boldsymbol{\varphi} \in \mathbb{R}^N$ iff there are no redundant base assets.*

Proof. The base assets are nonredundant iff their payoffs are linearly independent. That is,

$$\Pi_T^\varphi = \varphi_1 S_T^1 + \varphi_2 S_T^2 + \cdots + \varphi_N S_T^N \equiv 0 \text{ iff } \boldsymbol{\varphi} = \mathbf{0}.$$

If the only portfolio replicating the zero payoff vector is a zero portfolio, then there is no redundancy. Consider any two portfolios $\boldsymbol{\varphi}$ and $\boldsymbol{\psi}$ that are assumed to replicate the same payoff X. Then, $\Pi_T^\varphi = X = \Pi_T^\psi$. Moreover, by linearity, this is the case iff the difference portfolio $\boldsymbol{\varphi} - \boldsymbol{\psi}$ has a zero payoff, i.e.,

$$\Pi_T^\varphi = \Pi_T^\psi \iff \Pi_T^{\varphi-\psi} \equiv 0.$$

Thus, there are no redundant base assets iff $\boldsymbol{\varphi} - \boldsymbol{\psi} = \mathbf{0} \iff \boldsymbol{\varphi} = \boldsymbol{\psi}$, i.e., any attainable payoff X is replicated by a unique portfolio in the base assets iff there are no redundant base assets. $\square$

Corollary 5.4. *If there exists a redundant base asset, then every attainable payoff has infinitely many replicating portfolios.*

Proof. Suppose that there is a redundant base asset. Then, there exists a nonzero portfolio ψ_0 such that $\Pi_T^{\psi_0} = 0$. Therefore, for any portfolio φ replicating a payoff X, the portfolio $\varphi_\lambda := \varphi + \lambda\psi_0$ (with arbitrary constant $\lambda \in \mathbb{R}$) replicates X as well:

$$\Pi_T^{\varphi_\lambda} = \Pi_T^\varphi + \lambda\Pi_T^{\psi_0} = X + \lambda \cdot 0 = X.$$

Hence, we have an infinite family of portfolios (parametrized by $\lambda \in \mathbb{R}$) replicating the same payoff X. $\square$

5.2.2 Completeness of the Model

In this subsection, we investigate if any (arbitrary) payoff can be replicated by a portfolio in base assets. This is in essence the question of whether or not the market model is complete, as defined just below.

Definition 5.6. A market model is said to be *complete* if every payoff is attainable, i.e., if $\mathcal{L}(\Omega) = \mathcal{A}(\Omega)$ holds.

Since it is impossible to verify every payoff for attainability, we require some simple criterion for market completeness, as we now discuss.

Lemma 5.5. *The following statements are equivalent.*

(1) every payoff is attainable: $\mathcal{L}(\Omega) = \mathcal{A}(\Omega)$;

(2) every claim is attainable: $\mathcal{L}^+(\Omega) \subset \mathcal{A}(\Omega)$;

(3) every Arrow–Debreu security is attainable: $\mathcal{E}^j \in \mathcal{A}(\Omega)$, for all $j = 1, 2, \ldots, M$.

Proof. Let us prove that (1) implies (2), (2) implies (3), and (3) implies (1).

(1) $\implies$ (2): Obviously, $\mathcal{L}^+(\Omega) \subset \mathcal{L}(\Omega) = \mathcal{A}(\Omega)$.

(2) $\implies$ (3): This follows since, for every j, $\mathcal{E}^j \in \mathcal{L}^+(\Omega) \subset \mathcal{A}(\Omega)$.

(3) $\implies$ (1): According to Proposition 5.1, every payoff can be represented as a linear combination of AD securities, i.e., $X = \sum_{j=1}^M x_j\mathcal{E}^j$. By assumption, there exists a portfolio φ_j that replicates $\mathcal{E}^j$, for each $j = 1, 2, \ldots, M$. Then, the portfolio defined by $\varphi = \sum_{j=1}^M x_j\varphi_j$ replicates X, since

$$\Pi_T^\varphi = \sum_{j=1}^M x_j\Pi_T^{\varphi_j} = \sum_{j=1}^M x_j\mathcal{E}^j = X. \qquad \square$$

In conclusion, the market is complete iff the base security payoffs $S_T^1, S_T^2, \ldots, S_T^N$ span the vector space $\mathcal{L}(\Omega)$. That is, the N payoff vectors

$$\mathbf{v}_1 = \left[S_T^1(\omega^1), S_T^1(\omega^2), \ldots, S_T^1(\omega^M)\right]$$
$$\mathbf{v}_2 = \left[S_T^2(\omega^1), S_T^2(\omega^2), \ldots, S_T^2(\omega^M)\right]$$
$$\vdots$$
$$\mathbf{v}_N = \left[S_T^N(\omega^1), S_T^N(\omega^2), \ldots, S_T^N(\omega^M)\right]$$

span $\mathbb{R}^M$, the space of all 1-by-M real vectors. The dimension of $\mathcal{L}(\Omega)$ is M, hence for completeness to hold there has to be at least M base assets with linearly independent payoffs. Therefore, the market is complete iff rank$(\mathbf{D}) \geqslant M$. As was proved in Lemma 5.3, every attainable payoff is replicated uniquely iff rank$(\mathbf{D}) \geqslant N$. Since $\mathbf{D}$ is an N-by-M matrix, its rank does not exceed $\min\{N, M\}$. Therefore, the one-period model is complete and every payoff is replicated uniquely iff $\mathbf{D}$ is a square (i.e., $N = M$), nonsingular matrix. However, in a realistic asset price model, the number of market scenarios, M, is very large. On the other hand, the number of base assets is relatively small since it is otherwise impractical for an investor to operate with a large number of underlying securities. Thus, for a realistic one-period model we have that $M \gg N$, and hence such a model is incomplete.

Example 5.3. Consider the following 2-by-4 model with $N = 2$ base assets and $M = 4$ market scenarios:

$$\mathbf{S}_0 = \begin{bmatrix} 5 \\ 10 \end{bmatrix}, \quad \mathbf{D} = \begin{bmatrix} 6 & 6 & 4 & 2 \\ 12 & 8 & 8 & 12 \end{bmatrix}.$$

(a) Check if the model is complete and/or nonredundant.

(b) Is the payoff $X(\Omega) = \mathbf{x} = \begin{bmatrix} 0, 4, 0, -8 \end{bmatrix}$ attainable?

Solution. There are four states and only two base assets, $M = 4 > N = 2$, and hence the market is incomplete. Calculate the rank of the cash-flow matrix $\mathbf{D}$. Since $\det \begin{bmatrix} 6 & 6 \\ 12 & 8 \end{bmatrix} \neq 0$, we have that rank$(\mathbf{D}) = 2$, i.e., the payoff vectors (obtained from the two rows of $\mathbf{D}$), $\mathbf{v}_1 = [6, 6, 4, 2]$ and $\mathbf{v}_2 = [12, 8, 8, 12]$, are linearly independent. Therefore, there are no redundant base assets. To find a portfolio $\boldsymbol{\varphi} = \begin{bmatrix} \varphi_1, \varphi_2 \end{bmatrix}$ replicating X, we solve the following system of linear equations:

$$\boldsymbol{\varphi}\mathbf{D} = \mathbf{x} \iff \begin{cases} 6\varphi_1 + 12\varphi_2 = 0 \\ 6\varphi_1 + 8\varphi_2 = 4 \\ 4\varphi_1 + 8\varphi_2 = 0 \\ 2\varphi_1 + 12\varphi_2 = -8 \end{cases} \iff \begin{cases} \varphi_1 = 2 \\ \varphi_2 = -1 \end{cases}$$

Thus, X is attainable and is replicated by the unique portfolio $\boldsymbol{\varphi} = \begin{bmatrix} 2, -1 \end{bmatrix}$. The replicating portfolio value process is $\Pi_t = 2S_t^1 - S_t^2$, $t \in \{0, T\}$. Let us verify that $\Pi_T(\omega^j) = x_j$ for $j = 1, 2, 3, 4$:

$$\Pi_T(\omega^1) = 2 \cdot 6 - 12 = 0 = x_1 \qquad \checkmark$$
$$\Pi_T(\omega^2) = 2 \cdot 6 - 8 = 4 = x_2 \qquad \checkmark$$
$$\Pi_T(\omega^3) = 2 \cdot 4 - 8 = 0 = x_3 \qquad \checkmark$$
$$\Pi_T(\omega^4) = 2 \cdot 2 - 12 = -8 = x_4 \qquad \checkmark$$

$\square$

Let us summarize all criteria for investigating if a single-period market model is complete and/or has redundant base assets.

- If rank$(\mathbf{D}) < M$, then the market model is incomplete. Hence, not every payoff can be replicated by a portfolio in base assets.

- If rank$(\mathbf{D}) < N$, then there are redundant base assets. Hence, a zero claim can be replicated by a nonzero portfolio. Every attainable payoff has infinitely many replicating portfolios.

- If rank$(\mathbf{D}) = N = M$, then the market is complete and free of redundant base assets. Every payoff X is attainable, and the portfolio replicating a payoff $X(\Omega) = \mathbf{x}$ is uniquely given by $\boldsymbol{\varphi}_X = \mathbf{x}\mathbf{D}^{-1}$. Note that a square payoff matrix $\mathbf{D}$ has a full rank iff $\det(\mathbf{D}) \neq 0$.

5.3 No-Arbitrage Asset Pricing

5.3.1 The Law of One Price

Let us come back to the problem of calculating the initial price of a given security. A security with an attainable payoff can be priced by constructing a portfolio in base assets that replicates the payoff function. If there are no redundant securities, then such a replicating portfolio is unique for every attainable payoff. However, a payoff may be replicated nonuniquely. If this is the case, we expect that all replicating portfolios yield the same initial price. We say that *the Law of One Price* holds if for every attainable payoff all replicating portfolios have the same initial value, i.e., for any two portfolios $\varphi, \psi \in \mathbb{R}^N$ we have

$$\Pi_T^\varphi = \Pi_T^\psi \implies \Pi_0^\varphi = \Pi_0^\psi. \tag{5.5}$$

Thus, if the Law of One Price holds, then we can define the initial fair price $S_0 = \pi_0(S)$ of a security S with an attainable payoff S_T by

$$\pi_0(S) := \Pi_0^\psi \text{ for any } \psi \in \mathbb{R}^N \text{ such that } \Pi_T^\psi = S_T. \tag{5.6}$$

We will refer to this approach as *pricing via replication.*

Consider two securities with attainable payoffs, X and Y. Let their initial prices be $\pi_0(X)$ and $\pi_0(Y)$, respectively. We say that *linear pricing* holds if for every choice of constants a and b, the security with payoff $aX + bY$ has the initial price of $a\pi_0(X) + b\pi_0(Y)$. Clearly, the Law of One Price implies linear pricing. Indeed, let portfolios φ_X and φ_Y replicate X and Y, respectively. Then, the portfolio $a\varphi_X + b\varphi_Y$ replicates $aX + bY$. According to the Law of One Price, the initial price of the attainable payoff $aX + bY$ is

$$\pi_0(aX + bY) = \Pi_0[a\varphi_X + b\varphi_Y] = a\Pi_0[\varphi_X] + b\Pi_0[\varphi_Y] = a\pi_0(X) + b\pi_0(Y).$$

That is, the *pricing functional* $\pi_0 \colon \mathcal{A}(\Omega) \to \mathbb{R}$ is linear.

It is impossible to check the condition (5.5) for all replicating portfolios. The following lemma provides a simple criterion for verifying if the Law of One Price holds.

Lemma 5.6. *The Law of One Price holds iff,* $\forall \varphi \in \mathbb{R}^N$,

$$\Pi_T^\varphi = 0 \implies \Pi_0^\varphi = 0.$$

That is, the Law of One Price holds iff every portfolio φ that replicates the zero claim has zero initial value.

Proof.
The necessity part. Suppose that φ and ψ replicate the same payoff X:

$$\Pi_T^\varphi = \Pi_T^\psi = X.$$

Then, the portfolio $\theta = \varphi - \psi$ replicates the zero claim. By assumption, if $\Pi_T^\theta = 0$ then $\Pi_0^\theta = 0$. Therefore,

$$0 = \Pi_0^\theta = \Pi_0^{\varphi - \psi} = \Pi_0^\varphi - \Pi_0^\psi \implies \Pi_0^\varphi = \Pi_0^\psi.$$

The sufficiency part. Let the Law of One Price hold. Suppose that there exists a portfolio θ with $\Pi_0^\theta \neq 0$ and $\Pi_T^\theta = 0$. Take any nonzero attainable payoff X and a portfolio φ replicating the payoff of X. For $\lambda \in \mathbb{R}$, define $\varphi_\lambda = \varphi + \lambda\theta$. Then, $\forall \lambda \in \mathbb{R}$,

$$\Pi_T^{\varphi_\lambda} = \Pi_T^\varphi + \lambda\Pi_T^\theta = X + \lambda \cdot 0 = X.$$

However, the initial prices vary with λ:

$$\Pi_0^{\varphi_\lambda} = \Pi_0^{\varphi} + \lambda \Pi_0^{\theta} \neq \Pi_0^{\varphi}, \quad \text{for } \lambda \neq 0.$$

Thus, we can construct infinitely many portfolios replicating the same payoff and having different initial values. The supposition contradicts the Law of One Price. $\square$

Example 5.4. Verify if the Law of One Price holds for a 3-by-3 single-period model with

$$\mathbf{S}_0 = \begin{bmatrix} 10 \\ 10 \\ 10 \end{bmatrix} \quad \text{and} \quad \mathbf{D} = \begin{bmatrix} 5 & 10 & 15 \\ 5 & 10 & 15 \\ 15 & 10 & 5 \end{bmatrix}.$$

Solution. Find all portfolios replicating the zero claim. The general solution to $\varphi \mathbf{D} = \mathbf{0}$ is

$$\varphi = [-t, t, 0], \quad \text{where } t \in \mathbb{R}.$$

The intitial value Π_0^{φ} of any such portfolio φ replicating the zero claim is then

$$\varphi \mathbf{S}_0 = -10t + 10t = 0.$$

Since $\Pi_0^{\varphi} = 0$ for all $t \in \mathbb{R}$, the Law of One Price holds. $\square$

If there exist many portfolios replicating the same payoff X but having different initial values (i.e., the Law of One Price is violated), then it seems to be reasonable to define the fair initial price of X as the lowest cost for which we can replicate the target payoff. On the other hand, as follows from the next proposition, if the Law of One Price is violated, then for any attainable payoff there exists a replicating portfolio with arbitrarily low (or large) initial value. Thus, the fair price $\pi_0(X)$ cannot be defined meaningfully.

Proposition 5.7. *If the Law of One Price does not hold, then for every attainable payoff X and any real c_0 there exists a portfolio replicating X with the initial cost c_0.*

Proof. If the Law of One Price does not hold, then there exists θ so that $\Pi_T^{\theta} = 0$ and $\Pi_0^{\theta} \neq 0$. Let φ replicate payoff X. For any $c_0 \in \mathbb{R}$, set $\lambda = \frac{c_0 - \Pi_0^{\varphi}}{\Pi_0^{\theta}}$. Then $\varphi_\lambda = \varphi + \lambda \theta$ replicates X, i.e., $\Pi_T[\varphi_\lambda] = \Pi_T^{\varphi} + \lambda \Pi_T^{\theta} = \Pi_T^{\varphi} = X$ and $\Pi_0[\varphi_\lambda] = c_0$. $\square$

By comparing the results of Lemmas 5.3 and 5.6 we conclude that the nonredundancy condition implies the Law of One Price. However, the converse is generally not true. Indeed, let us consider a model with two states of the world and two base assets having the following cash-flow matrix and initial price vector:

$$\mathbf{D} = \begin{bmatrix} 1 & 2 \\ 1 & 2 \end{bmatrix} \quad \text{and} \quad \mathbf{S}_0 = \begin{bmatrix} 1 \\ 1 \end{bmatrix}.$$

Clearly, the two base assets S^1 and S^2 are redundant with the same payoff vector $[1, 2]$. Consider any portfolio $\varphi = [\varphi_1, \varphi_2]$ in S^1 and S^2 with zero terminal value:

$$\Pi_T^{\varphi} = \varphi_1 S_T^1 + \varphi_2 S_T^2 = 0 \iff \varphi_1 + \varphi_2 = 0.$$

Since $\Pi_0^{\varphi} = \varphi_1 S_0^1 + \varphi_2 S_0^2 = \varphi_1 + \varphi_2$, we have that $\Pi_T^{\varphi} = 0 \implies \Pi_0^{\varphi} = 0$. According to Lemma 5.6, the Law of One Price holds. Hence, this is a model with redundant assets and for which the Law of One Price holds. To conclude, market models for which the Law of One Price holds include models with and without redundant base assets. The next section ties together the Law of One Price with the concept of arbitrage in a market model.

5.3.2 Arbitrage

There exist two types of arbitrage opportunities. One type of arbitrage is an investment that gives a positive reward at time 0 and has no future cost at time T. Another type of arbitrage is an investment with zero initial value and nonnegative future cost that has a positive probability of yielding a strictly positive payoff at time T. Since any investment in a single-period model is a portfolio in base assets, we have the following formal definition of arbitrage.

Definition 5.7. An *arbitrage portfolio* is a portfolio φ such that one of the following two alternatives holds:

$$\Pi_0^\varphi = 0 \quad \text{and} \quad \Pi_T^\varphi > 0, \tag{5.7}$$

or

$$\Pi_0^\varphi < 0 \quad \text{and} \quad \Pi_T^\varphi \geqslant 0. \tag{5.8}$$

In other words, an arbitrage portfolio has zero initial value (i.e., there is no cost to set it up) and offers a potential gain with no potential liabilities at time $t = T$. Alternatively, an arbitrage opportunity is provided by a portfolio with a negative initial value and nonnegative terminal value (i.e., there are no potential liabilities). Conditions (5.7) and (5.8) can be written as inequalities

$$\varphi \, \mathbf{S}_0 \leqslant 0, \quad \varphi \, \mathbf{S}_T(\omega^j) \geqslant 0, \quad \text{for all} \quad j = 1, 2, \ldots, M, \tag{5.9}$$

where at least one inequality is strict. Note that the last inequality is equivalently written as $\varphi \, \mathbf{D} \geqslant 0$ by using the dividend (cash-flow) matrix.

If a market model admits no arbitrage, then any two portfolios replicating the same payoff must have the same initial value. Indeed, suppose that there exist two portfolios replicating the same payoff but having different initial values. By reversing positions in one portfolio (with a larger initial price) and combining it with the other portfolio, we can construct an arbitrage portfolio with a negative initial value and terminal value zero. Let us formally prove this fact.

Lemma 5.8. *No arbitrage implies the Law of One Price.*

Proof. Assume the absence of arbitrage opportunities. Recall that the Law of One Price holds when $\Pi_T^\varphi = 0 \implies \Pi_0^\varphi = 0$, for all $\varphi \in \mathbb{R}^N$. Suppose that there exists a portfolio ψ_0 such that $\Pi_T[\psi_0] = 0$ and $\Pi_0[\psi_0] \neq 0$. Take any attainable claim $X > 0$ (i.e., $X(\omega) \geqslant 0$ for all ω, and $X(\omega^*) > 0$ for some ω^*), and let φ_X replicate X. The portfolio $\varphi_\lambda := \varphi_X + \lambda \psi_0$ replicates X for any $\lambda \in \mathbb{R}$. Choose $\lambda = \lambda_0$ so that $\Pi_0[\varphi_{\lambda_0}] = 0$:

$$\Pi_0[\varphi_X + \lambda_0 \psi_0] = \Pi_0[\varphi_X] + \lambda_0 \Pi_0[\psi_0] = 0 \iff \lambda_0 = -\frac{\Pi_0[\varphi_X]}{\Pi_0[\psi_0]}.$$

Thus, φ_{λ_0} is an arbitrage portfolio, since $\Pi_0[\varphi_{\lambda_0}] = 0$ and $\Pi_T[\varphi_{\lambda_0}] = \Pi_T[\varphi_X] = X > 0$. This proves that violation of the Law of One Price implies the existence of an arbitrage portfolio. Hence, the equivalent contrapositive statement holds and the lemma is proven. $\square$

As follows from the Law of One Price, the initial price S_0 of a security with an attainable payoff S_T has to be equal to the initial cost of a replicating portfolio or else there exists arbitrage, as is given in (5.6). Indeed, if the security is priced at $S_0 \neq \Pi_0[\varphi_S]$, where φ_S replicates S_T, then an arbitrage portfolio combining the security S and base assets can be created. If $S_0 < \Pi_0[\varphi_S]$, then we buy the security S and form the portfolio $-\varphi_S$. If $S_0 > \Pi_0[\varphi_S]$, then we sell the asset and form the portfolio φ_S. In both cases, our proceeds at time 0 are positive, and the terminal value is zero. So the assumption $S_0 \neq \Pi_0[\varphi_S]$ leads to arbitrage. Thus, the price $S_0 = \pi_0(S) = \Pi_0[\varphi_S]$ given by (5.6) is called the *no-arbitrage price* of security S.

5.3.3   The First Fundamental Theorem of Asset Pricing

It can be difficult to find an arbitrage portfolio. Hence, it is reasonable to first investigate whether a model admits arbitrage. If the Law of One Price is violated, then, according to Lemma 5.8, there exists an arbitrage opportunity. However, we cannot make any conclusion if the Law of One Price holds. The *first fundamental theorem of asset pricing* (FTAP) provides a necessary and sufficient condition for the absence of arbitrage. Although we are proving the FTAP for a finite single-period economy, a very similar result is true for discrete-time multiperiod models and continuous-time models.

Theorem 5.9 (The first FTAP). *There are no arbitrage portfolios iff there exists a strictly positive solution* $\mathbf{\Psi} \in \mathbb{R}^M$ *to the linear system of equations*

$$\mathbf{D}\,\mathbf{\Psi} = \mathbf{S}_0. \qquad (5.10)$$

That is, there exists $\mathbf{\Psi} = \left[\Psi_1, \Psi_2, \cdots, \Psi_M\right]^\top \gg 0$ *such that*

$$\sum_{j=1}^{M} D_{ij}\,\Psi_j = S_0^i, \ i = 1, 2, \ldots, N,$$

where $D_{ij} \equiv S_T^i(\omega^j)$.

Note that (5.10) is a system of N equations in M unknowns, where $\mathbf{\Psi}$ is an $M \times 1$ column vector and $\mathbf{S}_0$ is an $N \times 1$ column vector. Before proving this theorem, let us illustrate it with the following example.

Example 5.5. Find an arbitrage portfolio (if any) for the model with three states and three assets specified as follows:

$$\mathbf{S}_0 = \begin{bmatrix} 1 \\ 1 \\ 1 \end{bmatrix}, \quad \mathbf{D} = \begin{bmatrix} 1 & 1 & 1 \\ 1 & 0 & 0 \\ 0 & 1 & 0 \end{bmatrix}.$$

Solution. Solving $\mathbf{D}\,\mathbf{\Psi} = \mathbf{S}_0$ gives the unique solution $\mathbf{\Psi} = [1, 1, -1]^\top$. Since not all Ψ_j's are positive, according to Theorem 5.9, there exists an arbitrage. We can form an arbitrage portfolio by replicating the Arrow–Debreu security $\mathcal{E}^3$ having payoff vector $\mathcal{E}^3(\Omega) = [\mathcal{E}^3(\omega^1), \mathcal{E}^3(\omega^2), \mathcal{E}^3(\omega^3)] = [0, 0, 1]$:

$$\boldsymbol{\varphi}\,\mathbf{D} = [0, 0, 1] \iff \begin{cases} \varphi_1 + \varphi_2 = 0 \\ \varphi_1 + \varphi_3 = 0 \\ \varphi_1 = 1 \end{cases} \iff \begin{cases} \varphi_1 = 1 \\ \varphi_2 = -1 \\ \varphi_3 = -1 \end{cases}$$

The replicating portfolio is $\boldsymbol{\varphi} = [1, -1, -1]$. Its initial value is negative:

$$\Pi_0^\varphi = \boldsymbol{\varphi}\,\mathbf{S}_0 = 1 - 1 - 1 = -1 < 0.$$

Since it's terminal value is $\Pi_T^\varphi = \mathcal{E}^3 \geqslant 0$, with $\mathcal{E}^3(\omega^3) > 0$, the solution $\boldsymbol{\varphi}$ is an arbitrage portfolio.   □

5.3.3.1 The First FTAP: Sufficiency Part

Proof. Suppose that $\boldsymbol{\Psi} >> 0$ solves (5.10). Then, the initial value of a portfolio $\boldsymbol{\varphi} \in \mathbb{R}^N$ is

$$\Pi_0^{\varphi} = \boldsymbol{\varphi}\,\mathbf{S}_0 = \boldsymbol{\varphi}\,(\mathbf{D}\,\boldsymbol{\Psi}) = (\boldsymbol{\varphi}\,\mathbf{D})\,\boldsymbol{\Psi} = \Pi_T^{\varphi}(\Omega)\,\boldsymbol{\Psi} = \sum_{j=1}^{M} \Pi_T^{\varphi}(\omega^j)\,\Psi_j. \qquad (5.11)$$

That is, the initial value of a portfolio is equal to a product of the terminal payoff vector and the state-price vector. Let $\boldsymbol{\varphi}^*$ be an arbitrage portfolio. Hence, $\Pi_0^{\varphi^*} \leqslant 0$, $\Pi_T^{\varphi^*} \geqslant 0$, and $\Pi_0^{\varphi^*} < 0$ or $\Pi_T^{\varphi^*}(\omega^k) > 0$ for some $k \in \{1,\dots,M\}$ holds. Applying (5.11) to $\boldsymbol{\varphi}^*$ gives

$$\Pi_0^{\varphi^*} = \sum_{j \neq k}^{M} \Pi_T^{\varphi^*}(\omega^j)\,\Psi_j + \Pi_T^{\varphi^*}(\omega^k)\,\Psi_k. \qquad (5.12)$$

In (5.12), either the left-hand side is negative and the right-hand side is nonnegative, or the left-hand side is zero and the right-hand side is positive. So, we have a contradiction. Hence, the existence of a strictly positive solution to (5.10) implies that there are no arbitrage portfolios. $\square$

5.3.3.2 The First FTAP: Necessity Part

Proof. The set $\mathbb{R}_+^{M+1} = \{\mathbf{x} \in \mathbb{R}^{M+1} \;:\; \mathbf{x} \geqslant 0\}$ is a closed convex cone in the vector space $\mathbb{R}^{M+1}$. That is,

- $\mathbb{R}_+^{M+1}$ contains its boundary points (it is a closed set),

- for all $\mathbf{x}, \mathbf{y} \in \mathbb{R}_+^{M+1}$ the segment $[\mathbf{x}, \mathbf{y}]$ of a straight line connecting $\mathbf{x}$ and $\mathbf{y}$ is contained in $\mathbb{R}_+^{M+1}$ (it is a convex set),

- for all $\mathbf{x} \in \mathbb{R}_+^{M+1}$ the ray $\{\lambda \mathbf{x} \;:\; \lambda \geqslant 0\}$ is contained in $\mathbb{R}_+^{M+1}$ (it is a cone).

Introduce another subset L of vectors in $\mathbb{R}^{M+1}$ defined by

$$L = \left\{ \left[-\boldsymbol{\theta}\mathbf{S}_0,\, \boldsymbol{\theta}\mathbf{S}_T(\omega^1),\dots,\boldsymbol{\theta}\mathbf{S}_T(\omega^M) \right] \;:\; \boldsymbol{\theta} \in \mathbb{R}^N \right\},$$

where $-\boldsymbol{\theta}\mathbf{S}_0 = -\sum_{i=1}^{N} \theta_i S_0^i$ and $\boldsymbol{\theta}\mathbf{S}_T(\omega^j) = \sum_{i=1}^{N} \theta_i S_T^i(\omega^j)$, $j = 1, 2, \dots, M$. Clearly, L is a linear (vector) subspace of $\mathbb{R}^{M+1}$, where for any $a, b \in \mathbb{R}$,

$$\mathbf{x}, \mathbf{y} \in L \implies a\mathbf{x} + b\mathbf{y} \in L.$$

Suppose that $\boldsymbol{\theta} \in \mathbb{R}^N$ is an arbitrage portfolio, i.e., $\boldsymbol{\theta}\mathbf{S}_0 < 0$ and $\boldsymbol{\theta}\mathbf{D} \geqslant 0$ (or $\boldsymbol{\theta}\mathbf{S}_0 \leqslant 0$ and $\boldsymbol{\theta}\mathbf{D} > 0$). Then there is a point in $L \cap \mathbb{R}_+^{M+1}$ corresponding to the arbitrage portfolio $\boldsymbol{\theta}$ and vice versa—any point in $L \cap \mathbb{R}_+^{M+1}$ corresponds to an arbitrage opportunity. Therefore, nonexistence of an arbitrage portfolio means that the subspace L and the cone $\mathbb{R}_+^{M+1}$ intersect only at the origin $\mathbf{0} = [0, 0, \dots, 0] \in \mathbb{R}^{M+1}$. The rest of the proof is based on the so-called separating hyperplane theorem, which, being applied to this situation, states the following. There exists a hyperplane $H \subset \mathbb{R}^{M+1}$ (i.e., a linear subspace of dimensions M) that separates $\mathbb{R}^{M+1}$ into two half-spaces H^+ and H^- such that

$$\mathbb{R}_+^{M+1} \subseteq H^+, \quad L \subseteq H^-, \quad H^+ \cap H^- = H.$$

In other words, the cone $\mathbb{R}_+^{M+1}$ lies on one side of H and the subspace L lies on the other side of H. The general equation for a hyperplane in $\mathbb{R}^{M+1}$ passing through the origin is

$$\boldsymbol{\lambda}\mathbf{x}^\top = 0 \iff \lambda_0 x_0 + \lambda_1 x_1 + \dots + \lambda_M x_M = 0,$$

where $\boldsymbol{\lambda} = [\lambda_0, \lambda_1, \ldots, \lambda_M]$ is a normal vector and $\boldsymbol{x} = [x_0, x_1, \ldots, x_M] \in \mathbb{R}^{M+1}$. The concept of separation can be expressed as follows: either $\boldsymbol{\lambda}\boldsymbol{x}^\top > \boldsymbol{\lambda}\boldsymbol{y}^\top$ or $\boldsymbol{\lambda}\boldsymbol{x}^\top < \boldsymbol{\lambda}\boldsymbol{y}^\top$ holds for all $\mathbf{x} \in \mathbb{R}_+^{M+1} \setminus \{\mathbf{0}\}$ and all $\mathbf{y} \in L$. In particular, the set $\{\boldsymbol{\lambda}\boldsymbol{y}^\top : \mathbf{y} \in L\}$ is bounded from above or below. This is possible iff $\boldsymbol{\lambda}\boldsymbol{y}^\top = 0$, i.e., L is contained in H. To show this, suppose that there exists $\mathbf{y} \in L$ such that $\boldsymbol{\lambda}\boldsymbol{y}^\top > 0$. Since L is a vector space, $\mathbf{y} \in L \implies a\mathbf{y} \in L$ for every $a \in \mathbb{R}$. Then the set $\{a\boldsymbol{\lambda}\boldsymbol{y}^\top : a \in \mathbb{R}\} = \mathbb{R}$ is unbounded. We arrive at a contradiction. On the other hand, since $\boldsymbol{\lambda}\boldsymbol{x}^\top > 0$ for every $\mathbf{x} \in \mathbb{R}_+^{M+1}$ (if $\boldsymbol{\lambda}\boldsymbol{x}^\top < 0$ then just replace $\boldsymbol{\lambda}$ by $-\boldsymbol{\lambda}$), all λ_i's are positive. Indeed, for each $j = 0, 1, \ldots, M$, take $\mathbf{x} = \mathbf{e}_j := [0, \ldots, 0, \underbrace{1}_{j\text{th}}, 0, \ldots, 0] \in \mathbb{R}^{M+1}$ to obtain that $\boldsymbol{\lambda}\mathbf{e}_j^\top = \lambda_j > 0$. Since $L \subset H$,

$$-\lambda_0 \boldsymbol{\theta} \mathbf{S}_0 + \sum_{j=1}^{M} \lambda_j \boldsymbol{\theta} \mathbf{S}_T(\omega^j) = 0$$

holds for every portfolio $\boldsymbol{\theta} \in \mathbb{R}^N$. Setting $\boldsymbol{\theta} = \mathbf{e}_i \in \mathbb{R}^N$, for each $i = 1, 2, \ldots, N$, we obtain

$$-\lambda_0 S_0^i + \sum_{j=1}^{M} \lambda_j S_T^i(\omega^j) = 0.$$

This implies that

$$-\lambda_0 \mathbf{S}_0 + \sum_{j=1}^{M} \lambda_j \mathbf{S}_T(\omega^j) = \mathbf{0} \iff \mathbf{S}_0 = \sum_{j=1}^{M} \underbrace{\frac{\lambda_j}{\lambda_0}}_{\equiv \Psi_j} \mathbf{S}_T(\omega^j).$$

In matrix form, we have

$$\mathbf{D}\,\boldsymbol{\Psi} = \mathbf{S}_0, \quad \text{where } \boldsymbol{\Psi}^\top = [\Psi_1, \Psi_2, \ldots, \Psi_M] = \left[\frac{\lambda_1}{\lambda_0}, \frac{\lambda_2}{\lambda_0}, \ldots, \frac{\lambda_M}{\lambda_0}\right] \gg 0. \qquad \square$$

Example 5.6. Consider the following 3-by-3 models with initial price vector $\mathbf{S}_0 = [1, 5, 10]^\top$ and payoff matrix

$$\text{(a) } \mathbf{D} = \begin{bmatrix} 1 & 1 & 1 \\ 1 & 6 & 15 \\ 12 & 8 & 6 \end{bmatrix} \quad \text{(b) } \mathbf{D} = \begin{bmatrix} 1 & 1 & 1 \\ 4 & 6 & 8 \\ 12 & 8 & 4 \end{bmatrix} \quad \text{(c) } \mathbf{D} = \begin{bmatrix} 1 & 1 & 1 \\ 3 & 5 & 5 \\ 10 & 10 & 15 \end{bmatrix}.$$

For each model, find an arbitrage opportunity, if any.

Solution. First, solve the matrix equation $\mathbf{D}\,\boldsymbol{\Psi} = \mathbf{S}_0$ for each matrix $\mathbf{D}$:

(a) $\boldsymbol{\Psi} = \left[\frac{8}{13}, \frac{2}{13}, \frac{3}{13}\right]^\top$;

(b) $\boldsymbol{\Psi} = \left[\frac{1}{2} + t, \frac{1}{2} - 2t, t\right]^\top$, where $t \in \mathbb{R}$;

(c) $\boldsymbol{\Psi} = [0, 1, 0]^\top$.

In case (a), the solution is unique and strictly positive. In case (b), we have a one-parameter family of solutions that are strictly positive iff $0 < t < \frac{1}{4}$. In case (c), the unique solution has non-positive components. Therefore, the model is arbitrage free only in cases (a) and (b) in which there exists at least one strictly positive $\boldsymbol{\Psi}$. Let us find an arbitrage portfolio in case (c). In doing so, we replicate the AD security $\mathcal{E}^1$ by solving the replication equation

$$\varphi\,\mathbf{D} = \mathcal{E}^1(\Omega), \quad \text{where } \mathcal{E}^1(\Omega) = [1, 0, 0].$$

The solution $\boldsymbol{\varphi} = [\frac{5}{2}, -\frac{1}{2}, 0]$ is an arbitrage portfolio since its initial value is zero,

$$\Pi_0^\varphi = \boldsymbol{\varphi}\,\mathbf{S}_0 = \frac{5}{2}\cdot 1 + \left(-\frac{1}{2}\right)\cdot 5 = 0,$$

its terminal value $\Pi_T^\varphi(\omega) = \mathcal{E}^1(\omega)$ is nonnegative for all $\omega \in \Omega$, and $\Pi_T^\varphi(\omega^1) = \mathcal{E}^1(\omega^1) = 1 > 0$. Alternatively, we can find another arbitrage portfolio $\boldsymbol{\psi} = [-2, 0, \frac{1}{5}]$ that replicates the AD security $\mathcal{E}^3$, as in the previous example. In this case we have $\Pi_0^\psi = 0$ and $\Pi_T^\psi = \mathcal{E}^3 \geqslant 0$ with $\Pi_T^\psi(\omega^3) = 1 > 0$. $\qquad\square$

5.3.3.3   A Geometric Interpretation of the First FTAP

We now give a geometric interpretation of Theorem 5.9. Define the set $C \subset \mathbb{R}^N$ by

$$C := \left\{ \sum_{j=1}^M x_j\,\mathbf{S}_T(\omega^j) \ : \ x_j \geqslant 0,\ 1 \leqslant j \leqslant M \right\}.$$

It is a convex closed cone, which is contained in the span of the M column vectors of the payoff matrix, $\mathbf{S}_T(\omega^j)$, $j = 1, 2, \ldots, M$. The absence of arbitrage means that the initial price vector $\mathbf{S}_0 \in \mathbb{R}^N$ lies in the interior of C. Indeed, suppose that $\mathbf{S}_0$ lies in the exterior or on the boundary of the cone C. The separating hyperplane theorem guarantees that there exists a hyperplane described by the equation

$$\boldsymbol{\theta}\mathbf{x}^\top = 0, \quad \mathbf{x} \in \mathbb{R}^N,$$

for some $\boldsymbol{\theta} \in \mathbb{R}^N$, which would separate the ray generated by $\mathbf{S}_0$ from the interior of the cone C. This fact implies the following inequalities:

$$\boldsymbol{\theta}\mathbf{S}_0 \leqslant 0 \text{ and } \boldsymbol{\theta}\mathbf{S}_T(\omega^j) \geqslant 0 \text{ for all } j = 1, 2, \ldots, M,$$

with $\boldsymbol{\theta}\mathbf{S}_T(\omega^k) > 0$ for some k. That is, $\boldsymbol{\theta}$ is an arbitrage portfolio.

For example, consider the 2-by-2 model with risk-free bond $S^1 \equiv B$ and risky stock $S^2 \equiv S$. The initial price vector and payoff matrix are, respectively, given by

$$\mathbf{S}_0 = \begin{bmatrix} (1+r)^{-1} \\ S_0 \end{bmatrix} \text{ and } \mathbf{D} = \begin{bmatrix} \mathbf{S}_T(\omega^1) \mid \mathbf{S}_T(\omega^2) \end{bmatrix} = \begin{bmatrix} 1 & 1 \\ S_0 u & S_0 d \end{bmatrix}.$$

The cone $C \subset \mathbb{R}^2$ is generated by strictly positive linear combinations of the vectors $\mathbf{S}_T(\omega^1)$ and $\mathbf{S}_T(\omega^2)$, i.e., according to (5.10) we have $\mathbf{S}_0 = \Psi_1\mathbf{S}_T(\omega^1) + \Psi_2\mathbf{S}_T(\omega^2)$ with $\Psi_1, \Psi_2 > 0$. The no-arbitrage condition means that the vector $\mathbf{S}_0$ lies inside the cone C. As follows from Figure 5.2, this is possible iff

$$S_0 d < (1+r)S_0 < S_0 u \iff d < 1 + r < u.$$

This criterion can be generalized on any model with $N \geqslant 2$ base assets and arbitrarily many states $M \geqslant 2$. The vector $\mathbf{S}_0$ lies inside the cone C iff

$$\min_{\omega \in \Omega} \frac{S_T^2(\omega)}{S_T^1(\omega)} < \frac{S_0^2}{S_0^1} < \max_{\omega \in \Omega} \frac{S_T^2(\omega)}{S_T^1(\omega)}.$$

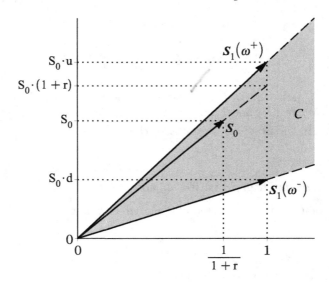

FIGURE 5.2: The no-arbitrage condition means that the initial price vector $\mathbf{S}_0$ has to lie inside the cone $C \subset \mathbb{R}_+^N$ generated by the M column vectors of the payoff matrix $\mathbf{D}$. This situation is illustrated in the case where $M = N = 2$.

5.3.4   Risk-Neutral Probabilities

An important component of Theorem 5.9 is the M-dimensional vector $\mathbf{\Psi} \gg 0$ which solves $\mathbf{D}\,\mathbf{\Psi} = \mathbf{S}_0$. As follows from (5.11), the initial value of a portfolio $\varphi \in \mathbb{R}^N$ is equal to the product of the terminal payoff vector and the state-price vector, $\Pi_0^\varphi = \Pi_T^\varphi(\Omega)\,\mathbf{\Psi}$. Therefore, the no-arbitrage initial value $\pi_0(X)$ of an attainable payoff X replicated by φ_X can be calculated as

$$\pi_0(X) = \Pi_0[\varphi_X] = \sum_{j=1}^{M} \Pi_T[\varphi_X](\omega^j)\,\Psi_j = \sum_{j=1}^{M} X(\omega^j)\,\Psi_j = \mathbf{X}\,\mathbf{\Psi}, \qquad (5.13)$$

where $\mathbf{X} = [X(\omega^1), \ldots, X(\omega^M)]$ is the payoff (row) vector in $\mathbb{R}^M$.

As is seen from (5.13), a replicating portfolio is not required to calculate the no-arbitrage initial value (i.e., the fair price) of a claim with an attainable payoff. We only need to find the vector $\mathbf{\Psi}$ and then take its inner product with the payoff vector to find $\pi_0(X)$. The solutions $\Psi_j \in (0, \infty)$, $j = 1, 2, \ldots, M$, are called *state prices*. To explain such a name, consider an Arrow–Debreu security $\mathcal{E}^j = \mathbb{I}_{\{\omega^j\}}$ having a nonzero payoff of one unit only for the state ω^j. According to (5.13), the initial price of $\mathcal{E}^j$ is

$$\pi_0(\mathcal{E}^j) = \sum_{k=1}^{M} \mathcal{E}^j(\omega^k)\,\Psi_k = \sum_{k=1}^{M} \mathbb{I}_{\{\omega^j\}}(\omega^k)\,\Psi_k = \Psi_j. \qquad (5.14)$$

Note that the j-th AD security has payoff given by the unit vector with unit j-th component, $\mathcal{E}^j(\Omega) = \hat{\mathbf{e}}_j$, where $\pi_0(\mathcal{E}^j) = \hat{\mathbf{e}}_j\,\mathbf{\Psi} = \Psi_j$. Thus, the financial meaning of the components $\Psi_j \in (0, \infty)$, $j = 1, 2, \ldots, M$, is that they are no-arbitrage prices of the Arrow–Debreu securities. In particular, Ψ_j can also be thought of as the price of an insurance contract that pays one unit of currency in case scenario ω^j occurs.

The pricing formula (5.13) can also be derived using (5.14) and the principle of linear pricing. According to Proposition 5.1, every payoff X can be represented as a portfolio in AD securities:

$$X = x_1 \mathcal{E}^1 + x_2 \mathcal{E}^2 + \cdots + x_M \mathcal{E}^M \quad \text{for some} \quad x_1, x_2, \ldots, x_M \in \mathbb{R}.$$

Thus, the initial value $\pi_0(X)$ is the same as that of the portfolio in the AD securities:

$$\pi_0(X) = x_1 \pi_0(\mathcal{E}^1) + x_2 \pi_0(\mathcal{E}^2) + \cdots + x_M \pi_0(\mathcal{E}^M)$$
$$= x_1 \Psi_1 + x_2 \Psi_2 + \cdots + x_M \Psi_M = \mathbf{X}\,\mathbf{\Psi}.$$

Consider a risk-less asset B that pays one unit of currency at the terminal time (i.e., B is a zero-coupon bond). According to (5.13), since $B_T(\omega^j) = 1$ for all $j = 1, \ldots, M$, the initial price of the payoff $B_T \equiv 1$ is

$$B_0 := \pi_0(B_T) = \sum_{j=1}^{M} \Psi_j.$$

Let us set $1 + r = \left(\sum_{j=1}^{M} \Psi_j \right)^{-1}$. Hence $B_0 = \frac{1}{1+r}$. The quantity r is in fact the single-period return on B:

$$\frac{B_T - B_0}{B_0} = \frac{B_T}{B_0} - 1 = \frac{1}{(1+r)^{-1}} - 1 = r.$$

Now we are ready to rewrite the security pricing formula (5.13) in a more familiar form. First, note that the positive state prices Ψ_j, $j = 1, 2, \ldots, M$, can be normalized to give us a new set of probabilities:

$$\tilde{p}_j := \frac{\Psi_j}{\sum_{i=1}^{M} \Psi_i}, \quad j = 1, 2, \ldots, M. \tag{5.15}$$

Clearly, all $\tilde{p}_j$'s are positive and sum to unity. These probabilities are called *risk-neutral probabilities* and they have no relation to the real-world (physical) probabilities $\{p_j = \mathbb{P}(\omega^j)\}$. The real-world probabilities can be estimated from historical data, whereas the risk-neutral probabilities are not observed. Let $\widetilde{\mathbb{P}}$ denote the *risk-neutral probability measure* defined by $\widetilde{\mathbb{P}}(\omega^j) = \tilde{p}_j$ for $j = 1, 2, \ldots, M$. The risk-neutral probability of any event $E \subseteq \Omega$ is then given by

$$\widetilde{\mathbb{P}}(E) = \sum_{\omega \in E} \widetilde{\mathbb{P}}(\omega) = \sum_{\omega^j \in E} \tilde{p}_j.$$

The mathematical expectation of a random variable $X \in \mathcal{L}(\Omega)$ with respect to $\widetilde{\mathbb{P}}$, denoted by $\widetilde{\mathbb{E}}[X]$, is defined by

$$\widetilde{\mathbb{E}}[X] = \sum_{j=1}^{M} X(\omega^j)\, \tilde{p}_j.$$

Using (5.15), the initial price in (5.13) can now be written as the mathematical expectation of the discounted payoff function with respect to the risk-neutral probability measure:

$$\pi_0(X) = \sum_{j=1}^{M} X(\omega^j)\, \Psi_j = \left(\sum_{k=1}^{M} \Psi_k \right) \sum_{j=1}^{M} X(\omega^j)\, \frac{\Psi_j}{\sum_{k=1}^{M} \Psi_k}$$

$$= \frac{1}{1+r} \sum_{j=1}^{M} X(\omega^j)\, \tilde{p}_j = \frac{1}{1+r} \widetilde{\mathbb{E}}[X]. \tag{5.16}$$

Here, $\frac{X}{1+r}$ is the discounted value of X. That is, it is the present (time-0) value of the payoff X paid at time T. If we apply (5.16) to a portfolio that only contains one unit of the i-th base asset S^i, then we have that

$$S_0^i = \frac{1}{1+r}\widetilde{\mathbb{E}}[S_T^i]. \tag{5.17}$$

Therefore, the risk-neutral expectation of the return on asset S^i is

$$\widetilde{\mathbb{E}}\left[\frac{S_T^i - S_0^i}{S_0^i}\right] = \frac{\widetilde{\mathbb{E}}[S_T^i]}{S_0^i} - 1 = (1+r) - 1 = r.$$

In other words, in the risk-neutral probability measure, the expected return of any base security is equal to the risk-free interest rate r on the riskless asset (e.g., bond) B. Since $B_0/B_T = (1+r)^{-1}$, we can rewrite (5.17) as

$$\frac{S_0^i}{B_0} = \widetilde{\mathbb{E}}\left[\frac{S_T^i}{B_T}\right] \tag{5.18}$$

for every base asset S^i, $i = 1, \ldots, N$. In other words, the discounted asset price processes $\left\{\overline{S}_t^i := \frac{S_t^i}{B_t}\right\}_{t \in \{0,T\}}$, $i = 1, 2, \ldots, N$, are martingales under the risk-neutral probability measure $\widetilde{\mathbb{P}}$. The subject of martingales will be introduced and covered in depth in the next chapter. It suffices to note here that the discounted price processes $\left\{\overline{S}_t^i\right\}_{t \in \{0,T\}}$ are examples of single-period stochastic processes. A single-period stochastic process, say $\{X_t\}_{t \in \{0,T\}}$, is a called a martingale with respect to the measure $\widetilde{\mathbb{P}}$ if $\widetilde{\mathbb{E}}[X_T] = X_0$. That is, our best prediction of the future time-T value (i.e., the expected future value with respect to the given measure) of the process is the same as the current time-0 value.

To summarize, if the single-period market model is arbitrage-free, then there exists a risk-neutral probability measure $\widetilde{\mathbb{P}}$. Conversely, assuming no arbitrage, we can construct a state-space vector from the risk-neutral probabilities as

$$\Psi_j = \frac{\tilde{p}_j}{1+r}, \quad j = 1, 2, \ldots, M.$$

Hence, the existence of $\widetilde{\mathbb{P}}$ implies no arbitrage and vice versa. Therefore, Theorem 5.9 can now be formulated as follows.

Theorem 5.10 (The first FTAP—the 2nd version). *There are no arbitrage portfolios in a single-period N-by-M model iff there exist probabilities $\{\widetilde{\mathbb{P}}(\omega^j) \equiv \tilde{p}_j > 0 \; : \; j = 1, 2, \ldots, M\}$ such that the discounted asset price processes $\left\{\overline{S}_t^i\right\}_{t \in \{0,T\}}$, $i = 1, 2, \ldots, N$, are all martingales with respect to the probability measure $\widetilde{\mathbb{P}}$. Such a probability measure is called the risk-neutral probability measure.*

Let us come back to the binomial model from Example 5.1 with two states and two assets: $S_t^1 = B_t$ and $S_t^2 = S_t$. We readily solve for the state-price vector as follows:

$$\begin{bmatrix} 1 & 1 \\ S_0 u & S_0 d \end{bmatrix} \begin{bmatrix} \Psi_1 \\ \Psi_2 \end{bmatrix} = \begin{bmatrix} (1+r)^{-1} \\ S_0 \end{bmatrix} \iff \begin{cases} \Psi_1 + \Psi_2 = \frac{1}{1+r} \\ u\,\Psi_1 + d\,\Psi_2 = 1 \end{cases}$$

$$\iff \begin{cases} \Psi_1 = \frac{(1+r)-d}{(1+r)(u-d)} \\ \Psi_2 = \frac{u-(1+r)}{(1+r)(u-d)} \end{cases} \tag{5.19}$$

The solution $\mathbf{\Psi} = [\Psi_1, \Psi_2]^\top$ in (5.19) is strictly positive iff $d < 1 + r < u$. If $1 + r \leqslant d$, then an arbitrage portfolio can be obtained by shorting the bond (for example, $\varphi_1 = -S_0(1 + r)$ and $\varphi_2 = 1$). If $1 + r \geqslant u$, then an arbitrage portfolio can be obtained by shorting the stock (for example, $\varphi_1 = S_0(1 + r)$ and $\varphi_2 = -1$). The reader may verify that Π_0^φ is zero and Π_T^φ is positive in both cases. Finally, we can calculate the risk-neutral probabilities for the binomial model:

$$\tilde{p} \equiv \tilde{p}_1 = \frac{\Psi_1}{\Psi_1 + \Psi_2} = \frac{(1 + r) - d}{u - d}, \quad 1 - \tilde{p} \equiv \tilde{p}_2 = \frac{\Psi_2}{\Psi_1 + \Psi_2} = \frac{u - (1 + r)}{u - d}.$$

The probabilities are strictly positive iff the no-arbitrage condition $d < 1 + r < u$ holds. Finally, note that the state prices are themselves given by the discounted (risk-neutral) expected value of the Arrow–Debreu payoffs:

$$\Psi_j = \frac{1}{1 + r} \widetilde{\mathrm{E}}[\mathcal{E}^j], \ j = 1, \dots, M.$$

5.3.5 The Second Fundamental Theorem of Asset Pricing

We are now ready to present the second fundamental theorem of asset pricing which states a sufficient and necessary condition of completeness of a market model. We give two versions of the theorem: first in terms of state prices and then in terms of risk-neutral probabilities.

Theorem 5.11 (The second FTAP). *Assuming absence of arbitrage, there exists a unique solution $\mathbf{\Psi} \gg 0$ to the state-price equation (5.10) iff the market is complete.*

Theorem 5.12 (The second FTAP—the 2nd version). *Assuming absence of arbitrage, there exists a unique set of risk-neutral probabilities $\{\tilde{p}_j > 0 : j = 1, 2, \dots, M\}$, as stated in the first FTAP above, iff the market is complete.*

Proof. The no-arbitrage assumption implies that there exists a strictly positive solution $\mathbf{\Psi}$ to the matrix equation $\mathbf{D}\,\mathbf{\Psi} = \mathbf{S}_0$. If the market is complete, then $\mathrm{rank}(\mathbf{D}) = M$. Hence the solution is unique. Conversely, let the solution $\mathbf{\Psi} \gg 0$ be unique. We will argue that the market is complete by contradiction. If the market is not complete, then $\mathrm{rank}(\mathbf{D}) < M$. Therefore, there exists a nontrivial solution $\boldsymbol{\lambda} \neq \mathbf{0}$ to the matrix equation $\mathbf{D}\,\boldsymbol{\lambda} = \mathbf{0}$. Since $\Psi_j > 0$ for all $1 \leqslant j \leqslant M$, there exists a sufficiently small constant $a \neq 0$ such that $\Psi_j + a\lambda_j > 0$ for all $1 \leqslant j \leqslant M$, i.e., $\mathbf{\Psi} + a\boldsymbol{\lambda} \gg 0$. Moreover, this vector solves

$$\mathbf{D}\,(\mathbf{\Psi} + a\boldsymbol{\lambda}) = \mathbf{D}\,\mathbf{\Psi} + a\mathbf{D}\,\boldsymbol{\lambda} = \mathbf{S}_0 + a\,\mathbf{0} = \mathbf{S}_0.$$

Thus, $\mathbf{\Psi} + a\boldsymbol{\lambda}$ is another state-price vector. This contradicts our hypothesis that the state-price vector is unique. □

Example 5.7. Verify whether or not the single-period model from Example 5.6 is complete.

Solution. In cases (a) and (b), there exists a positive solution to the state-price equation $\mathbf{D}\,\mathbf{\Psi} = \mathbf{S}_0$. Hence, there are no arbitrage opportunities. In case (a), the solution $\mathbf{\Psi}$ is unique; therefore, the model is complete. In case (b), there are infinitely many positive solutions parametrized by $t \in (0, \frac{1}{4})$; therefore, the model is incomplete. Indeed, in case (b) the determinant of the payoff matrix is zero; hence, the model is incomplete. For example, the payoff of an at-the-money call option on asset S^2 is unattainable since the linear system $\varphi\mathbf{D} = [0, 0, 2]$ is inconsistent, i.e., there is no replicating portfolio for this call option. □

5.3.6   Investment Portfolio Optimization

A no-arbitrage price of a derivative security is calculated by taking the risk-neutral expectation of the payoff function. So, a model that does not admit arbitrage opportunities has two probability measures, the actual (real-world) measure and the risk-neutral measure. We only use the latter in no-arbitrage pricing. As is demonstrated in previous chapters, the real-world measure is used in asset management and risk management. However, as is shown below, the risk-neutral measure also plays an important role when determining an optimal portfolio allocation.

Consider the following investment problem for an arbitrage-free complete model. Being given an initial capital W_0, we find a portfolio $\varphi = \mathbb{R}^N$ with initial value $\Pi_0^\varphi = W_0$ that maximizes the (real-world) expected utility of the terminal portfolio value, $\mathrm{E}[u(\Pi_T^\varphi)]$, where u is a utility function, i.e., $u(x)$ is generally a nondecreasing and concave function of $x \in \mathbb{R}$. That is, we need to find the solution to the following constrained optimization problem:

$$\text{maximize } \mathrm{E}[u(\Pi_T^\varphi)] = \sum_{j=1}^{M} u(\Pi_T^\varphi(\omega^j))\, p_j = \sum_{j=1}^{M} u\left(\sum_{i=1}^{N} \varphi_i D_{ij}\right) p_j \text{ w.r.t. } \varphi \in \mathbb{R}^N \quad (5.20)$$

$$\text{subject to } \Pi_0^\varphi = \sum_{i=1}^{N} \varphi_i S_0^i = W_0. \quad (5.21)$$

Since the market model is complete, the solution of the problem (5.20)–(5.21) can be split into two steps: first, we find the terminal value of the optimal portfolio Π_T^φ, i.e., an optimal payoff X that maximizes the expected utility $\mathrm{E}[u(X)]$; second, we find the portfolio $\varphi \equiv \varphi_X$ that replicates the given optimal payoff X. To obtain a constraint equation on X, we use the fact that the discounted value process is a martingale under the risk-neutral probability measure:

$$\frac{1}{1+r}\tilde{\mathrm{E}}\big[\Pi_T^\varphi\big] = \Pi_0^\varphi = W_0.$$

By assumed replication, we have $X = \Pi_T^\varphi$ and using state prices gives

$$\frac{1}{1+r}\tilde{\mathrm{E}}[X] = \frac{1}{1+r}\sum_{j=1}^{M} x_j\, \tilde{p}_j = \sum_{j=1}^{M} x_j \Psi_j = W_0.$$

The optimization problem (5.20)–(5.21) now takes the form

$$\mathrm{E}[u(X)] = \sum_{j=1}^{M} u(x_j)\, p_j \to \max_{\mathbf{X} \in \mathbb{R}^M} \quad (5.22)$$

$$\text{subject to } \mathbf{X}\,\Psi = \sum_{j=1}^{M} x_j \Psi_j = W_0. \quad (5.23)$$

Here, $\mathbf{X} = [x_1, x_2, \ldots, x_M]$ denotes the payoff vector for $X \in \mathcal{L}(\Omega)$, i.e., $X(\omega^j) = x_j$ for all $j = 1, 2, \ldots, M$. Let $\mathbf{X}^*$ be a solution to (5.22)–(5.23). Solving the matrix-vector equation $\Pi_T^{\varphi^*} = \varphi^* \mathbf{D} = \mathbf{X}^*$ gives the optimal portfolio $\varphi = \varphi^*$ replicating the optimal payoff vector $\mathbf{X}^*$. Under the completeness assumption, the solution φ^* exists and is unique.

Example 5.8. Consider a single-period model with two scenarios $\{\omega^+, \omega^-\}$ and two base assets, namely, a risky stock with prices $S_T(\omega^+) = 12$, $S_T(\omega^-) = 8$, $S_0 = 10$, and an at-the-money call option on the stock with initial price $C_0 = 1$. Suppose that the real-world probabilities are $p = \mathbb{P}(\omega^+) = \frac{1}{4}$ and $1 - p = \mathbb{P}(\omega^-) = \frac{3}{4}$. Find the optimal portfolio (φ_1, φ_2) with initial value $W_0 = 100$ that maximizes $\mathrm{E}\big[\sqrt{\Pi_T[(\varphi_1, \varphi_2)]}\big]$.

Solution. The strike price of the call option is $K = S_0$; the payoff is $C_T(\Omega) = (S_T(\Omega) - S_0)_+ = [2, 0]$. Hence, the initial price vector and payoff matrix are, respectively, given by

$$\mathbf{S}_0 = \begin{bmatrix} 10 \\ 1 \end{bmatrix}, \quad \mathbf{D} = \begin{bmatrix} 12 & 8 \\ 2 & 0 \end{bmatrix}.$$

Solving the matrix-vector equation, $\mathbf{D}\,\mathbf{\Psi} = \mathbf{S}_0$ for $\mathbf{\Psi} = [\Psi_1, \Psi_2]^\top$, gives the state prices $\Psi_1 = \Psi_2 = \frac{1}{2}$.

First, we find the payoff vector $\mathbf{X} = [x_1, x_2]$ that solves the optimization problem

$$E[\sqrt{X}] = \frac{1}{4}\sqrt{x_1} + \frac{3}{4}\sqrt{x_2} \to \max_{x_1, x_2}$$

$$\text{subject to } x_1\Psi_1 + x_2\Psi_2 = \frac{x_1 + x_2}{2} = 100.$$

The optimal value of x_1 occurs at a maximum of the function

$$f(x) = \frac{1}{4}\sqrt{x} + \frac{3}{4}\sqrt{200 - x}.$$

Equating the derivative $f'(x) = \frac{1}{8\sqrt{x}} - \frac{3}{8\sqrt{200-x}}$ to zero gives the solution $x = 20$. Since the second derivative $f''(x)$ is strictly negative for all $x \in (0, 200)$, the function f attains its maximum at $x = 20$. Thus, the terminal payoff of the optimal portfolio is given by

$$x_1 = 20 \quad \text{and} \quad x_2 = 180.$$

The Profit&Loss realized is $\Pi_T^{\varphi^*} - \Pi_0^{\varphi^*} = X - W_0$, which is equal to -80 in state ω^+ and 80 in state ω^-.

Second, we find portfolio $[\varphi_1, \varphi_2] = [\varphi_1^*, \varphi_2^*]$ by replicating the payoff $\mathbf{X} = [20, 180]$. Solving the replication equations gives:

$$\begin{cases} 12\varphi_1 + 2\varphi_2 = 20 \\ 8\varphi_1 + 0\varphi_2 = 180 \end{cases} \implies \begin{cases} \varphi_1 = \frac{45}{2} \\ \varphi_2 = -125 \end{cases}$$

So, the optimal allocation portfolio is $[\varphi_1^*, \varphi_2^*] = [\frac{45}{2}, -125]$. This means that we should sell (short) 125 call contracts and purchase 22.5 units of stock. $\square$

To find a general solution to the optimization problem (5.22)–(5.23), we apply the method of Lagrange multipliers. The Lagrangian is

$$L(\mathbf{X}, \lambda) = \sum_{j=1}^{M} u(x_j)\, p_j - \lambda \left(\sum_{j=1}^{M} x_j \Psi_j - W_0 \right).$$

Differentiating L w.r.t. $x_1, x_2, \ldots, x_M$, and λ, and equating the derivatives obtained to zero gives the following simultaneous equations:

$$\frac{\partial L}{\partial x_j} = u'(x_j)\, p_j - \lambda \Psi_j = 0, \quad j = 1, 2, \ldots, M, \tag{5.24}$$

$$\frac{\partial L}{\partial \lambda} = W_0 - \sum_{j=1}^{M} x_j \Psi_j = 0. \tag{5.25}$$

Suppose that the derivative of the utility function, u', is strictly monotone, i.e., it is a strictly increasing function everywhere it is finite. Additionally, we assume that the range of possible values of $u'(x)$ includes all positive reals. Examples of such utility functions include $\ln x$ and x^γ with $\gamma \in (0,1)$. Under the above assumptions, the equation $u'(x) = y$ has a unique solution for every $y \in (0,\infty)$. Thus, we can define the inverse function v for u', $v(y) := (u')^{-1}(y)$, with the property that $v(u'(y)) = u'(v(y)) = y$ for all $y \in (0,\infty)$.

Now, solving the equations in (5.24) individually gives (for any λ)

$$u'(x_j) = \frac{\lambda \Psi_j}{p_j} \implies x_j = v\left(\frac{\lambda \Psi_j}{p_j}\right), \quad j = 1, 2, \ldots, M. \tag{5.26}$$

So, we have a formula for the optimal payoff in terms of the multiplier λ. Substituting (5.26) into (5.25) gives

$$\sum_{j=1}^{M} v\left(\frac{\lambda \Psi_j}{p_j}\right) \Psi_j = W_0. \tag{5.27}$$

Solving (5.27) for λ and substituting the solution λ^* into (5.26) gives the optimal payoff vector components

$$x_j^* = v\left(\frac{\lambda^* \Psi_j}{p_j}\right), \quad j = 1, 2, \ldots, M. \tag{5.28}$$

Let us show that the solution X^* with values in (5.28) maximizes $\mathrm{E}[u(X)]$. That is, let us show that for every payoff $X \in \mathcal{L}(\Omega)$ chosen such that $\pi_0(X) = \mathbf{X} \mathbf{\Psi} = W_0$ we have

$$\mathrm{E}[u(X)] \leqslant \mathrm{E}[u(X^*)].$$

Consider the function $f(x) := u(x) - yx$ with a positive parameter y. Clearly, $x = v(y)$ maximizes f. Indeed, solving $f'(x) = u'(x) - y = 0$ gives $x = v(y)$; since $f''(x) = u''(x) \leqslant 0$ for all x, $x = v(y)$ is a point of maximum. Therefore, we have

$$u(x) - yx \leqslant u(v(y)) - yv(y) \quad \text{for all} \quad x. \tag{5.29}$$

Replacing x and y in (5.29) by x_j and $\frac{\lambda^* \Psi_j}{p_j}$, respectively, multiplying both parts of the inequality by p_j, and then using (5.28) gives

$$u(x_j)p_j - \lambda^* \Psi_j x_j \leqslant u\left(x_j^*\right) p_j - \lambda^* \Psi_j x_j^* \quad \text{for all } j = 1, 2, \ldots, M.$$

Adding all the above M inequalities up gives

$$\sum_{j=1}^{M} u(x_j)p_j - \lambda^* \sum_{j=1}^{M} \Psi_j x_j \leqslant \sum_{j=1}^{M} u\left(x_j^*\right) p_j - \lambda^* \sum_{j=1}^{M} \Psi_j x_j^*.$$

This is equivalent to

$$\mathrm{E}[u(X)] - \lambda \pi_0(X) \leqslant \mathrm{E}[u(X^*)] - \lambda \pi_0(X^*).$$

Since the payoffs X and X^* have the same initial value equal to W_0, we obtain that $\mathrm{E}[u(X)] \leqslant \mathrm{E}[u(X^*)]$. That is, X^* maximizes $\mathrm{E}[u(X)]$ and hence it is the payoff of the optimal portfolio. Finally, note that the difference $X^*(\omega^j) - W_0$ is the actual gain (if positive) or loss (if negative) realized in state ω^j. Under the no-arbitrage condition, the gain $X - W_0$ (if it is not identically zero) has to change its sign: to be negative in at least one state and positive in another state.

Example 5.9. Consider a 3-by-3 model with

$$\mathbf{D} = \begin{bmatrix} 1 & 1 & 1 \\ 1 & 6 & 15 \\ 12 & 8 & 6 \end{bmatrix} \quad \text{and} \quad \mathbf{S}_0 = \begin{bmatrix} 1 \\ 5 \\ 10 \end{bmatrix}.$$

Suppose that the real-world-probabilities are $p_1 = p_3 = \frac{1}{4}$ and $p_2 = \frac{1}{2}$. Find the optimal portfolio φ that maximizes $\mathrm{E}[\ln(\Pi_T[\varphi])]$ and has initial value $W_0 = 1$. Find the gain/loss realized.

Solution. By solving $\mathbf{D}\,\boldsymbol{\Psi} = \mathbf{S}_0$, we obtain the state-price vector $\boldsymbol{\Psi} = \left[\frac{8}{13}, \frac{2}{13}, \frac{3}{13}\right]^{\top}$. It is strictly positive and unique, hence the model is arbitrage-free and complete. The utility function $u(x) = \ln x$ satisfies the criteria stated above since the derivative $u'(x) = \frac{1}{x}$ varies from 0 to ∞ as $x \nearrow \infty$ and $x \searrow 0$, respectively. Substituting the inverse of the derivative, $v(y) = \frac{1}{y}$, into (5.28) gives the optimal payoff $\mathbf{X}^*$ in terms of λ^*:

$$x_j^* = \frac{p_j}{\lambda^* \Psi_j}, \; j = 1, 2, 3 \implies \mathbf{X}^* = \frac{1}{\lambda^*}\left[\frac{13}{32}, \frac{13}{4}, \frac{13}{12}\right].$$

Solving (5.27) with $W_0 = 1$ for λ gives

$$1 = \sum_{j=1}^{3} \frac{p_j}{\lambda \Psi_j} \Psi_j = \frac{1}{\lambda}\sum_{j=1}^{3} p_j = \frac{1}{\lambda} \implies \lambda^* = 1.$$

Therefore, the optimal payoff vector is $\mathbf{X}^* = \left[\frac{13}{32}, \frac{13}{4}, \frac{13}{12}\right]$. Solving the matrix-vector equation $\varphi\,\mathbf{D} = \mathbf{X}^*$ gives the optimal allocation portfolio:

$$\varphi^* = \left[\frac{853}{48}, -\frac{53}{96}, -\frac{269}{192}\right].$$

The gain of the investment is $\mathbf{X}^* - W_0 = \left[-\frac{19}{32}, \frac{9}{4}, \frac{1}{12}\right]$. So the return on the investment is only positive in states ω^2 and ω^3. $\quad\square$

5.4 Pricing in an Incomplete Market

5.4.1 A Trinomial Model of an Incomplete Market

The simplest single-period incomplete market model is a model with three states of the world, i.e., $\Omega = \{\omega^1, \omega^2, \omega^3\} \equiv \{\omega^+, \omega^0, \omega^-\}$, and two assets: B and S. The risk-free asset B provides the rate of return r and it pays \$1 at maturity:

$$B_0 = \frac{1}{1+r}, \quad B_T = 1.$$

The risky asset S admits three possible cash flows at time T:

$$S_T(\omega) = \begin{cases} S_0 u & \text{if } \omega = \omega^1, \\ S_0 m & \text{if } \omega = \omega^2, \\ S_0 d & \text{if } \omega = \omega^3, \end{cases}$$

where $S_0 > 0$ is the initial price and $0 < d < m < u$ are price factors. The initial price vector and the cash-flow matrix are, respectively, given by

$$\mathbf{S}_0 = \begin{bmatrix} (1+r)^{-1} \\ S_0 \end{bmatrix} \quad \text{and} \quad \mathbf{D} = \begin{bmatrix} B_T(\omega^1) & B_T(\omega^2) & B_T(\omega^3) \\ S_T(\omega^1) & S_T(\omega^2) & S_T(\omega^3) \end{bmatrix} = \begin{bmatrix} 1 & 1 & 1 \\ S_0 u & S_0 m & S_0 d \end{bmatrix}.$$

$\mathbf{D}$ has rank 2 since $\begin{vmatrix} 1 & 1 \\ S_0 u & S_0 m \end{vmatrix} = S_0(m - u) \neq 0$. The two payoff vectors $[1, 1, 1]$ and $[S_0 u, S_0 m, S_0 d]$ are clearly independent. Therefore, we conclude that there are no redundant base assets but the market is incomplete as the two vectors do not span $\mathbb{R}^3$. We wish to find all strictly positive solutions $\mathbf{\Psi}$ to the equation $\mathbf{D}\mathbf{\Psi} = \mathbf{S}_0$. This is a linear system of two equations and three unknowns. Generally, the solution represents a line in $\mathbb{R}^3$ corresponding to the intersection of two planes:

$$\begin{cases} \Psi_1 + \Psi_2 + \Psi_3 = (1+r)^{-1} \\ u\Psi_1 + m\Psi_2 + d\Psi_3 = 1 \end{cases} \iff \begin{cases} \Psi_1 = \dfrac{(1+r) - d}{(1+r)(u-d)} - \dfrac{m-d}{u-d} c \\ \Psi_2 = c \\ \Psi_3 = \dfrac{u - (1+r)}{(1+r)(u-d)} - \dfrac{u-m}{u-d} c \end{cases} \tag{5.30}$$

We set $c > 0$, giving $\Psi_2 > 0$. The limiting value of $\mathbf{\Psi}$ as $c \searrow 0$ is

$$\Psi_2 \searrow 0, \quad \Psi_1 \nearrow \frac{(1+r) - d}{(1+r)(u-d)}, \quad \text{and} \quad \Psi_3 \nearrow \frac{u - (1+r)}{(1+r)(u-d)}.$$

The limiting values of Ψ_1 and Ψ_3, as $c \searrow 0$, are strictly positive iff $d < 1 + r < u$ holds. Therefore, there exists a strictly positive solution $\mathbf{\Psi} = \begin{bmatrix} \Psi_1, \Psi_2, \Psi_3 \end{bmatrix}^\top$ iff $d < 1 + r < u$. Indeed, if $1 + r \leqslant d$, then $\Psi_1 < 0$ for any $\Psi_2 = c > 0$; if $u \leqslant 1 + r$, then $\Psi_3 < 0$ for any $\Psi_2 = c > 0$. From (5.30), we obtain all values of c such that $\Psi_j > 0$, $j = 1, 2, 3$:

$$\Psi_1 > 0 \iff c < \frac{(1+r) - d}{(1+r)(m-d)},$$

$$\Psi_2 > 0 \iff c > 0,$$

$$\Psi_3 > 0 \iff c < \frac{u - (1+r)}{(1+r)(u-m)}.$$

Thus, the solution $\mathbf{\Psi} = \mathbf{\Psi}(c)$ from (5.30) is strictly positive iff

$$c \in (0, c_{\max}), \quad \text{where} \quad c_{\max} := \frac{1}{1+r} \min\left\{ \frac{(1+r) - d}{m-d}, \frac{u - (1+r)}{u-m} \right\}.$$

Such a set of positive solutions, denoted by $\{\mathbf{\Psi}\}$, is an open segment of a line in $\mathbb{R}^3$. Indeed, according to (5.30), $\mathbf{\Psi}(c)$ is a linear function of c. Hence all solutions lie on a straight line. Since $0 \leqslant \Psi_j \leqslant (1+r)^{-1}$ for $j = 1, 2, 3$, the set $\{\mathbf{\Psi}\}$ is bounded. The coordinates of the two endpoints of $\{\mathbf{\Psi}\}$, denoted by $\mathbf{\Psi}(0)$ and $\mathbf{\Psi}(c_{\max})$ are obtained by setting $c = 0$ and $c = c_{\max}$ in (5.30), respectively:

$$\mathbf{\Psi}(0) = \begin{bmatrix} \dfrac{(1+r) - d}{(1+r)(u-d)}, & 0, & \dfrac{u - (1+r)}{(1+r)(u-d)} \end{bmatrix}^\top, \tag{5.31}$$

$$\mathbf{\Psi}(c_{\max}) = \begin{cases} \begin{bmatrix} \dfrac{(1+r) - m}{(1+r)(u-m)}, & \dfrac{u - (1+r)}{(1+r)(u-m)}, & 0 \end{bmatrix}^\top, & \text{if } m \leqslant 1 + r, \\[4mm] \begin{bmatrix} 0, & \dfrac{(1+r) - d}{(1+r)(m-d)}, & \dfrac{m - (1+r)}{(1+r)(m-d)} \end{bmatrix}^\top, & \text{if } m \geqslant 1 + r. \end{cases} \tag{5.32}$$

All the solutions of $\{\Psi\}$ can be parametrized as $\Psi(v\,c_{\max}) = (1-v)\Psi(0) + v\Psi(c_{\max})$, where $v = {}^{c}/_{c_{\max}} \in (0,1)$.

The binomial model (with two states) is recovered as a limiting case of the trinomial model as $\Psi_2 \to 0$. By setting $c = 0$ in (5.30), we obtain

$$\Psi_1 = \frac{(1+r)-d}{(1+r)(u-d)}, \quad \Psi_2 = 0, \quad \Psi_3 = \frac{u-(1+r)}{(1+r)(u-d)}.$$

The formulae of the state prices Ψ_1 and Ψ_3 are exactly the same as those of the binomial state prices in (5.19). Two other extreme cases are obtained by setting either Ψ_1 or Ψ_3 to zero:

- Let $\Psi_1 \searrow 0$. Then $\Psi_3 = \frac{1}{1+r}\frac{m-(1+r)}{m-d}$ and $\Psi_2 = \frac{1}{1+r}\frac{(1+r)-d}{m-d}$. If $d < 1+r < m$, then both Ψ_2 and Ψ_3 are positive.

- Let $\Psi_3 \searrow 0$. Then $\Psi_1 = \frac{1}{1+r}\frac{(1+r)-m}{u-m}$ and $\Psi_2 = \frac{1}{1+r}\frac{u-(1+r)}{u-m}$. If $m < 1+r < u$, then both Ψ_1 and Ψ_2 are positive.

Alternatively, one can recover the binomial model in the limiting case as $m \searrow d$ or $m \nearrow u$.

For every positive state-price vector $\Psi \in \{\Psi\}$ we can identify a respective risk-neutral probability measure with the probabilities $\tilde{p}_j = (1+r)\Psi_j$, $j = 1,2,3$. Therefore, any incomplete arbitrage-free model has infinitely many risk-neutral probability measures. For the trinomial model described above, the collection of risk-neutral probability measures (i.e., the probability mass functions) can be denoted as vectors $\tilde{\mathbf{p}} = [\tilde{p}_1, \tilde{p}_2, \tilde{p}_3]$, forming an open linear segment in $\mathbb{R}^3$, whose endpoints are

$$\tilde{\mathbf{p}}(0) = (1+r)\Psi(0) \quad \text{and} \quad \tilde{\mathbf{p}}(c_{\max}) = (1+r)\Psi(c_{\max}). \tag{5.33}$$

Hence, we have a collection of risk-neutral measures and risk-neutral probability vectors that we denote by $\{\widetilde{\mathbb{P}}\}$ and $\{\tilde{\mathbf{p}}\}$, respectively.

Example 5.10. Consider a single-period trinomial model with two assets B and $\overset{\circ}{S}$ where $S_0 = \$100$, $d = 0.8$, $m = 1.1$, $u = 1.4$, $r = 0.2$. Show that the model is arbitrage-free. Find all strictly positive state-price vectors and risk-neutral probability vectors.

Solution. The initial price vector and payoff matrix are, respectively,

$$\mathbf{S}_0 = \begin{bmatrix} \frac{5}{6} \\ 100 \end{bmatrix} \quad \text{and} \quad \mathbf{D} = \begin{bmatrix} 1 & 1 & 1 \\ 140 & 110 & 80 \end{bmatrix}.$$

Let us find the general solution to the state-price equation $\mathbf{D}\,\Psi = \mathbf{S}_0$:

$$\begin{bmatrix} 1 & 1 & 1 & \Big| & \frac{5}{6} \\ 140 & 110 & 80 & \Big| & 100 \end{bmatrix} \sim \begin{bmatrix} 1 & \frac{1}{2} & 0 & \Big| & \frac{5}{9} \\ 0 & \frac{1}{2} & 1 & \Big| & \frac{5}{18} \end{bmatrix} \implies \begin{cases} \Psi_1 = \frac{5}{9} - \frac{c}{2}, \\ \Psi_2 = c, \\ \Psi_3 = \frac{5}{18} - \frac{c}{2}. \end{cases}$$

The state-price vector

$$\Psi(c) = \left[\frac{5}{9} - \frac{c}{2},\; c,\; \frac{5}{18} - \frac{c}{2}\right]^{\top} \tag{5.34}$$

is strictly positive iff $0 < c < \frac{5}{9}$. We hence have a continuum of strictly positive state-price vectors and there is no arbitrage. Alternatively, we notice that the condition $d < 1+r < u$ is satisfied and hence the model is arbitrage-free.

Now, we find the risk-neutral probabilities. Applying (5.31)–(5.32) and (5.33) gives the following endpoints of the collection of risk-neutral probability vectors $\{\tilde{\mathbf{p}}\}$:

$$\tilde{\mathbf{p}}(0) = \left[\frac{(1+r)-d}{u-d}, 0, \frac{u-(1+r)}{u-d} \right]^{\top} = \left[\frac{1.2-0.8}{1.4-0.8}, 0, \frac{1.4-1.2}{1.4-0.8} \right]^{\top} = \left[\frac{2}{3}, 0, \frac{1}{3} \right]^{\top},$$

$$\tilde{\mathbf{p}}\left(\tfrac{5}{9}\right) = \left[\frac{(1+r)-m}{u-m}, \frac{u-(1+r)}{u-m}, 0 \right]^{\top} = \left[\frac{1.2-1.1}{1.4-1.1}, \frac{1.4-1.2}{1.4-1.1}, 0 \right]^{\top} = \left[\frac{1}{3}, \frac{2}{3}, 0 \right]^{\top}.$$

The collection $\{\tilde{\mathbf{p}}\}$ of risk-neutral probability vectors can be parametrized by a single variable as follows:

$$\tilde{\mathbf{p}}\left(\tfrac{5v}{9}\right) = (1-v)\tilde{\mathbf{p}}(0) + v\tilde{\mathbf{p}}\left(\tfrac{5}{9}\right) = \left[\frac{2-v}{3}, \frac{2v}{3}, \frac{1-v}{3} \right]^{\top}, \quad v \in (0,1). \qquad (5.35)$$

Alternatively, the solution (5.35) can be obtained by normalizing the state-price vector in (5.34) and applying the change of variables $c = \tfrac{5}{9}v$:

$$\frac{\boldsymbol{\Psi}(c)}{\sum_{j=1}^{3} \Psi_j(c)} = \left[\frac{2}{3} - \frac{3c}{5}, \frac{6c}{5}, \frac{1}{3} - \frac{3c}{5} \right]^{\top} = \left[\frac{2-v}{3}, \frac{2v}{3}, \frac{1-v}{3} \right]^{\top}, \quad v \in (0,1). \qquad \square$$

5.4.2 Pricing Unattainable Payoffs: The Bid-Ask Spread

Assume that the Law of One Price holds. Then the no-arbitrage initial price of any financial claim with an attainable payoff is unique and is given by the initial value of any portfolio in the base assets that replicates the payoff. We can use (5.13) to price an attainable payoff X in the trinomial model. Alternatively, the pricing formula in (5.16) is used with the risk-neutral probabilities $\tilde{p}_j = (1+r)\Psi_j$, $j = 1, 2, 3$. Any choice of state-price vector $\boldsymbol{\Psi}$, or the corresponding risk-neutral probability measure $\widetilde{\mathbb{P}}$, gives us the same value for the initial price $\pi_0(X)$.

Example 5.11. Consider the trinomial model from Example 5.10. Find the no-arbitrage initial price of a long forward contract F with the delivery (strike) price $K = \$100$.

Solution. The payoff of a long forward contract is attainable as it is replicated by a portfolio consisting of one long position in the stock and a loan in the amount of K:

$$F_T := S_T - K = S_T - K\,B_T.$$

That is, $F_T \equiv \Pi_T^{\varphi} = \varphi_1 B_T + \varphi_2 S_T$ where $[\varphi_1, \varphi_2] = [-K, 1]$. The initial price F_0 of the forward contract is unique and equal to $\boldsymbol{\varphi}\mathbf{S}_0$:

$$F_0 = S_0 - K\,B_0 = S_0 - \frac{K}{1+r} = 100 - \frac{100}{1.2} = \frac{50}{3} \cong \$16.67.$$

On the other hand, we know that the risk-neutral pricing formula (5.16) must give us the same initial value of the contract regardless of the choice of the risk-neutral probability measure $\widetilde{\mathbb{P}}(c)$ with $c \in (0, \frac{5}{9})$. Using the probabilities in (5.35) gives

$$F_0 = \frac{1}{1+r} E^{\widetilde{\mathbb{P}}(c)}[F_T] = \frac{1}{1+r} \sum_{j=1}^{3} \left(S_T(\omega^j) - K\right) \tilde{p}_j(c)$$

$$= \frac{1}{1.2} \left((80 - 100) \cdot \frac{1-v}{3} + (110 - 100) \cdot \frac{2v}{3} + (140 - 100) \cdot \frac{2-v}{3}\right)$$

$$= \frac{-20(1-v) + 20v + 40(2-v)}{3.6} = \frac{-20 + 20v + 20v + 80 - 40v}{3.6}$$

$$= \frac{60}{3.6} = \frac{50}{3} \cong \$16.67,$$

which is independent of c. Alternatively, using the pricing formula (5.13) and solution (5.34) gives the same unique price:

$$F_0 = \sum_{j=1}^{3} F_T(\omega^j) \, \Psi_j(c) = (-20) \cdot \left(\frac{5}{18} - \frac{c}{2}\right) + 10 \cdot c + 40 \cdot \left(\frac{5}{9} - \frac{c}{2}\right)$$

$$= (10 + 10 - 20)c + \frac{200 - 50}{9} = \frac{50}{3}. \qquad \square$$

To price an unattainable payoff, we can formally apply (5.13), or equivalently the asset pricing formula in (5.16). However, the no-arbitrage initial price of $X \notin \mathcal{A}(\Omega)$ is now not unique since there are infinitely many no-arbitrage state-price vectors Ψ, or equivalently infinitely many risk-neutral measures. There are several approaches to deal with incomplete market models. One approach is to complete the market by including extra tradeable securities into the market model for which the initial value is known. It should be evident that the additional security should not be a redundant one as this would not change the original space of attainable payoffs. For example, we can add in some other security such as an option on the underlying risky asset or stock with known market value. We now give an example of how this is accomplished.

Example 5.12. Consider the incomplete trinomial model from Example 5.10.

(a) Show that the payoff of the European call option with strike $K = \$100$ is unattainable.

(b) Assume that the initial price of the call option from (a) is $20. Add this option in the trinomial model as a third base asset. Show that the new model with three base assets is complete and arbitrage-free.

Solution.
(a) First, calculate the values of the European call payoff $X = (S_T - K)^+$:

$$X(\omega^1) = (S_T(\omega^1) - K)^+ = (140 - 100)^+ = 40,$$
$$X(\omega^2) = (S_T(\omega^2) - K)^+ = (110 - 100)^+ = 10,$$
$$X(\omega^3) = (S_T(\omega^3) - K)^+ = (80 - 100)^+ = 0.$$

This gives the call payoff vector $\mathbf{X} = [40, 10, 0]$. Second, we try to replicate X by a portfolio in B and S. That is, find (if possible) a portfolio vector $\varphi \in \mathbb{R}^2$ such that

$$\varphi \mathbf{D} = \mathbf{X} \iff \begin{cases} \varphi_1 + 140\varphi_2 = 40 \\ \varphi_1 + 110\varphi_2 = 10 \\ \varphi_1 + 80\varphi_2 = 0 \end{cases} \qquad (5.36)$$

The rank of the augmented coefficient matrix of the system in (5.36) is 3 and hence is not equal to $\text{rank}(\mathbf{D}) = 2$ since

$$\begin{vmatrix} 1 & 140 & 40 \\ 1 & 110 & 10 \\ 1 & 80 & 0 \end{vmatrix} = -600 \neq 0.$$

Therefore, the system of linear equations in (5.36) is inconsistent, i.e., it does not have a solution. Hence, the call option payoff is unattainable.

(b) Now assume that the European call is the third base asset. The initial price vector and payoff matrix of the new 3-by-3 model are, respectively,

$$\widehat{\mathbf{S}}_0 = \begin{bmatrix} \frac{5}{6} \\ 100 \\ 20 \end{bmatrix} \quad \text{and} \quad \widehat{\mathbf{D}} \equiv \widehat{\mathbf{S}}_T(\Omega) = \begin{bmatrix} 1 & 1 & 1 \\ 140 & 110 & 80 \\ 40 & 10 & 0 \end{bmatrix}.$$

The payoff matrix $\widehat{\mathbf{D}}$ has a full rank so the 3-by-3 model does not have redundant assets and is complete. To prove the absence of arbitrage, we need to show that the solution $\mathbf{\Psi}$ to the linear system $\widehat{\mathbf{D}}\,\mathbf{\Psi} = \widehat{\mathbf{S}}_0$ is strictly positive. Solve for $\mathbf{\Psi}$ as follows:

$$\begin{cases} \Psi_1 + \Psi_2 + \Psi_3 = \frac{5}{6} \\ 140\Psi_1 + 110\Psi_2 + 80\Psi_3 = 100 \\ 40\Psi_1 + 10\Psi_2 + 0\Psi_3 = 20 \end{cases} \iff \begin{cases} \Psi_1 = \frac{4}{9} \\ \Psi_2 = \frac{2}{9} \\ \Psi_3 = \frac{1}{6} \end{cases}$$

As we can see, the state-price vector $\mathbf{\Psi} = \left[\frac{4}{9}, \frac{2}{9}, \frac{1}{6}\right]^{\top}$ is strictly positive. Therefore, the 3-by-3 model is arbitrage-free. The risk-neutral probabilities are

$$\tilde{p}_1 = (1+r)\Psi_1 = \frac{8}{15}, \quad \tilde{p}_2 = (1+r)\Psi_2 = \frac{4}{15}, \quad \tilde{p}_3 = (1+r)\Psi_3 = \frac{1}{5}. \qquad \square$$

Another approach to pricing a security with an unattainable payoff X is to find two attainable claims with payoffs X^d and X^u so that $X^d(\omega) \leqslant X(\omega) \leqslant X^u(\omega)$ for all $\omega \in \Omega$. Then, the no-arbitrage price $\pi_0(X)$ is bounded by the initial prices of the two attainable claims:

$$\pi_0(X^d) < \pi_0(X) < \pi_0(X^u).$$

Indeed, let the portfolios φ^d and φ^u replicate X^d and X^u, respectively. Suppose that $\pi_0(X^d) \geqslant \pi_0(X)$, then there exists the following arbitrage opportunity. At time $t = 0$ we buy the security and sell the portfolio φ^d. Our proceeds, which we keep in cash, are nonnegative:

$$\Pi_0[\varphi^d] - \pi_0(X) = \pi_0(X^d) - \pi_0(X) \geqslant 0.$$

At time $t = T$ the net position is nonnegative in all states and strictly positive in some states, which we denote by:

$$X - \Pi_T[\varphi^d] = X - X^d > 0.$$

Indeed, since $X^d \leqslant X$, $X^d \in \mathcal{A}(\Omega)$, and $X \notin \mathcal{A}(\Omega)$, we have that $X^d(\omega) < X(\omega)$ in at least one state ω. Hence, this is an arbitrage portfolio with a positive payoff equal to $(\pi_0(X^d) - \pi_0(X)) + X - X^d$. Similarly, there would be an arbitrage opportunity if the security with payoff X were found to be selling in the market at a price $\pi(X) \geqslant \pi_0(X^u)$.

A *super-replicating portfolio* φ with the terminal value

$$\Pi_T^{\varphi} \geqslant X \qquad (5.37)$$

covers the liabilities of the writer of the security with payoff X. So the writer will choose φ with minimal initial cost Π_0^φ subject to the constraints (5.37). As a result, the writer obtains the following linear programming problem (the super-replication problem):

$$\Pi_0^\varphi \to \min_{\varphi \in \mathbb{R}^N} \text{ subject to } \Pi_T^\varphi \geqslant X. \tag{5.38}$$

Since the risk-free security is strictly positive, the linear programming problem is feasible. There always exists a trivial risk-free portfolio that satisfies the constraints (5.37). The problem (5.38) is equivalent to the minimization of $\pi_0(Y)$ over the set of dominating attainable claims for X:

$$\pi_0(Y) \to \min_{Y \in \mathcal{D}_X}, \text{ where } \mathcal{D}_X := \{Y \in \mathcal{A}(\Omega) : X \leqslant Y\}. \tag{5.39}$$

The set $\mathcal{D}_X \neq \emptyset$ includes all attainable payoffs Y that *dominate* the claim's payoff X.

The problem (5.38) or (5.39) can be solved graphically or numerically. Let us denote the optimal solution to (5.38) by φ^u. Then $\pi_0^u(X) := \Pi_0[\varphi^u]$ is an upper bound for a no-arbitrage initial price $\pi_0(X)$ for the derivative or claim with payoff X. If the writer were selling the security at a price higher than $\pi_0^u(X)$, then another agent could form an arbitrage portfolio.

Let us now look at the matter from the buyer's perspective. Let the portfolio $\varphi = \varphi^d$ solve the linear programming problem (the sub-replication problem):

$$\Pi_0^\varphi \to \max_{\varphi \in \mathbb{R}^N} \text{ subject to } \Pi_T^\varphi \leqslant X. \tag{5.40}$$

The problem (5.40) is equivalent to the maximization of $\pi_0(Y)$ over the set of attainable claims dominated by X:

$$\pi_0(Y) \to \max_{Y \in \mathcal{M}_X}, \text{ where } \mathcal{M}_X := \{Y \in \mathcal{A}(\Omega) : Y \leqslant X\}. \tag{5.41}$$

The set $\mathcal{M}_X \neq \emptyset$ includes all attainable payoffs Y that are *dominated* by the claim's payoff X. Denote the initial value of the portfolio φ^d by $\pi_0^d(X)$. The buyer of the derivative security with payoff X will not agree to pay more than $\pi_0^d(X)$ since it will be more beneficial to purchase the portfolio in base assets rather than the derivative. As a result we obtain a lower bound for a no-arbitrage price of $\pi_0(X)$. In summary, the initial price $\pi_0(X)$ for the claim with payoff X must satisfy the (above-mentioned) inequality relation

$$\pi_0^d(X) < \pi_0(X) < \pi_0^u(X). \tag{5.42}$$

Otherwise, an arbitrage opportunity will arise. The interval $\left(\pi_0^d(X), \pi_0^u(X)\right)$ is called the *bid-ask spread* for the (no-arbitrage) initial price of a derivative security with payoff X.

The ask price $\pi_0^u(X)$ and the bid price $\pi_0^d(X)$ can also be calculated by respectively taking the maximum and the minimum of the discounted expectation of the payoff function over the set of possible risk-neutral measures:

$$\pi_0^u(X) = \sup_{\widetilde{\mathbb{P}} \in \{\widetilde{\mathbb{P}}\}} \left\{ \frac{1}{1+r} \sum_{i=1}^{M} \tilde{p}_i X(\omega^i) \right\} = \sup_{\boldsymbol{\Psi} \in \{\boldsymbol{\Psi}\}} \mathbf{X}\,\boldsymbol{\Psi}, \tag{5.43}$$

$$\pi_0^d(X) = \inf_{\widetilde{\mathbb{P}} \in \{\widetilde{\mathbb{P}}\}} \left\{ \frac{1}{1+r} \sum_{i=1}^{M} \tilde{p}_i X(\omega^i) \right\} = \inf_{\boldsymbol{\Psi} \in \{\boldsymbol{\Psi}\}} \mathbf{X}\,\boldsymbol{\Psi}. \tag{5.44}$$

Here, $\{\boldsymbol{\Psi}\}$ and $\{\widetilde{\mathbb{P}}\}$ denote the collection of all state-price vectors and the corresponding collection of all risk-neutral probabilities measures, respectively.

Consider the case of the trinomial model. State-price vectors form a finite segment in $\mathbb{R}^3$. Thus, the range of possible no-arbitrage initial prices of a nonattainable payoff is also a finite segment. Since the price $\pi_0 = \mathbf{X}\,\mathbf{\Psi}$ is a linear function of the vector $\mathbf{\Psi}$, these bounds are attained at the endpoints (5.31)–(5.32) of the state-price vector set $\{\mathbf{\Psi}\}$.

Example 5.13. Consider the trinomial (incomplete) model from Example 5.10. Find the bid-ask spread for the initial price of the call option with strike $K = \$100$.

Solution. The first approach is to solve the super- and sub-replication problems given by (5.38) and (5.40), respectively,

$$
\begin{cases} \Pi_0[\varphi^d] \to \max_{\varphi^d \in \mathbb{R}^2} \\ \Pi_T[\varphi^d] \leqslant X \end{cases}
\iff
\begin{cases} \frac{5}{6}\varphi_1^d + 100\varphi_2^d \to \max_{[\varphi_1^d,\varphi_2^d]\in\mathbb{R}^2} \\ \varphi_1^d + 80\varphi_2^d \leqslant 0, \\ \varphi_1^d + 110\varphi_2^d \leqslant 10, \\ \varphi_1^d + 140\varphi_2^d \leqslant 40 \end{cases}
\tag{5.45}
$$

$$
\begin{cases} \Pi_0[\varphi^u] \to \min_{\varphi^u \in \mathbb{R}^2} \\ \Pi_T[\varphi^u] \geqslant X \end{cases}
\iff
\begin{cases} \frac{5}{6}\varphi_1^u + 100\varphi_2^u \to \min_{[\varphi_1^u,\varphi_2^u]\in\mathbb{R}^2} \\ \varphi_1^u + 80\varphi_2^u \geqslant 0, \\ \varphi_1^u + 110\varphi_2^u \geqslant 10 \\ \varphi_1^u + 140\varphi_2^u \geqslant 40 \end{cases}
\tag{5.46}
$$

The solutions to the linear programming problems (5.45) and (5.46) are, respectively,

$$
\varphi^d = \begin{bmatrix} -100, & 1 \end{bmatrix} \quad \text{and} \quad \varphi^u = \begin{bmatrix} -53\tfrac{1}{3}, & \tfrac{2}{3} \end{bmatrix}.
$$

The bid price π_0^d and ask price π_0^u of the call option are then given by

$$
\pi_0^d = \Pi_0[\varphi^d] = \varphi^d \mathbf{S}_0 = -100 \cdot \frac{5}{6} + 1 \cdot 100 = \frac{50}{3} \cong \$16.67,
$$

$$
\pi_0^u = \Pi_0[\varphi^u] = \varphi^u \mathbf{S}_0 = -53\frac{1}{3} \cdot \frac{5}{6} + \frac{2}{3} \cdot 100 = \frac{200}{9} \cong \$22.22.
$$

Thus, the bid-ask spread is $(16.67, 22.22)$.

The other approach is to find the bid-ask spread by computing the extreme values of the risk-neutral expectation of the discounted payoff. Since the mathematical expectation of a random variable from $\mathcal{L}(\Omega)$ is a linear function of the state probabilities, the maximum and minimum in (5.43) and (5.44) are attained at the endpoints of the domain $\{\widetilde{\mathbb{P}}\}$. Therefore, we just need to calculate the expected value

$$
\pi_0(c) = \frac{1}{1+r} \mathrm{E}^{\widetilde{\mathbb{P}}(c)}[(S_T - K)^+] = \frac{1}{1+r} \sum_{j=1}^{3} \tilde{p}_j(c)(S_T(\omega^j) - K)^+
$$

using risk-neutral probabilities given by (5.35) with extreme values $c = 0$ and $c = \frac{5}{9}$. The bid price is the smaller of the two prices calculated, and the ask price is the larger one. The above expectation is calculated explicitly in the measure $\widetilde{\mathbb{P}}(v)$:

$$
\pi_0(\tfrac{5v}{9}) = \frac{5}{6} \cdot \left(0 \cdot \frac{1-v}{3} + 10 \cdot \frac{2v}{3} + 40 \cdot \frac{2-v}{3} \right) = \frac{50(4-v)}{9}, \quad 0 \leqslant v \leqslant 1.
$$

The extreme values are $\pi_0(v = 0) = \frac{200}{9}$ and $\pi_0(v = 1) = \frac{50}{3} = \frac{150}{9}$. The bid price is

$\min\{\frac{200}{9}, \frac{50}{3}\} = \frac{50}{3}$, and the ask price is $\max\{\frac{200}{9}, \frac{50}{3}\} = \frac{200}{9}$. As required, both prices agree with the values obtained using the first approach. □

In the conclusion of this section, let us consider an example of a single-period model with four states of the world and two tradable assets.

Example 5.14. Let the initial price vector and payoff matrix be given by

$$\mathbf{S}_0 = \begin{bmatrix} 10 \\ 10 \end{bmatrix} \text{ and } \mathbf{D} = \begin{bmatrix} 11 & 11 & 11 & 11 \\ 6 & 8 & 12 & 14 \end{bmatrix}.$$

(a) Show that the market is arbitrage-free. Find the general solution for the state prices and illustrate it with a diagram.

(b) Find the bid-ask spread for the put option on the risky asset with strike price $K = 11$.

Solution. The general solution to the state-price equation $\mathbf{D}\,\boldsymbol{\Psi} = \mathbf{S}_0$ is a function of two free parameters x, y:

$$\begin{cases} \Psi_1 = 2x + 3y - \frac{15}{11} \\ \Psi_2 = \frac{25}{11} - 3x - 4y \\ \Psi_3 = x \\ \Psi_4 = y \end{cases}$$

The solution is strictly positive if

$$0 < x < \frac{25}{33} \quad \text{and} \quad \max\left\{\frac{5}{11} - \frac{2}{3}x, 0\right\} < y < \frac{24}{44} - \frac{3}{4}x \tag{5.47}$$

holds. The domain of admissible solutions, $D = \{[x, y] : \boldsymbol{\Psi}(x, y) \gg 0\}$, can be illustrated by a diagram on the (x, y)-plane (Figure 5.3).

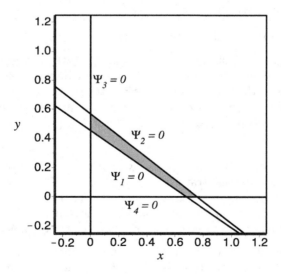

FIGURE 5.3: The domain of admissible solutions $\boldsymbol{\Psi}(x, y) \gg 0$.

The no-arbitrage initial price π_0 of the put option with strike price $K = 11$ is given by a product of the payoff vector

$$\mathbf{X} = (K - S(\Omega))^+ = [5, 3, 0, 0]$$

and the state-price vector $\mathbf{\Psi}$:

$$\pi_0 = \mathbf{X}\,\mathbf{\Psi} = 5\cdot\left(2x+3y-\frac{15}{11}\right)+3\cdot\left(\frac{25}{11}-3x-4y\right)=x+3y.$$

The bid and ask prices are, respectively, the minimum and maximum values attained by the function $x+3y$ in the domain D as is specified in (5.47). These values can be found by solving respective linear programming problems. Alternatively, we can use the fact that a linear function attains its extreme values at the boundary. The boundary of the domain D is piecewise linear. Hence, by applying the same principle one more time, we can conclude that the maximum and minimum values of $x+3y$ in D are attained at the corner points

$$\mathbf{q}_1 = \left[0,\frac{15}{33}\right],\quad \mathbf{q}_2 = \left[0,\frac{25}{44}\right],\quad \mathbf{q}_3 = \left[\frac{15}{22},0\right],\quad \mathbf{q}_4 = \left[\frac{25}{33},0\right].$$

Calculate the value of π_0 at each of these points:

$$\pi_0(\mathbf{q}_1) = \frac{15}{11},\quad \pi_0(\mathbf{q}_2) = \frac{75}{44},\quad \pi_0(\mathbf{q}_3) = \frac{15}{22},\quad \pi_0(\mathbf{q}_4) = \frac{25}{33}.$$

The smaller price is the bid price; the larger price is the ask price:

$$\min_{i=1,2,3,4}\pi_0(\mathbf{q}_i) = \frac{15}{22} \cong 0.68182,\qquad \max_{i=1,2,3,4}\pi_0(\mathbf{q}_i) = \frac{75}{55} \cong 1.70455.$$

Thus, the bid-ask spread for the put option is $(0.68182, 1.70455)$. $\square$

5.5 Change of Numéraire

5.5.1 The Concept of a Numéraire Asset

Suppose that there exists a risk-free asset $\{B_t\}_{t\in\{0,T\}}$ with a fixed rate of return r. Note that for a risk-free bond we assume nonnegative interest rates so that $r \geqslant 0$. As was shown in the previous section, the market model is arbitrage-free iff there exists a probability function $\widetilde{\mathbb{P}}$ called a *risk-neutral probability measure* or *martingale probability measure* such that the discounted price processes $\left\{\frac{S_t^i}{B_t}\right\}_{t\in\{0,T\}}$ for $i = 1,2,\ldots,N$ are martingales with respect to $\widetilde{\mathbb{P}}$.

As a process, the prices of base assets divided by the value of the risk-free account are martingales within a risk-neutral probability measure. There is nothing preventing us from also expressing prices of assets relative to another choice of available strictly positively valued security. The security may be any one of the base assets, or any strictly positively valued portfolio in the base assets, or even a strictly positively valued derivative security.

Definition 5.8. A *numéraire*, or *numéraire asset*, or *accounting unit*, denoted $\{g_t\}_{t\in\{0,T\}}$, is any strictly positive asset price process, i.e., with payoff $g_T \gg 0$ and initial price $g_0 > 0$.

Hence, for any numéraire, we have $g_t(\omega) > 0$ for all time $t \geqslant 0$ and all outcomes $\omega \in \Omega$. The numéraire is used for measuring the relative worth of any asset or a portfolio of assets. The ratio process $\{\frac{S_t}{g_t}\}_{t\geqslant 0}$ is called the value process of asset S discounted by g or relative to g. The ratio $\frac{S_t}{g_t}$ gives the number of units of the numéraire g that can be exchanged for one unit of asset S at time t. For example, in the binomial model both the bond B and stock S can be chosen as a numéraire.

5.5.2 Change of Numéraire in a Binomial Model

Consider a single-period binomial model with two states of the world $\Omega = \{\omega^+, \omega^-\}$ and two base securities B and S such that

$$B_0 = (1+r)^{-1}, \quad B_T = 1, \quad S_T(\omega) = (d\,\mathbb{I}_{\{\omega^-\}}(\omega) + u\,\mathbb{I}_{\{\omega^+\}}(\omega))S_0,$$

where $0 < d < u$. The market is arbitrage-free and complete if $d < 1+r < u$. Let $\widetilde{\mathbb{P}}$ be a risk-neutral probability measure with $\tilde{p} := \widetilde{\mathbb{P}}(\omega^+) = \frac{(1+r)-d}{u-d}$ and $1 - \tilde{p} = \widetilde{\mathbb{P}}(\omega^-) = \frac{u-(1+r)}{u-d}$. Under $\widetilde{\mathbb{P}}$, the base asset price processes discounted by the numéraire $g = B$ are martingales:

$$\widetilde{\mathrm{E}}\left[\frac{S_T}{B_T}\right] = \frac{S_0}{B_0}, \quad \widetilde{\mathrm{E}}\left[\frac{B_T}{B_T}\right] = 1 = \frac{B_0}{B_0}.$$

Moreover, any portfolio $[\Delta, \beta]$ with Δ shares in the stock S and β units of the bond B discounted by the value of asset B is a martingale:

$$\widetilde{\mathrm{E}}\left[\frac{\Pi_T}{B_T}\right] \equiv \widetilde{\mathrm{E}}\left[\frac{\Delta S_T + \beta B_T}{B_T}\right] = \Delta\widetilde{\mathrm{E}}\left[\frac{S_T}{B_T}\right] + \beta\widetilde{\mathrm{E}}\left[\frac{B_T}{B_T}\right] = \Delta\frac{S_0}{B_0} + \beta = \frac{\Delta S_0 + \beta B_0}{B_0} = \frac{\Pi_0}{B_0}.$$

In summary, there is no arbitrage in the binomial model iff there exists a martingale probability measure $\widetilde{\mathbb{P}}^{(B)}$ for the numéraire $g = B$. The no-arbitrage binomial model is complete, hence $\widetilde{\mathbb{P}}^{(B)}$ is unique. Is it possible to use the stock price as a numéraire? Is there a unique risk-neutral probability function $\widetilde{\mathbb{P}}^{(S)}$ for the numéraire $g = S$ under which the relative price processes for all base assets are martingales? Both questions are answered positively just below. There exists $\widetilde{\mathbb{P}}^{(B)}$ with bond B as numéraire iff there exists $\widetilde{\mathbb{P}}^{(S)}$ with stock S as numéraire. The value process of any attainable claim discounted by numéraire $g = B$ or $g = S$ is a martingale under $\widetilde{\mathbb{P}}^{(B)}$ or $\widetilde{\mathbb{P}}^{(S)}$, respectively. Thus, these probability measures are called *martingale measures*.

The martingale probability measures $\widetilde{\mathbb{P}}^{(B)}$ and $\widetilde{\mathbb{P}}^{(S)}$ are said to be *equivalent*. That is, any event E with zero probability under one measure has zero probability under the other equivalent measure, i.e., $\widetilde{\mathbb{P}}^{(B)}(E) = 0$ iff $\widetilde{\mathbb{P}}^{(S)}(E) = 0$ for all $E \subseteq \Omega$. So, two equivalent measures are consistent in their assignment of nonzero probability values to all events having nonzero probability, although the probability values assigned to the events will generally differ. Hence, both $\widetilde{\mathbb{P}}^{(B)}$ and $\widetilde{\mathbb{P}}^{(S)}$ are called *equivalent martingale measures*. The formal definition of equivalent martingale measures in the context of multiple base assets within the single-period setting is given in the next section. Note that the martingale measures $\widetilde{\mathbb{P}}^{(B)}$ and $\widetilde{\mathbb{P}}^{(S)}$ and the real-world measure $\mathbb{P}$ are equivalent to each other; however, $\mathbb{P}$ is not an equivalent martingale measure and hence cannot be used for computing no-arbitrage prices!

For the binomial case, we only need to show that there exists a probability $\tilde{p} = \widetilde{\mathbb{P}}^{(B)}(\omega^+) \in (0,1)$ iff there exists a probability $\tilde{q} = \widetilde{\mathbb{P}}^{(S)}(\omega^+) \in (0,1)$. Let us work out the martingale probability measure $\widetilde{\mathbb{P}}^{(S)}$. In this case the numéraire asset price at time t is the stock price, $g_t \equiv S_t$. Hence, the stock price process discounted by g_t is a constant and hence is automatically a martingale, i.e., $\frac{S_t}{g_t} \equiv \frac{S_t}{S_t} \equiv 1$, $t \in \{0, T\}$. The bond price B_T discounted by S_T is given by

$$\frac{B_T(\omega)}{S_T(\omega)} = \frac{(1+r)B_0}{S_T(\omega)} = \begin{cases} \frac{1+r}{d}\frac{B_0}{S_0} & \text{if } \omega = \omega^-, \\ \frac{1+r}{u}\frac{B_0}{S_0} & \text{if } \omega = \omega^+. \end{cases}$$

So we have that $\widetilde{\mathrm{E}}^{(S)}\left[\frac{B_T}{S_T}\right] = \frac{B_0}{S_0}$ iff

$$\frac{1+r}{u}\frac{B_0}{S_0}\tilde{q} + \frac{1+r}{d}\frac{B_0}{S_0}(1-\tilde{q}) = \frac{B_0}{S_0} \iff \frac{1+r}{u}\tilde{q} + \frac{1+r}{d}(1-\tilde{q}) = 1.$$

Here $\widetilde{\mathrm{E}}^{(S)}[\,\cdot\,]$ denotes the mathematical expectation under the probability measure $\widetilde{\mathbb{P}}^{(S)}$. Solving the above equation for $\tilde{q}$:

$$\left(\frac{1+r}{u} - \frac{1+r}{d}\right)\tilde{q} = 1 - \frac{1+r}{d} \iff \frac{(1+r)(u-d)}{ud}\tilde{q} = \frac{(1+r)-d}{d}$$

$$\iff \tilde{q} = \frac{(1+r)-d}{u-d}\,\frac{u}{1+r}$$

The probability measure $\widetilde{\mathbb{P}}^{(S)}$ is defined by the probabilities $\tilde{q}$ and $1 - \tilde{q}$, which can be expressed in terms of the probabilities for measure $\widetilde{\mathbb{P}}^{(B)}$, $\tilde{p} = \frac{(1+r)-d}{u-d}$ and $1 - \tilde{p}$, as follows:

$$\tilde{q} = \frac{(1+r)-d}{u-d} \cdot \frac{u}{1+r} = \tilde{p} \cdot \frac{u}{1+r}, \tag{5.48}$$

$$1 - \tilde{q} = \frac{u-(1+r)}{u-d} \cdot \frac{d}{1+r} = (1-\tilde{p}) \cdot \frac{d}{1+r}. \tag{5.49}$$

Clearly, $\tilde{q} \in (0,1)$ iff $d < 1 + r < u$. The latter condition is equivalent to $\tilde{p} \in (0,1)$. So assuming no-arbitrage, $\widetilde{\mathbb{P}}^{(B)}$ and $\widetilde{\mathbb{P}}^{(S)}$ are equivalent probability measures. Therefore, in the binomial model there exists an equivalent martingale measure (EMM) w.r.t. the stock, denoted by $\widetilde{\mathbb{P}}^{(S)}$, and an EMM w.r.t. the bond, denoted by $\widetilde{\mathbb{P}}^{(B)}$, iff the market admits no arbitrage. The probability measures are unique and hence the model is complete.

Every attainable asset price process that can be replicated by some portfolio value process Π_t^φ, $t \in \{0,T\}$, for some φ, discounted by the numéraire (the bond or stock) is then a martingale w.r.t. the equivalent martingale measure for the given numéraire. Thus, the unique initial price of any (derivative) asset can be calculated by using any appropriate choice of equivalent martingale measure. The initial price $\pi_0(X)$ of any attainable payoff X is equivalently given by

$$\pi_0(X) = g_0\,\widetilde{\mathrm{E}}^{(g)}\left[\frac{X}{g_T}\right] \quad \text{(for any choice of numéraire } g)$$

$$= B_0\,\widetilde{\mathrm{E}}^{(B)}\left[\frac{X}{B_T}\right] = \frac{1}{1+r}\left[X(\omega^+)\tilde{p} + X(\omega^-)(1-\tilde{p})\right] \quad \text{(for } g = B)$$

$$= S_0\,\widetilde{\mathrm{E}}^{(S)}\left[\frac{X}{S_T}\right] = \frac{X(\omega^+)}{u}\tilde{q} + \frac{X(\omega^-)}{d}(1-\tilde{q}) \quad \text{(for } g = S).$$

5.5.3 Change of Numéraire in a General Single Period Model

Consider the general case of a single-period model with M states of the world and N base assets. Any attainable generic asset g with a strictly positive price process can be used as a numéraire. Since g is attainable, its value g_t at any time $t \in \{0,T\}$ corresponds to the value of a replicating portfolio $\boldsymbol{\theta}^{(g)}$ in the base assets:

$$g_t = \Pi_t[\boldsymbol{\theta}^{(g)}] = \sum_{i=1}^{N} \theta_i^{(g)} S_t^i, \quad t \in \{0,T\}.$$

Since a numéraire asset must have strictly positive value, the portfolio in the base assets, $\boldsymbol{\theta}^{(g)} \in \mathbb{R}^N$, is chosen such that $g_0 > 0$ and $g_T(\omega) > 0$ for all $\omega \in \Omega$. For the ith base asset to qualify as a potential numéraire asset, we need only require that it has a strictly positive payoff (i.e., the respective row in the payoff matrix is strictly positive) and a positive initial price. It is convenient to define base asset price processes discounted by the numéraire asset price g_t as

$$\overline{S}_t^i := \frac{S_t^i}{g_t}, \quad t \in \{0, T\}, \quad i = 1, 2, \ldots, N.$$

$\overline{S}_t^i$ is the number of units of the numéraire that can be purchased at time t for the same amount needed to buy one unit of the ith base asset. Note that the discounted price of the numéraire asset itself is equal to 1. For a given portfolio $\varphi \in \mathbb{R}^N$ we can consider its *discounted value process relative to the numéraire g*:

$$\overline{\Pi}_t^\varphi := \frac{\Pi_t^\varphi}{g_t} = \sum_{i=1}^N \varphi_i \, \overline{S}_t^i.$$

Clearly, $\overline{\Pi}_t[\,\cdot\,]$ is a linear function, i.e., for any two portfolios $\varphi, \psi \in \mathbb{R}^N$ and constants $a, b \in \mathbb{R}$, we have

$$\overline{\Pi}_t[a\varphi + b\psi] = a\overline{\Pi}_t^\varphi + b\overline{\Pi}_t[\psi].$$

Definition 5.9. An *equivalent martingale measure* (EMM) $\widetilde{\mathbb{P}} \equiv \widetilde{\mathbb{P}}^{(g)}$ for a given numéraire asset $\{g_t\}_{t \in \{0,T\}}$ is a probability measure defined on the state space Ω such that the following conditions are fulfilled.

- The probability measure $\widetilde{\mathbb{P}}$ is equivalent to the real-world probability measure $\mathbb{P}$, i.e., $\widetilde{\mathbb{P}}(E) = 0$ iff $\mathbb{P}(E) = 0$ for all $E \subseteq \Omega$. Note that for finite or countable Ω, two probability measures, $\mathbb{P}$ and $\widetilde{\mathbb{P}}$, are equivalent when $\widetilde{\mathbb{P}}(\omega) = 0$ iff $\mathbb{P}(\omega) = 0$ for all $\omega \in \Omega$. Since $\mathbb{P}(\omega) > 0$ for every outcome ω, the measures $\mathbb{P}$ and $\widetilde{\mathbb{P}}$ are equivalent if $\widetilde{\mathbb{P}}(\omega) > 0$ for all $\omega \in \Omega$.

- The discounted base asset price process $\left\{\overline{S}_t^i := \frac{S_t^i}{g_t}\right\}_{t \in \{0,T\}}$ is a martingale w.r.t. $\widetilde{\mathbb{P}}$ for every base asset S^i, $i = 1, 2, \ldots, N$, i.e., for a single period model we have that

$$\widetilde{\mathrm{E}}^{(g)}[\overline{S}_T^i] = \widetilde{\mathrm{E}}^{(g)}\left[\frac{S_T^i}{g_T}\right] = \frac{S_0^i}{g_0} = \overline{S}_0^i. \tag{5.50}$$

Here, $\widetilde{\mathrm{E}}^{(g)}[\,\cdot\,]$ denotes the expectation w.r.t. $\widetilde{\mathbb{P}}^{(g)}$.

Suppose that a risk-free asset is used as a numéraire asset, i.e., $g_t = B_t$. Let r denote the risk-free rate of return on a risk-free numéraire: $r = \frac{g_T - g_0}{g_0}$. Then, the martingale condition (5.50) takes the form:

$$\frac{1}{1+r}\widetilde{\mathrm{E}}[S_T^i] = S_0^i.$$

We should note here that the formal definition of a martingale process for a multi-period model is presented in Chapter 6, as it will later come into play when we discuss asset pricing in a multi-period setting in Chapter 7. Within a single-period trading model, the next result tells us that the existence of an EMM for a given choice of numéraire is equivalent to saying that the value of any portfolio in the base assets discounted by the numéraire asset value is a martingale w.r.t. the EMM.

Lemma 5.13. *Consider a single-period model with M states and N base assets. Let g be a numéraire asset and $\widetilde{\mathbb{P}}^{(g)}$ be a probability measure equivalent to $\mathbb{P}$. Then $\widetilde{\mathbb{P}}^{(g)}$ is an equivalent martingale measure iff for every portfolio $\varphi \in \mathbb{R}^N$ the discounted value process, $\{\overline{\Pi}_t \equiv \overline{\Pi}_t^\varphi\}_{t \in \{0,T\}}$, is a martingale w.r.t. $\widetilde{\mathbb{P}}^{(g)}$.*

Proof. Let $\widetilde{\mathbb{P}}^{(g)}$ be an equivalent martingale measure. Then,

$$\widetilde{\mathrm{E}}^{(g)}\left[\overline{\Pi}_T\right] = \sum_{\omega \in \Omega} \overline{\Pi}_T(\omega)\widetilde{\mathbb{P}}^{(g)}(\omega) = \sum_{j=1}^{M} \sum_{i=1}^{N} \varphi_i \overline{S}_T^i(\omega^j)\widetilde{\mathbb{P}}^{(g)}(\omega^j)$$

$$= \sum_{i=1}^{N} \varphi_i \sum_{j=1}^{M} \overline{S}_T^i(\omega^j)\widetilde{\mathbb{P}}^{(g)}(\omega^j) = \sum_{i=1}^{N} \varphi_i \widetilde{\mathrm{E}}^{(g)}[\overline{S}_T^i] = \sum_{i=1}^{N} \varphi_i \overline{S}_0^i = \overline{\Pi}_0.$$

The converse follows by assumption. That is, for every $i = 1, 2, \ldots, N$ consider a portfolio that only contains one unit of the ith base asset, i.e., $\overline{\Pi}_t = \overline{S}_t^i$. Then, the process $\left\{\overline{S}_t^i\right\}_{t \in \{0,T\}}$ is a martingale w.r.t. $\widetilde{\mathbb{P}}^{(g)}$. $\qquad\square$

According to Lemma 5.13, the discounted value process of a financial claim with an attainable payoff X is a martingale w.r.t. $\widetilde{\mathbb{P}}^{(g)}$. Let $\boldsymbol{\varphi}_X \in \mathbb{R}^N$ be a replicating portfolio such that $\Pi_T[\boldsymbol{\varphi}_X] = X$. Then, the initial price of a claim with payoff X is calculated as follows:

$$\frac{\pi_0^{(g)}(X)}{g_0} = \frac{\Pi_0[\boldsymbol{\varphi}_X]}{g_0} = \widetilde{\mathrm{E}}^{(g)}\left[\frac{\Pi_T[\boldsymbol{\varphi}_X]}{g_T}\right] = \widetilde{\mathrm{E}}^{(g)}\left[\frac{X}{g_T}\right] \implies \pi_0^{(g)}(X) = g_0\widetilde{\mathrm{E}}^{(g)}\left[\frac{X}{g_T}\right]. \quad (5.51)$$

Since the initial value of a replicating portfolio is unique, the initial price $\pi_0^{(g)}(X) \equiv \pi_0(X)$ *does not depend on the choice of numéraire g.* If the payoff vector is positive, then the initial price is positive as well:

$$\mathbf{X} > 0 \implies \pi_0^{(g)}(X) > 0.$$

The functional $\pi_0^{(g)}$ in (5.51) is defined on the set of attainable payoffs $\mathcal{A}(\Omega)$ but it can be extended on $\mathcal{L}(\Omega)$ in the case of an incomplete market. For example, we may set the initial price of an unattainable claim $X \in \mathcal{L}(\Omega) \setminus \mathcal{A}(\Omega)$ equal to the ask price:

$$\pi_0^{(g),u}(X) := \inf_{Y \in \mathcal{D}_X} \pi_0^{(g)}(Y), \text{ where } \mathcal{D}_X = \{Y \in \mathcal{A}(\Omega) : X \leqslant Y\}.$$

As was proved above, the no-arbitrage condition is equivalent to the existence of a (strictly positive) state-price vector $\boldsymbol{\Psi}$ that solves $\mathbf{D}\boldsymbol{\Psi} = \mathbf{S}_0$. Consider a numéraire asset g and respective martingale measure $\widetilde{\mathbb{P}}^{(g)}$ with probabilities $\tilde{p}_j^{(g)} = \widetilde{\mathbb{P}}^{(g)}(\omega_j)$, $j = 1, 2, \ldots, M$. Let us find the relationship between $\widetilde{\mathbb{P}}^{(g)}$ and a state-price vector $\boldsymbol{\Psi}$. By the definition of an EMM, we have:

$$\widetilde{\mathrm{E}}^{(g)}\left[\frac{S_T^i}{g_T}\right] \equiv \sum_{j=1}^{M} \frac{S_T^i(\omega^j)}{g_T(\omega^j)}\tilde{p}_j^{(g)} = \frac{S_0^i}{g_0} \implies \sum_{j=1}^{M} \frac{g_0\tilde{p}_j^{(g)}}{g_T(\omega^j)}S_T^i(\omega^j) = S_0^i, \, i = 1, 2, \ldots, N.$$

Let us denote $\Psi_j = \frac{g_0\tilde{p}_j^{(g)}}{g_T(\omega^j)}$. Then, as before, we have

$$\sum_{j=1}^{M} S_T^i(\omega^j)\,\Psi_j = S_0^i, \, i = 1, 2, \ldots, N \iff \mathbf{D}\,\boldsymbol{\Psi} = \mathbf{S}_0.$$

Therefore, the existence of an EMM implies the existence of a strictly positive state-price vector. Since a numéraire asset has strictly positive prices, we have that $\Psi_j > 0$ if $\tilde{p}_j^{(g)} > 0$. Conversely, let $\Psi_j > 0$, $j = 1, 2, \ldots, M$, be the state prices. Define $\tilde{p}_j^{(g)} = \frac{g_T(\omega^j)}{g_0}\Psi_j$, $j =$

$1, 2, \ldots, M$. All $\tilde{p}_j^{(g)}$ are positive and they sum up to one if g is an attainable asset, i.e., if $g_t = \Pi_t[\boldsymbol{\theta}^{(g)}]$ for some portfolio $\boldsymbol{\theta}^{(g)} \in \mathbb{R}^N$. Indeed,

$$\sum_{j=1}^M \Psi_j g_T(\omega^j) = \sum_{j=1}^M \Psi_j \Pi_T[\boldsymbol{\theta}^{(g)}](\omega^j) = \sum_{j=1}^M \Psi_j \sum_{i=1}^N \theta_i^{(g)} S_T^i(\omega^j)$$

$$= \sum_{i=1}^N \theta_i^{(g)} \sum_{j=1}^M \Psi_j S_T^i(\omega^j) = \sum_{i=1}^N \theta_i^{(g)} S_0^i$$

$$= \Pi_0[\boldsymbol{\theta}^{(g)}] = g_0 \implies \sum_{j=1}^M \Psi_j \frac{g_T(\omega^j)}{g_0} = \sum_{j=1}^M \tilde{p}_j^{(g)} = 1.$$

Therefore, $\tilde{p}_j^{(g)}$, $j = 1, 2, \ldots, M$, are probabilities. As a result, we can restate the first and second fundamental theorems of asset pricing in terms of equivalent martingale measures as follows.

Theorem 5.14 (The first and second FTAPs—the 3rd version).

1. *There are no arbitrage opportunities iff there exists an equivalent martingale measure with respect to a given numéraire asset.*

2. *Assuming absence of arbitrage, there exists a unique equivalent martingale measure with respect to a given numéraire asset iff the market is complete.*

Example 5.15. Consider a 3-by-3 model from Example 5.12 with

$$\mathbf{S}_0 = \begin{bmatrix} \frac{5}{6} \\ 100 \\ 20 \end{bmatrix} \quad \text{and} \quad \mathbf{D} = \begin{bmatrix} 1 & 1 & 1 \\ 140 & 110 & 80 \\ 40 & 10 & 0 \end{bmatrix}.$$

Find the EMM $\widetilde{\mathbb{P}}^{(g)}$ using one of the base assets as numéraire.

Solution. To find the EMM $\widetilde{\mathbb{P}}^{(g)}$ we use the martingale condition (5.50) written for each base asset. As a result, we obtain a system of equations for the probabilities $\tilde{p}_j^{(g)} = \tilde{p}_j^{(i)} \equiv \widetilde{\mathbb{P}}^{(S^i)}(\omega^j)$, $j = 1, 2, 3$, for each choice of strictly positive numéraire asset $g = S^i$. For each choice of g we have:

$$\widetilde{\mathrm{E}}^{(g)}\left[\frac{S_T^1}{g_T}\right] = \frac{S_T^1(\omega^1)}{g_T(\omega^1)}\tilde{p}_1^{(g)} + \frac{S_T^1(\omega^2)}{g_T(\omega^2)}\tilde{p}_2^{(g)} + \frac{S_T^1(\omega^3)}{g_T(\omega^3)}\tilde{p}_3^{(g)} = \frac{S_0^1}{g_0} \tag{5.52}$$

$$\widetilde{\mathrm{E}}^{(g)}\left[\frac{S_T^2}{g_T}\right] = \frac{S_T^2(\omega^1)}{g_T(\omega^1)}\tilde{p}_1^{(g)} + \frac{S_T^2(\omega^2)}{g_T(\omega^2)}\tilde{p}_2^{(g)} + \frac{S_T^2(\omega^3)}{g_T(\omega^3)}\tilde{p}_3^{(g)} = \frac{S_0^2}{g_0} \tag{5.53}$$

$$\widetilde{\mathrm{E}}^{(g)}\left[\frac{S_T^3}{g_T}\right] = \frac{S_T^3(\omega^1)}{g_T(\omega^1)}\tilde{p}_1^{(g)} + \frac{S_T^3(\omega^2)}{g_T(\omega^2)}\tilde{p}_2^{(g)} + \frac{S_T^3(\omega^3)}{g_T(\omega^3)}\tilde{p}_3^{(g)} = \frac{S_0^3}{g_0} \tag{5.54}$$

First, consider the choice $g = S^1$:

$$\begin{cases} \tilde{p}_1^{(1)} + \tilde{p}_2^{(1)} + \tilde{p}_3^{(1)} = 1 \\ 140\tilde{p}_1^{(1)} + 110\tilde{p}_2^{(1)} + 80\tilde{p}_3^{(1)} = \frac{100}{5/6} \\ 40\tilde{p}_1^{(1)} + 10\tilde{p}_2^{(1)} + 0 \cdot \tilde{p}_3^{(1)} = \frac{20}{5/6} \end{cases} \implies \begin{cases} \tilde{p}_1^{(1)} = \frac{8}{15} \\ \tilde{p}_2^{(1)} = \frac{4}{15} \\ \tilde{p}_3^{(1)} = \frac{1}{5} \end{cases} \tag{5.55}$$

Second, consider the choice $g = S^2$:

$$\begin{cases} \frac{1}{140}\tilde{p}_1^{(2)} + \frac{1}{110}\tilde{p}_2^{(2)} + \frac{1}{80}\tilde{p}_3^{(2)} = \frac{5/6}{100} \\ \tilde{p}_1^{(2)} + \tilde{p}_2^{(2)} + \tilde{p}_3^{(2)} = 1 \\ \frac{40}{140}\tilde{p}_1^{(2)} + \frac{10}{110}\tilde{p}_2^{(2)} + \frac{0}{80}\tilde{p}_3^{(2)} = \frac{20}{100} \end{cases} \implies \begin{cases} \tilde{p}_1^{(2)} = \frac{28}{45} \\ \tilde{p}_2^{(2)} = \frac{11}{45} \\ \tilde{p}_3^{(2)} = \frac{2}{15} \end{cases} \tag{5.56}$$

Note that the third asset S^3 cannot serve as numéraire since its payoff vector (third column in $\mathbf{D}$) is not strictly positive. $\qquad\square$

Suppose that there are two choices, g and f, for the numéraire asset. For the two numéraire assets, let $\widetilde{\mathbb{P}}^{(g)}$ and $\widetilde{\mathbb{P}}^{(f)}$ be the respective equivalent martingale measures (EMMs) and $\widetilde{\mathbb{E}}^{(g)}[\,\cdot\,]$ and $\widetilde{\mathbb{E}}^{(f)}[\,\cdot\,]$ denote the respective mathematical expectations under the two measures. We are interested in the relationship between the two sets of probabilities $\tilde{p}_j^{(g)} := \widetilde{\mathbb{P}}^{(g)}(\omega^j)$ and $\tilde{p}_j^{(f)} := \widetilde{\mathbb{P}}^{(f)}(\omega^j)$, $j = 1, 2, \ldots, M$. For this purpose, consider any attainable claim having value $\Pi_t = \Pi_t^\varphi$, $t \in \{0, T\}$, $\varphi \in \mathbb{R}^N$. Applying (5.51) with EMMs $\widetilde{\mathbb{P}}^{(g)}$ and $\widetilde{\mathbb{P}}^{(f)}$ gives the no-arbitrage initial value of the claim

$$\Pi_0 = g_0 \widetilde{\mathbb{E}}^{(g)}\left[\frac{\Pi_T}{g_T}\right] = f_0 \widetilde{\mathbb{E}}^{(f)}\left[\frac{\Pi_T}{f_T}\right] \implies \widetilde{\mathbb{E}}^{(g)}\left[\frac{\Pi_T}{g_T}\right] = \widetilde{\mathbb{E}}^{(f)}\left[\frac{g_T/g_0}{f_T/f_0}\frac{\Pi_T}{g_T}\right]. \tag{5.57}$$

Let us define a new random variable

$$\varrho \equiv \varrho^{(f \to g)} := \frac{g_T/g_0}{f_T/f_0}. \tag{5.58}$$

The variable ϱ is called the *Radon–Nikodym derivative*. As follows from (5.57), it acts as a re-weighting factor that enters into the expected value of any random variable payoff X when changing probability measures in the expectation from one EMM into another EMM:

$$\widetilde{\mathbb{E}}^{(g)}[X] = \widetilde{\mathbb{E}}^{(f)}[\varrho X]. \tag{5.59}$$

Equivalently, and more explicitly,

$$\sum_{j=1}^{M} X(\omega^j)\tilde{p}_j^{(g)} = \sum_{j=1}^{M} \varrho(\omega^j)X(\omega^j)\tilde{p}_j^{(f)}. \tag{5.60}$$

We have the following basic properties of a Radon–Nikodym derivative.

Proposition 5.15 (Properties of a Radon–Nikodym derivative).

(1) The random variable $\varrho^{(f \to g)}$ is strictly positive;

(2) The expected value of $\varrho^{(f \to g)}$ under measure $\widetilde{\mathbb{P}}^{(f)}$ is one,

$$\widetilde{\mathbb{E}}^{(f)}\left[\varrho^{(f \to g)}\right] = 1.$$

Proof. The first property is obvious since for every ω, the value $\varrho(\omega)$ is a ratio of positive values, i.e., both f_t and g_t are strictly positive numéraire assets. The second property follows directly from (5.59) by setting $X \equiv 1$. $\qquad\square$

As follows from (5.60), the probabilities $\{\tilde{p}_j^{(g)}\}$ and $\{\tilde{p}_j^{(f)}\}$ relate to each other by

$$\tilde{p}_j^{(g)} = \varrho^{(f \to g)}(\omega^j)\,\tilde{p}_j^{(f)} = \frac{g_T(\omega^j)/g_0}{f_T(\omega^j)/f_0}\,\tilde{p}_j^{(f)}, \quad j = 1, 2, \ldots, M, \tag{5.61}$$

or, equivalently,

$$\varrho^{(f \to g)}(\omega) = \frac{\widetilde{P}^{(g)}(\omega)}{\widetilde{P}^{(f)}(\omega)} \text{ for all } \omega \in \Omega. \tag{5.62}$$

Hence, being given the risk-neutral probabilities for one EMM and the Radon–Nikodym derivative, we can calculate the state probabilities of the other EMM by simply using the relation in (5.61).

Note that the variable $\frac{1}{\varrho^{(f \to g)}} \equiv \varrho^{(g \to f)}$ is also a Radon–Nikodym derivative that allows for switching from $\widetilde{\mathbb{P}}^{(g)}$ to $\widetilde{\mathbb{P}}^{(f)}$. For any random variable (payoff) $X \in \mathcal{L}(\Omega)$ we have the following property, which is symmetric to that in (5.59):

$$\widetilde{\mathrm{E}}^{(f)}[X] = \widetilde{\mathrm{E}}^{(g)}\left[\frac{1}{\varrho^{(f \to g)}}\,X\right] = \widetilde{\mathrm{E}}^{(g)}\left[\varrho^{(g \to f)}\,X\right].$$

If the market is incomplete and arbitrage-free, then there are many equivalent $\widetilde{\mathbb{P}}^{(g)}$-martingale measures for a given numéraire g. Hence, $\tilde{p}_j^{(f)}$ and $\tilde{p}_j^{(g)}$ are not necessarily uniquely related. However, given a set of probabilities $\{\tilde{p}_j^{(f)}\}$ for a martingale measure $\widetilde{\mathbb{P}}^{(f)}$ there exists a set $\{\tilde{p}_j^{(g)}\}$ for a martingale measure $\widetilde{\mathbb{P}}^{(g)}$ using (5.61). The relationship and hence the martingale measure $\widetilde{\mathbb{P}}^{(g)}$ for any given numéraire g is unique iff the market is complete.

Example 5.16 (Example 5.15 continued). Find the Radon–Nikodym derivative $\varrho^{(S^1 \to S^2)}$.

Solution. The Radon–Nikodym derivative $\varrho \equiv \varrho^{(S^1 \to S^2)}$, for changing measures from $S^1 \to S^2$ as numéraire assets, is given by the ratio

$$\varrho(\omega) = \frac{S_T^2(\omega)/S_0^2}{S_T^1(\omega)/S_0^1} = \begin{cases} \frac{7}{6} & \omega = \omega^1 \\ \frac{11}{12} & \omega = \omega^2 \\ \frac{2}{3} & \omega = \omega^3. \end{cases}$$

This random variable allows us to express the probabilities $\tilde{p}_j^{(1)} \equiv \tilde{p}_j^{(S^1)}$ in terms of $\tilde{p}_j^{(2)} \equiv \tilde{p}_j^{(S^2)}$ and vice versa. For example, we have $\tilde{p}_j^{(2)} = \varrho(\omega^j)\,\tilde{p}_j^{(1)}$. Indeed, using the solutions (5.55) and (5.56) allows us to verify this:

$$\varrho(\omega^1)\,\tilde{p}_1^{(1)} = \frac{7}{6} \cdot \frac{8}{15} = \frac{28}{45} = \tilde{p}_1^{(2)}, \quad \checkmark$$

$$\varrho(\omega^2)\,\tilde{p}_2^{(1)} = \frac{11}{12} \cdot \frac{4}{15} = \frac{11}{45} = \tilde{p}_2^{(2)}, \quad \checkmark$$

$$\varrho(\omega^3)\,\tilde{p}_3^{(1)} = \frac{2}{3} \cdot \frac{1}{5} = \frac{2}{15} = \tilde{p}_3^{(2)}. \quad \checkmark$$

$\square$

More generally, a Radon–Nikodym derivative can be defined for *any pair of equivalent probability measures* that are not required to relate to particular numéraires, i.e., not necessarily related to EMMs. Let us consider a finite sample space Ω on which two equivalent probability measures $\mathbb{P}$ and $\widehat{\mathbb{P}}$ are defined. We assume that both $\mathbb{P}$ and $\widehat{\mathbb{P}}$ assign a positive probability to every element of the sample space, so we can calculate the quotient

$$\varrho(\omega) \equiv \varrho^{\mathbb{P} \to \widehat{\mathbb{P}}}(\omega) = \frac{\widehat{\mathbb{P}}(\omega)}{\mathbb{P}(\omega)}$$

for every $\omega \in \Omega$ (compare it with (5.62)). The random variable ϱ is called the *Radon–Nikodym derivative of $\widehat{\mathbb{P}}$ with respect to $\mathbb{P}$*. As proved above in Proposition 5.15 and in equation (5.59), the variable ϱ has the following properties:

(a) $\varrho > 0$ with probability 1;

(b) $\mathrm{E}[\varrho] = 1$;

(c) for any random variable Y, we have $\widehat{\mathrm{E}}[Y] = \mathrm{E}[\varrho Y]$.

Here, $\mathrm{E}[\,\cdot\,]$ and $\widehat{\mathrm{E}}[\,\cdot\,]$ denote the mathematical expectations under $\mathbb{P}$ and $\widehat{\mathbb{P}}$, respectively. Moreover, since $\varrho^{\widehat{\mathbb{P}} \to \mathbb{P}}(\omega) = \frac{\mathbb{P}(\omega)}{\widehat{\mathbb{P}}(\omega)}$, we have the simple relation $\varrho^{\widehat{\mathbb{P}} \to \mathbb{P}}(\omega) = \frac{1}{\varrho^{\mathbb{P} \to \widehat{\mathbb{P}}}(\omega)}$ between the two Radon–Nikodym derivatives for changing measures from $\mathbb{P} \to \widehat{\mathbb{P}}$ and vice versa from $\widehat{\mathbb{P}} \to \mathbb{P}$.

5.6 Exercises

Exercise 5.1. Consider a standard binomial model with the state space $\Omega = \{\omega^+, \omega^-\}$ and two base securities, risk-free bond B and risky stock S.

(a) Determine the portfolios $[\beta^\pm, \Delta^\pm]$ in the base assets B and S that replicate the payoffs of Arrow–Debreu securities $\mathcal{E}^\pm = \mathbb{I}_{\{\omega^\pm\}}$.

(b) Determine the no-arbitrage prices $\pi_0(\mathcal{E}^\pm)$.

(c) For an arbitrary payoff function X, we can write

$$X(\omega) = X(\omega^+)\mathcal{E}^+(\omega) + X(\omega^-)\mathcal{E}^-(\omega).$$

Since the replicating portfolio (β_X, Δ_X) is given by

$$[\beta_X, \Delta_X] = X(\omega^+)[\beta^+, \Delta^+] + X(\omega^-)[\beta^-, \Delta^-]$$

derive the formula for $[\beta_X, \Delta_X]$ using the result of (a).

(d) Using the results of (b) and (c), derive the formula for $\pi_0(X)$.

(e) Determine the no-arbitrage price $\pi_0(X)$ for the two payoffs $X = \max(S_T, K)$ and $X = \max(S_T - K, 0)$ with $K > 0$. [Hint: Consider three cases: 1) $\frac{K}{S_0} \leqslant d$; 2) $d < \frac{K}{S_0} < u$; 3) $u \leqslant \frac{K}{S_0}$.]

Exercise 5.2. Consider a standard binomial model. To hedge a short position in a claim C with payoff $[C^+, C^-]$, the investor forms a portfolio with Δ shares of stock. Obtain the optimal value of Δ that minimizes the variance of the terminal value $\Pi_T = -C_T + \Delta S_T$. What is the variance of Π_T when Δ is optimal?

Exercise 5.3. A stock currently trades at \$100. In three months its price will either be \$125 (in state ω^1), \$100 (in state ω^2) or \$85 (in state ω^3). You sell a call option on this stock with strike price \$95 and maturity time $T = 3$ months for \$11. You hedge your exposure by purchasing Δ shares, borrowing $100\Delta - 11$ in order to fund the stock purchase. The simple rate of interest is 12%.

(a) What will your profit/loss be in three months? That is, obtain the terminal value of your portfolio, $\Pi_T(\Delta)$, for every possible outcome ω^j. Leave your answer in terms of Δ.

(b) Is it possible for you to completely hedge your exposure (i.e., completely eliminate the risk)? Explain. [Hint: Show whether or not there exists a value for Δ s.t. $\Pi_T(\Delta)$ is constant.]

Exercise 5.4. Consider a trinomial single-period model with only a stock S and a bond B. Let $B_0 = 100$, $B_T = 105$, $S_0 = 10$, and

$$S_T(\omega) = \begin{cases} 14, & \text{if } \omega = \omega^1, \\ 12, & \text{if } \omega = \omega^2, \\ 8, & \text{if } \omega = \omega^3. \end{cases}$$

(a) Argue that the model is incomplete.

(b) Show that the payoff vector $\mathbf{X} = [3, 2, 0]$ is attainable and obtain a portfolio of base assets replicating this payoff.

(c) Determine all state price vectors and formulate the no-arbitrage condition.

(d) Determine the bid-ask price interval for a call option on the stock with strike $K = 10$.

(e) Suppose that the call option (from part d) is now base asset 3. If its initial price were $C_0 = \frac{3}{2}$, would the new 3-by-3 model be arbitrage free?

(f) Determine the equivalent martingale measure for the numéraire $g = S$ (provide the general solution).

Exercise 5.5. Assume a 2-by-2 economy with the state space $\Omega = \{\omega^1, \omega^2\}$ and two contingent assets with prices S_t^1 and S_t^2, $t \in \{0, T\}$. Assume $S_0^1 > 0$ and $S_0^2 > 0$. Let $S_T^i(\omega^j) = S_0^i R_j^i$ for some positive constants R_j^i, $i, j = 1, 2$.

(a) Derive a simple condition (in terms of R_j^i) for the completeness of this market model.

(b) Determine the risk-neutral probabilities $\tilde{p} = \widetilde{\mathbb{P}}(\omega^1)$ and $1 - \tilde{p} = \widetilde{\mathbb{P}}(\omega^2)$ such that the discounted price process $\frac{S_t^2}{S_t^1}$ is a martingale, i.e.

$$\widetilde{\mathbb{E}}\left[\frac{S_T^2}{S_T^1}\right] = \frac{S_0^2}{S_0^1}.$$

(c) Assume that the market is complete (i.e., the condition derived in (a) holds). Determine the portfolio (φ_1, φ_2) that replicates an arbitrary payoff X. Obtain the formula for $\pi_0(X)$.

Exercise 5.6 (A variant of the Law of One Price). Suppose that there are no arbitrage opportunities. Let X_t and Y_t, $t \in \{0, T\}$, be the prices of two securities such that $X_T(\omega) \geqslant Y_T(\omega)$ for all $\omega \in \Omega$ and $X_T(\omega^*) > Y_T(\omega^*)$ for at least one $\omega^* \in \Omega$. Prove that $X_0 > Y_0$.

Exercise 5.7. Consider a 3-by-3 market model with the following payoff matrix:

$$\mathbf{D} = \begin{bmatrix} 23 & 33 & 9 \\ 15 & 19 & 21 \\ 7 & 5 & 33 \end{bmatrix}.$$

(a) Show that the market is not complete.

(b) Determine any redundant base asset and represent its payoff vector as a linear combination of the payoffs of the other two assets.

Exercise 5.8. Consider a one-period binomial model. Assume that $B_0 = \$0.9$, $B_T = \$1$, $S_0 = \$100$, and that the two possible values of S_T are $\$90$ and $\$105$.

(a) Is this model arbitrage-free? Use the geometric interpretation of the first FTAP to verify this. Provide a diagram.

(b) If the model is arbitrage-free, obtain a state-price vector and risk-neutral probabilities. If the model admits arbitrage, obtain an arbitrage portfolio.

Exercise 5.9. Consider a one-period, arbitrage-free binomial model. Determine a replicating portfolio for a forward contract with strike price K. Determine the initial value of the forward contract.

Exercise 5.10. Three assets A, B, and C have market prices and payoffs as given in the table below:

Asset	Price	Payoff in state 1	Payoff in state 2
A	$70	$50	$100
B	$60	$30	$120
C	$80	$38	$112

(a) Construct a portfolio $[\varphi_A, \varphi_B]$ in assets A and B that replicates the payoff of asset C.

(b) Determine the present time-0 value of the replicating portfolio. Is there a possible arbitrage in this market? Explain.

(c) Determine whether or not assets A and B form the basis for a complete arbitrage-free two-state market. If so, obtain the risk-neutral probabilities.

Exercise 5.11. Consider a single-period economy with three states w^j, $j = 1, 2, 3$ and three base assets, a zero-coupon bond with interest rate $r = 0.1$, a stock S with spot price $S_0 = 45$, and a call option on the stock S with strike $K = 40$ and maturity T. Assume that the call option has current price $C_0 = C_0(S_0, K) = 10$ and that the stock can attain terminal values of 60, 50, and 30 in the respective states 1, 2, and 3.

(a) Define appropriate base asset price processes $\{S_t^i\}_{t \in \{0,T\}}$, $i = 1, 2, 3$, and provide the payoff matrix $\mathbf{D}$. Show that the market is complete.

(b) Using the three base assets, obtain the three replicating portfolios φ_j, $j = 1, 2, 3$ for the Arrow–Debreu securities that pay one unit of account in the respective jth state. From this, determine the risk-neutral probabilities $\tilde{p}_j$, $j = 1, 2, 3$.

(c) Determine a replicating portfolio in the three base assets for a put option struck at $K = 40$ and determine its initial price $P_0 = P_0(S_0, K)$.

(d) Re-price the put option from part (c) but this time simply employ the asset pricing formula (i.e., discounted expected value) using appropriate risk-neutral probabilities. Is this price the same as that obtained in part (c)? If so, why?

Exercise 5.12. Consider a 3-by-2 one-period model with three assets: a bond B and two stocks S^i, $i = 1, 2$. Assume that the bond sells for $1 at $t = 0$ and pays $R at $t = T$. For stock 1, we have $S_0^1 = 10$ and two possible prices of $12 and $8, respectively, at $t = T$. The initial price of stock 2 is $8, and S_T^2 has two possible prices of $Z and $4. Here, $Z \geqslant 0$ and $R \geqslant 1$.

(a) Show that there are redundant base assets. Represent stock 2 as a portfolio of the bond and stock 1.

(b) Verify that this model is complete for all choices of $Z \geqslant 0$ and $R \geqslant 1$.

(c) For what values of Z and R is the model arbitrage-free?

(d) Suppose that a call on stock 1 with a strike price of $9 has a price of $2. Determine Z and R. Determine a unique state-price vector and risk-neutral probabilities.

Exercise 5.13. Consider a 3-by-3 single-period model with

$$\mathbf{S}_0 = \begin{bmatrix} 10 \\ 10 \\ 10 \end{bmatrix} \text{ and } \mathbf{D} = \begin{bmatrix} 5 & 10 & 15 \\ 5 & 10 & 15 \\ 15 & 10 & 5 \end{bmatrix}.$$

Answer the following questions and provide your explanations.

(a) Are there redundant base securities?

(b) Is the model complete?

(c) Is a security with the payoff vector $\mathbf{X} = \begin{bmatrix} 1 & 2 & 3 \end{bmatrix}$ attainable? If it is so, obtain all the replicating portfolios.

Exercise 5.14. Consider a 3×3 single-period model with $\mathbf{S}_0 = \begin{bmatrix} 1 \\ 5 \\ 10 \end{bmatrix}$ and $\mathbf{D} = \begin{bmatrix} 1 & 2 & 1 \\ 2 & 6 & 15 \\ 12 & 8 & 6 \end{bmatrix}$.

(a) Verify that $\boldsymbol{\Psi} = \begin{bmatrix} \frac{34}{49}, & \frac{2}{49}, & \frac{11}{49} \end{bmatrix}^{\top}$ is a state price vector.

(b) Is the model arbitrage free? If yes, then why?

(c) Obtain the risk-free rate of return.

(d) Determine the risk-neutral probabilities.

(e) Determine the no-arbitrage price of a put option on asset S^3 with strike $K = 10$.

Exercise 5.15. Consider the following 4-by-4 Arrow–Debreu model:

$$\mathbf{S}_0 = \begin{bmatrix} 9/10 \\ 1 \\ 1/4 \\ 1/4 \end{bmatrix} \quad \text{and} \quad \mathbf{D} = \begin{bmatrix} 1 & 1 & 1 & 1 \\ 0 & 1 & 0 & 3 \\ 1 & 0 & 0 & 0 \\ 0 & 1 & 0 & 0 \end{bmatrix}$$

(a) Show that the model is arbitrage-free and complete.

(b) Determine a unique state-price vector and risk-neutral probabilities.

(c) Determine the initial price of a put option on S_T^2 with strike price $K = 2$.

(d) Determine the initial price of a call option on the maximum of S_T^2, S_T^3, and S_T^4 with strike price $K = 2$.

(e) Determine the initial price of a put option on the portfolio $\varphi = \begin{bmatrix} 0, & 1, & 1, & 1 \end{bmatrix}$ with strike price $K = 3$ (the payoff at $t = T$ is $(K - (S_T^2 + S_T^3 + S_T^4))^+)$.

Exercise 5.16. Consider a one-period model with three states and two assets, a risk-free bond and a stock. Assume that $B_0 = 100$, $B_T = 105$, $S_0 = 10$, and $S_T \in \{8, 9, 12\}$.

(a) Determine if the model is arbitrage-free and complete.

(b) Obtain the general solution for the state prices: a unique solution if the market is complete, or a range of solutions if the market is incomplete.

(c) Consider a call option on the stock with strike price $K = 10$. Determine the bid-ask spread at $t = 0$.

(d) Suppose the market price of the call option is zero. Construct an arbitrage portfolio.

Exercise 5.17. Determine if the following Arrow–Debreu models with $M = 4$ and $N = 3$ are arbitrage-free. If yes, find a state-price vector (just one). If not, find an arbitrage portfolio.

(a) $\mathbf{S_0} = \begin{bmatrix} 1 \\ 2 \\ 10 \end{bmatrix}$, $\mathbf{D} = \begin{bmatrix} 1 & 1 & 1 & 1 \\ 12 & 3 & 0 & 0 \\ 0 & 0 & 0 & 10 \end{bmatrix}$;

(b) $\mathbf{S_0} = \begin{bmatrix} 1 \\ 2 \\ 10 \end{bmatrix}$, $\mathbf{D} = \begin{bmatrix} 1 & 1 & 1 & 1 \\ 12 & 3 & 0 & 0 \\ 0 & 0 & 0 & 20 \end{bmatrix}$.

Exercise 5.18. Consider a trinomial single-period model with stock price S_t and zero-coupon bond price B_t, $t \in \{0, T\}$. Let $B_0 = \frac{1}{1+r}$, $B_T = 1$, and

$$\frac{S_T(\omega)}{S_0} = \begin{cases} u, & \text{if } \omega = \omega^1, \\ m, & \text{if } \omega = \omega^2, \\ d, & \text{if } \omega = \omega^3. \end{cases}$$

This model is incomplete. Determine the bid-ask spreads for the three Arrow–Debreu securities.

Exercise 5.19. Consider a single-period model with four states of the world and two tradable assets. Let the initial price vector and payoff matrix be given by

$$\mathbf{S_0} = \begin{bmatrix} 10 \\ 10 \end{bmatrix} \quad \text{and} \quad \mathbf{D} = \begin{bmatrix} 11 & 11 & 11 & 11 \\ 8 & 10 & 11 & 14 \end{bmatrix}.$$

(a) Show that the market is arbitrage-free. Obtain the general solution for the state prices and illustrate it with a diagram.

(b) Determine the bid-ask spread for the price of a put on the risky asset with strike price $K = 11$.

Exercise 5.20. Consider a single-period binomial model with two base assets. Asset #1 is risky with initial value $S_0^1 = 10$ and terminal values $S_T^1(\omega^1) = 8$ and $S_T^1(\omega^2) = 14$. Asset #2 is risk-free with initial value $S_0^2 = 10$ and rate of return $r = 20\%$. Assume that scenario ω^1 is twice more likely than scenario ω^2.

(a) Obtain the payoff matrix $\mathbf{D}$ and the initial price vector $\mathbf{S_0}$.

(b) Argue that the model is complete and arbitrage free (without computing state prices or risk-neutral probabilities).

(c) Obtain the real-world probabilities $p_1 = \mathbb{P}(\omega^1)$ and $p_2 = \mathbb{P}(\omega^2)$.

(d) Determine a portfolio $\varphi \in \mathbb{R}^2$ with initial value of 100 that maximizes $\mathrm{E}[\Pi_T^\varphi]$.

Exercise 5.21. Consider a 3-by-3 single-period model with

$$\mathbf{S_0} = \begin{bmatrix} 10 \\ 10 \\ 10 \end{bmatrix} \quad \text{and} \quad \mathbf{D} = \begin{bmatrix} 6 & 9 & 12 \\ 12 & 10 & 6 \\ 0 & 8 & 18 \end{bmatrix}.$$

The vector $\boldsymbol{\Psi} = \left[-\frac{5}{24} + \frac{11}{8}s, \ \frac{5}{4} - \frac{9}{4}s, \ s \right]^\top$, $s \in \mathbb{R}$, solves $\mathbf{D}\boldsymbol{\Psi} = \mathbf{S_0}$.

(a) Verify that the model is arbitrage free and obtain the no-arbitrage condition on the parameter s.

(b) Obtain the no-arbitrage initial price (or the bid-ask spread if the price is not unique) of a unit zero-coupon bond. If the no-arbitrage price is not unique, provide your explanations of this fact.

(c) Obtain the no-arbitrage initial price (or the bid-ask spread if the price is not unique) of a call option on the maximum price, $\max\{S_T^1, S_T^2, S_T^3\}$, with strike $K = 10$. If the no-arbitrage price is not unique, provide your explanations of this fact.

Exercise 5.22. Assume a two-state economy with two base assets S^1, S^2 such that

$$\mathbf{S}_0 = \begin{bmatrix} 10 \\ 15 \end{bmatrix} \quad \text{and} \quad \mathbf{D} = \begin{bmatrix} 5 & 15 \\ 20 & 10 \end{bmatrix}.$$

(a) Show that market is complete and there are no redundant base assets.

(b) Determine the equivalent martingale measure for the numéraire $g = S^1$.

(c) Determine the equivalent martingale measure for the numéraire $g = S^2$.

(d) Using first the EMM $\widetilde{\mathbb{P}}^{(S^1)}$ and then $\widetilde{\mathbb{P}}^{(S^2)}$, calculate and compare the risk-neutral intial price C_0 of the call option with payoff $C_T = (M_T - 15)^+$, where $M_T = \max\{S_T^1, S_T^2\}$.

Exercise 5.23. Consider a single-period economy with three states and two base assets, a zero-coupon bond B with interest rate $r = 0.1$ and a stock S with spot price $S_0 = 45$ and terminal values of 60, 50, and 30 in the respective states ω^1, ω^2, and ω^3.

(a) Determine the payoff matrix. Show that the market is incomplete.

(b) Show that the market is arbitrage-free. Provide the general formula for the state prices.

(c) Consider a call option on the stock with strike $K = 40$ and maturity T. Determine the bid-ask spread for its initial price C_0.

(d) Assume that the call option has initial price $C_0 = 10$. Consider this option as the third base asset (i.e., we make the market complete). Provide the upgraded payoff matrix. Show that the new market model is complete. Is it arbitrage-free?

(e) For the market with three base assets introduced in part (d), determine the unique fair prices of the Arrow–Debreu securities that pay one unit of currency in the respective states of the world. From this, determine the risk-neutral probabilities $\tilde{p}_j$, $j = 1, 2, 3$.

(f) For the market with three base assets introduced in part (d), consider a put option struck at $K = 40$. Determine its initial price P_0 by using the asset pricing formula with the risk-neutral probabilities from part (e).

Exercise 5.24. Consider the following 4-by-3 market model:

$$\mathbf{S}_0 = \begin{bmatrix} 1 \\ 1 \\ 1 \\ 2 \end{bmatrix} \quad \text{and} \quad \mathbf{D} = \begin{bmatrix} 1 & 0 & 1 \\ 2 & 1 & 0 \\ 0 & 3 & 1 \\ 2 & 0 & 2 \end{bmatrix}.$$

(a) Verify that this model is arbitrage-free and complete by finding a unique state-price vector.

(b) What is the risk-free rate r of return in this model? Replicate the risk-free asset.

(c) Obtain the price at $t = 0$ for the call option on the maximum of S^1, S^2, S^3, and S^4, with strike price $K = 2$.

(d) Obtain the price at $t = 0$ for the put option on the arithmetic average of S^1, S^2, S^3, and S^4, with strike price $K = 2$.

Exercise 5.25. Assume a two-state economy. Consider a market with two risky assets S_t^1 and S_t^2, $t \in \{0, T\}$, whose initial prices are $S_0^1 = 10$ and $S_0^2 = 25$ and terminal values are $S_T^1(\omega^1) = 5$, $S_T^1(\omega^2) = 15$, $S_T^2(\omega^1) = c$, $S_T^2(\omega^2) = c + 10$, respectively. Here c is some positive parameter.

(a) For what values of c is the market complete?

(b) Determine the equivalent martingale measure $\widetilde{\mathbb{P}}^{(g)}$ for the numéraire $g = S^1$. Provide the general solution (with arbitrary c).

(c) For what values of the parameter c is the market arbitrage-free?

Exercise 5.26. Show that if a market model is incomplete, then at least one Arrow–Debreu security is not attainable.

Exercise 5.27. Assume a two-state single period economy with two base assets, the zero-coupon bond with initial price B_0 and a stock with price S_t at time $t \in \{0, T\}$. Let $r > 0$ be the return on the bond. Denote $S_+ = S_T(\omega^1)$ and $S_- = S_T(\omega^2)$. Assume that $S_- < S_+$. Let $q_j = \widetilde{\mathbb{P}}^{(g)}(\omega^j)$, $j = 1, 2$, be the probabilities for the martingale measure with g as numéraire asset defined by the price process $g_t = \alpha S_t + \beta B_t$, $t \in \{0, T\}$. Assume the positions α, β are chosen such that $g_t(\omega) > 0$ for all t and all ω.

(a) Show that:
$$q_1 = \frac{\alpha S_+ + \beta(1 + r)B_0}{\alpha S_0 + \beta B_0} \left(\frac{S_0 - S_-(1 + r)^{-1}}{S_+ - S_-} \right)$$

and
$$q_2 = \frac{\alpha S_- + \beta(1 + r)B_0}{\alpha S_0 + \beta B_0} \left(\frac{S_+(1 + r)^{-1} - S_0}{S_+ - S_-} \right).$$

(b) Verify that $q_1 + q_2 = 1$.

(c) Show that $q_1, q_2 > 0$ from the no-arbitrage condition.

(d) Obtain the current price of the two Arrow–Debreu securities by explicitly taking expectations with respect to the $\widetilde{\mathbb{P}}^{(g)}$ measure.

Exercise 5.28. Assume a three-state single period economy with three base assets, the zero-coupon bond with price $S_t^1 = B_t$, a stock with price $S_t^2 = S_t$, and a call on the stock with price $S_t^3 = C_t = C_t(S_t, K)$, $t \in \{0, T\}$. Let $S_T(\omega^1) = S_+$, $S_T(\omega^2) = S_0$, $S_T(\omega^3) = S_-$ and assume that the call is struck at $K = S_0$ where $S_+ > S_0 > S_-$.

(a) Set $g = S$ as the numéraire asset price process and write down the linear system of three equations in the probabilities $q_j = \mathbb{P}^{(g)}(\omega^j)$, $j = 1, 2, 3$, corresponding to the martingale conditions for the relative base asset prices. Leave the equations expressed in terms of S_+, S_-, S_0, C_0, and the bond return r.

(b) Now assume $C_0 = 20/3$, $S_- = 20$, $S_0 = 40$, $S_+ = 60$, $r = 0$ and hence from part (a) determine the probabilities q_1, q_2, q_3.

(c) Assuming the same parameters as in part (b), obtain the initial price of a call with strike 30 and maturity T in this economy.

Exercise 5.29. Consider a single-period economy having three assets and two states with initial price vector and payoff matrix:

$$\mathbf{S}_0 = \begin{bmatrix} 1 \\ 10 \\ S_0^3 \end{bmatrix}, \quad \mathbf{D} = \begin{bmatrix} 1.05 & 1.05 \\ 20 & 5 \\ 10 & 0 \end{bmatrix}.$$

(a) Determine the state-price vector $\boldsymbol{\Psi}$ and the corresponding risk-neutral probabilities. Determine the no-arbitrage price S_0^3.

(b) Obtain the portfolio $[\varphi_1, \varphi_2, 0]$, consisting of only nonzero positions in the first two base assets, that replicates any arbitrary claim with cash-flow vector $[c_1, c_2]$ in the respective states $\{\omega^1, \omega^2\}$. Based on this, determine the initial price of such a claim. Express your answers in terms of c_1 and c_2.

(c) Determine the initial price V_0 of a chooser option with payoff

$$V_T = \max\{S_T^2, 3S_T^3\}.$$

Exercise 5.30. Consider a three-state single-period model with three base assets, a stock with the price S_t, a European call struck at K_1 with the price C_t^1, and a European call struck at K_2 with the price C_t^2, where $t \in \{0, T\}$. Both options expire at time T. Let $S_T(\omega^1) = S_+$, $S_T(\omega^2) = S_0$, $S_T(\omega^3) = S_-$, where $0 < S_- < S_0 < S_+$. Suppose that $K_1 = S_-$ and $K_2 = S_0$.

(a) Provide the payoff matrix $\mathbf{D}$ in terms of S_+, S_0, S_- only.

(b) Determine if the market is complete and free of redundant securities.

(c) Determine the three replicating portfolios $\boldsymbol{\varphi}^{(j)} = [\varphi_1^{(j)}, \varphi_2^{(j)}, \varphi_3^{(j)}]$ for the respective Arrow–Debreu securities $\mathcal{E}^j = \mathbb{I}_{\{\omega^j\}}$, $j = 1, 2, 3$.

(d) Without computing the prices C_0^1 and C_0^2, determine which one is larger. Explain.

(e) Suppose that $S_+ = 3$, $S_0 = 2$, $S_- = 1$, $C_0^1 = 1$. Determine the upper and lower bounds on the no-arbitrage price C_0^2.

Exercise 5.31. Consider a single-period economy with sample space $\Omega = \{\omega^1, \omega^2, \omega^3\}$. Assume an equivalent martingale measure $\widetilde{\mathbb{P}}$ with cash as numéraire is given by the state probabilities $\widetilde{\mathbb{P}}(\omega^j) = \tilde{p}_j = 1/3$, for all $j = 1, 2, 3$. Note that cash is an asset equivalent to a money market account (or bond) with no interest. Two assets X and Y in this economy have payoff vectors $X_T(\Omega) = [0, 1, 2]$ and $Y_T(\Omega) = [1, 2, 6]$.

(a) Let $[\varphi_X, \varphi_Y]$ be a portfolio where φ_X and φ_Y are fixed (and unknown) positions in assets X and Y, respectively. Determine the value of the ratio φ_X/φ_Y that makes the initial value of this portfolio equal zero. Hint: Use the $\widetilde{\mathbb{P}}$-martingale measure conditions.

(b) Determine the probabilities $\hat{p}_j := \widehat{\mathbb{P}}^{(g)}(\omega^j)$, $j = 1, 2, 3$, for the equivalent martingale measure with $g = Y$ as numéraire asset.

Exercise 5.32. Consider a trinomial model with the state space $\Omega = \{\omega^d, \omega^m, \omega^u\}$ and two base assets, risky asset S and risk-free asset B. To hedge the short position in a claim C with payoff $[C^d, C^m, C^u]$ where the values C^d, C^m, C^u are all different, the investor forms a portfolio with Δ shares of stock.

(a) Determine the optimal value of Δ that minimizes the variance of the terminal value $\Pi_T = -C_T + \Delta\, S_T$.

(b) Show that the variance of Π_T is always strictly positive, so the Δ-hedging cannot entirely eliminate the risk associated with such a risky claim in this trinomial model.

(c) Show that the optimal value of Δ can be expressed as a linear combination of optimal Δ's calculated for three binomial submodels with respective state spaces $\Omega_1 = \{\omega^d, \omega^u\}$, $\Omega_2 = \{\omega^m, \omega^u\}$, and $\Omega_3 = \{\omega^d, \omega^m\}$ (see Exercise 5.2).

Exercise 5.33. Consider a general N-by-M model with dividend matrix $\mathbf{D}$. Form a portfolio in base securities, $\varphi \in \mathbb{R}^N$, that hedges the short position in a claim with terminal payoff C_T. Show that the optimal portfolio vector φ that minimizes the variance of $-C_T + \Pi_T^\varphi$ is given by

$$\varphi = \mathbf{B}\,\boldsymbol{\Sigma}^{-1},$$

where $\mathbf{B} = \left[\, \mathrm{Cov}(C_T, S_T^i) \,\right]_{i=1,\ldots,N}$ and $\boldsymbol{\Sigma} = \left[\, \mathrm{Cov}(S_T^i, S_T^j) \,\right]_{i,j=1,\ldots,N}$.

6

Introduction to Discrete-Time Stochastic Calculus

The term *stochastic* means random. Because it is usually used together with the term *process*, it makes people think of something that changes in a random way over time. The term *calculus* refers to ways to calculate things or find objects that can be calculated (like derivatives in the differential calculus). *Stochastic Calculus* is the study of stochastic processes through a collection of special methods such as stochastic differential equations and stochastic integrals used to find probability distributions and to compute quantitative characteristics. Many fundamental concepts of stochastic calculus like filtration, conditioning, martingale, and stopping time are easy to introduce in a discrete-time setting. This chapter deals with discrete-time stochastic processes although many concepts will be defined in a general way.

6.1 A Multi-Period Binomial Probability Model

6.1.1 The Binomial Probability Space

6.1.1.1 A Sample Space

A *sample space* is a set of all possible outcomes of some experiment with an uncertain result. Examples of such *random experiments* include tossing coins and rolling dice. We shall denote a sample space by Ω. The elements $\omega \in \Omega$ are called *outcomes*. The goal of this section is the construction and characterization of a sample space that describes the multi-period binomial price model. Consider a risky asset such as a stock with initial price $S_0 > 0$. Suppose the stock price $\{S_k\}_{k=0,1,2,\ldots,N}$ follows a binomial tree model with T periods that has been introduced in Chapter 2. Recall that the stock price is defined by a recurrence relation

$$S_n = \begin{cases} u\, S_{n-1} & \text{with probability } p, \\ d\, S_{n-1} & \text{with probability } 1-p, \end{cases} \tag{6.1}$$

where $p \in (0,1)$, $k = 1, 2, \ldots, N$, and u and d are, respectively, the up-factor and down-factor satisfying $0 < d < u$. Let outcome ω_k describe the stock price dynamics in the kth period from time $k-1$ to time k as follows: $\omega_k = \mathsf{U}$ if the price moves up, and $\omega_k = \mathsf{D}$ if the price moves down. These two possible k-th market moves are represented as the respective events: $\{\omega_k = \mathsf{U}\}$ and $\{\omega_k = \mathsf{D}\}$. We shall call ω_k a *market move* in the kth period. During N periods, the stock price changes N times; thus its evolution is described by N market moves $\omega_1, \ldots, \omega_N$. Each possible market scenario, denoted by ω, is an N-step path in the binomial tree, and it can be represented by a string of D's and U's of length N:

$$\omega = (\omega_1, \omega_2, \ldots, \omega_N), \text{ where } \omega_k \in \{\mathsf{D}, \mathsf{U}\}, \ 1 \leqslant k \leqslant N.$$

The set of all strings of D's and U's of length N is a Cartesian product

$$\Omega_N = \prod_{k=1}^{N} \{\mathsf{D}, \mathsf{U}\} = \{(\omega_1, \omega_2, \ldots, \omega_N) \ : \ \omega_k \in \{\mathsf{D}, \mathsf{U}\}, \ 1 \leqslant k \leqslant N\}.$$

To simplify the notation, we shall write $\omega_1 \omega_2 \ldots \omega_N$ instead of $(\omega_1, \omega_2, \ldots, \omega_N)$. For instance, the sample spaces for $N =$1-, 2-, and 3-period models are, respectively,

$$\Omega_1 = \{\mathsf{D}, \mathsf{U}\};$$
$$\Omega_2 = \{\mathsf{DD}, \mathsf{DU}, \mathsf{UD}, \mathsf{UU}\};$$
$$\Omega_3 = \{\mathsf{DDD}, \mathsf{DDU}, \mathsf{DUD}, \mathsf{DUU}, \mathsf{UDD}, \mathsf{UDU}, \mathsf{UUD}, \mathsf{UUU}\}.$$

Note that Ω_N contains 2^N elements since there are N periods and two outcomes are possible in each period. The set Ω_N is called a *sample space* of the (N-period) binomial tree model whose elements ω^k, $1 \leqslant k \leqslant 2^N$, correspond to all possible market scenarios. In fact, each string from Ω_N can be equivalently viewed as a possible outcome of N tosses of a coin whose two sides are labelled U (a "head") and D (a "tail"), respectively. Thus, Ω_N is a sample space for the Bernoulli experiment with N independent tosses of a coin.

Any subset E of a sample space Ω is called an *event*. For instance, in the 3-period model

$$E = \{\mathsf{UUU}, \mathsf{UUD}, \mathsf{UDU}, \mathsf{UDD}\} \subset \Omega_3$$

represents the event that the market moves up in the first period. [Note: we shall use $\subset$ to mean subset where $A \subset B$ means every element in A is also in B.] We say that an event E *occurs* if the actual outcome ω belongs to E. We can consider a collection of all possible events (subsets including the empty set) of Ω. Such a collection is denoted 2^Ω and called the *power set* of Ω. If Ω is a finite set with $|\Omega|$ elements, then the power set 2^Ω contains $2^{|\Omega|}$ elements (events) since for every element of Ω there are two possibilities: to be a member of a particular event, or not to be. Therefore, the power set 2^{Ω_N} of the binomial sample space Ω_N contains 2^{2^N} elements.

6.1.1.2   Random Variables

Typically, the actual outcome $\omega \in \Omega \equiv \Omega_N$ of a random experiment is not observable, but some information about ω can be revealed via values of *random variables* that are (set) functions that map Ω to $\mathbb{R}$. For example, we introduce two special random variables on Ω:

$$\#\mathsf{D}(\omega) = \text{number of D's in } \omega, \tag{6.2}$$
$$\#\mathsf{U}(\omega) = \text{number of U's in } \omega, \tag{6.3}$$

$\forall \omega \in \Omega$. In some cases, we will need to calculate the number of D's and U's in the first k market moves. For this reason, we also define the random variables

$$\mathsf{D}_k(\omega) := \text{number of D's in } \omega_1 \omega_2 \ldots \omega_k = \#\mathsf{D}(\omega_1 \omega_2 \ldots \omega_k), \tag{6.4}$$
$$\mathsf{U}_k(\omega) := \text{number of U's in } \omega_1 \omega_2 \ldots \omega_k = \#\mathsf{U}(\omega_1 \omega_2 \ldots \omega_k). \tag{6.5}$$

for $1 \leqslant k \leqslant N$ and $\mathsf{D}_0 \equiv \mathsf{U}_0 \equiv 0$. There are two obvious properties:

1. $\mathsf{U}_k(\omega), \mathsf{D}_k(\omega) \in \{0, 1, \ldots, k\}$,

2. $\mathsf{U}_k(\omega) + \mathsf{D}_k(\omega) = k$,

for all $\omega \in \Omega_N$. That is, the random variables U_k and $\mathsf{D}_k = k - \mathsf{U}_k$ are linearly dependent.

In the recombining binomial tree model, the main objects are stock prices $\{S_k\}_{k=0,1,\ldots,N}$ defined via the recurrence relation in (6.1). These stock prices are functions of the market scenario $\omega = \omega_1\omega_2\ldots\omega_N$. For instance, there are two distinct one-step market moves with two distinct values for S_1; four distinct two-step market moves with a total of three possible values for S_2; eight distinct three-step market moves with a total of four possible values for S_3:

$$S_1(\omega) = \begin{cases} u\,S_0 & \text{if } \omega_1 = \mathsf{U} \\ d\,S_0 & \text{if } \omega_1 = \mathsf{D} \end{cases}$$

$$S_2(\omega) = \begin{cases} u\,S_1(\omega) & \text{if } \omega_2 = \mathsf{U} \\ d\,S_1(\omega) & \text{if } \omega_2 = \mathsf{D} \end{cases} = \begin{cases} u^2\,S_0 & \text{if } \omega_1\omega_2 = \mathsf{UU} \\ u\,d\,S_0 & \text{if } \omega_1\omega_2 \in \{\mathsf{DU},\mathsf{UD}\} \\ d^2\,S_0 & \text{if } \omega_1\omega_2 = \mathsf{DD} \end{cases}$$

$$S_3(\omega) = \begin{cases} u\,S_2(\omega) & \text{if } \omega_3 = \mathsf{U} \\ d\,S_2(\omega) & \text{if } \omega_3 = \mathsf{D} \end{cases} = \begin{cases} u^3\,S_0 & \text{if } \omega_1\omega_2\omega_3 = \mathsf{UUU} \\ u^2\,d\,S_0 & \text{if } \omega_1\omega_2\omega_3 \in \{\mathsf{DUU},\mathsf{UDU},\mathsf{UUD}\} \\ u\,d^2\,S_0 & \text{if } \omega_1\omega_2\omega_3 \in \{\mathsf{DDU},\mathsf{DUD},\mathsf{UDD}\} \\ d^3\,S_0 & \text{if } \omega_1\omega_2\omega_3 = \mathsf{DDD} \end{cases}$$

Clearly, distinct multi-step market moves can result in the same value for the stock price; e.g., there are eight possible outcomes $\omega_1\omega_2\omega_3$ for a three-step move, but only four different stock prices S_3 at time 3.

Using the random variables $\#\mathsf{D}$ and $\#\mathsf{U}$ allows us to rewrite the recurrence relation (6.1) in a compact form:

$$S_k(\omega) = S_{k-1}(\omega)u^{\#\mathsf{U}(\omega_k)}d^{\#\mathsf{D}(\omega_k)}, \quad 1 \leqslant k \leqslant N. \tag{6.6}$$

By recursively applying this formula n times, we have the following expression for the stock price at time n:

$$\begin{aligned} S_n(\omega) &= S_{n-1}(\omega)u^{\#\mathsf{U}(\omega_n)}d^{\#\mathsf{D}(\omega_n)} \\ &= S_{n-2}(\omega)u^{\#\mathsf{U}(\omega_{n-1})+\#\mathsf{U}(\omega_n)}d^{\#\mathsf{D}(\omega_{n-1})+\#\mathsf{D}(\omega_n)} \\ &\vdots \\ &= S_0 u^{\sum_{k=1}^n \#\mathsf{U}(\omega_k)}d^{\sum_{k=1}^n \#\mathsf{D}(\omega_k)} = S_0 u^{\#\mathsf{U}(\omega_1\omega_2\ldots\omega_n)}d^{\#\mathsf{D}(\omega_1\omega_2\ldots\omega_n)} \\ &= S_0 u^{\mathsf{U}_n(\omega)}d^{\mathsf{D}_n(\omega)} = S_0 u^{\mathsf{U}_n(\omega)}d^{n-\mathsf{U}_n(\omega)} \end{aligned} \tag{6.7}$$

for any $\omega = \omega_1\omega_2\ldots\omega_N \in \Omega_N$ and $1 \leqslant n \leqslant N$. This formula shows us that the stock price S_n depends only on the first n market moves $\omega_1, \omega_2, \ldots, \omega_n$. Since it is clear that all random variables in the binomial model are functions of outcome ω, we will omit ω in some cases to simplify the notation. For instance, (6.7) gives the expression for random variable S_n in terms of random variables $\mathsf{U}_n, \mathsf{D}_n$:

$$S_n = S_0 u^{\mathsf{U}_n}d^{\mathsf{D}_n}, \quad 1 \leqslant n \leqslant N.$$

As is seen from (6.7), the binomial price S_n can only take on a value in the set (support)

$$\{S_{n,k} := S_0 u^k d^{n-k}; \ k = 0, 1, \ldots, n\}.$$

Equation (6.7) can be used to represent S_n in terms of S_m, the stock value at an earlier time m, $0 \leqslant m \leqslant n$. Later we shall use this for conditioning on a particular value of the random variable S_m which has support $\{S_{m,\ell} = S_0 u^\ell d^{m-\ell}; \; \ell = 0, 1, \ldots, m\}$. Writing $S_n(\omega) = S_m(\omega)\frac{S_n(\omega)}{S_m(\omega)}$ and employing (6.7) gives

$$S_n(\omega) = S_m(\omega)\frac{S_0 u^{\mathsf{U}_n(\omega)} d^{\mathsf{D}_n(\omega)}}{S_0 u^{\mathsf{U}_m(\omega)} d^{\mathsf{D}_m(\omega)}} = S_m(\omega)\, u^{\mathsf{U}_n(\omega)-\mathsf{U}_m(\omega)} d^{\mathsf{D}_n(\omega)-\mathsf{D}_m(\omega)}$$

$$= S_m(\omega)\, u^{\#\mathsf{U}(\omega_{m+1}\omega_{m+2}\ldots\omega_n)} d^{\#\mathsf{D}(\omega_{m+1}\omega_{m+2}\ldots\omega_n)}. \tag{6.8}$$

Since

$$\#\mathsf{U}(\omega_{m+1}\ldots\omega_n) \in \{0, 1, \ldots, n-m\} \text{ and } \#\mathsf{D}(\omega_{m+1}\ldots\omega_n) = (n-m) - \#\mathsf{U}(\omega_{m+1}\ldots\omega_n),$$

the random variable S_n conditional on a given value $S_m = s_m$, at time $m < n$, has support

$$\{s_m u^k d^{n-m-k} \; : \; k = 0, 1, \ldots, n-m\}.$$

These points give the possible nodes (n, S_n) attainable at time n, after $(n-m)$ market moves on the binomial tree, given that we start at a node (m, S_m) at time m.

Suppose that Ω is a finite sample space. Then every real-valued random variable X on Ω takes on a finite number of possible values, say $\{x_1, \ldots, x_K\}$. Such a finite set $X(\Omega) = \{x_1, x_2, \ldots, x_K\}$ is called the *range or support* of X. For each x_j in the range of X we can form the event

$$\{X = x_j\} \equiv \{\omega \in \Omega \; : \; X(\omega) = x_j\} = X^{-1}(\{x_j\}).$$

In what follows we will also denote the inverse (pre-)image of a singleton set $X^{-1}(\{x\})$ by $X^{-1}(x)$ where $x \in \mathbb{R}$, e.g., $X^{-1}(\{x_j\}) \equiv X^{-1}(x_j)$. Since the range of a real-valued random variable (excluding the points $\pm\infty$) is a subset of $\mathbb{R}$, we can also consider events such as

$$\{X \leqslant x\} \equiv \{\omega \in \Omega \; : \; -\infty < X(\omega) \leqslant x\} = X^{-1}((-\infty, x]).$$

The following example introduces an important special kind of random variable that corresponds to the first (passage or hitting) time of the stock price for a given upper level. This kind of random variable has range $\mathbb{R} \cup \{\infty\}$ and is covered in more detail in Section 6.3.8.

Example 6.1. Suppose that $S_0 = 4$, $d = \frac{1}{2}$, and $u = 2$. Let random variable τ be given by

$$\tau(\omega) = \min\{n \in \{0, 1, 2, 3\} \; : \; S_n(\omega) \geqslant 6\}$$

if the set is nonempty and let $\tau(\omega) = \infty$ if the set is empty, i.e., we let $\tau(\omega) = \infty$ if the set $\min\{n \in \{0, 1, 2, 3\} \; : \; S_n(\omega) \geqslant 6\} = \emptyset$.

Determine the event $\{\tau \leqslant 3\} := \{\omega \; : \; \tau(\omega) \leqslant 3\}$ as a subset of the 3-period sample space Ω_3.

Solution. First note that $S_0 < 6$, hence $\tau \geqslant 1$. Since Ω_3 includes only eight outcomes, we can find $\tau(\omega)$ for each $\omega \in \Omega_3$. For instance, if $\omega = \mathsf{DUU}$, then $S_1(\omega) = S_0 d = 2 < 6$, $S_2(\omega) = S_0 ud = 4 < 6$, $S_3(\omega) = S_0 u^2 d = 8 > 6$. Hence, $\tau(\mathsf{DUU}) = 3$. For $\omega = \mathsf{UUU}$, $S_1(\omega) = S_0 u = 8 > 6$ so $\tau(\mathsf{UUU}) = 1$. In fact, for any outcome of the form $\omega = \mathsf{U}\omega_2\omega_3$ (where the first move is up) we have $S_1(\omega) = S_0 u = 8$ and hence $\tau(\mathsf{U}\omega_2\omega_3) = 1$. In this way, we have

$$\tau(\mathsf{UUU}) = \tau(\mathsf{UUD}) = \tau(\mathsf{UDU}) = \tau(\mathsf{UDD}) = 1,$$
$$\tau(\mathsf{DUU}) = 3, \; \tau(\mathsf{DUD}) = \infty, \; \tau(\mathsf{DDU}) = \infty, \; \tau(\mathsf{DDD}) = \infty.$$

By convention (as defined above), if $S_n(\omega^*) < 6$, for all $n = 0, 1, 2, 3$, then we set $\tau(\omega^*) = \infty$

for all such $\omega^* \in \Omega_3$; i.e., the process never reaches the value 6 for the given outcome. In summary,

$$\{\tau \leqslant 3\} = \{\mathsf{UUU}, \mathsf{UUD}, \mathsf{UDU}, \mathsf{UDD}, \mathsf{DUU}\}.$$

This is equivalent to $A_\mathsf{U} \cup \{\mathsf{DUU}\}$ where $A_\mathsf{U} = \{\omega \in \Omega_3 \ : \ \omega_1 = \mathsf{U}\} \equiv \{\omega_1 = \mathsf{U}\}$ is the event that the first move is up. Alternatively, by the equivalence of events $\{\tau \leqslant 3\} = \cup_{n=1}^{3}\{\tau \leqslant n\} = \cup_{n=1}^{3}\{S_n \geqslant 6\} = \{\max\{S_1, S_2, S_3\} \geqslant 6\}$, we can also use the representation:

$$\begin{aligned}
\{\tau \leqslant 3\} &= \{S_1 \geqslant 6 \text{ or } S_2 \geqslant 6 \text{ or } S_3 \geqslant 6\} = \{S_1 \geqslant 6\} \cup \{S_2 \geqslant 6\} \cup \{S_3 \geqslant 6\} \\
&= A_\mathsf{U} \cup \{\mathsf{UUU}, \mathsf{UUD}\} \cup \{\mathsf{UUU}, \mathsf{UUD}, \mathsf{UDU}, \mathsf{DUU}\} \\
&= \{\mathsf{UUU}, \mathsf{UUD}, \mathsf{UDU}, \mathsf{UDD}, \mathsf{DUU}\}. \qquad \square
\end{aligned}$$

6.1.1.3 Probability Measure

Definition 6.1. A real-valued function $\mathbb{P}$ defined on the set of all subsets $\mathcal{F} = 2^\Omega$ of a sample space Ω is called a *probability measure* on a sample space Ω if it satisfies the following properties:

1. $\mathbb{P}(E) \geqslant 0$ for every event $E \in \mathcal{F}$,

2. $\mathbb{P}(\Omega) = 1$,

3. $\mathbb{P}(\cup_{i \geqslant 1} E_i) = \sum_{i \geqslant 1} \mathbb{P}(E_i)$ for any countable collection $\{E_i\}_{i \geqslant 1}$, $E_i \in \mathcal{F}$, of pairwise disjoint events.

These properties are called the *Kolmogorov axioms*.

Recall that a family of sets is pairwise disjoint or mutually disjoint if every two sets in the family are disjoint, i.e., $i \neq j \implies E_i \cap E_j = \emptyset$. Suppose that Ω is a finite sample space. Then, Axiom 3, called the *countable additivity property*, can be simplified as follows:

3*. $\mathbb{P}(E_1 \cup E_2) = \mathbb{P}(E_1) + \mathbb{P}(E_2)$ for any two disjoint events $E_1, E_2 \in \mathcal{F}$.

Indeed, if Ω is finite, then 2^Ω is finite as well. So every collection of subsets of Ω consists of finitely many different events. If events $E_1, E_2, \ldots, E_n$ are mutually disjoint, then by applying Axiom 3* we have

$$\mathbb{P}(\cup_{i=1}^{n} E_i) = \mathbb{P}(\cup_{i=1}^{n-1} E_i) + \mathbb{P}(E_n) = \mathbb{P}(\cup_{i=1}^{n-2} E_i) + \mathbb{P}(E_{n-1}) + \mathbb{P}(E_n) = \cdots = \sum_{i=1}^{n} \mathbb{P}(E_i).$$

All outcomes (elements) of a finite sample space Ω can be enumerated:

$$\Omega = \{\omega^1, \omega^2, \ldots, \omega^M\}.$$

The probability (likelihood) of each outcome is specified by the numbers

$$p_i = \mathbb{P}(\{\omega^i\}) \equiv \mathbb{P}(\omega^j) \in [0, 1], \quad j = 1, 2 \ldots, M.$$

We assume that all probabilities p_i are strictly positive (if $p_j = 0$, then the outcome ω^j can be removed from Ω as an impossible one). Since $\mathbb{P}(\Omega) = 1$ by Axioms 2 and 3, we have

$$1 = \mathbb{P}(\Omega) = p_1 + p_2 + \cdots + p_M.$$

For the case with a finite sample space Ω, every nonempty event $E \in \mathcal{F} = 2^\Omega$ can be described by specifying all its elements:

$$E = \{\omega^{j_1}, \omega^{j_2}, \ldots, \omega^{j_k}\} \text{ with } 1 \leqslant j_1 < j_2 < \cdots < j_k \leqslant M. \tag{6.9}$$

Therefore, the probability of E can be calculated by using the additivity property:

$$\mathbb{P}(E) = \mathbb{P}\left(\cup_{\ell=1}^{k}\{\omega^{j_\ell}\}\right) = \sum_{\ell=1}^{k}\mathbb{P}(\{\omega^{j_\ell}\}) = \sum_{\ell=1}^{k}p_{j_\ell}. \tag{6.10}$$

Thus, for a finite Ω with $|\Omega| = M$ outcomes, the probability measure $\mathbb{P}: 2^{\Omega} \to [0,1]$ can be defined by (6.9) and (6.10), where the probabilities $p_j > 0$, $j = 1, 2, \ldots, M$, sum to one:

$$p_1 + p_2 + \cdots + p_M = 1.$$

Definition 6.2. A *finite probability space* is a triplet $(\Omega, \mathcal{F}, \mathbb{P})$, which consists of a finite nonempty sample space Ω and a probability measure $\mathbb{P}$ defined on the set of subsets $\mathcal{F} = 2^{\Omega}$.

Let us return to the sample space $\Omega = \Omega_N$ of the binomial tree model where $|\Omega_N| = M = 2^N$. Our goal is to define the probability measure $\mathbb{P}$. Since Ω_N is finite, it is sufficient to define $\mathbb{P}(\omega)$ for any single market scenario $\omega \in \Omega_N$ and then apply the above approach to compute $\mathbb{P}(E)$ for any $E \subset \Omega_N$. It is useful to consider two special complementary events, for any given $k \in \{1, 2, \ldots, N\}$:

$$\{\omega_k = \mathsf{U}\} \equiv \{\omega \in \Omega_N \; : \; \omega_k = \mathsf{U}\} = \{\text{all outcomes with } k\text{th move as up}\}$$

and

$$\{\omega_k = \mathsf{D}\} \equiv \{\omega \in \Omega_N \; : \; \omega_k = \mathsf{D}\} = \{\text{all outcomes with } k\text{th move as down}\}.$$

By assumed independence of successive moves, it follows trivially that the respective probabilities of these events are $\mathbb{P}(\omega_k = \mathsf{U}) = p > 0$ and $\mathbb{P}(\omega_k = \mathsf{D}) = 1 - p > 0$. Hence, we can write

$$\mathbb{P}(\{\omega_k = \omega_k^*\}) = p^{\#\mathsf{U}(\omega_k^*)}(1-p)^{\#\mathsf{D}(\omega_k^*)}, \quad \omega_k^* \in \{\mathsf{D}, \mathsf{U}\}.$$

Moreover, for every $k, m \in \{1, 2, \ldots, N\}$, $k \neq m$, the events $\{\omega_k = \omega_k^*\}$ and $\{\omega_m = \omega_m^*\}$, $\omega_k^*, \omega_m^* \in \{\mathsf{D}, \mathsf{U}\}$, are independent. Then, for any particular outcome $\omega^* = \omega_1^*\omega_2^* \cdots \omega_N^* \in \Omega_N$, we have

$$\mathbb{P}(\omega^*) = \mathbb{P}(\{\omega_1 = \omega_1^*\} \cap \{\omega_2 = \omega_2^*\} \cap \cdots \cap \{\omega_N = \omega_N^*\})$$

$$= \prod_{k=1}^{N}\mathbb{P}(\{\omega_k = \omega_k^*\}) = \prod_{k=1}^{N}p^{\#\mathsf{U}(\omega_k^*)}(1-p)^{\#\mathsf{D}(\omega_k^*)}$$

$$= p^{\#\mathsf{U}(\omega_1^*)+\cdots+\#\mathsf{U}(\omega_N^*)}(1-p)^{\#\mathsf{D}(\omega_1^*)+\cdots+\#\mathsf{D}(\omega_N^*)} = p^{\#\mathsf{U}(\omega^*)}(1-p)^{\#\mathsf{D}(\omega^*)}.$$

It is not difficult to show that, as is required by Axiom 2,

$$\mathbb{P}(\Omega_N) = \mathbb{P}(\{\omega \; : \; \omega \in \Omega_N\}) = \sum_{\omega \in \Omega_N}\mathbb{P}(\omega) = 1, \tag{6.11}$$

for any finite integer $N \geqslant 1$. One way is by induction. Indeed, if $N = 1$ then we simply have

$$\sum_{\omega_1 \in \{\mathsf{U},\mathsf{D}\}}\mathbb{P}(\omega_1) = \mathbb{P}(\mathsf{D}) + \mathbb{P}(\mathsf{U}) = p + (1-p) = 1.$$

Now assuming (6.11) holds for N, it follows that the formula holds true for $N+1$ as well:

$$\sum_{\omega \in \Omega_{N+1}}\mathbb{P}(\omega) = \sum_{\omega \in \Omega_N}\mathbb{P}(\omega\mathsf{D}) + \sum_{\omega \in \Omega_N}\mathbb{P}(\omega\mathsf{U})$$

$$= \sum_{\omega \in \Omega_N}\mathbb{P}(\omega)\mathbb{P}(\{\omega_{N+1} = \mathsf{D}\}) + \sum_{\omega \in \Omega_N}\mathbb{P}(\omega)\mathbb{P}(\{\omega_{N+1} = \mathsf{U}\})$$

$$= \sum_{\omega \in \Omega_N}\mathbb{P}(\omega)\cdot(1-p) + \sum_{\omega \in \Omega_N}\mathbb{P}(\omega)\cdot p = (1-p+p)\sum_{\omega \in \Omega_N}\mathbb{P}(\omega) = 1.$$

An alternative proof, as follows, is also instructive. Let

$$E_k = \{U_N = k\} \equiv \{\omega \in \Omega_N \;:\; U_N(\omega) = k\}$$
$$= \{\text{all outcomes with } k \text{ up moves in total}\}.$$

Then, the entire outcome space is simply a union of such exclusive events, $\Omega_N = \cup_{k=0}^N E_k$. As noted below, $\mathbb{P}(U_N = k) = \binom{N}{k}p^k(1-p)^{N-k}$. Hence, by Axiom 3 and the binomial expansion:

$$\mathbb{P}(\Omega_N) = \sum_{k=0}^N \mathbb{P}(E_k) = \sum_{k=0}^N \mathbb{P}(U_N = k) = \sum_{k=0}^N \binom{N}{k}p^k(1-p)^{N-k} = 1.$$

The probability of any event $E \subset \Omega_N$ is given by

$$\mathbb{P}(E) = \sum_{\omega \in E} \mathbb{P}(\omega) = \sum_{\omega \in E} p^{U_N(\omega)}(1-p)^{D_N(\omega)}.$$

Thus, $\mathbb{P}$ is a probability measure defined on $\mathcal{F}_N = 2^{\Omega_N}$. The triplet $(\Omega_N, \mathcal{F}_N, \mathbb{P})$ is a finite probability space called the *N-period binomial probability space*.

Example 6.2. Consider the random variable τ from Example 6.1. Suppose that $p = \frac{1}{4}$. Determine $\mathbb{P}(\{\tau \leqslant 3\})$.

Solution. From the solution in Example 6.1, we have

$$\mathbb{P}(\{\tau \leqslant 3\}) = \mathbb{P}(\{\mathsf{UUU, UUD, UDU, UDD, DUU}\})$$
$$= p^3 + p^2(1-p) + p^2(1-p) + p(1-p)^2 + p^2(1-p)$$
$$= p^3 + 3p^2(1-p) + p(1-p)^2$$
$$= \left(\frac{1}{4}\right)^3 + 3\left(\frac{1}{4}\right)^2\left(\frac{3}{4}\right) + \left(\frac{1}{4}\right)\left(\frac{3}{4}\right)^2 = \frac{19}{64}. \qquad \square$$

Example 6.3. Obtain the probability distributions of D_n and U_n for $n = 1, 2, \ldots, N$. For what values of p, do U_n and D_n have the same distribution, i.e., determine p such that $U_n \overset{d}{=} D_n$?

Solution. The variables D_n and U_n, respectively, give the number of D's and the number of U's in the first n ω_j's. So they only depend on $\omega_1, \omega_2, \ldots, \omega_n$. For every $n = 1, 2, \ldots, N$ and $k = 0, 1, \ldots, n$, there exist $\binom{n}{k}$ scenarios in Ω_n that have exactly k U's. The probability of each particular scenario with k U's is $p^k(1-p)^{n-k}$. Therefore, U_n has the binomial distribution $Bin(n, p)$ where

$$\mathbb{P}(U_n = k) = \binom{n}{k}p^k(1-p)^{n-k}.$$

Similarly, D_n has the binomial distribution $Bin(n, 1-p)$. Thus, $U_n \overset{d}{=} D_n$ iff $p = 1 - p$, i.e., iff $p = \frac{1}{2}$. $\qquad \square$

6.1.2 Random Processes

Definition 6.3. A *random process* (or a *stochastic process*) is a collection of random variables that are all defined on a common sample space Ω and indexed by $t \in \mathbf{T}$:

$$\{X_t(\omega)\}_{t \in \mathbf{T}}, \text{ such that } \forall t \in \mathbf{T}, \; X_t: \Omega \to \mathbb{R}.$$

T is an index set. Typically, **T** is a subset of the positive real half-line since t is usually referred to as time. The random process is said to be a *discrete-time* process if **T** $= \{0, 1, 2, \ldots\}$; it is said to be a *continuous-time* process if **T** is an interval, e.g., **T** $= [0, \infty)$. Note that **T** can be bounded or unbounded. For a given (fixed) outcome $\omega \in \Omega$, $X_t(\omega)$ considered as function of time $t \in$ **T** is called a *sample path* or a *trajectory* of the stochastic process.

Let us consider two important examples of discrete-time stochastic processes defined on the binomial sample space Ω_N.

6.1.2.1   Binomial Price Process and Path Probabilities

The binomial prices defined in (6.7) constitute a discrete-time stochastic process:

$$\{S_n\}_{n=0,1,\ldots,N}, \quad \text{where } S_n(\omega) = S_0 u^{\mathsf{U}_n(\omega)} d^{\mathsf{D}_n(\omega)}, \; 1 \leqslant n \leqslant N, \; \omega \in \Omega_N. \tag{6.12}$$

Moreover, the collection of random variables D_n and U_n that respectively count D's and U's in ω are also considered random processes:

$$\{\mathsf{U}_n\}_{n=0,1,\ldots,N} \text{ and } \{\mathsf{D}_n\}_{n=0,1,\ldots,N}.$$

As is shown in Example 6.3, U_n and D_n have the binomial probability distribution, $\mathsf{U}_n \sim Bin(n, p)$ and $\mathsf{D}_n \sim Bin(n, 1 - p)$, for any $n = 1, 2, \ldots, N$, with probability mass functions

$$\mathbb{P}(\mathsf{U}_n = k) = \binom{n}{k} p^k (1-p)^{n-k}, \quad \mathbb{P}(\mathsf{D}_n = k) = \binom{n}{k} p^{n-k} (1-p)^k, \quad 0 \leqslant k \leqslant n.$$

Therefore, the probability mass function (PMF) of S_n is given by

$$\mathbb{P}(S_n = S_{n,k}) = \binom{n}{k} p^k (1-p)^{n-k}, \quad S_{n,k} = S_0 u^k d^{n-k}, \quad 0 \leqslant k \leqslant n \leqslant N.$$

A path of a binomial price process passes through the nodes of a binomial tree: $\{(n, k) : 0 \leqslant k \leqslant n \leqslant N\}$. At the node (n, k), the process S_n takes on the value $S_{n,k}$. Let us calculate the *transition probability* that a path of the binomial price process will pass through the node (n, k) given that it passes through (m, ℓ) with $0 \leqslant m < n$, $0 \leqslant \ell \leqslant m$, $0 \leqslant k \leqslant n$, and $\ell \leqslant k$. To get from (m, ℓ) to (n, k), the process makes $k - \ell$ up moves and $(n - m) - (k - \ell)$ down moves. Hence, $S_{n,k} = S_{m,\ell} u^{k-\ell} d^{(n-m)-(k-\ell)}$. There are $\binom{n-m}{k-\ell} = \frac{(n-m)!}{(k-\ell)!(n-m-k+\ell)!}$ distinct paths in the binomial tree that connect the nodes (m, ℓ) and (n, k). From independence of one-step moves, we obtain that the probability for any one particular path is $p^{k-\ell}(1 - p)^{n-m-k+\ell}$. Since each particular path has the same probability of occurring, and each path represents a mutually exclusive event, we have

$$\mathbb{P}\left(S_n = S_{n,k} \mid S_m = S_{m,\ell}\right) = \binom{n-m}{k-\ell} p^{k-\ell} (1-p)^{n-m-k+\ell}.$$

This conditional probability can also be obtained by operating with events in terms of the ω_j's and by using the relation (6.8) as well as the independence of random variables $\frac{S_n}{S_m}$

and S_m:

$$\mathbb{P}\left(S_n = S_{n,k} \mid S_m = S_{m,\ell}\right) = \mathbb{P}\left(S_m \frac{S_n}{S_m} = S_{n,k} \middle| S_m = S_{m,\ell}\right)$$

$$= \mathbb{P}\left(\frac{S_n}{S_m} = \frac{S_{n,k}}{S_{m,\ell}}\right)$$

$$= \mathbb{P}\left(u^{\#\mathsf{U}(\omega_{m+1}\ldots\omega_n)} d^{(n-m)-\#\mathsf{U}(\omega_{m+1}\ldots\omega_n)} = u^{k-\ell} d^{(n-m)-(k-\ell)}\right)$$

$$= \mathbb{P}(\#\mathsf{U}(\omega_{m+1}\ldots\omega_n) = k-\ell) = \binom{n-m}{k-\ell} p^{k-\ell}(1-p)^{n-m-k+\ell}.$$

The last expression follows since $\#\mathsf{U}(\omega_{m+1}\ldots\omega_n) \sim Bin(n-m, p)$.

6.1.2.2 Random Walk

A random walk is a mathematical formalization of a process that consists of taking successive random steps of size 1 upwards or downwards. For $\omega = \omega_1\omega_2\cdots\omega_N \in \Omega_N$ we define random variables

$$X_n(\omega) = \begin{cases} 1 & \text{if } \omega_n = \mathsf{U}, \\ -1 & \text{if } \omega_n = \mathsf{D}, \end{cases} \quad n = 1, 2, \ldots, N.$$

Consider the stochastic process $\{M_n\}_{n=0,1,2,\ldots,N}$ defined as follows:

$$M_0 := 0, \quad M_n := \sum_{k=1}^{n} X_k, \ n = 1, 2, \ldots, N.$$

Like the time-n binomial price S_n, the variable M_n is a function of the first n market moves $\omega_1, \omega_2, \ldots, \omega_n$. $\{M_n\}_{n\geqslant0}$ is called a *random walk*. It admits a recurrence representation:

$$M_{n+1} = \sum_{k=1}^{n+1} X_k = M_n + X_{n+1} = \begin{cases} M_n + 1 & \text{with probability } p, \\ M_n - 1 & \text{with probability } 1-p. \end{cases}$$

If $p = \frac{1}{2}$, then $\{M_n\}_{n\geqslant0}$ is called a *symmetric random walk* since at every step it may go equally likely upwards or downwards.

Since $\mathsf{U}_n = \sum_{k=1}^{n}(X_k)^+$ and $\mathsf{D}_n = \sum_{k=1}^{n}(X_k)^-$ (and using $x = (x)^+ - (x)^-$, where $(x)^+ = \max\{x, 0\}$ and $(x)^- = \max\{-x, 0\}$), we can express M_n in terms of U_n and D_n as follows:

$$M_n = \sum_{k=1}^{n} X_k = \sum_{k=1}^{n}(X_k)^+ - \sum_{k=1}^{n}(X_k)^- = \mathsf{U}_n - \mathsf{D}_n.$$

This is clearly the difference between the number of upward and downward moves. Since $\mathsf{D}_n + \mathsf{U}_n = n$, we also write M_n in terms of only D_n or U_n:

$$M_n = n - 2\mathsf{D}_n = 2\mathsf{U}_n - n.$$

Thus, by equivalence of events $\{\mathsf{U}_n = k\} = \{M_n = 2k - n\}$, the PMF of M_n is given by

$$\mathbb{P}(M_n = 2k - n) = \binom{n}{k} p^k (1-p)^{n-k}, \quad k = 0, 1, \ldots, n.$$

Example 6.4. Construct the sample path of a particular random walk and find the path probability for $\omega^* = \mathsf{DUUDDUDD} \in \Omega_8$.

Solution. A random walk starts at the origin, i.e., $M_0 = 0$. Find recursively the sample values $M_n = M_n(\omega^*)$ for $n = 1, 2, \ldots, 8$:

$$M_1 = M_0 + X_1 = 0 - 1 = -1, \qquad M_2 = M_1 + X_2 = -1 + 1 = 0,$$
$$M_3 = M_2 + X_3 = 0 + 1 = 1, \qquad M_4 = M_3 + X_4 = 1 - 1 = 0,$$
$$M_5 = M_4 + X_5 = 0 - 1 = -1, \qquad M_6 = M_5 + X_5 = -1 + 1 = 0,$$
$$M_7 = M_6 + X_7 = 0 - 1 = -1, \qquad M_8 = M_7 + X_8 = -1 - 1 = -2.$$

Figure 6.1 demonstrates the sample paths of the process $\{M_n\}$ and $\{U_n\}$ for the outcome ω^*. Using $\#U(\omega^*) = 3$, the path probability is

$$\mathbb{P}(\omega^*) = p^{\#U(\omega^*)}(1-p)^{\#D(\omega^*)} = p^3(1-p)^5. \qquad \square$$

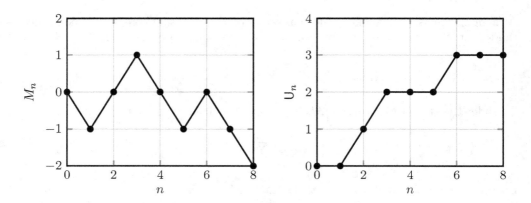

FIGURE 6.1: Sample paths of the random walk $\{M_n\}$ and the process $\{U_n\}$ in an eight-period model for $\omega = $ DUUDDUDD.

6.2 Information Flow

The pricing of derivative securities is based on contingent claims. We need a way to mathematically model the arriving information on which our future decisions can be based. In the binomial tree model, that information is the knowledge of all market moves between the initial and future dates. We hence introduce, in what follows, the concepts of partition, σ-algebra, and filtration.

6.2.1 Partitions and Their Refinements

Suppose that some random experiment is performed, and its actual outcome ω is unknown. However, we might be given some information that is enough to narrow down the possible value of ω. We can then create a list of events that are sure to contain the actual outcome, and other sets that are sure not to contain it. These sets are *resolved* by the available information. This leads us to the following definition.

Definition 6.4. A *partition* $\mathcal{P}$ of a sample space $\Omega \neq \emptyset$ is a (set) collection of mutually disjoint nonempty subsets whose union is Ω. That is,

$$\mathcal{P} = \{A_i\}_{i \geqslant 1} = \{A_1, A_2, \ldots\}, \text{ so that } \cup_{i \geqslant 1} A_i = \Omega, \text{ and } i \neq j \implies A_i \cap A_j = \emptyset.$$

The subsets A_i are called *atoms* of the partition $\mathcal{P}$.

If the information about the actual outcome ω is available in the form of a partition, then we are able to say which event from the collection has occurred. There are three trivial examples of partitions:

(a) $\mathcal{P}_0 = \{\Omega\}$;

(b) $\mathcal{P}_E = \{E, E^C\}$ where $E \cup E^C = \Omega$, $E \neq \emptyset$;

(c) $\mathcal{P}_\Omega = \{\{\omega\} : \omega \in \Omega\}$.

The partition $\mathcal{P}_0$ (i.e., no partition) represents the absence of any information regarding the actual outcome ω, and $\mathcal{P}_\Omega$ (i.e., greatest partition into all possible outcomes) represents the full information about ω. The partition $\mathcal{P}_E$ (i.e., partition into two mutually exclusive events, E and E^C) means that we are only able to say whether the event E has occurred or not.

Example 6.5. Construct a partition representing the available information in each case below.

(a) Roll two dice. Suppose that the sum of values on the facing-up sides is known.

(b) Toss three coins. Suppose that the total number of heads is known.

Solution.
(a) The sample space consists of 36 pairs on integers: $\Omega = \{(i, j) : 1 \leqslant i, j \leqslant 6\}$. The sum of values on the facing-up sides is an integer between 2 and 12. So the partition $\mathcal{P}$ contains 11 elements: $\mathcal{P} = \{A_2, A_3, \ldots, A_{12}\}$, where $A_k = \{(i, j) : 1 \leqslant i, j \leqslant 6 \text{ and } i + j = k\}$:

$$A_2 = \{(1, 1)\},$$
$$A_3 = \{(1, 2), (2, 1)\},$$
$$A_4 = \{(1, 3), (2, 2), (3, 1)\},$$
$$A_5 = \{(1, 4), (2, 3), (3, 2), (4, 1)\},$$
$$A_6 = \{(1, 5), (2, 4), (3, 3), (4, 2), (5, 1)\},$$
$$A_7 = \{(1, 6), (2, 5), (3, 4), (4, 3), (5, 2), (6, 1)\},$$
$$A_8 = \{(2, 6), (3, 5), (4, 4), (5, 3), (6, 2)\},$$
$$A_9 = \{(3, 6), (4, 5), (5, 4), (6, 3)\},$$
$$A_{10} = \{(4, 6), (5, 5), (6, 4)\},$$
$$A_{11} = \{(5, 6), (6, 5)\},$$
$$A_{12} = \{(6, 6)\},$$

where $\cup_i A_i = \Omega$ and $A_j \cap A_k = \emptyset$ if $j \neq k$.

(b) The coin toss sample space consists of eight combinations of H's and T's:

$$\Omega_3 = \{TTT, TTH, THT, THH, HTT, HTH, HHT, HHH\}.$$

The number of heads can be any integer between 0 and 3. Therefore, the partition $\mathcal{P} = \{A_0, A_1, A_2, A_3\}$ consists of four atoms:

$$A_0 = \{TTT\}, \ A_1 = \{TTH, THT, HTT\}, \ A_2 = \{THH, HTH, HHT\}, \ A_3 = \{HHH\}.$$

□

The next example illustrates a succession of three partitions where each new partition is obtained by subdividing (breaking down) the atoms in the previous one.

Example 6.6. Consider a three-period binomial model. Represent the available information in the form of a partition if (a) nothing is known; (b) the first market move ω_1 is known; (c) the first two market moves ω_1, ω_2 are known.

Solution.

(a) $\mathcal{P}_0 = \{\Omega_3\}$.

(b) $\mathcal{P}_1 = \{\{\mathsf{DDD, DDU, DUD, DUU}\}, \{\mathsf{UDD, UDU, UUD, UUU}\}\} := \{A_\mathsf{D}, A_\mathsf{U}\}$.

(c) $\mathcal{P}_2 = \{\{\mathsf{DDD, DDU}\}, \{\mathsf{DUD, DUU}\}, \{\mathsf{UDD, UDU}\}, \{\mathsf{UUD, UUU}\}\}$
    $:= \{A_\mathsf{DD}, A_\mathsf{DU}, A_\mathsf{UD}, A_\mathsf{UU}\}$.

Note: $A_\mathsf{D} \cup A_\mathsf{U} = \Omega_3$, $A_\mathsf{D} = A_\mathsf{DD} \cup A_\mathsf{DU}$, $A_\mathsf{U} = A_\mathsf{UD} \cup A_\mathsf{UU}$. Hence, the one atom in $\mathcal{P}_0$ is given by the union of the atoms in $\mathcal{P}_1$, and each atom in $\mathcal{P}_1$ is in turn given by a union of two atoms in $\mathcal{P}_2$.

□

6.2.1.1   Partition Generated by a Random Variable

Consider a random variable X defined on a sample space Ω. If the set Ω is finite, then any random variable on Ω is a discrete random variable with range $X(\Omega) \equiv \mathsf{S}_X = \{x_1, x_2, \ldots, x_K\}$. The partition *generated by* X is the collection

$$\mathcal{P}(X) = \{X^{-1}(x_1), X^{-1}(x_2), \ldots, X^{-1}(x_K)\},$$

i.e., with atoms $A_i = X^{-1}(x_i) \equiv \{\omega \in \Omega : X(\omega) = x_i\}$, $i = 1, \ldots, K$. Note that the random variable also has the representation in terms of indicator functions w.r.t. the atoms: $X = \sum_i x_i \, \mathbb{I}_{A_i}$. As an example, consider picking a card at random from a deck of playing cards with four suits $\clubsuit, \diamondsuit, \heartsuit, \spadesuit$. We can represent the suit of the card selected by an integer-valued random variable defined by $X(\{\heartsuit\}) = 1$, $X(\{\diamondsuit\}) = 2$, $X(\{\clubsuit\}) = 3$, $X(\{\spadesuit\}) = 4$. Then the partition generated by X consists of four subcollections of cards where each sub-collection contains cards of the same suit, i.e., $\mathcal{P}(X) = \{X^{-1}(1), X^{-1}(2), X^{-1}(3), X^{-1}(4)\} = \{\{\heartsuit\}, \{\diamondsuit\}, \{\clubsuit\}, \{\spadesuit\}\}$ where the set of all cards is $\Omega = \{\heartsuit\} \cup \{\diamondsuit\} \cup \{\clubsuit\} \cup \{\spadesuit\}$.

In the next example we work out the respective partitions generated by two random variables on the three-period binomial model.

Example 6.7. Consider a three-period binomial model. Find the partitions $\mathcal{P}(S_2)$ and $\mathcal{P}(Y)$ generated by the stock price at time 2 and the Bernoulli random variable $Y := \mathbb{I}_{\{\omega_2 = \mathsf{U}\}}$ that only takes on a nonzero value of 1 when the second market move is up.

Solution. The price $X \equiv S_2$ admits three possible values: $X \in \{S_0 d^2, S_0 ud, S_0 u^2\}$. The partition $\mathcal{P}(X) \equiv \mathcal{P}(S_2)$ has three atoms:

$$X^{-1}(S_0 d^2) = \{\omega_1 = \mathsf{D}, \omega_2 = \mathsf{D}\} = \{\mathsf{DDD, DDU}\} = A_\mathsf{DD},$$

$$X^{-1}(S_0 ud) = \{\omega_1 = \mathsf{D}, \omega_2 = \mathsf{U}\} \cup \{\omega_1 = \mathsf{U}, \omega_2 = \mathsf{D}\}$$
$$= \{\mathsf{DUD, DUU, UDD, UDU}\} = A_\mathsf{DU} \cup A_\mathsf{UD},$$

$$X^{-1}(S_0 u^2) = \{\omega_1 = \mathsf{U}, \omega_2 = \mathsf{U}\} = \{\mathsf{UUD, UUU}\} = A_\mathsf{UU}.$$

Since $Y \in \{0,1\}$, then $\mathcal{P}(Y)$ has two atoms:

$$Y^{-1}(0) = \{\omega_2 = \mathsf{D}\} = \{\mathsf{DDD}, \mathsf{DDU}, \mathsf{UDD}, \mathsf{UDU}\},$$
$$Y^{-1}(1) = \{\omega_2 = \mathsf{U}\} = \{\mathsf{DUD}, \mathsf{DUU}, \mathsf{UUD}, \mathsf{UUU}\}. \qquad \square$$

Consider a collection of discrete random variables, $X_k(\Omega) \to \mathbb{R}$, $k = 1, 2 \ldots, M$, defined on a common sample space Ω. Let each X_k have support $X_k(\Omega) = \mathsf{S}_k$ and hence the random vector $\mathbf{X} = (X_1, X_2, \ldots, X_M)$ has support $\mathsf{S}_{\mathbf{X}} \equiv \{(x_1, \ldots, x_M) : x_k \in \mathsf{S}_k, k = 1, \ldots, M\}$. We may then define a partition generated by the random vector $\mathbf{X}$ as the collection $\mathcal{P}(\mathbf{X}) \equiv \mathcal{P}(X_1, X_2, \ldots, X_M) = \{A_{\mathbf{x}}; \mathbf{x} \in \mathsf{S}_{\mathbf{X}}\}$ with each atom $A_{\mathbf{x}} \equiv A_{(x_1, \ldots, x_M)}$ given by

$$A_{(x_1, \ldots, x_M)} := \{X_1 = x_1, \ldots, X_M = x_M\} \equiv \{\omega \in \Omega : \mathbf{X}(\omega) = \mathbf{x}\}.$$

In the N-period binomial model, we allow the market to move N times (or toss a coin N times) to obtain Ω_N—the set of 2^N possible sequences $\omega = \omega_1 \ldots \omega_N$ of N-tuples of U's and D's. Consider the vector $(X_1, \ldots, X_n)$, $1 \leqslant n \leqslant N$ of i.i.d. Bernoulli random variables $X_i = \mathbb{I}_{\{\omega_i = \mathsf{U}\}}$, $1 \leqslant i \leqslant n$. That is, X_i takes on value 1 if the ith market move ω_i is up and zero if it's down. Hence, any given sequence $\omega_1^*, \ldots, \omega_n^*$ that specifies (resolves) the first n moves can be equivalently described in terms of the sequence of 0's and 1's, i.e., the event $\{\omega_1 = \omega_1^*, \ldots, \omega_n = \omega_n^*\}$ is equivalent to $\{X_1 = x_1^*, \ldots, X_n = x_n^*\}$ where $x_1^* = \mathbb{I}_{\{\omega_1^* = \mathsf{U}\}}, \ldots, x_n^* = \mathbb{I}_{\{\omega_n^* = \mathsf{U}\}}$. Hence, the atoms and the partitions that are generated by $(X_1, \ldots, X_n)$ and by the subsets of ω's that resolve the first n moves are equivalent. In particular, we have the atoms $A_{\omega_1^* \ldots \omega_n^*} \equiv A_{(x_1^*, \ldots, x_n^*)}$ of the partition $\mathcal{P}_n \equiv \mathcal{P}(X_1, X_2, \ldots, X_n)$:

$$A_{\omega_1^* \ldots \omega_n^*} := \{\omega = \omega_1 \ldots \omega_N \in \Omega_N : \omega_1 = \omega_1^*, \ldots, \omega_n = \omega_n^*\}, \qquad (6.13)$$

for every $\omega_j^* \in \{\mathsf{D}, \mathsf{U}\}$, $j = 1, \ldots, n$. That is, each such atom consists of fixing the first n letters in ω with all others $\omega_{n+1}, \ldots, \omega_N$ allowed to be U or D. Since $|\Omega_n| = 2^n$, the partition $\mathcal{P}_n$ contains 2^n atoms. For example, the first partition $\mathcal{P}_1 = \{A_\mathsf{D}, A_\mathsf{U}\} = \{A_{\omega_1} : \omega_1 \in \{\mathsf{U}, \mathsf{D}\}\}$ resolves the first market move, the second $\mathcal{P}_2 = \{A_{\mathsf{DD}}, A_{\mathsf{DU}}, A_{\mathsf{UD}}, A_{\mathsf{UU}}\} = \{A_{\omega_1 \omega_2} : \omega_1, \omega_2 \in \{\mathsf{U}, \mathsf{D}\}\}$ resolves the first two market moves, the third partition is given by $\mathcal{P}_3 = \{A_{\mathsf{DDD}}, A_{\mathsf{DDU}}, A_{\mathsf{DUD}}, A_{\mathsf{DUU}}, A_{\mathsf{UDD}}, A_{\mathsf{UDU}}, A_{\mathsf{UUD}}, A_{\mathsf{UUU}}\}$ and this resolves the first three moves, and so on. Finally, for $n = N$ the partition $\mathcal{P}_N$ contains 2^N atoms where each atom has only one element:

$$\mathcal{P}_N = \{\{\omega^1\}, \{\omega^2\}, \ldots, \{\omega^{2^N}\}\}.$$

The kth member $\{\omega^k\}$ of this collection is the singleton set $\{\omega_1^{(k)} \ldots \omega_N^{(k)}\}$ representing the complete N-period path #k among 2^N possible distinct paths. This partition represents the full information about the actual market scenario ω where all N moves are resolved.

Note that the partitions $\mathcal{P}_n$ can be generated by the first n stock prices as well:

$$\mathcal{P}_n = \mathcal{P}(S_1, S_2, \ldots, S_n), \quad 1 \leqslant n \leqslant N.$$

This is seen, for example, by noting that the value of ω_j can be revealed from the ratio $\frac{S_j}{S_{j-1}}$. In particular, knowing S_1 reveals a value $\omega_1 = \omega_1^*$. Knowing S_1 and S_2 then reveals the string $\omega_1 \omega_2 = \omega_1^* \omega_2^*$. Iterating gives us that knowing the values $S_1, S_2, \ldots, S_{n-1}$ is knowledge of $\omega_1^* \ldots \omega_{n-1}^*$ and combining this with knowing S_n gives $\omega_1^* \ldots \omega_n^*$. Hence, knowledge of the first n binomial prices is equivalent to knowing the first n market moves.

6.2.1.2 Refinements of Partitions

Let $\mathcal{P}$ and $\mathcal{Q}$ be two partitions of a sample space Ω. $\mathcal{Q}$ is said to be a *refinement* of $\mathcal{P}$, denoted $\mathcal{P} \preceq \mathcal{Q}$, if every atom of $\mathcal{P}$ is expressible as a union of atoms of $\mathcal{Q}$. So to speak, partition $\mathcal{Q}$

is obtained by breaking down the atoms of $\mathcal{P}$. By splitting any atom of a partition into two or more disjoint and exhaustive subsets, we can obtain a refinement of the original partition. For example, by slicing a pizza, one can obtain a sequence of refinements (see Figure 6.2).

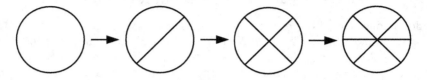

FIGURE 6.2: Slicing a pizza produces a new refinement with every new cut made.

Another example of a sequence of refinements is the following:

$$\text{body} \preceq \text{molecules} \preceq \text{atoms} \preceq \text{elementary particles}.$$

Given a sequence of random variables $\{X_k\}_{k\geqslant 1}$, we can generate a refinement as follows:

$$\mathcal{P}(X_1) \preceq \mathcal{P}(X_1, X_2) \preceq \mathcal{P}(X_1, X_2, X_3) \preceq \cdots$$

Clearly, every addition of a random variable can only increase the amount of available information.

A partition represents information about a stochastic process available to us at a particular moment. A refinement of a partition represents a transition from one information level to another level that is more detailed. As time passes, we learn more and more about the process and its history. The information is accumulated and catalogued using a sequence of partitions. A sequence of refinements,

$$\mathcal{P}_0 \preceq \mathcal{P}_1 \preceq \cdots \preceq \mathcal{P}_N,$$

is called an *information structure*. The collection of partitions $\{\mathcal{P}_n\}_{n=0,\ldots,N}$ where each $\mathcal{P}_n$ is defined by the atoms in (6.13) is an information structure for the N-period binomial model. The information in this case is based on market moves. In the beginning, the actual state of the model is unknown. This absence of information is represented by the partition $\mathcal{P}_0 = \{\Omega_N\}$. At the end of each period, a new portion of information (i.e., another market move) is revealed. At the end of period n we know the first n market moves $\omega_1, \omega_2, \ldots, \omega_n$. Clearly, for every $n = 1, 2, \ldots, N-1$, the partition $\mathcal{P}_n$ is a refinement of $\mathcal{P}_{n+1}$. At time $t = 1$, we have our first refinement $\mathcal{P}_0 \preceq \mathcal{P}_1 = \{A_\mathsf{D}, A_\mathsf{U}\}$; at time $t = 2$, we have $\mathcal{P}_1 \preceq \mathcal{P}_2 = \{A_\mathsf{DD}, A_\mathsf{DU}, A_\mathsf{UD}, A_\mathsf{UU}\}$, and so on. Therefore, the partitions form an information structure, i.e., $\mathcal{P}_0 \preceq \mathcal{P}_1 \preceq \cdots \preceq \mathcal{P}_N$. Figure 6.3 illustrates the information structure for Ω_3.

6.2.2   Sigma-Algebras

Suppose that we are given a partition $\mathcal{P}$ of a sample space Ω that represents the set of available information. We are able to say which event from collection $\mathcal{P}$ has occurred, but we might not know the actual outcome. Clearly, if we can say whether events A and/or B has occurred, then we can say whether events $A^{\complement}$, $A \cap B$, and $A \cup B$ occurred. Based on the available information, we can construct a collection of all "observable" events. That is, we are able to comment on the occurrence of every event from such a collection. The collection of all possible events of interest forms a collection of sets that is known as a σ-algebra.

Definition 6.5. A σ-algebra (or σ-field) $\mathcal{F}$ on a nonempty set Ω is a collection of subsets of Ω with the following properties:

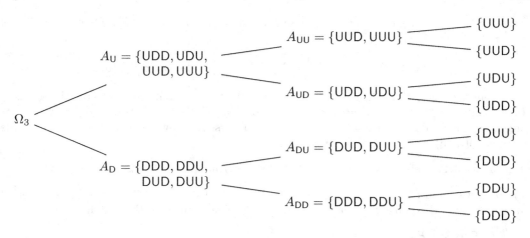

FIGURE 6.3: An information structure for Ω_3.

1. $\emptyset \in \mathcal{F}$;

2. $E \in \mathcal{F} \implies E^{\mathsf{C}} \in \mathcal{F}$;

3. for every countable collection $\{E_i\}_{i \geqslant 1} \in \mathcal{F}$, we have $\cup_{i \geqslant 1} E_i \in \mathcal{F}$.

The pair $(\Omega, \mathcal{F})$ is called a *measurable space*.

In other words, a σ-algebra on a set Ω is a nonempty collection $\mathcal{F}$ of subsets of Ω (including Ω itself) that is *closed under taking complementation and countable unions of its members*. As follows from Properties 2 and 3 and De Morgan's Laws, a σ-algebra is closed under countable intersections of its elements as well:

$$\{E_i\}_{i \geqslant 1} \in \mathcal{F} \implies \{E_i^{\mathsf{C}}\}_{i \geqslant 1} \in \mathcal{F} \implies \cup_{i \geqslant 1} E_i^{\mathsf{C}} \in \mathcal{F} \implies \left(\cap_{i \geqslant 1} E_i\right)^{\mathsf{C}} \in \mathcal{F} \implies \cap_{i \geqslant 1} E_i \in \mathcal{F}.$$

Notice that if Ω is a finite sample space, then it is sufficient to say that

3*. for every pair $E_1, E_2 \in \mathcal{F}$ we have $E_1 \cup E_2 \in \mathcal{F}$,

since there are only finitely many different subsets of a finite set. Note that a collection of subsets of Ω is called an *algebra* if it contains Ω and is closed w.r.t. the formation of complements and *finite* unions. Thus, in the case of a finite set Ω, every algebra on Ω is a σ-algebra on Ω.

The following are examples of σ-algebras.

(a) The minimal σ-algebra consisting only of the empty set and the set Ω: $\mathcal{F}_0 = \{\emptyset, \Omega\}$. It is also called a *trivial σ-algebra*.

(b) Let E be a subset of Ω such that $E \neq \emptyset$ and $E^{\mathsf{C}} \neq \emptyset$. Then there exists a smallest σ-algebra containing E: $\mathcal{F}_E = \{\emptyset, E, E^{\mathsf{C}}, \Omega\}$.

(c) The power set of Ω: $2^\Omega = \{E \ : \ E \subset \Omega\} \cup \emptyset$.

An interesting fact is that an intersection of any two σ-algebras on the same set Ω is again a σ-algebra on Ω, as is proved in Theorem 6.1. In contrast, a union of two σ-algebras is not necessarily a σ-algebra (and may not even be an algebra). As a simple example, consider $\Omega = \{a, b, c\}$ with three elements. Let $A = \{a\}$ and $B = \{b\}$. Then the union of $\mathcal{F}_A = \{\emptyset, A, A^C, \Omega\}$ and $\mathcal{F}_B = \{\emptyset, B, B^C, \Omega\}$ is

$$\mathcal{F}_A \cup \mathcal{F}_B = \{\emptyset, A, B, A^C, B^C, \Omega\} = \{\emptyset, \{a\}, \{b\}, \{b, c\}, \{a, c\}, \{a, b, c\}\}.$$

This collection is not an algebra (hence not a σ-algebra) since it does not contain $A \cup B = \{a, b\}$ and $\{a, b\}^C = \{c\}$. However, the intersection $\mathcal{F}_A \cap \mathcal{F}_B = \{\emptyset, \Omega\}$ is the trivial σ-algebra. The result just below tells us that any intersection over a collection of (countable or uncountable) σ-algebras is again a σ-algebra.

Theorem 6.1. *An intersection of a family of σ-algebras on Ω is a σ-algebra on Ω.*

Proof. Let $\mathcal{C}$ be a family of σ-algebras on Ω. Let $\mathcal{F}_{\mathcal{C}} = \bigcap \{\mathcal{F} : \mathcal{F} \in \mathcal{C}\}$ be the intersection of all members of $\mathcal{C}$. Using the fact that every $\mathcal{F} \in \mathcal{C}$ is a σ-algebra, then we have:

1. $\emptyset \in \mathcal{F}$ for all $\mathcal{F} \in \mathcal{C}$, so $\emptyset \in \mathcal{F}_{\mathcal{C}}$.

2. If $E \in \mathcal{F}_{\mathcal{C}}$, then $E \in \mathcal{F}$ and hence $E^C \in \mathcal{F}$ for all $\mathcal{F} \in \mathcal{C}$. Therefore, $E^C \in \mathcal{F}_{\mathcal{C}}$.

3. If $E_k \in \mathcal{F}_{\mathcal{C}}$, then $E_k \in \mathcal{F}$ for $k = 1, 2 \ldots$, and hence $\cup_{k \geqslant 1} E_k \in \mathcal{F}$ for all $\mathcal{F} \in \mathcal{C}$. Thus, $\cup_{k \geqslant 1} E_k \in \mathcal{F}_{\mathcal{C}}$. $\qquad\square$

Note that any intersection of σ-algebras is a nonempty collection since it includes at least two sets: $\emptyset$ and Ω. Theorem 6.1 now also allows us to construct the *smallest σ-algebra that includes a given collection* $\mathcal{A}$ of subsets of Ω. Let $\mathcal{C}(\mathcal{A})$ denote the family of all σ-algebras on Ω that contain $\mathcal{A}$. That is, let $\mathcal{C}(\mathcal{A}) \equiv \{\mathcal{F} : \mathcal{A} \subset \mathcal{F} \text{ and } \mathcal{F} \text{ is a } \sigma\text{-algebra on } \Omega\}$. Note that the power set 2^{Ω} is the collection of all subsets of Ω (including $\emptyset$ and Ω) and hence $\mathcal{A} \subset 2^{\Omega}$. Since 2^{Ω} is itself a σ-algebra, the collection $\mathcal{C}(\mathcal{A})$ is nonempty, i.e., $\mathcal{F} = 2^{\Omega}$ is one such $\mathcal{F}$ in $\mathcal{C}(\mathcal{A})$. The *σ-algebra generated by a collection* $\mathcal{A} \subset 2^{\Omega}$, denoted as $\sigma(\mathcal{A})$, is defined by the intersection:

$$\sigma(\mathcal{A}) := \bigcap \{\mathcal{F} : \mathcal{F} \in \mathcal{C}(\mathcal{A})\}.$$

By Theorem 6.1, $\sigma(\mathcal{A})$ is a σ-algebra, and by construction it is the smallest σ-algebra that contains all sets (events) from $\mathcal{A}$.

Example 6.8. Let A and B be nonempty subsets of Ω such that $A \cap B = \emptyset$. Find $\sigma(\{A, B\})$.

Solution. We can construct the smallest σ-algebra $\mathcal{F} = \sigma(\mathcal{A}) \equiv \sigma(\{A, B\})$ by requiring that the sets $A, B \in \mathcal{F}$. As well, their complements and intersections (or unions) must be in $\mathcal{F}$. Listing all the elements gives:

$$\sigma(\{A, B\}) = \{\emptyset, A, B, A^C, B^C, A \cup B, A^C \cap B^C, \Omega\}.$$

It is simple to check that $\sigma(\{A, B\})$ satisfies the properties in Definition 6.5 and that it is the smallest $\mathcal{F}$ such that $\{A, B\} \subset \mathcal{F}$. $\qquad\square$

A very important example of a σ-algebra that is central to measure theory and general probability theory is the *Borel σ-algebra* on $\mathbb{R}$, denoted $\mathcal{B}(\mathbb{R})$. Every member of the Borel σ-algebra, $B \in \mathcal{B}(\mathbb{R})$, is called a *Borel set*. The pair $(\mathbb{R}, \mathcal{B}(\mathbb{R}))$ is called a *Borel space*. The collection $\mathcal{B}(\mathbb{R})$ of all Borel sets in $\mathbb{R}$ is the smallest σ-algebra on $\mathbb{R}$ that contains all intervals (including zero length intervals or points) in $\mathbb{R}$. One way to form this collection is to start with all intervals and then add in all countable unions, countable intersections, and relative complements. $\mathcal{B}(\mathbb{R})$ is the σ-algebra generated by all the intervals in $\mathbb{R}$:

$$\mathcal{B}(\mathbb{R}) = \bigcap \{\mathcal{F} : \mathcal{F} \text{ is a } \sigma\text{-algebra containing all intervals in } \mathbb{R}\}.$$

To gain a better understanding of $\mathcal{B}(\mathbb{R})$ we now observe that it has equivalent representations as the σ-algebra generated by the collection of all open sets in $\mathbb{R}$, or all intervals of the form (a, b), or $[a, b]$, or (a, ∞), or $[a, \infty)$, or $(-\infty, b)$, or $(-\infty, b]$. That is, the Borel σ-algebra $\mathcal{B}(\mathbb{R})$ is equivalently given by each $\sigma(\mathcal{A}_i)$, $i = 1, \ldots, 6$, where

$$\mathcal{A}_1 = \{(a, b) : a, b \in \mathbb{R}\}, \quad \mathcal{A}_2 = \{[a, b] : a, b \in \mathbb{R}\}, \quad \mathcal{A}_3 = \{(a, \infty) : a \in \mathbb{R}\},$$
$$\mathcal{A}_4 = \{[a, \infty) : a \in \mathbb{R}\}, \quad \mathcal{A}_5 = \{(-\infty, b) : b \in \mathbb{R}\}, \quad \mathcal{A}_6 = \{(-\infty, b] : b \in \mathbb{R}\}.$$

In fact, any open set in $\mathbb{R}$ is a countable union of open intervals, and hence this gives $\mathcal{B}(\mathbb{R}) = \sigma(\mathcal{O})$, where $\mathcal{O}$ is the set of all open sets in $\mathbb{R}$. Other representations of $\mathcal{B}(\mathbb{R})$ are also possible. It is not difficult to prove that $\sigma(\mathcal{A}_1) = \cdots = \sigma(\mathcal{A}_6)$, i.e., are all equivalent representations of $\mathcal{B}(\mathbb{R})$. For instance, the equivalences $\sigma(\mathcal{A}_1) = \sigma(\mathcal{A}_2)$ and $\sigma(\mathcal{A}_1) = \sigma(\mathcal{A}_5)$ follow by the respective identities for countable intersections and unions over open intervals:

$$[a, b] = \bigcap_{n=1}^{\infty} \left(a - \frac{1}{n}, b + \frac{1}{n}\right) \quad \text{and} \quad (a, b) = \bigcup_{n=1}^{\infty} (-\infty, b) \setminus \left(-\infty, a + \frac{1}{n}\right)$$

where $A \setminus B \equiv A \cap B^{\complement}$. By closure of a σ-algebra with respect to taking countable unions and intersections, the sets on the right-hand side are all in the $\mathcal{B}(\mathbb{R})$. The other equivalences follow by employing similar identities.

In $\mathbb{R}^n$, $n \geqslant 2$, the Borel σ-algebra is denoted by $\mathcal{B}_n \equiv \mathcal{B}(\mathbb{R}^n)$. The Borel sets $B \in \mathcal{B}_n$ are formed as n-outer products of Borel sets in $\mathbb{R}$, i.e., $\mathcal{B}_n = \mathcal{B} \times \cdots \times \mathcal{B}$ where $\mathcal{B} \equiv \mathcal{B}(\mathbb{R})$. In particular, every Borel set $B \in \mathcal{B}_n$ has the form $B_1 \times \cdots \times B_n$, where $B_1, \ldots, B_n \in \mathcal{B}$. So these sets are n-outer products of all intervals (open, closed, semi-open, or points) $I_1, \ldots, I_n \subset \mathbb{R}$. For example, these sets include n-dimensional rectangles $B = [a_1, b_1] \times \cdots \times [a_n, b_n]$ where $a_1 < b_1, \ldots, a_n < b_n$, single points in $(a_1, \ldots, a_n) \in \mathbb{R}^n$, lines and hyperplanes, etc. For $n = 2$ dimensions, the Borel sets are outer products $I_1 \times I_2$ on the plane $\mathbb{R}^2$ where I_1, I_2 are any intervals.

The concept of Borel functions is important in the defining property of random variables and their expectation. A single variable real-valued function $g: \mathbb{R} \to \mathbb{R}$ is said to be *Borel measurable (or simply a Borel function)* on $\mathbb{R}$ if

$$g^{-1}(B) \equiv \{x \in \mathbb{R} : g(x) \in B\} \in \mathcal{B}(\mathbb{R}) \text{ for all } B \in \mathcal{B}(\mathbb{R}).$$

That is, the pre-image $g^{-1}(B)$ of any Borel set B in $\mathbb{R}$ is itself a Borel set in $\mathbb{R}$. Borel functions include practically all (and for our purposes all) types of real-valued functions. For example, this includes every continuous or piecewise continuous function, indicator functions of Borel sets, simple functions over Borel sets, the limit of a sequence of Borel functions, and the list goes on! It is actually quite challenging to come up with a function that is not a Borel function. Similarly, a real-valued function of two variables, $g : \mathbb{R}^2 \to \mathbb{R}$, is said to be a Borel function on $\mathbb{R}^2$ if $g^{-1}(B) \equiv \{(x, y) \in \mathbb{R}^2 : g(x, y) \in B\} \in \mathcal{B}_2$ for every $B \in \mathcal{B}$. For any dimension $n \geqslant 1$, $g : \mathbb{R}^n \to \mathbb{R}$ is a Borel function on $\mathbb{R}^n$ if $g^{-1}(B) \equiv \{(x_1, \ldots, x_n) \in \mathbb{R}^n : g(x_1, \ldots, x_n) \in B\} \in \mathcal{B}_n$ for all $B \in \mathcal{B}(\mathbb{R})$, i.e., the pre-image of g of any Borel set B in $\mathbb{R}$ is a Borel set in $\mathbb{R}^n$.

6.2.2.1 Construction of a Sigma-Algebra from a Partition

In the previous section we introduced the concept of a σ-algebra generated by a collection $\mathcal{A}$ of subsets of some given set (sample space) Ω. We can hence consider the special case in which the collection corresponds to a partition $\mathcal{P}$ of Ω, i.e., $\mathcal{A} = \mathcal{P}$ is a collection of atoms. In this situation the σ-algebra generated by the collection is denoted by $\sigma(\mathcal{P})$ and it is the smallest σ-algebra containing the collection $\mathcal{P}$. Since a partition consists of mutually disjoint and exhaustive sets in Ω, then $\sigma(\mathcal{P})$ corresponds to the power set $2^{\mathcal{P}}$ of $\mathcal{P}$. The result just below states this formally.

Theorem 6.2. *Consider a partition $\mathcal{P} = \{A_i\}_{i\in\mathcal{I}}$ of a set Ω, where $\mathcal{I}$ is a finite or countably infinite index set: $\mathcal{I} = \{1, 2, \ldots, M\}$ or $\mathcal{I} = \mathbb{N}$. The smallest σ-algebra, $\sigma(\mathcal{P})$, generated by $\mathcal{P}$ consists of sets of the form*

$$E = \bigcup_{i\in I} A_i, \tag{6.14}$$

where $I \subset \mathcal{I}$ is a set of indices including $I = \emptyset$ and $I = \mathcal{I}$.

Proof. Let $\mathcal{F}$ be a collection of sets of the form (6.14). First, by taking in (6.14) a set $I = \{i\}$ with only one index, we have that $\mathcal{F}$ includes the partition $\mathcal{P}$. Second, let us show that $\mathcal{F}$ is a σ-algebra. If we take $I = \emptyset$ and $I = \mathcal{I}$ then we obtain that $E = \emptyset$ and $E = \Omega$ are both in $\mathcal{F}$. If $E \in \mathcal{F}$, then $E = \cup_{i\in I} A_i$ for some $I \subset \mathcal{I}$. Hence $E^{\complement} = \cup_{i\in\mathcal{I}\setminus I} A_i \in \mathcal{F}$ since $\mathcal{I} \setminus I$ is again a set of indices. Finally, let $E_1, E_2, \ldots \in \mathcal{F}$. For each E_j, there exists a set of indices I_j. The union of all E_j's is again a set of the form (6.14):

$$\bigcup_{j\geqslant 1} E_j = \bigcup_{j\geqslant 1}\bigcup_{i\in I_j} A_i = \bigcup_{i\in\cup_{j\geqslant 1} I_j} A_i$$

where $\cup_{j\geqslant 1} I_j \subset \mathcal{I}$. Finally, removing any element from $\mathcal{F}$ leads to violating the property that a σ-algebra is closed under taking countable unions. Therefore, $\mathcal{F}$ is the smallest σ-algebra generated by the partition $\mathcal{P}$. $\square$

For instance, the σ-algebra generated by the partition $\mathcal{P} = \{E, E^{\complement}\}$ for some $E \subset \Omega$ is $\sigma(\mathcal{P}) = \{\emptyset, E, E^{\complement}, \Omega\} = 2^{\mathcal{P}}$. The σ-algebra constructed in Example 6.8 is generated by the partition $\mathcal{P} = \{A, B, A^{\complement} \cap B^{\complement}\}$ (since $A \cap B = \emptyset$). As we can see from (6.14), every element of $\sigma(\mathcal{P})$ is formed from atoms of $\mathcal{P}$. For a finite partition $\mathcal{P}$, the total number of possible combinations is $2^{|\mathcal{P}|}$. Hence, $|\sigma(\mathcal{P})| = 2^{|\mathcal{P}|}$ as must be the case since $\sigma(\mathcal{P}) = 2^{\mathcal{P}}$.

As an explicit construction, let us consider the N-period binomial model. We have seen that the partitions $\mathcal{P}_n, 0 \leqslant n \leqslant N$ are generated by the first n market moves with 2^n atoms in $\mathcal{P}_n$ given by (6.13):

$$\mathcal{P}_n = \mathcal{P}(\{A_{\omega_1\ldots\omega_n} : \omega_1, \ldots, \omega_n \in \{\mathsf{U}, \mathsf{D}\}\}). \tag{6.15}$$

We can now construct the σ-algebras $\mathcal{F}_n = \sigma(\mathcal{P}_n) = 2^{\mathcal{P}_n}, 0 \leqslant n \leqslant N$, as the power sets generated by such partitions. For time $n = 0$ (no market moves) we have the trivial case $\mathcal{F}_0 = \sigma(\mathcal{P}_0) = \sigma(\Omega_N) = \{\emptyset, \Omega_N\}$. For time $n = 1$ (first market move is resolved):

$$\mathcal{F}_1 = \sigma(\mathcal{P}_1) = \sigma(\{A_{\mathsf{U}}, A_{\mathsf{D}}\}) = \{\emptyset, A_{\mathsf{U}}, A_{\mathsf{D}}, \Omega_N\}.$$

For time $n = 2$ (first two market moves are resolved)

$$\mathcal{F}_2 = \sigma(\mathcal{P}_2) = \sigma(\{A_{\mathsf{UU}}, A_{\mathsf{UD}}, A_{\mathsf{DU}}, A_{\mathsf{DD}}\})$$
$$= \{\emptyset, A_{\mathsf{U}}, A_{\mathsf{D}}, A_{\mathsf{UU}}, A_{\mathsf{UD}}, A_{\mathsf{DU}}, A_{\mathsf{DD}}, A_{\mathsf{UU}}^{\complement}, A_{\mathsf{UD}}^{\complement}, A_{\mathsf{DU}}^{\complement}, A_{\mathsf{DD}}^{\complement},$$
$$A_{\mathsf{UU}} \cup A_{\mathsf{DU}}, A_{\mathsf{UU}} \cup A_{\mathsf{DD}}, A_{\mathsf{UD}} \cup A_{\mathsf{DU}}, A_{\mathsf{UD}} \cup A_{\mathsf{DD}}, \Omega_N\}.$$

This collection contains all the possible events of interest in which we can resolve the first two market moves (without any information on the third and subsequent moves). Note that $\mathcal{F}_1 \subset \mathcal{F}_2$, i.e., $\mathcal{F}_2$ includes all the events in $\mathcal{F}_1$. For example, A_{U} is the event that the first move is up, A_{UU} is the event that the first two moves are up, $A_{\mathsf{UU}}^{\complement}$ is the event that the first two moves are both not up, $A_{\mathsf{UU}} \cup A_{\mathsf{DD}}$ is the event that the first two moves are either both up or both down, $(A_{UU} \cup A_{DD})^{\complement} = A_{UD} \cup A_{DU}$ is the event that the first two moves are different, and so on.

Note that $\mathcal{F}_0$ has $2^{2^0} = 2$ events, $\mathcal{F}_1$ has $2^{2^1} = 4$ events, $\mathcal{F}_2$ has $2^{2^2} = 2^4 = 16$ events. For $n = 3$, then $\mathcal{F}_3 = \sigma(\mathcal{P}_3)$ has $2^{2^3} = 2^8 = 256$ events. For any $0 \leqslant n \geqslant N$, $\mathcal{F}_n = 2^{\mathcal{P}_n}$ is the power set of $\mathcal{P}_n$ having $2^{|\mathcal{P}_n|} = 2^{2^n}$ subsets (events) $E \subset \Omega_N$. At the terminal time $n = N$, all possible moves are resolved and $\mathcal{F}_N = 2^{\mathcal{P}_N}$ contains all the possible 2^{2^N} events of interest.

6.2.2.2 Sigma-Algebra Generated by a Random Variable

We recall that by observing the value of a random variable X, we may comment on the actual value of $\omega \in \Omega$. Formally, a real-valued[1] random variable $X : \Omega \to \mathbb{R}$ on a probability space $(\Omega, \mathcal{F}, \mathbb{P})$ is defined as a set function such that $X^{-1}((-\infty, b])$ is in $\mathcal{F}$, for all $b \in \mathbb{R}$:

$$\{X \in (-\infty, b]\} \equiv \{\omega \in \Omega \ : \ -\infty < X(\omega) \leqslant b\} \in \mathcal{F}.$$

Based on the properties of the Borel sets, we can re-state this definition using any type of interval or simply as $X^{-1}(B) \equiv \{X \in B\} \in \mathcal{F}$ for all $B \in \mathcal{B}(\mathbb{R})$. We therefore interpret a random variable to be an $\mathcal{F}$-*measurable* function. Later we shall precisely define the measurability of a random variable. We will see that X is said to be $\mathcal{F}$-*measurable* if $\sigma(X) \subset \mathcal{F}$ where $\sigma(X)$ denotes the σ-algebra generated by X, as defined just below.

At this point we need to introduce the concept of σ-algebras generated by random variables. All possible information extracted from X can be represented in the form of a σ-algebra, denoted $\sigma(X)$. We say that $\sigma(X)$ is generated by X. That is, $\sigma(X)$ should contain all events of interest associated to X. Such events should include all those of the form $\{X \leqslant b\}$, $\{X \geqslant a\}$, $\{a < X \leqslant b\}$, and so on. More generally, all events $\{X \in B\}$, $B \in \mathcal{B}(\mathbb{R})$, should be contained in $\sigma(X)$. This leads us to the following definition.

Definition 6.6. The σ-algebra, denoted by $\sigma(X)$, generated by a random variable $X : \Omega \to \mathbb{R}$ is the σ-algebra generated by the collection of all subsets of Ω of the form $\{X \in B\}$, $B \in \mathcal{B}(\mathbb{R})$.

Note that this tells us that generally $\sigma(X)$ is the smallest σ-algebra containing all subsets of Ω of the form $X^{-1}(J) \equiv \{X \in J\}$ for every interval J in $\mathbb{R}$. For a discrete random variable X, this therefore corresponds to simply defining $\sigma(X) = \sigma(\mathcal{P}(X))$, i.e., as the σ-algebra generated by the partition $\mathcal{P}(X)$. Indeed, let $\mathcal{I} = \{1, 2, \ldots, M\}$ or $\mathcal{I} = \mathbb{N}$ where the support of X is the countable set of numbers $\{x_1, x_2, \ldots\}$. Given any interval J, the set $X^{-1}(J) = \bigcup_{i \in \mathcal{J}} X^{-1}(x_i) = \bigcup_{i \in \mathcal{J}} A_i$ for some subset of indicies $\mathcal{J} \subset \mathcal{I}$. That is, every set $X^{-1}(J)$ is a union of a sub-collection of atoms $\{A_i\}_{i \in \mathcal{J}}$ in $\mathcal{P}(X)$. Hence, the σ-algebra generated by all the sets $X^{-1}(J)$ is the same as the σ-algebra generated by the partition $\mathcal{P}(X)$. So in this case $\sigma(X) = \sigma(\mathcal{P}(X)) = 2^{\mathcal{P}(X)}$, i.e., the power set of the partition $\mathcal{P}(X)$.

Example 6.9. Consider Example 6.7. Determine $\sigma(S_2)$ and $\sigma(Y)$.

Solution. Using the respective partitions $\mathcal{P}(S_2)$ and $\mathcal{P}(Y)$ in Example 6.7 we have:

$$\sigma(S_2) = \sigma(\mathcal{P}(S_2)) = \sigma(\{A_{DD}, A_{DU} \cup A_{UD}, A_{UU}\}) = 2^{\mathcal{P}(S_2)}$$

$$= \{\emptyset, A_{DD}, A_{UU}, A_{DU} \cup A_{UD}, A_{DD}^{\complement}, A_{UU}^{\complement}, A_{DD} \cup A_{UU}, \Omega_3\}$$

---

[1] More generally, some random variables can also take on infinite values ∞ or $-\infty$. These are defined on the extended real axis. That is, $X^{-1}(\infty) \equiv \{\omega \in \Omega \colon X(\omega) = \infty\}$ and $X^{-1}(-\infty) \equiv \{\omega \in \Omega \colon X(\omega) = -\infty\}$ are mutually exclusive events in $\mathcal{F}$ for which we can assign respective probabilities of occurrence, $\mathbb{P}(X = \infty)$ and $\mathbb{P}(X = -\infty)$. For example, we have already seen that a first passage time random variable for a process can be assigned a value of ∞.

and

$$\sigma(Y) = \sigma(\{\{\omega_2 = \mathsf{D}\}, \{\omega_2 = \mathsf{U}\}\}) = 2^{\mathcal{P}(Y)} = \{\emptyset, \{\omega_2 = \mathsf{D}\}, \{\omega_2 = \mathsf{U}\}, \Omega_3\}.$$

$\square$

Note that both σ-algebras $\sigma(S_2)$ and $\sigma(Y)$ are contained within $\mathcal{F}_2 \equiv \sigma(\mathcal{P}_2)$, which is the σ-algebra containing all the information (events) for the first two market moves. As discussed further below, we also have $\mathcal{F}_2 = \sigma(S_1, S_2)$, the σ-algebra generated by the pair of stock prices at times $n = 1$ and $n = 2$.

Consider now a sigma algebra generated by multiple (joint) random variables such as $X_1, \ldots, X_M$ or even by a countable or uncountable collection of random variables. In the case with M discrete random variables, $\sigma(X_1, \ldots, X_M) = \sigma(\mathcal{P}(X_1, \ldots, X_M))$. In the case where we have M random variables that are not necessarily discrete-valued, then $\sigma(X_1, \ldots, X_M)$ is the smallest σ-algebra containing all subsets of the form

$$\{\omega \in \Omega \ : \ X_i(\omega) \in B\} \text{ where } B \in \mathcal{B}(\mathbb{R}), \ 1 \leqslant i \leqslant M.$$

More generally, we consider a countable or uncountable index set Λ and define $\sigma(\{X_\lambda\}_{\lambda \in \Lambda})$ as the smallest σ-algebra generated by the collection of random variables $\{X_\lambda\}_{\lambda \in \Lambda}$. This σ-algebra consists of all events of the form $\{X_\lambda \in B\}$, $B \in \mathcal{B}(\mathbb{R})$, $\lambda \in \Lambda$.

6.2.3   Filtration

As time passes, we acquire more information about the actual state of a stochastic model. At any instant, the current level of information available to us is represented by a partition or a σ-algebra. The flow of information can be modeled by a sequence of partitions that we named an information structure. Since it is more convenient to work with σ-algebras, we can similarly define a sequence of σ-algebras that accumulates more and more knowledge as time increases. We have seen this explicitly in the binomial model where more and more information about events is available with each new market move, where $\mathcal{F}_1 \subset \mathcal{F}_2 \subset \ldots \mathcal{F}_N$. The collection of such σ-algebras is an example of what is called a *filtration*. The following is a formal definition in the continuous-time case.

Definition 6.7. A *filtration* $\mathbb{F}$ is a sequence of σ-algebras $\{\mathcal{F}_t\}_{0 \leqslant t \leqslant T}$ over a set Ω such that $\mathcal{F}_s \subset \mathcal{F}_t$ for all $0 \leqslant s \leqslant t \leqslant T$.

This definition implies that information can only increase (not decrease) with time. In the discrete-time case, a filtration is a finite sequence of σ-algebras $\{\mathcal{F}_n\}_{0 \leqslant n \leqslant N} \equiv \{\mathcal{F}_0, \mathcal{F}_1, \ldots, \mathcal{F}_N\}$ with the property $\mathcal{F}_{n-1} \subset \mathcal{F}_n$, for all $1 \leqslant n \leqslant N$, i.e.,

$$\mathcal{F}_0 \subset \mathcal{F}_1 \subset \cdots \subset \mathcal{F}_N.$$

6.2.3.1   Construction of a Filtration from an Information Structure

Consider an information structure $\mathcal{P}_0 \preceq \mathcal{P}_1 \preceq \cdots \preceq \mathcal{P}_N$. Define $\mathcal{F}_n = \sigma(\mathcal{P}_n)$, $0 \leqslant n \leqslant N$. Then, the σ-algebras $\mathcal{F}_0, \mathcal{F}_1, \ldots, \mathcal{F}_N$ form a filtration. It is sufficient to prove the following assertion.

Proposition 6.3. *Suppose that $\mathcal{P}$ and $\mathcal{Q}$ are two partitions of Ω such that $\mathcal{P} \preceq \mathcal{Q}$. Then $\sigma(\mathcal{P}) \subset \sigma(\mathcal{Q})$.*

Proof. First, enumerate all atoms in the partitions $\mathcal{P}$ and $\mathcal{Q}$:

$$\mathcal{P} = \{A_i\}_{i \geqslant 1}, \quad \mathcal{Q} = \{B_i\}_{i \geqslant 1}.$$

Take any $E \in \sigma(\mathcal{P})$. Since $\sigma(\mathcal{P}) = \sigma(\{A_i\}_{i \geqslant 1})$, there exists an index set $I_E \subset \mathbb{N}$ such that $E = \bigcup_{i \in I_E} A_i$. Moreover, $\mathcal{P} \preceq \mathcal{Q}$ implies that for every $A_i \in \mathcal{P}$ there is a countable collection of atoms of $\mathcal{Q}$, $\{B_j\}_{j \geqslant I_i}$, for some index set $I_i \subset \mathbb{N}$, such that $A_i = \bigcup_{j \in I_i} B_j$. Hence, event E can be represented as a union of atoms of $\mathcal{Q}$:

$$E = \bigcup_{i \in I_E} A_i = \bigcup_{i \in I_E} \bigcup_{j \in I_i} B_j.$$

This means that E is an element of $\sigma(\mathcal{Q})$ by definition. □

By Proposition 6.3, for all $1 \leqslant n \leqslant N$ we have $\mathcal{F}_{n-1} \subset \mathcal{F}_n$ since $\mathcal{P}_{n-1} \preceq \mathcal{P}_n$. Therefore, $\{\mathcal{F}_n\}_{0 \leqslant n \leqslant N}$ is a filtration. This type of filtration is a sequence of power sets on finer and finer partitions. Explicit examples are the σ-algebras $\mathcal{F}_n = \sigma(\mathcal{P}_n)$ with $\mathcal{P}_n$ in (6.15).

6.2.3.2 Construction of a Filtration from a Stochastic Process: Natural Filtration

A natural method to obtain a filtration is to generate σ-algebras from a sequence of random variables. Consider a stochastic process $\{X_t\}_{0 \leqslant t \leqslant T}$ with sample space Ω. Let

$$\mathcal{F}_t^X := \sigma(\{X_u : 0 \leqslant u \leqslant t\})$$

be the smallest σ-algebra generated by all paths up to time t: $\{X_u\}_{0 \leqslant u \leqslant t}$. So $\mathcal{F}_t^X$ contains *all information available from the observation of the process up to time t*. Note that, since time $t \in [0, T]$ is continuous, then $\mathcal{F}_t^X$ is the σ-algebra generated by the uncountable collection of random variables $\{X_u\}_{0 \leqslant u \leqslant t}$.

In the discrete-time case, $\mathcal{F}_n^X = \sigma(\{X_k : k = 0, \ldots, n\}) \equiv \sigma(X_0, X_1, \ldots, X_n)$. Clearly, $\mathcal{F}_s^X \subset \mathcal{F}_t^X$ for all $0 \leqslant s < t \leqslant T$. The filtration $\{F_t^X\}_{0 \leqslant t \leqslant T}$ is called a *natural filtration* of the process $\{X_t\}_{0 \leqslant t \leqslant T}$. As is demonstrated in the following example, σ-algebras generated *only from the time-t value X_t of a stochastic process* (rather than from all path values from 0 to t) may not form a filtration.

Example 6.10. Show that σ-algebras generated by binomial prices, $\{\mathcal{F}_n\}_{0 \leqslant n \leqslant 2}$ where $\mathcal{F}_n = \sigma(S_n)$, do not constitute a filtration.

Solution. Consider the N-period model for any $N \geqslant 2$. Since the range of S_1 is $\{S_0 d, S_0 u\}$, the partition $\mathcal{P}(S_1)$ consists of two atoms:

$$\{S_1 = S_0 d\} = \{\omega \in \Omega_N : \omega_1 = \mathsf{D}\} \equiv A_\mathsf{D}$$
$$\{S_1 = S_0 u\} = \{\omega \in \Omega_N : \omega_1 = \mathsf{U}\} \equiv A_\mathsf{U}.$$

Similarly, the partition $\mathcal{P}(S_2)$ contains three atoms:

$$\{S_2 = S_0 d^2\} = \{\omega_1 = \omega_2 = \mathsf{D}\} \equiv A_\mathsf{DD},$$
$$\{S_2 = S_0 ud\} = \{\omega_1 = \mathsf{D}, \omega_2 = \mathsf{U}\} \cup \{\omega_1 = \mathsf{U}, \omega_2 = \mathsf{D}\} \equiv A_\mathsf{DU} \cup A_\mathsf{UD},$$
$$\{S_2 = S_0 u^2\} = \{\omega_1 = \omega_2 = \mathsf{U}\} \equiv A_\mathsf{UU}.$$

As we can see, not all (in fact none) of the atoms of $\mathcal{P}(S_1)$ can be obtained as a union of atoms of $\mathcal{P}(S_2)$. Therefore, $\mathcal{P}(S_2)$ is not a refinement of $\mathcal{P}(S_1)$, and $\sigma(S_1) \not\subset \sigma(S_2)$. □

This example points out that by only observing the stock price at one point in time after two or more moves, we are missing information about all the previous history (i.e., all the previous moves) of the stock price path up to that time.

As a main example of a natural filtration, re-consider the binomial model on the sample space Ω_N. The natural filtration $\{F_n\}_{0 \leqslant n \leqslant N}$ can be obtained from the binomial price process $\{S_n\}$, the random walk $\{M_n\}$, the Bernoulli random variables $X_n = \mathbb{I}_{\{\omega_n = \mathsf{U}\}}$, or directly from the market moves $\{\omega_n\}$, $0 \leqslant n \leqslant N$:

$$\mathcal{F}_0 = \{\emptyset, \Omega_N\}$$
$$\mathcal{F}_n = \sigma(S_1, \ldots, S_n) = \sigma(M_1, \ldots, M_n) = \sigma(\omega_1, \ldots, \omega_n), \quad 1 \leqslant n \leqslant N,$$

where we are denoting $\sigma(\omega_1, \ldots, \omega_n) \equiv \sigma(\mathcal{P}_n)$ with $\mathcal{P}_n$ in (6.15). The σ-algebra $\mathcal{F}_n = 2^{\mathcal{P}_n}$, $1 \leqslant n \leqslant N$, is generated from the partition $\mathcal{P}_n$ having 2^n atoms of the form (6.13) and $\mathcal{F}_n$ has 2^{2^n} events. As discussed above, $\mathcal{F}_1 = \sigma(\{A_\mathsf{D}, A_\mathsf{U}\})$, $\mathcal{F}_2 = \sigma(\{A_\mathsf{DD}, A_\mathsf{DU}, A_\mathsf{UD}, A_\mathsf{UU}\})$, until $\mathcal{F}_N = 2^{\Omega_N}$.

6.2.4   Filtered Probability Space

Now it is time to revisit the definition of a probability space. In Definition 6.2, it is assumed that the sample space Ω is finite and the probability can be calculated for every possible event in Ω. In other words, the probability measure $\mathbb{P}$ was defined as a function that maps $\mathcal{F} = 2^\Omega$ to $[0, 1]$. It is a typical situation when the set of measurable events (for which we can compute their probability of occurrence) is smaller than a power set. However, such a set must be a σ-algebra to match the Kolmogorov axioms. The power set 2^Ω is one example of a σ-algebra, but it corresponds to the ultimate case where every event (i.e., every subset of Ω) can be measured by $\mathbb{P}$. So we need to add a σ-algebra $\mathcal{F}$ to the pair $(\Omega, \mathbb{P})$. The pair $(\Omega, \mathcal{F})$ by itself is called a *measurable space*. To complete the picture, we need a probability function for "measuring" events (which are subsets of Ω and simply elements in $\mathcal{F}$). In other words, a probability space includes a measuring tool (i.e., a probability measure $\mathbb{P}$) together with $(\Omega, \mathcal{F})$, where the collection of all measurable events are contained in the σ-algebra $\mathcal{F}$ on Ω. This then leads us to the following definition.

Definition 6.8. The triple $(\Omega, \mathcal{F}, \mathbb{P})$ is called a *probability space* where

- Ω is a sample space,

- $\mathcal{F}$ is a σ-algebra on Ω,

- $\mathbb{P}: \mathcal{F} \to [0, 1]$ is a probability measure that satisfies the Kolmogorov axioms in (6.1).

Note that a finite probability space is a particular case of a general probability space with $\mathcal{F} = 2^\Omega$. Our main working examples stem from the binomial probability spaces $(\Omega_N, \mathcal{F}_k, \mathbb{P}_k)$, where $\mathcal{F}_k = \sigma(\mathcal{P}_k)$ for some $0 \leqslant k \leqslant N$. That is, $\mathcal{F}_k$ contains all the information about the first k periods (or moves). For each $k = 0, \ldots, N$, we can define a probability measure $\mathbb{P}_k: \mathcal{F}_k \to [0, 1]$ by

$$\mathbb{P}_k(E) = \sum_{\omega_1, \ldots, \omega_k \in \{\mathsf{U}, \mathsf{D}\} \, : \, \omega \in E} p^{\mathsf{U}_k(\omega)} (1 - p)^{\mathsf{D}_k(\omega)}, \qquad (6.16)$$

for every nonempty $E \in \mathcal{F}_k$ and with $\mathbb{P}_k(\emptyset) = 0$. Note that the sum is over the first k ω's and is restricted to outcomes $\omega = \omega_1 \ldots \omega_k \, \omega_{k+1} \ldots \omega_N \in \Omega_N$ contained in E. It is readily seen that $\mathbb{P}_k$ is a probability measure on $(\Omega_N, \mathcal{F}_k)$ where $\mathbb{P}_k(\Omega_N) = 1$. Indeed, setting $E = \Omega_N$ removes the restriction $\omega \in E$ and the above sum evaluates to its maximum value of unity (note that this also corresponds to $\mathbb{P}_k(\Omega_k) = 1$). It is important to note that the measure $\mathbb{P}_k$ is the probability measure $\mathbb{P} \equiv \mathbb{P}_N$ restricted to events that are in $\mathcal{F}_k = \sigma(\mathcal{P}_k)$ which is the σ-algebra on Ω_N generated by the 2^k atoms, each denoted by $A_{\omega_1 \ldots \omega_k}$ (see (6.15)). To see this, observe that any $E \in \mathcal{F}_k$ is expressible as

$$E = E \cap \Omega_N = \bigcup_{\omega_1, \ldots, \omega_k \in \{U, D\}} (E \cap A_{\omega_1 \ldots \omega_k}) = \bigcup_{\omega_1, \ldots, \omega_k \in \{U, D\} \,:\, \omega \in E} A_{\omega_1 \ldots \omega_k}.$$

The last expression is a union over the restricted sub-collection of atoms in $\mathcal{P}_k$ that collectively define E. Taking the probability $\mathbb{P}(E)$ then recovers (6.16) by

$$\mathbb{P}(A_{\omega_1 \ldots \omega_k}) = \sum_{\omega_{k+1}, \ldots, \omega_N \in \{U, D\}} \mathbb{P}(\{\omega_1 \ldots \omega_k \omega_{k+1} \ldots \omega_N\})$$

$$= p^{U_k(\omega)} (1 - p)^{D_k(\omega)} \left(\sum_{\omega_{k+1}, \ldots, \omega_N \in \{U, D\}} p^{\#U(\omega_{k+1} \ldots \omega_N)} (1 - p)^{\#D(\omega_{k+1} \ldots \omega_N)} \right)$$

$$= p^{U_k(\omega)} (1 - p)^{D_k(\omega)}$$

where the quantity in brackets is $\mathbb{P}(\omega_{k+1} \in \{U, D\}) \times \cdots \times \mathbb{P}(\omega_N \in \{U, D\}) = 1 \times \cdots \times 1 = 1$.

A probability space is a natural framework for dealing with "static" random objects like random variables that do not change in time. To work with stochastic processes we need to introduce a "dynamic" version of a probability space.

Definition 6.9. A *filtered probability space* is a quadruple $(\Omega, \mathcal{F}, \mathbb{P}, \{\mathcal{F}_t\}_{t \geqslant 0})$ where the first three objects form a probability space $(\Omega, \mathcal{F}, \mathbb{P})$ and $\{\mathcal{F}_t\}_{t \geqslant 0}$ is a filtration such that $\mathcal{F}_t \subset \mathcal{F}$ for all $t \geqslant 0$.

Again, the main working model of this chapter is the binomial model with the filtered probability space $(\Omega_N, \mathcal{F}_N, \mathbb{P}_N, \{\mathcal{F}_n\}_{0 \leqslant n \leqslant N})$, where all the required ingredients are constructed as discussed above.

6.3 Conditional Expectation and Martingales

6.3.1 Measurability of Random Variables and Processes

Measurability is an important concept that arises in the theory of measure and integration. Recall that a measurable space is a pair $(\Omega, \mathcal{F})$ consisting of a nonempty set Ω and a σ-algebra $\mathcal{F}$ on Ω; such subsets are said to be *measurable*. A function between measurable spaces is said to be *measurable* if its pre-image on each measurable set is measurable.

Definition 6.10. Let $X \colon \Omega \to \mathbb{R}$ be a random variable on Ω, and let $\mathcal{F}$ be a σ-algebra of subsets of Ω. X is said to be $\mathcal{F}$-*measurable* if $\sigma(X) \subset \mathcal{F}$.

Alternatively, X is $\mathcal{F}$-measurable iff $\{\omega \in \Omega \,:\, X(\omega) \in B\} \in \mathcal{F}$ for all $B \in \mathcal{B}(\mathbb{R})$. A discrete random variable X is $\mathcal{F}$-measurable if $\{\omega \in \Omega \,:\, X(\omega) = x\} \in \mathcal{F}$ for all $x \in X(\Omega)$. The image $X(\Omega) = \{X(\omega) \in \mathbb{R} \,:\, \omega \in \Omega\}$ is the support or range of X. If $\sigma(X) \subset \mathcal{F}$, then all information about X is contained in $\mathcal{F}$. Therefore, $\mathcal{F}$-measurability of X means that we can calculate the value of X if we know what events of $\mathcal{F}$ have occurred. Suppose that a σ-algebra $\mathcal{F}$ is generated by a partition $\mathcal{P}$. To verify whether X is $\mathcal{F}$-measurable, it is enough to check if X is constant on every atom of $\mathcal{P}$.

Here are some examples of measurable random variables.

(a) Any random variable X is measurable relative to the σ-algebra $\sigma(X)$ generated by X.

(b) A constant random variable is $\mathcal{F}_0$-measurable, where $\mathcal{F}_0 = \{\emptyset, \Omega\}$ is a trivial σ-algebra.

(c) In the N-period binomial model, S_n and M_n are $\mathcal{F}_n$-measurable for all $0 \leqslant n \leqslant N$.

In what follows, it will always be assumed that every random variable defined on a probability space $(\Omega, \mathcal{F}, \mathbb{P})$ is $\mathcal{F}$-measurable. It is a reasonable assumption, since otherwise it may be not possible to calculate all probabilities relating to a random variable that is not $\mathcal{F}$-measurable. For example, consider a probability space $(\Omega, \mathcal{F}, \mathbb{P})$ corresponding to a single roll of a fair die: $\Omega = \{1, 2, 3, 4, 5, 6\}$. Suppose that the σ-algebra of events, $\mathcal{F}$, is generated by observing whether the facing-up value of the die is odd or even: $\mathcal{F} = \{\emptyset, \Omega, \{1, 3, 5\}, \{2, 4, 6\}\}$. Define $\mathbb{P}$ on $\mathcal{F}$ as usual: $\mathbb{P}(\emptyset) = 0$, $\mathbb{P}(\Omega) = 1$, $\mathbb{P}(\{1, 3, 5\}) = \mathbb{P}(\{2, 4, 6\}) = \frac{1}{2}$. It is easy to verify that the random variable $X(\omega) = \omega$ that gives us the value of the die is not $\mathcal{F}$-measurable. Therefore, it is impossible to compute probabilities $\mathbb{P}(X = k)$, $1 \leqslant k \leqslant 6$, as we are missing information on the probabilities for obtaining any given number on the die.

It is a common situation when one random variable is measurable relative to the σ-algebra generated by another random variable. Let Y be $\sigma(X)$-measurable. Since $\sigma(Y) \subset \sigma(X)$, we may expect that being given the value of X we can calculate the value of Y. Indeed, the Doob–Dynkin theorem states that if Y is $\sigma(X)$-measurable, then there exists a Borel function f such that $Y = f(X)$. Recall from above that $f \colon \mathbb{R} \to \mathbb{R}$ is a Borel function on $\mathbb{R}$ if $f^{-1}(B) \in \mathcal{B}(\mathbb{R})$ for all $B \in \mathcal{B}(\mathbb{R})$.

Example 6.11. Consider a two-period binomial model Ω_2 with σ-algebras $\mathcal{F}_1 = \sigma(\omega_1)$ and $\mathcal{F}_2 = \sigma(\omega_1, \omega_2)$. Define two random variables X and Y on Ω_2 as follows:

$$X(\mathsf{UU}) = X(\mathsf{DU}) = 1, \ X(\mathsf{UD}) = X(\mathsf{DD}) = -1,$$
$$Y(\mathsf{UU}) = Y(\mathsf{UD}) = 1, \ Y(\mathsf{DD}) = Y(\mathsf{DU}) = -1.$$

Show that Y is $\mathcal{F}_1$-measurable and X is not.

Solution. Clearly, X and Y are both $\mathcal{F}_2$-measurable since $\mathcal{F}_2$ contains all information about the outcome $\omega_1 \omega_2$. $\mathcal{F}_1 = \{\emptyset, A_\mathsf{D}, A_\mathsf{U}, \Omega_2\}$ is generated by ω_1. Since $Y^{-1}(1) = \{\mathsf{UD}, \mathsf{UU}\} = A_\mathsf{U} \in \mathcal{F}_1$ and $Y^{-1}(-1) = \{\mathsf{DD}, \mathsf{DU}\} = A_\mathsf{D} \in \mathcal{F}_1$, Y is $\mathcal{F}_1$-measurable. For X we observe that $X^{-1}(1) = \{\mathsf{DU}, \mathsf{UU}\} \notin \mathcal{F}_1$. Thus, X is not $\mathcal{F}_1$-measurable.
[Note: this is also readily seen by writing $X = \mathbb{I}_{\{\omega_2 = \mathsf{U}\}} - \mathbb{I}_{\{\omega_2 = \mathsf{D}\}}$, which is $\mathcal{F}_2$-measurable but not $\mathcal{F}_1$-measurable, and $Y = \mathbb{I}_{\{\omega_1 = \mathsf{U}\}} - \mathbb{I}_{\{\omega_1 = \mathsf{D}\}}$, which is $\mathcal{F}_1$-measurable.] □

Proposition 6.4. *Consider an N-period binomial model with the natural filtration*

$$\{\mathcal{F}_n = \sigma(\omega_1, \ldots, \omega_n)\}_{0 \leqslant n \leqslant N}.$$

The time-n binomial price S_n, $0 \leqslant n \leqslant N$, is $\mathcal{F}_k$-measurable iff $n \leqslant k$.

Proof. The random variable S_n is a function of $\omega_1, \ldots, \omega_n$. Therefore, S_n is $\mathcal{F}_n$-measurable. If $k > n$ then $\mathcal{F}_k \supset \mathcal{F}_n$. Thus S_n is $\mathcal{F}_k$-measurable for all $k \geqslant n$. Suppose that $0 \leqslant k < n$. The σ-algebra $\mathcal{F}_k$ is generated by a partition $\mathcal{P}_k$. Since S_n is not constant on atoms of $\mathcal{P}_k$, it is not $\mathcal{F}_k$-measurable. □

Similarly, one can show that the time-n value of the random walk process M_n, $0 \leqslant n \leqslant N$, is $\mathcal{F}_k$-measurable iff $n \leqslant k$ (see Exercise 6.12).

Definition 6.11. Let $\mathbb{F} = \{\mathcal{F}_t\}_{0 \leqslant t \leqslant T}$ be a filtration over Ω. A random process $\{X_t\}_{0 \leqslant t \leqslant T}$ on Ω is said to be *adapted* to filtration $\mathbb{F}$ if X_t is $\mathcal{F}_t$-measurable for every t.

Since, S_n and M_n are both $\mathcal{F}_n$-measurable, for all $0 \leqslant n \leqslant N$, then processes $\{S_n\}_{0 \leqslant n \leqslant N}$ and $\{M_n\}_{0 \leqslant n \leqslant N}$ are adapted to their respective natural filtrations $\{\sigma(S_0, S_1, \ldots, S_n)\}_{0 \leqslant n \leqslant N}$ and $\{\sigma(M_0, M_1, \ldots, M_n)\}_{0 \leqslant n \leqslant N}$. In general, every stochastic process $\{X_t\}_{0 \leqslant t \leqslant T}$ is adapted to its natural filtration $\{\mathcal{F}_t^X\}_{0 \leqslant t \leqslant T}$ since, by definition, $\mathcal{F}_t^X \equiv \sigma(\{X_u : 0 \leqslant u \leqslant t\})$ and this obviously implies $\sigma(X_t) \subset \mathcal{F}_t^X$, i.e., X_t is $\mathcal{F}_t^X$-measurable.

6.3.2 Conditional Expectations

In what follows we shall assume *discrete-valued random variables*, although the theory can be presented more generally to include (continuous or mixed) random variables on uncountable probability spaces within a single consistent framework using Lebesgue-Stieltjies integration. To further simplify the discussion, we now consider a finite probability space $(\Omega, \mathcal{F}, \mathbb{P})$ with $\Omega = \{\omega^1, \ldots, \omega^M\}$ and an $\mathcal{F}$-measurable random variable X on Ω. The discussion below follows identically for a countably infinite sample space $(M = \infty)$ where we simply have infinite summations. Since Ω is finite (or countably infinite in case $M = \infty$), $X : \Omega \to \mathbb{R}$ is a discrete random variable that admits the following representation:

$$X = \sum_{k=1}^{M} X(\omega^k)\, \mathbb{I}_{\{\omega^k\}}$$

with indicator random variable $\mathbb{I}_{\{\omega^k\}} = \mathbb{I}_{\{\omega = \omega^k\}}$, i.e., $\mathbb{I}_{\{\omega^k\}}(\omega) = 1$, if $\omega = \omega^k$, and 0 if $\omega \neq \omega^k$. This is one way to express the random variable X as a simple function by using a sum over all outcomes. Note that X may be a nonzero constant (or zero) on more than one outcome. The (unconditional) expectation of X (w.r.t. the probability measure $\mathbb{P}$) is

$$\mathrm{E}[X] = \sum_{k=1}^{M} X(\omega^k)\, \mathrm{E}[\mathbb{I}_{\{\omega^k\}}] = \sum_{k=1}^{M} X(\omega^k)\, \mathbb{P}(\omega^k) = \sum_{\omega \in \Omega} X(\omega)\, \mathbb{P}(\omega) \tag{6.17}$$

where $\mathrm{E}[\mathbb{I}_{\{\omega^k\}}] = \mathbb{P}(\{\omega^k\}) \equiv \mathbb{P}(\omega^k)$. Note that this formula can also be written in a more traditional form by using the disjoint sets defined by $A_x := \{\omega \in \Omega : X(\omega) = x\}$ for all nonzero x values in the finite support S_X of X, i.e., $X = \sum_{x \in \mathsf{S}_X} x\, \mathbb{I}_{A_x}$, giving

$$\mathrm{E}[X] = \sum_{x \in \mathsf{S}_X} x\, \mathrm{E}[\mathbb{I}_{A_x}] = \sum_{x \in \mathsf{S}_X} x\, \mathbb{P}(A_x) = \sum_{x \in \mathsf{S}_X} x\, \mathbb{P}(X = x).$$

The mathematical expectation of a random variable X represents the best prediction of X given no additional information about X or the actual outcome $\omega \in \Omega$. Being given some additional information about X or ω, we may obtain best predictions of X conditional on the available information. The simplest case is the conditional expectation of X given an event $B \in \mathcal{F}$. Such an expected value is denoted by $\mathrm{E}[X \mid B]$. In a more general case, the available information can be represented by a σ-algebra $\mathcal{G}$ of subsets of Ω. The best prediction of X given the information in $\mathcal{G}$ is a conditional expectation of X given $\mathcal{G}$, denoted by $\mathrm{E}[X \mid \mathcal{G}]$. In fact, such an expected value is *generally not a number but a $\mathcal{G}$-measurable random variable on Ω*. Later, we shall define different versions of a conditional expectation.

6.3.2.1 Conditioning on an Event

Recall that the *conditional probability* of event $A \in \mathcal{F}$ given event $B \in \mathcal{F}$ is defined by

$$\mathbb{P}(A \mid B) := \frac{\mathbb{P}(A \cap B)}{\mathbb{P}(B)},$$

provided $\mathbb{P}(B) \neq 0$. The *conditional expectation* of X given event $B \in \mathcal{F}$ is defined as

$$E[X \mid B] := \sum_{\omega \in \Omega} X(\omega)\, \mathbb{P}(\omega \mid B) \qquad (6.18)$$

where $\mathbb{P}(\omega \mid B) \equiv \mathbb{P}(\{\omega\} \mid B)$ is the conditional probability of a given outcome $\omega \in \Omega$ given B. Using the above definition, this conditional probability is expressible as

$$\mathbb{P}(\omega \mid B) = \frac{\mathbb{P}(\{\omega\} \cap B)}{\mathbb{P}(B)} = \frac{\mathbb{P}(\omega)}{\mathbb{P}(B)}\, \mathbb{I}_B(\omega),$$

where we use the property that $\{\omega\} \cap B = \emptyset$ if $\omega \notin B$, and $\{\omega\} \cap B = \{\omega\}$ if $\omega \in B$. Therefore, the formula in (6.18) can be rewritten as

$$
\begin{aligned}
E[X \mid B] &= \sum_{\omega \in \Omega} X(\omega) \frac{\mathbb{P}(\omega)}{\mathbb{P}(B)} \mathbb{I}_B(\omega) \\
&= \frac{1}{\mathbb{P}(B)} \sum_{\omega \in \Omega} X(\omega)\, \mathbb{I}_B(\omega)\, \mathbb{P}(\omega) \\
&= \frac{1}{\mathbb{P}(B)} \sum_{\omega \in B} X(\omega)\, \mathbb{P}(\omega) = \frac{E[X\, \mathbb{I}_B]}{\mathbb{P}(B)}.
\end{aligned}
\qquad (6.19)
$$

We know that the expectation of an indicator function of an event equals the probability of the event: $E[\mathbb{I}_A] = \mathbb{P}(A)$, for $A \in \mathcal{F}$. It follows by (6.19) that the same result holds for the conditional expected value:

$$E[\mathbb{I}_A \mid B] = \frac{E[\mathbb{I}_A\, \mathbb{I}_B]}{\mathbb{P}(B)} = \frac{E[\mathbb{I}_{A \cap B}]}{\mathbb{P}(B)} = \frac{\mathbb{P}(A \cap B)}{\mathbb{P}(B)} = \mathbb{P}(A \mid B).$$

Since the (unconditional) expectation is a linear functional, so is the conditional expectation.

Proposition 6.5. *Let $B \in \mathcal{F}$ be such that $\mathbb{P}(B) \neq 0$. For any reals c_i and random variables X_i, $i = 1, 2$, we have*

$$E[c_1 X_1 + c_2 X_2 \mid B] = c_1 E[X_1 \mid B] + c_2 E[X_2 \mid B].$$

Proof. Using (6.19) and the linearity of the operator E:

$$
\begin{aligned}
E[c_1 X_1 + c_2 X_2 \mid B] &= \frac{E[(c_1 X_1 + c_2 X_2)\, \mathbb{I}_B]}{\mathbb{P}(B)} = \frac{c_1 E[X_1 \mathbb{I}_B] + c_2 E[X_2 \mathbb{I}_B]}{\mathbb{P}(B)} \\
&= c_1 \frac{E[X_1\, \mathbb{I}_B]}{\mathbb{P}(B)} + c_2 \frac{E[X_2\, \mathbb{I}_B]}{\mathbb{P}(B)} = c_1 E[X_1 \mid B] + c_2 E[X_2 \mid B]. \qquad \square
\end{aligned}
$$

Example 6.12. Consider a four-period binomial model with $p = \frac{1}{2}$. Find $E[\mathsf{U}_4 \mid B]$ for $B = \{\text{at least two U's}\}$.

Solution. The event $B \subset \Omega_4$ can be characterized by listing all elements $\omega = \omega_1 \omega_2 \omega_3 \omega_4$ of its complement:

$$B^{\complement} = \{\omega \mid \mathsf{U}_4(\omega) \leqslant 1\} = \{\mathsf{DDDD}, \mathsf{DDDU}, \mathsf{DDUD}, \mathsf{DUDD}, \mathsf{UDDD}\}.$$

Note that all outcomes in Ω_4 are all equally likely since $p = 1 - p = \frac{1}{2}$; the probability of each particular scenario is $\left(\frac{1}{2}\right)^4 = \frac{1}{16}$. First, calculate $\mathbb{P}(B)$:

$$\mathbb{P}(B) = 1 - \mathbb{P}(B^{\complement}) = 1 - 5 \cdot \frac{1}{16} = \frac{11}{16}.$$

Second, apply (6.19) to calculate $E[U_4 \mid B]$:

$$E[U_4 \, \mathbb{I}_B] = \sum_{\omega \in B} U_4(\omega) \, \mathbb{P}(\omega) = \sum_{\omega : U_4(\omega) \geqslant 2} U_4(\omega) \, \mathbb{P}(\omega) = \sum_{k=2}^{4} k \cdot \mathbb{P}(U_4 = k)$$

$$= \frac{1}{16} \cdot \left(2 \cdot \binom{4}{2} + 3 \cdot \binom{4}{3} + 4 \cdot \binom{4}{4} \right) = \frac{1}{16} \cdot (2 \cdot 6 + 3 \cdot 4 + 4 \cdot 1) = \frac{28}{16} = \frac{7}{4}.$$

Thus, $E[U_4 \mid B] = E[U_4 \, \mathbb{I}_B] / \mathbb{P}(B) = \frac{7/4}{11/16} = \frac{28}{11}$. $\qquad\square$

The concept of the conditional expectation allows us to generalize the law of total probability. Let $\mathcal{P}$ be a partition of Ω, i.e., $\Omega = \cup_{A \in \mathcal{P}} A$, and let X be a random variable on Ω. Since $\mathbb{I}_\Omega = \sum_{A \in \mathcal{P}} \mathbb{I}_A$, then the expectation of X can be calculated as a sum of products of conditional expectations of X with respect to every atom $A \in \mathcal{P}$ and the probability of the event corresponding to every atom:

$$E[X] = E[X \, \mathbb{I}_\Omega] = \sum_{A \in \mathcal{P}} E[X \, \mathbb{I}_A] = \sum_{A \in \mathcal{P}} E[X \mid A] \, \mathbb{P}(A). \tag{6.20}$$

6.3.2.2 Conditioning on a Sigma-Algebra

Let X be a random variable of a probability space $(\Omega, \mathcal{F}, \mathbb{P})$ and let $\mathcal{G}$ be a sub-σ-algebra of $\mathcal{F}$, i.e., $\mathcal{G}$ is a σ-algebra on Ω and $\mathcal{G} \subset \mathcal{F}$. Hence, X is $\mathcal{F}$-measurable. If X were also $\mathcal{G}$-measurable then the information contained in $\mathcal{G}$ is sufficient to determine a value for X. In particular, if X were $\mathcal{G}$-measurable and $\mathcal{G}$ were generated by a given partition then X would be a constant on the atoms of the partition. However, more generally, when X is not $\mathcal{G}$-measurable the information in $\mathcal{G}$ can only be used to provide an estimate of X. Such an estimate is the expectation of X conditioned on information in $\mathcal{G}$ and forms the basis of the following definition, which generalizes the concept of a conditional expectation.

Definition 6.12. A random variable Y is called the *conditional expectation* of X given $\mathcal{G}$, denoted by $Y = E[X \mid \mathcal{G}]$, if

(i) Y is a $\mathcal{G}$-measurable random variable, i.e., $\sigma(Y) \subset \mathcal{G}$;

(ii) $E[X \, \mathbb{I}_B] = E[Y \, \mathbb{I}_B]$ for every event $B \in \mathcal{G}$.

We remark that property (i) tells us that $E[X \mid \mathcal{G}]$ is itself a *random variable that is constant on every atom of a partition that generates* $\mathcal{G}$. Property (ii) is sometimes referred to as the *partial averaging property*. That is, the expected value of X restricted to any given event B in $\mathcal{G}$ is the same as the expected value of $E[X \mid \mathcal{G}]$ restricted to the same event B in $\mathcal{G}$. Note that for $B = \Omega$, property (ii) implies the nested expectation identity $E[X] = E[E[X \mid \mathcal{G}]]$, i.e., the expected value of X is the same as the expected value of X that has been conditioned on any information set $\mathcal{G} \subset \mathcal{F}$.

Three simple examples of conditional expectations with respect to a sub-σ-algebra are as follows.

(a) If $\mathcal{G} = \mathcal{F}_0 \equiv \{\emptyset, \Omega\}$, then $E[X \mid \mathcal{G}] = E[X]$.

(b) For a constant random variable $X \equiv C$, $E[C \mid \mathcal{G}] = C$.

(c) If X is $\mathcal{G}$-measurable, then $E[X \mid \mathcal{G}] = X$.

Property (a) states that computing an expectation by conditioning on the trivial σ-algebra $\mathcal{F}_0$ is the same as taking the unconditional expectation, i.e., conditioning on no information.

Property (b) states that computing an expectation of a constant by conditioning on any σ-algebra is simply equal to the constant. Property (c) states that we can "pull out" the random variable X when computing its expectation conditional on a σ-algebra for which X is measurable.

To show that a given random variable Y is indeed $E[X \mid \mathcal{G}]$ we need to *verify that properties (i) and (ii) above hold for the given $\mathcal{G}$.* For (a), we let $Y = E[X]$ and $\mathcal{G} = \mathcal{F}_0$. Since Y is a constant, then (i) holds since $\sigma(Y) = \mathcal{F}_0$ implies that Y is $\mathcal{F}_0$-measurable. Property (ii) is now verified for every $B \in \mathcal{F}_0$, i.e., for $B = \emptyset$ and for $B = \Omega$. Since $\mathbb{I}_\emptyset = 0$, then $B = \emptyset$ gives $E[X \mathbb{I}_\emptyset] = 0 = E[Y \mathbb{I}_\emptyset]$. Since $\mathbb{I}_\Omega = 1$, then $B = \Omega$ gives $E[X \mathbb{I}_\Omega] = E[X] = Y = E[Y \mathbb{I}_\Omega]$. For (b), we let $Y = C$. Hence, property (i) holds since $\sigma(Y) = \sigma(Y^{-1}(C)) = \sigma(\Omega) = \mathcal{F}_0 \subset \mathcal{G}$ for any $\mathcal{G}$. Property (ii) holds in the obvious manner since $Y = X = C$. For case (c), we let $Y = X$. So properties (i) and (ii) follow automatically since X is $\mathcal{G}$-measurable implies Y is $\mathcal{G}$-measurable and the expectations in (ii) are equivalent.

Example 6.13. Consider a random variable X on $(\Omega, \mathcal{F}, \mathbb{P})$. Suppose that $\mathcal{G} = \sigma(A) \equiv \{\emptyset, A, A^{\mathsf{C}}, \Omega\}$ with nonempty $A \in \mathcal{F}$. Show that

$$E[X \mid \mathcal{G}](\omega) = E[X \mid A] \mathbb{I}_A(\omega) + E[X \mid A^{\mathsf{C}}] \mathbb{I}_{A^{\mathsf{C}}}(\omega) = \begin{cases} E[X \mid A] & \text{if } \omega \in A, \\ E[X \mid A^{\mathsf{C}}] & \text{if } \omega \in A^{\mathsf{C}}. \end{cases}$$

Solution. Let $Y = E[X \mid A] \mathbb{I}_A + E[X \mid A^{\mathsf{C}}] \mathbb{I}_{A^{\mathsf{C}}}$. Note that $E[X \mid A]$ and $E[X \mid A^{\mathsf{C}}]$ are two constants. Hence, if $E[X \mid A] \neq E[X \mid A^{\mathsf{C}}]$ then $\sigma(Y) = \sigma(\mathbb{I}_A) = \sigma(\mathbb{I}_{A^{\mathsf{C}}}) = \sigma(A) = \mathcal{G}$; otherwise Y is constant (since $\mathbb{I}_A + \mathbb{I}_{A^{\mathsf{C}}} = 1$) and $\sigma(Y) = \mathcal{F}_0$. In either case, Y is $\mathcal{G}$-measurable which verifies property (i). Here we verify property (ii) for every $B \in \mathcal{G}$. [Note, however, that it suffices to verify property (ii) for only $B = A$ and $B = A^{\mathsf{C}}$ as it then follows for $B = \emptyset$ and $B = \Omega$ by using identities: $\mathbb{I}_A + \mathbb{I}_{A^{\mathsf{C}}} = 1$, $\mathbb{I}_A \mathbb{I}_A = \mathbb{I}_A$ and $\mathbb{I}_A \mathbb{I}_{A^{\mathsf{C}}} = \mathbb{I}_\emptyset$.]

$B = \emptyset$: $E[Y \mathbb{I}_\emptyset] = 0 = E[X \mathbb{I}_\emptyset]$.

$B = \Omega$: Using $E[\mathbb{I}_A] = \mathbb{P}(A)$ and applying the total probability law in (6.20) gives

$$E[Y \mathbb{I}_\Omega] = E[Y] = E[X \mid A] \cdot \mathbb{P}(A) + E[X \mid A^{\mathsf{C}}] \cdot \mathbb{P}(A^{\mathsf{C}}) = E[X] = E[X \mathbb{I}_\Omega].$$

$B = A(\text{or } A^{\mathsf{C}})$: Since $Y \mathbb{I}_B = E[X \mid B] \mathbb{I}_B$, we have

$$E[Y \mathbb{I}_B] = E[X \mid B] \cdot \mathbb{P}(B) = E[X \mathbb{I}_B].$$

Hence, $E[X \mid \mathcal{G}] = E[X \mid A] \mathbb{I}_A + E[X \mid A^{\mathsf{C}}] \mathbb{I}_{A^{\mathsf{C}}}$. $\square$

Example 6.14. Consider a three-period binomial model. Compute $E[U_3 \mid \sigma(\omega_1)]$.

Solution. The σ-algebra $\sigma(\omega_1)$ is of the form considered in Example 6.13:

$$\sigma(\omega_1) = \{\emptyset, A_{\mathsf{D}}, A_{\mathsf{U}}, \Omega\}.$$

Therefore, $E[U_3 \mid \sigma(\omega_1)] = E[U_3 \mid \{\omega_1 = \mathsf{U}\}] \mathbb{I}_{A_{\mathsf{U}}} + E[U_3 \mid \{\omega_1 = \mathsf{D}\}] \mathbb{I}_{A_{\mathsf{D}}}$. We calculate the two expectations conditional on the respective events of the up move and the down move:

$$E[U_3 \mid \{\omega_1 = U\}] = \frac{E[U_3 \, \mathbb{I}_{\{\omega_1 = U\}}]}{\mathbb{P}(\omega_1 = U)}$$

$$= \frac{1}{p} \left\{ U_3(UDD)p(1-p)^2 + [U_3(UDU) + U_3(UUD)]p^2(1-p) + U_3(UUU)p^3 \right\}$$

$$= (1-p)^2 + 4p(1-p) + 3p^2 = 1 + 2p$$

$$E[U_3 \mid \{\omega_1 = D\}] = \frac{E[U_3 \, \mathbb{I}_{\{\omega_1 = D\}}]}{\mathbb{P}(\omega_1 = D)}$$

$$= \frac{1}{(1-p)} \left\{ U_3(DDD)(1-p)^3 + [U_3(DDU) + U_3(DUD)]p(1-p)^2 + U_3(DUU)(1-p)p^2 \right\}$$

$$= 2p(1-p) + 2p^2 = 2p.$$

Thus, $E[U_3 \mid \sigma(\omega_1)] = (1 + 2p) \, \mathbb{I}_{A_U} + 2p \, \mathbb{I}_{A_D}$. Note that we have the nested property:

$$E[E[U_3 \mid \sigma(\omega_1)]] = (1 + 2p) \, \mathbb{P}(A_U) + 2p \, \mathbb{P}(A_D) = (1 + 2p)p + 2p(1-p) = 3p = E[U_3]. \quad \square$$

The above definition of $E[X \mid \mathcal{G}]$ seems to be formal and nonconstructive. However, as demonstrated in Example 6.13, where $\mathcal{G}$ is the σ-algebra generated by a simple partition of any (nonempty) one event $A \in \mathcal{F}$, and its complement, it is possible to explicitly construct the random variable $E[X \mid \mathcal{G}]$ for some special cases of $\mathcal{G}$. Below we provide such an explicit construction of the conditional expectation for any σ-algebra generated by any *countable partition* of the sample space. The theorem is valid for any type of random variable (discrete or continuous).

Theorem 6.6. *Let X be a random variable on $(\Omega, \mathcal{F}, \mathbb{P})$ and $\mathcal{P} = \{A_i\}_{i \geqslant 1}$ be a countable partition of Ω with $\mathcal{G} = \sigma(\mathcal{P}) \subset \mathcal{F}$. Then,*

$$E[X \mid \mathcal{G}] = \sum_{i \geqslant 1} E[X \mid A_i] \, \mathbb{I}_{A_i}. \tag{6.21}$$

Proof. Let Y equal the right-hand side of (6.21). Note that Y has the form of a simple function, $Y = \sum_i y_i \, \mathbb{I}_{A_i}$, with constants $y_i = E[X \mid A_i]$. Hence, if all y_i's are distinct, the σ-algebra generated by Y is $\sigma(Y) = \sigma(\{Y^{-1}(y_i)\}_{i \geqslant 1}) = \sigma(\{A_i\}_{i \geqslant 1}) = \sigma(\mathcal{P}) = \mathcal{G}$ and Y is $\mathcal{G}$-measurable. Otherwise, if not all y_i's are distinct then some of the subsets $Y^{-1}(y_i)$ will be unions of atoms in $\mathcal{P}$, i.e., the partition $\{Y^{-1}(y_i)\}_{i \geqslant 1} \preceq \mathcal{P}$. In this case, $\sigma(Y)$ will be a proper subset of $\sigma(\mathcal{P})$ and we still have that Y is $\mathcal{G}$-measurable. Take any $B \in \mathcal{G}$. By definition, any element of $\mathcal{G} = \sigma(\mathcal{P})$ is a countable union of atoms:

$$B = \bigcup_{j \geqslant 1} A_{i_j} \text{ for some } i_1 < i_2 < \cdots$$

Since all atoms are mutually disjoint, the indicator of B is a sum of indicators of A_{i_j}:

$\mathbb{I}_B = \sum_{j \geqslant 1} \mathbb{I}_{A_{i_j}}$. It follows that $\mathrm{E}[Y \mathbb{I}_B] = \mathrm{E}[X \mathbb{I}_B]$:

$$
\begin{aligned}
\mathrm{E}[Y \mathbb{I}_B] &= \sum_{j \geqslant 1} \mathrm{E}\left[Y \mathbb{I}_{A_{i_j}}\right] \\
&= \sum_{j \geqslant 1} \mathrm{E}\left[\mathrm{E}[X \mid A_{i_j}] \cdot \mathbb{I}_{A_{i_j}}\right] = \sum_{j \geqslant 1} \mathrm{E}[X \mid A_{i_j}] \cdot \mathrm{E}[\mathbb{I}_{A_{i_j}}] \\
&= \sum_{j \geqslant 1} \mathrm{E}[X \mid A_{i_j}] \cdot \mathbb{P}(A_{i_j}) \\
&= \sum_{j \geqslant 1} \mathrm{E}[X \mathbb{I}_{A_{i_j}}] = \mathrm{E}\left[X \cdot \sum_{j \geqslant 1} \mathbb{I}_{A_{i_j}}\right] = \mathrm{E}[X \mathbb{I}_B].
\end{aligned}
$$

Here, we used the property that $Y \mathbb{I}_{A_k} = \mathrm{E}[X \mid A_k] \cdot \mathbb{I}_{A_k}$ for all $A_k \in \mathcal{P}$. $\square$

6.3.2.3 Conditioning on a Random Variable

Let X and Y be two random variables on the same probability space $(\Omega, \mathcal{F}, \mathbb{P})$. Formally, the conditional expectation of X given Y is equal to the conditional expectation of X given the σ-algebra generated by Y:

$$
\mathrm{E}[X \mid Y] := \mathrm{E}[X \mid \sigma(Y)].
$$

Equivalently, for every $\omega \in \Omega$,

$$
\mathrm{E}[X \mid Y](\omega) := \mathrm{E}[X \mid Y = Y(\omega)].
$$

Note that in this case the sub-σ-algebra, $\mathcal{G} = \sigma(Y) \subset \mathcal{F}$, contains all the information about Y only. The random variable $\mathrm{E}[X \mid Y]$ is $\sigma(Y)$-measurable. If the joint distribution of the pair (X, Y) is known, then one can obtain the conditional distribution of X given the value of Y. Then $\mathrm{E}[X \mid Y = y]$ is the expected value of X relative to the conditional distribution of X given $Y = y$. In any standard text on probability theory, the reader can find formulas of such conditional expectations for the cases with discrete or continuous random variables. For a discrete random variable, $Y \in Y(\Omega) \equiv \{y_k; k \geqslant 1\}$, the atoms $A_k := \{Y = y_k\} \equiv \{\omega \in \Omega : Y(\omega) = y_k\}$, $k \geqslant 1$, generate $\sigma(Y) = \sigma(\{A_k\}_{k \geqslant 1})$. Applying (6.21) gives the representation

$$
\mathrm{E}[X \mid Y] = \sum_{k \geqslant 1} \mathrm{E}[X \mid Y = y_k] \mathbb{I}_{\{Y = y_k\}}. \tag{6.22}
$$

That is, for every outcome $\omega \in \{Y = y_k\}$ the random variable $\mathrm{E}[X \mid Y]$ takes on a constant value given by the conditional expectation $\mathrm{E}[X \mid Y = y_k]$. Hence, given that Y has range $\{y_k; k \geqslant 1\}$ then $\mathrm{E}[X \mid Y]$ has range given by the set of values $\{\mathrm{E}[X \mid Y = y_k]; k \geqslant 1\}$.

Example 6.15. Consider the four-period binomial model. Define two random variables

$$
X(\omega) \equiv \text{\# of U's before the first D in } \omega \in \Omega_4 \text{ and } Y \equiv \mathsf{U}_4.
$$

1. Find $\mathrm{E}[X \mid Y]$.

2. Calculate $\mathbb{P}(\mathrm{E}[X \mid Y] \leqslant 2)$.

Solution. $\mathsf{U}_4 \in \{0, 1, 2, 3, 4\}$ and we apply (6.22) by finding the conditional expectations $\mathrm{E}[X \mid Y = k]$ for $0 \leqslant k \leqslant 4$:

$$
\mathrm{E}[X \mid Y = k] = \frac{1}{\mathbb{P}(Y = k)} \sum_{\omega : Y(\omega) = k} X(\omega) \mathbb{P}(\omega).
$$

$k = 0$: Since X is zero on the set $\{Y = 0\} = \{\text{DDDD}\}$, we have $\mathrm{E}[X \mid Y = 0] = 0$.

$k = 1$: The probability $\mathbb{P}(Y = 1) = \mathbb{P}(\{\text{UDDD}, \text{DUDD}, \text{DDUD}, \text{DDDU}\}) = 4p(1-p)^3$. Note that $X(\text{UDDD}) = 1$ and $X(\omega)$ is zero for other $\omega \in \{Y = 1\}$. Therefore,

$$\mathrm{E}[X \mid Y = 1] = \frac{1}{4p(1-p)^3} X(\text{UDDD}) \, \mathbb{P}(\text{UDDD}) = \frac{1}{4}.$$

$k = 2$: By listing all $\omega \in \Omega_4$ for which $Y(\omega) = 2$ and then calculating $X(\omega)$:

$$\mathrm{E}[X \mid Y = 2] = \frac{1}{\mathbb{P}(Y = 2)} \sum_{\omega \,:\, Y(\omega) = 2} X(\omega) \, \mathbb{P}(\omega) = \frac{1}{6p^2(1-p)^2} 4p^2(1-p)^2 = \frac{2}{3}.$$

$k = 3$: Similarly, we obtain $\mathrm{E}[X \mid Y = 3] = \frac{3}{2}$.

$k = 4$: Since $\{Y = 4\} = \{\text{UUUU}\}$ and $X(\text{UUUU}) = 4$, we have $\mathrm{E}[X \mid Y = 4] = 4$.

In summary,

$$E[X \mid Y](\omega) = \begin{cases} 0 & \text{if } Y(\omega) = 0 \\ \frac{1}{4} & \text{if } Y(\omega) = 1 \\ \frac{2}{3} & \text{if } Y(\omega) = 2 \\ \frac{3}{2} & \text{if } Y(\omega) = 3 \\ 4 & \text{if } Y(\omega) = 4 \end{cases}$$

This gives the random variable $E[X \mid Y] = \frac{1}{4}\mathbb{I}_{\{Y=1\}} + \frac{2}{3}\mathbb{I}_{\{Y=2\}} + \frac{3}{2}\mathbb{I}_{\{Y=3\}} + 4\,\mathbb{I}_{\{Y=4\}}$. Finally, we can compute the probability:

$$\mathbb{P}(\mathrm{E}[X \mid Y] \leqslant 2) = 1 - \mathbb{P}(\mathrm{E}[X \mid Y] > 2) = 1 - \mathbb{P}(\mathrm{E}[X \mid Y] = 4) = 1 - \mathbb{P}(Y = 4) = 1 - p^4. \quad \square$$

6.3.3 Properties of Conditional Expectations

Consider a probability space $(\Omega, \mathcal{F}, \mathbb{P})$. Suppose that X, X_1, and X_2 are $\mathcal{F}$-measurable random variables on Ω, and let $\mathcal{G}$ be a sub-σ-algebra of $\mathcal{F}$. We now derive some important identities that are satisfied by expectations conditional on such σ-algebras. Although the properties below are valid in more general cases, for simplicity of proofs we assume that $\mathcal{G}$ is generated by some countable partition. In particular, we will simply assume a finite number of atoms in the partition $\mathcal{P} = \{A_1, A_2, \ldots, A_K\}$ of the sample space Ω, i.e., $\mathcal{G} = \sigma(\mathcal{P})$.

6.3.3.1 Linearity

For any real c_1 and c_2,

$$\mathrm{E}[c_1 X_1 + c_2 X_2 \mid \mathcal{G}] = c_1 \mathrm{E}[X_1 \mid \mathcal{G}] + c_2 \mathrm{E}[X_2 \mid \mathcal{G}]. \tag{6.23}$$

Proof. By using the linearity of the expectation conditional on any event and by (6.21):

$$\mathrm{E}[c_1 X_1 + c_2 X_2 \mid \mathcal{G}] = \sum_{k=1}^{K} \mathrm{E}[c_1 X_1 + c_2 X_2 \mid A_k] \, \mathbb{I}_{A_k}$$

$$= c_1 \sum_{k=1}^{K} \mathrm{E}[X_1 \mid A_k] \, \mathbb{I}_{A_k} + c_2 \sum_{k=1}^{K} \mathrm{E}[X_2 \mid A_k] \, \mathbb{I}_{A_k}$$

$$= c_1 \mathrm{E}[X_1 \mid \mathcal{G}] + c_2 \mathrm{E}[X_2 \mid \mathcal{G}]. \qquad \square$$

6.3.3.2 Independence

First, let us define what it means for a pair of random variables and a pair of σ-algebras to be independent w.r.t. a given probability measure $\mathbb{P}$.

Definition 6.13. (Pairwise independence)

1. Two random variables X_1 and X_2 are said to be independent w.r.t. $\mathbb{P}$ if

$$\mathbb{P}(X_1 \in B_1, X_2 \in B_2) = \mathbb{P}(X_1 \in B_1)\,\mathbb{P}(X_2 \in B_2)$$

   for all $B_1, B_2 \in \mathcal{B}(\mathbb{R})$.

2. Two σ-algebras $\mathcal{G}_1, \mathcal{G}_2 \subset \mathcal{F}$ are said to be independent w.r.t. $\mathbb{P}$ if $\mathbb{P}(A_1 \cap A_2) = \mathbb{P}(A_1)\mathbb{P}(A_2)$ for all $A_1 \in \mathcal{G}_1$ and $A_2 \in \mathcal{G}_2$.

3. A random variable X and σ-algebra $\mathcal{G}$ are said to be independent w.r.t. $\mathbb{P}$ if one of the two equivalent conditions holds:

   (i) for every $A \in \mathcal{G}$, $\mathbb{I}_A$ and X are independent random variables w.r.t. $\mathbb{P}$;

   (ii) $\sigma(X)$ and $\mathcal{G}$ are independent σ-algebras w.r.t. $\mathbb{P}$.

As is seen, the independence between X and $\mathcal{G}$ means that we do not gain any information about X if we know $\mathcal{G}$ and vice versa.

Suppose that a random variable X and a σ-algebra $\mathcal{G}$ are independent. Then, the conditional expectation of X given $\mathcal{G}$ is equal to the unconditional expectation of X:

$$\mathrm{E}[X \mid \mathcal{G}] = \mathrm{E}[X]. \tag{6.24}$$

Proof. For every $A \in \mathcal{G}$, X and $\mathbb{I}_A$ are independent. Therefore,

$$\mathrm{E}[X\,\mathbb{I}_A] = \mathrm{E}[X]\,\mathrm{E}[\mathbb{I}_A] = \mathrm{E}[\mathrm{E}[X]\,\mathbb{I}_A].$$

Clearly, $\mathrm{E}[X]$ is $\mathcal{G}$-measurable since it is a constant. By definition of $\mathrm{E}[X \mid \mathcal{G}]$, the assertion is proved. $\qquad\square$

Since any random variable X is independent of the trivial σ-algebra $\mathcal{F}_0 = \{\emptyset, \Omega\}$, we have that $\mathrm{E}[X \mid \mathcal{F}_0] = \mathrm{E}[X]$. Let X_1 and X_2 be independent random variables. Then,

$$\mathrm{E}[X_1 \mid X_2] \equiv \mathrm{E}[X_1 \mid \sigma(X_2)] = \mathrm{E}[X_1] \text{ and } \mathrm{E}[X_2 \mid X_1] = \mathrm{E}[X_2].$$

For independent random variables, we then also have $\mathrm{E}[X_1 X_2] = \mathrm{E}[X_1]\mathrm{E}[X_2]$.

The above definition of independence extends naturally to an arbitrary (countable or uncountable) collection of random variables and σ-algebras. We present the definition for a countable collection of random variables and σ-algebras, as follows. Note that we simply generalize parts 1 and 2 in the above definition.

Definition 6.14 (Independence for a collection). Let n be a positive integer and consider a collection of random variables $X_1, X_2, \ldots, X_n$ defined on the probability space $(\Omega, \mathcal{F}, \mathbb{P})$ and a collection of sub-σ-algebras $\mathcal{G}_1, \mathcal{G}_2, \ldots, \mathcal{G}_n \subset \mathcal{F}$.

1. $X_1, X_2, \ldots, X_n$ are said to be (mutually) independent w.r.t. $\mathbb{P}$ if, for every subcollection of indices $1 \leqslant i_1 < i_2 < \cdots < i_k \leqslant n$, $1 \leqslant k \leqslant n$,

$$\mathbb{P}(X_{i_1} \in B_1, X_{i_2} \in B_2, \ldots, X_{i_k} \in B_k) = \mathbb{P}(X_{i_1} \in B_1) \cdot \mathbb{P}(X_{i_2} \in B_2) \cdots \mathbb{P}(X_{i_k} \in B_k)$$

   for all $B_1, B_2, \ldots, B_k \in \mathcal{B}(\mathbb{R})$.

2. $\mathcal{G}_1, \mathcal{G}_2, \ldots, \mathcal{G}_n$ are said to be (mutually) independent w.r.t. $\mathbb{P}$ if, for every subcollection of indices $1 \leqslant i_1 < i_2 < \cdots < i_k \leqslant n$, $1 \leqslant k \leqslant n$,

$$\mathbb{P}(A_{i_1} \cap A_{i_2} \cap \cdots \cap A_{i_k}) = \mathbb{P}(A_{i_1}) \cdot \mathbb{P}(A_{i_2}) \cdots \mathbb{P}(A_{i_k})$$

for all $A_{i_1} \in \mathcal{G}_{i_1}, A_{i_2} \in \mathcal{G}_{i_2}, \ldots, A_{i_k} \in \mathcal{G}_{i_k}$.

A countable sequence of random variables $X_1, X_2, \ldots$ and sub-σ-algebras $\mathcal{G}_1, \mathcal{G}_2, \ldots$, are (respectively) independent w.r.t. $\mathbb{P}$ if properties 1 and 2 hold for all $n \geqslant 1$.

We note that it suffices to take the Borel sets in property 1 to be of the form $B_i = (-\infty, x_i]$, $x_i \in \mathbb{R}$, since these semi-open infinite intervals also generate the Borel σ-algebra $\mathcal{B}(\mathbb{R})$.

6.3.3.3 Taking out What Is Known

Suppose that Y is $\mathcal{G}$-measurable, then in the conditional expectation $\mathrm{E}[XY \mid \mathcal{G}]$, the variable Y can be taken out of the expectation:

$$\mathrm{E}[XY \mid \mathcal{G}] = Y \, \mathrm{E}[X \mid \mathcal{G}]. \tag{6.25}$$

In particular, letting $X \equiv 1$ in (6.25) we obtain

$$\mathrm{E}[Y \mid \mathcal{G}] = Y \, \mathrm{E}[1 \mid \mathcal{G}] = Y.$$

Proof. To prove (6.25), let $\mathcal{G} = \sigma(\mathcal{P})$. Since Y is $\mathcal{G}$-measurable it is constant on every atom $A \in \mathcal{P}$. Thus, by (6.21) we have

$$\mathrm{E}[XY \mid \mathcal{G}] = \sum_{A \in \mathcal{P}} \mathrm{E}[XY \mid A]\, \mathbb{I}_A = \sum_{A \in \mathcal{P}} \frac{\mathrm{E}[XY\, \mathbb{I}_A]}{\mathbb{P}(A)}\, \mathbb{I}_A$$

$$= \sum_{A \in \mathcal{P}} Y \frac{\mathrm{E}[X\, \mathbb{I}_A]}{\mathbb{P}(A)}\, \mathbb{I}_A = Y \sum_{A \in \mathcal{P}} \mathrm{E}[X \mid A]\, \mathbb{I}_A = Y \, \mathrm{E}[X \mid \mathcal{G}]. \qquad \square$$

For an alternate proof of this result see Exercise (6.18). This result is readily extended as follows. Suppose that the σ-algebra $\mathcal{G}$ in (6.25) is generated by another random variable Z, and that Y is a function of Z: $\mathcal{G} = \sigma(Z)$ and $Y = f(Z)$. Clearly, Y is $\mathcal{G}$-measurable since $Y = f(Z)$ implies $\sigma(Y) \subset \sigma(Z)$. Therefore, we recover a familiar result

$$\mathrm{E}[X\, f(Z) \mid \sigma(Z)] \equiv \mathrm{E}[X\, f(Z) \mid Z] = f(Z)\, \mathrm{E}[X \mid Z].$$

Consider now the case of a function of two or more random variables in which some of the variables are $\mathcal{G}$-measurable and the rest are independent of $\mathcal{G}$. Then, we have a general version of "taking out what is known" whereby the expectation of the function, conditional on $\mathcal{G}$, is turned into an unconditional expectation. We will see later that such a property turns out to be quite useful. We now state and prove this result in the case of two random variables as follows. Note that here we give the proof for the case that the conditioning σ-algebra, $\mathcal{G}$, is generated by a countable partition of Ω, with discrete random variables. However, this result holds in the more general case of any type of random variable and (countable or uncountable) σ-algebra $\mathcal{G}$.

Proposition 6.7 (Independence Proposition for Two Random Variables). *Let X and Y be random variables on $(\Omega, \mathcal{F}, \mathbb{P})$. Suppose that X is $\mathcal{G}$-measurable and that Y is independent of the σ-algebra $\mathcal{G} \subset \mathcal{F}$. Let $h : \mathbb{R}^2 \to \mathbb{R}$ be a Borel function. Then*

$$\mathrm{E}[h(X, Y) \mid \mathcal{G}] = g(X),$$

with function $g : \mathbb{R} \to \mathbb{R}$ given by the unconditional expectation $g(x) := \mathrm{E}[h(x, Y)]$.

Proof. Let $\mathcal{G} = \sigma(\mathcal{P})$, $\mathcal{P} = \{A_i\}_{i \geqslant 1}$. Then, by (6.21) of Theorem 6.6:

$$\mathrm{E}[h(X,Y) \mid \mathcal{G}] = \sum_{i \geqslant 1} \mathrm{E}[h(X,Y) \mid A_i] \, \mathbb{I}_{A_i} = \sum_{i \geqslant 1} \frac{\mathrm{E}[h(X,Y) \, \mathbb{I}_{A_i}]}{\mathbb{P}(A_i)} \, \mathbb{I}_{A_i}.$$

Any given outcome $\omega \in \Omega$ must be in only one atom, say $\omega \in A_k$ for some value $k \geqslant 1$. Since X is $\mathcal{G}$-measurable, it is constant on a given atom of $\mathcal{G}$, i.e., say $X(\omega) = x_k$ and hence $h(X,Y)\mathbb{I}_{A_k} = h(x_k,Y)\mathbb{I}_{A_k}$. Moreover, the atoms are mutually exclusive so that $\mathbb{I}_{A_i}(\omega) = \delta_{ik}$. Note that Y and $\mathbb{I}_{A_k}$ are independent and hence $h(x_k,Y)$ and $\mathbb{I}_{A_k}$ are independent random variables. By combining these facts, the above random variable $\mathrm{E}[h(X,Y) \mid \mathcal{G}]$ evaluated on ω is

$$\mathrm{E}[h(X,Y) \mid \mathcal{G}](\omega) = \frac{\mathrm{E}[h(x_k,Y) \, \mathbb{I}_{A_k}]}{\mathbb{P}(A_k)} = \frac{\mathrm{E}[h(x_k,Y)] \, \mathrm{E}[\mathbb{I}_{A_k}]}{\mathbb{P}(A_k)} = \mathrm{E}[h(x_k,Y)].$$

By definition of g, the last expression is $g(x_k) = g(X(\omega)) = g(X)(\omega)$. Since the above argument holds for arbitrary ω, we have $\mathrm{E}[h(X,Y) \mid \mathcal{G}] = g(X)$.   □

We now remark on the way this result is used in practice and its essence. We begin by fixing the random variable X to some ordinary variable (parameter) x in $h(X,Y)$ and compute the function $g(x)$ as the unconditional expectation, $g(x) = \mathrm{E}[h(x,Y)]$. After having obtained $g(x)$, we put back the random variable X in place of x, giving the random variable $g(X)$, which is the same as $\mathrm{E}[h(X,Y) \mid \mathcal{G}]$. That is, conditioning on the information in $\mathcal{G}$ allows us to hold X as constant (as a given value x) and then the conditional expectation of $h(x,Y)$ given the information in $\mathcal{G}$ becomes an unconditional expectation since Y, and hence $h(x,Y)$, is independent of $\mathcal{G}$. Note that when $\mathcal{G} = \sigma(X)$ we obtain the well-known formula used for computing the expectation of a function of two random variables, say $h(X,Y)$, conditional on one of the random variables, say X, where Y is independent of X (i.e., Y is independent of $\sigma(X)$):

$$\mathrm{E}[h(X,Y) \mid \sigma(X)] \equiv \mathrm{E}[h(X,Y) \mid X] = g(X)$$

where $g(x) = \mathrm{E}[h(X,Y) \mid X](x) \equiv \mathrm{E}[h(X,Y) \mid X = x] = \mathrm{E}[h(x,Y)]$.

Proposition 6.7 extends to the multidimensional case in the obvious manner. Since this is an important result that is further used, we state it in a proposition as follows. A similar proof as given above for Proposition 6.7 in the case of $\mathcal{G}$ being generated by a countable partition, with discrete random variables, is left as an exercise for the reader. Theorem 6.8 is in fact valid for all types of random variables and σ-algebras.

Proposition 6.8 (Independence Proposition for Several Random Variables). *Consider two random vectors* $\mathbf{X} = (X_1, \ldots, X_m)$ *and* $\mathbf{Y} = (Y_1, \ldots, Y_n)$, *where all components of* $\mathbf{X}$ *are assumed* $\mathcal{G}$-measurable *and all components of* $\mathbf{Y}$ *are assumed independent of* $\mathcal{G}$. *Let* $h : \mathbb{R}^{m+n} \to \mathbb{R}$ *be a Borel function. Then,* $\mathrm{E}[h(\mathbf{X}, \mathbf{Y}) \mid \mathcal{G}] = g(\mathbf{X})$ *where* $g : \mathbb{R}^m \to \mathbb{R}$ *is defined by the unconditional expectation* $g(\mathbf{x}) := \mathrm{E}[h(\mathbf{x}, \mathbf{Y})]$, *i.e.,*

$$\mathrm{E}[h(X_1, \ldots, X_m, Y_1, \ldots, Y_n) \mid \mathcal{G}] = g(X_1, \ldots, X_m)$$

where $g(x_1, \ldots, x_m) := \mathrm{E}[h(x_1, \ldots, x_m, Y_1, \ldots, Y_n)]$.

As in the case of two variables, this result is applied by assigning ordinary variables $X_1 = x_1, \ldots, X_m = x_m$ in $h(X_1, \ldots, X_m, Y_1, \ldots, Y_n)$ and computing the function g as the unconditional expectation of the random variable $h(x_1, \ldots, x_m, Y_1, \ldots, Y_n)$ which is a function of random variables $Y_1, \ldots, Y_n$ with $x_1, \ldots, x_m$ as parameters assigned to the first m arguments of h. Then, $\mathrm{E}[h(\mathbf{X}, \mathbf{Y}) \mid \mathcal{G}]$ is the random variable $g(X_1, \ldots, X_m)$. An important case is when $\mathcal{G} = \sigma(\mathbf{X}) \equiv \sigma(X_1, \dots, X_m)$, i.e., conditioning on a random vector $\mathbf{X}$ which is independent of random vector $\mathbf{Y}$. We obtain the known formula for $\mathrm{E}[h(\mathbf{X}, \mathbf{Y}) \mid \sigma(\mathbf{X})] \equiv \mathrm{E}[h(\mathbf{X}, \mathbf{Y}) \mid \mathbf{X}]$:

$$\mathrm{E}[h(\mathbf{X}, \mathbf{Y}) \mid \mathbf{X}] = g(\mathbf{X}) \quad \text{where} \quad g(\mathbf{x}) = \mathrm{E}[h(\mathbf{X}, \mathbf{Y}) \mid \mathbf{X} = \mathbf{x}] = \mathrm{E}[h(\mathbf{x}, \mathbf{Y})].$$

6.3.3.4 Tower Property (Iterated Conditioning)

Let $\mathcal{H}$ be a sub-σ-algebra of $\mathcal{G}$, i.e., assume $\mathcal{H} \subset \mathcal{G} \subset \mathcal{F}$. Then,

$$\mathrm{E}[\mathrm{E}[X \mid \mathcal{G}] \mid \mathcal{H}] = \mathrm{E}[X \mid \mathcal{H}]. \tag{6.26}$$

This very useful property is proven in an elegant manner as follows, allowing the random variables to be discrete or continuous, and the σ-algebras $\mathcal{G}$ and $\mathcal{H}$ to be countable or uncountable.

Proof. Denote $Y = \mathrm{E}[X \mid \mathcal{G}]$ and $Z = \mathrm{E}[X \mid \mathcal{H}]$. By the definition of an expectation conditional on the respective σ-algebras $\mathcal{G}$ and $\mathcal{H}$,

$$\forall A \in \mathcal{G}: \quad \mathrm{E}[X \, \mathbb{I}_A] = \mathrm{E}[Y \, \mathbb{I}_A],$$
$$\forall A \in \mathcal{H}: \quad \mathrm{E}[X \, \mathbb{I}_A] = \mathrm{E}[Z \, \mathbb{I}_A].$$

Since $\mathcal{H} \subset \mathcal{G}$, $A \in \mathcal{H}$ implies $A \in \mathcal{G}$ and we hence have

$$\forall A \in \mathcal{H}: \quad \mathrm{E}[Y \, \mathbb{I}_A] = \mathrm{E}[X \, \mathbb{I}_A] = \mathrm{E}[Z \, \mathbb{I}_A].$$

Since Z is $\mathcal{H}$-measurable, then Z is $\mathrm{E}[Y \mid \mathcal{H}]$ and hence

$$\mathrm{E}[X \mid \mathcal{H}] = \mathrm{E}[\mathrm{E}[X \mid \mathcal{G}] \mid \mathcal{H}]. \qquad \square$$

By taking $\mathcal{H} = \mathcal{F}_0 = \{\emptyset, \Omega\}$, we recover the *law of double expectation*:

$$\mathrm{E}[\mathrm{E}[X \mid \mathcal{G}]] = \mathrm{E}[\mathrm{E}[X \mid \mathcal{G}] \mid \mathcal{F}_0] = \mathrm{E}[X \mid \mathcal{F}_0] = \mathrm{E}[X]. \tag{6.27}$$

For two random variables X and Y, where the inner expectation involves conditioning on Y (i.e., with $\mathcal{G} = \sigma(Y)$), this law takes the form:

$$\mathrm{E}[\mathrm{E}[X \mid Y]] = \mathrm{E}[X].$$

A more trivial property resembling (6.26) also follows by reversing the order of conditioning of $\mathcal{G}$ with $\mathcal{H}$. Note that, by definition, $\mathrm{E}[X \mid \mathcal{H}]$ is $\mathcal{H}$-measurable. Since $\mathcal{H} \subset \mathcal{G}$, then $\mathrm{E}[X \mid \mathcal{H}]$ is a $\mathcal{G}$-measurable random variable. Hence this corresponds to case (c) just after Definition 6.12 where $\mathrm{E}[X \mid \mathcal{H}]$ now plays the role of X. Thus, the tower property (6.26) also works if we swap $\mathcal{G}$ and $\mathcal{H}$:

$$\mathrm{E}[\mathrm{E}[X \mid \mathcal{H}] \mid \mathcal{G}] = \mathrm{E}[X \mid \mathcal{H}].$$

6.3.4 Conditioning in the Binomial Model

Now we can apply the above properties of conditional expectations to the *filtered* binomial probability space $(\Omega_N, \mathcal{F}, \mathbb{P}, \mathbb{F})$. Recall that

$$\Omega_N = \{\omega = \omega_1 \omega_2 \dots \omega_N : \omega_i \in \{\mathsf{D}, \mathsf{U}\}, 1 \leqslant i \leqslant N\},$$
$$\mathcal{F} = 2^{\Omega_N},$$
$$\mathbb{F} = \{\mathcal{F}_n = \sigma(\omega_1, \omega_2, \dots, \omega_n)\}_{n=0,1,2,\dots,N}, \quad \text{and} \quad \mathcal{F}_0 = \{\emptyset, \Omega_N\}, \ \mathcal{F}_N \equiv \mathcal{F},$$
$$\mathbb{P}(\omega) = p^{\#\mathsf{U}(\omega)} (1 - p)^{\#\mathsf{D}(\omega)}.$$

It proves very convenient in what follows to use the more compact notation

$$E_n[X] \equiv E[X \mid \mathcal{F}_n],$$

where X is an $\mathcal{F}$-measurable (i.e., $\mathcal{F}_N$-measurable) random variable on Ω_N. For $n = 0$, the conditional expectation reduces to an unconditional one: $E_0[X] = E[X \mid \mathcal{F}_0] = E[X]$. For the case $n = N$, the property in (6.25) gives us $E_N[X] = E[X \mid \mathcal{F}_N] = X$. Note that $E_n[X]$ is an $\mathcal{F}_n$-measurable random variable and $\mathcal{F}_{n-1} \subset \mathcal{F}_n$, for every $n = 1, \ldots, N$. The tower property (6.26) therefore takes the following compact form:

$$E_n[E_m[X]] = E_n[X], \text{ for } 0 \leqslant n \leqslant m \leqslant N. \tag{6.28}$$

Recall that $\mathcal{F}_n = \sigma(\mathcal{P}_n)$ is generated by 2^n atoms $A_{\omega_1^* \ldots \omega_n^*} \equiv \{\omega_1 = \omega_1^*, \ldots, \omega_n = \omega_n^*\}$, where each corresponds to fixing the first n market moves. Making use of (6.21) with $\mathcal{G} = \mathcal{F}_n$, $E_n[X]$ is then a simple random variable:

$$E_n[X] \equiv E[X \mid \mathcal{F}_n] = \sum_{\omega_1^*, \ldots, \omega_n^* \in \{U, D\}} E[X \mid A_{\omega_1^* \ldots \omega_n^*}] \, \mathbb{I}_{A_{\omega_1^* \ldots \omega_n^*}} . \tag{6.29}$$

Hence, for every particular outcome $\omega = \omega^* \equiv \omega_1^* \ldots \omega_N^* \in \Omega_N$, we see that the random variable $E_n[X]$ in (6.29) evaluates to a unique real number, $E_n[X](\omega^*) = E_n[X](A_{\omega_1^* \ldots \omega_n^*})$, which is a function of only the first n moves:

$$E_n[X](\omega^*) = E[X \mid A_{\omega_1^* \ldots \omega_n^*}] . \tag{6.30}$$

This dependence on *only the first n ω's in every outcome* is represented as follows:

$$E_n[X](\omega) = E_n[X](\omega_1 \omega_2 \ldots \omega_n) \equiv E_n[X](A_{\omega_1 \omega_2 \ldots \omega_n})$$

given any outcome $\omega = \omega_1 \ldots \omega_n \omega_{n+1} \ldots \omega_N$. That is, the random variable $E_n[X]$ is constant on each one of the 2^n atoms $A_{\omega_1 \ldots \omega_n}$ where the first n moves are fixed to a sequence $\omega_1, \ldots, \omega_n$. The unconditional expectation of X can be calculated using (6.17):

$$E[X] = \sum_{\omega \in \Omega_N} X(\omega) \mathbb{P}(\omega) \equiv \sum_{\omega_1, \ldots, \omega_N \in \{U, D\}} X(\omega_1 \ldots \omega_N) \mathbb{P}(A_{\omega_1 \ldots \omega_N}) \tag{6.31}$$

where $\mathbb{P}(A_{\omega_1 \ldots \omega_N}) = p^{\#U(\omega_1 \ldots \omega_N)} (1 - p)^{\#D(\omega_1 \ldots \omega_N)}$ From (6.30) and applying (6.19), we have a direct formula for calculating $E_n[X](\omega^*)$ for any given outcome $\omega^* = \omega_1^* \ldots \omega_N^*$:

$$E_n[X](\omega^*) = E_n[X](\omega_1^* \ldots \omega_n^*) = \frac{E[X \, \mathbb{I}_{A_{\omega_1^* \ldots \omega_n^*}}]}{\mathbb{P}(A_{\omega_1^* \ldots \omega_n^*})}. \tag{6.32}$$

By the definition of the unconditional expectation in (6.31), the numerator involves a sum over all $\omega = \omega_1 \ldots \omega_n \omega_{n+1} \ldots \omega_N$, but with indicator function fixing the first n values, i.e., $\mathbb{I}_{A_{\omega_1^* \ldots \omega_n^*}}(\omega) = 1$, if $\omega_1 = \omega_1^*, \ldots, \omega_n = \omega_n^*$ and equals zero otherwise. Using this fact, the expectation sum over $\omega_1, \ldots, \omega_n, \omega_{n+1}, \ldots, \omega_N \in \{D, U\}$ collapses to a sum over only the last $N - n$ market moves $\omega_{n+1}, \ldots, \omega_N \in \{D, U\}$, with the first n fixed to $\omega_1^* \ldots \omega_n^*$:

$$E[X \, \mathbb{I}_{A_{\omega_1^* \ldots \omega_n^*}}] = \sum_{\omega_{n+1}, \ldots, \omega_N \in \{D, U\}} X(\omega_1^* \ldots \omega_n^* \omega_{n+1} \ldots \omega_N) \, \mathbb{P}(A_{\omega_1^* \ldots \omega_n^* \omega_{n+1} \ldots \omega_N}). \tag{6.33}$$

By independence, $\mathbb{P}(A_{\omega_1^* \ldots \omega_n^* \omega_{n+1} \ldots \omega_N}) = \mathbb{P}(A_{\omega_1^* \ldots \omega_n^*}) \cdot \mathbb{P}(\omega_{n+1} \ldots \omega_N)$, with the sequence of moves $\omega_{n+1} \ldots \omega_N$ having probability $\mathbb{P}(\omega_{n+1} \ldots \omega_N) \equiv p^{\#U(\omega_{n+1}\ldots\omega_N)} (1-p)^{\#D(\omega_{n+1}\ldots\omega_N)}$. Hence, $\mathbb{P}(A_{\omega_1^* \ldots \omega_n^*})$ cancels out in (6.32), giving

$$E_n[X](\omega_1^* \ldots \omega_n^*) = \sum_{\omega_{n+1},\ldots,\omega_N \in \{D,U\}} X(\omega_1^* \ldots \omega_n^* \omega_{n+1} \ldots \omega_N) \, \mathbb{P}(\omega_{n+1} \ldots \omega_N). \qquad (6.34)$$

We can also remove the $*$ on all ω's and thereby state that for any $\omega = \omega_1 \ldots \omega_N \in \Omega_N$:

$$E_n[X](\omega_1 \ldots \omega_n) = \sum_{\omega_{n+1},\ldots,\omega_N \in \{D,U\}} X(\omega_1 \ldots \omega_n \omega_{n+1} \ldots \omega_N) \, \mathbb{P}(\omega_{n+1} \ldots \omega_N). \qquad (6.35)$$

It is sometimes the case that X is $\mathcal{F}_m$-measurable, where $0 \leqslant n < m < N$. The random variable X is now constant on the atoms $A_{\omega_1 \ldots \omega_m}$, i.e., $X(\omega_1 \ldots \omega_N) = X(\omega_1 \ldots \omega_m)$ depends only on the first m market moves. Making use of this within (6.35), combined with the fact that $\mathbb{P}(\omega_{n+1} \ldots \omega_N) = \mathbb{P}(\omega_{n+1} \ldots \omega_m) \cdot \mathbb{P}(\omega_{m+1} \ldots \omega_N)$, collapses the sum[2] over $\omega_{m+1} \ldots \omega_N$ to give

$$E_n[X](\omega_1 \ldots \omega_n) = \sum_{\omega_{n+1},\ldots,\omega_m \in \{D,U\}} X(\omega_1 \ldots \omega_n \omega_{n+1} \ldots \omega_m) \, \mathbb{P}(\omega_{n+1} \ldots \omega_m) \qquad (6.36)$$

where $\mathbb{P}(\omega_{n+1} \ldots \omega_m) = p^{\#U(\omega_{n+1}\ldots\omega_m)} (1-p)^{\#D(\omega_{n+1}\ldots\omega_m)}$.

Let us consider some simple examples of the above expectation properties applied to binomial prices S_n, $0 \leqslant n \leqslant N$. Here we define i.i.d. Bernoulli $Bin(1,p)$ random variables $B_m \stackrel{d}{=} \mathbb{I}_{\{\omega_m=U\}} = U_m - U_{m-1}$, for $1 \leqslant m \leqslant N$, where $U_0 \equiv 0$. Hence, B_n is a function of only the nth market move ω_n, i.e., it is $\mathcal{F}_n$-measurable and independent of $\mathcal{F}_{n-1} = \sigma(\omega_1, \omega_2, \ldots, \omega_{n-1}) = \sigma(B_1, \ldots, B_{n-1})$.

1. For $0 \leqslant n < N$, we have by (6.6):

$$\begin{aligned}
E_n[S_{n+1}] &= E_n\left[S_n \, u^{B_{n+1}} d^{1-B_{n+1}}\right] \\
&\quad \left(S_n \text{ is } \mathcal{F}_n\text{-measurable} \implies (6.25) \text{ is applied}\right) \\
&= S_n \, E_n\left[u^{B_{n+1}} d^{1-B_{n+1}}\right] \\
&\quad \left(B_{n+1} \text{ is independent of } \mathcal{F}_n \implies (6.24) \text{ is applied}\right) \\
&= S_n \, E\left[u^{B_{n+1}} d^{1-B_{n+1}}\right] \\
&= S_n \, (u \, p + d \, (1-p)). \qquad (6.37)
\end{aligned}$$

We can also derive this formula by applying (6.36), where $X = S_m$ and $m = n+1$. The sum is only over $\omega_{n+1} \in \{D, U\}$, giving

$$\begin{aligned}
E_n[S_{n+1}](\omega_1 \ldots \omega_n) &= S_{n+1}(\omega_1 \ldots \omega_n \, U) \, \mathbb{P}(U) + S_{n+1}(\omega_1 \ldots \omega_n \, D) \, \mathbb{P}(D) \\
&= up \, S_n(\omega_1 \ldots \omega_n) + d(1-p) \, S_n(\omega_1 \ldots \omega_n) \\
&= (up + d(1-p)) \, S_n(\omega_1 \ldots \omega_n).
\end{aligned}$$

This holds for every sequence $\omega_1 \ldots \omega_n$. Hence, $E_n[S_{n+1}] = (u \, p + d \, (1-p)) \, S_n$.

---

[2] Note that the probabilities over all market moves from time $m + 1$ to N sum to unity:

$$\sum_{\omega_{m+1},\ldots,\omega_N \in \{D,U\}} \mathbb{P}(\omega_{m+1} \ldots \omega_N) \equiv \sum_{\omega_{m+1},\ldots,\omega_N \in \{D,U\}} p^{\#U(\omega_{m+1},\ldots,\omega_N)} (1-p)^{\#D(\omega_{m+1},\ldots,\omega_N)} = 1.$$

2. For $0 \leqslant n < k \leqslant N$ we similarly have

$$\mathrm{E}_n[S_k] = \mathrm{E}_n \left[S_n \prod_{m=n+1}^{k} u^{\mathsf{B}_m} d^{1-\mathsf{B}_m} \right] = S_n \mathrm{E}_n \left[\prod_{m=n+1}^{k} u^{\mathsf{B}_m} d^{1-\mathsf{B}_m} \right]$$

$$(\mathsf{B}_{n+1}, \ldots, \mathsf{B}_k \text{ are all mutually independent and independent of } \mathcal{F}_n)$$

$$= S_n \prod_{m=n+1}^{k} \mathrm{E}\left[u^{\mathsf{B}_m} d^{1-\mathsf{B}_m} \right] = S_n \left(u\,p + d\,(1-p) \right)^{k-n}. \tag{6.38}$$

On the other hand, (6.38) can be derived from (6.37) by applying the tower property (6.28):

$$\mathrm{E}_n[S_k] = \mathrm{E}_n[\mathrm{E}_{n+1}[S_k]] = \cdots = \mathrm{E}_n[\mathrm{E}_{n+1}[\cdots \mathrm{E}_{k-2}[\mathrm{E}_{k-1}[S_k]]\cdots]]$$
$$= \mathrm{E}_n[\mathrm{E}_{n+1}[\cdots \mathrm{E}_{k-2}[S_{k-1}(up + d(1-p))]\cdots]] = \cdots$$
$$= S_n (up + d(1-p))^{k-n}.$$

Again, in a less elegant manner, we can re-derive the above formula by applying (6.36), where $X = S_k$ and $m = k$:

$$\mathrm{E}_n[S_k](\omega_1 \ldots \omega_n) = \sum_{\omega_{n+1} \ldots \omega_k \in \{\mathsf{D},\mathsf{U}\}} S_k(\omega_1 \ldots \omega_n \omega_{n+1} \ldots \omega_k)\, \mathbb{P}(\omega_{n+1} \ldots \omega_k)$$

$$= S_n(\omega_1 \ldots \omega_n) \sum_{\omega_{n+1} \ldots \omega_k \in \{\mathsf{D},\mathsf{U}\}} (up)^{\#\mathsf{U}(\omega_{n+1} \ldots \omega_k)} (d(1-p))^{\#\mathsf{D}(\omega_{n+1} \ldots \omega_k)}$$

$$= S_n(\omega_1 \ldots \omega_n) \sum_{m=0}^{k-n} \binom{k-n}{m} (up)^m (d(1-p))^{k-n-m}$$

$$= S_n(\omega_1 \ldots \omega_n)(up + d(1-p))^{k-n}$$

This holds for every sequence $\omega_1 \ldots \omega_n$. Hence, $\mathrm{E}_n[S_k] = S_n (up + d(1-p))^{k-n}$.

6.3.5 Binomial Model with Interdependent Market Moves

In the standard binomial model, dealt with so far, the main assumption is that the probabilities of an up and down market move are fixed for all times and are independent of previous moves. That is, we assume $\mathbb{P}(\omega_n = \mathsf{U}) = p$, $\mathbb{P}(\omega_n = \mathsf{D}) = 1 - p$ are fixed for all $n = 1, \ldots, N$. The market moves correspond to a standard Bernoulli experiment. Every particular atom $A_{\omega_1^* \ldots \omega_n^*}$ has probability $\mathbb{P}(A_{\omega_1^* \ldots \omega_n^*}) \equiv p^{\#\mathsf{U}(\omega_1^* \ldots \omega_n^*)} (1-p)^{\#\mathsf{D}(\omega_1^* \ldots \omega_n^*)}$. Recall that, by independence, we have

$$\mathbb{P}(A_{\omega_1^* \ldots \omega_N^*}) \equiv \mathbb{P}(\omega_1 = \omega_1^*, \ldots, \omega_n = \omega_n^*, \omega_{n+1} = \omega_{n+1}^*, \ldots, \omega_N = \omega_N^*)$$
$$= \mathbb{P}(\omega_1 = \omega_1^*, \ldots, \omega_n = \omega_n^*) \cdot \mathbb{P}(\omega_{n+1} = \omega_{n+1}^*, \ldots, \omega_N = \omega_N^*)$$
$$\equiv \mathbb{P}(A_{\omega_1^* \ldots \omega_n^*}) \cdot \mathbb{P}(\omega_{n+1} = \omega_{n+1}^*, \ldots, \omega_N = \omega_N^*),$$

and therefore all conditional probabilities simplify to

$$\frac{\mathbb{P}(A_{\omega_1^* \ldots \omega_N^*})}{\mathbb{P}(A_{\omega_1^* \ldots \omega_n^*})} \equiv \mathbb{P}(\omega_{n+1} = \omega_{n+1}^*, \ldots, \omega_N = \omega_N^* | \omega_1 = \omega_1^*, \ldots, \omega_n = \omega_n^*)$$

$$= \mathbb{P}(\omega_{n+1} = \omega_{n+1}^*, \ldots, \omega_N = \omega_N^*)$$
$$= p^{\#\mathsf{U}(\omega_{n+1}^* \ldots \omega_N^*)} (1-p)^{\#\mathsf{D}(\omega_{n+1}^* \ldots \omega_N^*)}.$$

We applied this formula and a similar one with N replaced by m, $0 \leqslant n \leqslant m \leqslant N$, to obtain the conditional expectation formulas in (6.35) and (6.36).

We now consider a more general binomial model where the probabilities of an up and down market move at a given time are *not* necessarily independent of previous market moves. The above simple formula for the conditional probabilities is not applicable. More generally, for any particular sequence $\omega_1^* \ldots \omega_N^*$, we have the conditional probabilities

$$p(\omega_{n+1}^* \ldots \omega_m^* | \omega_1^* \ldots \omega_n^*) := \mathbb{P}(\omega_{n+1} = \omega_{n+1}^*, \ldots, \omega_m = \omega_m^* | \omega_1 = \omega_1^*, \ldots, \omega_n = \omega_n^*)$$

$$\equiv \mathbb{P}(A_{\omega_1^* \ldots \omega_m^*} | A_{\omega_1^* \ldots \omega_n^*}) = \frac{\mathbb{P}(A_{\omega_1^* \ldots \omega_m^*})}{\mathbb{P}(A_{\omega_1^* \ldots \omega_n^*})}, \tag{6.39}$$

$1 \leqslant n < m \leqslant N$. Note that we may also equivalently write $p(\omega_{n+1} \ldots \omega_m | \omega_1 \ldots \omega_n) = \frac{\mathbb{P}(A_{\omega_1 \ldots \omega_m})}{\mathbb{P}(A_{\omega_1 \ldots \omega_n})} \equiv \frac{\mathbb{P}(\omega_1 \ldots \omega_m)}{\mathbb{P}(\omega_1 \ldots \omega_n)}$, where in the last expression we use shorthand to represent the atoms. Now, combining (6.32) and (6.33) gives the general version of (6.34) (or equivalently (6.35)) for any $\mathcal{F}_N$-measurable X:

$$E_n[X](\omega_1 \ldots \omega_n)$$
$$= \sum_{\omega_{n+1}, \ldots, \omega_N \in \{D, U\}} X(\omega_1 \ldots \omega_n \, \omega_{n+1} \ldots \omega_N) \, p(\omega_{n+1} \ldots \omega_N | \omega_1 \ldots \omega_n) \tag{6.40}$$

and (6.36) for any $\mathcal{F}_m$-measurable X now clearly reads

$$E_n[X](\omega_1 \ldots \omega_n)$$
$$= \sum_{\omega_{n+1}, \ldots, \omega_m \in \{D, U\}} X(\omega_1 \ldots \omega_n \, \omega_{n+1} \ldots \omega_m) \, p(\omega_{n+1} \ldots \omega_m | \omega_1 \ldots \omega_n). \tag{6.41}$$

Remark: For $n = N$, (6.40) is simply $E_N[X](\omega_1 \ldots \omega_N) = X(\omega_1 \ldots \omega_N)$ and for $n = m$, (6.41) is simply $E_m[X](\omega_1 \ldots \omega_m) = X(\omega_1 \ldots \omega_m)$. For $n = 0$, (6.41) corresponds to the unconditional expectation of X, with unconditional probabilities $p(\omega_{n+1} \ldots \omega_m | \omega_1 \ldots \omega_n) = \mathbb{P}(A_{\omega_1 \ldots \omega_m}) \equiv \mathbb{P}(\omega_1 \ldots \omega_m)$ and (6.40) is the unconditional expectation of X, with unconditional probabilities $p(\omega_{n+1} \ldots \omega_N | \omega_1 \ldots \omega_n) = \mathbb{P}(A_{\omega_1 \ldots \omega_N}) \equiv \mathbb{P}(\omega_1 \ldots \omega_N)$.

If we are given the conditional probabilities, $p(\omega_{n+1} \ldots \omega_m | \omega_1 \ldots \omega_n)$, $1 \leqslant n < m \leqslant N$, for all possible moves and given the probabilities of the first move, $\mathbb{P}(A_U) \equiv \mathbb{P}(\omega_1 = U) \equiv \mathbb{P}(U)$ and $\mathbb{P}(A_D) \equiv \mathbb{P}(\omega_1 = D) \equiv \mathbb{P}(D)$, then all event probabilities in the N-period binomial model can be calculated. In particular, we can compute $\mathbb{P}(A_{\omega_1 \ldots \omega_n})$, for all $n = 1, \ldots, N$. Indeed, we can apply successive conditioning in various ways. For example, we have

$$\mathbb{P}(A_{\omega_1 \omega_2}) = \frac{\mathbb{P}(A_{\omega_1 \omega_2})}{\mathbb{P}(A_{\omega_1})} \cdot \mathbb{P}(A_{\omega_1}) = p(\omega_2 | \omega_1) \, \mathbb{P}(\omega_1) \,,$$

$$\mathbb{P}(A_{\omega_1 \omega_2 \omega_3}) = \frac{\mathbb{P}(A_{\omega_1 \omega_2 \omega_3})}{\mathbb{P}(A_{\omega_1 \omega_2})} \cdot \mathbb{P}(A_{\omega_1 \omega_2}) = p(\omega_3 | \omega_1 \omega_2) \cdot \mathbb{P}(A_{\omega_1 \omega_2})$$

$$\overset{\text{or}}{=} \frac{\mathbb{P}(A_{\omega_1 \omega_2 \omega_3})}{\mathbb{P}(A_{\omega_1})} \cdot \mathbb{P}(A_{\omega_1}) = p(\omega_2 \omega_3 | \omega_1) \, \mathbb{P}(\omega_1) \,, \ldots,$$

$$\mathbb{P}(A_{\omega_1 \ldots \omega_n}) = p(\omega_n | \omega_1 \ldots \omega_{n-1}) p(\omega_{n-1} | \omega_1 \ldots \omega_{n-2}) \cdot \ldots \cdot p(\omega_3 | \omega_1 \omega_2) p(\omega_2 | \omega_1) \, \mathbb{P}(\omega_1) \,.$$

Of course, there are equivalent multiple ways to express $\mathbb{P}(A_{\omega_1 \ldots \omega_n})$ in terms of other conditional probabilities.

On the other hand, by using the total law of probabilities and the fact that successive atoms are refinements of one another, we can compute all the conditional probabilities given only knowledge of the probabilities $\mathbb{P}(A_{\omega_1 \ldots \omega_N})$. Indeed, assume we are only given all 2^N

probabilities $\mathbb{P}(\omega_1 \ldots \omega_N) \equiv \mathbb{P}(A_{\omega_1 \ldots \omega_N})$, where $\omega_n \in \{\mathsf{D}, \mathsf{U}\}$, $n = 1, \ldots, N$. Clearly, each of the 2^{N-1} atoms $A_{\omega_1 \ldots \omega_{N-1}}$, with first $N-1$ moves fixed, is a disjoint union of two atoms (where $\omega_N = \mathsf{U}$ or $\omega_N = \mathsf{D}$): $A_{\omega_1 \ldots \omega_{N-1}} = A_{\omega_1 \ldots \omega_{N-1}\mathsf{U}} \cup A_{\omega_1 \ldots \omega_{N-1}\mathsf{D}}$. Hence, we obtain the probabilities for all atoms $A_{\omega_1 \ldots \omega_{N-1}}$ as a sum:

$$\mathbb{P}(A_{\omega_1 \ldots \omega_{N-1}}) = \mathbb{P}(A_{\omega_1 \ldots \omega_{N-1}\mathsf{U}}) + \mathbb{P}(A_{\omega_1 \ldots \omega_{N-1}\mathsf{D}}) \equiv \sum_{\omega_N \in \{\mathsf{D}, \mathsf{U}\}} \mathbb{P}(A_{\omega_1 \ldots \omega_{N-1}\omega_N}).$$

Similarly, each of the 2^{N-2} atoms $A_{\omega_1 \ldots \omega_{N-2}}$, with first $N-2$ moves fixed, is a disjoint union of either two atoms, with first $N-1$ moves fixed, or four atoms, with first $N-2$ moves fixed. Their probabilities are given by either sum:

$$\mathbb{P}(A_{\omega_1 \ldots \omega_{N-2}}) = \sum_{\omega_{N-1} \in \{\mathsf{D}, \mathsf{U}\}} \mathbb{P}(A_{\omega_1 \ldots \omega_{N-2}\omega_{N-1}}) = \sum_{\omega_{N-1}, \omega_N \in \{\mathsf{D}, \mathsf{U}\}} \mathbb{P}(A_{\omega_1 \ldots \omega_{N-2}\omega_{N-1}\omega_N}).$$

Proceeding recursively gives the probabilities for all 2^n atoms $A_{\omega_1 \ldots \omega_n}$, for each $n = 1, \ldots, N-1$:

$$\mathbb{P}(A_{\omega_1 \ldots \omega_n}) = \sum_{\omega_{n+1} \in \{\mathsf{D}, \mathsf{U}\}} \mathbb{P}(A_{\omega_1 \ldots \omega_n \omega_{n+1}})$$

and

$$\mathbb{P}(A_{\omega_1 \ldots \omega_n}) = \sum_{\omega_{n+1}, \ldots, \omega_N \in \{\mathsf{D}, \mathsf{U}\}} \mathbb{P}(A_{\omega_1 \ldots \omega_n \omega_{n+1} \ldots \omega_N}).$$

Hence, all conditional probabilities, for all $1 \leqslant n \leqslant m \leqslant N$, are simply obtained as ratios, $p(\omega_{n+1} \ldots \omega_m | \omega_1 \ldots \omega_n) = \frac{\mathbb{P}(A_{\omega_1 \ldots \omega_m})}{\mathbb{P}(A_{\omega_1 \ldots \omega_n})}$.

Example 6.16. Consider a three-period binomial model with given probabilities

$$\mathbb{P}(\mathsf{UUU}) = \frac{1}{12}, \ \mathbb{P}(\mathsf{UUD}) = \frac{1}{6}, \ \mathbb{P}(\mathsf{UDU}) = \frac{1}{4}, \ \mathbb{P}(\mathsf{UDD}) = \frac{1}{12},$$

$$\mathbb{P}(\mathsf{DUU}) = \frac{1}{6}, \ \mathbb{P}(\mathsf{DUD}) = \frac{1}{24}, \ \mathbb{P}(\mathsf{DDU}) = \frac{1}{24}, \ \mathbb{P}(\mathsf{DDD}) = \frac{1}{6}.$$

(a) Compute the conditional probabilities: $p(\omega_3 | \omega_1, \omega_2)$ and $p(\omega_2, \omega_3 | \omega_1)$, for all $\omega_1 \omega_2 \omega_3 \in \Omega_3$.
(b) Compute the conditional expectations: $\mathsf{E}_2[U_3]$ and $\mathsf{E}_1[U_3]$.

Solution. (a) We have $p(\omega_3 | \omega_1, \omega_2) = \frac{\mathbb{P}(\omega_1 \omega_2 \omega_3)}{\mathbb{P}(\omega_1 \omega_2)}$. Hence, we begin by computing the probabilities $\mathbb{P}(\omega_1 \omega_2) = \mathbb{P}(\omega_1 \omega_2 \mathsf{U}) + \mathbb{P}(\omega_1 \omega_2 \mathsf{D})$:

$$\mathbb{P}(\mathsf{UU}) = \mathbb{P}(\mathsf{UUU}) + \mathbb{P}(\mathsf{UUD}) = \frac{1}{12} + \frac{1}{6} = \frac{1}{4}, \quad \mathbb{P}(\mathsf{UD}) = \mathbb{P}(\mathsf{UDU}) + \mathbb{P}(\mathsf{UDD}) = \frac{1}{4} + \frac{1}{12} = \frac{1}{3}$$

$$\mathbb{P}(\mathsf{DU}) = \mathbb{P}(\mathsf{DUU}) + \mathbb{P}(\mathsf{DUD}) = \frac{1}{6} + \frac{1}{24} = \frac{5}{24}, \quad \mathbb{P}(\mathsf{DD}) = \mathbb{P}(\mathsf{DDU}) + \mathbb{P}(\mathsf{DDD}) = \frac{1}{24} + \frac{1}{6} = \frac{5}{24}$$

Hence, for $\omega_3 = \mathsf{U}$:

$$p(\mathsf{U}|\mathsf{UU}) = \frac{\mathbb{P}(\mathsf{UUU})}{\mathbb{P}(\mathsf{UU})} = \frac{\frac{1}{12}}{\frac{1}{4}} = \frac{1}{3}, \quad p(\mathsf{U}|\mathsf{UD}) = \frac{\mathbb{P}(\mathsf{UDU})}{\mathbb{P}(\mathsf{UD})} = \frac{\frac{1}{4}}{\frac{1}{3}} = \frac{3}{4}$$

$$p(\mathsf{U}|\mathsf{DU}) = \frac{\mathbb{P}(\mathsf{DUU})}{\mathbb{P}(\mathsf{DU})} = \frac{\frac{1}{6}}{\frac{5}{24}} = \frac{4}{5}, \quad p(\mathsf{U}|\mathsf{DD}) = \frac{\mathbb{P}(\mathsf{DDU})}{\mathbb{P}(\mathsf{DD})} = \frac{\frac{1}{24}}{\frac{5}{24}} = \frac{1}{5}.$$

For $\omega_3 = \mathsf{D}$:

$$p(\mathsf{D}|\mathsf{UU}) = \frac{\mathbb{P}(\mathsf{UUD})}{\mathbb{P}(\mathsf{UU})} = \frac{\frac{1}{6}}{\frac{1}{4}} = \frac{2}{3}, \quad p(\mathsf{D}|\mathsf{UD}) = \frac{\mathbb{P}(\mathsf{UDD})}{\mathbb{P}(\mathsf{UD})} = \frac{\frac{1}{12}}{\frac{1}{3}} = \frac{1}{4}$$

$$p(\mathsf{D}|\mathsf{DU}) = \frac{\mathbb{P}(\mathsf{DUD})}{\mathbb{P}(\mathsf{DU})} = \frac{\frac{1}{24}}{\frac{5}{24}} = \frac{1}{5}, \quad p(\mathsf{D}|\mathsf{DD}) = \frac{\mathbb{P}(\mathsf{DDD})}{\mathbb{P}(\mathsf{DD})} = \frac{\frac{1}{6}}{\frac{5}{24}} = \frac{4}{5}.$$

We have $p(\omega_2, \omega_3 | \omega_1) = \frac{\mathbb{P}(\omega_1 \omega_2 \omega_3)}{\mathbb{P}(\omega_1)}$, where $\mathbb{P}(\omega_1)$ is given by

$$\mathbb{P}(\mathsf{U}) = \mathbb{P}(\mathsf{UU}) + \mathbb{P}(\mathsf{UD}) = \frac{1}{4} + \frac{1}{3} = \frac{7}{12}, \quad \mathbb{P}(\mathsf{D}) = \mathbb{P}(\mathsf{DU}) + \mathbb{P}(\mathsf{DD}) = \frac{5}{24} + \frac{5}{24} = \frac{5}{12}.$$

Hence, for $\omega_1 = \mathsf{U}$:

$$p(\mathsf{UU}|\mathsf{U}) = \frac{\mathbb{P}(\mathsf{UUU})}{\mathbb{P}(\mathsf{U})} = \frac{\frac{1}{12}}{\frac{7}{12}} = \frac{1}{7}, \quad p(\mathsf{UD}|\mathsf{U}) = \frac{\mathbb{P}(\mathsf{UUD})}{\mathbb{P}(\mathsf{U})} = \frac{\frac{1}{6}}{\frac{7}{12}} = \frac{2}{7}$$

$$p(\mathsf{DU}|\mathsf{U}) = \frac{\mathbb{P}(\mathsf{UDU})}{\mathbb{P}(\mathsf{U})} = \frac{\frac{1}{4}}{\frac{7}{12}} = \frac{3}{7}, \quad p(\mathsf{DD}|\mathsf{U}) = \frac{\mathbb{P}(\mathsf{UDD})}{\mathbb{P}(\mathsf{U})} = \frac{\frac{1}{12}}{\frac{7}{12}} = \frac{1}{7}.$$

For $\omega_1 = \mathsf{D}$:

$$p(\mathsf{UU}|\mathsf{D}) = \frac{\mathbb{P}(\mathsf{DUU})}{\mathbb{P}(\mathsf{D})} = \frac{\frac{1}{6}}{\frac{5}{12}} = \frac{2}{5}, \quad p(\mathsf{UD}|\mathsf{D}) = \frac{\mathbb{P}(\mathsf{DUD})}{\mathbb{P}(\mathsf{D})} = \frac{\frac{1}{24}}{\frac{5}{12}} = \frac{1}{10}$$

$$p(\mathsf{DU}|\mathsf{D}) = \frac{\mathbb{P}(\mathsf{DDU})}{\mathbb{P}(\mathsf{D})} = \frac{\frac{1}{24}}{\frac{5}{12}} = \frac{1}{10}, \quad p(\mathsf{DD}|\mathsf{D}) = \frac{\mathbb{P}(\mathsf{DDD})}{\mathbb{P}(\mathsf{D})} = \frac{\frac{1}{6}}{\frac{5}{12}} = \frac{2}{5}.$$

(b) We now make use of the conditional probabilities in (a) within (6.40) with $N = 3$. Computing $\mathrm{E}_1[\mathsf{U}_3](\omega_1)$, $\omega_1 \in \{\mathsf{U}, \mathsf{D}\}$ gives

$$\mathrm{E}_1[\mathsf{U}_3](\mathsf{U}) = \mathsf{U}_3(\mathsf{UUU})\,p(\mathsf{UU}|\mathsf{U}) + \mathsf{U}_3(\mathsf{UUD})\,p(\mathsf{UD}|\mathsf{U}) + \mathsf{U}_3(\mathsf{UDU})\,p(\mathsf{DU}|\mathsf{U}) + \mathsf{U}_3(\mathsf{UDD})\,p(\mathsf{DD}|\mathsf{U})$$

$$= 3 \cdot \frac{1}{7} + 2 \cdot \frac{2}{7} + 2 \cdot \frac{3}{7} + 1 \cdot \frac{1}{7} = 2$$

and

$$\mathrm{E}_1[\mathsf{U}_3](\mathsf{D}) = \mathsf{U}_3(\mathsf{DUU})\,p(\mathsf{UU}|\mathsf{D}) + \mathsf{U}_3(\mathsf{DUD})\,p(\mathsf{UD}|\mathsf{D}) + \mathsf{U}_3(\mathsf{DDU})\,p(\mathsf{DU}|\mathsf{D}) + \mathsf{U}_3(\mathsf{DDD})\,p(\mathsf{DD}|\mathsf{D})$$

$$= 2 \cdot \frac{2}{5} + 1 \cdot \frac{1}{10} + 1 \cdot \frac{1}{10} + 0 \cdot \frac{2}{5} = 1.$$

Hence, $\mathrm{E}_1[\mathsf{U}_3] = 2\,\mathbb{I}_{A_\mathsf{U}} + \mathbb{I}_{A_\mathsf{D}}$.

Computing $\mathrm{E}_2[\mathsf{U}_3](\omega_1 \omega_2)$, $\omega_1, \omega_2 \in \{\mathsf{U}, \mathsf{D}\}$ gives

$$\mathrm{E}_2[\mathsf{U}_3](\mathsf{UU}) = \mathsf{U}_3(\mathsf{UUU})\,p(\mathsf{U}|\mathsf{UU}) + \mathsf{U}_3(\mathsf{UUD})\,p(\mathsf{D}|\mathsf{UU}) = 3 \cdot \frac{1}{3} + 2 \cdot \frac{2}{3} = \frac{7}{3},$$

$$\mathrm{E}_2[\mathsf{U}_3](\mathsf{UD}) = \mathsf{U}_3(\mathsf{UDU})\,p(\mathsf{U}|\mathsf{UD}) + \mathsf{U}_3(\mathsf{UUD})\,p(\mathsf{D}|\mathsf{UD}) = 2 \cdot \frac{3}{4} + 1 \cdot \frac{1}{4} = \frac{7}{4},$$

$$\mathrm{E}_2[\mathsf{U}_3](\mathsf{DU}) = \mathsf{U}_3(\mathsf{DUU})\,p(\mathsf{U}|\mathsf{DU}) + \mathsf{U}_3(\mathsf{DUD})\,p(\mathsf{D}|\mathsf{DU}) = 2 \cdot \frac{4}{5} + 1 \cdot \frac{1}{5} = \frac{9}{5},$$

$$\mathrm{E}_2[\mathsf{U}_3](\mathsf{DD}) = \mathsf{U}_3(\mathsf{DDU})\,p(\mathsf{U}|\mathsf{DD}) + \mathsf{U}_3(\mathsf{DDD})\,p(\mathsf{D}|\mathsf{DD}) = 1 \cdot \frac{1}{5} + 0 \cdot \frac{4}{5} = \frac{1}{5}.$$

Hence, $\mathrm{E}_2[\mathsf{U}_3] = \frac{7}{3}\,\mathbb{I}_{A_\mathsf{UU}} + \frac{7}{4}\,\mathbb{I}_{A_\mathsf{UD}} + \frac{9}{5}\,\mathbb{I}_{A_\mathsf{DU}} + \frac{1}{5}\,\mathbb{I}_{A_\mathsf{DD}}.$ $\qquad\square$

6.3.6 Sub-, Super-, and True Martingales

Recall that a stochastic process $X \equiv \{X_t\}_{t \in \mathbf{T}}$ is a collection of random variables (indexed by a set $\mathbf{T} \subset [0, \infty)$) on a filtered probability space $(\Omega, \mathcal{F}, \mathbb{P}, \mathbb{F} = \{\mathcal{F}_t\}_{t \in \mathbf{T}})$ so that $\forall t \in \mathbf{T}$ X_t is $\mathcal{F}_t$-measurable, i.e., X is adapted to the filtration. Recall that X is a discrete-time process if $\mathbf{T} = \{0, 1, 2, \dots\}$; it is a continuous-time process if $\mathbf{T} = [0, \infty)$ or an interval in $[0, \infty)$. We now give a formal definition of important classes of processes that have many financial applications.

Definition 6.15. Consider a filtered probability space $(\Omega, \mathcal{F}, \mathbb{P}, \mathbb{F})$ and a stochastic process $\{M_t\}_{t \in \mathbf{T}}$ adapted to the filtration $\mathbb{F} = \{\mathcal{F}_t\}_{t \in \mathbf{T}}$ so that $\mathrm{E}[|M_t|] < \infty$ for every $t \in \mathbf{T}$ (in this case the process is said to be *integrable*). Then, $\{M_t\}_{t \in \mathbf{T}}$ is called:

(a) a *martingale* if $\mathrm{E}[M_{t+s} \mid \mathcal{F}_t] = M_t$ (it has no tendency to fall or to rise),

(b) a *sub-martingale* if $\mathrm{E}[M_{t+s} \mid \mathcal{F}_t] \geqslant M_t$ (it has no tendency to fall),

(c) a *super-martingale* if $\mathrm{E}[M_{t+s} \mid \mathcal{F}_t] \leqslant M_t$ (it has no tendency to rise),

for all $t, s \in \mathbf{T}$. In cases where the inequalities are strict, we say that the process is a *strict sub-martingale* or *strict super-martingale*, respectively. The processes that are really of interest to us are either discrete-time processes or continuous-time processes where time values are all real numbers $s, t \geqslant 0$ or $0 \leqslant s, t \leqslant T$ for some fixed time $T > 0$.

We note that in the strict sense of the above definition, which includes processes having a continuous state space, relations (a)–(c) are meant to hold true with probability one, i.e., we say *almost surely (a.s.)*. Moreover, a process is a martingale (or sub- or super-martingale) w.r.t. a given filtration *and* a given probability measure. A process may be a martingale w.r.t. a given filtration and measure, but it may not necessarily be a martingale if either the filtration or measure is changed. So, more precisely, we say that $\{M_t\}_{t \geqslant 0}$ is a $(\mathbb{P}, \mathbb{F})$-martingale, i.e., w.r.t. a measure $\mathbb{P}$ and filtration $\mathbb{F}$. In what follows in this chapter we simply fix the probability measure $\mathbb{P}$, with expectation E implied w.r.t. this measure $\mathbb{P}$, and then it suffices to say that the process is a martingale w.r.t. (or relative to) a given filtration $\mathbb{F}$.

When the filtration is fixed (e.g., to a specified natural filtration) and we change probability measures, then we shall refer to a process as a martingale w.r.t. a given measure, i.e., we say that the process $\{M_t\}_{t \geqslant 0}$ is a $\mathbb{P}$-martingale or a martingale under measure $\mathbb{P}$. In Chapter 5, we already saw an example of measure changes from the physical measure, say $\mathbb{P}$, into other risk-neutral measures, e.g., $\widetilde{\mathbb{P}}$, where we simply had a single-period stochastic process with discrete-time parameter having only an initial value $t = 0$ and a terminal value $T > 0$. In that case the conditional expectations w.r.t. information at time $t = 0$ are w.r.t. the trivial filtration $\mathcal{F}_0$, and hence the martingale condition is simply stated as an unconditional expectation. That is, $\{M_t\}_{t=0,T}$ is a $\mathbb{P}$-martingale if $\mathrm{E}[M_T] = M_0$, since $\mathrm{E}[M_T \mid \mathcal{F}_0] \equiv \mathrm{E}[M_T]$. For example, the risk-neutral measure $\widetilde{\mathbb{P}} \equiv \widetilde{\mathbb{P}}^{(B)}$ discussed in Chapter 5 was defined such that $\widetilde{\mathrm{E}}[S_T^i / B_T] = S_0^i / B_0$, for all discounted security price processes $\{S_t^i / B_t\}_{t=0,T}$, where the expectation is taken w.r.t. risk-neutral measure $\widetilde{\mathbb{P}}$. Hence, according to the above definition, each of these discounted price processes is a $\widetilde{\mathbb{P}}$-martingale.

In the discrete-time setting, the definition of a sub-, or super-martingale can be simplified thanks to the tower property. A discrete-time stochastic process $\{M_n\}_{n=0,1,2,\dots}$ is called a martingale w.r.t. a filtration $\{\mathcal{F}_n\}_{n \geqslant 0}$ if:

(i) M_n is $\mathcal{F}_n$-measurable for all $n \geqslant 0$ (the process is adapted to the filtration);
(ii) $\mathrm{E}[|M_n|] < \infty$ for all $n \geqslant 0$ (finite expectation of its absolute value);
(iii) $\mathrm{E}_n[M_{n+1}] = M_n$ for all $n \geqslant 0$.

Indeed, (iii) is equivalent to the expectation property in Definition 6.15 (a). This is seen by repeatedly applying the tower property (6.28) in reverse, for every single time step:

$$M_n = \mathrm{E}_n[M_{n+1}] = \mathrm{E}_n[\mathrm{E}_{n+1}[M_{n+2}]] = \mathrm{E}_n[M_{n+2}] = \ldots = \mathrm{E}_n[M_{n+m}]$$

for all $n, m = 0, 1, 2, \ldots$. This also leads to the following result, which tells us that the expectation of a martingale is constant over time. In other words, the expected value of a martingale is conserved.

Proposition 6.9. *Let $\{M_n\}_{n \geqslant 0}$ be a martingale. Then*

$$\mathrm{E}[M_n] = \mathrm{E}[M_0] \text{ for all } n = 1, 2, \ldots.$$

Proof. By the above application of the tower property we have $M_k = \mathrm{E}_k[M_{k+n}]$, for all $n, k \geqslant 0$. Hence, $\mathrm{E}[M_k] = \mathrm{E}[\mathrm{E}_k[M_{k+n}]] = \mathrm{E}[M_{k+n}]$. In particular, for $k = 0$, $\mathrm{E}[M_0] = \mathrm{E}[M_n]$. □

We note that in most cases M_0 is a known constant so that $M_0 = \mathrm{E}[M_n], n \geqslant 0$. The notion of a sub-, or super-martingale can be described in terms of fairness of a game. Suppose that we start playing a game with an initial capital W_0. Let W_n be the total winning after n rounds of the game. The game is considered to be fair if $W := \{W_n\}_{n=0,1,2\ldots}$ is a martingale; it is a favorable game if W is a sub-martingale; it is an unfavorable game if W is a super-martingale.

6.3.6.1 Examples

(a) Let X be an integrable random variable, i.e., $\mathrm{E}[|X|] < \infty$. Consider a filtration $\mathbb{F} = \{\mathcal{F}_t\}_{t \geqslant 0}$. Define $X_t := \mathrm{E}[X \mid \mathcal{F}_t]$ for $t \geqslant 0$. Then $\{X_t\}_{t \geqslant 0}$ is a martingale relative to $\mathbb{F}$. This is known as the so-called Doob-Lévy martingale and is valid for discrete as well as continuous time.

Solution. First, show that the process is integrable by using the conditional Jensen's inequality:

$$\mathrm{E}[|X_t|] = \mathrm{E}[|E[X \mid \mathcal{F}_t]|] \leqslant \mathrm{E}[\mathrm{E}[|X| \mid \mathcal{F}_t]] = \mathrm{E}[|X|] < \infty.$$

Applying the tower property with $\mathcal{F}_t \subset \mathcal{F}_{t+s}$ ($s, t \geqslant 0$), and using the definition of the process, gives

$$\mathrm{E}[X_{t+s} \mid \mathcal{F}_t] = \mathrm{E}[\mathrm{E}[X \mid \mathcal{F}_{t+s}]|\mathcal{F}_t] = \mathrm{E}[X \mid \mathcal{F}_t] = X_t.$$

Finally, note that $X_t := \mathrm{E}[X \mid \mathcal{F}_t]$ is $\mathcal{F}_t$-measurable. □

(b) Consider a sequence of integrable i.i.d. random variables, $\{Y_n\}_{n=1,2,\ldots}$, with $\mathrm{E}[Y_1] = \mu$ and $\mathrm{E}[|Y_1|] < \infty$. Define $\mathcal{F}_0 = \{\emptyset, \Omega\}$ and $\mathcal{F}_n = \sigma(Y_1, \ldots, Y_n)$ for $n = 1, 2, \ldots$ and hence form the natural filtration $\mathbb{F} = \{\mathcal{F}_n\}_{n \geqslant 0}$.

(i) Let $X_n := \sum_{k=1}^{n} Y_k$, $n = 1, 2, \ldots$ and $X_0 = 0$. The process $\{X_n\}_{n \geqslant 0}$ is a martingale w.r.t. $\mathbb{F}$ iff $\mu = 0$. It is a strict sub-martingale or a strict super-martingale iff $\mu > 0$ or $\mu < 0$, respectively.

Proof. Note that $\mathrm{E}[|X_n|] \leqslant \mathrm{E}[|Y_1|] + \cdots + \mathrm{E}[|Y_n|] < \infty$. Since X_n is a function of $Y_1, \ldots, Y_n$, X_n is $\mathcal{F}_n$-measurable for all $n \geqslant 0$. Noting that $X_{n+1} = X_n + Y_{n+1}$ and applying the linearity property of conditional expectations and the independence of Y_{n+1} and $\mathcal{F}_n$:

$$\mathrm{E}[X_{n+1} \mid \mathcal{F}_n] = \mathrm{E}[X_n \mid \mathcal{F}_n] + \mathrm{E}[Y_{n+1} \mid \mathcal{F}_n] = X_n + \mathrm{E}[Y_{n+1}] = X_n + \mu,$$

for $n \geqslant 0$. □

For example, the random walk process is given by

$$X_n = \sum_{k=1}^n Y_k, \; n \geqslant 1, X_0 = 0 \; \text{with} \; Y_k = \begin{cases} 1 & \text{with probability } p \\ -1 & \text{with probability } 1-p \end{cases}, \; k \geqslant 1.$$

In this case $\mu = \mathrm{E}[Y_1] = 2p - 1$. Therefore, the random walk is a martingale, strict sub-martingale, or strict super-martingale if $p = \frac{1}{2}$ (i.e., it is a symmetric random walk), $p > \frac{1}{2}$, or $p < \frac{1}{2}$, respectively. Similarly, the log-price process in the binomial model is defined by $X_0 = 0$ and

$$X_n := \ln \frac{S_n}{S_0} = \sum_{k=1}^n Y_k, \; n \geqslant 1, \; \text{with} \; Y_k = \begin{cases} \ln u & \text{with probability } p, \\ \ln d & \text{with probability } 1-p, \end{cases} \; k \geqslant 1.$$

Here $\mu = \mathrm{E}[Y_1] = p \ln u + (1-p) \ln d = 0$. If $0 < d < 1 < u$, then the log-price process is a martingale iff $p = \ln(\frac{1}{d}) / \ln(\frac{u}{d})$.

(ii) The process $X_n := \sum_{k=1}^n Y_k - \mu n, \; n \geqslant 1, \; X_0 = 0$, is a martingale for any μ.

(iii) Suppose that $\mathrm{E}[Y_1] = 0$ and $\mathrm{E}[Y_1^2] = \sigma^2 < \infty$. Then the process started at X_0 and given by $X_n := \left(\sum_{k=1}^n Y_k\right)^2 - \sigma^2 n, \; n \geqslant 1$, is a martingale. That is, squaring a symmetric random walk and subtracting its variance gives a martingale.

Proof. First, we express X_{n+1} in terms of X_n:

$$X_{n+1} = \left(\sum_{k=1}^n Y_k + Y_{n+1}\right)^2 - \sigma^2(n+1)$$

$$= \underbrace{\left(\sum_{k=1}^n Y_k\right)^2 - \sigma^2 n}_{=X_n} + 2Y_{n+1} \sum_{k=1}^n Y_k + Y_{n+1}^2 - \sigma^2.$$

Now, we compute the (single-step) conditional expectation $\mathrm{E}[X_{n+1} \mid \mathcal{F}_n]$ by making use of the assumed independence of all the Y_k random variables (as well as the independence of each Y_k and $\mathcal{F}_n$) and the fact that $\sum_{k=1}^n Y_k$ is $\mathcal{F}_n$-measurable since $\mathcal{F}_n = \sigma(Y_1, \ldots, Y_n)$:

$$\mathrm{E}[X_{n+1} \mid \mathcal{F}_n] = \mathrm{E}[X_n \mid \mathcal{F}_n] + 2\mathrm{E}[Y_{n+1} \sum_{k=1}^n Y_k \mid \mathcal{F}_n] + \mathrm{E}[Y_{n+1}^2 - \sigma^2 \mid \mathcal{F}_n]$$

$$= X_n + 2\underbrace{\mathrm{E}[Y_{n+1}]}_{=0} \sum_{k=1}^n Y_k + \underbrace{\mathrm{E}[Y_{n+1}^2 - \sigma^2]}_{=0} = X_n.$$

Moreover, $\mathrm{E}[|X_n|] \leqslant \sum_{k=1}^n \mathrm{E}[Y_k^2] + \sigma^2 n = (n+1)\sigma^2 < \infty$. Hence, the process is integrable and X_n is clearly $\mathcal{F}_n$-measurable. □

(c) The binomial price process $\{S_n\}_{n \geqslant 1}$ is given by

$$S_n = S_0 \prod_{k=1}^n Z_k, \; n \geqslant 1, \; \text{with} \; Z_k = \begin{cases} u & \text{with probability } p, \\ d & \text{with probability } 1-p, \end{cases} \; k \geqslant 1.$$

It is a martingale iff $up + d(1-p) = 1$. Indeed, since $S_{n+1} = S_n Z_{n+1}$,

$$\mathrm{E}_n[S_{n+1}] = S_n \mathrm{E}_n[Z_{n+1}] = S_n(up + d(1-p)) = S_n \; \text{iff} \; up + d(1-p) = 1.$$

Here we used the independence of Z_{n+1} and $\mathcal{F}_n$, i.e., $\mathrm{E}_n[Z_{n+1}] = \mathrm{E}[Z_{n+1}] = \mathrm{E}[Z_k]$, for all $k \geqslant 1$.

Suppose that $0 < d < 1 < u$. Then, $\{S_n\}_{n \geqslant 0}$ is a martingale iff $p = \frac{1-d}{u-d}$.

6.3.7 Classification of Stochastic Processes

The following classes of stochastic processes can be considered.

1. X is said to be a *process with independent increments* if $\forall n \in \mathbb{N}$ and $\forall t_1, t_2, \ldots, t_N \in \mathbf{T}$ so that $0 \leqslant t_1 < t_2 < \cdots < t_n$, the increments $X_{t_{i+1}} - X_{t_i}$, $1 \leqslant i \leqslant n-1$, are jointly independent random variables.

2. X is said to be a *process with stationary increments* if $\forall s, t \in \mathbf{T}$ with $s > t$ the probability distribution of the increment $X_s - X_t$ depends only on the length $s - t$ of the interval $[t, s]$ and not on s and t separately.

3. X is said to be a *martingale* with respect to the filtration $\mathbb{F}$ if it is adapted to $\mathbb{F}$, $\mathrm{E}[|X_t|] < \infty$ and $\mathrm{E}[X_s \mid \mathcal{F}_t] = X_t$ for every $t, s \in \mathbf{T}$ with $t \leqslant s$.

4. X is said to be a *Markov process or Markovian* if, given the value of X_t, $t \in \mathbf{T}$, the probability distribution of X_s, $s > t$, does not depend on the past history of values $\{X_u : u \in \mathbf{T}, u < t\}$. In other words,

$$\mathbb{P}(X_s \in B \mid \mathcal{F}_t^X) = \mathbb{P}(X_s \in B \mid X_t), \quad \forall B \in \mathcal{B}(\mathbb{R}).$$

This property can also be stated as $\mathbb{P}(X_s \leqslant x \mid \mathcal{F}_t^X) = \mathbb{P}(X_s \leqslant x \mid X_t)$, for any $x \in \mathbb{R}$. Equivalently, X is a Markov process if for any Borel function f and time index $t \in \mathbf{T}$ there exists a (Borel) function g (that may generally be defined in terms of f and s, t) such that

$$\mathrm{E}[f(X_s) \mid \mathcal{F}_t^X] = g(X_t), \quad \forall s \in \mathbf{T}, \ s \geqslant t,$$

i.e., $\mathrm{E}[f(X_s) \mid \mathcal{F}_t^X] = \mathrm{E}[f(X_s) \mid X_t]$, for $s \geqslant t$.

It should be noted here that the Markov property is defined by conditioning on the natural filtration, i.e., on F_t^X. However, if the above conditional expectation properties hold for any filtration, i.e., by conditioning on $\mathcal{F}_t$ instead of F_t^X, with X still assumed adapted to the filtration, then the process is Markov. This follows by the tower property since $F_t^X \subset F_t$ and $g(X_t)$ is F_t^X-measurable. Indeed, the condition $\mathrm{E}[f(X_s) \mid \mathcal{F}_t] = g(X_t)$ implies that $\mathrm{E}[f(X_s) \mid \mathcal{F}_t^X] = \mathrm{E}[\mathrm{E}[f(X_s) \mid \mathcal{F}_t] \mid \mathcal{F}_t^X] = \mathrm{E}[g(X_t) \mid \mathcal{F}_t^X] = g(X_t)$, $s \geqslant t$.

For a Markov process, its distribution at any future time is completely determined from the information contained only in the sub-σ-algebra $\sigma(X_t)$ generated by the process at the current time t. For a *discrete-time Markov process* $X = \{X_n\}_{n=0,1,\ldots}$, it follows that $\mathrm{E}[f(X_m) \mid \mathcal{F}_n] \equiv \mathrm{E}_n[f(X_m)] = g(X_n)$, for all $0 \leqslant n \leqslant m$. Hence, by the tower property, a discrete-time process is Markov iff the (one-step ahead) condition $\mathrm{E}_n[f(X_{n+1})] = g(X_n)$ holds for all $n = 0, 1, \ldots$. This gives us a very practical way to verify the Markov property when we make use of Proposition 6.7.

The above definition extends to the multidimensional case in the following obvious manner. Let $\mathbf{X} = \{\mathbf{X}_t := (X_t^1, \ldots, X_t^m)\}_{t \geqslant 0}$ be an $\mathbb{R}^m$-valued joint process, for integer $m \geqslant 1$. Then, we say that this vector process is Markovian or a Markov process if, given any Borel function $f : \mathbb{R}^m \to \mathbb{R}$, we have

$$\mathrm{E}[f(\mathbf{X}_s) \mid \mathcal{F}_t^{\mathbf{X}}] = g(\mathbf{X}_t), \forall s > t,$$

for some $g : \mathbb{R}^m \to \mathbb{R}$. If time is discrete then the vector process $\{\mathbf{X}_n := (X_n^1, \ldots, X_n^m)\}_{n \geqslant 0}$ is said to be Markovian if $\mathrm{E}_n[f(\mathbf{X}_{n+1})] = g(\mathbf{X}_n)$, for all $n = 0, 1, \ldots$.

Example 6.17. Show that the binomial log-price process $\{\ln S_n\}_{n \geqslant 0}$ is a Markov process with stationary and independent increments.

Solution. Denote $X_n := \ln S_n$. Then $X_{n+1} = \ln S_{n+1} = \ln S_n + Y_{n+1} = X_n + Y_{n+1}$, with i.i.d. random variables Y_k, $k \geqslant 1$, taking on value $Y_k = \ln u$ with probability p and value $\ln d$ with probability $1 - p$. The natural filtration is given by $\mathcal{F}_n = \sigma(S_1, \ldots, S_n) = \sigma(Y_1, \ldots, Y_n)$. Hence, Y_{n+1} and $\mathcal{F}_n$ are independent and X_n is $\mathcal{F}_n$-measurable, so we may use Proposition 6.7 (i.e., take $\mathcal{G} = \mathcal{F}_n, X = X_n, Y = Y_{n+1}, f(x, y) = h(x + y)$). That is, for any Borel function h:

$$E_n[h(X_{n+1})] = E_n[h(X_n + Y_{n+1})] = g(X_n)$$

where $g : \mathbb{R} \to \mathbb{R}$ is defined by $g(x) := E[h(x + Y_{n+1})]$. Hence, the log-price process $\{X_n\}_{n \geqslant 0}$ is Markov. Now, consider any two adjacent intervals $[n-1, n]$ and $[n, n+1]$. Then, $X_{n+1} - X_n = Y_{n+1}$ and $X_n - X_{n-1} = Y_n$ are independent and it clearly follows that process X has independent increments. Consider arbitrary intervals $[m + \ell, n + \ell]$, $\ell \geqslant 0$, of the same length $(n - m)$, with integers $n > m \geqslant 0$. The stationarity property now follows since $X_{n+\ell} - X_{m+\ell} = \sum_{k=m+\ell+1}^{n+\ell} Y_k$ is a sum of $(n - m)$ i.i.d. random variables having the same distribution as $Y_1 + \cdots + Y_{(n-m)}$. $\qquad\square$

Example 6.18. Let $\{Z_k\}_{k \in \mathbb{N}}$ be a sequence of i.i.d. random variables. Construct a piecewise-constant continuous-time process $\{X_t\}_{t \geqslant 0}$ as follows:

$$X_t = \sum_{k=1}^{\lfloor t \rfloor} Z_k, \ t > 0, \quad X_0 \equiv 0.$$

Show that X is a Markov process with independent increments. What condition guarantees that X is a martingale w.r.t. its natural filtration? [Note that increments of X are non-stationary in general.]

Solution. Take any two indices $t, s \in \mathbf{T}$ with $t < s$. The increment of X over $[t, s]$ is

$$X_s - X_t = \sum_{k=\lfloor t \rfloor + 1}^{\lfloor s \rfloor} Z_k.$$

So, $X_s - X_t$ is zero if $(t, s]$ does not contain integers, i.e., if $\lfloor t \rfloor = \lfloor s \rfloor$ then $X_s - X_t = \sum_{k=\lfloor t \rfloor + 1}^{\lfloor t \rfloor} Z_k \equiv 0$; otherwise it is a sum of Z_k, $k \in (t, s] \cap \mathbb{N}$. Therefore, for every selection of times $0 \leqslant t_1 < t_2 < \cdots < t_n$, all nonzero increments are sums of different selections of Z_k's and hence are jointly independent random variables. Moreover, X_s can be written as a sum of two independent random variables: X_t and $Y \equiv \sum_{k=\lfloor t \rfloor + 1}^{\lfloor s \rfloor} Z_k$. Note that X_t is clearly $\mathcal{F}_t^X$-measurable and Y is independent of $\sigma(Z_1, \ldots, Z_{\lfloor t \rfloor}) = \sigma(X_1, \ldots, X_{\lfloor t \rfloor}) = \mathcal{F}_{\lfloor t \rfloor}^X = \mathcal{F}_t^X$. Using Proposition 6.7 (set $\mathcal{G} = \mathcal{F}_t^X, X = X_t, f(x, y) = h(x+y)$) gives (for any Borel function h):

$$E[h(X_s) \mid \mathcal{F}_t^X] = E[h(X_t + Y) \mid \mathcal{F}_t^X] = g(X_t)$$

where $g : \mathbb{R} \to \mathbb{R}$ is defined by $g(x) := E[h(x + Y)]$. Hence the process is Markov. Putting $h(x) = x$, $g(x) = E[x + Y] = x + E[Y]$,

$$E[X_s \mid \mathcal{F}_t^X] = X_t + E[Y] = X_t + \sum_{k=\lfloor t \rfloor + 1}^{\lfloor s \rfloor} E[Z_k].$$

Hence, the process is a martingale w.r.t. its natural filtration iff $E[Z_k] = 0$. $\qquad\square$

6.3.8 Stopping Times

Let us fix a filtered probability space $(\Omega, \mathcal{F}, \mathbb{P}, \mathbb{F})$. There is a class of random variables that play an important role in financial modelling and derivative pricing. These random variables are known as "stopping times." The formal definition of a stopping time random variable (given just below) is quite general. That is, given any nonnegative number (time) $t \geqslant 0$, then a random variable $\mathcal{T}$ is a stopping time with respect to a filtration $\mathbb{F} = \{\mathcal{F}_t\}_{t \geqslant 0}$ if the event $\{\mathcal{T} \leqslant t\}$ is $\mathcal{F}_t$-measurable. For example, if we let $\mathcal{T}$ represent the first time that the share price of a stock has passed a certain level, then the information contained in $\mathcal{F}_t$ is enough to know whether or not $\mathcal{T} \leqslant t$, for any given value of time $t \geqslant 0$, i.e., the event that it took less than or equal to a given time t (or greater than time t) for the stock to pass a certain level is contained in $\mathcal{F}_t$. The event $\{\mathcal{T} \leqslant t\}$ and its complement $\{\mathcal{T} > t\}$ are in the σ-algebra $\mathcal{F}_t$. The following is a general definition.

Definition 6.16. A random variable $\mathcal{T} \colon \Omega \to [0, \infty) \cup \{\infty\}$ is called a stopping time w.r.t. a filtration $\{\mathcal{F}_t\}_{t \geqslant 0}$ if the event $\{\mathcal{T} \leqslant t\} \equiv \{\omega \in \Omega \ : \ \mathcal{T}(\omega) \leqslant t\} \in \mathcal{F}_t$ for every $t \geqslant 0$.

In the discrete-time setting, stopping times take on integer values and as a consequence there is an equivalent definition contained in the following proposition.

Proposition 6.10. *In the discrete-time case, $\mathcal{T} \colon \Omega \to \mathbb{N}_0 \cup \{\infty\}$, where $\mathbb{N}_0 = \{0, 1, 2, \ldots\}$, is a stopping time iff $\{\mathcal{T} = n\} \in \mathcal{F}_n$ for every $n \in \mathbb{N}_0$.*

Proof. Suppose that, $\forall n \in \mathbb{N}_0$, $\{\mathcal{T} \leqslant n\} \in \mathcal{F}_n$. If $n = 0$, then $\{\mathcal{T} \leqslant 0\} = \{\mathcal{T} = 0\} \in \mathcal{F}_0$. Now let $n \geqslant 1$. Then, $\{\mathcal{T} \leqslant n\} \in \mathcal{F}_n$ and $\{\mathcal{T} \leqslant n - 1\} \in \mathcal{F}_{n-1} \subset \mathcal{F}_n$. Since $\mathcal{F}_n$ is a σ-algebra, $\{\mathcal{T} = n\} = \{\mathcal{T} \leqslant n\} \setminus \{\mathcal{T} \leqslant n - 1\} \in \mathcal{F}_n$. Conversely, suppose that $\forall n \in \mathbb{N}_0$ $\{\mathcal{T} = n\} \in \mathcal{F}_n$. Therefore, $\{\mathcal{T} = k\} \in \mathcal{F}_k \subset \mathcal{F}_n$ for every $k = 0, 1, \ldots, n$. Again, by using the fact that $\mathcal{F}_n$ is a σ-algebra, we have

$$\{\mathcal{T} \leqslant n\} = \bigcup_{k=0}^{n} \{\mathcal{T} = k\} \in \mathcal{F}_n. \qquad \square$$

Denoting $x \wedge y := \min\{x, y\}$ and $x \vee y := \max\{x, y\}$, we have the following basic properties of any stopping times.

Proposition 6.11. *Suppose that $\mathcal{T}$, $\mathcal{T}_1$, and $\mathcal{T}_2$ are stopping times. Then, $\mathcal{T} \wedge m$, $m \in \mathbb{N}_0$, $\mathcal{T}_1 \wedge \mathcal{T}_2$, and $\mathcal{T}_1 \vee \mathcal{T}_2$ are stopping times as well.*

The proof of Proposition 6.11 is left as an exercise for the reader (see Exercises 6.38 and 6.39).

Some important examples of stopping times are so-called first passage times or hitting times of a process. For a discrete-time stochastic process $\{X_n\}_{n \geqslant 0}$ the *first hitting time to a level ℓ* is defined as the smallest nonnegative integer value of time such that the process attains the value ℓ:

$$\mathcal{T}_\ell := \min\{n \geqslant 0 \ : \ X_n = \ell\}. \tag{6.42}$$

We put $\mathcal{T}_\ell = \infty$ if the set $\{\mathcal{T}_\ell < \infty\} = \emptyset$. Hence, $\mathcal{T}_\ell$ is allowed to be infinite, i.e., $\mathcal{T}_\ell \colon \Omega \to \mathbb{N}_0 \cup \{\infty\}$. If $\mathbb{P}(\mathcal{T}_\ell < \infty) = 1$ then we say that $\mathcal{T}_\ell$ is (almost surely) finite.

The *first passage time up to level ℓ, $\mathcal{T}_\ell^+$*, and the *first passage time down to level ℓ, $\mathcal{T}_\ell^-$*, are both in $\mathbb{N}_0 \cup \{\infty\}$ and are defined by

$$\mathcal{T}_\ell^+ := \min\{n \geqslant 0 \ : \ X_n \geqslant \ell\} \quad \text{and} \quad \mathcal{T}_\ell^- := \min\{n \geqslant 0 \ : \ X_n \leqslant \ell\}.$$

We put $\mathcal{T}_\ell^+ = \infty$ if $\{\mathcal{T}_\ell^+ < \infty\} = \emptyset$ and, similarly, we put $\mathcal{T}_\ell^- = \infty$ if $\{\mathcal{T}_\ell^- < \infty\} = \emptyset$. For the respective $\pm$ cases, if $\mathbb{P}(\mathcal{T}_\ell^\pm < \infty) = 1$ then we say that $\mathcal{T}_\ell^\pm$ is (almost surely) finite.

We note also that if the process is defined only for finite integer times $0 \leqslant n \leqslant N$, for some fixed integer $N > 0$, then we set $\{\mathcal{T}_\ell^+ > N\} \equiv \{\mathcal{T}_\ell^+ = \infty\}$, $\{\mathcal{T}_\ell^- > N\} \equiv \{\mathcal{T}_\ell^- = \infty\}$ and $\{\mathcal{T}_\ell > N\} \equiv \{\mathcal{T}_\ell = \infty\}$. The meaning here is that if the process has not attained level ℓ by the maximum allowed finite time N then it will never attain (or takes an "infinite time" to attain) level ℓ.

Example 6.19. Consider the three-period binomial model with $S_0 = 1$, $u = 2$, and $d = \frac{1}{2}$. Show that $\mathcal{T} = \mathcal{T}_2^+$ and $\mathcal{T} = \mathcal{T}_{\frac{1}{4}}^-$ are stopping times w.r.t. the natural filtration $\{\mathcal{F}_n = \sigma(S_0, S_1, \ldots, S_n)\}_{n \geqslant 0}$ generated by the stock price process $S \equiv \{S_n\}_{n \geqslant 0}$.

Solution. It is sufficient to show that $\{\mathcal{T} = k\} \in \mathcal{F}_k$ for each $k = 0, 1, 2, 3$, as follows:

$$\{\mathcal{T}_2^+ = 0\} = \emptyset \in \mathcal{F}_0, \qquad\qquad \{\mathcal{T}_{\frac{1}{4}}^- = 0\} = \emptyset \in \mathcal{F}_0,$$

$$\{\mathcal{T}_2^+ = 1\} = A_\mathsf{U} \in \mathcal{F}_1, \qquad\qquad \{\mathcal{T}_{\frac{1}{4}}^- = 1\} = \emptyset \in \mathcal{F}_1,$$

$$\{\mathcal{T}_2^+ = 2\} = \emptyset \in \mathcal{F}_2, \qquad\qquad \{\mathcal{T}_{\frac{1}{4}}^- = 2\} = A_\mathsf{DD} \in \mathcal{F}_2,$$

$$\{\mathcal{T}_2^+ = 3\} = \{\mathsf{DUU}\} \in \mathcal{F}_3, \qquad\qquad \{\mathcal{T}_{\frac{1}{4}}^- = 3\} = \emptyset \in \mathcal{F}_3. \qquad \square$$

Figure 6.4 illustrates the values of the random variables $\mathcal{T}_b^+$ and $\mathcal{T}_a^-$, where $a < b$, for two (paths) outcomes on the binomial stock price model with six periods.

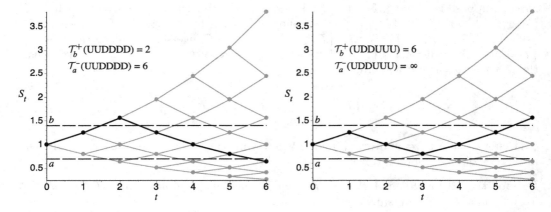

FIGURE 6.4: First passage time values $\mathcal{T}_b^+(\omega)$ and $\mathcal{T}_a^-(\omega)$ in the case of the six-period binomial model ($N = 6$) with parameters $u = 1.25$ and $d = 1/u = 0.8$, for upper level $b = 1.4$ and lower level $a = 0.7$, for two particular sample paths: (left) $\omega = \mathsf{UUDDDD}$ and (right) $\omega = \mathsf{UDDUUU}$.

First passage times are examples of stopping times. The precise claim and its proof, for the case of a discrete-time process, is given just below. Before proving the result, we define two other important and related random variables. One is the *running (or sampled) maximum*

$$M_n := \max\{X_k : k = 0, 1, \ldots, n\}$$

and the other is the *running (or sampled) minimum*

$$m_n := \min\{X_k : k = 0, 1, \ldots, n\}$$

of the process $\{X_k\}_{k=0,1,\ldots}$ observed from time zero to time $n \geqslant 0$. These are now related to the first passage times by noting that $\{\mathcal{T}_\ell^+ \leqslant n\}$ is the event that the process takes at most n time steps to reach or go above level ℓ and $\{\mathcal{T}_\ell^- \leqslant n\}$ is the event that it takes at most n time steps to reach or go below level ℓ. Hence, we have the equivalence of events:

$$\{\mathcal{T}_\ell^+ \leqslant n\} = \{X_0 \geqslant \ell\} \cup \{X_1 \geqslant \ell\} \cup \ldots \cup \{X_n \geqslant \ell\} = \{\max(X_0, \ldots, X_n) \geqslant \ell\} \equiv \{M_n \geqslant \ell\}$$

and

$$\{\mathcal{T}_\ell^- \leqslant n\} = \{X_0 \leqslant \ell\} \cup \{X_1 \leqslant \ell\} \cup \ldots \cup \{X_n \leqslant \ell\} = \{\min(X_0, \ldots, X_n) \leqslant \ell\} \equiv \{m_n \leqslant \ell\}.$$

The next result shows that first passage times (to a given level) of a discrete-time stochastic process are in fact stopping times with respect to the natural filtration of the process.

Proposition 6.12. *The first passage times $\mathcal{T}_\ell^\pm$ for a real-valued discrete-time process $\{X_n\}_{n \geqslant 0}$ are stopping times w.r.t. its natural filtration $\{\mathcal{F}_n := \sigma(X_0, X_1, \ldots, X_n)\}_{n \geqslant 0}$.*

Proof. Note that $\{X_k \geqslant \ell\} \in \mathcal{F}_n$ and $\{X_k \leqslant \ell\} \in \mathcal{F}_n$, since X_k is $\mathcal{F}_n$-measurable for every $0 \leqslant k \leqslant n$. Now, fix any integer $n \geqslant 0$. By the above equivalence of events we immediately have $\{\mathcal{T}_\ell^+ \leqslant n\} = \cup_{k=0}^n \{X_k \geqslant \ell\} \in \mathcal{F}_n$ and $\{\mathcal{T}_\ell^- \leqslant n\} = \cup_{k=0}^n \{X_k \leqslant \ell\} \in \mathcal{F}_n$ by closure under countable unions of sets in $\mathcal{F}_n$. $\square$

The above definitions extend to *continuous-time* real-valued processes $X \equiv \{X(t)\}_{t \geqslant 0}$. The *first hitting time to a level* $\ell \in \mathbb{R}$ is defined as the smallest nonnegative real value of time such that the process attains the value ℓ :

$$\mathcal{T}_\ell := \inf\{t \geqslant 0 : X(t) = \ell\}. \tag{6.43}$$

We put $\mathcal{T}_\ell = \infty$ if the set $\{\mathcal{T}_\ell < \infty\} = \emptyset$, that is, if there are no paths ω for which $\mathcal{T}_\ell < \infty$. More importantly, if $\mathbb{P}(\mathcal{T}_\ell < \infty) = 0$, then we say that the first hitting time to ℓ is (almost surely or with probability one) infinite. The first hitting time $\mathcal{T}_\ell$ is hence a random variable with mapping $\mathcal{T}_\ell : \Omega \to [0, \infty) \cup \{\infty\}$.

The *first passage time up*, $\mathcal{T}_\ell^+$, *and the first passage time down*, $\mathcal{T}_\ell^-$, to a level $\ell \in \mathbb{R}$, are defined as follows:

$$\mathcal{T}_\ell^+ := \inf\{t \geqslant 0 : X(t) \geqslant \ell\} \quad \text{and} \quad \mathcal{T}_\ell^- := \inf\{t \geqslant 0 : X(t) \leqslant \ell\} \tag{6.44}$$

where we set $\mathcal{T}_\ell^+ = \infty$ if $\{t \geqslant 0 : X(t) \geqslant \ell\} = \emptyset$ and, similarly, we set $\mathcal{T}_\ell^- = \infty$ if $\{t \geqslant 0 : X(t) \leqslant \ell\} = \emptyset$. Let $X(0) = x_0$ be the initial value of the process and assume the process has continuous paths (such as Brownian motion). Then, we clearly have $\mathcal{T}_\ell^+ = \mathcal{T}_\ell$ if $x_0 \leqslant \ell$ (hitting or passing above level ℓ) and $\mathcal{T}_\ell^- = \mathcal{T}_\ell$ if $x_0 \geqslant \ell$ (hitting or passing below level ℓ). That is, for processes with continuous paths the first hitting time corresponds to the appropriate first passage time.

Some continuous-time processes are defined only up to a finite time T, i.e., $\{X(t)\}_{0 \leqslant t \leqslant T}$. Then, for each of the above respective first passage times we mean $\{\mathcal{T}_\ell = \infty\} \equiv \{\mathcal{T}_\ell > T\}$ or $\{\mathcal{T}_\ell^+ = \infty\} \equiv \{\mathcal{T}_\ell^+ > T\}$ or $\{\mathcal{T}_\ell^- = \infty\} \equiv \{\mathcal{T}_\ell^- > T\}$. We therefore observe that the first passage times are the smallest nonnegative (and possibly infinite) values of time such that the process hits or goes, respectively, above or below the given level ℓ. In the trivial respective cases where the process starts at $X(0) = x_0 \geqslant \ell$ (or, respectively, $x_0 \leqslant \ell$) then $\mathcal{T}_\ell^+ = 0$ (or, respectively, $\mathcal{T}_\ell^- = 0$). By combining the two types of first passage times, we can also consider the *first exit time* $\mathcal{T}_{(a,b)}$ from an interval (a, b), assuming the process begins at an interior point $x_0 \in (a, b)$ and $\mathcal{T}_{(a,b)} = 0$ in the case $x_0 \notin (a, b)$. It follows by definition that $\mathcal{T}_{(a,b)} = \mathcal{T}_a^- \wedge \mathcal{T}_b^+ = \inf\{t \geqslant 0 : X(t) \leqslant a \text{ or } X(t) \geqslant b\}$.

Generally, the sampled maximum of a process X, up to a time $t \geqslant 0$, is defined by

$$M(t) := \sup\{X(u) : 0 \leqslant u \leqslant t\} \tag{6.45}$$

and its sampled minimum is defined by

$$m(t) := \inf\{X(u) : 0 \leqslant u \leqslant t\}. \tag{6.46}$$

Note that the process is sampled (observed) continuously in time from time zero to time $t \geqslant 0$. As in the above discrete-time case, we have a simple relationship between the first passage times and these two extreme values of the process. That is, the event that the process takes more than a given time t to first pass above level ℓ is the same as the event that its observed maximum value up to time t is less than ℓ. Similarly, the event that the process takes more than a time t to first pass below level ℓ is the event that its observed minimum value up to time t is greater than ℓ. Precisely in terms of events, we have the equivalences:

$$\{\mathcal{T}_\ell^+ > t\} = \{X(u) < \ell, \text{ for all } u \leqslant t\} \equiv \{M(t) < \ell\} \tag{6.47}$$

and

$$\{\mathcal{T}_\ell^- > t\} = \{X(u) > \ell, \text{ for all } u \leqslant t\} \equiv \{m(t) > \ell\}. \tag{6.48}$$

Based on these identity relations we can readily prove that, in the general case of a continuous-time process, the first passage or hitting times $\mathcal{T}_\ell^\pm$, and hence $\mathcal{T}_\ell$, are stopping times w.r.t. the natural filtration of the process.

Proposition 6.13. *The first passage times $\mathcal{T}_\ell^\pm$ defined for a real-valued process $\{X(t)\}_{t \geqslant 0}$ are stopping times w.r.t. the natural filtration $\mathbb{F}^X = \{\mathcal{F}_t^X\}_{t \geqslant 0}$.*

Proof. We first realize that the event $\{M(t) < \ell\} \equiv \cap_{0 \leqslant u \leqslant t}\{X(u) < \ell\} \in \mathcal{F}_t^X$ by the very definition of the natural filtration $\mathbb{F}^X$, i.e., $\mathcal{F}_t^X$ contains all events $\{X(u) \in B\}$, for any Borel set B and real times $0 \leqslant u \leqslant t$. [Note: We can equivalently write $\cap_{0 \leqslant u \leqslant t}\{X(u) < \ell\}$ as a countable intersection using only rational time values $\cap_{0 \leqslant u \leqslant t : u \in \mathbb{Q}}\{X(u) < \ell\}$, which is a set in $\mathcal{F}_t^X$ by closure under countable unions of sets $\{X(u) < \ell\} \in \mathcal{F}_t^X$.]. Then, $\{\mathcal{T}_\ell^+ > t\} = \{M(t) < \ell\}$ gives $\{\mathcal{T}_\ell^+ > t\} \in \mathcal{F}_t^X$ and hence $\{\mathcal{T}_\ell^+ > t\}^{\complement} = \{\mathcal{T}_\ell^+ \leqslant t\} \in \mathcal{F}_t^X$. The proof that $\{\mathcal{T}_\ell^- \leqslant t\} \in \mathcal{F}_t^X$ follows in the same manner and we leave it to the reader. $\qquad\square$

According to its name, a stopping time is used to stop a stochastic process. Let process $\{X(t)\}_{t \geqslant 0}$ be adapted to a filtration $\mathbb{F} = \{\mathcal{F}_t\}_{t \geqslant 0}$, and $\mathcal{T}$ be a stopping time w.r.t. $\mathbb{F}$. The process $\{X(t \wedge \mathcal{T})\}_{t \geqslant 0}$ is called a *stopped process* and is defined by

$$X(t \wedge \mathcal{T})(\omega) \equiv X(t \wedge \mathcal{T}(\omega), \omega) := \begin{cases} X(t, \omega) & \text{if } t < \mathcal{T}(\omega) \\ X(\mathcal{T}(\omega), \omega) & \text{if } t \geqslant \mathcal{T}(\omega) \end{cases}$$

for every $\omega \in \Omega$. We can represent this compactly in terms of indicator functions on events: $X(t \wedge \mathcal{T}) = X(t)\mathbb{I}_{\{t < \mathcal{T}\}} + X(\mathcal{T})\mathbb{I}_{\{t \geqslant \mathcal{T}\}}$. Note that stopping the process does not mean that time itself stops. What stopping really means is that, for all times $t \geqslant \mathcal{T}$, the process is kept constant (i.e., it is fixed, pinned down, or frozen) at the value $X(\mathcal{T}, \omega)$ taken at the stopping time $\mathcal{T} = \mathcal{T}(\omega)$ for any given realization (path) $\omega \in \Omega$.

As an example, consider the six-period binomial model with parameters $S_0 = 1, u = 1.25, d = 1/u = 0.8$, as shown in Figure 6.4. For the choice of stopping time $\mathcal{T} = \mathcal{T}_b^+$ we have $\mathcal{T}(\omega) \equiv \mathcal{T}_b^+(\omega) = 2$ for all $\omega \in A_{\mathsf{UU}}$, i.e., the first passage time above level $b = 1.4$ is $t = 2$ time steps for all 16 stock price trajectories (scenarios) with the first two upward moves. In particular, $\mathcal{T}(A_{\mathsf{UU}}) = 2$, i.e., $\mathcal{T}(\mathsf{UUDDDD}) = \mathcal{T}(\mathsf{UUUDDD}) = \mathcal{T}(\mathsf{UUDUDD}) = \ldots = \mathcal{T}(\mathsf{UUUUUU}) = 2$. For each $\omega \in A_{\mathsf{UU}}$, the stopped stock price process, denoted by $\hat{S}_t := S_{t \wedge \mathcal{T}}$, has the path $\hat{S}_0(\omega) = S_0 = 1, \hat{S}_1(\omega) = S_1 = 1.25$ and $\hat{S}_t(\omega) = S_2 = 1.5625$ for $t = 2, 3, 4, 5, 6$.

Clearly, the process X and the stopping time are both adapted to $\mathbb{F}$, the stopped process is adapted to $\mathbb{F}$. Moreover, assuming standard conditions, if the original process is a martingale w.r.t. $\mathbb{F}$, then the stopped process is also a martingale w.r.t. $\mathbb{F}$. Below we state this result and give the proof for a discrete-time process $\{X_t\}_{t=0,1,\ldots}$. However, the theorem also holds for continuous-time processes $\{X(t)\}_{t\geq 0}$.

Proposition 6.14. *Let $\mathcal{T}$ be a stopping time w.r.t. a filtration $\mathbb{F} = \{\mathcal{F}_t\}_{t=0,1,\ldots}$ and let $\{X_t\}_{t=0,1,\ldots}$ be a stochastic process adapted to $\mathbb{F}$. If the process $\{X_t\}_{t=0,1,\ldots}$ is a martingale (or supermartingale, or submartingale) w.r.t. $\mathbb{F}$, then the stopped process $\{X_{t\wedge\mathcal{T}}\}_{t=0,1,\ldots}$ is also a martingale (or supermartingale, or submartingale, respectively).*

Proof. Let $\{X_t\}_{t=0,1,\ldots}$ be a martingale w.r.t. filtration $\mathbb{F}$, then

$$\mathrm{E}[X_t \mid \mathcal{F}_{t-1}] = X_{t-1}, \quad t = 1, 2, \ldots$$

For the stopped process we have $X_{t\wedge\mathcal{T}} = \mathbb{I}_{\{\mathcal{T}<t\}}X_\mathcal{T} + \mathbb{I}_{\{\mathcal{T}\geq t\}}X_t$. Since the event $\{\mathcal{T} < t\} = \cup_{n=1}^{t-1}\{\mathcal{T} = n\}$, i.e., this is a union of mutually exclusive (disjoint) events where $\mathbb{I}_{\{\mathcal{T}<t\}} = \sum_{n=1}^{t-1}\mathbb{I}_{\{\mathcal{T}=n\}}$, then

$$\mathrm{E}[X_{t\wedge\mathcal{T}} \mid \mathcal{F}_{t-1}] = \sum_{n=1}^{t-1}\mathrm{E}[\mathbb{I}_{\{\mathcal{T}=n\}}X_n \mid \mathcal{F}_{t-1}] + \mathrm{E}[\mathbb{I}_{\{\mathcal{T}\geq t\}}X_t \mid \mathcal{F}_{t-1}]$$

$$= \sum_{n=1}^{t-1}\mathbb{I}_{\{\mathcal{T}=n\}}X_n + \mathrm{E}[\mathbb{I}_{\{\mathcal{T}\geq t\}}X_t \mid \mathcal{F}_{t-1}]$$

(since $\mathbb{I}_{\{\mathcal{T}=n\}}X_n$ is $\mathcal{F}_{t-1}$-measurable for $0 \leq n \leq t - 1$)

$$= \sum_{n=1}^{t-1}\mathbb{I}_{\{\mathcal{T}=n\}}X_n + \mathbb{I}_{\{\mathcal{T}\geq t\}}\mathrm{E}[X_t \mid \mathcal{F}_{t-1}]$$

(since $\mathbb{I}_{\{\mathcal{T}\geq t\}} = 1 - \mathbb{I}_{\{\mathcal{T}\leq t-1\}}$ is $\mathcal{F}_{t-1}$-measurable)

$$= \sum_{n=1}^{t-1}\mathbb{I}_{\{\mathcal{T}=n\}}X_n + \mathbb{I}_{\{\mathcal{T}\geq t\}}X_{t-1}$$

(upon using the above martingale property)

$$= \sum_{n=1}^{t-2}\mathbb{I}_{\{\mathcal{T}=n\}}X_n + \mathbb{I}_{\{\mathcal{T}=t-1\}}X_{t-1} + \mathbb{I}_{\{\mathcal{T}\geq t\}}X_{t-1}$$

$$= \sum_{n=1}^{t-2}\mathbb{I}_{\{\mathcal{T}=n\}}X_n + \mathbb{I}_{\{\mathcal{T}\geq t-1\}}X_{t-1} = X_{(t-1)\wedge\mathcal{T}}.$$

The last line follows since we can write $\sum_{n=1}^{t-2}\mathbb{I}_{\{\mathcal{T}=n\}}X_n = \mathbb{I}_{\{\mathcal{T}\leq t-2\}}X_\mathcal{T} = \mathbb{I}_{\{\mathcal{T}<t-1\}}X_\mathcal{T}$ and then $\mathbb{I}_{\{\mathcal{T}<t-1\}}X_\mathcal{T} + \mathbb{I}_{\{\mathcal{T}\geq t-1\}}X_{t-1} = X_{(t-1)\wedge\mathcal{T}}$. Hence, the stopped process has the martingale conditional expectation property. Also, $X_{t\wedge\mathcal{T}}$ is $\mathcal{F}_t$-measurable since $\mathcal{T}$ is $\mathcal{F}_t$-measurable and X_t is $\mathcal{F}_t$-measurable for every $t \geq 0$. Lastly, $X_{t\wedge\mathcal{T}}$ is integrable, i.e., $\mathrm{E}[|X_{t\wedge\mathcal{T}}|] < \infty$, since $t \wedge \mathcal{T} \leq t$ and $\mathrm{E}[|X_t|] < \infty$ for all $t \geq 0$. The proof for a supermartingale (or submartingale) is very similar and left to the reader. □

Let us analyze one application of how this theorem can be used in practice. In Game Theory, a martingale is a betting strategy such that the gambler doubles the bet after every loss. By doing this, the gambler guarantees that the first win would recover all previous losses plus it would give a profit equal to the original stake. Let us consider the following game of chance. Flip a coin repeatedly to generate a sequence $\omega \in \Omega_\infty := \prod_{i=1}^\infty \{H, T\}$. Win an amount equal to the stake on heads (with probability $p > 0$) and lose on tails (with probability $1 - p > 0$). Consider the "martingale strategy": if the gambler loses, then he/she doubles the stake and plays again; if he/she wins, then he/she takes the winnings and quits playing. Suppose that the initial stake is $\alpha_1 = \$1$. The stake at time $t \geqslant 2$ is

$$\alpha_t(\omega) = \alpha_t(\omega_1, \ldots, \omega_{t-1}) = \begin{cases} 2^{t-1} & \text{if } \omega_1 = \cdots = \omega_{t-1} = T, \\ 0 & \text{otherwise.} \end{cases}$$

Note that the stake at time t (t-th game) is predicated on the previous games up to time $t - 1$, i.e., α_t is $\mathcal{F}_{t-1}$-measurable. For a given outcome ω, the Profit & Loss, or wealth, at time t is

$$V_t(\omega) = \sum_{k=1}^t \alpha_k(\omega) X_k(\omega), \quad \text{where } X_k(\omega) = \begin{cases} 1 & \text{if } \omega_k = H, \\ -1 & \text{if } \omega_k = T. \end{cases}$$

The time to quit playing, $\mathcal{T} = \min\{t : \omega_t = H\}$, is a stopping time w.r.t. the natural filtration generated by the coin flips. Note that $\mathbb{P}(\mathcal{T} < \infty) = 1$ since, by the continuity of the probability measure,

$$\mathbb{P}(\mathcal{T} = \infty) = \lim_{n \to \infty} \mathbb{P}(\mathcal{T} > n) = \lim_{n \to \infty} \mathbb{P}(\omega_1 = T, \ldots, \omega_n = T) = \lim_{n \to \infty} (1 - p)^n = 0.$$

At time $\mathcal{T}$, the total winning is $\$1$:

$$V_\mathcal{T} = -1 - 2 - 4 - \cdots - 2^{\mathcal{T}-2} + 2^{\mathcal{T}-1} = 1.$$

Let us find the average total loss at time $\mathcal{T} - 1$, i.e., one step before winning:

$$\mathbb{E}[V_{\mathcal{T}-1}] = \mathbb{E}[1 - 2^{\mathcal{T}-1}] = 1 - \sum_{n=1}^\infty 2^{n-1} \mathbb{P}(\mathcal{T} = n)$$

$$\left(\text{using } \mathbb{P}(\mathcal{T} = n) = \mathbb{P}(\omega_1 = T, \ldots, \omega_{n-1} = T, \omega_n = H) = p(1-p)^{n-1}\right)$$

$$= 1 - \sum_{n=1}^\infty 2^{n-1} p(1-p)^{n-1} = 1 - p \sum_{m=0}^\infty (2 - 2p)^m = \begin{cases} 1 - \frac{p}{2p-1} & \text{if } p > \frac{1}{2} \\ -\infty & \text{if } p \leqslant \frac{1}{2} \end{cases}$$

This game of chance creates a paradox. On the one hand, with probability 1 we stop playing after a finite number of games and our total winning is $\$1$. On the other hand, the average total loss is infinite (when $p \leqslant \frac{1}{2}$) if we quit playing one step before winning. Note that the process $\{V_t\}_{t=0,1,\ldots}$ is a martingale when $p = \frac{1}{2}$, so it is a fair game but with unbounded loss.

The above proposition leads to the important result known as the Optional Sampling Theorem, as now follows.

Theorem 6.15 (Optional Sampling Theorem). *Let $\mathcal{T}$ be a stopping time w.r.t. a filtration $\mathbb{F} = \{\mathcal{F}_t\}_{t=0,1,\ldots}$ and let $\{X_t\}_{t=0,1,\ldots}$ be a martingale w.r.t. $\mathbb{F}$. If the conditions (i) $\mathbb{E}[\sup_{t \geqslant 0} |X_{t \wedge \mathcal{T}}|] < \infty$ and (ii) $\mathbb{P}(\mathcal{T} < \infty) = 1$ hold, then $X_\mathcal{T}$ is integrable and we have $\mathbb{E}[X_\mathcal{T}] = \mathbb{E}[X_0]$.*

Proof. Since $\mathbb{P}(X_\mathcal{T} < \infty) = 1$, we have $\sum_{n=0}^\infty \mathbb{I}_{\{\mathcal{T}=n\}} = 1$ and hence

$$X_\mathcal{T} = X_\mathcal{T} \sum_{n=0}^\infty \mathbb{I}_{\{\mathcal{T}=n\}} = \sum_{n=0}^\infty X_n \mathbb{I}_{\{\mathcal{T}=n\}} = \sum_{n=0}^\infty X_{n \wedge \mathcal{T}} \mathbb{I}_{\{\mathcal{T}=n\}}.$$

This gives $\mathbb{E}[|X_\mathcal{T}|] \leqslant \mathbb{E}[\sup_{t \geqslant 0} |X_{t \wedge \mathcal{T}}|] < \infty$, where the assumed boundedness condition was used. Note that the martingale property in Proposition 6.14 implies the constant expectation, $E[X_{n \wedge \mathcal{T}}] = E[X_0]$ for all integers $n \geqslant 0$, i.e., $\lim_{n \to \infty} E[X_{n \wedge \mathcal{T}}] = E[X_0]$. It is now sufficient to show that $\lim_{n \to \infty} E[X_{n \wedge \mathcal{T}}] = E[X_{\mathcal{T}}]$ since this implies $E[X_{\mathcal{T}}] = E[X_0]$. In particular,

$$|E[X_{n \wedge \mathcal{T}}] - E[X_{\mathcal{T}}]| \leqslant E[(X_{n \wedge \mathcal{T}} - X_{\mathcal{T}})\mathbb{I}_{\{\mathcal{T} > n\}}] \leqslant 2E[Y\mathbb{I}_{\{\mathcal{T} > n\}}]$$

where $Y \equiv \sup_{n \geqslant 0} |X_{n \wedge \mathcal{T}}|$. Finally, based on conditions (i)-(ii), $\lim_{n \to \infty} E[Y\mathbb{I}_{\{\mathcal{T} > n\}}] = 0$ (which we do not prove here). Hence, $\lim_{n \to \infty} E[X_{n \wedge \mathcal{T}}] = E[X_{\mathcal{T}}]$ and this proves that $E[X_{\mathcal{T}}] = E[X_0]$. □

The above theorem is also referred to as (Doob's) Optional Stopping Theorem and can be restated with differing assumptions. In particular, the same conclusion obtains with assumptions (i)-(ii) replaced by (i) $\mathbb{P}(\mathcal{T} < \infty) = 1$, (ii) $E[|X_{\mathcal{T}}|] < \infty$ and (iii) $\lim_{n \to \infty} E[X_n \mathbb{I}_{\{\mathcal{T} > n\}}] = 0$. The theorem has several applications for computing certain expectations and probabilities associated with stopping times, such as first passage (or hitting) times for stochastic processes. In particular, we can derive some important facts about random walks. The following is a simple example illustrating the application of the theorem to a one-dimensional random walk. More examples of the application of the theorem are left as exercises at the end of this chapter.

Example 6.20. Consider the random walk $\{M_n\}_{n=0,1,\ldots}$ started at zero, $M_0 = 0$, as defined at the start of Section 6.1.2.2. Let $p = \frac{1}{2}$ (i.e., symmetric random walk) and define the first exit time

$$\mathcal{T} := \min\{n \geqslant 0 : M_n = a \text{ or } M_n = b\} \tag{6.49}$$

where $a < 0$ and $b > 0$ are integers. Determine the probability that the process reaches level a before b.

Solution. Note that the random walk will reach either finite level a or b in a finite number of time steps, i.e., $\mathbb{P}(\mathcal{T} < \infty) = 1$. Moreover, the process is clearly bounded with $\sup_{n \geqslant 0} |X_{n \wedge \mathcal{T}}| \leqslant \max\{-a, b\}$, i.e., $E[\sup_{t \geqslant 0} |X_{t \wedge \mathcal{T}}|] < \infty$. Hence, conditions (i)-(ii) of Theorem 6.15 hold, giving $E[M_{\mathcal{T}}] = E[M_0] = 0$. Since the events of reaching either level before the other are mutually exclusive, we have $\mathbb{P}(M_{\mathcal{T}} = b) = 1 - \mathbb{P}(M_{\mathcal{T}} = a)$ where $\mathbb{P}(M_{\mathcal{T}} = a)$ is the probability of reaching a before b and $\mathbb{P}(M_{\mathcal{T}} = b)$ is the probability of reaching b before a. Hence,

$$E[M_{\mathcal{T}}] = 0 = a \cdot \mathbb{P}(M_{\mathcal{T}} = a) + b \cdot \mathbb{P}(M_{\mathcal{T}} = b) \implies \mathbb{P}(M_{\mathcal{T}} = a) = \frac{b}{b-a} = \frac{b}{|a|+b}$$

and $\mathbb{P}(M_{\mathcal{T}} = b) = \frac{|a|}{|a|+b}$. □

6.4   Exercises

Exercise 6.1. Define events (a)–(c) as subsets of the coin toss sample space

$$\Omega_3 = \{TTT, TTH, THT, THH, HTT, HTH, HHT, HHH\}$$

and compute their probabilities in terms of $\mathbb{P}(H) = p \in (0, 1)$.

(a) The second toss results in a head.

(b) A tail comes before a head.

(c) No heads.

Exercise 6.2. Define an indicator function $\mathbb{I}_A$ of event A as follows:

$$\mathbb{I}_A(\omega) = \begin{cases} 0 & \text{if } \omega \notin A, \\ 1 & \text{if } \omega \in A. \end{cases}$$

Prove the following properties:

(a) $\mathbb{I}_{A \cap B} = \mathbb{I}_A \cdot \mathbb{I}_B$;

(b) $\mathbb{I}_{A \cup B} = \mathbb{I}_A + \mathbb{I}_B - \mathbb{I}_A \cdot \mathbb{I}_B$;

(c) $\mathbb{I}_{A^c} = 1 - \mathbb{I}_A$.

Exercise 6.3. Express the indicator function of $(A \backslash B) \cup (B \backslash A)$ in terms of the indicator functions $\mathbb{I}_A$ and $\mathbb{I}_B$.

Exercise 6.4. Consider a three-period binomial model with $S_0 = 4$, $u = 2$, $d = 0.5$, and $p = \frac{1}{4}$.

(a) Construct a binomial tree.

(b) Determine the probability distribution of the stock price S_3. Find its expected value.

(c) Determine the probability distributions of the geometric average $\sqrt[3]{S_1 S_2 S_3}$. Find its expected value.

Exercise 6.5. Consider a binomial model with $S_0 = 4$, $u = 2$, $d = 0.5$, and $p = \frac{1}{4}$. Let $\mathcal{T} = \mathcal{T}_b^+ := \min\{n = 0, 1, 2, \ldots : S_n \geqslant 6\}$ be the first passage time above level $b = 6$. Construct the following events as subsets of the sample space Ω_3 and find their probabilities:

(a) $\{\mathcal{T} \leqslant 3\}$;   (b) $\{\mathcal{T} = 3\}$;   (c) $\{\mathcal{T} > 3\}$.

Exercise 6.6. Consider the six-period model with parameters given in Figure 6.4.

(a) Determine the events $\{\mathcal{T}_b^+ = k\}$ for $k = 0, 1, 2, 3, 4, 5, 6$ and $k = \infty$ as subsets of Ω_6.

(b) Assume $p = 1/4$ is the probability of an upward move and determine the probability mass function (i.e., distribution) of $\mathcal{T}_b^+$.

(c) Let $\hat{S}_t := S_{t \wedge \mathcal{T}}$, where $\mathcal{T} = \mathcal{T}_b^+$. Determine the probability mass function of $\hat{S}_t$ for $t = 2, 3, 4, 5, 6$.

Exercise 6.7. Roll two standard dice. For each case, determine a partition $\mathcal{P}$ of the sample space Ω of all outcomes which resolves the following information:

(a) we only know whether each face value is even or odd numbered;

(b) we only know whether or not the two face values are equal to each other;

(c) we only know whether or not the two face values differ by one.

Exercise 6.8. Obtain a σ-algebra over Ω_3 generated by:

(a) $D_3 = \#$ of D's in $\omega \in \Omega_3$;

(b) the joint pair of random variables $X_1 := \ln\left(\frac{S_2}{S_1}\right)$ and $X_2 := \ln\left(\frac{S_3}{S_2}\right)$, where S_n, $n = 0, 1, \ldots$ are the stock prices in the standard binomial model with factors $d < u$.

Exercise 6.9. Let $\mathcal{F}_n =$ be a σ-algebra generated by n coin tosses. Find the smallest n such that the following event belongs to $\mathcal{F}_n$:

(a) $A = \{$the first occurrence of heads is preceded by no more than 10 tails$\}$;

(b) $B = \{$there is at least one head in the sequence $\omega_1, \omega_2, \ldots\}$;

(c) $C = \{$the first 100 tosses produce the same outcome$\}$;

(d) $D = \{$there are no more than two heads and two tails among the first five tosses$\}$.

Exercise 6.10. Show that in the three-period binomial model the smallest σ-algebras respectively generated by the random vector (S_1, S_2, S_3) and by the product $S_1 \cdot S_2 \cdot S_3$ are not the same. What can you say about $\sigma(S_1, S_2)$ and $\sigma(S_1 \cdot S_2)$?

Exercise 6.11. Let $\{M_n\}_{n \geqslant 0}$ be a simple random walk on Ω_3. Show that the sequence $(\sigma(M_1), \sigma(M_1 + M_2), \sigma(M_1 + M_2 + M_3))$ is not a filtration.

Exercise 6.12. Consider the N-period binomial model. Let $\{\mathcal{F}_n\}_{0 \leqslant n \leqslant N}$ be the natural filtration. Show that the time-k value M_k of the random walk process, $\{M_n, 0 \leqslant n \leqslant N\}$, is $\mathcal{F}_n$-measurable iff $n \geqslant k$.

Exercise 6.13. Consider a binomial tree model with $S_0 = 1$, $u = \frac{1}{d} = 2$, and $p = \frac{1}{2}$.

(a) Obtain the partition generated by S_2, i.e., $\mathcal{P}(S_2)$.

(b) Evaluate the conditional expectation of $E[S_3 \mid S_2] \equiv E[S_3 \mid \sigma(S_2)]$.

Exercise 6.14. Consider the N-period binomial tree model with stock price process $\{S_n\}_{n \geqslant 0}$. Let $S_0 = 1$ and $u = \frac{1}{d} = e$ hold (where $e = 2.71828\ldots$ is the base of the natural logarithm). Assume the probability of an up move $\mathbb{P}(\omega_j = \mathsf{U}) = p$, for all $j = 1, \ldots, N$.

(a) Show that the log-price $L_n := \ln S_n$ is given by $L_n = 2U_n - n$ for $n = 0, 1, \ldots, N$ (where U_n is the number of upward market moves during the first n periods).

(b) Determine the $\mathbb{P}$-measure expectation $E[L_n]$ for $n = 1, 2, \ldots, N$.

(c) Derive an expression for the joint probability $\mathbb{P}(S_1 < S_2 < \ldots < S_{N-1} < S_N)$.

(d) Determine the atoms in the partition generated by the first passage time

$$\tau = \min\{n \in \{0, 1, 2, 3\} : S_n \geqslant e\}.$$

Exercise 6.15. Consider the standard multi-period binomial tree model with the stock price process $\{S_n\}_{n\geqslant 0}$ defined on the filtered probability space $(\Omega, \mathcal{F}, \mathbb{P}, \mathbb{F} = \{\mathcal{F}_n\}_{n\geqslant 0})$, with $\mathbb{F}$ as the natural filtration. Let p be the probability of an up move, u be the up factor and d be the down factor. Derive expressions for the following conditional expectations (for $n \geqslant 1$):

(a) $\mathrm{E}[f(S_{n+1}) \mid \mathsf{D}_n]$, where f is any Borel function and D_n is the number of down moves until time n.

(b) $\mathrm{E}[S_1 S_2 \cdots S_n S_{n+1} \mid \mathcal{F}_n]$.

(c) $\mathrm{E}[S_n \mid \mathsf{U}_n \geqslant n - 1]$, where U_n is the number of up moves until time n.

Exercise 6.16. Consider the binomial probability space $(\Omega_4, \mathcal{F}, \mathbb{P})$ for any $\mathbb{P}(\mathsf{U}) = p \in (0, 1)$. Compute the conditional expectation

$$\mathrm{E}[\mathsf{D}_4 \mid \{\text{at least two } \mathsf{U}\text{'s}\}].$$

Exercise 6.17. Let $\mathbb{P} = \{A_1, A_2, \ldots, A_M\}$ be a disjoint partition of a sample space Ω. Prove the law of total probability

$$\mathbb{P}(B) = \sum_{m=1}^{M} \mathbb{P}(B \mid A_m)\, \mathbb{P}(A_m), \quad B \subset \Omega.$$

Exercise 6.18. Prove (6.25) by using the fact that (since it is $\mathcal{G}$-measurable) Y is a simple random variable of the form $Y = \sum_{A' \in \mathcal{P}} \mathrm{E}[Y \mid A']\, \mathbb{I}_{A'}$. Combine this with the identity $\mathbb{I}_A \mathbb{I}_{A'} = \mathbb{I}_A$ if $A = A'$, and $\mathbb{I}_A \cdot \mathbb{I}_{A'} = 0$ if $A \neq A'$, for any two atoms $A, A' \in \mathcal{P}$.

Exercise 6.19. Consider the binomial probability space $(\Omega_N, \mathcal{F}, \mathbb{P}_N)$. Let $\mathcal{F} = \sigma(\mathsf{U}_N)$ be the smallest σ-algebra generated by the number of U's. Determine the following conditional expectations w.r.t. σ-algebra $\mathcal{F}$:

(a) $\mathrm{E}[\mathsf{D}_N \mid \mathcal{F}]$;

(b) $\mathrm{E}[\mathsf{UbD} \mid \mathcal{F}]$, where $\mathsf{UbD}(\omega)$ is the number of U's before the first D in $\omega \in \Omega_N$. Assume $N \geqslant 2$.

Exercise 6.20. Consider the binomial model on $(\Omega_N, \mathcal{F}, \mathbb{P}_N, \{\mathcal{F}_n\}_{0\leqslant n\leqslant N})$. Derive formulas for conditional expectations $\mathrm{E}_n[\mathsf{U}_m]$ and $\mathrm{E}_n[\mathsf{D}_m]$ for arbitrary n and m, $0 \leqslant n, m \leqslant N$.

Exercise 6.21. Consider a standard binomial tree model for the stock price process $\{S_n\}_{n\geqslant 0}$, where u and d are up and down factors, and p is the up-move probability. Let $\{\mathcal{F}_n\}_{n\geqslant 0}$ be the natural filtration for the stock price process.

(a) Derive a formula for the conditional expectation $\mathrm{E}\left[(S_m)^2 \mid \mathcal{F}_n\right]$ where $0 \leqslant n \leqslant m$.

(b) Determine p such that $\{X_n := (S_n)^2\}_{n\geqslant 0}$ is a martingale w.r.t. $\{\mathcal{F}_n\}_{n\geqslant 0}$. Assume $d < 1 < u$.

Exercise 6.22. Consider a three-period binomial tree model with initial price $S_0 > 0$, up move factor $u = \frac{1}{d} > 1$, down move factor $d > 0$, and probability $p \in (0, 1)$ for an up move.

(a) State the support and probability distribution of S_3.

(b) Determine the number of elements in the σ-algebra generated by S_3, i.e., $|\sigma(S_3)|$.

(c) Express the events $\{S_3 > S_1\}$ and $\{S_3 = S_1\}$ as subsets of Ω_3.

(d) Compute the probability $\mathbb{P}(S_3 \geqslant S_1)$ as a function of p.

(e) Compute the conditional probability $\mathbb{P}(S_3 > S_1 | S_1 = uS_0)$ as a function of p.

Exercise 6.23. Consider the three-period *binomial model for the stock price process* $\{S_n\}_{0 \leqslant n \leqslant 3}$ on $(\Omega_3, \mathcal{F}, \mathbb{P})$, with probability of an up move $\mathbb{P}(\mathsf{U}) = p$ and with up and down factors $u = 2$ and $d = 1/2$, respectively. Fix $S_0 = 1$ and define $M_n = \max\{S_0, S_1, \ldots, S_n\}$ as the sampled maximum of the stock price at times $n = 0, 1, 2, 3$.

(a) Determine the support (range) of *each* random variable M_1, M_2 and M_3.

(b) Determine the probability distribution (p.m.f.) of M_3.

(c) Determine the conditional expectations $\mathrm{E}[M_2 \mid \sigma(S_1)]$ and $\mathrm{E}[M_3 \mid \sigma(S_1)]$.

(d) For *each* $n = 0, 1, 2, 3$, determine the partition $\mathcal{P}_n$ (i.e., the atoms) such that $\sigma(M_n) = \sigma(\mathcal{P}_n)$, i.e., $\sigma(\mathcal{P}_0) = \sigma(M_0), \sigma(\mathcal{P}_1) = \sigma(M_1), \ldots, \sigma(\mathcal{P}_3) = \sigma(M_3)$. State your results in terms of subsets of Ω_3. Is $\{\mathcal{F}_n := \sigma(M_n)\}_{0 \leqslant n \leqslant 3}$ a filtration? Why or why not?

Exercise 6.24. Let X be any integrable random variable on a probability space $(\Omega, \mathcal{F}, \mathbb{P})$ (that is, $\mathrm{E}[|X|] < \infty$) and let process $\{X_n\}_{n=0,1,\ldots} \in S$ be defined by $X_n := \mathrm{E}[X \mid \mathcal{F}_n]$, $n = 0, 1, \ldots$, where $\mathbb{F}^X = \{\mathcal{F}_n := \sigma(X_0, X_1, \ldots, X_n)\}_{n=0,1,\ldots}$. Let $\ell \in S$ and define $\mathcal{T}_\ell := \min\{n \in \{0, 1 \ldots\} : X_n \geqslant \ell\}$.
Prove that $\{X_{n \wedge \mathcal{T}_\ell}\}_{n=0,1,\ldots}$ is a martingale w.r.t. $\mathbb{F}^X$.

Exercise 6.25. Let $X_0, X_1, X_2, \ldots$ be a sequence of i.i.d. random variables with common moments $\mathrm{E}[X_1] = 0$ and $\mathrm{E}[X_1^2] = b^2$. Let filtration $\mathbb{F}^X = \{\mathcal{F}_n\}_{n \geqslant 0}$ be defined by $\mathcal{F}_n := \sigma(X_0, X_1, \ldots, X_n)$. Prove that the following processes are martingales w.r.t. $\mathbb{F}^X$:

(a) $Y_n = \sum\limits_{k=0}^{n} X_k$, $n = 0, 1, 2, \ldots$;

(b) $Z_n = \left(\sum\limits_{k=0}^{n} X_k \right)^2 - b^2 \, n$, $n = 0, 1, 2, \ldots$.

Exercise 6.26. Let $\{M_n\}_{n \geqslant 0}$ be a symmetric simple random walk. Show that the process $\{Y_n\}_{n \geqslant 0}$ defined by
$$Y_n = (-1)^n \cos(\pi M_n)$$
is a martingale w.r.t. the natural filtration $\mathbb{F}^Y = \{\mathcal{F}_n := \sigma(Y_0, Y_1, \ldots, Y_n)\}_{n=0,1,\ldots}$.
[Hint: Make use of the identity $\cos(a + b) = \cos a \cos b - \sin a \sin b$ when proving the martingale expectation property.]

Exercise 6.27. Let $\{M_n\}_{n \geqslant 0}$ be a symmetric simple random walk. Fix a real parameter θ. Prove that the process $S_n = e^{\theta M_n} \left(\frac{2}{e^\theta + e^{-\theta}} \right)^n$, $n = 0, 1, 2, \ldots$, is a martingale w.r.t. the natural filtration $\mathbb{F}^S = \{\mathcal{F}_n := \sigma(S_0, S_1, \ldots, S_n)\}_{n=0,1,\ldots}$.

Exercise 6.28. Let $\{M_n\}_{n \geqslant 0}$ be a symmetric simple random walk. Prove that the process $M_n^2 - n$, $n = 0, 1, 2, \ldots$ is a martingale w.r.t. the natural filtration generated by the random walk, $\{\mathcal{F}_n := \sigma(M_0, M_1, \ldots, M_n)\}_{n=0,1,\ldots}$.

Exercise 6.29. Using a similar procedure as in Example 6.17, show that the random walk $\{M_n\}_{n \geqslant 0}$ is a Markov process with stationary and independent increments.

Exercise 6.30. Let $\{Z_n\}_{n \geqslant 0}$ be any sequence of square integrable random variables, i.e., assume $\mathrm{E}[Z_n^2] < \infty$ for all $n \geqslant 0$. Show that if the process $\{Z_n\}_{n \geqslant 0}$ is a martingale w.r.t. a given filtration $\{\mathcal{F}_n\}_{n \geqslant 0}$, then the (squared) process $\{Z_n^2\}_{n \geqslant 0}$ is a submartingale w.r.t. $\{\mathcal{F}_n\}_{n \geqslant 0}$.

Exercise 6.31. Let $\{M_n\}_{n \geqslant 0}$ be an *asymmetric* simple random walk, i.e., $M_0 = 0$ and $M_n = \sum_{k=1}^{n} X_k$, $n \geqslant 1$, where $\{X_k\}_{k \geqslant 0}$ is a sequence of i.i.d. random variables such that $X_k = 1$ with probability $p \neq \frac{1}{2}$ and $X_k = -1$ with probability $q = 1 - p$, for $k \geqslant 1$, $p \in (0,1)$. Show that $\{Z_n := M_n - n(p-q); \ n = 0,1,2,\ldots\}$ is a martingale w.r.t. its natural filtration.

Exercise 6.32. Let $\{Y_k\}_{k \geqslant 0}$ be any sequence of random variables and assume that $X_n := \sum_{k=0}^{n} Y_k$, $n \geqslant 0$, is a martingale w.r.t. filtration $\{\mathcal{F}_n = \sigma(X_0, \ldots, X_n)\}_{n \geqslant 0}$. Show that $\mathrm{E}[Y_i Y_j] = 0$, i.e., that Y_i and Y_j, for all $j \neq i$, are uncorrelated.
[Hint: Consider the expectation of nonoverlapping increments of the process and then apply the tower and martingale properties.]

Exercise 6.33. Let $\{Y_k\}_{k=1,\ldots}$ be a sequence of i.i.d. random variables with common moment-generating function $\phi(\lambda) = \mathrm{E}[e^{\lambda Y_k}]$ defined on an interval $\lambda \in (-\delta, \delta)$, for some $\delta > 0$. Show that the process $\{X_n\}_{n=0,1,\ldots}$ defined by $X_0 = 1$ and

$$X_n = [\phi(\lambda)]^{-n} \prod_{k=1}^{n} e^{\lambda Y_k}, \ n = 1, 2, \ldots,$$

is a martingale w.r.t. its natural filtration.

Exercise 6.34. Let $f(x) > 0$ and $g(x)$ be two probability density functions defined on some interval $I \subset \mathbb{R}$ and $\{Y_k\}_{k=0,1,\ldots}$ be a sequence of i.i.d. continuous random variables on I with common probability density function $f(x) > 0$. Show that the sequence $\{X_n\}_{n=0,1,\ldots}$ of likelihood ratios defined by

$$X_n = \prod_{k=0}^{n} \frac{g(Y_k)}{f(Y_k)}, \ n = 0, 1, \ldots,$$

is a martingale w.r.t. its natural filtration.

Exercise 6.35. Let $\{Y_k\}_{k=0,1,\ldots} \in I$, $I \subset \mathbb{R}$, be a Markov process defined by the (one-step) transition function

$$P(y|x) := \mathbb{P}(Y_{n+1} \leqslant y | Y_n = x)$$

and transition density $p(y|x) = \frac{\partial}{\partial y} P(y|x)$, $x, y \in I$. Assume $\mathrm{E}[|f(Y_n)|] < \infty$, for all $n \geqslant 0$, and that

$$\lambda f(x) = \int_I f(y) p(y|x) dy,$$

$\forall x \in I$, for some real parameter λ and a given function $f : I \to \mathbb{R}$. Show that the process defined by $X_n := \lambda^{-n} f(Y_n)$, $n = 0, 1, \ldots$, is a martingale w.r.t. its natural filtration.

Exercise 6.36. Consider the multi-period binomial model. Let $\overline{S}_n := \dfrac{S_n}{(1+r)^n}$, $n \geqslant 0$, be a discounted stock price process with the interest rate $r \geqslant 0$. Find the up-move probability p such that $\{\overline{S}_n\}_{n \geqslant 0}$ is a martingale w.r.t. the natural filtration generated by the stock price process.

Exercise 6.37. Assume a standard binomial stock price process $\{S_t\}_{t=0,1,\ldots}$. In each case below, show (or give a brief explanation of) whether or not the defined random variable $\mathcal{T}$ is a stopping time with respect to the natural filtration generated by the stock price process:

(a) $\mathcal{T} := \min\{t \geqslant 0 \ : \ S_t > S_0\}$,

(b) $\mathcal{T} := \min\{t \geqslant 0 \; : \; S_t > S_{t+1}\}$,

(c) $\mathcal{T} := \min\{t \geqslant 1 \; : \; S_t > S_{t-1}\}$,

(d) $\mathcal{T} := \min\{t \geqslant 1 \; : \; D_t > 1\}$, where D_t is the number of down moves until time t.

Exercise 6.38. Show that if $\mathcal{T}$ is a stopping time w.r.t. some filtration then so is $\mathcal{T} \wedge m := \min\{\mathcal{T}, m\}$ for any constant $m \geqslant 0$.

Exercise 6.39. Show that if $\mathcal{T}_1$ and $\mathcal{T}_2$ are stopping times w.r.t. some filtration then so are $\mathcal{T}_1 \wedge \mathcal{T}_2 := \min\{\mathcal{T}_1, \mathcal{T}_2\}$ and $\mathcal{T}_1 \vee \mathcal{T}_2 := \max\{\mathcal{T}_1, \mathcal{T}_2\}$.

Exercise 6.40. Consider the asymmetric random walk $\{M_n\}_{n \geqslant 0}$, as defined in Exercise 6.31. Let $\mathcal{T}$ be the first exit time, as defined in (6.49) of Example 6.20.

(a) Apply Theorem 6.15 to the process $\{Z_n\}_{n \geqslant 0}$, defined in Exercise 6.31, to show that

$$\mathrm{E}[M_{\mathcal{T}}] = \mu \mathrm{E}[\mathcal{T}]$$

where $\mu = p - q$.

(b) Show that the process $X_n := \left(\frac{q}{p}\right)^{M_n}$, $n \geqslant 0$, is a martingale w.r.t. its natural filtration.

(c) Apply Theorem 6.15 to the process in (b) and derive a formula for $\mathbb{P}(M_{\mathcal{T}} = a)$, the probability that the asymmetric random walk reaches integer level $a < 0$ before integer level $b > 0$.

(d) Based on (a) and (c), obtain a formula for $\mathrm{E}[\mathcal{T}]$, assuming $q \neq p$.

7

Replication and Pricing in the Binomial Tree Model

7.1 The Standard Binomial Tree Model

By combining the probabilistic framework in Chapter 6 with the main formal concepts of derivative asset pricing presented for the single-period model in Chapter 5, we are now ready to formally discuss derivative asset pricing within the multi-period binomial tree model. Let us begin by recalling (from Chapter 6) the salient features of the *standard T-period (recombining) binomial tree model* on the space $(\Omega, \mathbb{P}, \mathcal{F}, \mathbb{F})$ with two assets, namely, a risky stock S and a risk-free asset B, such as a bank account. The model is specified as follows.

- The time is discrete: $t \in \{0, 1, 2, \ldots, T\}$.

- There are 2^T possible market scenarios:

$$\Omega \equiv \Omega_T = \{\omega = \omega_1 \omega_2 \cdots \omega_T \ : \ \omega_t \in \{\mathsf{D}, \mathsf{U}\}, \ t = 1, 2, \ldots, T\},$$

  where each market outcome (scenario) $\omega \in \Omega_T$ can be represented by a path in a multi-period recombining binomial tree.

- The set of events is the power set $\mathcal{F} = 2^\Omega$.

- The (physical) probability function $\mathbb{P}: \mathcal{F} \to [0, 1]$ is given by

$$\mathbb{P}(E) = \sum_{\omega \in E} \mathbb{P}(\omega), \ E \in \mathcal{F}, \ \text{where} \ \mathbb{P}(\omega) \equiv \mathbb{P}(\{\omega\}) = p^{\#\mathsf{U}(\omega)}(1-p)^{\#\mathsf{D}(\omega)}, \qquad (7.1)$$

  and $p \in (0, 1)$ is the (physical) probability of the event $\{\omega_t = \mathsf{U}\} = \{\omega \in \Omega \ : \ \omega_t = \mathsf{U}\}$ for every $t = 1, 2, \ldots, T$. Note that the market moves are assumed to be mutually independent in the standard binomial model.

- The flow of information is described by the filtration $\mathbb{F} = \{\mathcal{F}_t\}_{0 \leqslant t \leqslant T}$, where $\mathcal{F}_0 = \emptyset$ and $\mathcal{F}_t$ is generated by the first t market moves $\omega_1, \ldots, \omega_t$ for every $t = 1, 2, \ldots, T$, i.e., $\mathcal{F}_t = \sigma(\mathcal{P}_t)$, where the partition $\mathcal{P}_t$ is a collection of atoms of the form

$$A_{\omega_1^*, \omega_2^*, \ldots, \omega_t^*} = \{\omega \in \Omega \ : \ \omega_n = \omega_n^* \text{ for all } n = 1, 2, \ldots, t\}, \quad \omega_1^*, \omega_2^*, \ldots, \omega_t^* \in \{\mathsf{D}, \mathsf{U}\}$$

  (in particular, $\mathcal{F}_T \equiv \mathcal{F} = 2^\Omega$).

- The stock price process, $\{S_t\}_{0 \leqslant t \leqslant T}$, is adapted to the filtration $\mathbb{F}$, i.e., S_t is $\mathcal{F}_t$-measurable with $S_t(\omega) = S_t(\omega_1 \ldots \omega_t)$ for all t, $\omega \in \Omega_T$. In the case of zero dividends, the stock price is given by the recurrence

$$S_{t+1}(\omega) = Y_{t+1}(\omega)S_t(\omega), \quad t = 0, 1, \ldots, T-1, \qquad (7.2)$$

  where $Y_{t+1}(\omega) = Y_{t+1}(\omega_{t+1}) := u\mathbb{I}_{\{\omega_{t+1}=\mathsf{U}\}} + d\mathbb{I}_{\{\omega_{t+1}=\mathsf{D}\}} \equiv u^{\#\mathsf{U}(\omega_{t+1})}d^{\#\mathsf{D}(\omega_{t+1})}$, i.e.,

$Y_{t+1}(\mathsf{U}) = u, Y_{t+1}(\mathsf{D}) = d$. Equivalently, we have the stock price in terms of all the market moves up to time t:

$$S_t(\omega) = S_0 \prod_{k=1}^{t} Y_k(\omega_k) = S_0 u^{\mathsf{U}_t(\omega)} d^{\mathsf{D}_t(\omega)}, \quad t = 0, 1, \ldots, T, \tag{7.3}$$

where D_t and U_t, defined in (6.4)-(6.5), count the respective number of downward and upward market moves in the sequence $\omega_1 \ldots \omega_t$; d and u are, respectively, downward and upward market movement factors (assumed constant) satisfying $0 < d < u$. The initial price of the stock, $S_0 > 0$, is known (not random). The random variables $\{Y_k\}_{k=1,\ldots,T}$ are i.i.d. with Y_{t+1} being $\mathcal{F}_t$-independent.

In the case of a dividend paying stock, equation (7.2) is modified to

$$S_{t+1}(\omega) = (1 - q_{t+1}(\omega))Y_{t+1}(\omega)S_t(\omega) \tag{7.4}$$

and (7.3) modified to

$$S_t(\omega) = S_0 \prod_{k=1}^{t} (1 - q_k(\omega))Y_k(\omega_k) = S_0 c_t(\omega) \prod_{k=1}^{t} Y_k(\omega_k) = S_0 c_t(\omega) u^{\mathsf{U}_t(\omega)} d^{\mathsf{D}_t(\omega)}. \tag{7.5}$$

The adapted process defined by $c_0 = 1$ and

$$c_t := (1 - q_1) \times \cdots \times (1 - q_t), \tag{7.6}$$

$1 \leqslant t \leqslant T$, where $q_k(\omega) = q_k(\omega_1 \ldots \omega_k) \in (0, 1)$, $k = 1, \ldots, T$, is the time-t cumulative after-dividend factor. This factor accounts for the drop in the stock price due to all dividends paid in proportion to the stock price at the end of every time interval until time t. At time $t+1$ of each time interval $[t, t+1]$, a stock dividend in the amount of $q_{t+1}Y_{t+1}S_t$ is paid to the holder. Hence, to avoid arbitrage, the stock price must drop by this same amount. Hence, the stock price S_{t+1} is the value at time $t + 1$ due to the market moves (not including dividends) minus the dividend amount, i.e., $S_{t+1} = Y_{t+1}S_t - q_{t+1}Y_{t+1}S_t$. This is the recurrence relation in (7.4). We note that in a more general (non-recombining) binomial model (as discussed in Chapter 8) we can allow the dividend factors to be stochastic where $c_t(\omega) = c_t(\omega_1 \ldots \omega_t) = \prod_{k=1}^{t}(1 - q_k(\omega_1 \ldots \omega_k))$, $c_{t+1} = (1 - q_{t+1})c_t$. The relations simplify in the obvious manner when q_k are constant. Moreover, the zero dividend case is recovered when all $q_k = 0$. For a standard binomial model in which the stock price tree is recombining, we can take $q_k(\omega) = q_k(\omega_k)$ with all q_k as i.i.d., where the dividends have a common value for an upward move and a downward move: $q_1(\mathsf{U}) = \cdots = q_T(\mathsf{U})$ and $q_1(\mathsf{D}) = \cdots = q_T(\mathsf{D})$. A simpler case of a recombining stock price tree is when each q_t is nonrandom (but possibly dependent on time t).

- The deterministic value process, $\{B_t\}_{0 \leqslant t \leqslant T}$, for the risk-free asset (bank account) is given by

$$B_t = B_0(1 + r)^t, \quad t = 0, 1, 2, \ldots, T, \tag{7.7}$$

where $r > 0$ is a one-period return. Note that a unit zero-coupon bond paying \$1 at time T, has initial value $(1 + r)^{-T}$.

We recall the discussion in Section 5.5.2 of Chapter 5. A single-period binomial-tree model admits no (static) arbitrage portfolios in base assets iff there exists an equivalent martingale measure (EMM) $\widetilde{\mathbb{P}}^{(g)}$ for numéraire $g \in \{B, S\}$. An EMM $\widetilde{\mathbb{P}}^{(g)}$ is defined so that it is equivalent to the real-world (physical) measure $\mathbb{P}$ and all base asset price processes discounted by a numéraire asset price process g are $\widetilde{\mathbb{P}}^{(g)}$-martingales. Let us extend this idea to the multi-period case. The probability function in (7.1) is specified by a single probability $p = \mathbb{P}(\mathsf{U})$ of an upward move over a single period. The respective risk-neutral probabilities of an upward move, $\tilde{p}^{(g)} = \widetilde{\mathbb{P}}^{(g)}(\mathsf{U})$, for either choice $g = B$ or $g = S$, are (recall equation (5.48)):

$$\tilde{p} \equiv \tilde{p}^{(B)} = \frac{1 + r - d}{u - d} \quad \text{or} \quad \tilde{q} \equiv \tilde{p}^{(S)} = \tilde{p} \cdot \frac{u}{1 + r} = \frac{1 + r - d}{u - d} \cdot \frac{u}{1 + r}. \tag{7.8}$$

In either case, the probability $\tilde{p}^{(g)} \in (0, 1)$ exists iff

$$d < 1 + r < u. \tag{7.9}$$

In the standard multi-period binomial model, the respective single-period $\widetilde{\mathbb{P}}^{(g)}$-measure probabilities in (7.8) arise as the defining martingale property of the discounted (ratio) processes $\frac{B_t}{g_t}$ and $\frac{S_t}{g_t}$, $0 \leqslant t \leqslant T$, where the stock price S_t is taken to follow (7.2). Hence, throughout this chapter we define the respective $\tilde{p}^{(B)}$ and $\tilde{p}^{(S)}$-measure probabilities of an upward move, $\tilde{p}^{(B)}$ and $\tilde{p}^{(S)}$, by (7.8). Hence, by replacing in (7.1) the (physical) probability p of an upward move by a risk-neutral probability $\tilde{p}^{(g)}$, given by (7.8) in either case with $g = B$ or $g = S$, we construct a risk-neutral probability measure $\widetilde{\mathbb{P}}^{(g)}$ defined on the σ-algebra $\mathcal{F} = 2^{\Omega}$. As is proved below, (non-dividend paying) base asset price processes discounted by a (non-dividend paying) numéraire g are $\widetilde{\mathbb{P}}^{(g)}$-martingales iff the condition in (7.9) holds. The probability measures $\widetilde{\mathbb{P}}^{(B)}$ and $\widetilde{\mathbb{P}}^{(S)}$ are equivalent martingale measures (EMMs) for the standard multi-period binomial model.

Theorem 7.1. *Assuming the stock pays no dividends, the discounted base asset price processes*

$$\left\{ \overline{S}_t := \frac{S_t}{g_t} \right\}_{0 \leqslant t \leqslant T} \quad \text{and} \quad \left\{ \overline{B}_t := \frac{B_t}{g_t} \right\}_{0 \leqslant t \leqslant T}$$

are $\widetilde{\mathbb{P}}^{(g)}$-martingales for $g \in \{B, S\}$, i.e.,

$$\widetilde{\mathrm{E}}_t^{(g)}[\overline{S}_{t+1}] = \overline{S}_t \quad \text{and} \quad \widetilde{\mathrm{E}}_t^{(g)}[\overline{B}_{t+1}] = \overline{B}_t$$

holds for all $t = 0, 1, \ldots, T - 1$, iff (7.9) holds. If the stock pays dividends, according to (7.4), then

$$\left\{ \frac{S_t}{c_t B_t} \right\}_{0 \leqslant t \leqslant T} \quad \text{and} \quad \left\{ \frac{c_t B_t}{S_t} \right\}_{0 \leqslant t \leqslant T}$$

are $\widetilde{\mathbb{P}}^{(B)}$- and $\widetilde{\mathbb{P}}^{(S)}$-martingales, respectively, iff (7.9) holds.

Proof. Assume no dividends on the stock. Let us consider the case with $g = B$ (the other case with $g = S$ is treated similarly). The discounted risk-free asset price process $\overline{B}_t$ is equal to 1 for all times t and hence the process $\{\overline{B}_t\}$ is a martingale. We now show that the process $\{\overline{S}_t\}$ is a martingale under $\widetilde{\mathbb{P}} \equiv \widetilde{\mathbb{P}}^{(B)}$. Fix arbitrarily $t \in \{0, 1, \ldots, T - 1\}$. Using (7.2), we have

$$\widetilde{\mathrm{E}}_t[\overline{S}_{t+1}] \equiv \widetilde{\mathrm{E}}_t\left[\frac{S_{t+1}}{B_{t+1}}\right] = \widetilde{\mathrm{E}}_t\left[\frac{Y_{t+1} S_t}{(1 + r) B_t}\right] = \frac{S_t}{B_t} \widetilde{\mathrm{E}}\left[\frac{Y_{t+1}}{1 + r}\right] = \frac{S_t}{B_t} \equiv \overline{S}_t.$$

We used the fact that Y_{t+1} is $\mathcal{F}_t$-independent and S_t is $\mathcal{F}_t$-measurable. The reader can verify that $\widetilde{\mathrm{E}}[Y_{t+1}] = 1 + r$ by employing the above probabilities $(\tilde{p}, 1 - \tilde{p})$ and the definition of Y_{t+1}. The martingale condition is fulfilled iff $\tilde{p} = \frac{1 + r - d}{u - d}$. The condition $\tilde{p} \in (0, 1)$ is equivalent to (7.9). The proof of the respective martingale properties for the case that the stock pays dividends is left as an exercise (see Exercise 7.11). $\qquad\square$

Note that the above $\widetilde{\mathbb{P}}^{(B)}$-martingale property of the discounted (non-dividend) stock price process is also written equivalently as:

$$S_t(\omega_1 \ldots \omega_t) = \frac{1}{1+r} \left[\tilde{p} S_{t+1}(\omega_1 \ldots \omega_t \mathsf{U}) + (1 - \tilde{p}) S_{t+1}(\omega_1 \ldots \omega_t \mathsf{D}) \right] \qquad (7.10)$$

where (7.2) gives $S_{t+1}(\omega_1 \ldots \omega_t \mathsf{U}) = u S_t(\omega_1 \ldots \omega_t)$ and $S_{t+1}(\omega_1 \ldots \omega_t \mathsf{D}) = d S_t(\omega_1 \ldots \omega_t)$, for every sequence of market moves $\omega_1 \ldots \omega_t$, $t \in \{0, 1, \ldots, T-1\}$. In the dividend paying case, we have the $\widetilde{\mathbb{P}}^{(B)}$-martingale property of $\left\{ \frac{S_t}{c_t B_t} \right\}_{0 \leqslant t \leqslant T}$ stated equivalently as:

$$S_t(\omega_1 \ldots \omega_t) = \frac{1}{1+r} \left[\tilde{p} \frac{S_{t+1}(\omega_1 \ldots \omega_t \mathsf{U})}{1 - q_{t+1}(\omega_1 \ldots \omega_t \mathsf{U})} + (1 - \tilde{p}) \frac{S_{t+1}(\omega_1 \ldots \omega_t \mathsf{D})}{1 - q_{t+1}(\omega_1 \ldots \omega_t \mathsf{D})} \right] \qquad (7.11)$$

where (7.4) gives $\frac{S_{t+1}(\omega_1 \ldots \omega_t \mathsf{U})}{1 - q_{t+1}(\omega_1 \ldots \omega_t \mathsf{U})} = u S_t(\omega_1 \ldots \omega_t)$ and $\frac{S_{t+1}(\omega_1 \ldots \omega_t \mathsf{D})}{1 - q_{t+1}(\omega_1 \ldots \omega_t \mathsf{D})} = d S_t(\omega_1 \ldots \omega_t)$.

The above result tells us that an EMM $\widetilde{\mathbb{P}}^{(g)}$ for $g \in \{B, S\}$ exists iff (7.9) holds. Thus, according to the fundamental theorem of asset pricing (proved for the single-period case), there are no arbitrage portfolios iff $d < 1 + r < u$, i.e., the return on the risk-free asset is strictly between the downward and upward returns on the stock: $d - 1 < r < u - 1$. [We remark, from the discussion just below (7.6), that an investment (portfolio) consisting of only one share (unit) of the stock at time t, say $\Pi_t = S_t$, will have a time-$(t+1)$ net value given by $\Pi_{t+1} = Y_{t+1} S_t$ regardless of whether the stock pays a dividend since the (after-dividend) drop in the share price of the stock is exactly compensated by the dividend payout to the investor. The single-period return on the stock investment within $[t, t+1]$ is then $\frac{\Pi_{t+1}}{\Pi_t} - 1 = Y_{t+1} - 1$, which is either $Y_{t+1}(\mathsf{U}) - 1 = u - 1$ or $Y_{t+1}(\mathsf{D}) - 1 = d - 1$.]

In the next sections, we will introduce the notion of an arbitrage portfolio strategy and will prove the fundamental theorems in the multi-period case. It turns out, the condition (7.9) guarantees the absence of arbitrage strategy in the binomial tree model as well.

7.2   Self-Financing Strategies and Their Value Processes

Consider an investor who begins with an initial capital to be invested in base securities. Suppose that injecting or withdrawing funds is not allowed in the future time, although the investor can modify the investment portfolio by changing the positions in base assets. For example, the investor may sell some stock shares and invest the proceeds without risk. As a result, a sequence of investment portfolios in the base assets indexed by time is constructed. Recall from Section 2.2.5 in Chapter 2 that such a sequence of portfolios that does not allow for injecting or withdrawing funds is called a *self-financing strategy*. A self-financing strategy allows the investor to create a portfolio with a target probability distribution or hedge a cash flow during a period of time. Self-financing strategies are important for the no-arbitrage pricing of derivative securities when combined with replication in the multi-period setting where trading (i.e., portfolio re-balancing) in the base assets is allowed at times $t = 0, 1, \ldots, T$. The simplest example of a self-financing strategy is a static portfolio in the base assets that does not change in time.

For the binomial model there are only two base assets, namely, a risky stock S and a risk-free bank account B. Thus, any investment (or trading) strategy Φ is a sequence of portfolios in the two base assets:

$$\Phi = \{\varphi_t\}_{0 \leqslant t \leqslant T-1}, \text{ where } \varphi_t = (\beta_t, \Delta_t).$$

Throughout we shall use β_t and Δ_t to denote the respective positions in assets B and S at time t. For each $t = 0, 1, \ldots, T - 1$, the portfolio φ_t is formed at time t and held until time $t + 1$, i.e., $\varphi_t = (\beta_t, \Delta_t)$ is the portfolio held in the time period $[t, t + 1)$. At time $t + 1$ the investor can re-balance the portfolio to form the new portfolio $\varphi_{t+1} = (\beta_{t+1}, \Delta_{t+1})$, which is held in the time period $[t + 1, t + 2)$, and so on. At each trading time, we will insist that the re-balancing of the positions must satisfy the self-financing condition.

Let us begin with time $t = 0$. Assume zero stock dividend. The investor begins with a given initial capital or wealth Π_0 which completely finances the initial portfolio with positions $\varphi_0 = (\beta_0, \Delta_0)$ in the base securities, i.e., with acquisition value

$$\Pi_0 = \Pi_0[\varphi_0] := \Delta_0 S_0 + \beta_0 B_0.$$

We use the notation $\Pi_t[\varphi]$ to denote the time-t value of a portfolio $\varphi = (\beta, \Delta)$ in the base assets B and S:

$$\Pi_t[\varphi] \equiv \Pi_t[(\beta, \Delta)] := \Delta S_t + \beta B_t.$$

By the above self-financing of the initial portfolio, the initial position Δ_0 in the stock determines the initial investment in the risk-free asset:

$$\beta_0 = \frac{\Pi_0 - \Delta_0 S_0}{B_0} \implies \Pi_0 = \Delta_0 S_0 + \left(\frac{\Pi_0 - \Delta_0 S_0}{B_0} \right) B_0.$$

The investor holds this portfolio until a time just prior to time $t = 1$, and at time $t = 1$ the investor liquidates it to form a new one. The liquidation value is the value of the portfolio with the positions being those at time 0 but with the prices of the base assets being those at the present time $t = 1$:

$$\Pi_1 := \Pi_1[\varphi_0] = \Delta_0 S_1 + \beta_0 B_1 = \Delta_0 S_1 + (\Pi_0 - \Delta_0 S_0)\frac{B_1}{B_0} = \Delta_0 S_1 + (\Pi_0 - \Delta_0 S_0)(1 + r).$$

These proceeds are used entirely to finance the formation of a new portfolio $\varphi_1 = (\beta_1, \Delta_1)$ with the same acquisition value $\Pi_1 = \Pi_1[\varphi_1] = \Delta_1 S_1 + \beta_1 B_1$. This is the self-financing condition applied at time $t = 1$, giving

$$\beta_1 = \frac{\Pi_1 - \Delta_1 S_1}{B_1} \implies \Pi_1 = \Delta_1 S_1 + \left(\frac{\Pi_1 - \Delta_1 S_1}{B_1} \right) B_1.$$

The position Δ_1 is determined based on the information available at time 1, i.e., $\Delta_1 = \Delta_1(\omega_1)$. By repeating the same procedure at every time step, we obtain general formulae for the equivalent liquidation and acquisition values Π_t, $t = 1, \ldots, T$, for any self-financing strategy:

$$\Pi_t := \Pi_t[\varphi_{t-1}] = \Delta_{t-1} S_t + (\Pi_{t-1} - \Delta_{t-1} S_{t-1})(1 + r), \tag{7.12}$$

$$= \Pi_t[\varphi_t] = \Delta_t S_t + \underbrace{\left(\frac{\Pi_t - \Delta_t S_t}{B_t} \right)}_{=\beta_t} B_t, \tag{7.13}$$

where $\varphi_t = (\beta_t, \Delta_t)$ is a portfolio in the base assets B and S formed at time $t = 0, 1, \ldots, T - 1$. Since the liquidation value and the acquisition value of a self-financing strategy are the same at every date $t \geqslant 0$, we will only speak of the value Π_t of a self-financing strategy.

At every time step $t \geqslant 0$, the positions Δ_t and β_t are determined based on the market information available at time t. In other words, Δ_t and β_t depend on the first t market moves and we express this as

$$\Delta_t(\omega) = \Delta_t(\omega_1 \ldots \omega_t), \quad \beta_t(\omega) = \beta_t(\omega_1 \ldots \omega_t).$$

Therefore, the portfolio process $\{\varphi_t\}_{0 \leqslant t \leqslant T-1}$ is adapted to the natural filtration $\mathbb{F}$, i.e., Δ_t and β_t are $\mathcal{F}_t$-measurable random variables for all $t \geqslant 0$.

Clearly, any self-financing strategy in the binomial tree model is fully described by the process $\{\Delta_t\}_{0 \leqslant t \leqslant T-1}$ of stock positions (i.e., the delta positions) and the initial value Π_0. The positions β_t can be calculated with the use of the self-financing condition:

$$\beta_t = \frac{\Pi_t - \Delta_t S_t}{B_t}. \tag{7.14}$$

Since the delta process is adapted to the natural filtration $\mathbb{F}$, the value process is expected to be adapted to $\mathbb{F}$, as is proved in the next proposition.

Proposition 7.2. *Let* $\{\Delta_t\}_{0 \leqslant t \leqslant T-1}$ *be a process adapted to the natural filtration* $\mathbb{F} = \{\mathcal{F}_t\}_{0 \leqslant t \leqslant T}$ *of a standard binomial tree model with non-dividend paying stock, and let* Π_0 *be the initial known capital. Then the value process* $\{\Pi_t\}_{0 \leqslant t \leqslant T}$ *defined recursively by the* **wealth equation**

$$\Pi_{t+1} = \Delta_t S_{t+1} + (\Pi_t - \Delta_t S_t)(1+r), \quad 0 \leqslant t \leqslant T-1, \tag{7.15}$$

where $S_{t+1} = Y_{t+1}S_t$, *is adapted to the natural filtration as well, i.e.,* $\Pi_t(\omega) = \Pi_t(\omega_1 \omega_2 \ldots \omega_t)$, *for all* $\omega \in \Omega$, $0 \leqslant t \leqslant T$.

Proof. Let us prove the assertion by induction. The initial value Π_0 is an $\mathcal{F}_0$-measurable constant. For any $0 \leqslant t \leqslant T-1$, the right-hand side of (7.15), giving Π_{t+1}, is $\mathcal{F}_{t+1}$-measurable since it is a linear combination of $\mathcal{F}_{t+1}$-measurable variables Y_{t+1}, S_t, Δ_t, and Π_t which is assumed to be $\mathcal{F}_t$-measurable. Hence, Π_t is $\mathcal{F}_t$-measurable implies Π_{t+1} is $\mathcal{F}_{t+1}$-measurable for all $0 \leqslant t \leqslant T-1$. $\qquad \square$

Note: In the case that the stock pays dividends according to (7.4), the wealth equation has the same form as in (7.15):

$$\begin{aligned} \Pi_{t+1} &= \Delta_t S_{t+1} + (\Pi_t - \Delta_t S_t)(1+r) + \Delta_t q_{t+1} Y_{t+1} S_t \\ &= \Delta_t Y_{t+1} S_t + (\Pi_t - \Delta_t S_t)(1+r). \end{aligned} \tag{7.16}$$

Note that the stock price process in (7.16) follows (7.4), whereas it follows (7.2) in (7.15). The simple argument for the validity of (7.16) is left as an exercise for the reader (see Exercise 7.11).

Example 7.1. Consider a three-period recombining binomial tree model with $S_0 = 8$, $B_0 = 1$, $u = \frac{3}{2}$, $d = \frac{1}{2}$, and $r = \frac{1}{4}$. Assume zero dividends on the stock. Determine the terminal value of a self-financing strategy with the initial value $\Pi_0 = 10$ and stock positions Δ_t given as the number of upward moves in the first t market movements for each $t = 0, 1, 2$.

Solution. The recombining binomial tree for the stock price process is given in Figure 7.1. Now construct the self-financing strategy and calculate its value step by step going forward in time. Note that the positions $\Delta_0 = 0$, $\Delta_1 = \#\mathsf{U}(\omega_1)$, $\Delta_2 = \#\mathsf{U}(\omega_1\omega_2)$ are $\mathcal{F}_0$-, $\mathcal{F}_1$-, and $\mathcal{F}_2$-measurable, respectively. The self-financing investment strategy is therefore adapted to the natural filtration. Construct the strategy for each $t = 0, 1, 2$ as follows.

$t = 0$: Since $\Delta_0 = 0$, at time 0 all the capital is invested in the bank account with $\beta_0 = \frac{\Pi_0}{B_0} = 10$.

$t = 1$: There are no shares of stock in our investment portfolio so far. Therefore, its value is independent of ω_1 and by the above wealth equation:

$$\Pi_1 = \beta_0 B_1 = \Pi_0 \cdot (1+r) = 12.5.$$

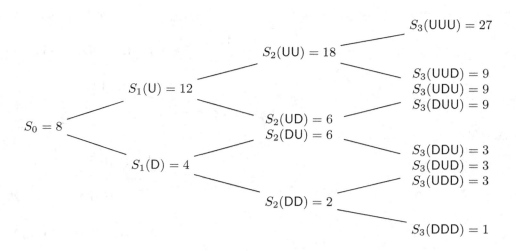

FIGURE 7.1: A three-period recombining binomial tree.

If $\omega_1 = D$, then $\Delta_1(D) = 0$ and $\beta_1(D) = \beta_0 = 10$. If $\omega_1 = U$, then $\Delta_1(U) = 1$ and

$$\beta_1(U) = \frac{\Pi_1(U) - S_1(U)\Delta_1(U)}{B_1} = \frac{12.5 - 12 \cdot 1}{1.25} = 0.4.$$

$t = 2$: The self-financed portfolio has value $\Pi_2(\omega_1\omega_2) = \beta_1(\omega_1)B_2 + \Delta_1(\omega_1)S_2(\omega_1\omega_2)$ for each of four scenarios: $\omega_1\omega_2 \in \{DD, DU, UD, UU\}$. Let us calculate the respective liquidation values of Δ_2 and β_2 in each case.

- If $\omega_1\omega_2 = DD$, then

$$\Pi_2(DD) = \beta_1(D)B_2 + \Delta_1(D)S_2(DD) = 10 \cdot 1.25^2 + 0 \cdot 2 = 15.625,$$
$$\Delta_2(DD) = \#U(DD) = 0,$$
$$\beta_2(DD) = \frac{\Pi_2(DD) - S_2(DD)\Delta_2(DD)}{B_2} = \frac{15.625 - 2 \cdot 0}{1.25^2} = 10.$$

- If $\omega_1\omega_2 = DU$, then

$$\Pi_2(DU) = \beta_1(D)B_2 + \Delta_1(D)S_2(DU) = 10 \cdot 1.25^2 + 0 \cdot 6 = 15.625,$$
$$\Delta_2(DU) = \#U(DU) = 1,$$
$$\beta_2(DU) = \frac{\Pi_2(DU) - S_2(DU)\Delta_2(DU)}{B_2} = \frac{15.625 - 6 \cdot 1}{1.25^2} = 6.16.$$

- If $\omega_1\omega_2 = UD$, then

$$\Pi_2(UD) = \beta_1(U)B_2 + \Delta_1(U)S_2(UD) = 0.4 \cdot 1.25^2 + 1 \cdot 6 = 6.625,$$
$$\Delta_2(UD) = \#U(UD) = 1,$$
$$\beta_2(UD) = \frac{\Pi_2(UD) - S_2(UD)\Delta_2(UD)}{B_2} = \frac{6.625 - 6 \cdot 1}{1.25^2} = 0.4.$$

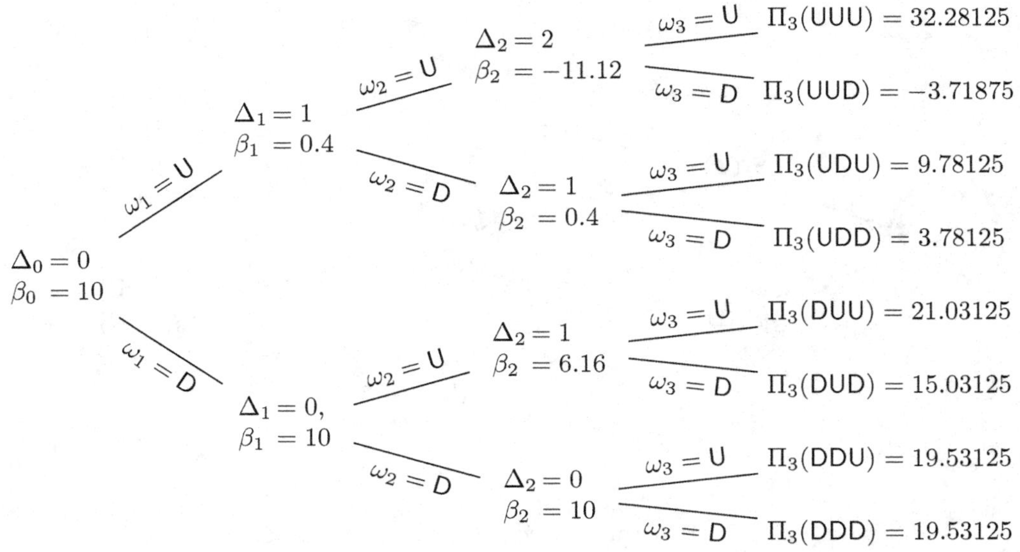

FIGURE 7.2: A self-financing strategy and its value process constructed in Example 7.1.

- If $\omega_1\omega_2 = \mathsf{UU}$, then

$$\Pi_2(\mathsf{UU}) = \beta_1(\mathsf{U})B_2 + \Delta_1(\mathsf{U})S_2(\mathsf{UU}) = 0.4 \cdot 1.25^2 + 1 \cdot 18 = 18.625,$$
$$\Delta_2(\mathsf{UU}) = \#\mathsf{U}(\mathsf{UU}) = 2,$$
$$\beta_2(\mathsf{UU}) = \frac{\Pi_2(\mathsf{UU}) - S_2(\mathsf{UU})\Delta_2(\mathsf{UU})}{B_2} = \frac{18.625 - 18 \cdot 2}{1.25^2} = -11.12.$$

The terminal value Π_3 of the strategy is calculated by

$$\Pi_3(\omega_1\omega_2\omega_3) = \beta_2(\omega_1\omega_2)B_3 + \Delta_2(\omega_1\omega_2)S_3(\omega_1\omega_2\omega_3), \quad \omega_1, \omega_2, \omega_3 \in \{\mathsf{D}, \mathsf{U}\}.$$

The full details of this three-period strategy, for every scenario, are summarized in Figure 7.2. $\qquad\qquad\qquad\qquad\qquad\qquad\qquad\qquad\qquad\qquad\qquad\qquad\qquad\qquad\qquad\Box$

7.2.1 Equivalent Martingale Measures for the Binomial Model

As mentioned at the start of this chapter, we fix the filtration to be the one generated by the market moves (i.e., the scenarios in the binomial tree); hence, the filtration also corresponds to the natural filtration generated by the stock price process. Since our main interest is in pricing derivative assets, the martingale property will be relevant under risk-neutral probability measures. We already referred to these measures as so-called equivalent martingale measures (EMMs). In the next chapter we shall define an EMM for the general case of N basic securities within the multi-period setting. At this point we need only to deal with the binomial multi-period model with two basic securities: the risk-free bank account B (or money market account, or zero-coupon bond) and the risky stock S. Hence, given a numéraire asset g, a corresponding EMM $\widetilde{\mathbb{P}} \equiv \widetilde{\mathbb{P}}^{(g)}$ is defined such that the discounted value processes of the risk-free asset and the stock are both $\widetilde{\mathbb{P}}^{(g)}$-martingales (see Theorem 7.1).

Recall Chapter 5, where we proved that for a single-period (Arrow–Debreu) model the value process of any static portfolio in the base securities, discounted by a numéraire asset $g \in \{B, S\}$, is a $\widetilde{\mathbb{P}}^{(g)}$-martingale, i.e., the discounted portfolio value process is a martingale under the measure $\widetilde{\mathbb{P}}^{(g)}$. We are now ready to extend such a result for the binomial tree model to the multi-period setting in the case of dynamic self-financing portfolio strategies.

Theorem 7.3. *Suppose that an equivalent martingale measure $\widetilde{\mathbb{P}} \equiv \widetilde{\mathbb{P}}^{(g)}$ for a numéraire asset g (i.e., $g = B$ or $g = S$) exists. Let $\{\Delta_t\}_{0 \leqslant t \leqslant T-1}$ be a process adapted to the natural filtration of the binomial tree model. Let $\{\Pi_t\}_{0 \leqslant t \leqslant T}$ be generated recursively by the wealth equation (7.15) with stock price process given by (7.2). Then the value process discounted by g, $\left\{\overline{\Pi}_t := \frac{\Pi_t}{g_t}\right\}_{0 \leqslant t \leqslant T}$, is a $\widetilde{\mathbb{P}}$-martingale. If the stock price process satisfies (7.4), with wealth equation (7.16), then $\left\{\frac{c_t \Pi_t}{S_t}\right\}_{0 \leqslant t \leqslant T}$ is a $\widetilde{\mathbb{P}}^{(S)}$-martingale and $\left\{\frac{\Pi_t}{B_t}\right\}_{0 \leqslant t \leqslant T}$ is a $\widetilde{\mathbb{P}}^{(B)}$-martingale.*[1]

Proof. We prove the case with no dividends. The case with dividends is proven similarly and left to the reader (see Exercise 7.11). Note that, for every $0 \leqslant t \leqslant T$, the value $\overline{\Pi}_t$ is clearly $\mathcal{F}_t$-measurable and integrable, i.e., $\widetilde{\mathbb{E}}[|\overline{\Pi}_t|] < \infty$. It then suffices to prove the single-step martingale expectation property under the $\widetilde{\mathbb{P}}$-measure: $\widetilde{\mathbb{E}}[\overline{\Pi}_{t+1} \mid \mathcal{F}_t] \equiv \widetilde{\mathbb{E}}_t[\overline{\Pi}_{t+1}] = \overline{\Pi}_t$, for every $0 \leqslant t \leqslant T-1$. We now use the fact that the discounted stock price and risk-free asset price processes, $\overline{S}_t := \frac{S_t}{g_t}$ and $\overline{B}_t := \frac{B_t}{g_t}$, are $\widetilde{\mathbb{P}}$-martingales and that $\Delta_t, \overline{S}_t, \overline{B}_t, \overline{\Pi}_t$ are $\mathcal{F}_t$-measurable. From the wealth equation (7.15) with $B_{t+1}/B_t = 1 + r$ we have:

$$
\begin{aligned}
\widetilde{\mathbb{E}}_t[\overline{\Pi}_{t+1}] &\equiv \widetilde{\mathbb{E}}_t\left[\frac{\Pi_{t+1}}{g_{t+1}}\right] = \widetilde{\mathbb{E}}_t\left[\frac{\Delta_t S_{t+1} + (\Pi_t - \Delta_t S_t)(B_{t+1}/B_t)}{g_{t+1}}\right] \\
&= \widetilde{\mathbb{E}}_t\left[\Delta_t \frac{S_{t+1}}{g_{t+1}} + \left(\frac{\Pi_t}{g_t} - \Delta_t \frac{S_t}{g_t}\right)\frac{B_{t+1}/g_{t+1}}{B_t/g_t}\right] \\
&= \widetilde{\mathbb{E}}_t\left[\Delta_t \overline{S}_{t+1} + (\overline{\Pi}_t - \Delta_t \overline{S}_t)\frac{\overline{B}_{t+1}}{\overline{B}_t}\right] \\
&= \Delta_t \underbrace{\widetilde{\mathbb{E}}_t[\overline{S}_{t+1}]}_{=\overline{S}_t} + \left(\frac{\overline{\Pi}_t - \Delta_t \overline{S}_t}{\overline{B}_t}\right)\underbrace{\widetilde{\mathbb{E}}_t[\overline{B}_{t+1}]}_{=\overline{B}_t} \\
&= \Delta_t \overline{S}_t + \overline{\Pi}_t - \Delta_t \overline{S}_t = \overline{\Pi}_t.
\end{aligned}
$$

Hence, the process $\{\overline{\Pi}_t\}$ is a $\widetilde{\mathbb{P}}$-martingale. $\square$

We note that the above theorem is applicable to the more general (non-standard) binomial model where the time-t upward and downward moves, as well as the interest rate, are stochastic (i.e., depend upon the sequence of market moves up to time t). These models are considered in Chapter 8.

Within the standard binomial model, we assume the interest rate r to be fixed where the risk-free bank account is given by (7.7). Hence, by taking the bank account as numéraire asset, we obtain the following well-known result for the risk-neutral expectation (and risk-neutral growth rate) of the self-financing portfolio value process.

---

[1] This theorem can be summarized by stating that $\left\{\frac{\Pi_t}{\hat{g}_t}\right\}_{0 \leqslant t \leqslant T}$ is a $\widetilde{\mathbb{P}}^{(g)}$-martingale, where $\hat{g}_t := \frac{g_t}{c_t^g}$ with c_t^g being the time-t after-dividend factor for the numéraire asset g. For $g = B$, we have $c_t^g = c_t^B \equiv 1$ since the bank account pays no dividend, i.e., $\hat{B}_t \equiv B_t$. For $g = S$, we have $c_t^g = c_t^S \equiv c_t$ given by (7.6) so $\hat{g}_t \equiv \hat{S}_t = S_t/c_t$. The latter simplifies to $\hat{S}_t = S_t$ when $c_t = 1$ in the case of zero stock dividend.

Corollary 7.4. *Let* $\widetilde{\mathbb{P}} \equiv \widetilde{\mathbb{P}}^{(B)}$ *be the EMM for the numéraire* $g = B$, *i.e.,* $\widetilde{\mathrm{E}}_t[\frac{\Pi_{t+1}}{B_{t+1}}] = \frac{\Pi_t}{B_t}$ *and* $\widetilde{\mathrm{E}}_t[\Pi_{t+1}] = (1+r)\Pi_t$ *since* $B_{t+1} = (1+r)B_t$ *for all* $t \geqslant 0$. *Then,*

$$\widetilde{\mathrm{E}}_s[\Pi_t] = (1+r)^{(t-s)}\Pi_s \quad \text{for all } 0 \leqslant s \leqslant t \leqslant T,$$

and hence $\widetilde{\mathrm{E}}_0[\Pi_t] = (1+r)^t \Pi_0$ *for all* $t \geqslant 0$.

Example 7.2. Verify that the self-financed portfolio value process in Example 7.1, discounted by the risk-free asset price process, is a martingale under the measure $\widetilde{\mathbb{P}} \equiv \widetilde{\mathbb{P}}^{(B)}$.

Solution. The risk-neutral probabilities for the up and down moves are

$$\tilde{p} = \widetilde{\mathbb{P}}(\omega_i = \mathsf{U}) = \frac{1.25 - 0.5}{1.5 - 0.5} = 0.75 \quad \text{and} \quad \widetilde{\mathbb{P}}(\omega_i = \mathsf{D}) = 1 - \tilde{p} = 0.25.$$

It is sufficient to verify that $\Pi_t = \frac{1}{1+r}\widetilde{\mathrm{E}}_t[\Pi_{t+1}]$ for $t = 0, 1, 2$.

$t = 0$:
$$\Pi_0 = 10 \overset{?}{=} \frac{1}{1+r}\widetilde{\mathrm{E}}_0[\Pi_1] = \frac{\Pi_1(\mathsf{U}) \cdot \tilde{p} + \Pi_1(\mathsf{D}) \cdot (1-\tilde{p})}{1+r}$$
$$= \frac{12.5 \cdot 0.75 + 12.5 \cdot 0.25}{1.25} = 10 \qquad \checkmark$$

$t = 1$:
$$\Pi_1(\mathsf{D}) = 12.5 \overset{?}{=} \frac{1}{1+r}\widetilde{\mathrm{E}}_1[\Pi_2](\mathsf{D}) = \frac{\Pi_2(\mathsf{DU}) \cdot \tilde{p} + \Pi_2(\mathsf{DD}) \cdot (1-\tilde{p})}{1+r}$$
$$= \frac{15.625 \cdot 0.75 + 15.625 \cdot 0.25}{1.25} = 12.5 \qquad \checkmark$$

$$\Pi_1(\mathsf{U}) = 12.5 \overset{?}{=} \frac{1}{1+r}\widetilde{\mathrm{E}}_1[\Pi_2](\mathsf{U}) = \frac{\Pi_2(\mathsf{UU}) \cdot \tilde{p} + \Pi_2(\mathsf{UD}) \cdot (1-\tilde{p})}{1+r}$$
$$= \frac{18.625 \cdot 0.75 + 6.625 \cdot 0.25}{1.25} = 12.5 \qquad \checkmark$$

$t = 2$:
$$\Pi_2(\mathsf{DD}) = 15.625 \overset{?}{=} \frac{1}{1+r}\widetilde{\mathrm{E}}_2[\Pi_3](\mathsf{DD}) = \frac{\Pi_3(\mathsf{DDU}) \cdot \tilde{p} + \Pi_3(\mathsf{DDD}) \cdot (1-\tilde{p})}{1+r}$$
$$= \frac{19.53125 \cdot 0.75 + 19.53125 \cdot 0.25}{1.25} = 15.625 \qquad \checkmark$$

$$\Pi_2(\mathsf{DU}) = 15.625 \overset{?}{=} \frac{1}{1+r}\widetilde{\mathrm{E}}_2[\Pi_3](\mathsf{DU}) = \frac{\Pi_3(\mathsf{DUU}) \cdot \tilde{p} + \Pi_3(\mathsf{DUD}) \cdot (1-\tilde{p})}{1+r}$$
$$= \frac{21.03125 \cdot 0.75 + 15.03125 \cdot 0.25}{1.25} = 15.625 \qquad \checkmark$$

$$\Pi_2(\mathsf{UD}) = 6.625 \overset{?}{=} \frac{1}{1+r}\widetilde{\mathrm{E}}_2[\Pi_3](\mathsf{UD}) = \frac{\Pi_3(\mathsf{UDU}) \cdot \tilde{p} + \Pi_3(\mathsf{UDD}) \cdot (1-\tilde{p})}{1+r}$$
$$= \frac{9.78125 \cdot 0.75 + 3.78125 \cdot 0.25}{1.25} = 6.625 \qquad \checkmark$$

$$\Pi_2(\mathsf{UU}) = 18.625 \overset{?}{=} \frac{1}{1+r}\widetilde{\mathrm{E}}_2[\Pi_3](\mathsf{UU}) = \frac{\Pi_3(\mathsf{UUU}) \cdot \tilde{p} + \Pi_3(\mathsf{UUD}) \cdot (1-\tilde{p})}{1+r}$$
$$= \frac{32.28125 \cdot 0.75 - 3.78125 \cdot 0.25}{1.25} = 18.625 \qquad \checkmark$$

$\square$

Another corollary of Theorem 7.3 is the first fundamental theorem of asset pricing (FTAP). It states that there are no arbitrage opportunities iff there exists an EMM $\widetilde{\mathbb{P}}^{(g)}$ for a numéraire g. So far, such an assertion has been proved for static portfolios in the one-period or multi-period setting. However, one can create an arbitrage opportunity by manipulating with a portfolio in base assets without injecting or withdrawing funds. In other words, a specially constructed self-financing strategy can be an arbitrage. Let us prove that there are no self-financing arbitrage strategies in a binomial tree model iff there exists an EMM. First, we need to have a formal definition of an arbitrage in a multi-period trading model as follows.

Definition 7.1. An *arbitrage strategy* in a multi-period model is a self-financing strategy with nonnegative value process and zero initial value Π_0 such that $\mathbb{P}(\Pi_t > 0) > 0$ holds at some time $t \in \{1, 2, \ldots, T\}$.

Suppose that a binomial tree model admits no arbitrage. In particular, there are no static (i.e., with constant positions in base assets as in any single-period setting) arbitrage portfolios. Then, as was proved before, there exists an EMM. To complete the proof of the FTAP for the binomial tree model we only need to prove the following converse statement.

Lemma 7.5. *Suppose that an EMM $\widetilde{\mathbb{P}} = \widetilde{\mathbb{P}}^{(g)}$ for a numéraire g exists. Then there are no self-financing arbitrage strategies.*

Proof. Suppose that a self-financed arbitrage strategy with initial value zero and delta positions $\{\Delta_t\}_{t \geq 0}$ exists. The value process for such a strategy must satisfy the wealth equation (7.15), or (7.16) in the case of stock dividends, such that (by the arbitrage assumption) $\Pi_t(\omega) \geq 0$ for all $\omega \in \Omega$ and $t \geq 0$. Moreover, arbitrage implies that there exists a time $m \in \{1, 2, \ldots, T\}$ and market scenario $\omega^* \in \Omega$ such that $\Pi_m(\omega^*) > 0$. Therefore,

$$\widetilde{\mathbb{E}}_0[\overline{\Pi}_m] = \sum_{\omega \in \Omega} \overline{\Pi}_m(\omega) \, \widetilde{\mathbb{P}}(\omega) \geq \frac{\Pi_m(\omega^*)}{g(\omega^*)} \, \widetilde{\mathbb{P}}(\omega^*) > 0.$$

This contradicts Theorem 7.3, since $\widetilde{\mathbb{E}}[\overline{\Pi}_m] = \overline{\Pi}_0 = \Pi_0/g_0 = 0$. $\qquad\square$

7.3 Dynamic Replication in the Binomial Tree Model

7.3.1 Dynamic Replication of Payoffs

A key idea in modern finance is the replication of a financial claim with the use of portfolios in other (base) assets. The two most important applications of replication is the no-arbitrage pricing of derivative securities and hedging their liabilities. Consider a derivative security with maturity at time T and payoff function $X : \Omega_T \to \mathbb{R}$. Recall that $\mathcal{L}(\Omega_T)$ denotes the collection of all payoff functions on the sample space Ω_T. The structure of a payoff can be quite general. It can depend on the whole path ω, or on a quantity calculated along the stock price path such as an average of the stock prices over some time window, or only on the terminal stock price S_T. For example, the payoff of a non-path-dependent derivative, such as a standard European call or put option, with exercise only at maturity T, is a function of only the terminal stock price S_T and is given by $X(\omega) = \Lambda(S_T(\omega))$ with function $\Lambda : \mathbb{R}_+ \to \mathbb{R}$, where $\mathbb{R}_+ := [0, \infty)$. Hence, in what follows we allow for generally path-dependent payoffs where the random variable X is $\mathcal{F}_T$-measurable, i.e., the payoff is determined by possibly the entire sequence of market moves $\omega \equiv \omega_1 \ldots \omega_T \in \Omega_T$.

Let $\{V_t\}_{0 \leq t \leq T}$ denote the *price process of the derivative security*, i.e., V_t is the price of the derivative security at time t. At maturity time, the derivative price is given by the payoff function $V_T = X$, where $X : \Omega_T \to \mathbb{R}$. The writer of a derivative needs to calculate the no-arbitrage current price, V_0, of the contract. As we saw in great detail in Chapters 2 and 5, in the single-period model, the price of a derivative security must equal the initial value of a portfolio replicating the derivative payoff at maturity. In the multi-period setting, such a replication can only be done dynamically with the use of self-financing strategies. It is possible to construct such a strategy that replicates the whole derivative price process from time 0 until maturity T. Knowledge of the derivative price process is required for hedging the writer's liabilities.

Definition 7.2. A self-financing strategy $\{\varphi_t\}_{0 \leqslant t \leqslant T-1}$ is said to *replicate* the payoff X at maturity T if its value at maturity T, given by $\Pi_T = \Pi_T[\varphi_{T-1}]$, equals the payoff value for all possible scenarios, i.e.,

$$\Pi_T(\omega) = X(\omega) \text{ for all } \omega \in \Omega.$$

According to the law of one price, the initial value Π_0 of a strategy that replicates the payoff X of a derivative maturing at time T must equal the initial value V_0 of the derivative, or else an arbitrage opportunity exists. Moreover, we shall prove that in the absence of arbitrage opportunities, the price of the derivative, V_t is equal to the value $\Pi_t \equiv \Pi_t[\varphi_t]$ of the replicating strategy at every time t.

To construct a self-financing strategy that replicates a derivative with payoff X, we proceed as follows. First, we construct a no-arbitrage derivative price process $\{V_t\}_{0 \leqslant t \leqslant T}$ recursively backward in time starting from maturity time T. Second, we obtain a sequence of (delta) positions in the stock, $\{\Delta_t\}_{0 \leqslant t \leqslant T-1}$, corresponding to the price process. Finally, we show that the process $\{\Delta_t\}$ is nothing more than a replicating strategy for the derivative price process so that the value process $\{\Pi_t\}_{0 \leqslant t \leqslant T}$ generated by the strategy coincides with the derivative price process at *all intermediate dates and at maturity*, i.e., $\Pi_t(\omega) = V_t(\omega)$ for all $0 \leqslant t \leqslant T$ and all scenarios $\omega \in \Omega$. Note that $\Pi_T(\omega) = V_T(\omega) = X(\omega)$ holds at maturity by the definition of a replicating strategy. Before we prove a general result, let us study how this procedure works for the simple one- and two-period cases in the following example. We generally allow for stock dividends while using the wealth equation (7.16).

Example 7.3 (The cases with $T = 1$ and $T = 2$). Assume that a binomial tree model admits no-arbitrage, i.e., $d < 1 + r < u$ holds. Let $\widetilde{\mathbb{P}}$ be the usual risk-neutral measure (for the numéraire asset $g = B$). Construct the replicating strategy and find the no-arbitrage prices for an arbitrary derivative with maturity time (a) $T = 1$ and (b) $T = 2$.

Solution.
Case with $T = 1$. In the one-period case, the replicating portfolio is static and formed at time 0, as we already saw in Chapter 2. The initial value Π_0 is invested in Δ_0 shares of stock, leaving us with the risk-free asset position with value $\Pi_0 - \Delta_0 S_0$. At time 1, the portfolio value is $\Pi_1(\omega) = \Delta_0 S_0 Y_1(\omega) + (\Pi_0 - \Delta_0 S_0)(1+r)$. Let us choose $\Pi_0 = V_0$ and Δ_0 such that $\Pi_1(\omega) = V_1(\omega)$ for $\omega \in \Omega_1 = \{\mathsf{D}, \mathsf{U}\}$. Since there are two possibilities for ω_1, we obtain the linear system of two equations in two unknowns Δ_0 and V_0:

$$\begin{cases} \Pi_1(\mathsf{U}) = V_1(\mathsf{U}) \\ \Pi_1(\mathsf{D}) = V_1(\mathsf{D}) \end{cases} \iff \begin{cases} \Delta_0 S_0 Y_1(\mathsf{U}) + (V_0 - \Delta_0 S_0)(1+r) = V_1(\mathsf{U}) \\ \Delta_0 S_0 Y_1(\mathsf{D}) + (V_0 - \Delta_0 S_0)(1+r) = V_1(\mathsf{D}) \end{cases}$$
$$\iff \begin{cases} (Y_1(\mathsf{U}) - (1+r))\Delta_0 S_0 + (1+r)V_0 = V_1(\mathsf{U}) \\ (Y_1(\mathsf{D}) - (1+r))\Delta_0 S_0 + (1+r)V_0 = V_1(\mathsf{D}) \end{cases} \qquad (7.17)$$

This system has a unique solution with Δ_0 obtained by simply subtracting the first and second equations in (7.17):

$$\Delta_0 = \frac{V_1(\mathsf{U}) - V_1(\mathsf{D})}{S_0(Y_1(\mathsf{U}) - Y_1(\mathsf{D}))} = \frac{V_1(\mathsf{U}) - V_1(\mathsf{D})}{S_0(u - d)}. \qquad (7.18)$$

Substituting this expression for Δ_0 into either of the equations in (7.17) gives the current price of the derivative in terms of known (payoff) values $V_1(\mathsf{U})$ and $V_1(\mathsf{D})$:

$$V_0 = (1+r)^{-1}\left[V_1(\mathsf{U}) - (Y_1(\mathsf{U}) - (1+r))\Delta_0 S_0\right] \tag{7.19}$$

$$= (1+r)^{-1}\left[\frac{V_1(\mathsf{U})(Y_1(\mathsf{U}) - Y_1(\mathsf{D})) - (Y_1(\mathsf{U}) - (1+r))(V_1(\mathsf{U}) - V_1(\mathsf{D}))}{Y_1(\mathsf{U}) - Y_1(\mathsf{D})}\right]$$

$$= (1+r)^{-1}\left[\frac{(1+r) - Y_1(\mathsf{D})}{Y_1(\mathsf{U}) - Y_1(\mathsf{D})}V_1(\mathsf{U}) + \frac{Y_1(\mathsf{U}) - (1+r)}{Y_1(\mathsf{U}) - Y_1(\mathsf{D})}V_1(\mathsf{D})\right]$$

$$= (1+r)^{-1}\left[\frac{(1+r) - d}{u - d}V_1(\mathsf{U}) + \frac{u - (1+r)}{u - d}V_1(\mathsf{D})\right]$$

$$= (1+r)^{-1}\left[\tilde{p}V_1(\mathsf{U}) + (1 - \tilde{p})V_1(\mathsf{D})\right]$$

$$= (1+r)^{-1}\widetilde{\mathbb{E}}_0[V_1].$$

Note that the last expression corresponds to the discounted expected value of the derivative payoff with risk-neutral probabilities with asset B as numéraire:

$$\tilde{p} = \frac{(1+r)S_0 - S_0 d}{S_0 u - S_0 d} = \frac{(1+r) - d}{u - d}, \quad 1 - \tilde{p} = \frac{S_0 u - (1+r)S_0}{S_0 u - S_0 d} = \frac{u - (1+r)}{u - d}.$$

Case with $T = 2$. Let us construct a replicating strategy $\{\Delta_t\}_{t=0,1}$ with the time-2 value satisfying $\Pi_2(\omega) = V_2(\omega)$ for all $\omega \in \Omega_2 = \{\mathsf{DD}, \mathsf{DU}, \mathsf{UD}, \mathsf{UU}\}$. At time 0, the strategy is already specified above, i.e., take a position Δ_0 in shares of stock and a position in the risk-free bank account with value $\Pi_0 - \Delta_0 S_0$. This gives Δ_0 and V_0, respectively, as in (7.18) and (7.19). At time 1, we liquidate the old portfolio and form a new portfolio having value $\Pi_1(\omega_1) = V_1(\omega_1)$ and with new position $\Delta_1(\omega_1)$ in the stock for any given market scenario with first move ω_1. At time 2, the value of this new portfolio, as given by the wealth equation for the second period, becomes

$$\Pi_2(\omega_1\omega_2) = \Delta_1(\omega_1)S_1(\omega_1)Y_2(\omega_2) + (V_1(\omega_1) - \Delta_1(\omega_1)S_1(\omega_1))(1+r).$$

By replication, $\Pi_2(\omega_1\omega_2)$ must equal $V_2(\omega_1\omega_2)$ for all four possible market scenarios $\omega_1\omega_2$. Hence, we obtain a system of four linear equations which are grouped into two pairs of equations. The first pair (for $\omega_1 = \mathsf{U}$) gives two equations in two unknowns $\Delta_1(\mathsf{U}), V_1(\mathsf{U})$:

$$\begin{cases} V_2(\mathsf{UU}) = \Delta_1(\mathsf{U})S_1(\mathsf{U})Y_2(\mathsf{U}) + (V_1(\mathsf{U}) - \Delta_1(\mathsf{U})S_1(\mathsf{U}))(1+r) \\ V_2(\mathsf{UD}) = \Delta_1(\mathsf{U})S_1(\mathsf{U})Y_2(\mathsf{D}) + (V_1(\mathsf{U}) - \Delta_1(\mathsf{U})S_1(\mathsf{U}))(1+r) \end{cases} \tag{7.20}$$

and the second pair (for $\omega_1 = \mathsf{D}$) in the two unknowns $\Delta_1(\mathsf{D}), V_1(\mathsf{D})$:

$$\begin{cases} V_2(\mathsf{DU}) = \Delta_1(\mathsf{D})S_1(\mathsf{D})Y_2(\mathsf{U}) + (V_1(\mathsf{D}) - \Delta_1(\mathsf{D})S_1(\mathsf{D}))(1+r) \\ V_2(\mathsf{DD}) = \Delta_1(\mathsf{D})S_1(\mathsf{D})Y_2(\mathsf{D}) + (V_1(\mathsf{D}) - \Delta_1(\mathsf{D})S_1(\mathsf{D}))(1+r) \end{cases} \tag{7.21}$$

At this point it is important to observe that both of these pairs of equations are of the same form as (7.17) and hence are solved in the same manner. In particular, solving (7.20) gives

$$\Delta_1(\mathsf{U}) = \frac{V_2(\mathsf{UU}) - V_2(\mathsf{UD})}{S_1(\mathsf{U})(Y_2(\mathsf{U}) - Y_2(\mathsf{D}))} = \frac{V_2(\mathsf{UU}) - V_2(\mathsf{UD})}{S_1(\mathsf{U})(u - d)} \tag{7.22}$$

and

$$V_1(\mathsf{U}) = (1+r)^{-1}\left[\frac{S_1(\mathsf{U})((1+r) - Y_2(\mathsf{D}))}{S_1(\mathsf{U})(Y_2(\mathsf{U}) - Y_2(\mathsf{D}))}V_2(\mathsf{UU}) + \frac{S_1(\mathsf{U})(Y_2(\mathsf{U}) - (1+r))}{S_1(\mathsf{U})(Y_2(\mathsf{U}) - Y_2(\mathsf{D}))}V_2(\mathsf{UD})\right]$$

$$= (1+r)^{-1}\left[\frac{(1+r) - d}{u - d}V_2(\mathsf{UU}) + \frac{u - (1+r)}{u - d}V_2(\mathsf{UD})\right]$$

$$= (1+r)^{-1}\left[\tilde{p}V_2(\mathsf{UU})) + (1 - \tilde{p})V_2(\mathsf{UD})\right]$$

$$= (1+r)^{-1}\widetilde{\mathbb{E}}_1[V_2](\mathsf{U}). \tag{7.23}$$

Similarly, solving (7.21) gives

$$\Delta_1(\mathsf{D}) = \frac{V_2(\mathsf{DU}) - V_2(\mathsf{DD})}{S_1(\mathsf{D})(Y_2(\mathsf{U}) - Y_2(\mathsf{D}))} = \frac{V_2(\mathsf{DU}) - V_2(\mathsf{DD})}{S_1(\mathsf{D})(u - d)} \tag{7.24}$$

and

$$\begin{aligned} V_1(\mathsf{D}) &= (1+r)^{-1}\left[\frac{S_1(\mathsf{D})((1+r) - Y_2(\mathsf{D}))}{S_1(\mathsf{D})(Y_2(\mathsf{U}) - Y_2(\mathsf{D}))}V_2(\mathsf{DU}) + \frac{S_1(\mathsf{D})(Y_2(\mathsf{U}) - (1+r))}{S_1(\mathsf{D})(Y_2(\mathsf{U}) - Y_2(\mathsf{D}))}V_2(\mathsf{DD})\right] \\ &= (1+r)^{-1}\left[\tilde{p}V_2(\mathsf{DU})) + (1-\tilde{p})V_2(\mathsf{DD})\right] \\ &= (1+r)^{-1}\widetilde{\mathsf{E}}_1[V_2](\mathsf{D}). \end{aligned} \tag{7.25}$$

Combining both expressions in (7.23) and (7.25) for the derivative prices at $t = 1$ gives

$$V_1(\omega_1) = (1+r)^{-1}\left[\tilde{p}V_2(\omega_1\mathsf{U}) + (1-\tilde{p})V_2(\omega_1\mathsf{D})\right]$$
$$\implies V_1 = (1+r)^{-1}\widetilde{\mathsf{E}}_1[V_2].$$

Similarly, we can combine (7.22) and (7.24) to give

$$\Delta_1(\omega_1) = \frac{V_2(\omega_1\mathsf{U}) - V_2(\omega_1\mathsf{D})}{S_1(\omega_1)(u - d)}.$$

We hence see that the derivative prices at times $t = 0$ and $t = 1$ are given by the discounted (risk-neutral) expected value of the derivative price at time $t + 1$. Hence, by the tower property, the initial price of the derivative is expressible as an expectation of the payoff at time $t = T = 2$, discounted back by two periods:

$$V_0 = (1+r)^{-1}\widetilde{\mathsf{E}}_0[V_1] = (1+r)^{-2}\widetilde{\mathsf{E}}_0\left[\widetilde{\mathsf{E}}_1[V_2]\right] = (1+r)^{-2}\widetilde{\mathsf{E}}_0[V_2].$$

$\square$

Now, we consider the general case for any finite number of $T \geqslant 1$ periods. Assume the absence of arbitrage opportunities. Let us define recursively backward in time the price process $\{V_t\}_{0 \leqslant t \leqslant T}$ of a derivative with maturity time T and given payoff function $X \in \mathcal{L}(\Omega_T)$. Fix arbitrarily time $t \in \{0, 1, \ldots, T-1\}$ and market moves $\omega_1, \omega_2, \ldots, \omega_t \in \{\mathsf{D}, \mathsf{U}\}$. The stock price S_{t+1} conditional on $\omega_1, \omega_2, \ldots, \omega_t$ follows a binomial single-period sub-tree with risk-neutral probabilities $\widetilde{\mathbb{P}}(\{\omega_{t+1} = \mathsf{U}\}) = \tilde{p}$, $\widetilde{\mathbb{P}}(\{\omega_{t+1} = \mathsf{D}\}) = 1 - \tilde{p}$, as given above. Applying the single-time-step discounted expectation formula to the sub-tree originated at the stock price $S_t(\omega_1\omega_2 \ldots \omega_t)$ gives

$$V_t(\omega_1\omega_2 \ldots \omega_t) = \frac{1}{1+r}\left(\tilde{p}V_{t+1}(\omega_1\omega_2 \ldots \omega_t\mathsf{U}) + (1-\tilde{p})V_{t+1}(\omega_1\omega_2 \ldots \omega_t\mathsf{D})\right) \tag{7.26}$$

$$\equiv \frac{1}{1+r}\widetilde{\mathsf{E}}_t[V_{t+1}](\omega_1\omega_2 \ldots \omega_t), \quad t = 0, 1, \ldots, T-1, \tag{7.27}$$

At maturity, we set the derivative price equal to the payoff

$$V_T(\omega_1\omega_2 \ldots \omega_T) = X(\omega_1\omega_2 \ldots \omega_T). \tag{7.28}$$

Define the strategy $\{\Delta_t\}_{0 \leqslant t \leqslant T-1}$ as follows:

$$\Delta_t(\omega_1\omega_2 \ldots \omega_t) = \frac{V_{t+1}(\omega_1\omega_2 \ldots \omega_t\mathsf{U}) - V_{t+1}(\omega_1\omega_2 \ldots \omega_t\mathsf{D})}{(u - d)S_t(\omega_1\omega_2 \ldots \omega_t)}, \quad t = 0, 1, \ldots, T-1. \tag{7.29}$$

Note that when the stock price satisfies (7.2) we can equally write the expression in (7.29) as: $\Delta_t(\omega_1\omega_2\ldots\omega_t) = \frac{V_{t+1}(\omega_1\omega_2\ldots\omega_t\mathsf{U}) - V_{t+1}(\omega_1\omega_2\ldots\omega_t\mathsf{D})}{S_{t+1}(\omega_1\omega_2\ldots\omega_t\mathsf{U}) - S_{t+1}(\omega_1\omega_2\ldots\omega_t\mathsf{D})}$.

Let us now prove that the derivative price process $\{V_t\}_{0\leqslant t\leqslant T}$ and the value process for the self-financing strategy $\{\Delta_t\}_{0\leqslant t\leqslant T-1}$ have the same value at all times $t = 0, 1, \ldots, T$, i.e., the portfolio value process with (delta hedging) strategy $\{\Delta_t\}_{0\leqslant t\leqslant T-1}$ replicates the derivative price process. We shall assume the general case that allows for a dividend-paying stock, wherein (7.16) and (7.4) hold (which are equivalent to (7.15) and (7.2) when all stock dividends are zero). Note that (7.27) is a statement of the fact that $\left\{\frac{V_t}{B_t}\right\}_{0\leqslant t\leqslant T}$ is a $\widetilde{\mathbb{P}}^{(B)}$-martingale where $\Pi_t = V_t$ (see Theorem 7.3).

Theorem 7.6. *Consider a derivative security with $\mathcal{F}_T$-measurable payoff X, at maturity T. Define the derivative price process $\{V_t\}_{0\leqslant t\leqslant T}$ by (7.26)–(7.28) and the self-financing portfolio strategy $\{\Delta_t\}_{0\leqslant t\leqslant T-1}$ by (7.29). Set $\Pi_0 = V_0$ and construct recursively forward in time the value process $\{\Pi_t\}_{0\leqslant t\leqslant T}$ via the wealth equation. Then, the strategy replicates the derivative price process at every time, i.e.,*

$$\Pi_t(\omega_1\omega_2\ldots\omega_t) = V_t(\omega_1\omega_2\ldots\omega_t) \tag{7.30}$$

holds for all $t = 0, 1, \ldots, T$ and all market moves $\omega_1, \omega_2, \ldots, \omega_t \in \{D, U\}$.

Proof. We prove the assertion by induction. For $t = 0$, the equality $\Pi_0 = V_0$ follows trivially by the definition of the initial price. Now assume that (7.30) holds for some time t and show that it holds for time $t+1$. Fix an arbitrary sequence of moves $\omega_1\omega_2\ldots\omega_t$. By the induction assumption, $\Pi_t(\omega_1\omega_2\ldots\omega_t) = V_t(\omega_1\omega_2\ldots\omega_t)$, and since $\omega_{t+1} = U$ or D, we need to prove that

$$\Pi_{t+1}(\omega_1\ldots\omega_t\mathsf{D}) = V_{t+1}(\omega_1\ldots\omega_t\mathsf{D}) \text{ and } \Pi_{t+1}(\omega_1\ldots\omega_t\mathsf{U}) = V_{t+1}(\omega_1\ldots\omega_t\mathsf{U}).$$

Let us consider the case with $\omega_{t+1} = \mathsf{D}$, where $Y_{t+1}(\omega_{t+1}) = Y_{t+1}(\mathsf{D}) = d$. The case with $\omega_{t+1} = \mathsf{U}$ follows similarly where $Y_{t+1}(\omega_{t+1}) = Y_{t+1}(\mathsf{U}) = u$. The wealth equation (7.16) (or (7.15)) gives

$$
\begin{aligned}
\Pi_{t+1}(\omega_1\ldots\omega_t\mathsf{D}) &= \Delta_t(\omega_1\ldots\omega_t)S_t(\omega_1\ldots\omega_t)d \\
&\quad + (\Pi_t(\omega_1\ldots\omega_t) - \Delta_t(\omega_1\ldots\omega_t)S_t(\omega_1\ldots\omega_t))(1+r) \\
&= \Delta_t(\omega_1\ldots\omega_t)S_t(\omega_1\ldots\omega_t)(d - (1+r)) + (1+r)\Pi_t(\omega_1\ldots\omega_t). \tag{7.31}
\end{aligned}
$$

Subsituting the expression in (7.29) for $\Delta_t(\omega_1\ldots\omega_t)$ and using $\tilde{p} = (1+r-d)/(u-d)$ and $\Pi_t = V_t$ gives

$$\Pi_{t+1}(\omega_1\ldots\omega_t\mathsf{D}) = -\tilde{p}\left(V_{t+1}(\omega_1\ldots\omega_t\mathsf{U}) - V_{t+1}(\omega_1\ldots\omega_t\mathsf{D})\right) + (1+r)V_t(\omega_1\ldots\omega_t)$$

(and by using equation (7.26))

$$
\begin{aligned}
&= \tilde{p}V_{t+1}(\omega_1\ldots\omega_t\mathsf{D}) - \tilde{p}V_{t+1}(\omega_1\ldots\omega_t\mathsf{U}) \\
&\quad + \tilde{p}V_{t+1}(\omega_1\ldots\omega_t\mathsf{U}) + (1-\tilde{p})V_{t+1}(\omega_1\ldots\omega_t\mathsf{D}) \\
&= V_{t+1}(\omega_1\ldots\omega_t\mathsf{D}). \qquad \square
\end{aligned}
$$

Remarks.

1. Applying the tower property of conditional expectations to (7.27) gives us the (multi-step ahead) risk-neutral pricing formula for any derivative with payoff $V_T = X \in \mathcal{L}(\Omega)$ at maturity T:

$$V_t(\omega_1\omega_2\ldots\omega_t) = \frac{1}{(1+r)^{T-t}}\widetilde{\mathbb{E}}_t[V_T](\omega_1\omega_2\ldots\omega_t), \quad 0 \leqslant t \leqslant T, \tag{7.32}$$

or, in short, as an $\mathcal{F}_t$-measurable random variable:

$$V_t = \frac{1}{(1+r)^{T-t}}\widetilde{\mathrm{E}}_t[V_T], \quad 0 \leqslant t \leqslant T, \tag{7.33}$$

where $\widetilde{\mathbb{P}} = \widetilde{\mathbb{P}}^{(B)}$. In particular, the initial derivative price is

$$V_0 = \frac{1}{(1+r)^T}\widetilde{\mathrm{E}}_0[V_T] = \frac{1}{(1+r)^T}\sum_{\omega \in \Omega_T} V_T(\omega)\widetilde{\mathbb{P}}(\omega). \tag{7.34}$$

Equations (7.32)-(7.34) were derived for the general case with stock dividends.

2. Theorem 7.6 is a particular case of a more general result presented below in Section 7.3.2, where here we take $X_t = 0$ for all $t = 0, 1, \ldots, T-1$ and $X_T = X \neq 0$, i.e., the derivative has a payoff (i.e., a cash flow) only at maturity T.

3. A corollary of Theorems 7.3 and 7.6 (giving $\Pi_t = V_t$) is the property that, in the measure $\widetilde{\mathbb{P}}^{(g)}$ with $g \in \{B, S\}$, the discounted derivative price process defined by $\left\{\overline{V}_t := \frac{V_t}{\hat{g}_t}\right\}_{0 \leqslant t \leqslant T}$, where $\hat{g}_t := \frac{g_t}{c_t^g}$ is the numéraire asset price divided by the after-dividend factor for the numéraire asset g, is a $\widetilde{\mathbb{P}}^{(g)}$-martingale:

$$\widetilde{\mathrm{E}}_s^{(g)}\left[\overline{V}_t\right] = \overline{V}_s \text{ for all } 0 \leqslant s \leqslant t \leqslant T.$$

For $g = B$, we simply have $\hat{g}_t \equiv g_t = B_t$ since the bank account has no dividend ($c_t^B \equiv 1$). In this case the growth rate of the portfolio value in the measure $\widetilde{\mathbb{P}}^{(B)} \equiv \widetilde{\mathbb{P}}$ is given by the interest rate, i.e., $B_t = (1+r)^{t-s}B_s$ gives $\widetilde{\mathrm{E}}_s[V_t] = (1+r)^{t-s}V_s$ For the stock as numéraire we have $\hat{g}_t = \hat{S}_t = S_t/c_t$, where $c_t^S \equiv c_t$ is given by (7.6).

According to Remark #3, the backward recurrence pricing formula in (7.26) (or (7.27)) can be written more generally with an arbitrary choice of numéraire asset g. Given any numéraire asset g that pays no dividends, we have the single-step recurrence pricing formula:

$$V_t(\omega_1\omega_2\ldots\omega_t) = g_t(\omega_1\omega_2\ldots\omega_t)\widetilde{\mathrm{E}}_t^{(g)}\left[\frac{V_{t+1}}{g_{t+1}}\right](\omega_1\omega_2\ldots\omega_t),$$

$$= g_t(\omega_1\omega_2\ldots\omega_t)\left[\widetilde{p}^{(g)}\frac{V_{t+1}(\omega_1\omega_2\ldots\omega_t\mathsf{U})}{g_{t+1}(\omega_1\omega_2\ldots\omega_t\mathsf{U})} + (1-\widetilde{p}^{(g)})\frac{V_{t+1}(\omega_1\omega_2\ldots\omega_t\mathsf{D})}{g_{t+1}(\omega_1\omega_2\ldots\omega_t\mathsf{D})}\right], \tag{7.35}$$

where $\widetilde{p}^{(g)} = \widetilde{\mathbb{P}}^{(g)}(\mathsf{U})$, $1 - \widetilde{p}^{(g)} = \widetilde{\mathbb{P}}^{(g)}(\mathsf{D})$. Equivalently,

$$\frac{V_t}{g_t} = \widetilde{\mathrm{E}}_t^{(g)}\left[\frac{V_{t+1}}{g_{t+1}}\right], \tag{7.36}$$

where $\widetilde{\mathrm{E}}_t^{(g)}$ denotes the expectation conditional on $\mathcal{F}_t$ and w.r.t. the risk-neutral probability measure $\widetilde{\mathbb{P}}^{(g)}$. Applying the tower property to (7.36) gives

$$\frac{V_t}{g_t} = \widetilde{\mathrm{E}}_t^{(g)}\left[\frac{V_T}{g_T}\right] \implies V_t = g_t\widetilde{\mathrm{E}}_t^{(g)}\left[\frac{V_T}{g_T}\right] \text{ for all } t \in \{0, 1, \ldots, T\}. \tag{7.37}$$

That is, the $\mathcal{F}_t$-measurable time-t derivative price V_t discounted by the numéraire value g_t is given by the mathematical expectation (in the risk-neutral measure $\widetilde{\mathbb{P}}^{(g)}$) of the discounted payoff V_T/g_T. Note how this more general result recovers equation (7.32) when we choose the bank account as numéraire, i.e., when $g_t = B_t$ we obtain the more familiar discount factor $\frac{g_t}{g_T} = \frac{B_t}{B_T} = (1+r)^{-(T-t)}$. We remark that g_t in (7.35)-(7.37) can be any strictly positive portfolio (positive linear combination) in the time-t base asset values B_t and S_t, where the measure $\widetilde{\mathbb{P}}^{(g)}$ is defined such that $\frac{B_t}{g_t}$ and $\frac{S_t}{g_t}$ are $\widetilde{\mathbb{P}}^{(g)}$-martingales.

If the numéraire asset g pays dividends, as mentioned in Remark #3 above, then the measure $\widetilde{\mathbb{P}}^{(g)}$ is defined such that $\frac{B_t}{\hat{g}_t}$ and $\frac{S_t}{\hat{g}_t}$ are $\widetilde{\mathbb{P}}^{(g)}$-martingales. Hence, in the backward recurrence pricing equations (7.35)-(7.37), we replace g_t by $\hat{g}_t := \frac{g_t}{c_t^g}$, which accounts for the time-t after-dividend factor c_t^g for the numéraire asset g, for all $t = 0, 1, \ldots, T$. In particular, (7.36) and (7.37) are replaced by

$$V_t = \hat{g}_t \widetilde{\mathrm{E}}_t^{(g)} \left[\frac{V_{t+1}}{\hat{g}_{t+1}} \right] \quad \text{and} \quad V_t = \hat{g}_t \widetilde{\mathrm{E}}_t^{(g)} \left[\frac{V_T}{\hat{g}_T} \right]. \tag{7.38}$$

For example, in the case of the stock as a dividend paying numéraire asset we have $\hat{g}_t = \hat{S}_t = \frac{S_t}{c_t}$ giving

$$V_t = \hat{S}_t \widetilde{\mathrm{E}}_t^{(S)} \left[\frac{V_{t+1}}{\hat{S}_{t+1}} \right] = S_t \widetilde{\mathrm{E}}_t^{(S)} \left[(1 - q_{t+1}) \frac{V_{t+1}}{S_{t+1}} \right], \tag{7.39}$$

and

$$V_t = \hat{S}_t \widetilde{\mathrm{E}}_t^{(S)} \left[\frac{V_T}{\hat{S}_T} \right] = S_t \widetilde{\mathrm{E}}_t^{(S)} \left[\prod_{k=t+1}^{T} (1 - q_k) \frac{V_T}{S_T} \right], \tag{7.40}$$

where we used (7.6). The measure $\widetilde{\mathbb{P}}^{(S)}$ is defined as above with $\widetilde{\mathbb{P}}^{(S)}(\mathsf{U}) = \frac{1+r-d}{u-d} \cdot \frac{u}{1+r}$. If the stock pays no dividends then all $q_k \equiv 0$ and (7.39)-(7.40) are simply

$$V_t = S_t \widetilde{\mathrm{E}}_t^{(S)} \left[\frac{V_{t+1}}{S_{t+1}} \right] \quad \text{and} \quad V_t = S_t \widetilde{\mathrm{E}}_t^{(S)} \left[\frac{V_T}{S_T} \right]. \tag{7.41}$$

We recall from Chapter 6 the explicit summation formula for computing any of the above conditional expectations within the standard binomial model with independent market moves. In particular, (7.32) can be written using equation (6.35) of Chapter 6 where $t = n$, $T = N$, and $X \equiv V_T$:

$$V_t(\omega_1 \ldots \omega_t) = (1+r)^{-(T-t)} \sum_{\omega_{t+1}, \ldots, \omega_T \in \{\mathsf{D}, \mathsf{U}\}} \cdots \sum \tilde{p}^{\#\mathsf{U}(\omega_{t+1} \ldots \omega_T)} (1 - \tilde{p})^{\#\mathsf{D}(\omega_{t+1} \ldots \omega_T)} V_T(\omega), \tag{7.42}$$

where $\tilde{p} \equiv \widetilde{\mathbb{P}}^{(B)}(\mathsf{U}) = \frac{1+r-d}{u-d}$ and $\omega \equiv \omega_1 \ldots \omega_t \omega_{t+1} \ldots \omega_T$, $0 \leqslant t \leqslant T$. If, instead, we choose the stock as numéraire, i.e., $g_t = S_t$, then (7.40) gives us yet another *equivalent formula* for the derivative price:

$$V_t(\omega_1 \ldots \omega_t) = S_t(\omega_1 \ldots \omega_t) \sum_{\omega_{t+1}, \ldots, \omega_T \in \{\mathsf{D}, \mathsf{U}\}} \cdots \sum \frac{\tilde{q}^{\#\mathsf{U}(\omega_{t+1} \ldots \omega_T)} (1 - \tilde{q})^{\#\mathsf{D}(\omega_{t+1} \ldots \omega_T)}}{(1 - q_{t+1}(\omega))^{-1} \times \ldots \times (1 - q_T(\omega))^{-1}} \frac{V_T(\omega)}{S_T(\omega)}$$

$$= \sum_{\omega_{t+1}, \ldots, \omega_T \in \{\mathsf{D}, \mathsf{U}\}} \cdots \sum \left(\frac{\tilde{q}}{u} \right)^{\#\mathsf{U}(\omega_{t+1} \ldots \omega_T)} \left(\frac{1 - \tilde{q}}{d} \right)^{\#\mathsf{D}(\omega_{t+1} \ldots \omega_T)} V_T(\omega), \tag{7.43}$$

where now $\tilde{q} \equiv \widetilde{\mathbb{P}}^{(S)}(\mathsf{U}) = \tilde{p} \frac{u}{1+r}$. Note that here we generally allowed for a dividend-paying stock. The simplification in the last equation line, showing explicitly the equivalence of (7.42) and (7.43), arises by using the stock price representation in (7.5):

$$S_T(\omega) = S_t(\omega_1 \ldots \omega_t) \left[\prod_{k=t+1}^{T} (1 - q_k(\omega)) \right] u^{\#\mathsf{U}(\omega_{t+1} \ldots \omega_T)} d^{\#\mathsf{D}(\omega_{t+1} \ldots \omega_T)}.$$

In summary, we have derived two methods for calculating derivative prices. The first method involves the use of a single-time-step recurrence formula, e.g., (7.26) or (7.36), which is used to calculate the prices one by one recursively backward in time. The second method employs a multi-step pricing formula, e.g., (7.37), (7.42), or (7.43). It allows us to calculate the derivative price at any intermediate date by averaging the values of the payoff function weighted by the risk-neutral probabilities of all the $(T-t)$-step paths $\omega_{t+1} \ldots \omega_T$.

The replicating strategy $\{\Delta_t\}_{t \geqslant 0}$ is also called a *delta hedging strategy*. It allows the writer of a derivative contract to hedge perfectly the contract until the expiration date. Suppose that the writer sells one derivative contract for V_0 dollars and uses the proceeds to form a replicating portfolio, (β_0, Δ_0), in the bank account and the stock. As a result, the total value of the initial investment portfolio of one short derivative, Δ_0 shares of stock, and β_0 units of the risk-free asset is zero. At the end of each period, the investor changes (i.e., re-adjusts) the positions in the stock using (7.29) and (7.14). According to Theorem 7.6, the total value of the portfolio remains equal to zero. At maturity, the investor closes the positions in the stock and bank account to pay out the premium (if any) to the holder of the derivative contract. The proceeds cover the payoff in full. The whole situation is a zero sum game without any risk involved.

The next question arising naturally is whether the binomial tree model is *complete*, that is, every claim can be replicated. Recall that a single-period model is said to be complete if every payoff can be replicated by a portfolio in base assets.

Definition 7.3. A multi-period model is said to be *complete* if every $\mathcal{F}_T$-measurable payoff $X: \Omega_T \to \mathbb{R}$ can be replicated by a self-financing trading strategy in base assets.

Theorem 7.6 states that every payoff can be replicated by a self-financing strategy. Thus, the binomial tree model is indeed *complete*. To illustrate this important result, we will find the price process and replication strategy for a path-dependent derivative in the following example.

Example 7.4. Consider a binomial tree model from Example 7.1. Determine the prices and replicating strategy for a *lookback call option* with payoff

$$X = S_3 - \min_{0 \leqslant t \leqslant 3} S_t.$$

Solution. First, we shall construct a tree of possible scenarios. To calculate the payoff function at maturity $T = 3$ we need to know the minimum price

$$m_3(\omega) = \min_{0 \leqslant t \leqslant 3} S_t(\omega)$$

for all market scenarios $\omega \in \Omega_3$. The sampled minimum $m_t = \min_{0 \leqslant n \leqslant t} S_n$, $t = 0, 1, 2, 3$, can be calculated simultaneously with stock prices using the formula $m_t = \min\{m_{t-1}, S_t\}$, $t = 1, 2, 3$, where $m_0 = S_0$. Since the value of m_t depends on the historical path of the stock price process (i.e., a market scenario ω), the result can be represented using a nonrecombining scenario tree, as given in Figure 7.3. As is clearly seen from Figure 7.3, the tree cannot be reduced to a recombining one since $m_2(\text{DU}) \neq m_2(\text{UD})$, $m_3(\text{UUD}) \neq m_3(\text{UDU}) \neq m_3(\text{DUU})$, etc..

To calculate the derivative prices V_t, $t = 0, 1, 2$, we first evaluate the payoff function

$$V_3(\omega) = S_3(\omega) - m_3(\omega)$$

for all eight possible scenarios $\omega \in \Omega_3$. After that, using the backward recursion (7.26) with

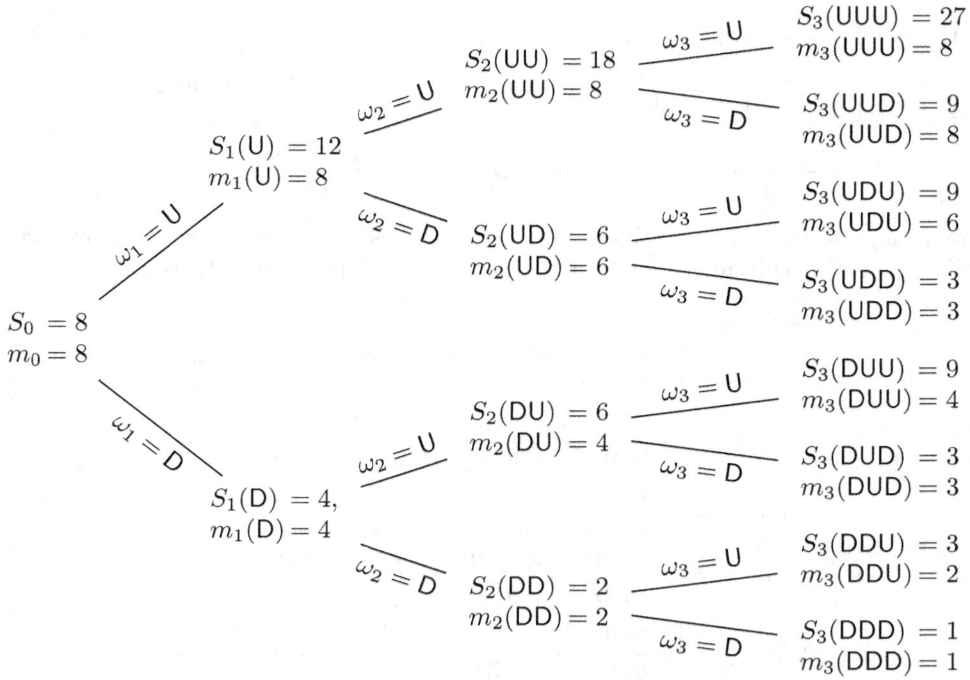

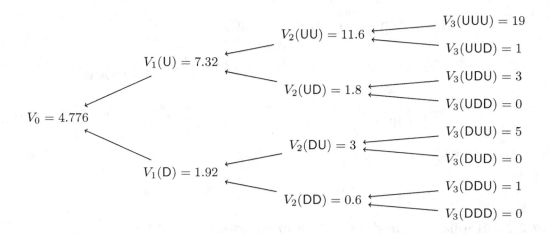

FIGURE 7.3: A binomial tree with stock prices S and sampled minimum values m calculated for each node. There are $|\Omega_3| = 8$ distinct paths from time 0 to 3.

$V_2(UU) = 11.6$ $V_3(UUU) = 19$

$V_1(U) = 7.32$

$V_3(UUD) = 1$

$V_3(UDU) = 3$

$V_2(UD) = 1.8$

$V_3(UDD) = 0$

$V_0 = 4.776$

$V_3(DUU) = 5$

$V_2(DU) = 3$

$V_3(DUD) = 0$

$V_1(D) = 1.92$

$V_3(DDU) = 1$

$V_2(DD) = 0.6$

$V_3(DDD) = 0$

FIGURE 7.4: A scenario tree with derivative prices calculated for each node of a three-period nonrecombining tree.

$\tilde{p} = \frac{3}{4}$ and $r = \frac{1}{4}$, we compute derivative prices as follows:

$$V_2(\omega_1\omega_2) = \frac{1}{1+r}\left(\tilde{p}V_3(\omega_1\omega_2\mathsf{U}) + (1-\tilde{p})V_3(\omega_1\omega_2\mathsf{D})\right) = \frac{3V_3(\omega_1\omega_2\mathsf{U}) + V_3(\omega_1\omega_2\mathsf{D})}{5},$$

$$V_1(\omega_1) = \frac{1}{1+r}\left(\tilde{p}V_2(\omega_1\mathsf{U}) + (1-\tilde{p})V_2(\omega_1\mathsf{D})\right) = \frac{3V_2(\omega_1\mathsf{U}) + V_2(\omega_1\mathsf{D})}{5},$$

$$V_0 = \frac{1}{1+r}\left(\tilde{p}V_1(\mathsf{U}) + (1-\tilde{p})V_1(\mathsf{D})\right) = \frac{3V_1(\mathsf{U}) + V_1(\mathsf{D})}{5},$$

where $\omega_1, \omega_2 \in \{\mathsf{D}, \mathsf{U}\}$. The results of our computations are summarized in Figure 7.4.

We can now also obtain the replicating strategy using the delta hedging formula in (7.29) for Δ_t, $t = 0, 1, 2$:

$$\Delta_0 = \frac{V_1(\mathsf{U}) - V_1(\mathsf{D})}{S_1(\mathsf{U}) - S_1(\mathsf{D})} = \frac{7.32 - 1.92}{12 - 4} = \frac{27}{40} = 0.675,$$

$$\Delta_1(\mathsf{U}) = \frac{V_1(\mathsf{UU}) - V_1(\mathsf{UD})}{S_1(\mathsf{UU}) - S_1(\mathsf{UD})} = \frac{11.6 - 1.8}{18 - 6} = \frac{49}{60} \cong 0.816,$$

$$\Delta_1(\mathsf{D}) = \frac{V_1(\mathsf{DU}) - V_1(\mathsf{DD})}{S_1(\mathsf{DU}) - S_1(\mathsf{DD})} = \frac{3 - 0.6}{6 - 2} = \frac{3}{5} = 0.6,$$

$$\Delta_2(\mathsf{UU}) = \frac{V_3(\mathsf{UUU}) - V_3(\mathsf{UUD})}{S_3(\mathsf{UUU}) - S_3(\mathsf{UUD})} = \frac{19 - 1}{27 - 9} = 1,$$

$$\Delta_2(\mathsf{UD}) = \frac{V_3(\mathsf{UDU}) - V_3(\mathsf{UDD})}{S_3(\mathsf{UDU}) - S_3(\mathsf{UDD})} = \frac{3 - 0}{9 - 3} = 0.5,$$

$$\Delta_2(\mathsf{DU}) = \frac{V_3(\mathsf{DUU}) - V_3(\mathsf{DUD})}{S_3(\mathsf{DUU}) - S_3(\mathsf{DUD})} = \frac{5 - 0}{9 - 3} = \frac{5}{6},$$

$$\Delta_2(\mathsf{DD}) = \frac{V_3(\mathsf{DDU}) - V_3(\mathsf{DDD})}{S_3(\mathsf{DDU}) - S_3(\mathsf{DDD})} = \frac{1 - 0}{3 - 1} = 0.5.$$

The alternative method for computing the derivative prices is to use (7.42). For example, the initial price can be computed by the following discounted expectation:

$$V_0 = (1+r)^{-3}\widetilde{\mathbb{E}}[V_3] = (1+r)^{-3}\sum_{\omega \in \Omega_3}\widetilde{\mathbb{P}}(\omega)V_3(\omega)$$

$$= \left(\frac{5}{4}\right)^{-3}\sum_{\omega_1,\omega_2,\omega_3 \in \{\mathsf{D},\mathsf{U}\}}\left(\frac{3}{4}\right)^{\#\mathsf{U}(\omega_1\omega_2\omega_3)}\left(\frac{1}{4}\right)^{\#\mathsf{D}(\omega_1\omega_2\omega_3)}V_3(\omega_1\omega_2\omega_3)$$

$$= \left(\frac{4}{5}\right)^3\left\{19\cdot\left(\frac{3}{4}\right)^3 + (5+3)\cdot\left(\frac{3}{4}\right)^2\frac{1}{4} + 1\cdot\left[\left(\frac{3}{4}\right)^2\frac{1}{4} + \left(\frac{1}{4}\right)^2\frac{3}{4}\right]\right\}$$

$$= 4.776,$$

$\square$

7.3.2 Replication and Valuation of Random Cash Flows

Consider a financial contract whose payoff is spread over all T periods. Let the time-t payment be X_t for all $t \in \{0, 1, \dots, T\}$. In general X_t is an $\mathcal{F}_t$-measurable random variable. In other words, X_t is a function of the first t market moves $\omega_1, \omega_2, \dots, \omega_t$. So we deal with a stream of random cash flows $\{X_t\}_{0 \leqslant t \leqslant T}$ adapted to the filtration $\mathbb{F}$. We allow these payments to be negative as well as positive. If a payment is positive (negative), then the holder, who is long on the contract, receives (makes) the payment.

Such a stream of cash flows can be replicated by a self-financing strategy in the base securities S and B. Start with the capital $\Pi_0 = V_0$. At time $t = 0$, we make the payment X_0 and form a portfolio (β_0, Δ_0) with the total cost of $\Pi_0 - X_0$. At time $t = 1$ its value is

$$\Pi_1 = \Delta_0 S_0 Y_1 + \beta_0 B_1 = \Delta_0 S_0 Y_1 + (\Pi_0 - X_0 - \Delta_0 S_0)(1 + r).$$

At every time $t+1 \in \{1, \dots, T\}$, we liquidate the portfolio (β_t, Δ_t) whose total value (before the payment X_t is made) is governed by the following wealth equation:

$$\Pi_{t+1} = \Delta_t S_t Y_{t+1} + (\Pi_t - X_t - \Delta_t S_t)(1 + r). \tag{7.44}$$

Note that this reduces to (7.16) (or (7.15)) when we have zero cash flows $X_t \equiv 0$.

Theorem 7.7. *Consider a derivative security with the payoff process $\{X_t\}_{0 \leqslant t \leqslant T}$ adapted to the filtration $\mathbb{F}$. Define the derivative price process $\{V_t\}_{0 \leqslant t \leqslant T}$ by the following backward-in-time recursion:*

$$V_T(\omega_1, \dots, \omega_T) = X_T(\omega_1, \dots, \omega_T), \tag{7.45}$$
$$V_t(\omega_1, \dots, \omega_t) = X_t(\omega_1, \dots, \omega_t)$$
$$+ \frac{1}{1+r}\big(\tilde{p}V_{t+1}(\omega_1, \dots, \omega_t, U) + (1 - \tilde{p})V_{t+1}(\omega_1, \dots, \omega_t, D)\big), \tag{7.46}$$
$$t = 0, 1, \dots, T - 1;$$

and the strategy $\{\Delta_t\}_{0 \leqslant t \leqslant T-1}$ by (7.29). Then the value process $\{\Pi_t\}_{0 \leqslant t \leqslant T}$ given by the wealth equation (7.44) coincides with the derivative price process:

$$\Pi_t(\omega_1 \omega_2 \dots \omega_t) = V_t(\omega_1 \omega_2 \dots \omega_t)$$

for all $0 \leqslant t \leqslant T$ and all $\omega_1, \omega_2, \dots, \omega_t \in \{D, U\}$. In other words, $\{\Delta_t\}_{0 \leqslant t \leqslant T-1}$ is a replicating strategy for the derivative security with the payoff process $\{X_t\}_{0 \leqslant t \leqslant T}$.

Proof. The proof is analogous to that of Theorem 7.6 and is left as an exercise for the reader. □

Equation (7.46) can be written in a compact form using the risk-neutral mathematical expectation:

$$V_t = X_t + (1 + r)^{-1}\widetilde{\mathbb{E}}_t[V_{t+1}], \quad t = 0, 1, \dots, T - 1.$$

Applying this formula successively and using the tower property gives

$$V_t = X_t + \widetilde{\mathbb{E}}_t\left[\sum_{s=t+1}^{T}(1 + r)^{-(s-t)}X_s\right]. \tag{7.47}$$

In other words, the value V_t of the stream of cash flows at time t is given by the sum of the time-t risk-neutral values of all present and future payments. Here, we assume the risk-neutral measure $\widetilde{\mathbb{P}} \equiv \widetilde{\mathbb{P}}^{(B)}$.

7.4 Pricing and Hedging Non-Path-Dependent Derivatives

Equation (7.42) allows us to calculate the current derivative price by averaging the discounted payoff values calculated for all market scenarios. However, it is time and memory consuming to tabulate the payoff function for 2^T scenarios. For example, a naive implementation of (7.42) for a 100-period binomial tree leads to a large amount of computation with $2^{100} \cong 1.268 \times 10^{30}$ market scenarios!

We know that in a recombining binomial tree many market scenarios lead to the same stock prices. For example, in a standard binomial model with stock prices given by (7.3), there are only $t + 1$ different time-t stock prices:

$$S_t \in \{S_{t,n}; n = 0, 1, \ldots, t\}, \text{ where } S_{t,n} := S_0 u^n d^{t-n},$$

for each $t = 0, 1, \ldots, T$. Consider a non-path-dependent derivative. At maturity, the payoff is a function of the time-T stock price, $X = \Lambda(S_T)$ (recall that Λ is used here to denote a known payoff function of the maturity price of the underlying asset). Thus, we expect that the derivative price is a function of the spot price for any time t preceding the maturity: $V_t = V_t(S_t)$. The number of possible values of S_t is $t + 1$, which is significantly less than 2^t. Therefore, we may organize the computation in a more efficient manner by computing derivative values on the recombining binomial tree.

To prove that the time-t derivative price V_t is only a function of the current spot S_t, we use the Markov property of the stock price process within the risk-neutral expectation pricing formula. Recall the definition of a discrete-time Markov process from Chapter 6. From the tower property, we have that if the Markov property holds for one period of time then it holds for any multiple time periods. Consider a non-dividend paying stock. Clearly, using (7.2), the stock price process $\{S_t\}_{t\geqslant 0}$ is a Markov process since, given any (Borel) function g,

$$\widetilde{\mathrm{E}}_t[g(S_{t+1})] = \widetilde{\mathrm{E}}_t[g(S_t Y_{t+1})] = f(S_t), \text{ where } f(x) = \widetilde{\mathrm{E}}[g(xY_{t+1})] = \tilde{p}g(xu) + (1 - \tilde{p})g(xd).$$

Here we used Proposition 6.7 where S_t is $\mathcal{F}_t$-measurable and Y_{t+1} is independent of $\mathcal{F}_t$. [Note that the Markov property holds in any equivalent probability measure.] Therefore, from the multi-step Markov property of the stock price process applied to the risk-neutral expectation of the discounted payoff function $\Lambda(S_T)$, the time-t derivative price must be given by a function of t, S_t, say $f(t, S_t)$:

$$V_t = \widetilde{\mathrm{E}}_t\left[(1 + r)^{-(T-t)}\Lambda(S_T)\right] = f(t, S_t),$$

for any fixed $T \geqslant t$. This is an equation stating the equivalence of two random variables. On any given $\omega \in \Omega$ we have $V_t(\omega) = f(t, S_t(\omega))$. Given a numerical value, S, for the stock price for a given outcome ω, i.e., $S_t(\omega) = S$, then the derivative price function $f(t, S)$ is an ordinary function of ordinary variables t, S. Below we shall also write $V_t(S)$ to denote the time-t derivative price for $S_t = S$. Note that the random variable $V_t(S_t)$ is $\sigma(S_t)$-measurable (recall: $\sigma(S_t) \subset \mathcal{F}_t$). Hence, $V_t(S_t)$ is constant on any atom $\{S_t = S_{t,n}\} \equiv \{\mathsf{U}_t(\omega) = n\}$, i.e., the $(t + 1)$ derivative prices take on values $V_t(S_{t,n})$, for each $n = 0, 1, \ldots, t$.

The Markov property gives the time-t (random) derivative prices, $V_t = V_t(S_t)$, for all t, as functions of the random stock price S_t. Hence, the single-time-step recurrence formula directly gives (recall (4.13) in Chapter 4):

$$V_t(S_t) = \frac{1}{1+r}\widetilde{\mathrm{E}}_t\left[V_{t+1}(S_{t+1})\right] = \frac{1}{1+r}\left[\tilde{p}V_{t+1}(uS_t) + (1 - \tilde{p})V_{t+1}(dS_t)\right]. \qquad (7.48)$$

The backward-time recurrence formula for the derivative value at each node of the binomial tree, $S_t = S_{t,n}$, is then:

$$V_t(S_{t,n}) = \frac{1}{1+r}\left[\tilde{p}V_{t+1}(uS_{t,n}) + (1 - \tilde{p})V_{t+1}(dS_{t,n})\right], \quad 0 \leqslant t \leqslant T - 1, \qquad (7.49)$$

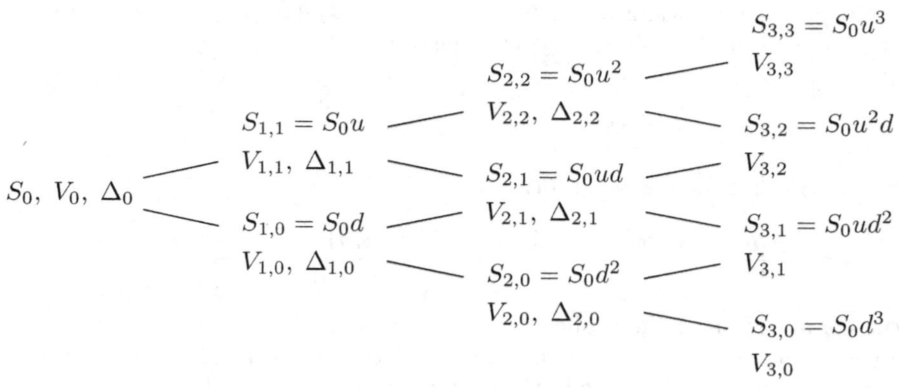

$$S_{3,3} = S_0 u^3$$
$$V_{3,3}$$

$$S_{2,2} = S_0 u^2$$
$$V_{2,2}, \ \Delta_{2,2}$$

$$S_{1,1} = S_0 u$$
$$V_{1,1}, \ \Delta_{1,1}$$
$$S_{3,2} = S_0 u^2 d$$
$$V_{3,2}$$

$$S_{2,1} = S_0 u d$$

$$S_0, \ V_0, \ \Delta_0$$

$$S_{1,0} = S_0 d$$
$$V_{2,1}, \ \Delta_{2,1}$$
$$S_{3,1} = S_0 u d^2$$
$$V_{1,0}, \ \Delta_{1,0}$$
$$V_{3,1}$$

$$S_{2,0} = S_0 d^2$$
$$V_{2,0}, \ \Delta_{2,0}$$
$$S_{3,0} = S_0 d^3$$
$$V_{3,0}$$

FIGURE 7.5: A recombining binomial tree with stock prices S, derivative prices V, and replication stock positions Δ given for each node.

and for $t = T$, $V_T(S_{T,n}) \equiv \Lambda(S_{T,n})$. Hence, for a non-dividend paying stock, this recovers the backward recurrence formula in equation (4.13) of Chapter 4.

On the other hand, we can derive a direct formula for $V_t(S_t)$ at $t = 0, 1, \ldots, T-1$, given as the discounted expectation of the payoff function. Applying (7.32) to a European-style derivative with payoff $\Lambda(S_T)$:

$$V_t(S_t) = (1+r)^{-(T-t)} \widetilde{\mathbb{E}}_t \left[V_T(S_T) \right] = (1+r)^{-(T-t)} \widetilde{\mathbb{E}}_t \left[\Lambda \left(S_t \frac{S_T}{S_t} \right) \right]. \tag{7.50}$$

Assume the stock price is given by (7.3). Hence, the ratio $\frac{S_T}{S_t}$ can be written as

$$\frac{S_T}{S_t} = u^{H_{T-t}} d^{T-t-H_{T-t}}, \tag{7.51}$$

where we conveniently define $H_{T-t}(\omega) := \#\mathsf{U}(\omega_{t+1}, \ldots, \omega_T) = \mathsf{U}_T(\omega) - \mathsf{U}_t(\omega)$, with U_t defined in (6.3) of Chapter 6. The variable H_{T-t} counts the number of upward moves in $\omega_{t+1}, \ldots, \omega_T$. Hence, $H_{T-t} \sim Bin(T-t, \tilde{p})$ under $\widetilde{\mathbb{P}}$. The random variable H_{T-t} (and hence $\frac{S_T}{S_t}$) is independent of $\mathcal{F}_t$ and S_t is $\mathcal{F}_t$-measurable. Therefore, by using Proposition 6.7, we reduce (7.50) to an unconditional expectation which is evaluated explicitly:

$$V_t(S_t) = \frac{1}{(1+r)^{T-t}} \widetilde{\mathbb{E}}_t \left[\Lambda \left(S_t u^{H_{T-t}} d^{(T-t)-H_{T-t}} \right) \right]$$

$$= \frac{1}{(1+r)^{T-t}} \widetilde{\mathbb{E}} \left[\Lambda \left(x u^{H_{T-t}} d^{(T-t)-H_{T-t}} \right) \right] \Bigg|_{x=S_t}$$

$$= \frac{1}{(1+r)^{T-t}} \sum_{n=0}^{T-t} \Lambda \left(S_t u^n d^{(T-t)-n} \right) \binom{T-t}{n} \tilde{p}^n (1-\tilde{p})^{(T-t)-n}. \tag{7.52}$$

This shows that V_t is a function of S_t. By setting $t = 0$, we recover the risk-neutral pricing formula (4.18) in Chapter 4 for the initial price V_0 of a European-style derivative.

Using (7.49), we can compute derivative prices for each node of the recombining binomial tree. As a result, we obtain the *derivative pricing function*, $V_t(S)$, which is a function of calendar (actual) time $t \geqslant 0$ and stock price (spot) $S > 0$. It is defined for every node of the binomial tree and hence can be represented in a tree form as well (see Figure 7.5).

The random stock positions of the replication strategy $\{\Delta_t\}_{0\leqslant t\leqslant T-1}$ given by (7.29) are now also $\sigma(S_t)$-measurable (since they are functions of S_t):

$$\Delta_t(S_t) = \frac{V_{t+1}(S_t u) - V_{t+1}(S_t d)}{(u-d)S_t}. \tag{7.53}$$

Equation (7.14) takes the following form:

$$\beta_t(S_t) = \frac{V_t(S_t) - \Delta_t(S_t)S_t}{B_t} = \frac{V_t(S_t) - (V_{t+1}(S_t u) - V_{t+1}(S_t d))/(u-d)}{B_t}. \tag{7.54}$$

Hence the computation of the replication strategy reduces to the calculation of stock positions for each node of the recombining binomial tree. As a result we obtain a binomial tree where three quantities are computed for each node (t, n), where $t = 0, 1, \ldots, T$ and $n = 0, 1, \ldots, t$: the stock price $S_{t,n} := S_0 u^n d^{t-n}$, the derivative price $V_{t,n} := V_t(S_{t,n})$, and the delta hedge (replicating position in the stock) $\Delta_{t,n} := \Delta_t(S_{t,n})$ (see Figure 7.5). We also have the position in the bank account at each node by (7.54) where $\beta_{t,n} := \beta_t(S_{t,n})$.

The recurrence formulae (7.49) and (7.53) can now be written in a more compact form:

$$V_{t,n} = \frac{1}{1+r} \left[\tilde{p} V_{t+1,n+1} + (1 - \tilde{p}) V_{t+1,n} \right], \tag{7.55}$$

$$\Delta_{t,n} = \frac{V_{t+1,n+1} - V_{t+1,n}}{S_{t,n}(u-d)}, \tag{7.56}$$

$0 \leqslant t \leqslant T-1$, $0 \leqslant n \leqslant t$. The terminal condition is $V_{T,n} = \Lambda(S_{T,n})$, $0 \leqslant n \leqslant T$.

Example 7.5. Consider the three-period recombining binomial tree model from Example 7.1 with $S_0 = 8$, $B_0 = 1$, $u = \frac{3}{2}$, $d = \frac{1}{2}$, and $r = \frac{1}{4}$. Determine the option prices and replication strategy for a standard European call with maturity time $T = 3$ and strike price $K = 8$ (i.e., the call option is at the money).

Solution. We first calculate the payoff values and then the call option prices $C_t(S_{t,n})$ at $t = 0, 1, 2$ using the backward-in-time recursion (7.49), which takes the following form:

$$C_t(S_{t,n}) = \frac{1}{1+r} \left(\tilde{p} C_{t+1}(S_{t,n}u) + (1 - \tilde{p}) C_{t+1}(S_{t,n}d) \right) = \frac{3C_{t+1}(S_{t,n}u) + C_{t+1}(S_{t,n}d)}{5},$$

where $S_{t,n} = S_0 u^n d^{t-n} = 8 \cdot 1.5^n \cdot 0.5^{t-n}$ for $n = 0, 1, \ldots, t$.

$t = 3$: At maturity $T = 3$, calculate the payoff values $C_3(S_3) = (S_3 - 8)^+$, where $S_3 \in \{1, 3, 9, 27\}$:

$$C_3(1) = C_3(3) = 0, \quad C_3(9) = 1, \quad C_3(27) = 19.$$

$t = 2$: Calculate the option prices at time $t = 2$:

$$C_2(2) = \frac{3C_3(3) + C_3(1)}{5} = 0, \quad C_2(6) = \frac{3C_3(9) + C_3(3)}{5} = 0.6,$$

$$C_2(18) = \frac{3C_3(27) + C_3(9)}{5} = 11.6.$$

$t = 1$: Calculate the option prices at time $t = 1$:

$$C_1(4) = \frac{3C_2(6) + C_2(2)}{5} = 0.36, \quad C_1(12) = \frac{3C_2(18) + C_2(6)}{5} = 7.08.$$

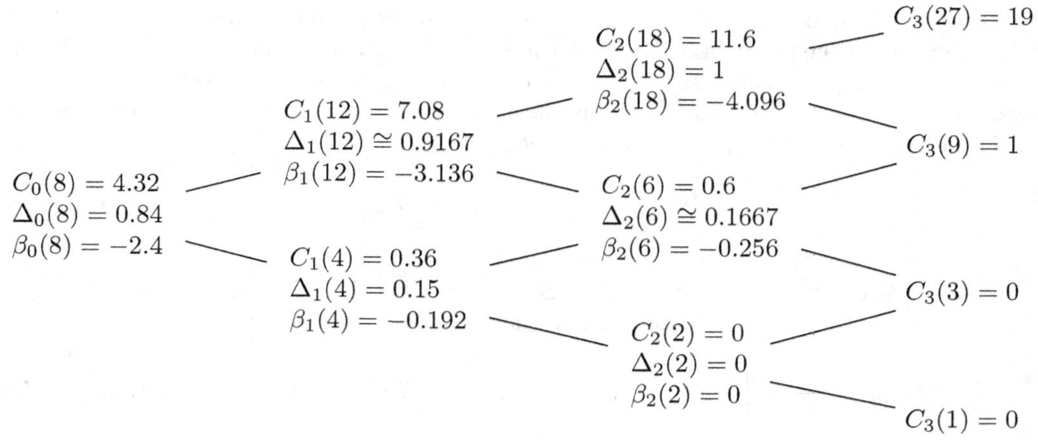

$$C_2(18) = 11.6$$
$$\Delta_2(18) = 1$$
$$\beta_2(18) = -4.096$$

$$C_3(27) = 19$$

$$C_1(12) = 7.08$$
$$\Delta_1(12) \cong 0.9167$$
$$\beta_1(12) = -3.136$$

$$C_3(9) = 1$$

$$C_0(8) = 4.32$$
$$\Delta_0(8) = 0.84$$
$$\beta_0(8) = -2.4$$

$$C_2(6) = 0.6$$
$$\Delta_2(6) \cong 0.1667$$
$$\beta_2(6) = -0.256$$

$$C_1(4) = 0.36$$
$$\Delta_1(4) = 0.15$$
$$\beta_1(4) = -0.192$$

$$C_3(3) = 0$$

$$C_2(2) = 0$$
$$\Delta_2(2) = 0$$
$$\beta_2(2) = 0$$

$$C_3(1) = 0$$

FIGURE 7.6: Call option prices and replication portfolio positions β and Δ calculated for each node of the recombining binomial tree.

$t = 0$: The initial option price is

$$C_0(8) = \frac{3C_1(12) + C_1(4)}{5} = 4.32.$$

Now, compute the replication strategy $\{(\beta_t, \Delta_t)\}_{t=0,1,2}$ using the formulae (7.53) and (7.54):

$$\Delta_0(8) = \frac{C_1(12) - C_1(4)}{12 - 4} = 0.84, \qquad \beta_0(8) = \frac{C_0(8) - 8\Delta_0(8)}{1} = -2.4,$$

$$\Delta_1(4) = \frac{C_2(6) - C_2(2)}{6 - 2} = 0.15, \qquad \beta_1(4) = \frac{C_1(4) - 4\Delta_1(4)}{1.25} = -0.192,$$

$$\Delta_1(12) = \frac{C_2(18) - C_2(6)}{18 - 6} = \frac{11}{12} \cong 0.9167, \quad \beta_1(12) = \frac{C_1(12) - 12\Delta_1(12)}{1.25} = -3.136,$$

$$\Delta_2(2) = \frac{C_3(3) - C_3(1)}{3 - 1} = 0, \qquad \beta_2(2) = \frac{C_2(2) - 2\Delta_2(2)}{1.25^2} = 0,$$

$$\Delta_2(6) = \frac{C_3(9) - C_3(3)}{9 - 3} = \frac{1}{6} \cong 0.1667, \qquad \beta_2(6) = \frac{C_2(6) - 6\Delta_2(6)}{1.25^2} = -0.256,$$

$$\Delta_2(18) = \frac{C_3(27) - C_3(9)}{27 - 9} = 1, \qquad \beta_2(18) = \frac{C_2(18) - 18\Delta_2(18)}{1.25^2} = -4.096.$$

The results are summarized in Figure 7.6. □

Equations (7.49)-(7.52) are readily extended to the case of a dividend-paying stock. The Markov property of the stock price process is maintained if we assume q_{t+1} in (7.4) are $\mathcal{F}_t$-independent, i.e., $q_{t+1}(\omega) = q_{t+1}(\omega_{t+1})$, where

$$S_{t+1}(\omega) = (1 - q_{t+1}(\omega_{t+1}))Y_{t+1}(\omega_{t+1})S_t(\omega). \tag{7.57}$$

Using Proposition 6.7, where S_t is $\mathcal{F}_t$-measurable and Y_{t+1} is $\mathcal{F}_t$-independent, gives the analogue of (7.48)

$$V_t(S_t) = \frac{1}{1+r} \left[\tilde{p}V_{t+1}(u(1 - q_{t+1}(\mathsf{U}))S_t) + (1 - \tilde{p})V_{t+1}(u(1 - q_{t+1}(\mathsf{D}))S_t) \right]. \tag{7.58}$$

Note that in this backward recurrence formula the dividend factors can generally depend on time. In order to obtain a recombining binomial stock price tree, we now make a further simplification to the case where the factors are i.i.d. for all time. That is, we let $q_{t+1}(\mathsf{U}) \equiv q_u$ and $q_{t+1}(\mathsf{D}) \equiv q_d$ be the respective fixed dividend factors for an upward and downward market move. The relation in (7.51) has the same form, but now with effective upward and downward factors $\hat{u} := u(1 - q_u)$ and $\hat{d} := d(1 - q_d)$:

$$\frac{S_T}{S_t} = \hat{u}^{H_{T-t}}\, \hat{d}^{T-t-H_{T-t}}. \tag{7.59}$$

The nodes of the tree are now given by $S_{t,n} := S_0\hat{u}^n\hat{d}^{t-n}$, $n = 0, 1, \ldots, t$; $t = 0, 1, \ldots, T$. The analogue of the backward-time recurrence formula (7.49) is then:

$$V_t(S_{t,n}) = \frac{1}{1+r}\left[\tilde{p}V_{t+1}(\hat{u}S_{t,n}) + (1-\tilde{p})V_{t+1}(\hat{d}S_{t,n})\right], \quad 0 \leqslant t \leqslant T-1, \tag{7.60}$$

$V_T(S_{T,n}) \equiv \Lambda(S_{T,n})$. Since $\hat{u}S_{t,n} = S_{t+1,n+1}$ and $\hat{d}S_{t,n} = S_{t+1,n}$, the derivative prices at the nodes, $V_{t,n} := V_t(S_{t,n})$, satisfy (7.55) and the hedge positions satisfy (7.56).

From (7.59), we also see that the pricing formula in (7.52) is now:

$$V_t(S_t) = \frac{1}{(1+r)^{T-t}}\sum_{n=0}^{T-t}\Lambda\left(S_t\hat{u}^n\hat{d}^{(T-t)-n}\right)\binom{T-t}{n}\tilde{p}^n(1-\tilde{p})^{(T-t)-n}. \tag{7.61}$$

7.5   Pricing Formulae for Standard European Options

The formula (7.52) (or (7.61) if we include stock dividends) allows us to price any non-path-dependent derivative (on a recombining stock price tree) with a given payoff function Λ which can include nonlinear functions of the stock price. However, the option price formula (7.52) (or (7.61)) can be written in a simpler form for the standard European call and put payoff functions. In Chapter 4 we presented the binomial pricing formulae (4.22) and (4.23) for initial (time-0) prices of the European call and put options. Those formulae are given in terms of the binomial CDF $\mathcal{B}$. Let us generalize that result and derive a closed-form formula of the time-t derivative price (with arbitrary calendar time t) for a class of derivatives with piecewise linear payoffs.

According to the pricing formula (7.50), the derivative price is given by the risk-neutral expectation of the payoff function. First, we observe that the European call, put, and chooser option payoff functions $\Lambda(S)$, which are respectively given by $(S-K)^+$, $(K-S)^+$ and $S \wedge K$, can be represented as a linear combination of (elemental) functions from the list

$$\{S, K, S\,\mathbb{I}_{\{S \leqslant K\}}, K\,\mathbb{I}_{\{S \leqslant K\}}\}.$$

Indeed, the call and put payoff functions admit the following representations:

$$(S - K)^+ = (S - K)(1 - \mathbb{I}_{\{S \leqslant K\}}) = S - K - S\,\mathbb{I}_{\{S \leqslant K\}} + K\,\mathbb{I}_{\{S \leqslant K\}},$$
$$(K - S)^+ = (K - S)\,\mathbb{I}_{\{S \leqslant K\}} = K\,\mathbb{I}_{\{S \leqslant K\}} - S\,\mathbb{I}_{\{S \leqslant K\}}.$$

Let us first consider the case with zero stock dividends. By the linearity of the mathematical expectation and the $\widetilde{\mathbb{P}}^{(B)}$-martingale property of the discounted stock price process, i.e., $S_t = \frac{1}{(1+r)^{T-t}}\widetilde{E}_t[S_T]$, we have the following formulae for time-t prices of the standard European call and put options:

$$C_t(S_t) = \frac{1}{(1+r)^{T-t}}\widetilde{E}_t[(S_T - K)^+]$$

$$= S_t - \frac{K}{(1+r)^{T-t}} - \frac{1}{(1+r)^{T-t}}\widetilde{E}_t[S_T \,\mathbb{I}_{\{S_T \leqslant K\}}] + \frac{K}{(1+r)^{T-t}}\widetilde{E}_t[\mathbb{I}_{\{S_T \leqslant K\}}], \quad (7.62)$$

$$P_t(S_t) = \frac{1}{(1+r)^{T-t}}\widetilde{E}_t[(K - S_T)^+]$$

$$= \frac{K}{(1+r)^{T-t}}\widetilde{E}_t[\mathbb{I}_{\{S_T \leqslant K\}}] - \frac{1}{(1+r)^{T-t}}\widetilde{E}_t[S_T \,\mathbb{I}_{\{S_T \leqslant K\}}], \quad 0 \leqslant t \leqslant T. \quad (7.63)$$

As is seen, the evaluation of European call and put option prices reduces to the computation of the conditional expectations $\widetilde{E}_t[\mathbb{I}_{\{S_T \leqslant K\}}]$ and $\widetilde{E}_t[S_T \,\mathbb{I}_{\{S_T \leqslant K\}}]$. The expectation of the indicator function, $\mathbb{I}_{\{S_T \leqslant K\}}$, is equal to the conditional probability of the event $\{S_T \leqslant K\}$:

$$\widetilde{E}_t[\mathbb{I}_{\{S_T \leqslant K\}}] = \widetilde{\mathbb{P}}(S_T \leqslant K \mid \mathcal{F}_t) = \widetilde{\mathbb{P}}\left(\frac{S_T}{S_t} \leqslant \frac{K}{S_t} \,\Big|\, S_t\right). \quad (7.64)$$

In the last expression we divided both sides of the inequality by S_t. Recall the ratio S_T/S_t expressed in (7.51) in terms of a binomial random variable $H_{T-t} \sim Bin(T-t, \tilde{p})$, which is independent of S_t. Expressing the inequality in terms of H_{T-t}:

$$\frac{S_T}{S_t} \leqslant \frac{K}{S_t} \iff u^{H_{T-t}} d^{T-t-H_{T-t}} \leqslant \frac{K}{S_t} \iff \left(\frac{u}{d}\right)^{H_{T-t}} \leqslant \frac{K}{S_t} d^{-(T-t)} \iff$$

$$H_{T-t} \ln\left(\frac{u}{d}\right) \leqslant \ln\left(\frac{K}{S_t}\right) - (T-t)\ln d \iff H_{T-t} \leqslant m_{T-t}(S_t),$$

with random variable $m_{T-t}(S_t) \equiv m_{T-t}(S_t, K)$, for any fixed strike $K \geqslant 0$, given by

$$m_{T-t}(S_t) = \max\left\{m \,:\, 0 \leqslant m \leqslant T-t;\ S_t\, u^m\, d^{(T-t)-m} \leqslant K\right\} = \left\lfloor \frac{\ln\left(\frac{K}{S_t}\right) - (T-t)\ln d}{\ln\left(\frac{u}{d}\right)} \right\rfloor$$

for $K \in [S_t d^{T-t}, S_t u^{T-t}]$. Here $\lfloor x \rfloor$ is the floor (integer part) of x. For values of $K < S_t d^{T-t}$, i.e., for strike values less than the minimum value of S_T conditional on S_t, we define $m_{T-t}(S_t) \equiv -\infty$. If $K > S_t u^{T-t}$ holds, i.e., the strike price is greater than the maximum value of S_T conditional on S_t, then we define $m_{T-t}(S_t) \equiv T-t$. Now, by independence of H_{T-t} and S_t and the above equivalence of events, (7.64) evaluates to

$$\widetilde{\mathbb{P}}(H_{T-t} \leqslant m_{T-t}(S_t)|S_t) = \widetilde{\mathbb{P}}(H_{T-t} \leqslant m_{T-t}(S))|_{S=S_t} = \mathcal{B}(m_{T-t}(S_t); T-t, \tilde{p}), \quad (7.65)$$

where $\mathcal{B}(m; n, p)$ is the CDF of $\mathsf{X} \sim Bin(n, p)$ given by

$$\mathcal{B}(m; n, p) = \mathbb{P}(\mathsf{X} \leqslant m) = \sum_{k=0}^{m} \binom{n}{k} p^k (1-p)^{n-k}.$$

Note that if $K < S_t d^{T-t}$ holds, then $\widetilde{E}_t[\mathbb{I}_{\{S_T \leqslant K\}}] = \mathcal{B}(-\infty; T-t, \tilde{p}) = 0$; if $K \geqslant S_t u^{T-t}$ holds, then $\widetilde{E}_t[\mathbb{I}_{\{S_T \leqslant K\}}] = \mathcal{B}(T-t; T-t, \tilde{p}) = 1$.

Similarly, we calculate the conditional expectation of a product of the stock price S_T and indicator function $\mathbb{I}_{\{S_T \leqslant K\}}$ as follows:

$$
\begin{aligned}
\widetilde{\mathrm{E}}_t[S_T \, \mathbb{I}_{\{S_T \leqslant K\}}] &= \widetilde{\mathrm{E}}_t[S_t \, u^{H_{T-t}} \, d^{T-t-H_{T-t}} \, \mathbb{I}_{\{H_{T-t} \leqslant m_{T-t}(S_t)\}}] \\
&= S_t \, \widetilde{\mathrm{E}}_t[u^{H_{T-t}} \, d^{T-t-H_{T-t}} \, \mathbb{I}_{\{H_{T-t} \leqslant m_{T-t}(S_t)\}}] \\
&= S_t \sum_{k=0}^{m_{T-t}(S_t)} \binom{T-t}{k} (u\,\tilde{p})^k \, (d\,(1-\tilde{p}))^{T-t-k}.
\end{aligned} \tag{7.66}
$$

The summation in (7.66) can also be written as a binomial CDF. Consider the probability

$$
\tilde{q} = \frac{u}{1+r} \cdot \tilde{p},
$$

which is the risk-neutral probability $\tilde{q} = \widetilde{\mathbb{P}}^{(S)}(\mathsf{U})$ of an upward market move in the EMM $\widetilde{\mathbb{P}}^{(S)}$. The probability $\widetilde{\mathbb{P}}^{(S)}(\mathsf{D})$ of a downward market move is

$$
1 - \tilde{q} = \frac{d}{1+r}(1-\tilde{p}).
$$

Thus, we have $u\tilde{p} = (1+r)\tilde{q}$ and $d(1-\tilde{p}) = (1+r)(1-\tilde{q})$. Hence,

$$
u^k \, \tilde{p}^k \, d^{(T-t)-k} \, (1-\tilde{p})^{(T-t)-k} = (1+r)^{T-t} \, \tilde{q}^k \, (1-\tilde{q})^{(T-t)-k}.
$$

Substituting this into (7.66) gives

$$
\begin{aligned}
\widetilde{\mathrm{E}}_t[S_T \, \mathbb{I}_{\{S_T \leqslant K\}}] &= S_t \, (1+r)^{T-t} \sum_{k=0}^{m_{T-t}(S_t)} \binom{T-t}{k} \tilde{q}^k \, (1-\tilde{q})^{(T-t)-k} \\
&= S_t \, (1+r)^{T-t} \mathcal{B}(m_{T-t}(S_t); T-t, \tilde{q}).
\end{aligned} \tag{7.67}
$$

By combining (7.62)-(7.63) with (7.65) and (7.67), and collecting terms in S_t and strike K, we obtain the time-t binomial pricing formula for the European call and put options:

$$
C_t(S_t) = S_t \left(1 - \mathcal{B}(m_{T-t}(S_t); T-t, \tilde{q})\right) - \frac{K}{(1+r)^{T-t}} \left(1 - \mathcal{B}(m_{T-t}(S_t); T-t, \tilde{p})\right), \tag{7.68}
$$

$$
P_t(S_t) = \frac{K}{(1+r)^{T-t}} \mathcal{B}(m_{T-t}(S_t); T-t, \tilde{p}) - S_t \, \mathcal{B}(m_{T-t}(S_t); T-t, \tilde{q}), \tag{7.69}
$$

with risk-neutral probabilities $\tilde{p}, \tilde{q}$ and $m_{T-t}(S_t) \equiv m_{T-t}(S_t, K)$ defined above. Note the option price in (7.68) and (7.69) satisfy the *put-call parity* relation for zero stock dividends:

$$
C_t(S_t) - P_t(S_t) = S_t - \frac{K}{(1+r)^{T-t}}, \quad 0 \leqslant t \leqslant T. \tag{7.70}
$$

That is, a portfolio of a long call and a short put is equivalent to a long forward contract with the same strike price and expiry date. As a result, the put (call) option value can be computed by combining the put-call parity and the call (put) pricing formula.

We point out an alternative and more direct way to evaluate the expectation in (7.66) by choosing $g_t = S_t$ within (7.41) with $\widetilde{\mathbb{P}}^{(g)} \equiv \widetilde{\mathbb{P}}^{(S)}$. Consider a derivative contract with payoff $V_T = S_T \, \mathbb{I}_{\{S_T \leqslant K\}}$. Then, since $V_T/S_T = \mathbb{I}_{\{S_T \leqslant K\}}$, its time-$t$ value is

$$
V_t = V_t(S_t) = S_t \widetilde{\mathrm{E}}_t^{(S)} \left[\mathbb{I}_{\{S_T \leqslant K\}}\right] = S_t \widetilde{\mathbb{P}}^{(S)} (S_T \leqslant K \mid S_t) = S_t \mathcal{B}(m_{T-t}(S_t); T-t, \tilde{q}).
$$

Here we simply made use of the formula in (7.65) for the conditional probability of the same event in (7.64), but now with $\tilde{q}$ replacing $\tilde{p}$, since we are in the measure $\widetilde{\mathbb{P}}^{(S)}$. Note that this is equivalent to (7.67) since $V_t(S_t)$ is also equal to $(1+r)^{-(T-t)}\widetilde{\mathbb{E}}_t[S_T \, \mathbb{I}_{\{S_T \leqslant K\}}]$.

Let us now consider the case where the stock pays dividends according to (7.59). Recall that now $\{\frac{S_t}{B_t}\}_{0 \leqslant t \leqslant T}$ is no longer a $\widetilde{\mathbb{P}}^{(B)}$-martingale, rather the process $\{\frac{S_t}{c_t B_t}\}_{0 \leqslant t \leqslant T}$ is a $\widetilde{\mathbb{P}}^{(B)}$-martingale. As part of the formulae for pricing the call and put options, we compute $(1+r)^{-(T-t)}\widetilde{\mathbb{E}}_t[S_T]$. Writing $S_T = S_t \frac{S_T}{S_t}$, and using the fact that $\frac{S_T}{S_t}$ is $\mathcal{F}_t$-independent and S_t is $\mathcal{F}_t$-measurable and the independence of each k-th term in the product,

$$\widetilde{\mathbb{E}}_t[S_T] = S_t \widetilde{\mathbb{E}}\left[\frac{S_T}{S_t}\right] = S_t \prod_{k=t+1}^{T} \widetilde{\mathbb{E}}\left[(1-q_k)Y_k\right] = S_t \left(\widehat{u}\widetilde{p} + \widehat{d}(1-\widetilde{p})\right)^{T-t}.$$

Hence, the time-t value of a contract having payoff S_T is

$$(1+r)^{-(T-t)}\widetilde{\mathbb{E}}_t[S_T] = S_t \left(\frac{\widehat{u}\widetilde{p} + \widehat{d}(1-\widetilde{p})}{1+r}\right)^{T-t}. \tag{7.71}$$

[The reader can also check that the same expression follows directly via (7.61) with choice $\Lambda(x) = x$, i.e., purely with the stock itself as payoff.] Note that this expression simplifies to S_t in case the dividends are zero: $q_u = q_d = 0 \implies \widehat{u} = u, \widehat{d} = d \implies \widehat{u}\widetilde{p} + \widehat{d}(1-\widetilde{p}) = u\widetilde{p} + d(1-\widetilde{p}) = 1 + r$. If we consider the common special case with constant dividend $q_u = q_d \equiv q \in [0,1)$ (so $\widehat{u} = u(1-q), \widehat{d} = d(1-q)$), then (7.71) reduces to

$$(1+r)^{-(T-t)}\widetilde{\mathbb{E}}_t[S_T] = S_t(1-q)^{T-t}. \tag{7.72}$$

Since (7.59) has the same form as (7.51), the conditional expectaton in (7.64) is obtained by simply replacing u by $\widehat{u}$ and d by $\widehat{d}$ in (7.65):

$$\widetilde{\mathbb{E}}_t[\mathbb{I}_{\{S_T \leqslant K\}}] = \mathcal{B}(\widehat{m}_{T-t}(S_t); T-t, \tilde{p}), \tag{7.73}$$

where, for any fixed strike $K \geqslant 0$,

$$\widehat{m}_{T-t}(S_t) \equiv \widehat{m}_{T-t}(S_t, K) = \left\lfloor \frac{\ln\left(\frac{K}{S_t}\right) - (T-t)\ln\widehat{d}}{\ln\left(\widehat{u}/\widehat{d}\right)} \right\rfloor$$

if $K \in [S_t\widehat{d}^{T-t}, S_t\widehat{u}^{T-t}]$; $\widehat{m}_{T-t}(S_t) = -\infty$ if $K < S_t\widehat{d}^{T-t}$; $\widehat{m}_{T-t}(S_t) = T-t$ if $K > S_t\widehat{u}^{T-t}$. Note that $\tilde{p} = \frac{1+r-d}{u-d}$ remains the same.

The expression for the conditional expectation in (7.66) follows by using (7.59):

$$\widetilde{\mathbb{E}}_t[S_T \, \mathbb{I}_{\{S_T \leqslant K\}}] = S_t \, \widetilde{\mathbb{E}}_t[\widehat{u}^{H_{T-t}} \, \widehat{d}^{T-t-H_{T-t}} \, \mathbb{I}_{\{H_{T-t} \leqslant \widehat{m}_{T-t}(S_t)\}}]$$

$$= S_t \sum_{k=0}^{\widehat{m}_{T-t}(S_t)} \binom{T-t}{k} (\widehat{u}\,\tilde{p})^k \, (\widehat{d}(1-\tilde{p}))^{T-t-k}$$

$$= (1+r)^{T-t} S_t \sum_{k=0}^{\widehat{m}_{T-t}(S_t)} \binom{T-t}{k} \tilde{q}^k \, (1-\tilde{q})^{T-t-k}(1-q_u)^k(1-q_d)^{T-t-k}. \tag{7.74}$$

The last expression was obtained using $u\tilde{p} = (1+r)\tilde{q}$ and $d(1-\tilde{p}) = (1+r)(1-\tilde{q})$. Hence, we have

$$(1+r)^{-(T-t)}\widetilde{E}_t[S_T \, \mathbb{I}_{\{S_T \leqslant K\}}] = S_t \sum_{k=0}^{\widehat{m}_{T-t}(S_t)} \binom{T-t}{k} \tilde{q}^k \, (1-\tilde{q})^{T-t-k}(1-q_u)^k(1-q_d)^{T-t-k}. \tag{7.75}$$

We point out again, as we have above, that this result is directly derived by choosing $g_t = S_t$ as numéraire. Since the stock now pays a dividend, where $\widehat{g}_t = S_t/c_t$ and $\widehat{g}_T = S_T/c_T$, we have the time-t value of the payoff $V_T := S_T \, \mathbb{I}_{\{S_T \leqslant K\}}$ equivalently given by the risk-neutral pricing formula within the $\widetilde{\mathbb{P}}^{(S)}$-measure:

$$V_t = V_t(S_t) = \widehat{g}_t \widetilde{E}_t^{(S)}\left[\frac{S_T}{\widehat{g}_T}\mathbb{I}_{\{S_T \leqslant K\}}\right] = S_t \widetilde{E}_t^{(S)}\left[\frac{c_T}{c_t}\mathbb{I}_{\{S_T \leqslant K\}}\right]$$

$$= S_t \widetilde{E}_t^{(S)}\left[(1-q_u)^{H_{T-t}}(1-q_d)^{T-t-H_{T-t}}\mathbb{I}_{\{H_{T-t} \leqslant \widehat{m}_{T-t}(S_t)\}}\right].$$

In the last line we used $\frac{c_T}{c_t} = (1-q_u)^{H_{T-t}}(1-q_d)^{T-t-H_{T-t}}$. The summation in (7.75) now follows by using the binomial p.m.f. of $H_{T-t} \sim Bin(T-t, \tilde{q})$, under measure $\widetilde{\mathbb{P}}^{(S)}$, and where S_t is $\mathcal{F}_t$-measurable and independent of H_{T-t}. The expression simplifies in the obvious manner in case the dividends are constant. That is, when $q_u = q_d \equiv q$ we have

$$(1+r)^{-(T-t)}\widetilde{E}_t[S_T \, \mathbb{I}_{\{S_T \leqslant K\}}] = S_t(1-q)^{T-t}\mathcal{B}(\widehat{m}_{T-t}(S_t); T-t, \tilde{q}). \tag{7.76}$$

Combining (7.71), (7.73), and (7.75) and collecting terms in S_t and K, gives us the time-t binomial pricing formula for the European call and put options in the case that the stock pays dividends according to (7.59). In the case of constant dividends the prices of the call and put options take the compact form:

$$C_t(S_t) = S_t(1-q)^{T-t}\left(1 - \mathcal{B}(\widehat{m}_{T-t}(S_t); T-t, \tilde{q})\right) - \frac{K}{(1+r)^{T-t}}\left(1 - \mathcal{B}(\widehat{m}_{T-t}(S_t); T-t, \tilde{p})\right), \tag{7.77}$$

$$P_t(S_t) = \frac{K}{(1+r)^{T-t}}\mathcal{B}(\widehat{m}_{T-t}(S_t); T-t, \tilde{p}) - S_t(1-q)^{T-t}\mathcal{B}(\widehat{m}_{T-t}(S_t); T-t, \tilde{q}), \tag{7.78}$$

where

$$\widehat{m}_{T-t}(S_t) = m_{T-t}(S_t(1-q)^{T-t}) = \left\lfloor \frac{\ln\left(\frac{K}{S_t(1-q)^{T-t}}\right) - (T-t)\ln d}{\ln(\frac{u}{d})} \right\rfloor$$

if $K \in [S_t(1-q)^{T-t}d^{T-t}, S_t(1-q)^{T-t}u^{T-t}]$; $\widehat{m}_{T-t}(S_t) = -\infty$ if $K < S_t(1-q)^{T-t}d^{T-t}$; $\widehat{m}_{T-t}(S_t) = T-t$ if $K > S_t(1-q)^{T-t}u^{T-t}$.

We make note of a useful symmetry relation between the option prices in the case of zero dividends and a constant proportional dividend q. In particular, we observe that the call and put prices in (7.77)-(7.78) are related to those for zero stock dividends, (7.68)-(7.69), by simply replacing $S_t \to S_t(1-q)^{T-t}$ in (7.68)-(7.69). In fact, this symmetry relation is readily shown to hold for any standard European option. Indeed, taking $q_u = q_d \equiv q \in [0,1)$, i.e., $\widehat{u} = u(1-q), \widehat{d} = d(1-q)$, within (7.61) gives

$$V_t(S_t) = \frac{1}{(1+r)^{T-t}} \sum_{n=0}^{T-t} \Lambda\left(S_t(1-q)^{T-t} u^n d^{T-t-n}\right)\binom{T-t}{n}\tilde{p}^n(1-\tilde{p})^{T-t-n}. \tag{7.79}$$

which is equal to the (non-dividend stock) European option price expression in (7.52) after replacing $S_t \to S_t(1-q)^{T-t}$. If we denote the derivative price in (7.79) by $V_t^q(S_t)$, for given constant dividend q, then we can represent this symmetry relation precisely as

$$V_t^q(S_t) = V_t^0\big(S_t(1-q)^{T-t}\big) \tag{7.80}$$

where $V_t^0(S)$ is the derivative pricing function for zero dividend $q = 0$, as given in (7.52), for the same given payoff function Λ.

The put-call symmetry in the case of constant dividends follows from the above call and put price expressions, but also simply from (7.70) with $S_t \to S_t(1-q)^{T-t}$ (i.e., by applying (7.80)):

$$C_t(S_t) - P_t(S_t) = S_t(1-q)^{T-t} - \frac{K}{(1+r)^{T-t}}, \quad 0 \leqslant t \leqslant T. \tag{7.81}$$

For any non-path-dependent European derivative, we also have a similar symmetry relation for the delta hedge positions. Let $\Delta_t^q(S_t)$ denote the delta hedge position for the derivative having price $V_t^q(S_t)$ in (7.79) for any $q \in [0,1)$. Then, by using the symmetry relation in (7.80), for time $t+1$, within the formula for the hedge position gives

$$
\begin{aligned}
\Delta_t^q(S_t) &= \frac{V_{t+1}^q(\widehat{u}S_t) - V_{t+1}^q(\widehat{d}S_t)}{(u-d)S_t} = \frac{V_{t+1}^0\big((1-q)^{T-t-1}\widehat{u}S_t\big) - V_{t+1}^0\big((1-q)^{T-t-1}\widehat{d}S_t\big)}{(u-d)S_t} \\
&= \frac{V_{t+1}^0\big(u\,(1-q)^{T-t}S_t\big) - V_{t+1}^0\big(d\,(1-q)^{T-t}S_t\big)}{(u-d)S_t} \\
&= (1-q)^{T-t}\Delta_t^0\big((1-q)^{T-t}S_t\big) \tag{7.82}
\end{aligned}
$$

where $\Delta_t^0(S) = \frac{V_{t+1}^0(uS) - V_{t+1}^0(dS)}{(u-d)S}$ is the time-t hedge position for zero dividend ($q = 0$) and spot S.

7.6 Pricing and Hedging Path-Dependent Derivatives

As the name implies, the payoff function of a path-dependent derivative depends on some quantity calculated along a trajectory (sample path) of the underlying security price process. The options or derivatives with path-dependent payoffs are called *exotic* since their features are more complex than commonly traded "vanilla" derivatives such as standard European and American put and call options. Examples of such path-dependent quantities include the observed maximum and minimum values of a security price process, the arithmetic and geometric averages, etc. By themselves, such quantities are generally not Markovian. However, when coupled with the underlying security price process, the combination forms a Markov vector process. By the Markov property, it then follows that we can price such path-dependent European-style derivatives by a backward recurrence method involving only derivative prices at the relevant nodes corresponding to the joint values of the underlying (stock) and whatever path-dependent quantities make up the payoff. The backward recurrence pricing formula involves derivative prices on a lattice with nodes specified by the stock price and path-dependent values.

7.6.1 Average Asset Prices and Asian Options: Recursive Evaluation

Asian options are typical examples of exotic options where the payoff functions depend on some form of averaging of the underlying asset price over the life of the option. By considering different types of averaging, such as arithmetic or geometric, one can generate different types of options. So the payoff of an Asian option maturing at time T depends on the arithmetic average, A_T, or the geometric average, G_T, of the prices of the underlying asset S. The sampled averages A_t and G_t, observed from time 0 to t, are, respectively, defined by

$$A_t := \frac{1}{t+1} \sum_{n=0}^{t} S_n, \quad G_t := \left(\prod_{n=0}^{t} S_n \right)^{\frac{1}{t+1}}, \quad 0 \leqslant t \leqslant T. \tag{7.83}$$

As follows from (7.83), at time 0 we have $A_0 = G_0 = S_0$.

In what follows we assume the stock price satisfies (7.2), i.e., we assume zero stock dividends within the standard binomial (recombining) tree model. Let us consider the case of the arithmetic average. Clearly, the average price process $\{A_t\}_{t \geqslant 0}$ is, by itself, not Markovian since we cannot calculate A_t by using only A_{t-1}. However, the vector process $\{(S_t, A_t)\}_{t \geqslant 0}$ is Markovian. The evolution of this process from time $t-1$ to time t for $t \geqslant 1$ is given by the following equation:

$$\begin{bmatrix} S_{t-1} \\ A_{t-1} \end{bmatrix} \rightarrow \begin{bmatrix} S_t \\ A_t \end{bmatrix} = \begin{bmatrix} Y_t S_{t-1} \\ \frac{1}{t+1}(tA_{t-1} + Y_t S_{t-1}) \end{bmatrix}, \tag{7.84}$$

where $Y_t = u\mathbb{I}_{\{\omega_t = \mathsf{U}\}} + d\mathbb{I}_{\{\omega_t = \mathsf{D}\}}$ depends only on ω_t. Hence, given knowledge of the vector process (S_{t-1}, A_{t-1}) at time $t-1$, we obtain the vector process (S_t, A_t) at time t given only the information of the outcome ω_t. That is, the pair S_{t-1}, A_{t-1} is $\mathcal{F}_{t-1}$-measurable and Y_t is independent of $\mathcal{F}_{t-1}$ so that we may use the appropriate multidimensional version of the Independence Proposition, i.e., Proposition 6.8 (with random vector (S_{t-1}, A_{t-1}) and random variable Y_t). Hence, given any (Borel) function $f : \mathbb{R}^2 \to \mathbb{R}$,

$$\mathrm{E}_{t-1}[f(S_t, A_t)] = \mathrm{E}_{t-1}\left[f\left(Y_t S_{t-1}, \frac{1}{t+1}(tA_{t-1} + Y_t S_{t-1}) \right) \right] = g(S_{t-1}, A_{t-1}), \tag{7.85}$$

where

$$g(x, y) = \mathrm{E}\left[f\left(Y_t x, \frac{1}{t+1}(ty + xY_t) \right) \right] = pf\left(ux, \frac{1}{t+1}(ty + xu) \right) + (1-p)f\left(dx, \frac{1}{t+1}(ty + xd) \right).$$

This proves that the above vector process is Markovian.

Hence, based on the above joint Markov property and the single-step discounted expectation formula (7.26), the no-arbitrage price process $V_t = V_t(S_t, A_t)$, $t = 0, 1, \ldots, T$, of a European style path-dependent option having payoff $\Lambda_T = \Lambda(S_T, A_T)$ is given by the backward recurrence formula:

$$\begin{aligned} V_{t-1}(S_{t-1}, A_{t-1}) &= \frac{1}{1+r} \widetilde{\mathrm{E}}_{t-1}[V_t(S_t, A_t)] \\ &= \frac{\tilde{p}}{1+r} V_t\left(uS_{t-1}, \frac{tA_{t-1} + uS_{t-1}}{t+1} \right) + \frac{1-\tilde{p}}{1+r} V_t\left(dS_{t-1}, \frac{tA_{t-1} + dS_{t-1}}{t+1} \right), \end{aligned} \tag{7.86}$$

$t = 1, \ldots, T$, where $V_T(S_T, A_T) = \Lambda(S_T, A_T)$. Here we used the fact that the $\mathcal{F}_t$-measurable prices are only functions of S_t and A_t, i.e., on every sequence of moves $\omega_1, \ldots, \omega_t$ the prices are given by $V_t(\omega_1, \ldots, \omega_t) = V_t(S_t(\omega_1, \ldots, \omega_t), A_t(\omega_1, \ldots, \omega_t))$. From (7.86) the derivative prices at each node $(S_{t-1}, A_{t-1}) = (S, A)$ at time $t-1$ can be computed by backward recurrence:

$$V_{t-1}(S, A) = \frac{\tilde{p}}{1+r} V_t\left(uS, \frac{tA + uS}{t+1} \right) + \frac{1-\tilde{p}}{1+r} V_t\left(dS, \frac{tA + dS}{t+1} \right), \tag{7.87}$$

for all $t = 1, \ldots, T$ and where $V_T(S, A) = \Lambda(S, A)$ for each node $(S_T, A_T) = (S, A)$ at maturity T. Since $V_t = V_t(S_t, A_t)$, the time-$(t-1)$ hedge positions at each node $(S_{t-1}, A_{t-1}) = (S, A)$, given by (7.29), are now equivalently given by

$$\Delta_{t-1}(S, A) = \frac{V_t\left(uS, \frac{tA+uS}{t+1}\right) - V_t\left(dS, \frac{tA+dS}{t+1}\right)}{S(u-d)}. \tag{7.88}$$

Note that all the stock price nodes on the tree are as discussed previously, i.e., S_t has support values $S_{t,n} = S_0 u^n d^{t-n}$, $n = 0, 1, \ldots, t$, for each $t = 0, 1, \ldots, T$. As seen in the above equations, each node (S, A) at time $t-1$ gives rise to a pair of nodes $\left(uS, \frac{tA+uS}{t+1}\right)$ and $\left(dS, \frac{tA+dS}{t+1}\right)$ at time t. Note that the above equations for pricing and hedging are of course valid for special cases where the payoff may depend only the average price, i.e., when $\Lambda_T = \Lambda(A_T)$.

For the geometric average case we have

$$G_t = ((S_0 \cdots S_{t-1}) \cdot S_t)^{\frac{1}{t+1}} = ((G_{t-1})^t \cdot S_t)^{\frac{1}{t+1}}$$

$$= ((G_{t-1})^t \cdot S_{t-1} Y_t)^{\frac{1}{t+1}} = ((G_{t-1})^t S_{t-1})^{\frac{1}{t+1}} Y_t^{\frac{1}{t+1}}. \tag{7.89}$$

By itself, the process $\{G_t\}_{t \geq 0}$ is not Markovian. However, the vector (G_t, S_t) is determined by (G_{t-1}, S_{t-1}) and Y_t where G_{t-1} and S_{t-1} are $\mathcal{F}_{t-1}$-measurable and Y_t is independent of $\mathcal{F}_{t-1}$. Hence, by the same steps as above, it can be shown that the vector process $\{(S_t, G_t)\}_{t \geq 0}$ is Markovian. Hence, we have similar formulas analogous to (7.86)-(7.94). We leave the details as an exercise for the reader. See Exercise 7.7.

In general, the payoff function of an Asian-style option exercised at maturity time $t = T$ is a function of the terminal price and the average price of the underlying asset: $\Lambda(S_T, A_T)$ (or $\Lambda(S_T, G_T)$ for geometric averaging). There are two main examples of Asian options: a floating price (denoted *AFP*) and a floating strike (denoted *AFS*). Sometimes, these options are also called "average price option" and "average strike option", respectively. The payoff functions of Asian calls (denoted C) and puts (denoted P) at maturity time T are as follows:

$$\begin{array}{ll} \Lambda_C^{AFP}(S_T, A_T) = (A_T - K)^+, & \Lambda_P^{AFP}(S_T, A_T) = (K - A_T)^+, \\ \Lambda_C^{AFS}(S_T, A_T) = (S_T - A_T)^+, & \Lambda_P^{AFS}(S_T, A_T) = (A_T - S_T)^+. \end{array} \tag{7.90}$$

Note that for the case of geometric averaging we replace A_T by G_T. Here K is a fixed strike price for the average price options. Note that the payoff of a floating price option is obtained by replacing the terminal asset price S_T with the average value A_T in the payoff function of the respective standard European option. The average strike options do not have a fixed strike price. Their payoffs can also be deduced from the standard European options by replacing the fixed strike with the average value A_T (or G_T).

Example 7.6. Consider the three-period model in Example 7.1 with $S_0 = 8$, $u = \frac{3}{2}$, $d = \frac{1}{2}$, $r = \frac{1}{4}$. For all relevant nodes (S_t, A_t), determine the no-arbitrage derivative price and the delta hedging position for the arithmetic average Asian call option with payoff $\Lambda_3 = (A_3 - K)^+$ and fixed strike $K = 8$.

Solution. By applying the arithmetic averaging formula in (7.83) to the stock prices on the binomial tree, the nodes (S_t, A_t), $t = 0, 1, 2, 3$, are obtained as follows:

$$(S_0, A_0) = (8, 8),$$

$$(S_1, A_1) \in \{(4, 6), (12, 10)\},$$

$$(S_2, A_2) \in \left\{(2, \frac{14}{3}), (6, 6), (6, \frac{26}{3}), (18, \frac{38}{3})\right\},$$

$$(S_3, A_3) \in \left\{(1, \frac{15}{4}), (3, \frac{17}{4}), (3, \frac{21}{4}), (3, \frac{29}{4}), (9, \frac{25}{4}), (9, \frac{35}{4}), (9, \frac{47}{4}), (27, \frac{65}{4})\right\}.$$

We have the single-step discounted probabilities $\frac{\tilde{p}}{1+r} = \frac{3}{5}$, $\frac{1-\tilde{p}}{1+r} = \frac{1}{5}$. Equation (7.87) reads

$$V_{t-1}(S, A) = \frac{3}{5}V_t\left(\frac{3}{2}S, \frac{tA + \frac{3}{2}S}{t+1}\right) + \frac{1}{5}V_t\left(\frac{1}{2}S, \frac{tA + \frac{1}{2}S}{t+1}\right),$$

$t = 1, 2, 3$. The option prices at maturity are $V_3(S, A) = (A - 8)^+$:

$$V_3(1, \frac{15}{4}) = V_3(3, \frac{17}{4}) = V_3(3, \frac{21}{4}) = V_3(3, \frac{29}{4}) = V_3(9, \frac{25}{4}) = 0;$$

$$V_3(9, \frac{35}{4}) = \frac{3}{4}, \ V_3(9, \frac{47}{4}) = \frac{15}{4}, \ V_3(27, \frac{65}{4}) = \frac{33}{4}.$$

Using the above backward recurrence formula for $t = 3, 2, 1$ gives:

$$V_2(2, \frac{14}{3}) = \frac{3}{5}V_3(3, \frac{17}{4}) + \frac{1}{5}V_3(1, \frac{15}{4}) = \frac{3}{5} \cdot 0 + \frac{1}{5} \cdot 0 = 0,$$

$$V_2(6, 6) = \frac{3}{5}V_3(3, \frac{29}{4}) + \frac{1}{5}V_3(3, \frac{21}{4}) = \frac{3}{5} \cdot 0 + \frac{1}{5} \cdot 0 = 0,$$

$$V_2(6, \frac{26}{3}) = \frac{3}{5}V_3(9, \frac{35}{4}) + \frac{1}{5}V_3(3, \frac{29}{4}) = \frac{3}{5} \cdot \frac{3}{4} + \frac{1}{5} \cdot 0 = \frac{9}{20} = 0.45,$$

$$V_2(18, \frac{38}{3}) = \frac{3}{5}V_3(27, \frac{65}{4}) + \frac{1}{5}V_3(9, \frac{47}{4}) = \frac{3}{5} \cdot \frac{33}{4} + \frac{1}{5} \cdot \frac{15}{4} = \frac{57}{10} = 5.7,$$

$$V_1(4, 6) = \frac{3}{5}V_2(6, 6) + \frac{1}{5}V_2(2, \frac{14}{3}) = \frac{3}{5} \cdot 0 + \frac{1}{5} \cdot 0 = 0,$$

$$V_1(12, 10) = \frac{3}{5}V_2(18, \frac{38}{3}) + \frac{1}{5}V_2(6, \frac{26}{3}) = \frac{3}{5} \cdot \frac{57}{10} + \frac{1}{5} \cdot \frac{9}{20} = \frac{351}{100} = 3.51,$$

$$V_0 \equiv V_0(8, 8) = \frac{3}{5}V_1(12, 10) + \frac{1}{5}V_1(4, 6) = \frac{3}{5} \cdot \frac{351}{100} + \frac{1}{5} \cdot 0 = \frac{1053}{500} = 2.106.$$

Equation (7.94), with $u - d = 1$, reads

$$\Delta_{t-1}(S, A) = \frac{V_t\left(\frac{3}{2}S, \frac{tA + \frac{3}{2}S}{t+1}\right) - V_t\left(\frac{1}{2}S, \frac{tA + \frac{1}{2}S}{t+1}\right)}{S}.$$

Using this equation for $t = 3, 2, 1$ gives:

$$\Delta_2(2, \frac{14}{3}) = \frac{V_3(3, \frac{17}{4}) - V_3(1, \frac{15}{4})}{2} = 0,$$

$$\Delta_2(6, 6) = \frac{V_3(3, \frac{29}{4}) - V_3(3, \frac{21}{4})}{6} = 0,$$

$$\Delta_2(6, \frac{26}{3}) = \frac{V_3(9, \frac{35}{4}) - V_3(3, \frac{29}{4})}{6} = \frac{3}{24} = 0.125,$$

$$\Delta_2(18, \frac{38}{3}) = \frac{V_3(27, \frac{65}{4}) - V_3(9, \frac{47}{4})}{18} = \frac{1}{4} = 0.25,$$

$$\Delta_1(4, 6) = \frac{V_2(6, 6) - V_2(2, \frac{14}{3})}{4} = 0,$$

$$\Delta_1(12, 10) = \frac{V_2(18, \frac{38}{3}) - V_2(6, \frac{26}{3})}{12} = \frac{21}{48} = 0.4375,$$

$$\Delta_0 \equiv \Delta_0(8, 8) = \frac{V_1(12, 10) - V_1(4, 6)}{8} = \frac{351}{800} = 0.43875.$$

□

Consider now the problem of pricing the above Asian options when the stock pays dividends. Recall that if we assume the stock price satisfies (7.57), i.e., $S_t = (1 - q_t)Y_t S_{t-1}$, $q_t(\omega) = q_t(\omega_t)$, then we have

$$\begin{bmatrix} S_t \\ A_t \end{bmatrix} = \begin{bmatrix} (1 - q_t)Y_t S_{t-1} \\ \frac{1}{t+1}(tA_{t-1} + (1 - q_t)Y_t S_{t-1}) \end{bmatrix}. \tag{7.91}$$

The stock price process is again Markovian since q_t is $\mathcal{F}_{t-1}$-independent. This property also led to (7.58). By Proposition 6.8, $\{(S_t, A_t)\}_{t \geqslant 0}$ is also again Markovian. Hence, we have the backward recurrence pricing formula:

$$V_{t-1}(S_{t-1}, A_{t-1}) = \frac{\tilde{p}}{1+r}V_t\left((1 - q_t(\mathsf{U}))uS_{t-1}, \frac{tA_{t-1} + (1 - q_t(\mathsf{U}))uS_{t-1}}{t+1}\right)$$
$$+ \frac{1 - \tilde{p}}{1+r}V_t\left((1 - q_t(\mathsf{D}))dS_{t-1}, \frac{tA_{t-1} + (1 - q_t(\mathsf{D}))dS_{t-1}}{t+1}\right), \tag{7.92}$$

$t = 1, \ldots, T$, where $V_T(S_T, A_T) = \Lambda(S_T, A_T)$. In considering a recombining binomial stock price tree, we again make a further simplification to the case where the dividend factors are assumed i.i.d. for all time, i.e., we set $q_{t+1}(\mathsf{U}) \equiv q_u$ and $q_{t+1}(\mathsf{D}) \equiv q_d$. The stock prices satisfy (7.59) and the derivative prices at each node $(S_{t-1}, A_{t-1}) = (S, A)$ at time $t - 1$ are computed by backward recurrence:

$$V_{t-1}(S, A) = \frac{\tilde{p}}{1+r}V_t\left(\hat{u}S, \frac{tA + \hat{u}S}{t+1}\right) + \frac{1 - \tilde{p}}{1+r}V_t\left(\hat{d}S, \frac{tA + \hat{d}S}{t+1}\right), \tag{7.93}$$

$t = 1, \ldots, T$; $V_T(S, A) = \Lambda(S, A)$ for each node $(S_T, A_T) = (S, A)$ at maturity T. The time-$(t - 1)$ hedge positions at each node $(S_{t-1}, A_{t-1}) = (S, A)$, given by (7.29), are now equivalently given by

$$\Delta_{t-1}(S, A) = \frac{V_t\left(\hat{u}S, \frac{tA + \hat{u}S}{t+1}\right) - V_t\left(\hat{d}S, \frac{tA + \hat{d}S}{t+1}\right)}{S(u - d)}. \tag{7.94}$$

The stock price nodes are as discussed previously where S_t takes on values $S_{t,n} = S_0\hat{u}^n\hat{d}^{t-n}$, $n = 0, 1, \ldots, t$, $t = 0, 1, \ldots, T$, where $\hat{u} = u(1 - q_u)$ and $\hat{d} = d(1 - q_d)$. In the case of a constant dividend we again simply take $q_d \equiv q_u \equiv q \in [0, 1)$, $\hat{u} = u(1 - q)$, $\hat{d} = d(1 - q)$. The analogues of (7.91)-(7.94) for the case of geometric averaging follow in a similar manner (See Exercise 7.7).

7.6.2 Extreme Asset Prices and Lookback Options

Lookback options are another example of path-dependent options that we consider. Their payoffs depend on the maximum or minimum value of the underlying asset price attained during the life of the option. The option allows the holder to "look back" over time to determine the payoff. Recall from Chapter 6 that, in the discrete time setting, the maximum and minimum prices of asset S are, respectively, defined by:

$$M_t = \max_{n=0,\ldots,t} S_n, \quad m_t = \min_{n=0,\ldots,t} S_n, \quad t = 0, 1, \ldots, T. \tag{7.95}$$

As in the case with the average process, the sampled maximum process, $\{M_t\}_{0 \leqslant t \leqslant T}$, and sampled minimum process, $\{m_t\}_{0 \leqslant t \leqslant T}$, are not Markovian by themselves.

Let's assume the stock price satisfies (7.2), i.e., with zero stock dividends. To update the extreme value for a single-period transition, we use the following recurrences:

$$M_t = \max\{M_{t-1}, S_t\} = \max\{M_{t-1}, Y_t S_{t-1}\}, \tag{7.96}$$

$$m_t = \min\{m_{t-1}, S_t\} = \min\{m_{t-1}, Y_t S_{t-1}\}, \tag{7.97}$$

for $t = 1, 2, \ldots, T$, where $S_{t-1}, M_{t-1}, m_{t-1}$ are $\mathcal{F}_{t-1}$-measurable random variables and Y_t is independent of $\mathcal{F}_{t-1}$. Hence, the pair of vector processes $\{(S_t, M_t)\}_{t \geqslant 0}$ and $\{(S_t, m_t)\}_{t \geqslant 0}$ are Markovian. When the stock pays a dividend, according to (7.4), equations (7.96)-(7.97) are modified as

$$M_t = \max\{M_{t-1}, (1 - q_t)Y_t S_{t-1}\}, \quad m_t = \min\{m_{t-1}, (1 - q_t)Y_t S_{t-1}\}. \tag{7.98}$$

The reader can verify that $\{(S_t, M_t)\}_{t \geqslant 0}$ and $\{(S_t, m_t)\}_{t \geqslant 0}$ are Markovian, assuming $q_t(\omega) = q_t(\omega_t)$.

There exist two kinds of lookback options: with floating strike and with floating price. For the *floating strike lookback* (denoted LFS), the option's strike price is floating and determined at maturity. The payoff of an LFS option is the maximum difference between the market asset's price at maturity and the floating strike. An LFS call gives its holder the right to buy at the lowest price recorded during the option's life. An LFS put gives the right to sell at the highest price recorded during the option's life. The payoffs to the holder are given by

$$\Lambda_C^{LFS} = (S_T - m_T)^+ = S_T - m_T, \quad \Lambda_P^{LFS} = (M_T - S_T)^+ = M_T - S_T, \tag{7.99}$$

respectively, for the LFS call and the LFS put. At maturity, the LFS options are never out of the money since $m_T \leqslant S_T \leqslant M_T$ by definition of the extreme values.

As for the floating price lookback (denoted LFP) options, their payoffs are the maximum differences between the optimal (maximum or minimum) underlying asset price and fixed strike. The payoff functions are given by

$$\Lambda_C^{LFP} = (M_T - K)^+, \quad \Lambda_P^{LFP} = (K - m_T)^+, \tag{7.100}$$

respectively, for the lookback call and the lookback put: In other words, the LFP options are structured so that the call (put) option has payoffs given by the underlying asset price at its highest (lowest) realized level during the lifetime of the option. Note that LFP options have the possibility of expiring worthless.

7.6.3   Recursive Evaluation of Lookback Options

The risk-neutral pricing formula (7.32) allows us to compute the price of any path-dependent derivative, including Asian-style options and many others. First, we simply need to calculate the path-dependent payoff for each possible path in the nonrecombing binomial tree (recall that there are 2^T possible paths). Second, we compute the sum of payoff values multiplied by respective risk-neutral probabilities. Third, we multiply the result obtained by a discounting factor. Equivalently, we can always apply the single-step discounted expectation formula in (7.26).

As we already saw in the case of non-path-dependent (standard) European-style options and average Asian options, the (completely pathwise) recurrence formula in (7.26) is simplified by invoking the Markov property. This gives rise to backward recurrence formulas for the derivative prices at each node. Moreover, the computational cost may be reduced when calculating the derivative prices recursively backward in time on the set of nodes for every allowable value of the spot price and path-dependent quantity. This also applies to any lookback option where the payoff $\Lambda(S_T, m_T)$ is a function of only S_T and m_T or only m_T. This includes any of the above mentioned lookback options on the minimum. Since $\{(S_t, m_t)\}_{t \geqslant 0}$ is a Markov process, we have that the option price at any time t is given as a function of only random variables S_t and m_t, i.e., $V_t = V_t(S_t, m_t)$ where $V_t(\omega_1, \ldots, \omega_t) = V_t(S_t(\omega_1, \ldots, \omega_t), m_t(\omega_1, \ldots, \omega_t))$. Then, in the case of zero stock dividends, using (7.97) and the joint Markov property gives

$$
\begin{aligned}
V_t(S_t, m_t) &= \frac{1}{1+r} \widetilde{E}_t \left[V_{t+1}(S_{t+1}, m_{t+1}) \right] . \\
&= \frac{1}{1+r} \widetilde{E}_t \left[V_{t+1}(S_t Y_{t+1}, m_t \wedge (S_t Y_{t+1})) \right] \\
&= \frac{1}{1+r} \left[\tilde{p} V_{t+1}(S_t u, m_t \wedge (S_t u)) + (1 - \tilde{p}) V_{t+1}(S_t d, m_t \wedge (S_t d)) \right] .
\end{aligned}
$$

We use the standard notation $a \wedge b := \min\{a, b\}$ and $a \vee b := \max\{a, b\}$. Hence, the derivative price at each node $(S_t, m_t) = (S, m)$ can be computed by employing the backward recurrence relation

$$
V_t(S, m) = \frac{1}{1+r} \left[\tilde{p} V_{t+1}(Su, m \wedge (Su)) + (1 - \tilde{p}) V_{t+1}(Sd, m \wedge (Sd)) \right], \tag{7.101}
$$

for all $t = 1, \ldots, T - 1$ and for $t = T$: $V_T(S, m) = \Lambda(S, m)$. Note that if $u > 1$, then we may simply replace the argument $m \wedge (Su)$ by m.

By a very similar analysis, due to the fact that $\{(S_t, M_t)\}_{t \geqslant 0}$ is Markov, we can derive a backward recurrence pricing formula for a lookback option whose payoff $\Lambda(S_T, M_T)$ is a function of only S_T and the maximum M_T (or only M_T). In particular, by using (7.96) and the joint Markov property, the derivative price at each node $(S_t, M_t) = (S, M)$ satisfies

$$
V_t(S, M) = \frac{1}{1+r} \left[\tilde{p} V_{t+1}(Su, M \vee (Su)) + (1 - \tilde{p}) V_{t+1}(Sd, M \vee (Sd)) \right], \tag{7.102}
$$

for all $t = 1, \ldots, T - 1$ and $V_T(S, M) = \Lambda(S, M)$. If $d < 1$, then the argument $M \vee (Sd) = M$.

In the more general case of a lookback option, the payoff function can depend on both the terminal maximum and the minimum values of the stock, and possibly the terminal stock price, i.e., $\Lambda(S_T, M_T, m_T)$. Then, since the triplet $\{(S_t, M_t, m_t)\}_{t \geqslant 0}$ is a vector Markov process, it follows that the option price process is a function $V_t = V_t(S_t, M_t, m_t)$. That is, the option price at any time $t = 0, 1, \ldots, T$ is given by the (ordinary) function $V_t(S, M, m)$ at each node $S_t = S$, $M_t = M$ and $m_t = m$. The above analysis leads to the backward recurrence formula:

$$
\begin{aligned}
V_t(S, M, m) = \frac{1}{1+r} \big[&\tilde{p} V_{t+1}(Su, M \vee (Su), m \wedge (Su)) \\
&+ (1 - \tilde{p}) V_{t+1}(Sd, M \vee (Sd), m \wedge (Sd)) \big],
\end{aligned} \tag{7.103}
$$

for $t = 0, 1, \ldots, T - 1$ and $V_T(S, M, m) = \Lambda(S, M, m)$. Again, if $d < 1$, then $M \vee (Sd) = M$ and if $u > 1$, then $m \wedge (Su) = m$.

The delta hedging strategy for the above path-dependent European-style options can also be computed based on the fact that option prices are given as functions whose arguments correspond to the nodal values. Consider the above first type of lookback options where $V_t = V_t(S_t, m_t)$. Then, $\Delta_t = \Delta_t(S_t, m_t)$ in (7.29). Hence, the corresponding hedging position in the stock at time t, given $S_t = S$, $m_t = m$, is given by

$$
\Delta_t(S, m) = \frac{V_{t+1}(Su, m \wedge (Su)) - V_{t+1}(Sd, m \wedge (Sd))}{S(u - d)}. \tag{7.104}
$$

Similarly, the delta hedging position for a lookback option with price $V_t = V_t(S_t, M_t)$ at time t, given $S_t = S, M_t = M$, is

$$\Delta_t(S, M) = \frac{V_{t+1}(Su, M \vee (Su)) - V_{t+1}(Sd, M \vee (Sd))}{S(u - d)}. \tag{7.105}$$

For the more general lookback options where $V_t = V_t(S_t, M_t, m_t)$, the hedging position at time t, given $S_t = S, M_t = M, m_t = m$, is

$$\Delta_t(S, M, m) = \frac{V_{t+1}(Su, M \vee (Su), m \wedge (Su)) - V_{t+1}(Sd, M \vee (Sd), m \wedge (Sd))}{S(u - d)}. \tag{7.106}$$

When the stock pays dividends we use (7.98). In particular, for the constant dividend case we have the analoguous equations where we replace $u \to \widehat{u}, d \to \widehat{d}$ in the recurrence relations (7.101)-(7.103) and in the numerators of expressions in (7.104)-(7.106). Again, we note that the risk-neutral probability $\widetilde{p} = \frac{1+r-d}{u-d}$ is the same in all equations.

Example 7.7. Consider the three-period model in Example 7.1 with $S_0 = 8, u = \frac{3}{2}, d = \frac{1}{2}$, $r = \frac{1}{4}$. Determine the no-arbitrage prices at all the relevant nodes for the floating strike lookback call option with payoff as in Example 7.4.

Solution. The nodes (S, m) at each time $t = 0, 1, 2, 3$ are displayed in Figure 7.3. Beginning with time $t = 3$, we have seven distinct pairs of values:

$$(S_3, m_3) = (S, m) \in \{(27, 8), (9, 8), (9, 6), (9, 4), (3, 3), (3, 2), (1, 1)\}.$$

The derivative price at those nodes is simply the payoff value, $V_3(S, m) = \Lambda(S, m) = S - m$:

$$V_3(27, 8) = 19, \ V_3(9, 8) = 1, \ V_3(9, 6) = 3, \ V_3(9, 4) = 5$$
$$V_3(3, 3) = 0, \ V_3(3, 2) = 1, \ V_3(1, 1) = 0.$$

The nodes at time $t = 2$ are $(S_2, m_2) = (S, m) \in \{(18, 8), (6, 6), (6, 4), (2, 2)\}$. Equation (7.101) with $\widetilde{p} = \frac{3}{4}$ now reads

$$V_t(S, m) = \frac{3}{5} V_{t+1}\left(\frac{3S}{2}, m \wedge \frac{3S}{2}\right) + \frac{1}{5} V_{t+1}\left(\frac{S}{2}, m \wedge \frac{S}{2}\right).$$

Applying this equation for $t = 2$ to all four nodes and using the above payoff values gives

$$V_2(18, 8) = \frac{3}{5} V_3(27, 8) + \frac{1}{5} V_3(9, 8) = 11.60; \ V_2(6, 6) = \frac{3}{5} V_3(9, 6) + \frac{1}{5} V_3(3, 3) = 1.80;$$
$$V_2(6, 4) = \frac{3}{5} V_3(9, 4) + \frac{1}{5} V_3(3, 3) = 3.00; \ V_2(2, 2) = \frac{3}{5} V_3(3, 2) + \frac{1}{5} V_3(1, 1) = 0.60.$$

Applying again the recurrence formula for $t = 1$ to the two nodes $(S_1, m_1) = (S, m) \in \{(12, 8), (4, 4)\}$:

$$V_1(12, 8) = \frac{3}{5} V_2(18, 8) + \frac{1}{5} V_2(6, 6) = 7.32; \ V_1(4, 4) = \frac{3}{5} V_2(6, 4) + \frac{1}{5} V_2(2, 2) = 1.92.$$

Finally, the price at current time $t = 0$ is calculated for $(S_0, m_0) = (S_0, S_0) = (8, 8)$:

$$V_0 \equiv V_0(8, 8) = \frac{3}{5} V_1(12, 8) + \frac{1}{5} V_1(4, 4) = 4.776.$$

Note that this agrees with the price V_0 in Example 7.4 which we previously computed by using two other related methods. □

Example 7.8. Consider again the three-period model in Example 7.1 with $S_0 = 8$, $u = \frac{3}{2}$, $d = \frac{1}{2}$, $r = \frac{1}{4}$. Determine the no-arbitrage prices and the delta hedging positions at all relevant nodes for the lookback option with payoff $\Lambda_3 = M_3 - m_3$.

Solution. The nodes $(S, M, m) \in \{(S_t, M_t, m_t)\}$ at time $t = 1, 2, 3$ are:

$$(S_3, M_3, m_3) \in \{(27, 27, 8), (9, 18, 8), (9, 12, 6), (9, 9, 4), (3, 12, 3), (3, 8, 3), (3, 8, 2), (1, 8, 1)\},$$
$$(S_2, M_2, m_2) \in \{(18, 18, 8), (6, 12, 6), (6, 8, 4), (2, 8, 2)\},$$
$$(S_1, M_1, m_1) \in \{(12, 12, 8), (4, 8, 4)\},$$

where $(S_0, M_0, m_0) = (8, 8, 8)$ is the initial node. The prices at maturity $T = 3$ are given by the payoff, $V_3(S, M, m) = M - m$:

$$V_3(27, 27, 8) = 19, \ V_3(9, 18, 8) = 10, \ V_3(9, 12, 6) = 6, \ V_3(9, 9, 4) = 5$$
$$V_3(3, 12, 3) = 9, \ V_3(3, 8, 3) = 5, \ V_3(3, 8, 2) = 6, \ V_3(1, 8, 1) = 7.$$

The backward recurrence pricing formula (7.103) reads

$$V_t(S, M, m) = \frac{3}{5} V_{t+1}\left(\frac{3S}{2}, M \vee \frac{3S}{2}, m \wedge \frac{3S}{2}\right) + \frac{1}{5} V_{t+1}\left(\frac{S}{2}, M \vee \frac{S}{2}, m \wedge \frac{S}{2}\right).$$

Applying this formula for all nodes at the respective times $t = 2, 1, 0$ gives:

$$V_2(18, 18, 8) = \frac{3}{5} V_3(27, 27, 8) + \frac{1}{5} V_3(9, 18, 8) = 13.40,$$

$$V_2(6, 12, 6) = \frac{3}{5} V_3(9, 12, 6) + \frac{1}{5} V_3(3, 12, 3) = 5.40,$$

$$V_2(6, 8, 4) = \frac{3}{5} V_3(9, 9, 4) + \frac{1}{5} V_3(3, 8, 3) = 4.00,$$

$$V_2(2, 8, 2) = \frac{3}{5} V_3(3, 8, 2) + \frac{1}{5} V_3(1, 8, 1) = 5.00,$$

$$V_1(12, 12, 8) = \frac{3}{5} V_2(18, 18, 8) + \frac{1}{5} V_2(6, 12, 6) = 9.12,$$

$$V_1(4, 8, 4) = \frac{3}{5} V_2(6, 8, 4) + \frac{1}{5} V_2(2, 8, 2) = 3.40,$$

$$V_0 \equiv V_0(8, 8, 8) = \frac{3}{5} V_1(12, 12, 8) + \frac{1}{5} V_1(4, 8, 4) = 6.152.$$

Equation (7.106) for the hedging positions is

$$\Delta_t(S, M, m) = \frac{V_{t+1}\left(\frac{3S}{2}, M \vee \frac{3S}{2}, m \wedge \frac{3S}{2}\right) - V_{t+1}\left(\frac{S}{2}, M \vee \frac{S}{2}, m \wedge \frac{S}{2}\right)}{S}.$$

Applying this for all nodes at the respective times $t = 2, 1, 0$ gives:

$$\Delta_2(18, 18, 8) = \frac{V_3(27, 27, 8) - V_3(9, 18, 8)}{18} = \frac{19 - 10}{18} = 0.5,$$

$$\Delta_2(6, 12, 6) = \frac{V_3(9, 12, 6) - V_3(3, 12, 3)}{6} = \frac{6 - 9}{6} = -0.5,$$

$$\Delta_2(6, 8, 4) = \frac{V_3(9, 9, 4) - V_3(3, 8, 3)}{6} = \frac{5 - 5}{6} = 0,$$

$$\Delta_2(2, 8, 2) = \frac{V_3(3, 8, 2) - V_3(1, 8, 1)}{2} = \frac{6 - 7}{6} = -\frac{1}{6} \approx -0.1667,$$

$$\Delta_1(12, 12, 8) = \frac{V_2(18, 18, 8) - V_2(6, 12, 6)}{12} = \frac{13.40 - 5.40}{12} = \frac{2}{3} \approx 0.6667$$

$$\Delta_1(4, 8, 4) = \frac{V_2(6, 8, 4) - V_2(2, 8, 2)}{1} = \frac{4 - 5}{1} = -1,$$

$$\Delta_0 \equiv \Delta_0(8, 8, 8) = \frac{V_1(12, 12, 8) - V_1(4, 8, 4)}{8} = \frac{9.12 - 3.40}{8} = 0.715.$$

□

7.6.3.1   Pricing Lookback Options on a Two-Dimensional Lattice

Let us construct a complete computational scheme for pricing a lookback derivative whose payoff $\Lambda(S, m)$ depends on the terminal and minimum asset prices. The case when the payoff is a function of the maximum asset price can be considered similarly. For simplicity, we shall assume the case with zero stock dividends and where $ud = 1$ (i.e., we deal with a symmetric stock price lattice) to reduce the range of possible values of S_t and m_t. At time $t \in \{0, 1, \ldots, T - 1\}$, the stock price can take one of $t + 1$ values:

$$S_t \in \{S_{t,n} = S_0 u^{2n-t} \; : \; n = 0, 1, \ldots, t\}.$$

The minimum price process $\{m_t\}_{t \geqslant 0}$ is nonincreasing. Thus, the value of m_t does not exceed S_0 at any time t. Let us find the range of values of m_t. At time $t = 0$, there is only one value of m_0, namely, $m_0 = S_0$. At time $t = 1$, there are two possible values: $m_1 \in \{S_0, S_0 u^{-1}\}$. At time $t = 2$, there are three possible values: $m_2 \in \{S_0, S_0 u^{-1}, S_0 u^{-2}\}$. By induction, we can show that m_t has $t + 1$ possible values:

$$m_t \in \{S_0 u^{k-t} \; : \; k = 0, 1, \ldots, t\}.$$

Thus, at time t, the random vector $[S_t, m_t]$ may have at most $(t + 1)^2$ values:

$$[S_t, m_t] \in \{[S_0 u^{2n-t}, S_0 u^{k-t}] \; : \; n = 0, 1, \ldots, t, \; k = 0, 1, \ldots, t\}. \tag{7.107}$$

The time-t lookback derivative value is a function of S_t and m_t, i.e., $V_t = V_t(S_t, m_t)$. Let $V_{t,n,k}$ denote the time-t derivative value at the node $[S_t, m_t] = [S_0 u^{2n-t}, S_0 u^{k-t}]$. Using the backward recurrence formula (7.103), we construct a scheme for computing the lookback derivative prices on a two-dimensional lattice of values of S_t and m_t. Since not all combinations in (7.107) are possible, we first find the range for the minimum price m_t conditional on $S_t = S_{t,n} = S_0 u^{2n-t}$. The node $(t, S_{t,n})$ of a binomial tree is attained by a path $\omega_1 \omega_2 \ldots \omega_t$ with n upward moves and $t - n$ downward moves. Considering all such paths, we find that the lowest value of m_t is attained on the path with $\omega_1 = \cdots = \omega_{t-n} = \mathsf{D}$ and $\omega_{t-n+1} = \cdots = \omega_t = \mathsf{U}$. It is equal to $S_0 u^{n-t}$. The largest value of m_t attained on a path with n upward moves is equal to $\min\{S_0, S_0 u^{2n-t}\} = S_0 u^{\min\{t, 2n\}-t}$. It is achieved on the path with $\omega_1 = \cdots = \omega_n = \mathsf{U}$ and $\omega_{n+1} = \cdots = \omega_t = \mathsf{D}$. Thus, there are $\min\{t - n, n\} + 1$ possible values of m_t given that $S_t = S_{t,n}$:

$$m_t \in \{S_0 u^{k-t} \; : \; n \leqslant k \leqslant \min\{t, 2n\}\}.$$

We now derive a backward-in-time recursion scheme for evaluation of the lookback derivative prices for every attainable value of $[S_t, m_t]$. At maturity, the derivative values are equal to the respective values of the payoff function. For $t < T$, using the one-period risk-neutral pricing formula (7.101) with $S_t = S_0 u^{2n-t}$ and $m_t = S_0 u^{k-t}$ (where $0 \leqslant n \leqslant t$

and $n \leqslant k \leqslant \min\{t, 2n\}$) gives

$$V_t(S_0 u^{2n-t}, S_0 u^{k-t}) = \frac{1}{1+r} \left[\tilde{p} \, V_{t+1}(S_0 u^{2n-t+1}, S_0 u^{k-t} \wedge S_0 u^{2n-t+1}) \right.$$

$$\left. + (1 - \tilde{p}) \, V_{t+1}(S_0 u^{2n-t-1}, S_0 u^{k-t} \wedge S_0 u^{2n-t-1}) \right]$$

$$= \frac{1}{1+r} \left[\tilde{p} \, V_{t+1}(S_0 u^{2(n+1)-(t+1)}, S_0 u^{(k+1)-(t+1)}) \right.$$

$$\left. + (1 - \tilde{p}) \, V_{t+1}(S_0 u^{2n-(t+1)}, S_0 u^{\min\{k+1,2n\}-(t+1)}) \right].$$

Therefore, we obtain the following recursion scheme:

$$V_{T,n,k} = \Lambda\left(S_0 u^{2n-T}, S_0 u^{k-T}\right), \quad 0 \leqslant n \leqslant T, \; n \leqslant k \leqslant \min\{T, 2n\}, \tag{7.108}$$

$$V_{t,n,k} = \frac{1}{1+r} \left[\tilde{p} \, V_{t+1,n+1,k+1} + (1 - \tilde{p}) \, V_{t+1,n,\min\{k+1,2n\}} \right], \quad 0 \leqslant t \leqslant T, \tag{7.109}$$

$$0 \leqslant n \leqslant t, \; n \leqslant k \leqslant \min\{t, 2n\}.$$

To compare the scheme (7.108)-(7.109) with the general approach (7.27)-(7.28), we find and compare the total number of derivative values to be calculated by using each method. The general method is implemented on a nonrecombining binomial tree with $1 + 2 + \cdots + 2^T = \mathcal{O}(2^{T+1})$ nodes. The scheme (7.108)-(7.109) needs to compute derivative prices for $n_t := \sum_{n=0}^{t}(\min\{n, t-n\} + 1)$ distinct pairs of values of $[S_t, m_t]$ for each $t = 0, 1, \ldots, T$. If t is even, then $n_t = \frac{(t+2)^2}{4}$. If t is odd, then $n_t = \frac{(t+1)(t+3)}{4}$. Therefore, there are $\mathcal{O}(\frac{1}{12}T^3)$ values to be calculated. For large values of T, the scheme (7.108)-(7.109) with a polynomial computational cost is much more efficient than the general approach whose cost is an exponential function of T. Note that pricing of a non-path-dependent option on a recombining binomial tree with T periods requires $\mathcal{O}(\frac{1}{2}T^2)$ arithmetic operations.

7.7 American Options

Recall that an American call (put) option gives the right to buy (to sell) the underlying asset for the strike price agreed in advance at any time from the time the option is written (which is time $t = 0$) to the expiry time $t = T$. The holder of an American option may exercise the option at any time up to and including the expiry date. In the discrete time setting, the option can only be exercised at times $0, 1, \ldots, T$.

7.7.1 Writer's Perspective: Pricing and Hedging

Let us review the pricing of an American derivative in a recombining binomial tree model. Let $\Lambda(S_t)$ be a non-path-dependent European-style payoff to the holder of an American derivative at time $t \in \{0, 1, \ldots, T\}$. For example, the call and put payoffs are $\Lambda(S_t) = (S_t - K)^+$ and $\Lambda(S_t) = (K - S_t)^+$, respectively, if the option is exercised at time $0 \leqslant t \leqslant T$. Let $V_t(S_t)$ denote the time-t value of the non-path-dependent American derivative for spot S_t that has not been exercised yet. At expiry time T, the value of the derivative is

$$V_T(S_T) = \max\{\Lambda(S_T), 0\}. \tag{7.110}$$

Given the American derivative has not been exercised before time $t \in \{0, \ldots, T-2, T-1\}$, the holder has the choice to exercise the derivative immediately at time t with payoff $\Lambda(S_t)$

or wait until the next time moment $t+1$ when the derivative will be worth $V_{t+1}(S_{t+1})$. The value $\Lambda(S_t)$ is called the *intrinsic value* (at time t). Assuming zero stock dividends, the time-t value of the latter alternative, called the *continuation value*, is given by a one-step derivative pricing formula:

$$\frac{1}{1+r}\widetilde{\mathbb{E}}_t\left[V_{t+1}(S_{t+1})\right] = \frac{1}{1+r}\left(\tilde{p}V_{t+1}(S_t u) + (1-\tilde{p})V_{t+1}(S_t d)\right)$$

where $\tilde{p} = \frac{1+r-d}{u-d}$. Since the holder of the option may choose either alternative, the time-t value of the derivative is the maximum of the intrinsic value and continuation value:

$$V_t(S_t) = \max\left\{\Lambda(S_t), \frac{1}{1+r}\left(\tilde{p}V_{t+1}(S_t u) + (1-\tilde{p})V_{t+1}(S_t d)\right)\right\}, \qquad (7.111)$$

for $t = 0, 1, 2, \ldots, T-1$. Equations (7.110)-(7.111) allow us to evaluate American derivative prices on a recombining binomial tree starting from the expiry date and then proceeding backward in time. In the case of stock dividends we replace $uS_t \to \hat{u}S_t, dS_t \to \hat{d}S_t$ in the right-hand side of the above equations, where $\hat{u}, \hat{d}$ are as previously discussed. As we have seen in the European case, option prices can be used to construct a delta hedging (self-financing) portfolio process that perfectly replicates the European option prices. Below we study whether an American option can be replicated by a self-financing strategy. We begin by considering an example.

Example 7.9. Consider the three-period recombining binomial tree model in Example 7.1 with $S_0 = 8$, $u = \frac{3}{2}$, $d = \frac{1}{2}$, and $r = \frac{1}{4}$. Determine and compare the prices of the standard European and American put options with common strike $K = 8$ and expiration $T = 3$.

Solution. The risk-neutral probabilities are $\tilde{p} = 0.75$ and $1 - \tilde{p} = 0.25$. For every stock price node, we apply recurrence formulae (7.49) and (7.111) to evaluate the respective prices of the European and American put options. Let P_t^E and P_t^A, respectively, denote the time-t prices of the European and American put options for $t = 0, 1, 2, 3$. At the expiration date, the option values must equal the payoff function:

$$P_3^E(S_3) = P_3^A(S_3) = (8 - S_3)^+, \qquad S_3 \in \{1, 3, 9, 27\}.$$

To determine the price for the times before expiration, use the following recurrences:

$$P_t^E(S_t) = \frac{1}{1.25}\cdot\left(\frac{3}{4}\cdot P_{t+1}^E\left(S_t\cdot{}^3/_2\right) + \frac{1}{4}\cdot P_{t+1}^E\left(S_t\cdot{}^1/_2\right)\right)$$

$$= \frac{3\cdot P_{t+1}^E\left(1.5\cdot S_t\right) + P_{t+1}^E\left(0.5\cdot S_t\right)}{5},$$

$$P_t^A(S_t) = \max\left\{(8-S_t)^+, \frac{3\cdot P_{t+1}^A\left(1.5\cdot S_t\right) + P_{t+1}^A\left(0.5\cdot S_t\right)}{5}\right\},$$

where $S_t = 8\cdot 1.5^n\cdot 0.5^{t-n}$, $\quad n = 0, 1, \ldots, t$, $\quad t = 2, 1, 0.$

As a result, we obtain the option prices given in Figure 7.7. The values of the American put are boxed at those nodes of the tree where the intrinsic value is greater than or equal to the continuation value:

$$P_2^A(6) = \max\left\{8 - 6, \frac{3\cdot 0 + 5}{5}\right\} = 2 \vee 1 = 2,$$

$$P_2^A(2) = \max\left\{8 - 2, \frac{3\cdot 5 + 7}{5}\right\} = 6 \vee 4.4 = 6,$$

$$P_1^A(4) = \max\left\{8 - 4, \frac{3\cdot 2 + 6}{5}\right\} = 4 \vee 2.4 = 4.$$

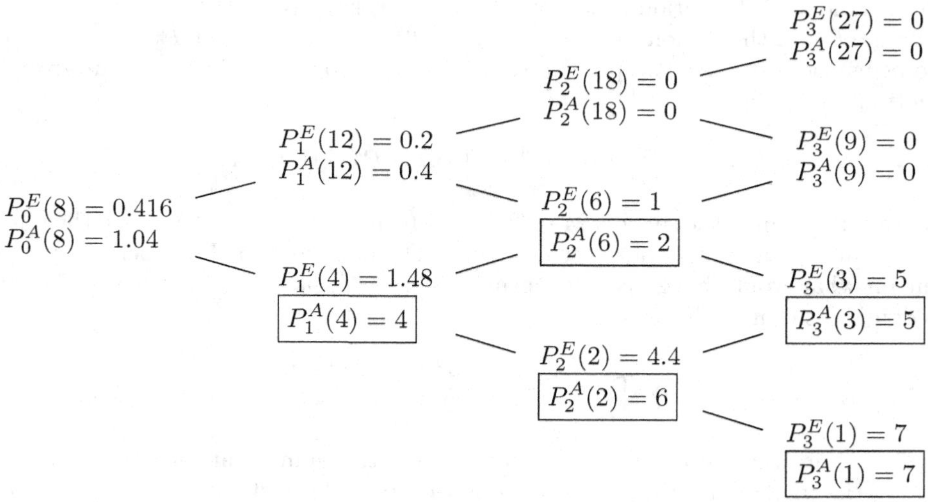

$$P_3^E(27) = 0$$
$$P_3^A(27) = 0$$

$$P_2^E(18) = 0$$
$$P_2^A(18) = 0$$

$$P_1^E(12) = 0.2$$
$$P_1^A(12) = 0.4$$

$$P_3^E(9) = 0$$
$$P_3^A(9) = 0$$

$$P_0^E(8) = 0.416$$
$$P_0^A(8) = 1.04$$

$$P_2^E(6) = 1$$
$$\boxed{P_2^A(6) = 2}$$

$$P_1^E(4) = 1.48$$
$$\boxed{P_1^A(4) = 4}$$

$$P_3^E(3) = 5$$
$$\boxed{P_3^A(3) = 5}$$

$$P_2^E(2) = 4.4$$
$$\boxed{P_2^A(2) = 6}$$

$$P_3^E(1) = 7$$
$$\boxed{P_3^A(1) = 7}$$

FIGURE 7.7: Prices of the standard European put (P^E) and American put (P^A) options.

So the holder of the American put should exercise the option early when the continuation value is strictly less than the intrinsic value, which happens when $S_1 = 4$, or $S_2 = 6$, or $S_2 = 2$. The price of the American option is strictly larger than that of the European put for those cases. Therefore, the initial price of the American put is strictly larger than that of the European put. □

Example 7.10. Construct a hedging strategy for the American put in Example 7.9.

Solution. $t = 0$: The initial value, Π_0, of the hedging strategy is the same as that of the option price: $\Pi_0 = P_0^A(8) = 1.04$. We calculate the number of shares of stock, Δ_0, and the position in the risk-free account, $\beta_0 B_0$, so that the hedging portfolio replicates the option prices at time 1. From (7.29), obtain the hedging position in the stock:

$$\Delta_0 = \frac{P_1^A(12) - P_1^A(4)}{12 - 4} = \frac{0.4 - 4}{8} = -0.45.$$

From the self-financing condition, the amount in the bank account is given by $\beta_0 B_0 = \Pi_0 - \Delta_0 S_0 = 1.04 + 0.45 \cdot 8 = 4.64$.

$t = 1$: At time $t = 1$, the stock price is either at $S_1 = S_1(\mathsf{U}) = 12$ ($\omega_1 = \mathsf{U}$) or at $S_1 = S_1(\mathsf{D}) = 4$ ($\omega_1 = \mathsf{D}$). Consider the case with $\omega_1 = \mathsf{U}$. There are two possible values of $P_2^A(S_2(\mathsf{U}\omega_2))$. The portfolio that hedges against these two possibilities has

$$\Delta_1(\mathsf{U}) = \frac{P_2^A(S_2(\mathsf{UU})) - P_2^A(S_2(\mathsf{UD}))}{(u - d)S_1(\mathsf{U})} = \frac{P_2^A(18) - P_2^A(6)}{18 - 6} = \frac{0 - 2}{12} = -\frac{1}{6}$$

shares of stock. In case $\omega_1 = \mathsf{D}$, the value of the hedging portfolio is

$$\Pi_1(\mathsf{D}) = (1 + r)\beta_0 B_0 + \Delta_0 S_1(\mathsf{D}) = 4.64 \cdot 1.25 - 0.45 \cdot 4 = 4.$$

The holder may exercise the option at time 1. If this is the case, then as writer we liquidate the hedging portfolio and deliver the premium $(K - S_1(\mathsf{D}))^+ = (8 - 4)^+ = 4$

to the holder. If the option is not exercised (i.e., is kept alive) then we continue hedging. At time $t = 2$, the option may be worth $P_2^A(6) = 2$ if $\omega_2 = \mathsf{U}$ or $P_2^A(2) = 6$ if $\omega_2 = \mathsf{D}$. To hedge against these two possibilities, we need a portfolio whose value is given by the continuation value a time $t = 1$:

$$\frac{4}{5}\left(0.75 \cdot P_2^A(6) + 0.25 \cdot P_2^A(2)\right) = 2.4\,.$$

So we can consume the surplus $\$4 - \$2.4 = \$1.6$ and continue hedging with the remaining $\$2.4$. This means that, under scenario $\omega_1 = \mathsf{D}$, the holder of the American put missed out on what would have been an optimal exercise opportunity at time 1. The writer's hedging position in this case is

$$\Delta_1(\mathsf{D}) = \frac{P_2^A(6) - P_2^A(2)}{6 - 2} = \frac{2 - 6}{4} = -1.$$

$t = 2$: There are three possibilities to consider. First, assume that $S_2 = 18$. The option is out of the money and the payoff value is zero regardless of the value of ω_3. The value of the hedging portfolio is zero. The position in stock changes to $\Delta_2(18) = 0$. Second, consider the case that $S_2 = 6$. Again, if the holder exercises the option, then we liquidate the hedging portfolio, which is worth $\$2$, and use the proceeds to pay out the premium. Otherwise, we consume $\$2 - \$1 = \$1$ and use the remaining $\$1$ to hedge against two possibilities: $P_3^A(9) = 0$ and $P_3^A(3) = 5$. We set

$$\Delta_2(S_2 = 6) = \frac{P_3^A(9) - P_3^A(3)}{9 - 3} = \frac{0 - 5}{6} = -\frac{5}{6}.$$

The last case is when $S_2 = 2$. The value of the hedging portfolio is $\$6$, which is sufficient to cover our liabilities if the holder decides to exercise. If the holder declines to exercise the option, then we consume $\$6 - \$4.4 = \$1.6$ and use the remaining $\$4.4$ to construct a hedging portfolio with $\Delta_2(S_2 = 2) = \frac{5-7}{3-1} = -1$ shares of stock. $\qquad\square$

As we can see, an American derivative can be replicated by means of a delta hedging portfolio strategy and a consumption process. If the holder of the derivative does not exercise at the optimal time, then all excess value due to delayed optimal exercise is consumed by the writer. The above example is a special case of a general algorithm for no-arbitrage pricing and hedging of a (path-independent) American derivative for a binomial model with parameters $0 < d < 1 + r < u$. That is, assuming a given payoff $\Lambda(S_T)$, let the prices $V_t(S_t), t = T, T - 1, \ldots, 1, 0$, satisfy the recurrence relations in (7.110) and (7.111). Moreover, let the hedging position $\Delta_t = \Delta_t(S_t)$ in the stock be given by

$$\Delta_t = \frac{V_{t+1}(S_t u) - V_{t+1}(S_t d)}{S_t(u - d)}, \tag{7.112}$$

and the corresponding consumption be given by the difference of the derivative value and the continuation value,

$$C_t = V_t(S_t) - \frac{1}{1 + r}\left(\tilde{p}V_{t+1}(S_t u) + (1 - \tilde{p})V_{t+1}(S_t d)\right), \tag{7.113}$$

for $t = 0, 1, 2, \ldots, T - 1$. If we begin with a portfolio having initial capital $\Pi_0 = V_0(S_0)$ and having time-t value Π_t given recursively forward in time by the (modified version of the wealth equation (7.16) due to consumption),

$$\Pi_{t+1} = \Delta_t Y_{t+1} S_t + (1 + r)(\Pi_t - C_t - \Delta_t S_t), \tag{7.114}$$

then the portfolio process will replicate the American derivative value for any scenario, i.e.,

$$\Pi_t(\omega) = V_t(S_t(\omega))$$

for every $\omega \in \Omega_T$ and $t = 0, \ldots, T$. It should be noted that equation (7.111) guarantees that the consumption is nonnegative, i.e., $C_t \geqslant 0$ for every $t \geqslant 0$. Moreover, the derivative prices are always as valuable as the intrinsic (early payoff) value, $V_t^A(S_t) \geqslant \Lambda(S_t)$, and by replication this also implies $\Pi_t \geqslant \Lambda(S_t)$, for all $t \geqslant 0$.

The above assertion of replication can be proven using similar steps as in the proof of Theorem 7.6. By assumption, the claim is obviously true for $t = 0$. We now give a compact version of a proof of replication that may be instructive. By induction, we assume $\Pi_t = V_t^A(S_t)$ and show $\Pi_{t+1} = V_{t+1}^A(S_{t+1})$ for $t = 0, 1, \ldots, T-1$, as follows. Equation (7.113) gives

$$(1+r)(\Pi_t - C_t) = (1+r)(V_t(S_t) - C_t) = \tilde{p}V_{t+1}(S_t u) + (1 - \tilde{p})V_{t+1}(S_t d) \,.$$

Then, using this expression and (7.112) within (7.114) gives

$$\begin{aligned}
\Pi_{t+1} &= \Delta_t(Y_{t+1}S_t - (1+r)S_t) + (1+r)(\Pi_t - C_t) \\
&= (V_{t+1}(S_t u) - V_{t+1}(S_t d)) \cdot \frac{Y_{t+1} - (1+r)}{(u-d)} + (1+r)(V_t(S_t) - C_t) \\
&= (V_{t+1}(S_t u) - V_{t+1}(S_t d)) \cdot \left((1-\tilde{p})\mathbb{I}_{\{\omega_{t+1}=\mathsf{U}\}} - \tilde{p}\mathbb{I}_{\{\omega_{t+1}=\mathsf{D}\}} \right) \\
&\quad + \tilde{p}V_{t+1}(S_t u) + (1-\tilde{p})V_{t+1}(S_t d) \\
&= V_{t+1}(S_t u)\,\mathbb{I}_{\{\omega_{t+1}=\mathsf{U}\}} + V_{t+1}(S_t d)\,\mathbb{I}_{\{\omega_{t+1}=\mathsf{D}\}} \\
&= V_{t+1}(S_{t+1}) \,.
\end{aligned}$$

Here we made use of $Y_{t+1} = u\,\mathbb{I}_{\{\omega_{t+1}=\mathsf{U}\}} + d\,\mathbb{I}_{\{\omega_{t+1}=\mathsf{D}\}}$ and the fact that $\mathbb{I}_{\{\omega_{t+1}=\mathsf{U}\}} + \mathbb{I}_{\{\omega_{t+1}=\mathsf{D}\}} = 1$. In the case of stock dividends we replace $uS_t \to \widehat{u}S_t, dS_t \to \widehat{d}S_t$ in the numerator of the right-hand side of (7.112) and in (7.113), where $\widehat{u}, \widehat{d}$ are as previously discussed. The above replication follows with the same replacements and (7.114) still holds.

For *path-dependent American derivatives* the above equations for the pricing algorithm and replication strategy (by delta hedging and consumption) are the same in structure within the standard binomial model. The payoff $\Lambda(S_T)$ is replaced by Λ_T, which is any $\mathcal{F}_T$-measurable random variable (not necessarily given as a function of S_T) that evaluates to a number $\Lambda_T(\omega)$ for every $\omega \in \Omega_T$. There is a sequence of $\mathcal{F}_t$-measurable random variables, $\{\Lambda_t\}_{t=0,1,\ldots,T}$, representing the intrinsic values at times $t = 0, 1, \ldots, T$. The derivative prices V_t, at time t, are $\mathcal{F}_t$-measurable random variables that evaluate to a number $V_t(\omega) = V_t(\omega_1 \ldots \omega_t)$ for every sequence $\omega_1 \ldots \omega_t$, $t = 0, 1, \ldots, T$, i.e., V_t are constant on every atom $A_{\omega_1 \ldots \omega_t}$. Equation (7.111) is then a special case of the following general backward recurrence relation for any path-dependent American derivative:

$$V_t(\omega_1 \ldots \omega_t) = \max\left\{ \Lambda_t(\omega_1 \ldots \omega_t), \frac{1}{1+r} \left[\tilde{p}V_{t+1}(\omega_1 \ldots \omega_t \mathsf{U}) + (1-\tilde{p})V_{t+1}(\omega_1 \ldots \omega_t \mathsf{D})\right] \right\}, \tag{7.115}$$

for $t = 0, 1, \ldots, T-1$ and $V_T(\omega_1 \ldots \omega_T) = \max\{\Lambda_T(\omega_1 \ldots \omega_T), 0\}$. The above delta hedging and consumption relations, (7.112) and (7.113), are replaced in the obvious manner by

$$\Delta_t(\omega_1 \ldots \omega_t) = \frac{V_{t+1}(\omega_1 \ldots \omega_t \mathsf{U}) - V_{t+1}(\omega_1 \ldots \omega_t \mathsf{D})}{(u-d)S_t(\omega_1 \ldots \omega_t)}, \tag{7.116}$$

and the consumption value is the difference between the option value and the continuation value:

$$C_t(\omega_1 \ldots \omega_t) = V_t(\omega_1 \ldots \omega_t) - \frac{1}{1+r} \left[\tilde{p} V_{t+1}(\omega_1 \ldots \omega_t U) + (1 - \tilde{p}) V_{t+1}(\omega_1 \ldots \omega_t D) \right]. \quad (7.117)$$

Finally, the wealth equation is defined exactly as in (7.114), which leads to replication, i.e., $\Pi_t(\omega_1 \ldots \omega_t) = V_t(\omega_1 \ldots \omega_t)$ for all scenarios $\omega_1 \ldots \omega_t$ and times $t = 0, 1, \ldots, T$.

In most path-dependent American-style derivatives the payoff is a given function of the stock price and some other observable variable(s). For American-style Asian options, the payoff and hence the intrinsic value is a function of the stock price and the average price: $\Lambda_t = \Lambda(S_t, A_t)$ or $\Lambda_t = \Lambda(S_t, G_t)$, $t = 0, 1, \ldots, T$. The derivative price process, continuation and consumption value processes, and delta hedging process are then all functions of the pair of random variables (S_t, A_t) or (S_t, G_t). In particular, for an arithmetic average Asian option we have $\Lambda_t = \Lambda(S_t, A_t)$. By the joint Markov property of $\{(S_t, A_t)\}_{t \geqslant 0}$, the above equations lead to $V_t = V_t(S_t, A_t)$, $C_t = C_t(S_t, A_t)$, $\Delta_t = \Delta_t(S_t, A_t)$. The backward recurrence pricing equation (7.115), in the case of zero stock dividends, is now equivalently recast as:

$$V_{t-1}(S, A) = \max \left\{ \Lambda(S, A), \frac{\tilde{p}}{1+r} V_t \left(uS, \frac{tA + uS}{t+1} \right) + \frac{1 - \tilde{p}}{1+r} V_t \left(dS, \frac{tA + dS}{t+1} \right) \right\},$$
$$(7.118)$$

for each node (S, A) at time $t = 1, \ldots, T$, and $V_T(S, A) = \Lambda(S, A)$ for each node (S, A) at maturity T. Note that the continuation values at time $t-1$ and node (S, A) are given by the second expression within the above maximum function. Since $V_t = V_t(S_t, A_t)$, the hedging process $\Delta_t = \Delta_t(S_t, A_t)$, $t = 0, \ldots, T-1$, is again given by (7.94). The consumption values at each node and time are now expressible as

$$C_{t-1}(S, A) = V_{t-1}(S, A) - \left[\frac{\tilde{p}}{1+r} V_t \left(uS, \frac{tA + uS}{t+1} \right) + \frac{1 - \tilde{p}}{1+r} V_t \left(dS, \frac{tA + dS}{t+1} \right) \right], \quad (7.119)$$

$t = 1, \ldots, T$. In the case of stock dividends we replace $uS \to \hat{u}S, dS \to \hat{d}S$ in (7.118)-(7.119) where $\hat{u}, \hat{d}$ are the effective up and down factors defined earlier.

Example 7.11. Consider the three-period binomial model in Example 7.6. For all relevant nodes (S_t, A_t), determine the no-arbitrage derivative price, consumption value and the delta hedging position for the American-style arithmetic average Asian call option with payoff $\Lambda_3 = (A_3 - 8)^+$.

Solution. All nodes (S_t, A_t), $t = 0, 1, 2, 3$, option prices at maturity (payoff values) $V_3(S, A) = (A - 8)^+$ and continuation values at each node (S, A) at time $t = 2$ (which are equal to the corresponding European option values $V_2^E(S, A)$) are all given in the solution to Example 7.6. Hence, we do not repeat them here. Equation (7.118) reads

$$V_{t-1}(S, A) = \max \left\{ (A - 8)^+, \frac{3}{5} V_t \left(\frac{3}{2} S, \frac{tA + \frac{3}{2}S}{t+1} \right) + \frac{1}{5} V_t \left(\frac{1}{2} S, \frac{tA + \frac{1}{2}S}{t+1} \right) \right\},$$

$t = 1, 2, 3$. Applying the backward recurrence for $t = 2$ (and using the continuation values $V_2^E(S, A)$) at each time-2 node gives:

$$V_2(2, \tfrac{14}{3}) = \max \left\{ (\tfrac{14}{3} - 8)^+, 0 \right\} = 0, \qquad V_2(6, 6) = \max \left\{ (6 - 8)^+, 0 \right\} = 0,$$

$$V_2(6, \tfrac{26}{3}) = \max \left\{ (\tfrac{26}{3} - 8)^+, 0.45 \right\} = \tfrac{2}{3}, \quad V_2(18, \tfrac{38}{3}) = \max \left\{ (\tfrac{38}{3} - 8)^+, 5.7 \right\} = 5.7.$$

We see that the consumption values are $C_2(2, \tfrac{14}{3}) = C_2(6, 6) = C_2(18, \tfrac{38}{3}) = 0$ and $C_2(6, \tfrac{26}{3}) = \tfrac{2}{3} - 0.45 = \tfrac{13}{60} \approx 0.21667$.

Applying the backward recurrence for $t = 1$ gives:

$$V_1(4,6) = \max\left\{(6-8)^+, \frac{3}{5}\cdot 0 + \frac{1}{5}\cdot 0\right\} = 0$$

$$V_1(12,10) = \max\left\{(10-8)^+, \frac{3}{5}\cdot 5.7 + \frac{1}{5}\cdot\frac{2}{3}\right\} = \frac{533}{150} \approx 3.55333,$$

with consumption values $C_1(4,6) = C_1(12,10) = 0$. For $t = 0$:

$$V_0 \equiv V_0(8,8) = \max\left\{(8-8)^+, \frac{3}{5}\cdot\frac{533}{150} + \frac{1}{5}\cdot 0\right\} = \frac{533}{250} = 2.132,$$

with consumption value $C_0(8,8) = 0$. Applying the delta hedging formula in (7.94),

$$\Delta_{t-1}(S, A) = \frac{V_t\left(\frac{3}{2}S, \frac{tA+\frac{3}{2}S}{t+1}\right) - V_t\left(\frac{1}{2}S, \frac{tA+\frac{1}{2}S}{t+1}\right)}{S},$$

gives the same values for $\Delta_2(S, A)$ as in the solution to Example 7.6. For $t = 1, 0$:

$$\Delta_1(4,6) = \frac{V_2(6,6) - V_2(2,\frac{14}{3})}{4} = \frac{0-0}{4} = 0,$$

$$\Delta_1(12,10) = \frac{V_2(18,\frac{38}{3}) - V_2(6,\frac{26}{3})}{12} = \frac{5.7 - \frac{2}{3}}{12} = \frac{151}{360} = 0.419444445,$$

$$\Delta_0 \equiv \Delta_0(8,8) = \frac{V_1(12,10) - V_1(4,6)}{8} = \frac{\frac{533}{150} - 0}{8} = \frac{533}{1200} \approx 0.444167.$$

$\square$

Due to the joint Markov property of $\{(S_t, M_t, m_t)\}_{t\geqslant 0}$, $\{(S_t, M_t)\}_{t\geqslant 0}$ and $\{(S_t, m_t)\}_{t\geqslant 0}$, an analogous simplification occurs for all American-style lookback options having payoffs $\Lambda_T = \Lambda(S_T, M_T, m_T)$, or $\Lambda(S_T, M_T)$ or $\Lambda(S_T, m_T)$, i.e., with respective intrinsic values $\Lambda_t = \Lambda(S_t, M_t, m_t)$, or $\Lambda(S_t, M_t)$ or $\Lambda(S_t, m_t)$, $t = 0, 1, \ldots, T$. Let us consider, for example, the case where $\Lambda_T = \Lambda(S_T, M_T)$. We then have $V_t = V_t(S_t, M_t)$, $C_t = C_t(S_t, M_t)$, $\Delta_t = \Delta_t(S_t, M_t)$. The backward recurrence pricing equation (7.102) for the European style derivative is now replaced by its American style version:

$$V_t(S, M) = \max\left\{\Lambda(S, M), \frac{\tilde{p}}{1+r}V_{t+1}(Su, M \vee (Su)) + \frac{1-\tilde{p}}{1+r}V_{t+1}(Sd, M \vee (Sd))\right\},$$
$$(7.120)$$

for all $t = 1, \ldots, T-1$ and $V_T(S, M) = \Lambda(S, M)$. The consumption values are

$$C_t(S, M) = V_t(S, M) - \left[\frac{\tilde{p}}{1+r}V_{t+1}(Su, M \vee (Su)) + \frac{1-\tilde{p}}{1+r}V_{t+1}(Sd, M \vee (Sd))\right], \quad (7.121)$$

$t = 0, 1, \ldots, T-1$. The delta hedging positions $\Delta_t(S, M)$ are calculated by the formula in (7.105). The analogues of (7.120)-(7.121) for the case where $\Lambda_T = \Lambda(S_T, M_T, m_T)$ or $\Lambda_T = \Lambda(S_T, m_T)$ are obvious and follow directly by their European counterparts, as discussed in Section 7.6.3. The respective delta hedging positions, $\Delta_t(S, M, m)$ or $\Delta_t(S, m)$, are calculated using the respective formulas in (7.106) and (7.104). In the case of stock dividends we replace $uS \to \hat{u}S, dS \to \hat{d}S$ in (7.120)-(7.121) and within the numerator expressions for the hedge positions.

Example 7.12. Consider the three-period model as in Example 7.8 with $S_0 = 8$, $u = \frac{3}{2}$, $d = \frac{1}{2}$, $r = \frac{1}{4}$. For all relevant nodes (S_t, M_t, m_t), determine the no-arbitrage derivative price, consumption value and the delta hedging position for the American lookback option with payoff $\Lambda_3 = M_3 - m_3$.

Solution. All nodes (S_t, M_t, m_t), $t = 0, 1, 2, 3$, option prices at maturity $V_3(S, M, m) = M - m$ and continuation values at each node (S, M, m) at time $t = 2$ (given by the European option values $V_2^E(S, M, m)$) are all given in the solution to Example 7.8. Hence, we do not repeat them here. The backward recurrence pricing formula reads

$$V_t(S, M, m) = \max\left\{M - m, \frac{3}{5}V_{t+1}\left(\frac{3S}{2}, M \vee \frac{3S}{2}, m \wedge \frac{3S}{2}\right) + \frac{1}{5}V_{t+1}\left(\frac{S}{2}, M \vee \frac{S}{2}, m \wedge \frac{S}{2}\right)\right\}.$$

Applying this formula for all nodes at time $t = 2$ while using the continuation values $V_2^E(S, M, m)$ from Example 7.8 gives:

$$V_2(18, 18, 8) = \max\{18 - 8, 13.40\} = 13.4, \quad V_2(6, 12, 6) = \max\{12 - 6, 5.40\} = 6,$$
$$V_2(6, 8, 4) = \max\{8 - 4, 4\} = 4, \qquad\qquad V_2(2, 8, 2) = \max\{8 - 2, 5\} = 6.$$

The consumption values at time $t = 2$ are: $C_2(18, 18, 8) = 0$, $C_2(6, 12, 6) = 6 - 5.4 = 0.6$, $C_2(6, 8, 4) = 0$, $C_2(2, 8, 2) = 6 - 5 = 1$. Applying the backward pricing formula for $t = 1$ gives

$$V_1(12, 12, 8) = \max\{12 - 8, \frac{3}{5}V_2(18, 18, 8) + \frac{1}{5}V_2(6, 12, 6)\}$$
$$= \max\{4, 0.6 \cdot 13.4 + 0.2 \cdot 6\} = \max\{4, 9.24\} = 9.24,$$
$$V_1(4, 8, 4) = \max\{8 - 4, \frac{3}{5}V_2(6, 8, 4) + \frac{1}{5}V_2(2, 8, 2)\}$$
$$= \max\{4, 0.6 \cdot 4 + 0.2 \cdot 6\} = \max\{4, 3.6\} = 4.$$

The consumption values for $t = 1$ are: $C_1(12, 12, 8) = 0$ and $C_1(4, 8, 4) = 4 - 3.6 = 0.4$. Applying the backward pricing formula for $t = 0$ gives the initial price

$$V_0 \equiv V_0(8, 8, 8) = \max\{0, 0.6 \cdot 9.24 + 0.2 \cdot 4\} = 6.344$$

with consumption value $C_0 \equiv C_0(8, 8, 8) = 0$.

The hedging positions are calculated using (7.106) (note: $u - d = 1$):

$$\Delta_t(S, M, m) = \frac{V_{t+1}\left(\frac{3S}{2}, M \vee \frac{3S}{2}, m \wedge \frac{3S}{2}\right) - V_{t+1}\left(\frac{S}{2}, M \vee \frac{S}{2}, m \wedge \frac{S}{2}\right)}{S}. \tag{7.122}$$

Note that $\Delta_2(18, 18, 8)$, $\Delta_2(6, 12, 6)$, $\Delta_2(6, 8, 4)$ and $\Delta_2(2, 8, 2)$ are the same as the European values already calculated in Example 7.8. For times $t = 1, 0$ we have:

$$\Delta_1(12, 12, 8) = \frac{V_2(18, 18, 8) - V_2(6, 12, 6)}{12} = \frac{13.4 - 6}{12} \approx 0.61667,$$
$$\Delta_1(4, 8, 4) = \frac{V_2(6, 8, 4) - V_2(2, 8, 2)}{4} = \frac{4 - 6}{4} = -0.5,$$
$$\Delta_0 \equiv \Delta_0(8, 8, 8) = \frac{V_1(12, 12, 8) - V_1(4, 8, 4)}{8} = \frac{9.24 - 4}{8} = 0.655.$$

$\square$

7.7.2   Buyer's Perspective: Optimal Exercise

Let us assume we are dealing with a non-path-dependent American derivative. The analysis and equations for path-dependent American derivatives extend in the obvious manner, i.e., as pointed out in the previous section, early exercise occurs whenever the intrinsic value

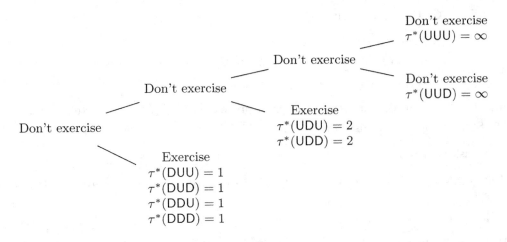

FIGURE 7.8: Exercise rule τ^* of Example 7.9.

equals or surpasses the continuation value. An American derivative (with assumed payoff function $\Lambda(S)$) can be exercised by its holder at any time $t \in \{0, 1, 2 \ldots, T\}$ before the expiry date T. To decide when to exercise the derivative, the holder needs to follow a certain strategy, which is simply a rule that tells if the derivative should be exercised at a particular time t based on the information revealed up to that point. An exercise strategy can be described by a function that maps the set of scenarios Ω to the set of dates. Given the outcome $\omega \in \Omega$, the holder exercises at time $t = \tau(\omega)$ and receives the payoff $\Lambda\left(S_{\tau(\omega)}(\omega)\right)$. In a discrete-time model, we have $\tau \colon \Omega \to \{0, 1, \ldots, T, \infty\}$. If $\tau = \infty$, then the derivative should not be exercised (e.g., the option is out of the money). Since the decision to exercise at time t (i.e., $\tau(\omega) = t$) only depends on the information revealed up to that time, the function τ is adapted to the filtration $\mathbb{F} = \{\mathcal{F}_t\}_{0 \leqslant t \leqslant T}$. That is, for all $0 \leqslant t \leqslant T$ we have $\{\omega \in \Omega \colon \tau(\omega) \leqslant t\} \in \mathcal{F}_t$. In other words, τ is a stopping time. Let $\mathcal{S}_{t,T}$ be the set of all stopping times τ such that $\tau \colon \Omega \to \{t, t+1, \ldots, T, \infty\}$, where $t = 0, 1, \ldots, T$. In particular, $\mathcal{S}_{0,T}$ contains every stopping time in the T-period model.

Let us come back to Example 7.9. As follows from the solution, the holder should exercise the option as soon as the intrinsic value exceeds the continuation value. In particular, if the stock price goes down at time $t = 1$ (i.e., $\omega_1 = \mathsf{D}$), then the option should be exercised at time $t = 1$. If the stock price goes up at time $t = 1$ (i.e., $\omega_1 = \mathsf{U}$), then there is no advantage to exercising at time $t = 1$ and the holder should wait until time $t = 2$. If the stock price goes down at time $t = 2$ (i.e., the scenario observed so far is $\omega_1\omega_2 = \mathsf{UD}$), then it is beneficial to exercise the option. If the stock price goes up, then the option is out of the money and there is no advantage to an early exercise of this American put. The exercising rule obtained is summarized in Figure 7.8, i.e., $\tau^*(A_\mathsf{D}) = 1, \tau^*(A_\mathsf{UD}) = 2, \tau^*(A_\mathsf{UU}) = \infty$ where $A_\mathsf{D} = \{\mathsf{DUU}, \mathsf{DUD}, \mathsf{DDU}, \mathsf{DDD}\}$, $A_\mathsf{UD} = \{\mathsf{UDU}, \mathsf{UDD}\}$, $A_\mathsf{UU} = \{\mathsf{UUU}, \mathsf{UUD}\}$.

Recall from Chapter 6 that, given a stochastic process $\{X_t\}_{t \geqslant 0}$ and stopping time τ, we can define a stopped process $\widehat{X}_t := X_{t \wedge \tau}$, i.e.,

$$\widehat{X}_t(\omega) := X_{t \wedge \tau(\omega)}(\omega) \equiv X_{\min\{t, \tau(\omega)\}}(\omega), \quad t \geqslant 0, \ \omega \in \Omega. \tag{7.123}$$

For example, consider the three-period binomial model with choice $\tau = \tau^*$. Then, for $t = 0$

we have $\widehat{X}_0 = X_0$. For $t \geqslant 1$ we have the following:

$$\omega \in A_\mathsf{D}: \quad \widehat{X}_t(\omega) = X_{t \wedge 1}(\omega) = X_1(\omega),$$

$$\omega \in A_\mathsf{UD}: \widehat{X}_t(\omega) = X_{t \wedge 2}(\omega) = X_1(\omega)\mathbb{I}_{\{t=1\}} + X_2(\omega)\mathbb{I}_{\{t \geqslant 2\}},$$

$$\omega \in A_\mathsf{UU}: \widehat{X}_t(\omega) = X_{t \wedge \infty}(\omega) = X_t(\omega).$$

Example 7.13. Consider the American price process $\{P^A_t\}_{0 \leqslant t \leqslant 3}$ constructed in Example 7.9. Show that (a) the discounted process $\left\{\overline{P}^A_t := \frac{P^A_t}{(1+r)^t}\right\}_{0 \leqslant t \leqslant 3}$ is a supermartingale under the risk-neutral measure $\widetilde{\mathbb{P}}$; (b) the discounted stopped process $\left\{\overline{P}^A_{t \wedge \tau^*}\right\}_{0 \leqslant t \leqslant 3}$, where the stopping time τ^* is the optimal exercising strategy presented in Figure 7.8, is a $\widetilde{\mathbb{P}}$-martingale .

Solution. First, let us construct the discounted derivative price process $\left\{\overline{P}^A_t\right\}_{0 \leqslant t \leqslant 3}$. Each American put time-t value in Figure 7.7 is divided by $(1+r)^t = 1.25^t$. The result is presented in Figure 7.9.

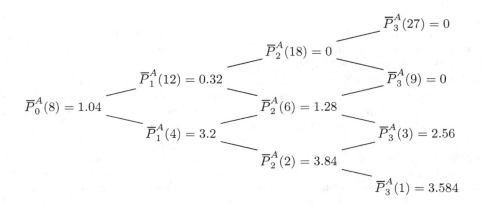

FIGURE 7.9: Values of the discounted price process $\{\overline{P}^A_t\}$ for the standard American put.

To verify whether the American put price process is a $\widetilde{\mathbb{P}}$-supermartingale, we only need to check if the inequality $\overline{P}^A_t(S_t) \geqslant \widetilde{\mathbb{E}}[\overline{P}^A_{t+1}(S_{t+1}) \mid S_t] \equiv 0.75 \cdot \overline{P}^A_{t+1}(\frac{3}{2}S_t) + 0.25 \cdot \overline{P}^A_{t+1}(\frac{1}{2}S_t)$ holds for each $t = 0, 1, 2$ and every possible price S_t. If we have a strict inequality in at least one case, then the process is a supermartingale but not a martingale. Checking this condition for each time step we have:

$t = 0: \quad \overline{P}^A_0(8) = 1.04 \overset{?}{\geqslant} \widetilde{\mathbb{E}}[\overline{P}^A_1] = 0.32 \cdot 0.75 + 3.2 \cdot 0.25 = 1.04$ ✓

$t = 1: \quad \overline{P}^A_1(4) = 3.2 \overset{?}{\geqslant} \widetilde{\mathbb{E}}[\overline{P}^A_2 \mid S_1 = 4] = 1.28 \cdot 0.75 + 3.84 \cdot 0.25 = 1.92$ ✓

$\qquad \overline{P}^A_1(12) = 0.32 \overset{?}{\geqslant} \widetilde{\mathbb{E}}[\overline{P}^A_2 \mid S_1 = 12] = 0 \cdot 0.75 + 1.28 \cdot 0.25 = 0.32$ ✓

$t = 2: \quad \overline{P}^A_2(2) = 3.84 \overset{?}{\geqslant} \widetilde{\mathbb{E}}[\overline{P}^A_3 \mid S_2 = 2] = 2.56 \cdot 0.75 + 3.584 \cdot 0.25 = 2.816$ ✓

$\qquad \overline{P}^A_2(6) = 1.28 \overset{?}{\geqslant} \widetilde{\mathbb{E}}[\overline{P}^A_3 \mid S_2 = 6] = 0 \cdot 0.75 + 2.56 \cdot 0.25 = 0.64$ ✓

$\qquad \overline{P}^A_2(18) = 0 \overset{?}{\geqslant} \widetilde{\mathbb{E}}[\overline{P}^A_3 \mid S_2 = 18] = 0 \cdot 0.75 + 0 \cdot 0.25 = 0$ ✓

Since every conditional expectation satisfies the above inequality, the discounted American put price process is indeed a supermartingale. Finally, consider the stopped process defined by $\overline{P}_t^* := \overline{P}_{t\wedge\tau^*}^A$ where the optimal stopping time τ^* is given in Figure 7.8. The dynamics of the stopped process can be represented by a scenario tree of all market moves, as in Figure 7.10. It is not difficult to verify that this stopped process is a $\widetilde{\mathbb{P}}$-martingale. Indeed, for every choice of market moves $\omega_1, \omega_2 \in \{\mathsf{D}, \mathsf{U}\}$ we have that

$$\overline{P}_0^* = \widetilde{\mathbb{E}}_0[\overline{P}_1^*] = \tilde{p}\overline{P}_1^*(\mathsf{U}) + (1-\tilde{p})\overline{P}_1^*(\mathsf{D}) = 0.75\overline{P}_1^*(\mathsf{U}) + 0.25\overline{P}_1^*(\mathsf{D}),$$
$$\overline{P}_1^*(\omega_1) = \widetilde{\mathbb{E}}_1[\overline{P}_2^*](\omega_1) = \tilde{p}\overline{P}_2^*(\omega_1\mathsf{U}) + (1-\tilde{p})\overline{P}_2^*(\omega_1\mathsf{D}),$$
$$\overline{P}_2^*(\omega_1\omega_2) = \widetilde{\mathbb{E}}_2[\overline{P}_3^*](\omega_1\omega_2) = \tilde{p}\overline{P}_3^*(\omega_1\omega_2\mathsf{U}) + (1-\tilde{p})\overline{P}_3^*(\omega_1\omega_2\mathsf{D}).$$ □

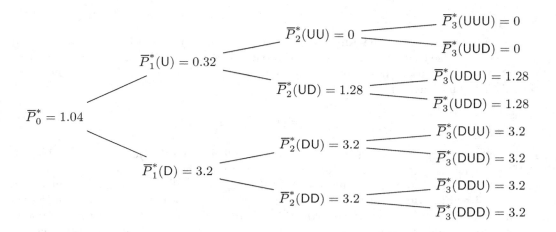

FIGURE 7.10: A full scenario tree for the three-period binomial model with values of the stopped derivative price process $\{\overline{P}_t^* := \overline{P}_{t\wedge\tau^*}^A\}$ for the standard American put.

As is seen from the above example, not exercising an American option at the optimal time is equivalent to the option being an unfavorable game for its holder. An interesting fact is that stopping at the optimal time may turn a supermartingale into a martingale. In general, a stopped (super-)martingale is a (super-)martingale, as was proved in the Optional Sampling Theorem from Chapter 6.

Now let us come back to the problem of pricing of an American derivative at time $t = 0$ based on an optimal exercise strategy. Suppose that the buyer follows a certain exercise strategy $\tau \in \mathcal{S}_{0,T}$, which is not necessarily an optimal one. If $\tau(\omega) \leqslant T$ for scenario $\omega \in \Omega$, then the holder exercises the derivative at time $\tau(\omega)$ to realize the payoff $\Lambda_{\tau(\omega)}(\omega)$ or $\Lambda(S_{\tau(\omega)}(\omega))$ for the non-path-dependent case. If $\tau(\omega) = \infty$, then the derivative is not exercised at all and the terminal payoff is zero. Therefore, the holder's payoff at time $t = T\wedge\tau$ is a product of the payoff function and the indicator function of the event $\{\tau \leqslant T\}$. Generally, for any $\mathcal{F}_t$-measurable intrinsic value Λ_t the payoff at time τ is $\mathbb{I}_{\{\tau\leqslant T\}}\Lambda_\tau$. If the payoff is only dependent on the stock price (i.e., non-path-dependent American options where $\Lambda_t = \Lambda(S_t)$), then the time-τ payoff will be given by $\mathbb{I}_{\{\tau\leqslant T\}}\Lambda(S_\tau)$. To find the time-0 risk-neutral value, denoted by $v_0(\tau)$, of this payoff under the exercise rule τ, we can apply the risk-neutral pricing formula for the random cash flow $\{\mathbb{I}_{\{\tau=t\}}\Lambda_t\}_{0\leqslant t\leqslant T}$ (see Section 7.3.2):

$$v_0(\tau) = \widetilde{\mathbb{E}}_0\left[\sum_{t=0}^{T}\mathbb{I}_{\{\tau=t\}}\frac{\Lambda_t}{(1+r)^t}\right] = \widetilde{\mathbb{E}}_0\left[\mathbb{I}_{\{\tau\leqslant T\}}\frac{\Lambda_\tau}{(1+r)^\tau}\right].$$

Here, we use the fact that $\mathbb{I}_{\{\tau\leqslant T\}} = \mathbb{I}_{\{\tau=0\}} + \mathbb{I}_{\{\tau=1\}} + \cdots + \mathbb{I}_{\{\tau=T\}}$.

At time $t = 0$ it is unknown what strategy is the optimal one. Moreover, the writer of the derivative has no control on what exercise strategy will be used by the buyer, who can choose any one from the set $\mathcal{S}_{0,T}$. Therefore, it is reasonable to set the fair price V_0 of the American derivative equal to the maximum over all values of $v_0(\tau)$ for all possible exercise strategies:

$$V_0 = \max_{\tau \in \mathcal{S}_{0,T}} v_0(\tau) = \max_{\tau \in \mathcal{S}_{0,T}} \widetilde{\mathrm{E}}_0 \left[\mathbb{I}_{\{\tau \leqslant T\}} \frac{\Lambda_\tau}{(1+r)^\tau} \right]. \tag{7.124}$$

The optimal exercise strategy (the optimal stopping time random variable) is defined as the stopping time $\tau^* \in \mathcal{S}_{0,T}$ which maximizes the expected final payoff under the risk-neutral measure. Hence, the price of the American claim is given in terms of an optimal stopping time τ^* by

$$V_0 = \widetilde{\mathrm{E}}_0 \left[\mathbb{I}_{\{\tau^* \leqslant T\}} \frac{\Lambda_{\tau^*}}{(1+r)^{\tau^*}} \right]. \tag{7.125}$$

Let us see that selling the American derivative for an amount other than V_0 leads to arbitrage. Suppose that the holder of the derivative does not withdraw the proceeds at time $t = \tau^*$ from the market, but allows them to grow at the risk-free rate r from time τ^* to time T. The final payoff is

$$\frac{B_T}{B_{\tau^*}} \mathbb{I}_{\{\tau^* \leqslant T\}} \Lambda_{\tau^*} = (1+r)^{T-\tau} \mathbb{I}_{\{\tau^* \leqslant T\}} \Lambda_{\tau^*}.$$

Let a self-financing strategy $\boldsymbol{\Phi}^*$ replicate this payoff at time T. Hence the initial value $\Pi_0[\boldsymbol{\Phi}^*]$ is equal to V_0 in (7.124). Note that in reality neither the seller of the American derivative nor the buyer knows in advance what strategy will be optimal and what the final payoff will be. Hence, at time $t = 0$ such a replicating strategy $\boldsymbol{\Phi}^*$ is unknown. As we saw in Example 7.10, an American derivative can be replicated by two processes, namely, a portfolio process and a consumption process. The worst case scenario for the seller of the derivative is when the buyer follows the optimal stopping strategy and he or she will exercise at the optimal time, leaving nothing to the writer to consume for free.

Now, suppose that the American derivative can be purchased for an amount less than $V_0 = \Pi_0[\boldsymbol{\Phi}^*]$. Then an arbitrage opportunity can be created by purchasing the cheaper derivative and selling short the more expensive strategy $\boldsymbol{\Phi}^*$. If the initial price of the American derivative is larger than V_0, then an arbitrage opportunity is available to the seller of the option, who can invest the proceeds in $\boldsymbol{\Phi}^*$. So we may conclude that the no-arbitrage price of the American derivative is equal to $\Pi_0[\boldsymbol{\Phi}^*] = V_0$.

The next example gives an explicit implementation of equations (7.124) and (7.125).

Example 7.14. Find values $v_0(\tau)$ for each exercise rule $\tau \in \mathcal{S}_{0,3}$ for the American put option from Example 7.9. Find an optimal stopping time τ^* such that $v_0(\tau^*)$ is a maximum value. Compare $v_0(\tau^*)$ with the no-arbitrage price P_0^A calculated in Example 7.9. Moreover, compare the optimal stopping time τ^* with the exercise rule given in Figure 7.8.

Solution. Without loss of generality we can assume that the option has to be exercised before or at the expiry time but with a possibly zero payoff to the holder if the option is out of the money or it is not optimal to exercise the option. In this case it is sufficient to only consider finite stopping times of the form $\tau \colon \Omega \to \{0, 1, 2, 3\}$, i.e., $\tau \in \widehat{\mathcal{S}}_{0,3}$, while calculating the maximum in (7.124). Then, (7.124) simplifies as follows:

$$P_0^A = \max_{\tau \in \widehat{\mathcal{S}}_{0,3}} \widetilde{\mathrm{E}}_0 \left[\frac{\Lambda(S_\tau)}{(1+r)^\tau} \right] = \max_{\tau \in \widehat{\mathcal{S}}_{0,3}} \widetilde{\mathrm{E}} \left[\frac{(8 - S_\tau)^+}{(1+r)^\tau} \right].$$

Using the definition of a stopping time, we can find all elements of $\widehat{\mathcal{S}}_{0,T}$. For example,

TABLE 7.1: Stopping times of $\widehat{\mathcal{S}}_{0,3}$.

ω	$\hat\tau_1$	$\hat\tau_2$	$\hat\tau_3$	$\hat\tau_4$	$\hat\tau_5$	$\hat\tau_6$	$\hat\tau_7$	$\hat\tau_8$	$\hat\tau_9$	$\hat\tau_{10}$	$\hat\tau_{11}$	$\hat\tau_{12}$	$\hat\tau_{13}$	$\hat\tau_{14}$	$\hat\tau_{15}$	$\hat\tau_{16}$	$\hat\tau_{17}$	$\hat\tau_{18}$	$\hat\tau_{19}$	$\hat\tau_{20}$	$\hat\tau_{21}$	$\hat\tau_{22}$	$\hat\tau_{23}$	$\hat\tau_{24}$	$\hat\tau_{25}$	$\hat\tau_{26}$
UUU	0	1	1	1	1	1	2	2	3	3	2	2	2	2	3	2	2	2	3	3	3	3	3	3	2	3
UUD	0	1	1	1	1	1	2	2	3	3	2	2	2	2	3	2	2	2	3	3	3	3	3	3	2	3
UDU	0	1	1	1	1	1	2	3	2	3	2	2	2	3	2	2	3	3	2	2	3	3	3	2	3	3
UDD	0	1	1	1	1	1	2	3	2	3	2	2	2	3	2	2	3	3	2	2	3	3	3	2	3	3
DUU	0	1	2	2	3	3	1	1	1	1	2	2	3	2	2	3	3	2	2	3	2	3	2	3	3	3
DUD	0	1	2	2	3	3	1	1	1	1	2	2	3	2	2	3	3	2	2	3	2	3	2	3	3	3
DDU	0	1	2	3	2	3	1	1	1	1	2	3	2	2	2	3	2	3	3	2	2	2	3	3	3	3
DDD	0	1	2	3	2	3	1	1	1	1	2	3	2	2	2	3	2	3	3	2	2	2	3	3	3	3

TABLE 7.2: The mathematical expectation $\widetilde{\mathrm{E}}\left[(1+r)^{-\tau}\Lambda(S_\tau)\right]$ is calculated for each stopping time $\tau = \hat\tau_k$ of Table 7.1.

$\hat\tau_1$	$\hat\tau_2$	$\hat\tau_3$	$\hat\tau_4$	$\hat\tau_5$	$\hat\tau_6$	$\hat\tau_7$	$\hat\tau_8$	$\hat\tau_9$	$\hat\tau_{10}$	$\hat\tau_{11}$	$\hat\tau_{12}$	$\hat\tau_{13}$
0	0.8	0.48	0.416	0.36	0.296	1.04	0.92	1.04	0.92	0.72	0.656	0.6

$\hat\tau_{14}$	$\hat\tau_{15}$	$\hat\tau_{16}$	$\hat\tau_{17}$	$\hat\tau_{18}$	$\hat\tau_{19}$	$\hat\tau_{20}$	$\hat\tau_{21}$	$\hat\tau_{22}$	$\hat\tau_{23}$	$\hat\tau_{24}$	$\hat\tau_{25}$	$\hat\tau_{26}$
0.6	0.72	0.536	0.48	0.536	0.656	0.6	0.6	0.48	0.536	0.536	0.416	0.416

Table 7.1 defines all 26 finite stopping times of $\widehat{\mathcal{S}}_{0,3}$ for a three-period model. We denote these random variables by $\hat\tau_k, k = 1, \ldots, 26$.

For each stopping time $\tau = \hat\tau_k$ of Table 7.1, we can calculate the risk-neutral expectation of the discounted payoff as follows:

$$
\widetilde{\mathrm{E}}_0\left[\frac{\Lambda(S_\tau)}{(1+r)^\tau}\right] = \sum_{\omega\in\Omega} \frac{\Lambda(S_{\tau(\omega)}(\omega))}{(1+r)^{\tau(\omega)}} \tilde{p}^{\#\mathsf{U}(\omega)}(1-\tilde{p})^{\#\mathsf{D}(\omega)}
$$
$$
= \sum_{\omega\in\Omega_3} 1.25^{-\tau(\omega)} \cdot \left(8 - S_{\tau(\omega)}(\omega)\right)^+ \cdot 0.75^{\#\mathsf{U}(\omega)} \cdot 0.25^{\#\mathsf{D}(\omega)}.
$$

For example, for $\tau = \hat\tau_8$:

$$
\widetilde{\mathrm{E}}_0\left[\frac{(8-S_\tau)^+}{1.25^\tau}\right] = \sum_{\omega\in\Omega_3} 1.25^{-\hat\tau_8(\omega)}\left(8 - S_0 u^{\#\mathsf{U}_{\hat\tau_8(\omega)}(\omega)} d^{\#\mathsf{D}_{\hat\tau_8(\omega)}(\omega)}\right) \cdot 0.75^{\#\mathsf{U}(\omega)} \cdot 0.25^{\#\mathsf{D}(\omega)}
$$

$$
= \underbrace{1.25^{-2} \cdot (8-18)^+ \cdot \left(0.75^3 + 0.75^2 \cdot 0.25\right)}_{\hat\tau_8(\omega)=2 \text{ for } \omega\in A_{\mathsf{UU}}}
$$
$$
+ \underbrace{1.25^{-3} \cdot \left((8-9)^+ \cdot 0.75^2 \cdot 0.25 + (8-3)^+ \cdot 0.75 \cdot 0.25^2\right)}_{\hat\tau_8(\omega)=3 \text{ for } \omega\in A_{\mathsf{UD}}}
$$
$$
+ \underbrace{1.25^{-1} \cdot (8-4)^+ \cdot \left(0.75^2 \cdot 0.25 + 2\cdot 0.75 \cdot 0.25^2 + 0.25^3\right)}_{\hat\tau_8(\omega)=1 \text{ for } \omega\in A_{\mathsf{D}}}
$$
$$
= 0 + 0 + 0 + 0.12 + 0.45 + 0.15 + 0.15 + 0.05 = 0.92.
$$

The results are presented in Table 7.2. As is seen, the maximum is equal to $P_0^A = 1.04$ and it is attained for stopping times $\tau = \hat\tau_7$ and $\hat\tau_9$. The exercise rules $\hat\tau_7$ and $\hat\tau_9$ of Table 7.1 and the optimal exercise rule τ^* of Figure 7.8 are all equivalent since

$$
\mathbb{I}_{\{\tau^*(\omega)\leqslant 3\}}\left(8 - S_{\tau^*(\omega)}(\omega)\right)^+ = \left(8 - \Lambda(S_{\hat\tau_7(\omega)}(\omega))\right)^+ = \left(8 - S_{\hat\tau_9(\omega)}(\omega)\right)^+
$$

holds for all $\omega \in \Omega_3$. $\qquad\square$

In the above three-period example, we considered all conceivable stopping times (i.e., stopping rules) and respectively computed the set of all discounted expected values of the put payoff. The fair value of the American put was then given by taking the maximum over all such computed values. Underlying all of this is a method (i.e., a rule) for choosing the optimal exercise time. As depicted in Figure 7.8, the rule corresponds to defining an optimal stopping time, τ^*. It turns out that this is given by the random variable:

$$\tau^* = \min\{t \in \{0,1,2,3\} \ : \ P_t^A = \Lambda_t\} = \min\{t \in \{0,1,2,3\} \ : \ P_t^A(S_t) = (8 - S_t)^+\}.$$

We note that, in the case that this set is empty, we define the minimum to be ∞. That is, we put $\tau^* = \infty$ if $P_t^A(S_t(\omega)) > (8 - S_t(\omega))^+$ for all $t \geqslant 0, \omega \in \Omega_3$. Given a scenario ω, then $\tau^*(\omega)$ is the nonnegative integer corresponding to the first time at which the American price equals the intrinsic value, i.e., the smallest integer $t \in \{0,1,2,3\}$ such that $P_t^A(S_t(\omega)) = (8 - S_t(\omega))^+$:

$$\tau^*(\omega) = \min\{t \in \{0,1,2,3\} \ : \ P_t^A(S_t(\omega)) = (8 - S_t(\omega))^+\}.$$

It is simple to verify that this produces the optimal stopping time. For instance, take the outcome $\omega \in A_{UU}$. For $t = 0$, $P_0^A(S_0) = P_0^A(8) = 1.04 > (8 - 8)^+ = 0$. For $t = 1$, $P_1^A(S_1(\omega)) = P_1^A(S_1(U)) = P_1^A(12) = 0.4 > (8 - 12)^+ = 0$. For $t = 2$, $P_2^A(S_2(\omega)) = P_2^A(S_2(UU)) = P_2^A(18) = 0 = (8 - 18)^+ = (8 - S_2(UU))^+$. Hence, $\tau^*(\omega) = 2$. Similarly, the reader can verify that $\tau^*(\omega) = 2$, for $\omega \in A_{UD}$, and $\tau^*(\omega) = 1$, for $\omega \in A_D$. Therefore, the above method generates an optimal stopping time where $\tau^* = \hat{\tau}_7$.

The following proposition now justifies the above for a generally path-dependent American option on the standard T-period binomial model. Here, it suffices to consider an optimal stopping $\tau^* \in \mathcal{S}_{0,T}$ which corresponds to the optimal stopping time in (7.125) for the time-0 no-arbitrage price the American option. Again, we note that in equation (7.126) below, if $\min\{t \in \{0,1,\ldots,T\} \ : \ V_t(\omega) = \Lambda_t(\omega)\} = \emptyset$ then we put $\tau^*(\omega) = \infty$ for any such $\omega \in \Omega_T$.

Proposition 7.8. *The stopping time defined by*

$$\tau^* = \min\{t \in \{0,1,\ldots,T\} \ ; \ V_t = \Lambda_t\} \tag{7.126}$$

maximizes the expectation on the right-hand side of (7.124). That is, the time-0 American option price is given by V_0^A in (7.125) with optimal stopping time τ^* given by (7.126).

Proof. The proof is rather immediate once we use the fact that the stopped discounted American option value process defined by $\left\{ \frac{V_{t \wedge \tau^*}}{(1+r)^{t \wedge \tau^*}} \right\}_{t=0,1,\ldots,T}$ is a $\widetilde{\mathbb{P}}$-martingale. This is shown by considering the backward recurrence relation in (7.115). In particular, arbitrarily fix the first t market moves and consider the two possible mutually exclusive events whose union is $A_{\omega_1,\ldots,\omega_t}$: (i) $\{\omega \in A_{\omega_1,\ldots,\omega_t} \ : \ \tau^*(\omega) > t\}$ and (ii) $\{\omega \in A_{\omega_1,\ldots,\omega_t} \ : \ \tau^*(\omega) \leqslant t\}$.

For case (i): $\tau^*(\omega) \wedge t = t$ and hence the American option price is given by the continuation value at time t. The latter is greater than the intrinsic value, i.e., $V_t^A(\omega) > \Lambda_t(\omega)$. Therefore, equation (7.115) implies

$$V_{\tau^* \wedge t}(\omega) \equiv V_{\tau^*(\omega) \wedge t}(\omega) = V_t(\omega) = \frac{1}{1+r}\widetilde{\mathbb{E}}_t\left[V_{t+1}\right](\omega) = \frac{1}{1+r}\widetilde{\mathbb{E}}_t\left[V_{\tau^* \wedge (t+1)}\right](\omega)$$

where we used $\tau^*(\omega) \wedge (t+1) = t+1$ since $\tau^*(\omega) \geqslant t+1$. Moreover, we then have $(1+r)^{\tau^*(\omega) \wedge (t+1)}/(1+r)^{\tau^*(\omega) \wedge t} = (1+r)^{t+1}/(1+r)^t = 1+r$, which implies the martingale expectation property:

$$\frac{V_{\tau^* \wedge t}}{(1+r)^{\tau^* \wedge t}}(\omega) = \widetilde{\mathbb{E}}_t\left[\frac{V_{\tau^* \wedge (t+1)}}{(1+r)^{\tau^* \wedge (t+1)}}\right](\omega).$$

For case (i) we have $\tau^*(\omega) \wedge t = \tau^*(\omega)$, giving

$$\frac{V_{\tau^*(\omega) \wedge t}(\omega)}{(1+r)^{\tau^*(\omega) \wedge t}} = \frac{V_{\tau^*(\omega)}(\omega)}{(1+r)^{\tau^*(\omega)}} = \frac{\Lambda_{\tau^*(\omega)}(\omega)}{(1+r)^{\tau^*(\omega)}} = \frac{\Lambda_{\tau^*}}{(1+r)^{\tau^*}}(\omega)$$

for all $t > \tau^*(\omega)$. Hence, for $t > \tau^*(\omega)$ the discounted process is a constant and therefore satisfies the martingale property.

By using the $\widetilde{\mathbb{P}}$-martingale property of the discounted stopped price process T-steps forward, and the fact that $V_{\tau^*} = \Lambda_{\tau^*}$, we finally obtain the required pricing formula:

$$V_0 = \widetilde{\mathrm{E}}_0 \left[\frac{V_{\tau^* \wedge T}}{(1+r)^{\tau^* \wedge T}} \right] = \widetilde{\mathrm{E}}_0 \left[\mathbb{I}_{\{\tau^* \leqslant T\}} \frac{V_{\tau^*}}{(1+r)^{\tau^*}} \right] = \widetilde{\mathrm{E}}_0 \left[\mathbb{I}_{\{\tau^* \leqslant T\}} \frac{\Lambda_{\tau^*}}{(1+r)^{\tau^*}} \right].$$

Here we also used the fact that $1 = \mathbb{I}_{\{\tau^* \leqslant T\}} + \mathbb{I}_{\{\tau^* = \infty\}}$ where $\mathbb{I}_{\{\tau^* = \infty\}} V_{\tau^* \wedge T} = \mathbb{I}_{\{\tau^* = \infty\}} V_T = 0$ since the event of never exercising the option amounts to zero payoff. $\qquad\square$

In a similar way, we can define the value process $\{v_t\}_{0 \leqslant t \leqslant T}$. Fix $t \in \{0, 1, \ldots, T\}$. Assuming that the derivative has not been exercised before time t, the holder may exercise it any time from t until T. Let the holder follow an exercise strategy $\tau \in \mathcal{S}_{t,T}$. The no-arbitrage time-$t$ value (relative to the rule τ) is given by the risk-neutral expectation of the discounted payoff conditional on the information available at time t: $v_t(\tau) = \widetilde{\mathrm{E}}_t \left[\mathbb{I}_{\{\tau \leqslant T\}} \frac{\Lambda_\tau}{(1+r)^{\tau-t}} \right]$. The fair price of the American derivative, V_t, which is independent of the exercise rule used by the holder of the derivative, is given by the largest possible value $V_t(\tau)$:

$$V_t = \max_{\tau \in \mathcal{S}_{t,T}} \widetilde{\mathrm{E}}_t \left[\mathbb{I}_{\{\tau \leqslant T\}} \frac{\Lambda_\tau}{(1+r)^{\tau-t}} \right] = \widetilde{\mathrm{E}}_t \left[\mathbb{I}_{\{\tau^* \leqslant T\}} \frac{\Lambda_{\tau^*}}{(1+r)^{\tau^*}} \right], \tag{7.127}$$

where $\tau^* \in \mathcal{S}_{t,T}$ is an optimal exercise time, which maximizes the expected final payoff in (7.127). Below is a concrete example of how the early exercise rule τ^* can be determined and how (7.127) is directly applied in the case of a path-dependent American option.

Example 7.15. Consider the American-style Asian call on the three-period binomial model in Example 7.11. Determine the early exercise rule τ^* and obtain the no-arbitrage initial price of the option using (7.127) for $t = 0$, i.e, (7.125).

Solution. Note that in this case the time-t intrinsic value random variable $\Lambda_t = (A_t - 8)^+$. The early exercise rule in this case corresponds to exercising early at time $t = 2$ if the node $(6, \frac{26}{3})$ is attained and at time $t = 3$ for nodes $(9, \frac{47}{4})$ and $(27, \frac{65}{4})$; otherwise there is no exercise. Hence, the optimal exercise rule is:

$$\tau^*(\omega) = \begin{cases} 2 & \text{if } \omega \in A_{\mathsf{UD}} \\ 3 & \text{if } \omega \in \{\mathsf{UUU}, \mathsf{UUD}\} \end{cases}$$

and $\tau^* = \infty$ for all other $\omega \in \Omega_3$. Using the random variable τ^* within (7.125) now gives:

$$V_0 = \widetilde{\mathrm{E}}_0 \left[\mathbb{I}_{\{\tau^* \leqslant 3\}} \frac{\Lambda_{\tau^*}}{(1+r)^{\tau^*}} \right]$$

$$= \frac{\widetilde{\mathbb{P}}(\mathsf{UD})}{(1+r)^2} (A_2(\mathsf{UD}) - 8)^+ + \frac{\widetilde{\mathbb{P}}(\mathsf{UUU})}{(1+r)^3} (A_3(\mathsf{UUU}) - 8)^+ + \frac{\widetilde{\mathbb{P}}(\mathsf{UUD})}{(1+r)^3} (A_3(\mathsf{UUD}) - 8)^+$$

$$= \frac{3}{5} \cdot \frac{1}{5} \cdot \left(\frac{26}{3} - 8 \right)^+ + \left(\frac{3}{5} \right)^3 \cdot \left(\frac{65}{4} - 8 \right)^+ + \left(\frac{3}{5} \right)^2 \cdot \frac{1}{5} \cdot \left(\frac{47}{4} - 8 \right)^+$$

$$= \frac{3}{5} \cdot \frac{1}{5} \cdot \frac{2}{3} + \left(\frac{3}{5} \right)^3 \cdot \frac{33}{4} + \left(\frac{3}{5} \right)^2 \cdot \frac{1}{5} \cdot \frac{15}{4} = 2.132.$$

As required, this agrees with the price obtained in Example 7.11 where the backward recurrence pricing formula was employed instead.   □

Example 7.16. Consider the American-style lookback option within the three-period binomial model in Example 7.12. Determine the early exercise rule τ^* and obtain the no-arbitrage initial price of the option using (7.127) for $t = 0$, i.e, (7.125).

Solution. The time-t intrinsic value random variable is $\Lambda_t = M_t - m_t$. The early exercise rule in this case corresponds to exercising early at time $t = 1$ if the first market move is down to node $(4, 8, 4)$, at time $t = 2$ if the first move is up and the second move is down to node $(6, 12, 6)$; at time $t = 3$ if either node $(9, 18, 8)$ or $(27, 27, 8)$ is attained; otherwise there is no exercise. Hence, the optimal exercise rule is:

$$\tau^*(\omega) = \begin{cases} 1 & \text{if } \omega \in A_\mathsf{D} \\ 2 & \text{if } \omega \in A_\mathsf{UD} \\ 3 & \text{if } \omega \in \{\mathsf{UUU}, \mathsf{UUD}\} \end{cases}$$

and $\tau^* = \infty$ for all other $\omega \in \Omega_3$. Using the random variable τ^* within (7.125):

$$\begin{aligned} V_0 &= \widetilde{\mathrm{E}}_0 \left[\mathbb{I}_{\{\tau^* \leqslant 3\}} \frac{\Lambda_{\tau^*}}{(1+r)^{\tau^*}} \right] \\ &= \frac{\widetilde{\mathbb{P}}(\mathsf{D})}{(1+r)}(M_1(\mathsf{D}) - m_1(\mathsf{D}) + \frac{\widetilde{\mathbb{P}}(\mathsf{UD})}{(1+r)^2}(M_2(\mathsf{UD}) - m_2(\mathsf{UD})) \\ &\quad + \frac{\widetilde{\mathbb{P}}(\mathsf{UUU})}{(1+r)^3}(M_3(\mathsf{UUU}) - m_3(\mathsf{UUU})) + \frac{\widetilde{\mathbb{P}}(\mathsf{UUD})}{(1+r)^3}(M_3(\mathsf{UUD}) - m_3(\mathsf{UUD})) \\ &= \frac{1}{5} \cdot (8-4) + \frac{3}{5} \cdot \frac{1}{5} \cdot (12-6) + \left(\frac{3}{5}\right)^3 \cdot (27-8) + \left(\frac{3}{5}\right)^2 \cdot \frac{1}{5} \cdot (18-8) \\ &= 0.8 + 0.72 + 0.72 + 4.104 = 6.344. \end{aligned}$$

This agrees with the price obtained in Example 7.12.   □

7.7.3   Early-Exercise Boundary for Non-Path-Dependent Options

An American option should be exercised according to the optimal stopping rule τ^* given in (7.126), i.e., as soon as the value $V_t = V_t(S_t)$ becomes equal to $\Lambda_t(S_t)$, for maximal gain. As a result, the binomial lattice is divided into two parts: a continuation domain for which the option is not exercised,

$$\mathcal{D}_C = \{(t, S_{t,n}) \ : \ 0 \leqslant t \leqslant T, \ 0 \leqslant n \leqslant t, \ V_t(S_{t,n}) > \Lambda_t(S_{t,n})\}, \tag{7.128}$$

and a stopping domain whereby the option is exercised early,

$$\mathcal{D}_S = \{(t, S_{t,n}) \ : \ 0 \leqslant t \leqslant T, \ 0 \leqslant n \leqslant t, \ V_t(S_{t,n}) = \Lambda_t(S_{t,n})\}. \tag{7.129}$$

The structure of the stopping domain may be quite complicated, and it is defined by the payoff function Λ and whether the underlying asset pays dividends. However, for the standard monotonic piecewise call and put payoffs, the stopping domain turns out to be simply connected. The early-exercise boundary, which separates the continuation and stopping domains, is a connected path in the binomial tree given by

$$\{(t, S) \ : \ 0 \leqslant t \leqslant T, \ S = S_t^*\},$$

where
$$S_t^* = \min\{S_{t,n} \; : \; C_t^A(S_{t,n}) = (S_{t,n} - K)^+, \; 0 \leqslant n \leqslant t\} \tag{7.130}$$

for a call and
$$S_t^* = \max\{S_{t,n} \; : \; P_t^A(S_{t,n}) = (K - S_{t,n})^+, \; 0 \leqslant n \leqslant t\} \tag{7.131}$$

for a put. Note that since the American option value is always nonnegative, the superscript + signs are omitted in (7.130) and (7.131).

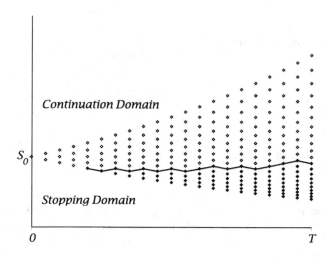

FIGURE 7.11: The early-exercise boundary for an American put option. The black nodes correspond to the stopping domain.

A typical early exercise boundary for an American put option is shown in Figure 7.11. The stopping domain is marked with black nodes. As follows from the definition of the stopping domain, the optimal exercise time τ^* defined in (7.126) is also the first hitting time of the stopping domain $\mathcal{D}_S$. That is, τ^* is the first time when the stock price process reaches the early-exercise boundary S^*:

$$\tau^* = \min\{t \; : \; 0 \leqslant t \leqslant T, \; S_t = S_t^*\}.$$

7.7.4 Pricing American Options: The Case with Dividends

Using a no-arbitrage argument, we have proved earlier that the price of a standard American call (with intrinsic value $\Lambda_t = (S_t - K)^+$, $0 \leqslant t \leqslant T$) is the same as that of the standard European call paying $(S_T - K)^+$ at expiration T. Note that the situation is different for a dividend-paying stock. If this is the case, then, as is demonstrated below, the American call can have a larger initial value than the European call.

Recall a dividend-paying stock that follows the binomial model between the dividend payment times. We assume that at each time $t = 1, 2, \ldots, T$, the dividend payment at time t denoted D_t is equal to $q_t S_t^*$ for some $q_t \in [0,1)$, where $S_t^* = Y_t S_{t-1}$ denotes the asset price just prior to the dividend payment. The no-arbitrage asset price just after the dividend payment is
$$S_t = S_t^* - D_t = S_t^* - q_t S_t^* = (1 - q_t)S_t^* = (1 - q_t)Y_t S_{t-1}.$$

This is equation (7.4), where $Y_t(\omega_t) = u^{\#U(\omega_t)} d^{\#D(\omega_t)}$, i.e., Y_t is independent of S_{t-1}. Applying (7.4) sequentially gives (7.5). We recall that for nonrandom dividend factors (or

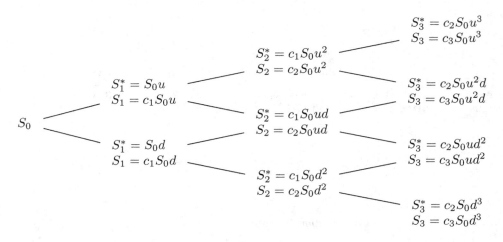

FIGURE 7.12: A three-period recombining tree for the asset price process with dividend payments $D_t = q_t S_t^*$. Here $c_t = (1 - q_1)(1 - q_2) \cdots (1 - q_t)$ is not random.

when $q_t(\omega) = q_t(\omega_t)$ with fixed $q_t(\mathsf{U}) = q_u$, $q_t(\mathsf{D}) = q_d$), the evolution of the asset price process can be represented by a recombining binomial tree (see a three-period tree presented in Figure 7.12). For more general $\mathcal{F}_t$-measurable dividend factors q_t, we have a non-recombining binomial tree.

In the general case, where the dividend payments, $\{D_t\}_{1 \leqslant t \leqslant T}$, are not given as a percentage of the asset price, the recurrence formula for the asset price takes the form:

$$S_t(\omega) = S_t^*(\omega) - D_t(\omega) = S_{t-1} Y_t(\omega_t) - D_t(\omega), \quad 1 \leqslant t \leqslant T, \quad \omega \in \Omega_T.$$

Generally, we have $D_t(\omega) = D_t(\omega_1 \ldots \omega_t)$ with $\mathcal{F}_t$-measurable D_t. In many applications the dividends are simply deterministic functions of time. In general, the price dynamics of such an asset cannot be represented by a recombining binomial tree. In the case of rare dividend payments, we have a partly recombining tree.

Example 7.17. Consider the three-period binomial tree model in Example 7.1 with $S_0 = 8$, $u = \frac{3}{2}$, $d = \frac{1}{2}$, and $r = \frac{1}{4}$. Assume the stock pays dividends at times $t = 2$ and $t = 3$ and each dividend payment is 10% of the stock price. That is, $q_1 = 0$ and $q_2 = q_3 = 0.1$.

(a) Construct the binomial tree with prior- and post-dividend stock prices.

(b) Determine and compare the prices of the standard European and American call options with common strike $K = 6$ and expiration $T = 3$.

(c) Determine the optimal exercise policy τ^* and verify that the initial no-arbitrage price of the American call is given by the expectation in (7.125)

Solution. The recombining binomial tree for the stock price process is given in Figure 7.13. The risk-neutral probabilities are $\tilde{p} = 0.75$ and $1 - \tilde{p} = 0.25$; $\tilde{p}/(1+r) = 3/5$, $(1-\tilde{p})/(1+r) = 1/5$. At each node of the tree, we apply recurrence formulae (7.49) and (7.111) to evaluate the respective prices of the European and American call options. Let C_t^E and C_t^A, respectively, denote the time-t prices of the European and American put options for $t = 0, 1, 2, 3$. At the expiration date, the option value must equal the payoff function: $C_3^E(S) = C_3^A(S) = (S-6)^+$

for $S \in \{0.81, 2.43, 7.29, 21.87\}$. To determine the price for the times before expiration, use the following recurrences:

$$C_t^E(S) = \frac{3 \cdot C_{t+1}^E(1.5 \cdot S) + C_{t+1}^E(0.5 \cdot S)}{5},$$

$$C_t^A(S) = \max\left\{(S-6)^+, \frac{3 \cdot C_{t+1}^A(1.5 \cdot S) + C_{t+1}^A(0.5 \cdot S)}{5}\right\}, \quad t = 2, 1, 0.$$

As a result, we obtain the option prices given in Figure 7.14. The values of the American put are boxed at those nodes of the tree where the intrinsic value is greater than or equal to the continuation value. Therefore, the American call option should be exercised at time $t = 2$ when the stock price is $S_2 = 16.2$, or at time $t = 3$ when the price S_3 exceeds the strike $K = 6$. Note that the initial price of the American call, C_0^A, is strictly larger than that of the European call, C_0^E. Generally, we have $C_t^A(S) \geqslant C_t^E(S)$, for all $t \geqslant 0$. $\qquad \square$

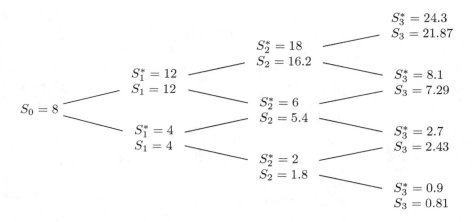

FIGURE 7.13: A three-period recombining binomial tree for the stock paying the dividend of $0.1 S_t^*$ at times $t = 1, 2$.

By applying the early exercise boundary rule, or as also seen from Figure 7.14, the optimal exercise rule for the American call is given by

$$\tau^*(\omega) = \begin{cases} 2 & \text{if } \omega \in A_{\mathsf{UU}} \\ 3 & \text{if } \omega \in \{\mathsf{UDU}, \mathsf{DUU}\} \end{cases}$$

and $\tau^* = \infty$ for all other $\omega \in \Omega_3$. By using this stopping time random variable we now verify that (7.125) reproduces the above American call price:

$$C_0^A = \tilde{\mathbb{E}}_0\left[\mathbb{I}_{\{\tau^* \leqslant 3\}} \frac{\Lambda_{\tau^*}}{(1+r)^{\tau^*}}\right]$$

$$= \frac{\tilde{\mathbb{P}}(\mathsf{UU})}{(1+r)^2}\Lambda(S_2(\mathsf{UU})) + \frac{\tilde{\mathbb{P}}(\mathsf{UDU})}{(1+r)^3}\Lambda(S_3(\mathsf{UDU})) + \frac{\tilde{\mathbb{P}}(\mathsf{DUU})}{(1+r)^3}\Lambda(S_3(\mathsf{DUU}))$$

$$= (\tilde{p}/(1+r))^2 \cdot \Lambda(16.2) + 2(\tilde{p}/(1+r))^2 \cdot ((1-\tilde{p})/(1+r)) \cdot \Lambda(7.29)$$

$$= (0.6)^2(10.2) + 2(0.6)^2(0.2)(1.29)$$

$$= 3.85776.$$

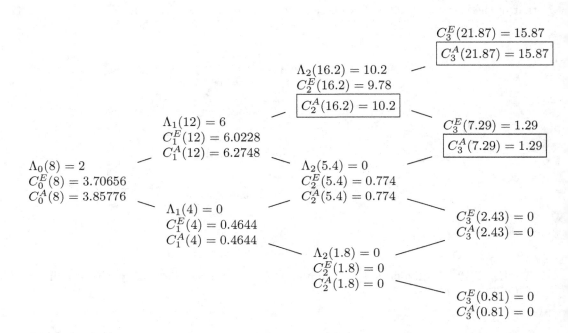

FIGURE 7.14: Prices of the standard European call, C_t^E, and American call, C_t^A, on the dividend-paying stock.

7.8 Exercises

Exercise 7.1. Consider a T-period binomial tree model with stock price $S_{t,n} = S_0 u^n d^{t-n}$ at each node (t,n) of the binomial tree for every $n = 0, 1, \ldots, t$ and every $t = 0, 1, \ldots, T$.

(a) Let $v, t \in \{0, 1, \ldots, T\}$ be two dates such that $v < t$. Determine the joint probability $\mathbb{P}(S_t = S_{t,n}, S_v = S_{v,m})$, $n = 0, 1, \ldots, t$, $m = 0, 1, \ldots, v$.

(b) Determine the conditional probability $\mathbb{P}(S_t = S_{t,n} \mid S_v = S_{v,m})$ using the result of (a).

Exercise 7.2. Consider a three-period binomial model with $S_0 = 100$, $u = 1.15$, and $d = 0.9$.

(a) Determine the distribution of S_3 conditional on $S_1 = 115$.

(b) Determine the distribution of S_3 conditional on $S_1 = 90$.

Exercise 7.3. Consider a three-period binomial tree model with $S_0 = 8$, $u = 1.5$, $d = 0.5$, and $r = 0.25$. Consider a European put option that expires at $T = 3$ and provides payoff $(10 - S_3)^+$. Let $P_{t,n}$ be the price of this option at time t given that $S_t = S_{t,n}$.

(a) Compute recursively backward in time the prices $P_{t,n}$, $t = 3, 2, 1, 0$, $n = 0, 1, \ldots, t$. Present the prices calculated in a recombining binomial tree diagram.

(b) Construct a hedging strategy $\{\Delta_t\}_{t=0,1,2}$. Represent it using a recombining binomial tree diagram.

Exercise 7.4. Complete the proof of Theorem 7.6 by showing that

$$\Pi_{t+1}(\omega_1 \ldots \omega_t \mathsf{U}) = V_{t+1}(\omega_1 \ldots \omega_t \mathsf{U}).$$

Exercise 7.5. Prove Theorem 7.7.

Exercise 7.6. Determine the initial no-arbitrage value of a contract that pays \$1 at the first time t when $\#\mathsf{U}(\omega_1, \omega_2, \ldots, \omega_t) = n$ for some nonrandom $n \in \{1, 2, \ldots, T\}$.

Exercise 7.7. Consider an arbitrary T-period binomial tree model. For $t \geqslant 0$, define $G_t = (S_0 \cdot S_1 \cdots S_t)^{\frac{1}{t+1}}$ to be the geometric average of stock prices between times zero and t. Consider an Asian call option that expires at time T and provides payoff $(G_T - K)^+$ with strike price K. Let $V_t(S, G)$ denote the price of this option at time $t \geqslant 0$ given that $S_t = S$ and $G_t = G$. In particular, $V_T(S, G) = (G - K)^+$.

(a) Develop a one-step pricing formula for $V_t = V_t(S, G)$ in terms of V_{t+1}, S, and G.

(b) Develop a formula for $\Delta_t(S, G)$, the number of shares of stock held in a replicating portfolio during the period $[t, t+1)$, given that $S_t = S$ and $G_t = G$.

Exercise 7.8. Consider a three-period binomial model with $S_0 = 8$, $u = 1.5$, $d = 0.5$, and $r = 0.25$. Consider an Asian call option that expires at time $T = 3$ and provides payoff $(G_3 - 8)^+$.

(a) Compute the prices V_t for each $t = 3, 2, 1, 0$, using a backward-time recursion. Represent these prices on a nonrecombining tree diagram.

(b) Construct a hedging strategy $\{\Delta_t\}_{t=0,1,2}$. Represent it on a nonrecombining tree diagram.

Exercise 7.9. Consider a binomial-tree arbitrage-free model with T periods and arithmetic floating-price (Asian) call and put options with respective payoff functions

$$C_T = (A_T - K)^+ \quad \text{and} \quad P_T = (K - A_T)^+,$$

where $K > 0$ is a strike price and $A_T = \frac{1}{T+1} \sum_{n=0}^{T} S_n$.

(a) Determine the risk-neutral conditional expectation $\widetilde{\mathbb{E}}_0[A_T]$.

(b) Using the result of (a), determine the put-call parity for the Asian options at time 0. That is, express the difference of initial prices $C_0 - P_0$ in terms of S_0, K, T, and r.

Exercise 7.10. Consider a T-period arbitrage-free binomial model and floating-price lookback European call and put options with fixed strike $K > 0$ defined by the respective payoffs

$$C_T = \left(M_T - K\right)^+ \quad \text{and} \quad P_T = \left(K - M_T\right)^+,$$

where $M_t := \max_{0 \leqslant n \leqslant t} S_n$ for $0 \leqslant t \leqslant T$.

(a) Prove (or argue) whether the floating-price lookback European call and put option prices C_t and P_t at arbitrary time $0 \leqslant t \leqslant T$ are functions of the pair of values (M_t, S_t) or whether they are just functions of M_t.

(b) Assume $ud = 1$ and $T = 3$. Determine the no-arbitrage present values C_0 and P_0. Leave your answers as expressions in terms of the parameters S_0, r, u, d.

Exercise 7.11. Assume the stock pays nonzero dividends according to (7.4).

(a) Show that the process $\left\{ \frac{S_t}{B_t} \right\}_{0 \leqslant t \leqslant T}$ is not a martingale but a strict supermartingale under the $\widetilde{\mathbb{P}}$-measure, $\widetilde{\mathbb{P}} \equiv \widetilde{\mathbb{P}}^{(B)}$, i.e., show that $\widetilde{\mathbb{E}}_t \left[\frac{S_{t+1}}{B_{t+1}} \right] < \frac{S_t}{B_t}$.

(b) Prove that the process $\left(\frac{S_t}{c_t B_t} \right)_{0 \leqslant t \leqslant T}$ is a $\widetilde{\mathbb{P}}^{(B)}$-martingale, as stated in Theorem 7.1.

(c) Show that $(1+r)\widetilde{\mathbb{E}}_t^{(S)} \left[\frac{1}{Y_{t+1}} \right] = 1$, where the expectation is under the $\widetilde{\mathbb{P}}^{(S)}$-measure. Use this fact to prove that $\left\{ \frac{c_t B_t}{S_t} \right\}_{0 \leqslant t \leqslant T}$ is a $\widetilde{\mathbb{P}}^{(S)}$-martingale, as stated in Theorem 7.1.

(d) Show (argue) that the wealth equation for the replicating portfolio Π_t takes the form in equation (7.16).

(e) Prove that the discounted wealth process $\left\{ \frac{\Pi_t}{B_t} \right\}_{0 \leqslant t \leqslant T}$ is a $\widetilde{\mathbb{P}}$-martingale, $\widetilde{\mathbb{P}} \equiv \widetilde{\mathbb{P}}^{(B)}$.

(f) Prove that $\left\{ \frac{c_t \Pi_t}{S_t} \right\}_{0 \leqslant t \leqslant T}$ is a $\widetilde{\mathbb{P}}^{(S)}$-martingale.

Exercise 7.12. Consider the T-period binomial model with stock price $S_{t,n} = S_0 u^n d^{t-n}$ at each node (t, n) with $t = 0, 1, \ldots, T$ and $n = 0, 1, \ldots, t$. Let $V_{t,n} = V_t(S_{t,n})$ denote the (no-arbitrage) price of any standard (non-path-dependent) European option at the node (t, n); let the relative derivative prices be defined by $\overline{V}_{t,n} := \frac{V_{t,n}}{S_{t,n}}$. Determine α and β (explicitly in terms of the model parameters) in the recurrence relation:

$$\overline{V}_{t,n} = \alpha \overline{V}_{t+1,n+1} + \beta \overline{V}_{t+1,n}.$$

Exercise 7.13. Consider a standard T-period binomial model and an Arrow–Debreu derivative security that pays one dollar if the *single particular* sequence $\bar{\omega} := \bar{\omega}_1 \cdots \bar{\omega}_T$ of T market moves occurs and pays zero otherwise. Hence, this is a derivative with payoff given by the indicator function random variable $V_T = \mathcal{E}_T := \mathbb{I}_{\{\omega = \bar{\omega}\}}$:

$$\mathcal{E}_T(\omega_1 \omega_2 \cdots \omega_T) = \begin{cases} 1 & \text{if } \omega_1 = \bar{\omega}_1, \omega_2 = \bar{\omega}_2, \ldots, \omega_T = \bar{\omega}_T, \\ 0 & \text{otherwise.} \end{cases}$$

(a) Derive a formula for the no-arbitrage derivative value, V_t, for time $t = 0, 1, \ldots, T$.

(b) Consider the self-financed replicating portfolio strategy for this derivative. Derive a formula for the delta hedging position in the stock, $\Delta_t(\omega_1 \omega_2 \cdots \omega_t)$, for time $t = 0, 1, \ldots, T-1$.

Exercise 7.14 (The Law of One Price for the Binomial Tree Model). Consider two securities with respective discrete-time price processes $\{X_t\}_{0 \leqslant t \leqslant T}$ and $\{Y_t\}_{0 \leqslant t \leqslant T}$ adapted to the binomial filtration $\mathbb{F} = \{\mathcal{F}_t\}_{0 \leqslant t \leqslant T}$. Suppose that $X_T(\omega) = Y_T(\omega)$ for all $\omega \in \Omega_T$. Prove that $X_t(\omega_1 \ldots \omega_t) = Y_t(\omega_1 \ldots \omega_t)$ for all $0 \leqslant t \leqslant T$ and $\omega_1, \ldots, \omega_t \in \{\mathsf{D}, \mathsf{U}\}$, or else an arbitrage opportunity exists.

Exercise 7.15. Consider a T-period binomial arbitrage-free model and two chooser options expiring at time T with respective payoffs

$$\text{(a) } \Lambda(S) = S \vee K, \quad \text{(b) } \Lambda(S) = S \wedge K.$$

For each payoff, derive the time-t ($t \leqslant T$) risk-neutral option pricing formula as a function of the (random) time-t stock value S_t. Express your answer in terms of the binomial cumulative probability mass function

$$\mathcal{B}(m; n, p) = \sum_{k=0}^{m} \binom{n}{k} p^k (1-p)^{n-k}, \quad 0 \leqslant m \leqslant n.$$

Exercise 7.16. Consider a T-period binomial arbitrage-free model and a European-style derivative expiring at time T, namely,

(a) an asset-or-nothing call option with payoff

$$\Lambda(S) = \begin{cases} S & \text{if } S > K, \\ 0 & \text{if } S \leqslant K, \end{cases}$$

(b) a bear spread with payoff

$$\Lambda(S) = \begin{cases} K_2 - K_1 & \text{if } S \leqslant K_1, \\ K_2 - S & \text{if } K_1 < S \leqslant K_2, \\ 0 & \text{if } S \geqslant K_2, \end{cases}$$

(c) a strangle with payoff

$$\Lambda(S) = \begin{cases} K_1 - S & \text{if } S \leqslant K_1, \\ 0 & \text{if } K_1 < S < K_2, \\ S - K_2 & \text{if } S \geqslant K_2. \end{cases}$$

Assume that $K_1 < S_0 < K_2$ in (b) and (c). For each payoff (a-c), derive the time-t ($t \leqslant T$) risk-neutral option pricing formula as a function of the (random) time-t stock value S_t. Express your answer in terms of the binomial cumulative probability mass function as in the previous question.

Exercise 7.17. Let $\sigma > 0$ and $r_\infty \geqslant 0$ be, respectively, an annual volatility and annual rate of interest compounded continuously. Consider the following parameterization of an N-period binomial model with the length of each period equal to $\delta t = \frac{1}{N}$:

$$u = e^{\sigma \sqrt{\delta t}}, \quad d = \frac{1}{u} = e^{-\sigma \sqrt{\delta t}}, \quad 1 + r = e^{r_\infty \delta t}.$$

Show that the binomial model is arbitrage free iff $N > \left(\frac{r_\infty}{\sigma}\right)^2$ holds.

Exercise 7.18. Consider a three-month call option on a dividend-paying stock with spot price $S_0 = \$50$ and strike price $K = \$51$. A stock dividend of $\$2.50$ is due *only at the end of the first month*. Use a three-period binomial model with annual interest rate $r_\infty = 5\%$ and annual volatility $\sigma = 12\%$ parametrized as in Exercise 7.17. Determine the no-arbitrage initial values of

(a) the European derivative,

(b) the American derivative.

Exercise 7.19. Repeat calculations (a) and (b) for the three-month call option in Exercise 7.19. Assume the same model and parameters. However, now assume that the stock dividend of $\$2.50$ is paid at the end of *every month* and the strike is $K = \$42.5$.

Exercise 7.20. In a three-period binomial model with $S_0 = 8$, $u = 1.5$, $d = 0.5$, and $1 + r = 1.25$, consider the American straddle with expiry $T = 3$ and payoff function $\Lambda(S) = (8 - S)^+ + (S - 8)^+$. For all nodes in the binomial tree, determine the derivative price, the delta hedging position and the value of the consumption process. Also, determine the optimal stopping time (early exercise) rule for this American option.

Exercise 7.21. In a three-period binomial model with parameters $S_0 = 8$, $u = 1.5$, $d = 0.5$, and $1 + r = 1.25$, consider the American style arithmetic average Asian put expiring at $T = 3$ and whose intrinsic value Λ_t at each time $t = 0, 1, 2, 3$ is

$$\Lambda_t = (4 - A_t)^+, \quad A_t := \frac{1}{t+1} \sum_{n=0}^{t} S_n.$$

For all nodes (S_t, A_t), $t = 0, 1, 2, 3$, determine the derivative price, the delta hedging position and the value of the consumption process. Also, determine the optimal stopping time (early exercise) rule for this American option.

Exercise 7.22. Consider the American derivative in Example 7.21. Verify that the initial arbitrage-free price of the option is given by the expectation in (7.125).

Exercise 7.23. In the three-period binomial model with $S_0 = 8$, $u = 1.5$, $d = 0.5$, and $1 + r = 1.25$, consider the American strangle that expires at $T = 3$ and has intrinsic value (for $0 \leqslant t \leqslant 3$)

$$\Lambda(S_t) = \begin{cases} 4 - S_t & , S_t \leqslant 4, \\ 0 & , 4 < S_t < 12, \\ S_t - 12 & , S_t \geqslant 12. \end{cases}$$

(a) Determine the derivative price process and the consumption strategy. Draw a binomial tree showing your findings.

(b) Determine the initial hedging portfolio (Δ_0, β_0) for the strangle.

(c) Determine the optimal exercise policy τ^*.

(d) Verify that the initial arbitrage-free price of this American strangle is given by the expectation in (7.125).

Exercise 7.24. In the three-period binomial model with $S_0 = 8$, $u = 1.25$, $d = 0.75$, and $1 + r = 1.2$, consider the American butterfly option that expires at $T = 3$ and has intrinsic value (for $0 \leqslant t \leqslant 3$)

$$\Lambda(S_t) = \begin{cases} (S_t - 5.5)^+ & , S_t \leqslant 8, \\ (10.5 - S_t)^+ & , S_t \geqslant 8. \end{cases}$$

(a) Determine the derivative price process and the consumption strategy. Draw a binomial tree showing your findings.

(b) Determine the initial hedging portfolio (Δ_0, β_0) for this option.

(c) Determine the optimal exercise policy τ^*.

(d) Verify that the initial arbitrage-free price of this option is given by the expectation in (7.125).

Exercise 7.25. Consider a standard three-period binomial model with parameters $S_0 = 8$, $u = 1.5$, $d = 0.5$, $1 + r = 1.25$. For $t = 0, 1, 2, 3$, we define the arithmetic average of the stock price, $A_t := \frac{1}{t+1} \sum_{k=0}^{t} S_k$. Consider a European style floating-strike Asian call option that expires at time $T = 3$ and provides payoff $(S_3 - A_3)^+$. Let $V_t(S, A)$ be the no-arbitrage price of this option at time t, given the spot values $S_t = S$ and $A_t = A$.

(a) Provide the general backward recurrence formula for $V_t(S, A)$ and thereby compute the option prices for all spot values and times $t = 0, 1, 2, 3$.

(b) Provide a general formula for computing $\Delta_t(S, A)$, the number of shares of stock held in a replicating portfolio for the option during the period $[t, t+1)$, as a function of the spot values S, A.

(c) Construct a dynamic hedging strategy $(\Delta_t, \beta_t B_t)_{t=0,1,2}$ to hedge one unit of the American option, with Δ_t shares of stock and $\beta_t B_t$ dollars in the bank account. Represent the strategy using a non-recombining binomial tree diagram.

]

8

General Multi-Asset Multi-Period Model

In this chapter we construct a general multi-asset discrete-time model with a finite state space. Most of the results of Chapters 5 and 7 will be generalized so that the single-period and binomial tree models become special cases of a more general framework.

8.1 Main Elements of the Model

Let us begin by describing the main components of a general multi-period model defined on a filtered probability space $(\Omega, \mathcal{F}, \mathbb{P}, \mathbb{F})$.

- There are $T + 1$ trading dates, $t \in \{0, 1, 2, \ldots, T\}$. The collection of trading dates is denoted by $\mathbf{T}$.

- A finite state space $\Omega = \{\omega^1, \omega^2, \ldots, \omega^M\}$ represents all possible final states (or scenarios) of the world after T periods.

- A filtration $\mathbb{F} = \{\mathcal{F}_t\}_{t \in \mathbf{T}}$ describes the arrival of information about the market with the passage of time:
$$\Omega = \mathcal{F}_0 \subseteq \mathcal{F}_1 \subseteq \mathcal{F}_2 \subseteq \cdots \subseteq \mathcal{F}_T \equiv \mathcal{F} = 2^\Omega.$$
  For each date $t \in \mathbf{T}$, the respective partition $\mathcal{P}_t$ of Ω that generates $\mathcal{F}_t$ is given by
$$\mathcal{P}_t = \{A_t^1, A_t^2, \ldots, A_t^{k_t}\}. \tag{8.1}$$
  Here A_t^j, $j = 1, 2, \ldots, k_t$, are atoms, and k_t is the number of atoms in the partition $\mathcal{P}_t$. The partitions form an information structure, i.e., $\mathcal{P}_{t-1} \preceq \mathcal{P}_t$ and for all $t = 1, 2, \ldots, T$. Therefore, the numbers k_t form an increasing sequence so that
$$1 = k_0 \leqslant k_1 \leqslant \cdots \leqslant k_T = M.$$
  Note that $\mathcal{P}_0 = \{\Omega\}$ and $\mathcal{P}_T = \{\{\omega^1\}, \ldots, \{\omega^M\}\}$.

- A probability measure $\mathbb{P}$ on $(\Omega, \mathcal{F})$ describes the "real-world" probabilities for the possible states of the world, i.e., $p_j = \mathbb{P}(\omega^j) > 0$ is the probability that the economy will be revealed to be in state $\omega^j \in \Omega$ at time T (i.e., after T periods).

The information structure $\{\mathcal{P}_t\}_{t \in \mathbf{T}}$ can be represented by a nonrecombining tree, where each node is an atom and two atoms are connected by an edge if one of the atoms is a subset of the other. We call it the *information tree*. Any final (T-period) outcome $\omega^j \in \Omega$ is contained in the unique sequence of atoms, one from each partition $\mathcal{P}_t$:

$$\{\omega^j\} = A_T^j \subseteq A_{T-1}^{j_{T-1}} \subseteq \cdots \subseteq A_1^{j_1} \subseteq A_0^1 = \Omega,$$

with indices $j_t \in \{1, 2, \ldots, k_t\}$ and with each j_{t-1} depending on j_t, where $j_T = j \in$

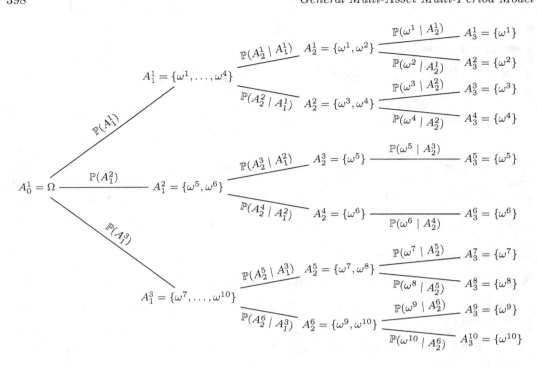

FIGURE 8.1: An example of an information tree with conditional (one-step) probabilities of going from one atom to another.

$\{1, \ldots, M\}$. That is, every (time-t) atom $A_t^{j_t}$ is a subset of some time-$(t-1)$ atom $A_{t-1}^{j_{t-1}}$. That is, the time-$(t-1)$ atoms are obtained by breaking down the time-t atoms for all $t = 0, 1, \ldots, T-1$. The sequence of atoms

$$A_0^1 \to A_1^{j_1} \to \cdots \to A_{T-1}^{j_{T-1}} \to A_T^j$$

form a T-period path in the information tree that goes from atom $A_0^1 = \Omega$ to atom $A_T^j = \{\omega^j\}$. Such a path uniquely represents ω^j. A particular example of a three-period information tree is given in Figure 8.1, where $k_1 = 3$ (i.e., 3 atoms at time $t = 1$), $k_2 = 6$ (i.e., 6 atoms at time $t = 2$), $k_3 = 10$ (i.e., 10 atoms at time $t = T = 3$).

The probability $\mathbb{P}(\omega^j) \equiv \mathbb{P}(\{\omega^j\})$ can be computed as a product of conditional probabilities:

$$\mathbb{P}(\omega^j) = \mathbb{P}(\{\omega^j\} \mid A_{T-1}^{j_{T-1}}) \cdot \mathbb{P}(A_{T-1}^{j_{T-1}})$$

$$= \mathbb{P}(\{\omega^j\} \mid A_{T-1}^{j_{T-1}}) \cdot \mathbb{P}(A_{T-1}^{j_{T-1}} \mid A_{T-2}^{j_{T-2}}) \cdot \mathbb{P}(A_{T-2}^{j_{T-2}})$$

$$\vdots$$

$$= \mathbb{P}(\{\omega^j\} \mid A_{T-1}^{j_{T-1}}) \cdot \mathbb{P}(A_{T-1}^{j_{T-1}} \mid A_{T-2}^{j_{T-2}}) \cdots \mathbb{P}(A_2^{j_2} \mid A_1^{j_1}) \cdot \mathbb{P}(A_1^{j_1}). \qquad (8.2)$$

Therefore, we may refer to $\mathbb{P}(\omega^j)$ as a *path probability*. Equation (8.2) gives us another method of constructing the probability measure $\mathbb{P}$. We only need to define the conditional probability $\mathbb{P}(A_{t+1}^{j_{t+1}} \mid A_t^{j_t})$ for every pair of atoms $A_t^{j_t} \in \mathcal{P}_t$ and $A_{t+1}^{j_{t+1}} \in \mathcal{P}_{t+1}$ with the property $A_{t+1}^{j_{t+1}} \subseteq A_t^{j_t}$.

For calculating derivative prices at an intermediate time $t \in \mathbf{T}$ we need conditional expectations of the form $\mathrm{E}_t[X] \equiv \mathrm{E}[X \mid \mathcal{F}_t]$ (where $X \colon \Omega \to \mathbb{R}$ is a random variable). The σ-algebra $\mathcal{F}_t = \sigma(\mathcal{P}_t)$ is generated by the partition $\mathcal{P}_t$ with k_t atoms, as given in (8.1). Therefore, the conditional expectation $\mathrm{E}_t[X]$ is a random variable that is constant on the k_t atoms of $\mathcal{P}_t$; it is given by applying (6.21):

$$\mathrm{E}_t[X] = \sum_{i=1}^{k_t} \mathrm{E}[X \mid A_t^i]\, \mathbb{I}_{A_t^i}\,.$$

The conditional expectation of X given atom (event) A_t^i is calculated using (6.19):

$$\mathrm{E}[X \mid A_t^i] = \frac{\mathrm{E}[X\,\mathbb{I}_{A_t^i}]}{\mathbb{P}(A_t^i)} = \frac{1}{\mathbb{P}(A_t^i)} \sum_{\omega \in A_t^i} X(\omega)\,\mathbb{P}(\omega)\,, \quad i = 1, 2, \ldots, k_t\,.$$

Note that the atoms making up each partition $\mathcal{P}_t$ are disjoint. Hence, for each state (outcome) contained in the i-th time-t atom, $\omega^j \in A_t^i$, there is a unique sequence of atoms from time t to T where

$$\{\omega^j\} = A_T^j \subseteq A_{T-1}^{j_{T-1}} \subseteq \cdots \subseteq A_{t+1}^{j_{t+1}} \subseteq A_t^i\,.$$

The probability of state ω^j can be calculated as follows:

$$\mathbb{P}(\omega^j) = \mathbb{P}(\{\omega^j\} \mid A_{T-1}^{j_{T-1}})\, \mathbb{P}(A_{T-1}^{j_{T-1}} \mid A_{T-2}^{j_{T-2}}) \cdots \mathbb{P}(A_{t+1}^{j_{t+1}} \mid A_t^i)\, \mathbb{P}(A_t^i)\,.$$

The above formula for the conditional expectation hence simplifies to obtain

$$\mathrm{E}[X \mid A_t^i] = \sum_{\omega \in A_t^i} X(\omega)\, \mathbb{P}(\{\omega\} \mid A_{T-1}^{j_{T-1}})\, \mathbb{P}(A_{T-1}^{j_{T-1}} \mid A_{T-2}^{j_{T-2}}) \cdots \mathbb{P}(A_{t+1}^{j_{t+1}} \mid A_t^i)\,.$$

Let us consider some examples of multi-asset multi-period models.

Example 8.1. Clearly, the standard binomial lattice model (with independent market moves) fits the definition of a general multi-period model. The state space Ω is the collection of $M = 2^T$ distinct paths in the recombining binomial tree with T periods, where each path is coded by a sequence of D's and U's (of length T). That is, $\omega = \omega_1 \omega_2 \ldots \omega_T$, where each $\omega_t \in \{\mathsf{D}, \mathsf{U}\}$ is the time-t market move. The partition $\mathcal{P}_t$ consists of $k_t = 2^t$ atoms. Each atom of $\mathcal{P}_t$ contains 2^{T-t} paths (outcomes), all having identical first t market moves. The probability measure $\mathbb{P}$ on Ω is defined by the probability $\mathbb{P}(\mathsf{U}) = p \in (0, 1)$ of a single upward move. The probability of any scenario $\omega \in \Omega$ is found to be

$$\mathbb{P}(\omega) = p^{\#\mathsf{U}(\omega)} (1 - p)^{\#\mathsf{D}(\omega)}\,.$$

Select an atom $A_t^i \in \mathcal{P}_t$, $i \in \{1, \ldots, 2^t\}$, so that $\omega \in A_t^i$ iff $\omega_1 = \bar{\omega}_1, \ldots, \omega_t = \bar{\omega}_t$ for some $\bar{\omega}_1, \ldots, \bar{\omega}_t \in \{\mathsf{D}, \mathsf{U}\}$. That is, $A_t^i = A_{\bar{\omega}_1 \ldots \bar{\omega}_t}$. This atom is a subset of $A_{\bar{\omega}_1 \ldots \bar{\omega}_{t-1}} \in \mathcal{P}_{t-1}$. The conditional probability of $A_{\bar{\omega}_1 \ldots \bar{\omega}_t}$ given $A_{\bar{\omega}_1 \ldots \bar{\omega}_{t-1}}$ is

$$\mathbb{P}(A_{\bar{\omega}_1 \ldots \bar{\omega}_t} \mid A_{\bar{\omega}_1 \ldots \bar{\omega}_{t-1}}) = p^{\#\mathsf{U}(\bar{\omega}_t)} (1 - p)^{\#\mathsf{D}(\bar{\omega}_t)}\,.$$

Therefore, the probability of an individual scenario $\omega \in \Omega$ is computed as

$$\mathbb{P}(\omega) = p^{\#\mathsf{U}(\omega)} (1 - p)^{\#\mathsf{D}(\omega)}\,.$$

The more general binomial model with interdependent moves, as described in Section 6.3.5, is yet another example of a multi-period model. The conditional probabilities are computed according to (6.39).

Example 8.2. A generalization of the binomial tree model is a *multinomial tree model*, where at each time there exist $n \geqslant 2$ possible continuations (i.e., market moves) denoted by M_j for $j = 1, 2, \ldots, n$. Note that for a binomial model ($n = 2$) we use $\mathsf{D} \equiv \mathsf{M}_1$ and $\mathsf{U} \equiv \mathsf{M}_2$. A scenario (outcome) in this model is a path in a nonrecombining n-nomial tree with T periods. Each path can be coded by a sequence of length T formed from the symbols $\mathsf{M}_1, \mathsf{M}_2, \ldots, \mathsf{M}_n$:

$$\Omega = \big\{ \omega_1 \omega_2 \ldots \omega_T \; : \; \omega_1, \omega_2, \ldots, \omega_T \in \{\mathsf{M}_1, \mathsf{M}_2, \ldots, \mathsf{M}_n\} \big\}.$$

Clearly, there are n^t atoms representing all t-period moves for $t = 0, 1, \ldots, T$. Every atom $A_{\bar{\omega}_1 \ldots \bar{\omega}_t}$ represents a sequence of the first t moves where $\bar{\omega}_k \in \{\mathsf{M}_1, \mathsf{M}_2, \ldots, \mathsf{M}_n\}$, $k = 1, \ldots, t$. The state space Ω has $M = n^T$ elements (i.e., n^T possible outcomes). For the simplest model, with all independent moves, the probability measure $\mathbb{P}$ is defined by the collection of probabilities $p_i \in (0, 1)$, $i = 1, 2, \ldots, n$, where $p_i = \mathbb{P}(\mathsf{M}_i)$ is a (fixed) probability of the i-th single-step move M_i. Let $\#\mathsf{M}_i(\omega)$ give the number of M_i in the sequence ω. The probability of any outcome $\omega \in \Omega$ is computed as

$$\mathbb{P}(\omega) = p_1^{\#\mathsf{M}_1(\omega)} p_2^{\#\mathsf{M}_2(\omega)} \cdots p_n^{\#\mathsf{M}_n(\omega)},$$

where $p_1 + p_2 + \cdots + p_n = 1$.

In the more general multinomial model the moves $\omega_1, \omega_2, \ldots, \omega_T$ may be interdependent. The conditional probabilities are then computed according to (6.39) with the appropriate atoms described above.

Example 8.3. Consider a two-period trinomial model. Assume that by the end of the first period the economy can end up in one of three possible states. Moreover, for each of these possibilities, there are three alternative continuations throughout the second period. Therefore, in total there are nine possible states of the world ($M = 9$) at time $T = 2$. The filtration $\mathbb{F} = \{\mathcal{F}_t\}_{t \in \{0,1,2\}}$ is then generated by the information structure $\mathcal{P}_0 \preceq \mathcal{P}_1 \preceq \mathcal{P}_2$ with the following partitions:

$$\mathcal{P}_0 = \big\{ A_0^1 = \{\omega^1, \omega^2, \ldots, \omega^9\} = \Omega \big\},$$
$$\mathcal{P}_1 = \big\{ A_1^1 = \{\omega^1, \omega^2, \omega^3\}, A_1^2 = \{\omega^4, \omega^5, \omega^6\}, A_1^3 = \{\omega^7, \omega^8, \omega^9\} \big\},$$
$$\mathcal{P}_2 = \big\{ A_2^1 = \{\omega^1\}, A_2^2 = \{\omega^2\}, \ldots, A_2^9 = \{\omega^9\} \big\}.$$

This trinomial model is a special case of the above multinomial model where $n = 3$, $T = 2$, $M = n^T = 3^2 = 9$.

8.2 Assets, Portfolios, and Strategies

8.2.1 Payoffs and Assets

Consider a financial contract maturing at time $T_m \in \{1, 2, \ldots, T\}$. The *payoff* of such a contract (to its holder) is a real-valued function defined on the set of states of the world. Such a function is contingent on the information available at time T_m and is therefore represented by an $\mathcal{F}_{T_m}$-measurable random variable $X : \Omega \to \mathbb{R}$. Trivial examples include constant payoffs.

Since the σ-algebra $\mathcal{F}_{T_m}$ is generated by the partition $\mathcal{P}_{T_m}$, the $\mathcal{F}_{T_m}$-measurability of X means that the payoff X is constant on the k_{T_m} atoms $A_{T_m}^1, A_{T_m}^2, \ldots, A_{T_m}^{k_{T_m}}$ of $\mathcal{P}_{T_m}$. Thus, payoff X can be represented by the vector

$$X(\mathcal{P}_{T_m}) = \left[X(A_{T_m}^1), X(A_{T_m}^2), \ldots, X(A_{T_m}^{k_{T_m}}) \right] \in \mathbb{R}^{k_{T_m}}.$$

A linear combination of $\mathcal{F}_{T_m}$-measurable random variables (i.e., payoffs maturing at the same time T_m) is again an $\mathcal{F}_{T_m}$-measurable random variable (i.e., a payoff maturing at time T_m). Thus, the set of payoffs maturing at time T_m is a vector subspace of the space $\mathcal{L}(\Omega)$. Recall that $\mathcal{L}(\Omega)$ consists of all payoff functions defined on Ω. We denote such a subspace by $\mathcal{L}_{T_m}(\Omega)$.

A European-style asset (also called a security) S maturing at time $T_m \leqslant T$ is described by

- a nonnegative terminal payoff function $S_{T_m} : \Omega \to \mathbb{R}_+$ at time T_m,

- nonnegative prices $S_t : \Omega \to \mathbb{R}_+$ with $t \in \{0, 1, \ldots, T_m - 1\}$.

Since the asset price S_t is contingent on the information available at time t, it is an $\mathcal{F}_t$-measurable random variable for all t. In other words, the asset S is described by a nonnegative price process $\{S_t\}_{0 \leqslant t \leqslant T_m}$ adapted to the filtration $\{\mathcal{F}_t\}_{0 \leqslant t \leqslant T_m}$.

The main building blocks of our model are $N + 1$ base (tradable) assets denoted by $S^0, S^1, S^2, \ldots, S^N$. Here, $\{S_t^0\}_{0 \leqslant t \leqslant T}$ is a strongly positive asset price process. It usually refers to a bank account or money market account. We will adopt a dual notation for such an asset: $S^0 \equiv B$. The other N assets with nonnegative price processes $\{S_t^i\}_{0 \leqslant t \leqslant T}$, $i = 1, 2, \ldots, N$, usually refer to equity stocks.

In analogy with the binomial model of Chapter 7, we can model the price dynamics of each risky assets (or stocks) in terms of market moves. In the case of zero dividends on each stock we have the single time-step dynamics (see the relation in (7.2)):

$$S_{t+1}^i(\omega) = Y_{t+1}^i(\omega) S_t^i(\omega), \quad t = 0, 1, \ldots, T-1, \tag{8.3}$$

for each stock $i = 1, \ldots, N$, where each Y_{t+1}^i is $\mathcal{F}_{t+1}$-measurable and each S_t^i is $\mathcal{F}_t$-measurable. Given a pair of atoms $A_t^k \in \mathbb{P}_t$ and $A_{t+1}^j \in \mathbb{P}_{t+1}$ with $A_{t+1}^j \subset A_t^k$, we have

$$S_{t+1}^i(A_{t+1}^j) = Y_{t+1}^i(A_{t+1}^j) S_t^i(A_t^k), \quad t = 0, 1, \ldots, T-1. \tag{8.4}$$

For each i, Y_{t+1}^i is strictly positive and generalizes the "upward" and "downward" moves within time period $[t, t+1]$ in the standard binomial model. Generally, $\{Y_{t+1}^i\}_{0 \leqslant t \leqslant T}$ are not i.i.d. and each Y_{t+1}^i is not necessarily $\mathcal{F}_t$-independent. Iterating (8.4) gives

$$S_t^i(A_t^{j_t}) = S_0^i \prod_{k=1}^t Y_k^i(A_k^{j_k}), \tag{8.5}$$

for any $t \geqslant 1$, where $A_t^{j_t} \subset A_{t-1}^{j_{t-1}} \subset \cdots \subset A_1^{j_1} \subset A_0$.

We can model the price dynamics of dividend-paying stocks by introducing a $\mathcal{F}_{t+1}$-measurable factor $q_{t+1}^i \in [0, 1)$ for the i-th stock in period $[t, t+1]$. In this case, (8.3)-(8.5) are modified to read:

$$S_{t+1}^i(\omega) = (1 - q_{t+1}^i(\omega)) Y_{t+1}^i(\omega) S_t^i(\omega), \tag{8.6}$$

$$S_{t+1}^i(A_{t+1}^j) = (1 - q_{t+1}^i(A_{t+1}^j)) Y_{t+1}^i(A_{t+1}^j) S_t^i(A_t^k), \tag{8.7}$$

$$S_t^i(A_t^{j_t}) = S_0^i \prod_{k=1}^t (1 - q_k^i(A_k^{j_k})) Y_k^i(A_k^{j_k}). \tag{8.8}$$

This relation reproduces (7.5) in Chapter 7 in the special case of the binomial model.

The adapted processes defined by $c_0^i = 1$ and

$$c_t^i := (1 - q_1^i) \times \cdots \times (1 - q_t^i), \tag{8.9}$$

$1 \leqslant t \leqslant T$, $i = 1, \ldots, N$, are time-t cumulative after-dividend factors for each respective stock. In particular, on every time-t atom $A_t^{j_t}$:

$$c_t^i(A_t^{j_t}) := \prod_{k=1}^{t} (1 - q_k^i(A_k^{j_k})), \tag{8.10}$$

where $A_k^{j_k} \subset A_{k-1}^{j_{k-1}}$.

Let us define the short rate process $\{r_t\}_{1 \leqslant t \leqslant T}$ as a sequence of one-period returns on B:

$$r_t := \frac{B_t - B_{t-1}}{B_{t-1}} \text{ for all } t \in \{1, 2, \ldots, T\}. \tag{8.11}$$

The condition that $r_t(\omega) > -1$ for all $\omega \in \Omega$ and for all $t \in \{1, 2, \ldots, T\}$ guarantees that the price process $\{B_t\}_{0 \leqslant t \leqslant T}$ is strictly positive (provided that $B_0 > 0$). Additionally, we assume that the short rate process is $\mathbb{F}$-*predictable* [1], i.e., r_t is a $\mathcal{F}_{t-1}$-measurable random variable for all $t \in \{1, 2, \ldots, T\}$. So the interest rate r_t, for the period $[t-1, t)$, is known at time $t-1$, i.e., it is fixed on each time-$(t-1)$ atom A_{t-1}^j with known (nonrandom $\mathcal{F}_0$-measurable) return r_1 in the first period $[0, 1]$. Note that B_0 dollars invested in the bank account over the time interval $[0, t]$ has value $B_t = B_0(1 + r_1) \times \cdots \times (1 + r_t)$, for $t = 1, \ldots, T$. Therefore, the bank account process is $\mathbb{F}$-predictable as well.

The bank account is used here as a numéraire asset. The discounted base asset prices are obtained by dividing the nondiscounted (original) prices by the price of the numéraire asset at the same time:

$$\overline{S}_t^i := \frac{S_t^i}{B_t} \text{ for } i \in \{0, 1, 2, \ldots, N\} \text{ and } t \in \{0, 1, \ldots, T\}.$$

Notice that $\overline{S}_t^0 \equiv 1$ for all t and $\overline{S}_0^i = S_0^i$ for all i.

Example 8.4. Assume the two-period trinomial model from Example 8.3. Let us now consider three assets in this model, namely, a risk-free cash account B with initial value $B_0 = 1$ and fixed interest rate $r_t = r \equiv 0$ and two stocks S^1 and S^2 with respective initial values $S_0^1 = 50$ and $S_0^2 = 100$. The time-1 and time-2 values of the stocks are specified in the following table.

ω	$S_1^1(\omega)$	$S_1^2(\omega)$	$S_2^1(\omega)$	$S_2^2(\omega)$
ω^1	50	115	45	130
ω^2	50	115	45	120
ω^3	50	115	60	95
ω^4	40	115	45	135
ω^5	40	115	35	120
ω^6	40	115	40	105
ω^7	60	70	60	100
ω^8	60	70	55	40
ω^9	60	70	70	70

---

[1]We are assuming that the interest rate is fixed to it's initial value within each time period $(t-1, t]$. According to our definition of r_t for the return on interest, it follows that r_t is known at time $t-1$. We also alternatively use the notation $R_{t-1} := \frac{B_t - B_{t-1}}{B_{t-1}}$, where $R_{t-1} \equiv r_t$. Hence, R_0 is a constant (known nonrandom $\mathcal{F}_0$-measurable rate for initial period $[0,1]$) and R_t is $\mathcal{F}_t$-measurable for $t \in \{1, 2, \ldots, T-1\}$. Accordingly, the bond prices are given by $B_t = B_0(1 + R_0) \times \cdots \times (1 + R_{t-1})$, for $t = 1, \ldots, T$.

The asset price processes $\{S_t^1\}_{t\in\{0,1,2\}}$ and $\{S_t^2\}_{t\in\{0,1,2\}}$ are adapted to the filtration since for every $t \in \{0,1,2\}$ and every $i \in \{1,2\}$ the price S_t^i is constant on atoms of the partition $\mathcal{P}_t$, as given in Example 8.3. For example, $S_1^1(\omega) = 50$ and $S_1^2(\omega) = 115$ for all $\omega \in \{\omega^1, \omega^2, \omega^3\} = A_1^1 \in \mathcal{P}_1$.

8.2.2 Static and Dynamic Portfolios

A portfolio is a combination of positions in several (or all) base assets. If the positions do not change as time passes by, then we speak of a static portfolio. Let β denote the position in B, where $\beta < 0$ corresponds to a loan and $\beta > 0$ is an investment; let Δ^i be the position in S^i for $i = 1, 2, \ldots, N$, where $\Delta^i < 0$ corresponds to shorting the asset and $\Delta^i > 0$ is a long position. Such a static portfolio is represented by the vector

$$\varphi = \left[\beta, \Delta^1, \Delta^2, \ldots, \Delta^N\right] \in \mathbb{R}^{N+1},$$

As opposed to static portfolios in a single-period model, a portfolio held in a multi-period environment can be re-balanced at intermediate dates (as seen in the binomial market model of Chapter 7). The portfolio holder can change or even liquidate some positions and open others. A sequence of re-balanced portfolios in the base assets indexed by time, $\{\varphi_t\}_{0 \leqslant t \leqslant T_m-1}$, is called a *portfolio strategy* maturing at time $T_m \in \{1, 2, \ldots, T\}$. The portfolio φ_t is formed at time t and held from time t to $t+1$. At time $t+1$, φ_t is liquidated and a new portfolio φ_{t+1} is set up. This procedure is repeated for every $t \in \{0, 1, \ldots, T_m - 1\}$. At time T_m, the final portfolio φ_{T_m-1} is liquidated and the proceeds are consumed.

The re-balancing of a portfolio at any date is contingent on the information available at that time. Therefore, the re-balancing is done in a way such that φ_t is $\mathcal{F}_t$-measurable for each $t \in \{0, 1, \ldots, T_m - 1\}$. Hence, a portfolio strategy is a stochastic vector process adapted to the filtration $\{\mathcal{F}_t\}_{0 \leqslant t \leqslant T_m-1}$.

Recall the notion of the value (wealth) of a portfolio. The time-t value of a static portfolio φ in the base assets is defined as

$$\Pi_t[\varphi] = \beta B_t + \Delta^1 S_t^1 + \Delta^2 S_t^2 + \cdots + \Delta^N S_t^N.$$

For a portfolio strategy $\boldsymbol{\Phi} = \{\varphi_t\}_{0 \leqslant t \leqslant T_m-1}$ maturing at time T_m, in which the portfolio $\varphi_t = \left[\beta_t, \Delta_t^1, \Delta_t^2, \ldots, \Delta_t^N\right]$ is held during the period $[t, t+1]$, we define its *acquisition value* $\Pi_t[\varphi_t]$ and its *liquidation value* $\Pi_{t+1}[\varphi_t]$, respectively, by

$$\Pi_t[\varphi_t] = \beta_t B_t + \Delta_t^1 S_t^1 + \Delta_t^2 S_t^2 + \cdots + \Delta_t^N S_t^N,$$
$$\Pi_{t+1}[\varphi_t] = \beta_t B_{t+1} + \Delta_t^1 S_{t+1}^1 + \Delta_t^2 S_{t+1}^2 + \cdots + \Delta_t^N S_{t+1}^N, \quad t \in \{0, 1, \ldots, T_m - 1\}.$$

Note that we are assuming zero dividends on every base asset S^i. For example, the base assets may be non-dividend paying stocks. At each date $t \in \{1, 2, \ldots, T_m-1\}$, the portfolio φ_{t-1} is liquidated with proceeds of value $\Pi_t[\varphi_{t-1}]$ just before the new portfolio φ_t is acquired for value $\Pi_t[\varphi_t]$. This process is referred to as *portfolio re-balancing*.

8.2.3 Self-Financing Strategies

The difference $\Pi_t[\varphi_t] - \Pi_t[\varphi_{t-1}]$ may be positive, meaning that some funds are withdrawn, or negative, meaning that some funds are injected. *Self-financing strategies* do not allow withdrawing or injecting funds at intermediate dates. At each time $t \in \{1, 2, \ldots, T_m-1\}$, the portfolio value just before re-balancing is exactly the same as the value after re-balancing: $\Pi_t[\varphi_t] = \Pi_t[\varphi_{t-1}]$. A self-financing strategy finances each re-balancing on its own without withdrawing or injecting funds. A trivial example of a self-financing strategy is a static portfolio.

Since for a self-financing strategy there is no need to distinguish between its acquisition and liquidation values, we will only speak of the portfolio value of a self-financing strategy $\boldsymbol{\Phi}$ given by

$$\Pi_t \equiv \Pi_t^{\Phi} = \beta_{t-1} B_t + \Delta_{t-1}^1 S_t^1 + \Delta_{t-1}^2 S_t^2 + \cdots + \Delta_{t-1}^N S_t^N, \quad t \in \{1, 2, \ldots, T_m\}.$$

The initial value of a self-financing strategy $\boldsymbol{\Phi}$ given by

$$\Pi_0 = \beta_0 B_0 + \Delta_0^1 S_0^1 + \Delta_0^2 S_0^2 + \cdots + \Delta_0^N S_0^N$$

is referred to as the *initial cost* (of setting up the strategy). The terminal value Π_{T_m} at maturity T_m is referred to as the *payoff* of $\boldsymbol{\Phi}$.

The self-financing condition can be written as follows:

$$\Pi_{t+1}[\boldsymbol{\varphi}_{t+1}] - \Pi_{t+1}[\boldsymbol{\varphi}_t] = B_{t+1}(\beta_{t+1} - \beta_t) + \sum_{i=1}^N S_{t+1}^i (\Delta_{t+1}^i - \Delta_t^i)$$

$$= \delta\beta_t \cdot B_{t+1} + \sum_{i=1}^N \delta\Delta_t^i \cdot S_{t+1}^i = 0 \qquad (8.12)$$

for $t \in \{0, 1, \ldots, T_m - 1\}$, where $\delta\beta_t := \beta_{t+1} - \beta_t$ and $\delta\Delta_t^i := \Delta_{t+1}^i - \Delta_t^i$ are single-period changes in the strategy positions. Let the one-period changes in the portfolio value and in the asset prices be given by

$$\delta\Pi_t := \Pi_{t+1} - \Pi_t, \quad \delta B_t := B_{t+1} - B_t, \quad \delta S_t^i := S_{t+1}^i - S_t^i,$$

respectively. By using the self-financing condition in (8.12), the change in the re-balanced portfolio value from time t to time $t+1$ is

$$\delta\Pi_t = \left(B_{t+1}\beta_{t+1} + \sum_{i=1}^N S_{t+1}^i \Delta_{t+1}^i \right) - \left(B_t\beta_t + \sum_{i=1}^N S_t^i \Delta_t^i \right)$$

$$= B_{t+1}(\beta_t + \delta\beta_t) + \sum_{i=1}^N S_{t+1}^i (\Delta_t^i + \delta\Delta_t^i) - \left(B_t\beta_t + \sum_{i=1}^N S_t^i \Delta_t^i \right)$$

$$= \beta_t \, \delta B_t + \sum_{i=1}^N \Delta_t^i \, \delta S_t^i.$$

As is seen, the self-financing condition is equivalent to the statement that the changes in portfolio values are due only to changes in base asset prices.

The self-financing condition imposes restrictions on the positions in the base assets. At any time t, the portfolio value Π_t and the delta-positions of a self-financing strategy determine the position in the bank account:

$$\Pi_t = \beta_t B_t + \sum_{i=1}^N \Delta_t^i S_t^i \implies \beta_t = \frac{\Pi_t - \sum_{i=1}^N \Delta_t^i S_t^i}{B_t}.$$

Moreover, we can show that the value process $\{\Pi_t\}_{0 \leqslant t \leqslant T_m}$ is governed by a wealth equation which is similar to (7.15) derived in the case with only two base assets:

$$\Pi_t = \sum_{i=1}^N \Delta_{t-1}^i S_t^i + \left(\Pi_{t-1} - \sum_{i=1}^N \Delta_{t-1}^i S_{t-1}^i \right)(1 + r_t), \quad 1 \leqslant t \leqslant T_m. \qquad (8.13)$$

As a result, a self-financing strategy, whose value dynamics follow (8.13), can be described by the delta strategy $\{\Delta_t\}_{0 \leqslant t \leqslant T_m - 1}$ and the initial cost Π_0.

Recall that a linear combination of two portfolios, which is merely a linear combination of two vectors, is also a portfolio. The linear combination of two strategies $\boldsymbol{\Phi} = \{\boldsymbol{\varphi}_t\}_{t \geqslant 0}$ and $\boldsymbol{\Psi} = \{\boldsymbol{\psi}_t\}_{t \geqslant 0}$ maturing at the same time T_m is defined by taking a linear combination of the portfolios $\boldsymbol{\varphi}_t$ and $\boldsymbol{\psi}_t$ at each time $t \in \{0, 1, \ldots, T_m - 1\}$. Any linear combination $a\boldsymbol{\Phi} + b\boldsymbol{\Psi}$ with $a, b \in \mathbb{R}$, of two self-financing strategies $\boldsymbol{\Phi}$ and $\boldsymbol{\Psi}$ maturing at the same time T_m is again a self-financing strategy with the same maturity. The proof follows from the fact the acquisition and liquidation values are linear functions of the positions. For example, the liquidation time-t value of the strategy $a\boldsymbol{\Phi} + b\boldsymbol{\Psi}$ is

$$\Pi_t[a\boldsymbol{\varphi}_{t-1} + b\boldsymbol{\psi}_{t-1}] = B_t \left(a\beta^{\boldsymbol{\Phi}}_{t-1} + b\beta^{\boldsymbol{\Psi}}_{t-1} \right) + \sum_{i=1}^{N} S^i_t \left(a\Delta^{\boldsymbol{\Phi}}_{t-1} + b\Delta^{\boldsymbol{\Psi}}_{t-1} \right)$$

$$= a \left(B_t \beta^{\boldsymbol{\Phi}}_{t-1} + \sum_{i=1}^{N} S^i_t \Delta^{\boldsymbol{\Phi}}_{t-1} \right) + b \left(B_t \beta^{\boldsymbol{\Psi}}_{t-1} + \sum_{i=1}^{N} S^i_t \Delta^{\boldsymbol{\Psi}}_{t-1} \right)$$

$$= a\Pi_t[\boldsymbol{\varphi}_{t-1}] + b\Pi_t[\boldsymbol{\psi}_{t-1}].$$

Applying the self-financing condition (8.12) to the strategy $a\boldsymbol{\Phi} + b\boldsymbol{\Psi}$ gives

$$\Pi_t[a\boldsymbol{\varphi}_t + b\boldsymbol{\psi}_t] - \Pi_t[a\boldsymbol{\varphi}_{t-1} + b\boldsymbol{\psi}_{t-1}] = a \left(\Pi_t[\boldsymbol{\varphi}_t] - \Pi_t[\boldsymbol{\varphi}_{t-1}] \right) + b \left(\Pi_t[\boldsymbol{\psi}_t] - \Pi_t[\boldsymbol{\psi}_{t-1}] \right)$$

$$= a \cdot 0 + b \cdot 0 = 0.$$

That is, any linear combination of two self-financing strategies, $a\boldsymbol{\Phi} + b\boldsymbol{\Psi}$, is a self-financing strategy.

Suppose that we are given a self-financing strategy $\boldsymbol{\Phi} = \{\boldsymbol{\varphi}_t\}_{0 \leqslant t \leqslant T_m - 1}$ maturing at time T_m. The terminal value (at time T_m) of this strategy is $\Pi^{\boldsymbol{\Phi}}_{T_m}$. We can lock in this value by liquidating the portfolio $\boldsymbol{\varphi}_{T_m - 1}$ at time T_m and investing all proceeds in the bank account with strictly positive returns. The new strategy $\boldsymbol{\Phi}' = \{\boldsymbol{\varphi}'_t\}_{0 \leqslant t \leqslant T - 1}$ is thus defined by

$$\boldsymbol{\varphi}'_t := \begin{cases} \boldsymbol{\varphi}_t & \text{if } 0 \leqslant t \leqslant T_m - 1, \\ \left[\overline{\Pi}, 0, \cdots, 0 \right] & \text{if } T_m \leqslant t \leqslant T, \end{cases} \tag{8.14}$$

where $\overline{\Pi} := \overline{\Pi}^{\boldsymbol{\Phi}}_{T_m} = \frac{\Pi^{\boldsymbol{\Phi}}_{T_m}}{B_{T_m}}$ is the discounted value of the terminal value of the strategy $\boldsymbol{\Phi}$. We refer to the trading strategy $\boldsymbol{\Phi}'$ as the strategy obtained by *locking the portfolio value* of $\boldsymbol{\Phi}$ at maturity. This new strategy is now maturing at time T. Thus, any trading strategy can be converted into a strategy maturing at the terminal time T.

8.2.4 Replication of Payoffs

Consider a financial contract maturing at time $T_m \leqslant T$ with payoff V_{T_m}. As usual, our objective is to find a fair initial price of such a contract. In the multi-period world, this goal can be achieved by constructing a dynamic portfolio strategy that replicates the payoff function. The value of the contract is then given by the cost of setting up the replication strategy.

Definition 8.1. A payoff V_{T_m} maturing at time T_m is said to be *attainable* if there exists a self-financing strategy $\boldsymbol{\Phi} = \{\boldsymbol{\varphi}_t\}_{0 \leqslant t \leqslant T_m - 1}$ maturing at time T_m such that its terminal value coincides with the payoff for all possible market scenarios: $\Pi^{\boldsymbol{\Phi}}_{T_m} = V_{T_m}$. Any trading strategy with this property is called a *replicating* or *hedging strategy* for the payoff V_{T_m}. A market model is said to be *complete* if every payoff for every maturity is attainable.

The initial fair value V_0 of a contract with payoff V_{T_m} maturing at time T_m is given by the cost of setting up a strategy $\boldsymbol{\Phi}$ that replicates V_{T_m}:

$$V_0 = \Pi_0^{\boldsymbol{\Phi}}, \text{ where } \Pi_{T_m}^{\boldsymbol{\Phi}} = X_{T_m}. \tag{8.15}$$

If there exist several replicating strategies (all having the same terminal value), then their initial values are expected to be the same. In other words, the Law of One Price is expected to hold for each maturity T_m. Otherwise, the definition (8.15) is meaningless.

For a multi-period model, the Law of One Price is stated as follows: any two self-financing strategies maturing at the same time and having the same terminal value have the same initial value. If the Law of One Price does not hold, then for each attainable payoff there exists a replicating strategy with an arbitrary pre-specified initial value. The proof of this statement is identical to that for single-period models considered in Chapter 5 (see Proposition 5.7). As in the single-period case, we can show that the Law of One Price holds iff each strategy replicating a zero payoff has initial value zero (see Lemma 5.6).

8.3   Fundamental Theorems of Asset Pricing

8.3.1   Arbitrage Strategies

Recall that an *admissible strategy* is a self-financing strategy with nonnegative values for all dates from time zero until the maturity. An *arbitrage strategy* is an admissible strategy with zero initial value and positive terminal value. In other words, a self-financing strategy $\boldsymbol{\Phi} = \{\boldsymbol{\varphi}_t\}_{0 \leqslant t \leqslant T_m-1}$ maturing at time T_m is an arbitrage opportunity if the conditions

$$\Pi_0^{\boldsymbol{\Phi}} = 0, \quad \Pi_t^{\boldsymbol{\Phi}} \geqslant 0 \text{ for all } t \in \{1, 2, \dots, T_m\}, \text{ and } \mathbb{P}(\Pi_{T_m}^{\boldsymbol{\Phi}} > 0) > 0$$

all hold.

When we say that the market model admits no arbitrage opportunity, we mean that the model admits no arbitrage strategies for any maturity $T_m \in \{1, 2, \dots, T\}$. The same convention is assumed when we say that the Law of One Price holds. However, it is sufficient to only consider strategies maturing at time T, as is shown in the following lemma.

Lemma 8.1. *Consider a multi-period, finite-state model.*

(a) The Law of One price holds for all maturities iff it holds for strategies maturing at time T.

(b) There are no arbitrage strategies for all maturities iff the market admits no arbitrage opportunities for the maturity T.

Proof. The proof of the sufficiency part is trivial. Let us prove the necessity part. Being given a strategy with maturity $T_m \leqslant T$, we can construct another strategy with maturity T by locking the portfolio value at time T_m.

(a) As was noticed above, the Law of One Price holds for maturity $T_m \leqslant T$ iff each strategy replicating the zero claim at time T_m has zero initial value. Let $\boldsymbol{\Phi} = \{\boldsymbol{\varphi}_t\}_{0 \leqslant t \leqslant T_m-1}$ be a strategy replicating the zero payoff $X_{T_m} \equiv 0$. Then the strategy $\boldsymbol{\Phi}' = \{\boldsymbol{\varphi}_t'\}_{0 \leqslant t \leqslant T-1}$ defined by

$$\boldsymbol{\varphi}_t' := \begin{cases} \boldsymbol{\varphi}_t & \text{if } 0 \leqslant t \leqslant T_m - 1 \\ \mathbf{0} & \text{if } T_m \leqslant t \leqslant T \end{cases}$$

is self-financing and replicates the zero claim with maturity T. If the Law of One Price holds for maturity T, then $\Pi_0[\mathbf{\Phi}] = \Pi_0[\mathbf{\Phi}'] = 0$. Therefore, the Law of One Price holds for maturity T_m.

(b) Assume that there are no arbitrage strategies with maturity T. Let us prove that there are no arbitrage opportunities for any time $T_m \leqslant T$. Indeed, if $\mathbf{\Phi} = \{\varphi_t\}_{0 \leqslant t \leqslant T_m - 1}$ is an arbitrage strategy for time T_m, then the strategy $\mathbf{\Phi}'$ constructed by (8.14) is an arbitrage opportunity for time T, which contradicts the assumption. $\qquad\square$

8.3.2 Enhancing the Law of One Price

In Chapter 5, we proved that the no-arbitrage condition implies that the Law of One Price holds for static portfolios. Let us show that if no arbitrage strategy exists then the Law of One Price holds in a stronger sense: the strategies replicating the same payoff have the same value process. In other words, if two self-financing strategies $\mathbf{\Phi}$ and $\mathbf{\Psi}$ have the same terminal value, i.e., $\Pi_{T_m}^{\mathbf{\Phi}} = \Pi_{T_m}^{\mathbf{\Psi}}$, then not only their initial values $\Pi_0^{\mathbf{\Phi}}$ and $\Pi_0^{\mathbf{\Psi}}$ are the same, but at every intermediate date $t \in \{0, 1, \ldots, T_m - 1\}$ the values $\Pi_t^{\mathbf{\Phi}}$ and $\Pi_t^{\mathbf{\Psi}}$ are also the same. First, we shall prove that any self-financing strategy with a nonnegative payoff is admissible provided that there are no arbitrage opportunities.

Lemma 8.2. *Assume the absence of arbitrage. Let $\mathbf{\Phi} = \{\varphi_t\}_{0 \leqslant t \leqslant T_m - 1}$ be a self-financing strategy maturing at time T_m.*

(i) If $\Pi_{T_m}^{\mathbf{\Phi}} \geqslant 0$ holds, then $\Pi_t^{\mathbf{\Phi}} \geqslant 0$ holds for all $t \in \{0, 1, 2, \ldots, T_m\}$.

(ii) If $\Pi_{T_m}^{\mathbf{\Phi}} = 0$ holds (i.e., $\mathbf{\Phi}$ replicates the zero claim), then $\Pi_t^{\mathbf{\Phi}} = 0$ holds for all $t \in \{0, 1, 2, \ldots, T_m\}$.

Proof. Suppose that statement (i) is not true. Let t_* be the largest time $t \in \{0, 1, \ldots, T_m - 1\}$ such that $\Pi_t^{\mathbf{\Phi}} \geqslant 0$ does not hold. That is, there exists a state of the world $\omega_* \in \Omega$ for which $\Pi_{t_*}^{\mathbf{\Phi}}(\omega_*) < 0$. We also have that $\Pi_t^{\mathbf{\Phi}} \geqslant 0$ for all $t_* < t \leqslant T_m$. Let $A_* \in \mathcal{P}_{t_*}$ be the atom that contains the state ω_*. Since $\Pi_{t_*}^{\mathbf{\Phi}}$ is $\mathcal{F}_{t_*}$-measurable, it is constant (and negative) on the atom A_*. Define a new self-financing strategy $\mathbf{\Psi} = \{\psi_t\}_{0 \leqslant t \leqslant T_m - 1}$ as follows:

$$\psi_t(\omega) = \begin{cases} 0 & \text{if } 0 \leqslant t < t_* \text{ or } \omega \notin A_*, \\ \varphi_t(\omega) - \varphi_*(\omega) & \text{if } t_* \leqslant t < T_m \text{ and } \omega \in A_*, \end{cases}$$

where $\varphi_* = \left[\frac{\Pi_{t_*}^{\mathbf{\Phi}}}{B_{t_*}}, 0, \ldots, 0\right]$. The value process of the strategy $\mathbf{\Psi}$ is zero for $t < t_*$. For $t > t_*$, the value of the strategy is

$$\Pi_t[\psi_t] = \mathbb{I}_{A_*} \cdot \left(\Pi_t^{\mathbf{\Phi}} - \frac{\Pi_{t_*}^{\mathbf{\Phi}}}{B_{t_*}} B_t\right).$$

At time t_*, the self-financing condition holds:

$$\Pi_{t_*}[\psi_{t_*-1}] = 0 = \Pi_{t_*}[\psi_{t_*}] = \mathbb{I}_{A_*} \cdot \left(\Pi_{t_*}^{\mathbf{\Phi}} - \frac{\Pi_{t_*}^{\mathbf{\Phi}}}{B_{t_*}} B_{t_*}\right).$$

The value $\Pi_t^{\mathbf{\Psi}}$ is nonnegative for all $t > t_*$ since $\Pi_t^{\mathbf{\Phi}} \geqslant 0$ and $-\frac{\Pi_{t_*}^{\mathbf{\Phi}}}{B_{t_*}} B_t = \frac{|\Pi_{t_*}^{\mathbf{\Phi}}|}{B_{t_*}} B_t > 0$. Moreover, $\Pi_t^{\mathbf{\Psi}}(\omega) > 0$ for all $\omega \in A_*$ and all $t > t_*$. Therefore, $\mathbf{\Psi}$ is an arbitrage strategy for maturity T_m, contradicting the no-arbitrage assumption. Hence, our supposition is wrong and statement (i) holds.

Now, assume that $\Pi^{\Phi}_{T_m} = 0$. By statement (i), which has been just proved, we have that

$$\Pi^{\Phi}_{T_m} \geqslant 0 \implies \Pi^{\Phi}_t \geqslant 0 \text{ for all } t \in \mathbf{T}_m,$$

where $\mathbf{T}_m := \{0, 1, \ldots, T_m\}$. On the other hand $\Pi^{-\Phi}_{T_m} = 0$, hence

$$\Pi^{-\Phi}_{T_m} \geqslant 0 \implies \Pi^{-\Phi}_t \geqslant 0 \text{ for all } t \in \mathbf{T}_m \implies \Pi^{\Phi}_t \leqslant 0 \text{ for all } t \in \mathbf{T}_m.$$

Therefore, $\Pi^{\Phi}_t = 0$ for all $t \in \mathbf{T}_m$ and statement (ii) holds.   □

As follows from Lemma 8.2, the initial no-arbitrage price of a positive payoff is strictly positive. Moreover, any intermediate portfolio value Π_t replicating such a payoff is strictly positive at some atoms of the partition $\mathcal{P}_t$, otherwise an arbitrage opportunity exists. Recall that $X \in \mathrm{L}(\Omega)$ is said to be positive if $X(\omega) \geqslant 0$ for all $\omega \in \Omega$ and $X(\omega^*) > 0$ for at least one $\omega^* \in \Omega$.

A stronger version of the Law of One Price follows from Lemma 8.2.

Corollary 8.3. *Assume the absence of arbitrage. Suppose that two self-financing strategies Φ and Ψ both maturing at T_m have the same terminal value, i.e., $\Pi^{\Phi}_{T_m} = \Pi^{\Psi}_{T_m}$. Then $\Pi^{\Phi}_t = \Pi^{\Psi}_t$ holds for all $0 \leqslant t \leqslant T_m$.*

Proof. The new self-financing strategy $\Phi - \Psi$ replicates the zero claim maturing at T_m. As follows from statement (ii) of Lemma 8.2, all intermediate-time values of $\Phi - \Psi$ are zero. By using the linearity of the value function, we obtain

$$\Pi_t\left[\Phi - \Psi\right] = 0 \implies \Pi_t\left[\Phi\right] - \Pi_t\left[\Psi\right] = 0 \implies \Pi_t\left[\Phi\right] = \Pi_t\left[\Psi\right],$$

for all $0 \leqslant t \leqslant T_m$.   □

8.3.3 Equivalent Martingale Measures

Recall the concept of a *numéraire asset*. In a discrete-time model, a *numéraire* is any tradable asset g with a strictly positive price process, that is, $g_0 > 0$ and $\mathbb{P}(g_t > 0) = 1$ for all $t \in \{1, 2, \ldots, T\}$. The bank account B hence satisfies the above criteria and it will be our primary choice for a numéraire asset. In what follows we are assuming that all base assets, as well as the numéraire asset, pay no dividends.

Definition 8.2. A probability measure $\widetilde{\mathbb{P}}^{(g)}$ on the state space Ω is called an equivalent martingale measure (EMM) or a risk-neutral measure relative to numéraire g if

(1) $\widetilde{\mathbb{P}}^{(g)}$ is equivalent to the real-world probability distribution $\mathbb{P}$ (that is, $\widetilde{\mathbb{P}}^{(g)}(\omega) > 0$ for all ω from a finite state space Ω);

(2) for each base asset the discounted price process is a $\widetilde{\mathbb{P}}^{(g)}$-martingale, that is,

$$\widetilde{\mathrm{E}}^{(g)}\left[\frac{S^i_{t+1}}{g_{t+1}} \,\middle|\, \mathcal{F}_t\right] = \frac{S^i_t}{g_t},$$

for all $i \in \{0, 1, 2, \ldots, N\}$ and all $t \in \{0, 1, \ldots, T-1\}$.

In the case of dividend-paying base assets, the EMM $\widetilde{\mathbb{P}}^{(g)}$ is defined such that every ratio $\frac{\hat{S}^i_t}{\hat{g}_t}$ is a martingale under $\widetilde{\mathbb{P}}^{(g)}$, where $\hat{S}^i_t \equiv \frac{S^i_t}{c^i_t}$, $\hat{g}_t \equiv \frac{g_t}{c^g_t}$ with after-dividend factors c^i_t and c^g_t for the i-th base asset and numéraire g, respectively.

Our primary choice for a numéraire asset will be the bank account B. Since B_t discounted by B_t is identically one, a probability measure $\widetilde{\mathbb{P}} \equiv \widetilde{\mathbb{P}}^{(B)}$ is a martingale measure if the discounted base asset price processes $\{\overline{S}^i_t := \frac{S^i_t}{B_t}\}_{0 \leqslant t \leqslant T}$, $i = 1, 2, \ldots, N$, are all $\widetilde{\mathbb{P}}$-martingales. Let us review some properties of an EMM which were discussed in Chapters 5 and 7.

Lemma 8.4. *A probability measure $\widetilde{\mathbb{P}}^{(g)}$ is an EMM with a numéraire (non-dividend paying) asset g iff for any self-financing strategy, the value process discounted by g is a $\widetilde{\mathbb{P}}^{(g)}$ — martingale as well.*

Proof. Consider a self-financing strategy $\{\varphi_t = [\beta_t, \Delta_t^1, \ldots, \Delta_t^N] \, ; 0 \leqslant t \leqslant T - 1\}$, and let $\widetilde{\mathbb{P}}^{(g)}$ be an EMM for a given numéraire asset g. Define the discounted portfolio value process:

$$\overline{\Pi}_t := \frac{\Pi_t}{g_t}, \quad 0 \leqslant t \leqslant T.$$

Now, compute the risk-neutral expectation of $\overline{\Pi}_{t+1}$ conditional on $\mathcal{F}_t$, while using linearity, the fact that φ_t is $\mathcal{F}_t$-measurable, the $\widetilde{\mathbb{P}}^{(g)}$-martingale property of every asset discounted by numéraire g and the self-financing condition:

$$
\begin{aligned}
\widetilde{\mathbb{E}}_t^{(g)}[\overline{\Pi}_{t+1}] &= \widetilde{\mathbb{E}}_t^{(g)} \left[\beta_t \overline{B}_{t+1} + \Delta_t^1 \overline{S}_{t+1}^1 + \cdots + \Delta_t^N \overline{S}_{t+1}^N \right] \\
&= \beta_t \widetilde{\mathbb{E}}_t^{(g)} \left[\overline{B}_{t+1} \right] + \Delta_t^1 \widetilde{\mathbb{E}}_t^{(g)} \left[\overline{S}_{t+1}^1 \right] + \cdots + \Delta_t^N \widetilde{\mathbb{E}}_t^{(g)} \left[\overline{S}_{t+1}^N \right] \\
&= \beta_t \overline{B}_t + \Delta_t^1 \overline{S}_t^1 + \cdots + \Delta_t^N \overline{S}_t^N \\
&= \overline{\Pi}_t.
\end{aligned}
$$

That is, $\widetilde{\mathbb{E}}_t[\overline{\Pi}_{t+1}] = \overline{\Pi}_t$ for any $t \in \{0, 1, \ldots, T - 1\}$, which is the martingale condition for the discounted value process. The converse is obvious, since any base asset forms a static portfolio with only one nonzero position. Since a static portfolio is a particular case of a self-financing strategy, for each base asset the discounted price process is a $\mathbb{P}$-martingale. $\square$

The martingale property of self-financing replicating strategies is the key to pricing derivatives. Suppose that the EMM $\widetilde{\mathbb{P}}^{(g)}$ exists. The time-t no-arbitrage value V_t of an attainable payoff V_{T_m} has to be equal to the time-t cost of a strategy Φ that replicates V_{T_m}:

$$V_t = \Pi_t^{\Phi} \text{ for any time } t \text{ with } 0 \leqslant t \leqslant T_m \leqslant T \tag{8.16}$$

Then, thanks to the fact that the discounted portfolio value process $\{\overline{\Pi}_t^{\Phi}\}_{0 \leqslant t \leqslant T}$ is a $\widetilde{\mathbb{P}}^{(g)}$-martingale and (by assumed replication) $\Pi_t^{\Phi} = V_t$ holds for all $0 \leqslant t \leqslant T_m$, the price process is given by the risk-neutral conditional expectation of the payoff function V_{T_m}:

$$V_t = g_t \widetilde{\mathbb{E}}_t^{(g)} \left[\frac{\Pi_{T_m}^{\Phi}}{g_{T_m}} \right] = g_t \widetilde{\mathbb{E}}_t^{(g)} \left[\frac{V_{T_m}}{g_{T_m}} \right] \text{ or, equivalently, } \frac{V_t}{g_t} = \widetilde{\mathbb{E}}_t^{(g)} \left[\frac{V_{T_m}}{g_{T_m}} \right]. \tag{8.17}$$

In particular, we have the single-time-step recurrence pricing formula:

$$V_t = g_t \widetilde{\mathbb{E}}_t^{(g)} \left[\frac{V_{t+1}}{g_{t+1}} \right], \tag{8.18}$$

with time-t derivative price, V_t, expressed as an $\mathcal{F}_t$-measurable random variable. That is, the discounted derivative price process $\left\{ \overline{V}_t := \frac{V_t}{g_t} \right\}_{0 \leqslant t \leqslant T}$ is a $\widetilde{\mathbb{P}}^{(g)}$-martingale. This corresponds to the pricing formula (7.37) in Chapter 7 which we derived within the binomial market model with two base assets. In the case of dividend paying assets the above pricing formulas are valid if we replace g_t with $\hat{g}_t \equiv \frac{g_t}{c_t^g}$ for all t, where c_t^g is the time-t cumulative after-dividend factor for the numéraire. To apply the risk-neutral pricing formula (8.17) or (8.18) we only need to construct the EMM $\widetilde{\mathbb{P}}^{(g)}$. In the next section we show how this problem is reduced to the solution of an algebraic system of linear equations.

8.3.4   Calculation of Martingale Measures

Since the state space Ω is finite, the computation of the martingale measure $\widetilde{\mathbb{P}} \equiv \widetilde{\mathbb{P}}^{(B)}$ (relative to the numéraire $g = B$) reduces to computing $\widetilde{\mathbb{P}}(\omega)$ for each $\omega \in \Omega$. As was pointed out in the beginning of this chapter, any state of the world $\omega^j \in \Omega$ lies in a unique sequence of atoms, one from each partition $\mathcal{P}_t$:

$$\{\omega^j\} = A_T^j \subseteq A_{T-1}^{j_{T-1}} \subseteq \cdots \subseteq A_1^{j_1} \subseteq A_0^1 \equiv \Omega.$$

By using (8.2), we can calculate the risk-neutral probability $\widetilde{\mathbb{P}}(\omega^j)$ as a product of conditional probabilities:

$$\widetilde{\mathbb{P}}(\omega^j) = \widetilde{\mathbb{P}}(\{\omega^j\} \mid A_{T-1}^{j_{T-1}}) \cdot \widetilde{\mathbb{P}}(A_{T-1}^{j_{T-1}} \mid A_{T-2}^{j_{T-2}}) \cdots \widetilde{\mathbb{P}}(A_2^{j_2} \mid A_1^{j_1}) \cdot \widetilde{\mathbb{P}}(A_1^{j_1}).$$

Thus, our goal is to calculate the conditional probability $\widetilde{\mathbb{P}}(A_{t+1} \mid A_t)$ for every pair of atoms $A_t \in \mathcal{P}_t$ and $A_{t+1} \in \mathcal{P}_{t+1}$ with $A_{t+1} \subseteq A_t$. These are essentially one-step transition probabilities in the risk-neutral measure $\widetilde{\mathbb{P}}$. Fix arbitrarily time $t \in \{0, 1, \ldots, T-1\}$ and atom $A_t \in \mathcal{P}_t$. The partition $\mathcal{P}_{t+1}$ is obtained by refining the partition $\mathcal{P}_t$. Suppose that the atom A_t is a union of $\ell \geqslant 1$ atoms

$$A_{t+1}^1, A_{t+1}^2, \ldots, A_{t+1}^\ell \in \mathcal{P}_{t+1}. \tag{8.19}$$

Let us find ℓ conditional probabilities $\widetilde{\mathbb{P}}(A_{t+1}^j \mid A_t)$, $j = 1, 2, \ldots, \ell$, such that for each (non-dividend-paying) base asset S^i the discounted price process satisfies the martingale condition

$$\widetilde{\mathrm{E}}_t\left[\bar{S}_{t+1}^i\right] = \bar{S}_t^i, \ t \geqslant 0.$$

[Note that in the case of dividend-paying assets we replace each $\bar{S}_t^i$ by $\hat{\bar{S}}_t^i \equiv \bar{S}_t^i / c_t^i$ where c_t^i is the time-t cumulative after-dividend factor for the i-th base asset.] Since $\bar{S}_t^i$ and $\widetilde{\mathrm{E}}_t\left[\bar{S}_{t+1}^i\right]$ are both $\mathcal{F}_t$-measurable random variables, they are constant on atoms of $\mathcal{P}_t$. Therefore, for the atom A_t, we can write $N + 1$ equations, one for each base asset:

$$\bar{S}_t^i(A_t) = \widetilde{\mathrm{E}}_t\left[\bar{S}_{t+1}^i\right](A_t) = \widetilde{\mathrm{E}}\left[\bar{S}_{t+1}^i \mid A_t\right] = \sum_{j=1}^\ell \bar{S}_{t+1}^i(A_{t+1}^j)\, \widetilde{\mathbb{P}}(A_{t+1}^j \mid A_t), \tag{8.20}$$

$i = 0, 1, 2, \ldots, N$. For $i = 0$, equation (8.20) reduces to

$$1 = \sum_{j=1}^\ell \widetilde{\mathbb{P}}(A_{t+1}^j \mid A_t), \tag{8.21}$$

since $\bar{S}_t^0 \equiv 1$. The above equation means that $\widetilde{\mathbb{P}}(\,\cdot\,\mid A_t)$ is a probability measure on the state space $\widehat{\Omega} = \left\{\widehat{\omega}^j := A_{t+1}^j;\ j = 1, 2, \ldots, \ell\right\}$. Thus, for each atom A_t at time t, we obtain a linear system of $N + 1$ equations with ℓ unknowns $q_j = \widetilde{\mathbb{P}}(A_{t+1}^j \mid A_t)$, $j = 1, 2, \ldots, \ell$,

$$\begin{cases} \sum_{j=1}^\ell \bar{S}_{t+1}^1(A_{t+1}^j) q_j = \bar{S}_t^1(A_t), \\ \ \vdots \\ \sum_{j=1}^\ell \bar{S}_{t+1}^N(A_{t+1}^j) q_j = \bar{S}_t^N(A_t), \\ \sum_{j=1}^\ell q_j = 1. \end{cases} \tag{8.22}$$

Such a system needs to be solved for every atom $A_t = A_t^k$, $k = 1, \ldots, k_t$, in each partition $\mathcal{P}_t = \mathcal{P}_0, \mathcal{P}_1, \ldots, \mathcal{P}_{T-1}$. In total, there are $1 + k_1 + k_2 + \cdots + k_{T-1}$ systems like (8.22) where k_t is the number of atoms in $\mathcal{P}_t$. Solving all these systems gives us all the conditional probabilities of the form $\widetilde{\mathbb{P}}(A_{t+1} \mid A_t)$. Note that for $t = 0$, we have only one atom $A_0 \equiv A_0^1 \equiv \Omega$; hence,

$$q_j = \widetilde{\mathbb{P}}(A_{t+1}^j \mid A_t) = \widetilde{\mathbb{P}}(A_1^j \mid \Omega) = \widetilde{\mathbb{P}}(A_1^j), \quad j = 1 \ldots, k_1.$$

As a result, we can calculate $\widetilde{\mathbb{P}}(\omega)$ for every state ω (and hence the risk-neutral probability of any event) using equation (8.2). Let us illustrate this method of computation of a martingale measure by the following example.

Example 8.5. Consider the two-period trinomial model with three base assets as given in Examples 8.3 and 8.4. Determine the EMM $\widetilde{\mathbb{P}} \equiv \widetilde{\mathbb{P}}^{(B)}$.

Solution. Note that $T = 2$ (two periods) and $N = 2$ ($N + 1 = 3$ base assets including the bank account). Since the interest rate is zero, discounting does not change the stock prices. First, let us determine the probabilities $q_j := \widetilde{\mathbb{P}}(A_1^j)$, $j = 1, 2, 3$, for atoms of $\mathcal{P}_1$ by solving the following system of three equations (8.22):

$$\begin{cases} 50q_1 + 40q_2 + 60q_3 = 50 \\ 115q_1 + 115q_2 + 70q_3 = 100 \\ q_1 + q_2 + q_3 = 1 \end{cases} \iff \begin{bmatrix} q_1 \\ q_2 \\ q_3 \end{bmatrix} = \begin{bmatrix} 1/3 \\ 1/3 \\ 1/3 \end{bmatrix}.$$

Here we have used the stock prices in the first two columns of the table in Example 8.4: $S_1^1(A_1^1) = 50$, $S_1^1(A_1^2) = 40$, $S_1^1(A_1^3) = 60$ and $S_1^2(A_1^1) = S_1^2(A_1^2) = 115$, $S_1^2(A_1^3) = 70$. At time $t = 1$, there are three trinomial sub-models, hence we need to solve three systems of linear equations to obtain the conditional probabilities $q_j^k := \widetilde{\mathbb{P}}(A_2^j \mid A_1^k)$ for $j = 1, 2, \ldots, 9$ and $k = 1, 2, 3$. By using the last two columns of the table in Example 8.4, equation (8.22) gives (note: $A_2^j = \omega_j, j = 1, \ldots, 9$):

$$\begin{cases} 45q_1^1 + 45q_2^1 + 60q_3^1 = 50 \\ 130q_1^1 + 120q_2^1 + 95q_3^1 = 115 \\ q_1^1 + q_2^1 + q_3^1 = 1 \end{cases} \iff \begin{bmatrix} q_1^1 \\ q_2^1 \\ q_3^1 \end{bmatrix} = \begin{bmatrix} 1/3 \\ 1/3 \\ 1/3 \end{bmatrix},$$

$$\begin{cases} 45q_4^2 + 35q_5^2 + 40q_6^2 = 40 \\ 135q_4^2 + 120q_5^2 + 105q_6^2 = 115 \\ q_4^2 + q_5^2 + q_6^2 = 1 \end{cases} \iff \begin{bmatrix} q_4^2 \\ q_5^2 \\ q_6^2 \end{bmatrix} = \begin{bmatrix} 2/9 \\ 2/9 \\ 5/9 \end{bmatrix},$$

$$\begin{cases} 60q_7^3 + 55q_8^3 + 70q_9^3 = 60 \\ 100q_7^3 + 40q_8^3 + 70q_9^3 = 70 \\ q_7^3 + q_8^3 + q_9^3 = 1 \end{cases} \iff \begin{bmatrix} q_7^3 \\ q_8^3 \\ q_9^3 \end{bmatrix} = \begin{bmatrix} 2/5 \\ 2/5 \\ 1/5 \end{bmatrix}.$$

All the above systems have strictly positive solutions, i.e., all conditional (transition) probabilities are strictly positive. Hence, the EMM exists and the model is arbitrage-free. The state probabilities can be computed as follows:

$$\tilde{p}_j \equiv \widetilde{\mathbb{P}}(\omega^j) = \widetilde{\mathbb{P}}(A_2^j \mid A_1^k) \cdot \widetilde{\mathbb{P}}(A_1^k) = q_j^k \cdot q_k,$$

where $k = 1$ if $j = 1, 2, 3$, $k = 2$ if $j = 4, 5, 6$, and $k = 3$ if $j = 7, 8, 9$. Hence, we obtain

$$\tilde{p}_1 = \frac{1}{9}, \ \tilde{p}_2 = \frac{1}{9}, \ \tilde{p}_3 = \frac{1}{9}, \ \tilde{p}_4 = \frac{2}{27}, \ \tilde{p}_5 = \frac{2}{27}, \ \tilde{p}_6 = \frac{5}{27}, \ \tilde{p}_7 = \frac{2}{15}, \ \tilde{p}_8 = \frac{2}{15}, \ \tilde{p}_9 = \frac{1}{15}.$$

The final solution is illustrated in Figure 8.2. $\qquad\qquad\square$

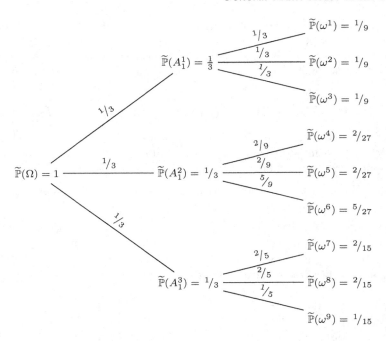

FIGURE 8.2: A trinomial tree with risk-neutral transition probabilities and state probabilities.

Example 8.6. Consider the two-period trinomial model with three base assets as given in Examples 8.3 and 8.4. Determine the no-arbitrage prices V_0 and V_1 of a basket call option maturing at time $T = 2$ with the payoff $V_2(\omega) = (\max\{2S_2^1(\omega), S_2^2(\omega)\} - 120)^+$, $\omega \in \Omega$.

Solution. Firstly, let us calculate the payoff function $V_2(\omega)$ for all possible scenarios ω as follows:

ω	ω^1	ω^2	ω^3	ω^4	ω^5	ω^6	ω^7	ω^8	ω^9
$2S_2^1(\omega)$	90	90	120	90	70	80	120	110	140
$S_2^2(\omega)$	130	120	95	135	120	105	100	40	70
$\max\{2S_2^1(\omega), S_2^2(\omega)\}$	130	120	120	135	120	105	120	110	140
$V_2(\omega)$	10	0	0	15	0	0	0	0	20

Applying the pricing formula (8.17) to the payoff V_2, while using $g_t = B_t \equiv 1$ (with assumed zero interest rate) gives

$$V_0 = B_0 \sum_{j=1}^{9} \frac{V_2(\omega^j)}{B_2(\omega^j)} \widetilde{p}_j = \sum_{j=1}^{9} V_2(\omega^j) \widetilde{p}_j$$

$$= \frac{1}{9} \cdot 10 + \frac{1}{9} \cdot 0 + \frac{1}{9} \cdot 0 + \frac{2}{27} \cdot 15 + \frac{2}{27} \cdot 0 + \frac{5}{27} \cdot 0 + \frac{2}{15} \cdot 0 + \frac{2}{15} \cdot 0 + \frac{1}{15} \cdot 20$$

$$= \frac{32}{9} \cong 3.556.$$

The time $t = 1$ price V_1 is an $\mathcal{F}_1$-measurable random variable. It is constant on the atoms A_1^i, $i = 1, 2, 3$, of $\mathcal{P}_1$. Using (8.17) for one backward step $t + 1 = 2$ to $t = 1$ and the fact that $B_1 \equiv B_2 \equiv 1$ gives

$$V_1(A_1^i) = B_1(A_1^i) \sum_{j=1}^{9} \frac{V_2(\omega^j)}{B_2(\omega^j)} \widetilde{\mathbb{P}}(\{\omega^j\} \mid A_1^i) = \sum_{j=1}^{9} V_2(\omega^j) \widetilde{\mathbb{P}}(\{\omega^j\} \mid A_1^i) \text{ for } i = 1, 2, 3.$$

The above sum is restricted to values of $j = 1, 2, 3$ if $i = 1$; $j = 4, 5, 6$ if $i = 2$, $j = 7, 8, 9$ if $i = 3$. Applying this formula for each $i = 1, 2, 3$ gives

$$V_1(A_1^1) = 10 \cdot \frac{1}{3} + 0 \cdot \frac{1}{3} + 0 \cdot \frac{1}{3} = \frac{10}{3} \cong 3.333,$$

$$V_1(A_1^2) = 15 \cdot \frac{2}{9} + 0 \cdot \frac{2}{9} + 0 \cdot \frac{5}{9} = \frac{10}{3} \cong 3.333,$$

$$V_1(A_1^3) = 0 \cdot \frac{2}{5} + 0 \cdot \frac{2}{5} + 20 \cdot \frac{1}{5} = 4.$$

Note that V_0 can also be obtained by using (8.17) for $t + 1 = 1$ to $t = 0$:

$$V_0 = \frac{1}{3} \cdot V_1(A_1^1) + \frac{1}{3} \cdot V_1(A_1^2) + \frac{1}{3} \cdot V_1(A_1^3) = \frac{1}{3} \cdot \frac{10}{3} + \frac{1}{3} \cdot \frac{10}{3} + \frac{1}{3} \cdot 4 = \frac{32}{9}.$$

$\square$

8.3.5 The First and Second FTAP

The first and second fundamental theorems of asset pricing (FTAP) for a multi-period model are formulated in exactly the same way as those for a single-period model. The absence of arbitrage is equivalent to the existence of an equivalent martingale measure. Moreover, under the no-arbitrage assumption, the martingale measure is unique iff the model is complete, i.e., every payoff can be replicated by a self-financing strategy. For simplicity of presentation, we assume that the bank account B serves as a numéraire asset, and the (equivalent) martingale measure $\widetilde{\mathbb{P}} \equiv \widetilde{\mathbb{P}}^{(B)}$ is constructed relative to such a numéraire.

In Chapter 5, we proved the first and second FTAP for a single-period model. Let us prove the theorems by reducing the multi-period model to a single-period case. As was discussed in the beginning of this section, a multi-period model with a finite number of states can be viewed as a union of a finite number of single-period sub-models, where the states of the world are atoms of the partitions $\mathcal{P}_0, \mathcal{P}_1, \dots, \mathcal{P}_T$. Let us consider a sub-model originated at some atom $A_t \in \mathcal{P}_t$ with $t \in \{0, 1, \dots, T-1\}$. Suppose that the atom A_t is a union of ℓ atoms as listed in (8.19). The multi-period model can be represented by a multinomial tree (see Figure 8.1), where every scenario $\omega \in A_t$ can be viewed as a path that goes through the atom A_t to one of the atoms from the list in (8.19). Such a single-step transition from time t to $t + 1$ can be described by a single-period model with the state space

$$\widehat{\Omega} = \{\widehat{\omega}^1 := A_{t+1}^1, \ \widehat{\omega}^2 := A_{t+1}^2, \ \dots, \ \widehat{\omega}^\ell := A_{t+1}^\ell\}. \tag{8.23}$$

The state probabilities are given by the conditional probabilities of going from A_t to one of its ℓ successors $A_{t+1}^1, A_{t+1}^2, \dots, A_{t+1}^\ell$:

$$\widehat{\mathbb{P}}(\widehat{\omega}^j) := \mathbb{P}\left(A_{t+1}^j \mid A_t\right) = \frac{\mathbb{P}(A_{t+1}^j)}{\mathbb{P}(A_t)}, \quad j = 1, 2, \dots, \ell, \tag{8.24}$$

since $A_{t+1}^j \cap A_t = A_{t+1}^j$. The single-period sub-model obtained is illustrated in Figure 8.3.

To complete the construction of a single-period model, we construct an initial price vector $\widehat{\mathbf{S}}_0$ and payoff matrix $\widehat{\mathbf{D}}$. Since the base asset price process is adapted to the filtration $\mathbb{F}$, the

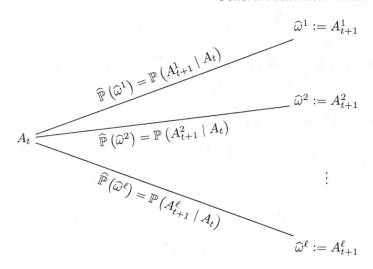

FIGURE 8.3: A single-period sub-model.

prices $B_t, S_t^1, \ldots, S_t^N$ are constant on an atom $A_t \in \mathcal{P}_t$, and the prices $B_{t+1}, S_{t+1}^1, \ldots, S_{t+1}^N$ are constant on each atom $A_{t+1}^1, \ldots, A_{t+1}^\ell$. Therefore, the initial price vector and payoff matrix are given by values of the base assets respectively calculated on A_t and the atoms of (8.19):

$$\widehat{\mathbf{S}}_0 = \left[B_t(A_t),\, S_t^1(A_t),\, \cdots,\, S_t^N(A_t) \right]^\top, \tag{8.25}$$

$$\widehat{\mathbf{D}} = \begin{bmatrix} B_{t+1}(A_{t+1}^1) & B_{t+1}(A_{t+1}^2) & \cdots & B_{t+1}(A_{t+1}^\ell) \\ S_{t+1}^1(A_{t+1}^1) & S_{t+1}^1(A_{t+1}^2) & \cdots & S_{t+1}^1(A_{t+1}^\ell) \\ \vdots & \vdots & \ddots & \vdots \\ S_{t+1}^N(A_{t+1}^1) & S_{t+1}^N(A_{t+1}^2) & \cdots & S_{t+1}^N(A_{t+1}^\ell) \end{bmatrix}. \tag{8.26}$$

As a result, we obtain a single-period model with the state space $\widehat{\Omega}$, probability distribution function $\widehat{\mathbb{P}}$, initial price vector $\widehat{\mathbf{S}}_0$, and payoff matrix $\widehat{\mathbf{D}}$. Recall that there are $k_0 = 1$ sub-models at time 0, k_1 sub-models at time 1, and so on. In total, we have $k_0 + k_1 + \cdots + k_{T-1}$ single-period sub-models.

The proof of the FTAP is split into several steps. First, we prove that there are no arbitrage strategies in the multi-period model iff there are no arbitrage portfolios in each single-period sub-model (Lemma 8.5). We also prove a similar result about the completeness of models (Lemma 8.6). Second, we prove that there exists a (unique) martingale measure for the multi-period model iff each sub-model has a (unique) martingale measure (Lemma 8.7).

Lemma 8.5. *There are no arbitrage opportunities in a multi-period model iff each single-period sub-model is arbitrage-free.*

Proof. Let us prove that if each single-period sub-model is arbitrage-free, then there are no arbitrage strategies. Suppose there exists an admissible arbitrage strategy $\boldsymbol{\Phi}$ maturing at time T_m. Let $t \geqslant 0$ be the largest time such that $\Pi_t^{\boldsymbol{\Phi}} \equiv 0$ and $\Pi_{t+1}^{\boldsymbol{\Phi}} > 0$. Such time t exists since the initial value $\Pi_0^{\boldsymbol{\Phi}}$ is zero and the terminal value $\Pi_{T_m}^{\boldsymbol{\Phi}}$ is positive. By construction, the $(t+1)$-time value of the strategy is positive for some outcome $\omega^j \in \Omega$, that is, $\Pi_{t+1}^{\boldsymbol{\Phi}}(\omega^j) > 0$. Let ω^j be contained in a chain of atoms:

$$\{\omega^j\} \subseteq A_{T-1}^{j_{T-1}} \subseteq \cdots \subseteq A_{t+1}^{j_{t+1}} \subseteq A_t^{j_t} \subseteq \cdots \subseteq A_0^1 = \Omega.$$

The value process is adapted to the filtration $\mathbb{F}$. Therefore, the strategy values Π_t^{Φ} and Π_{t+1}^{Φ} are constant on $A_t^{j_t}$ and $A_{t+1}^{j_{t+1}}$, respectively. Consider the single-period sub-model originated at atom $A_t^{j_t}$ and the portfolio $\psi := \varphi_t(\omega^j)$. Since $\Pi_t[\varphi_t] \equiv 0$, the initial value of ψ is zero. As $\Pi_{t+1}[\varphi_t](\omega^j) > 0$, the terminal value of ψ is strictly positive on the atom $A_{t+1}^{j_{t+1}}$. Therefore, ψ is an arbitrage portfolio in the sub-model originated at $A_t^{j_t}$. We arrive at a contradiction. Hence, the multi-period model is arbitrage-free. The converse is obvious since every static portfolio in a single-period sub-model is a special case of a dynamic strategy. Indeed, consider a single-period sub-model originated at some atom $A_t^j \in \mathcal{P}_t$ and a portfolio ψ in base assets with initial value zero. Construct a strategy maturing at time $t+1$ as follows: $\varphi_s \equiv \mathbf{0}$ for all $s \in \{0, 1, \ldots, t-1\}$, $\varphi_t(\omega) = \psi$ if $\omega \in A_t^j$ and $\varphi_t(\omega) = \mathbf{0}$ otherwise. The strategy is self-financing. If the portfolio ψ is an arbitrage opportunity, then $\Pi_{t+1}[\varphi_t] > 0$ or, in other words, it is an arbitrage strategy that contradicts the assumption. $\square$

Lemma 8.6. *A multi-period model is complete iff each single-period sub-model is complete.*

Proof. The proof is left as an exercise for the reader. $\square$

Lemma 8.7. *There exists a (unique) martingale measure for a multi-period model iff each single-period sub-model has a (unique) martingale measure.*

Proof. Being given the risk-neutral probability distribution function for each one-period sub-model, we have the conditional risk-neutral probabilities of the form $\widetilde{\mathbb{P}}(A_{t+1} \mid A_t)$ for all $t \in \{0, 1, \ldots, T-1\}$ and for any two atoms $A_t \in \mathcal{P}_t$, $A_{t+1} \in \mathcal{P}_{t+1}$ such that $A_{t+1} \subseteq A_t$. Using the path probability formula (8.2), we can then calculate the risk-neutral probability $\widetilde{\mathbb{P}}(\omega)$ for each outcome $\omega \in \Omega$. If the risk-neutral probability measure $\widetilde{\mathbb{P}}$ is known, the state probabilities for each sub-model can be computed by (8.24) where $\widetilde{\mathbb{P}}$ is used in place of $\mathbb{P}$. As usual, in the case with a finite probability space $(\Omega, \mathbb{P})$, the probability of event $E \subseteq \Omega$ is calculated as $\mathbb{P}(E) = \sum_{\omega \in E} \mathbb{P}(\omega)$. Clearly, the probability distribution is unique for each sub-model iff the probability distribution $\widetilde{\mathbb{P}}$ defined on Ω is unique. Moreover, the base asset price processes discounted by the numéraire asset B, $\{\overline{S}_t^i = \frac{S_t^i}{B_t}\}_{0 \leqslant t \leqslant T}$, are all martingales iff

$$\widetilde{\mathrm{E}}_t\left[\overline{S}_{t+1}^i\right](A_t) = \overline{S}_t^i(A_t) \tag{8.27}$$

holds for every time $t \in \{0, 1, \ldots, T-1\}$, all $i = 1, 2, \ldots, N$, and each atom $A_t \in \mathcal{P}_t$. Equation (8.27) is nothing but a single-period martingale condition. $\square$

8.3.6 Pricing and Hedging Derivatives

The pricing formula (8.17) allows us to calculate the price process for any derivative with an attainable payoff V_{T_m} at maturity T_m. Since for every time t the derivative price V_t is equal to the conditional expectation of V_{T_m} given the σ-algebra $\mathcal{F}_t$, this price is $\mathcal{F}_t$-measurable (hence, the derivative price process $\{V_t\}_{0 \leqslant t \leqslant T_m}$ is $\mathbb{F}$-adapted). By construction, any $\mathcal{F}_t$-measurable random variable is constant on atoms of the partition $\mathcal{P}_t$, where $\mathcal{F}_t = \sigma(\mathcal{P}_t)$. So we only need to calculate the derivative price $V_t(A_t)$ for every atom $A_t \in \mathcal{P}_t$. Suppose that A_t is a union of ℓ atoms from $\mathcal{P}_{t+1}$ as listed in (8.19). Since the discounted derivative price process is a $\widetilde{\mathbb{P}}$-martingale, $\overline{V}_t(A_t)$ is equal to the risk-neutral expectation of $\overline{V}_{t+1}$,

$$\overline{V}_t(A_t) = \widetilde{\mathrm{E}}_t\left[\overline{V}_{t+1}\right](A_t) = \widetilde{\mathrm{E}}\left[\overline{V}_{t+1} \mid A_t\right] = \sum_{j=1}^{\ell} \overline{V}_{t+1}(A_{t+1}^j)\,\widetilde{\mathbb{P}}(A_{t+1}^j \mid A_t)$$

where $\overline{V}_t = V_t/B_t$, $B_t/B_{t+1} = 1/(1+r_{t+1})$. Note that here we are using the bank account as numéraire asset. Therefore, the derivative prices can be computed using a backward-in-time recursion as follows:

$$V_t(A_t) = \frac{1}{1 + r_{t+1}(A_t)} \sum_{j=1}^{\ell} V_{t+1}(A_{t+1}^j)\, \widetilde{\mathbb{P}}(A_{t+1}^j \mid A_t) \tag{8.28}$$

for all $A_t \in \mathcal{P}_t$, $t = 0, 1, \ldots, T_m - 1$. Note that r_{t+1} is constant on any $A_t \in \mathcal{P}_t$. At maturity T_m, the $\mathcal{F}_{T_m}$-measurable payoff $V_{T_m}(A_{T_m})$ is known for every atom $A_{T_m} \in \mathcal{P}_{T_m}$. Using the above equation, we calculate the derivative prices $V_t(A_t)$ for all $A_t \in \mathcal{P}_t$ and for $t = T_m - 1, T_m - 2, \ldots, 0$. In the end, we find the initial price V_0. In total, to obtain a complete derivative price process, we need to calculate $1 + k_1 + \cdots + k_{T_m-1}$ conditional expectations. The same approach was used to price derivatives in the standard binomial tree model. Indeed, (8.28) reduces to (7.26) where each atom $A_t = A_{\omega_1 \ldots \omega_t}$ is a union of two atoms, e.g., $A_{t+1}^1 = A_{\omega_1 \ldots \omega_t U}$ and $A_{t+1}^2 = A_{\omega_1 \ldots \omega_t D}$.

Knowing the derivative prices $V_t(A_t)$ for all $t \in \{0, 1, \ldots, T_m\}$ and all $A_t \in \mathcal{P}_t$, we can find the self-financing portfolio strategy $\mathbf{\Phi} = \{\varphi_t\}_{0 \leqslant t \leqslant T_m-1}$ hedging the derivative. Suppose that the market model is arbitrage free and complete. Fix arbitrarily time $t \in \{0, 1, \ldots, T_m - 1\}$ and an atom $A_t \in \mathcal{P}_t$. The portfolio $\varphi_t(A_t)$ can be found by solving a single-period replication problem. Consider the one-period sub-model originating at the atom A_t. The state space $\widehat{\Omega}$, initial price vector $\widehat{\mathbf{S}}_0$, and payoff matrix $\widehat{\mathbf{D}}$ are given in (8.23), (8.25), and (8.26), respectively. Determine a portfolio in the base assets replicating the payoff vector $V_{t+1}(\widehat{\Omega}) = \left[V_{t+1}(A_{t+1}^1), V_{t+1}(A_{t+1}^2), \ldots, V_{t+1}(A_{t+1}^\ell)\right]$. The portfolio vector is a solution to the matrix-vector equation $\boldsymbol{\psi}\widehat{\mathbf{D}} = V_{t+1}(\widehat{\Omega})$. Set $\varphi_t(A_t) = \boldsymbol{\psi}$, i.e., equal to the replicating portfolio obtained. In particular, for each given time-t atom A_t we have $\varphi_t(A_t) = [\beta_t(A_t), \Delta_t^1(A_t), \ldots, \Delta_t^N(A_t)]$ where $\beta_t(A_t)$ is the position in the bank account and $\Delta_t^i(A_t)$ is the (delta hedge) position in the i-th asset (stock). At each time-t atom A_t the replicating portfolio positions solve the matrix equation:

$$\begin{bmatrix} B_{t+1}(A_{t+1}^1) & S_{t+1}^1(A_{t+1}^1) & \cdots & S_{t+1}^N(A_{t+1}^1) \\ B_{t+1}(A_{t+1}^2) & S_{t+1}^1(A_{t+1}^2) & \cdots & S_{t+1}^N(A_{t+1}^2) \\ \vdots & \vdots & \ddots & \vdots \\ B_{t+1}(A_{t+1}^\ell) & S_{t+1}^1(A_{t+1}^\ell) & \cdots & S_{t+1}^N(A_{t+1}^\ell) \end{bmatrix} \begin{bmatrix} \beta_t(A_t) \\ \Delta_t^1(A_t) \\ \vdots \\ \Delta_t^N(A_t) \end{bmatrix} = \begin{bmatrix} V_{t+1}(A_{t+1}^1) \\ V_{t+1}(A_{t+1}^2) \\ \vdots \\ V_{t+1}(A_{t+1}^\ell) \end{bmatrix}. \tag{8.29}$$

By construction, the initial cost to set up this portfolio is equal to

$$V_t(A_t) = \varphi_t(A_t)\widehat{\mathbf{S}}_0 = \beta_t(A_t)B_t(A_t) + \Delta_t^1(A_t)S_t^1(A_t) + \cdots + \Delta_t^N(A_t)S_t^N(A_t).$$

Repeat these steps for all t and all for all atoms $A_t \in \mathcal{P}_t$ to obtain a self-financing portfolio strategy $\mathbf{\Phi}$ that replicates the derivative process, i.e., $V_t^{\mathbf{\Phi}} = V_t$ for all $t = 0, 1, \ldots, T_m$.

8.3.7 Pricing under the Markov Property

As was discussed in Chapter 7, the Markov property can be used to simplify (or re-organize) the pricing and hedging of standard as well as path-dependent derivatives. In particular, let us consider a standard European multi-stock derivative whose value at maturity T_m, $0 \leqslant T_m \leqslant T$, is a given (payoff) function of N stock prices, $V_{T_m} = \Lambda(S_{T_m}^1, \ldots, S_{T_m}^N)$. Assume the joint multi-stock process $(S_t^1, \ldots, S_t^N)_{t=0,1,\ldots,T}$ is Markov. That is, given a Borel function $f : \mathbb{R}^N \to \mathbb{R}$ and the natural filtration $\{\mathcal{F}_t\}_{t \geqslant 0}$ for the multi-stock process, we have

$$\mathrm{E}\left[f(S_{t+1}^1, \ldots, S_{t+1}^N)|\mathcal{F}_t\right] = h(S_t^1, \ldots, S_t^N) \tag{8.30}$$

where $h(S_t^1, \ldots, S_t^N) := \mathrm{E}\left[h(S_{t+1}^1, \ldots, S_{t+1}^N)|S_t^1, \ldots, S_t^N\right]$ is the expectation conditional on the stock prices at time t. Note that $h(S_t^1, \ldots, S_t^N)$ is $\sigma(S_t^1, \ldots, S_t^N)$-measurable. Assume a (non-dividend paying) numéraire price process $g_t = g_t(S_t^1, \ldots, S_t^N)$. Hence, a direct application of the Markov property within the recurrence formula in (8.18) for $t = T_m - 1$, where $V_{t+1} \equiv V_{T_m} = \Lambda(S_{T_m}^1, \ldots, S_{T_m}^N)$, gives $V_{T_m-1} = V_{T_m-1}(S_{T_m-1}^1, \ldots, S_{T_m-1}^N)$. Similarly, by repeated application we obtain $V_t = V_t(S_t^1, \ldots, S_t^N)$, $0 \leqslant t \leqslant T_m$, i.e., the $\sigma(S_t^1, \ldots, S_t^N)$-measurable prices V_t as functions of $S_t^1, \ldots, S_t^N$. In particular, (8.18) gives

$$V_t(S_t^1, \ldots, S_t^N) = g_t(S_t^1, \ldots, S_t^N)\widetilde{\mathrm{E}}^{(g)}\left[\frac{V_{t+1}(S_{t+1}^1, \ldots, S_{t+1}^N)}{g_{t+1}(S_{t+1}^1, \ldots, S_{t+1}^N)}\Bigg| S_t^1, \ldots, S_t^N\right], \qquad (8.31)$$

for each $t = 0, 1, \ldots, T_m - 1$. At each time $t = 0, 1, \ldots, T_m$ the stock prices have a given support: $S_t^1 \in \{S_{t,1}^1, \ldots, S_{t,\kappa_1(t)}^1\}$, $S_t^2 \in \{S_{t,1}^2, \ldots, S_{t,\kappa_2(t)}^2\}, \ldots, S_t^N \in \{S_{t,1}^N, \ldots, S_{t,\kappa_N(t)}^N\}$, where $\kappa_i(t)$ is the number of time-t distinct values (nodes) of S_t^i, $i = 1, \ldots, N$. The unique derivative prices on each node in the multi-stock tree are therefore given by

$$V_t(S_{t,j_1}^1, \ldots, S_{t,j_N}^N)$$
$$= g_t(S_{t,j_1}^1, \ldots, S_{t,j_N}^N) \sum_{j_1', \ldots, j_N'} \widetilde{p}_{(t;j_1,\ldots,j_N)\to(t+1;j_1',\ldots,j_N')}^{(g)} \frac{V_{t+1}(S_{t+1,j_1'}^1, \ldots, S_{t+1,j_N'}^N)}{g_{t+1}(S_{t+1,j_1'}^1, \ldots, S_{t+1,j_N'}^N)} \qquad (8.32)$$

where

$$\widetilde{p}_{(t;j_1,\ldots,j_N)\to(t+1;j_1',\ldots,j_N')}^{(g)}$$
$$:= \widetilde{\mathbb{P}}^{(g)}\left(S_{t+1}^1 = S_{t+1,j_1'}^1, \ldots, S_{t+1}^N = S_{t+1,j_N'}^N \mid S_t^1 = S_{t,j_1}^1, \ldots, S_t^N = S_{t,j_N}^N\right)$$

are transition probabilities in the EMM $\widetilde{\mathbb{P}}^{(g)}$.

If we choose the bank account as numéraire, then $g_t/g_{t+1} = B_t/B_{t+1} = (1 + r_{t+1})^{-1} = (1 + R_t)^{-1}$, $r_{t+1} := R_t$. If we assume the single-period interest rates are given functions of the stock prices, $R_t = R_t(S_t^1, \ldots, S_t^N)$, then

$$V_t(S_{t,j_1}^1, \ldots, S_{t,j_N}^N)$$
$$= \frac{1}{1 + R_t(S_{t,j_1}^1, \ldots, S_{t,j_N}^N)} \sum_{j_1', \ldots, j_N'} \widetilde{p}_{(t;j_1,\ldots,j_N)\to(t+1;j_1',\ldots,j_N')} V_{t+1}(S_{t+1,j_1'}^1, \ldots, S_{t+1,j_N'}^N)$$
$$(8.33)$$

where $\widetilde{p}_{(t;j_1,\ldots,j_N)\to(t+1;j_1',\ldots,j_N')}$ are transition probabilities in the EMM $\widetilde{\mathbb{P}} \equiv \widetilde{\mathbb{P}}^{(B)}$. Note that in the case of constant interest rates, $R_t(S_t^1, \ldots, S_t^N) \equiv r$ in (8.33). The number of nonzero transition probabilities (i.e., the number of terms in the above sums) for a given time step $t \to t+1$ is a relatively small number. For example, in the case of a re-combining binomial tree for a single stock ($N = 1$) we have each time-t node (t, j), with price $S_{t,j}$, giving rise to only two nodes labelled by $(t+1, j)$ and $(t+1, j+1)$, with prices $S_{t+1,j}$ and $S_{t+1,j+1}$. In a trinomial model we have three single-step nonzero transition probabilities for each node, and so on.

The above backward recurrence relations can be used when the interest rate is either constant or is stochastic as a function of the stock prices. More generally, we may have a payoff $V_{T_m} = \Lambda(S_{T_m}^1, \ldots, S_{T_m}^N, R_{T_m})$ as a function of the stock prices and possibly the interest rate. The interest rate is now assumed stochastic where $(S_t^1, \ldots, S_t^N, R_t)_{t=0,1,\ldots,T}$ is a joint Markov process. The derivative price process, $V_t = V_t(S_t^1, \ldots, S_t^N, R_t)$, is now a function of the stock prices and the interest rate. In the place of (8.33), the derivative prices at each node $(t; j_1, \ldots, j_N, j)$, where $S_t^1 = S_{t,j_1}^1, \ldots, S_t^N = S_{t,j_N}^N, R_t = R_{t,j}$, now satisfy the backward recurrence relation

$$V_t(S_{t,j_1}^1, \ldots, S_{t,j_N}^N, R_{t,j}) = \frac{1}{1 + R_{t,j}} \sum_{j_1', \ldots, j_N'} V_{t+1}(S_{t+1,j_1'}^1, \ldots, S_{t+1,j_N'}^N, R_{t+1,j'})$$

$$\cdot \widetilde{p}_{(t;j_1, \ldots, j_N, j) \to (t+1; j_1', \ldots, j_N', j')} \qquad (8.34)$$

where

$$\widetilde{p}_{(t;j_1, \ldots, j_N, j) \to (t+1; j_1', \ldots, j_N', j')}$$
$$:= \widetilde{\mathbb{P}}(S_{t+1}^1 = S_{t+1,j_1'}^1, \ldots, S_{t+1}^N = S_{t+1,j_N'}^N, R_{t+1} = R_{t+1,j'} \mid S_t^1 = S_{t,j_1}^1, \ldots, S_t^N = S_{t,j_N}^N, R_t = R_{t,j})$$

are single-time-step transition probabilities in the EMM $\widetilde{\mathbb{P}} \equiv \widetilde{\mathbb{P}}^{(B)}$. At each time step the joint indices are restricted to a subset of the allowed nodes, $j_1 \in \{1, \ldots, \kappa_1(t)\}, j_2 \in \{1, \ldots, \kappa_2(t)\}, \ldots, j_N \in \{1, \ldots, \kappa_N(t)\}, j \in \{1, \ldots, \kappa_0(t)\}$.

The following is a simple example of an implementation of (8.33) to the pricing of a two-stock option on a two-stock trinomial price tree.

Example 8.7. Consider the two-period trinomial model with three base assets as given in Examples 8.3 and 8.4, e.g., see Example 8.6. Determine the no-arbitrage prices V_t for $t = 0, 1, 2$ at all the two-stock nodes for an exchange option having payoff $V_T = V_T(S_T^1, S_T^2) = (S_T^2 - S_T^1)^+$ at maturity $T = 2$.

Solution. We will determine the prices by using (8.33). The interest rate is assumed to be zero, $R_t \equiv 0$. Hence,

$$V_t(S_{t,i}^1, S_{t,j}^2) = \sum_{i',j'} \widetilde{p}_{(t;i,j) \to (t+1;i',j')} V_{t+1}(S_{t+1,i'}^1, S_{t+1,j'}^2).$$

The two-stock nodal prices $\{(S_t^1, S_t^2)\}_{t=0,1,2}$ together with the risk-neutral transition probabilities $\widetilde{p}_{(t;i,j) \to (t+1;i',j')}$ are given in Figure 8.4. We can label the nodes at times $t = 1, 2$ as: $(S_{1,1}^1, S_{1,1}^2) = (50, 115)$, $(S_{1,2}^1, S_{1,1}^2) = (40, 115)$, $(S_{1,3}^1, S_{1,2}^2) = (60, 70)$; $(S_{2,1}^1, S_{2,1}^2) = (45, 130)$, $(S_{2,1}^1, S_{2,2}^2) = (45, 120)$, $(S_{2,2}^1, S_{2,3}^2) = (60, 95)$, $(S_{2,1}^1, S_{2,4}^2) = (45, 135)$, $(S_{2,3}^1, S_{2,2}^2) = (35, 120)$, $(S_{2,4}^1, S_{2,5}^2) = (40, 105)$, $(S_{2,2}^1, S_{2,6}^2) = (60, 100)$, $(S_{2,5}^1, S_{2,7}^2) = (55, 40)$, $(S_{2,6}^1, S_{2,7}^2) = (70, 70)$. We have

$$\widetilde{p}_{(0;0,0) \to (1;1,1)} \equiv \widetilde{\mathbb{P}}(S_1^1 = 50, S_1^2 = 115 \mid S_0^1 = 50, S_0^2 = 100) = \widetilde{\mathbb{P}}(A_1^1 \mid A_0^1) = \widetilde{\mathbb{P}}(A_1^1) = 1/3,$$

$$\widetilde{p}_{(0;0,0) \to (1;2,1)} \equiv \widetilde{\mathbb{P}}(S_1^1 = 40, S_1^2 = 115 \mid S_0^1 = 50, S_0^2 = 100) = \widetilde{\mathbb{P}}(A_1^2 \mid A_0^1) = \widetilde{\mathbb{P}}(A_1^2) = 1/3,$$

$$\widetilde{p}_{(0;0,0) \to (1;3,2)} \equiv \widetilde{\mathbb{P}}(S_1^1 = 60, S_1^2 = 70 \mid S_0^1 = 50, S_0^2 = 100) = \widetilde{\mathbb{P}}(A_1^3 \mid A_0^1) = \widetilde{\mathbb{P}}(A_1^3) = 1/3,$$

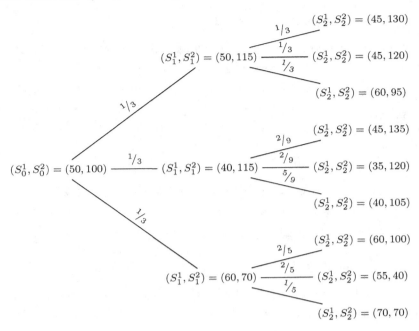

FIGURE 8.4: The two-stock trinomial tree with risk-neutral transition probabilities.

for time $0 \to 1$ transitions and

$$\widetilde{p}_{(1;1,1)\to(2;1,1)} \equiv \widetilde{\mathbb{P}}(S_2^1 = 45, S_2^2 = 130 | S_1^1 = 50, S_1^2 = 115) = \widetilde{\mathbb{P}}(A_2^1 | A_1^1) = 1/3,$$

$$\widetilde{p}_{(1;1,1)\to(2;1,2)} \equiv \widetilde{\mathbb{P}}(S_2^1 = 45, S_2^2 = 120 | S_1^1 = 50, S_1^2 = 115) = \widetilde{\mathbb{P}}(A_2^2 | A_1^1) = 1/3,$$

$$\widetilde{p}_{(1;1,1)\to(2;2,3)} \equiv \widetilde{\mathbb{P}}(S_2^1 = 60, S_2^2 = 95 | S_1^1 = 50, S_1^2 = 115) = \widetilde{\mathbb{P}}(A_2^3 | A_1^1) = 1/3,$$

$$\widetilde{p}_{(1;2,1)\to(2;1,4)} \equiv \widetilde{\mathbb{P}}(S_2^1 = 45, S_2^2 = 135 | S_1^1 = 40, S_1^2 = 115) = \widetilde{\mathbb{P}}(A_2^4 | A_1^2) = 2/9,$$

$$\widetilde{p}_{(1;2,1)\to(2;3,2)} \equiv \widetilde{\mathbb{P}}(S_2^1 = 35, S_2^2 = 120 | S_1^1 = 40, S_1^2 = 115) = \widetilde{\mathbb{P}}(A_2^5 | A_1^2) = 2/9,$$

$$\widetilde{p}_{(1;2,1)\to(2;4,5)} \equiv \widetilde{\mathbb{P}}(S_2^1 = 40, S_2^2 = 105 | S_1^1 = 40, S_1^2 = 115) = \widetilde{\mathbb{P}}(A_2^6 | A_1^2) = 5/9,$$

$$\widetilde{p}_{(1;3,2)\to(2;2,6)} \equiv \widetilde{\mathbb{P}}(S_2^1 = 60, S_2^2 = 100 | S_1^1 = 60, S_1^2 = 70) = \widetilde{\mathbb{P}}(A_2^7 | A_1^3) = 2/5,$$

$$\widetilde{p}_{(1;3,2)\to(2;5,7)} \equiv \widetilde{\mathbb{P}}(S_2^1 = 55, S_2^2 = 40 | S_1^1 = 60, S_1^2 = 70) = \widetilde{\mathbb{P}}(A_2^8 | A_1^3) = 2/5,$$

$$\widetilde{p}_{(1;3,2)\to(2;6,7)} \equiv \widetilde{\mathbb{P}}(S_2^1 = 70, S_2^2 = 70 | S_1^1 = 60, S_1^2 = 70) = \widetilde{\mathbb{P}}(A_2^9 | A_1^3) = 1/5,$$

for time $1 \to 2$ transitions. The prices at time $t = 2$ are given by the payoff function, $V_2(S_2^1, S_2^2) = (S_2^2 - S_2^1)^+$: $V_2(45, 130) = 85$, $V_2(45, 120) = 75$, $V_2(60, 95) = 35$, $V_2(45, 135) = 90$, $V_2(35, 120) = 85$, $V_2(40, 105) = 65$, $V_2(60, 100) = 40$, $V_2(55, 40) = 0$, $V_2(70, 70) = 0$.

Using these prices, we obtain the time $t = 1$ prices:

$$
\begin{aligned}
V_1(50, 115) &= \widetilde{p}_{(1;1,1)\to(2;1,1)} \cdot V_2(45, 130) + \widetilde{p}_{(1;1,1)\to(2;1,2)} \cdot V_2(45, 120) \\
&\quad + \widetilde{p}_{(1;1,1)\to(2;2,3)} \cdot V_2(60, 95) \\
&= \frac{1}{3}\,[85 + 75 + 35] = 65\,,
\end{aligned}
$$

$$
\begin{aligned}
V_1(40, 115) &= \widetilde{p}_{(1;2,1)\to(2;1,4)} \cdot V_2(45, 135) + \widetilde{p}_{(1;2,1)\to(2;3,2)} \cdot V_2(35, 120) \\
&\quad + \widetilde{p}_{(1;2,1)\to(2;4,5)} V_2(40, 105) \\
&= \frac{2}{9} \cdot 90 + \frac{2}{9} \cdot 85 + \frac{5}{9} \cdot 65 = 75\,,
\end{aligned}
$$

$$
\begin{aligned}
V_1(60, 70) &= \widetilde{p}_{(1;3,2)\to(2;2,6)} \cdot V_2(60, 100) + \widetilde{p}_{(1;3,2)\to(2;5,7)} \cdot V_2(55, 40) \\
&\quad + \widetilde{p}_{(1;3,2)\to(2;6,7)} V_2(70, 70) \\
&= \frac{2}{5} \cdot 40 + \frac{2}{5} \cdot 0 + \frac{1}{5} \cdot 0 = 16\,.
\end{aligned}
$$

The price at the initial node is

$$
\begin{aligned}
V_0(50, 100) &= \widetilde{p}_{(0;0,0)\to(1;1,1)} \cdot V_1(50, 115) + \widetilde{p}_{(0;0,0)\to(1;2,1)} \cdot V_1(40, 115) \\
&\quad + \widetilde{p}_{(0;0,0)\to(1;3,2)} \cdot V_1(60, 70) \\
&= \frac{1}{3}\,[65 + 75 + 16] = 52\,.
\end{aligned}
$$

□

Transition probabilities for any scalar or vector process adapted to the filtration generated by the elementary atoms in Section 8.3.4 are readily expressed in terms of conditional and unconditional probabilities associated to the atoms. In particular, consider a scalar process $\{Y_t\}_{t=0,1,\dots,T}$, which may be a single asset (stock) price or a function of the underlying asset prices. Let the set of values $Y_{t,k}$, $k = 1, \dots, N_t$ denote the support of Y_t for each $t = 0, 1, \dots, T$. Since the process is assumed adapted, we can represent the event $\{Y_t = Y_{t,k}\}$ as a union of time-t atoms:

$$
\{Y_t = Y_{t,k}\} = \bigcup_{i \in I_t(k)} A_t^i\,, \quad \text{where } Y_t(A_t^i) = Y_{t,k},\ \forall\, i \in I_t(k)\,.
$$

Here $I_t(k)$ represents the index set in $1, \dots, k_t$ (i.e., the subset of time-t atoms) such that Y_t has constant value $Y_{t,k}$. Similarly, at time $t + 1$ we have $\{Y_{t+1} = Y_{t+1,k'}\} = \bigcup_{j \in I_{t+1}(k')} A_{t+1}^j$ where $I_{t+1}(k')$ is the index set in $1, \dots, k_{t+1}$ such that $Y_{t+1}(A_{t+1}^j) = Y_{t+1,k'}$. Hence, under any EMM $\widetilde{\mathbb{P}}$, we can express the single-time-step transition probabilities of the Y-process in terms of conditional probabilities associated with the elementary atoms at time t and $t + 1$:

$$
\begin{aligned}
\widetilde{\mathbb{P}}(Y_{t+1} = Y_{t+1,k'} \mid Y_t = Y_{t,k}) &= \frac{\widetilde{\mathbb{P}}(Y_{t+1} = Y_{t+1,k'}, Y_t = Y_{t,k})}{\widetilde{\mathbb{P}}(Y_t = Y_{t,k})} = \frac{\displaystyle\sum_{i \in I_t(k)} \sum_{j \in I_{t+1}(k')} \widetilde{\mathbb{P}}(A_{t+1}^j \cap A_t^i)}{\displaystyle\sum_{i \in I_t(k)} \widetilde{\mathbb{P}}(A_t^i)} \\
&= \frac{\displaystyle\sum_{i \in I_t(k)} \widetilde{\mathbb{P}}(A_t^i) \sum_{j \in I_{t+1}(k')} \widetilde{\mathbb{P}}(A_{t+1}^j \mid A_t^i)}{\displaystyle\sum_{i \in I_t(k)} \widetilde{\mathbb{P}}(A_t^i)}\,.
\end{aligned}
\tag{8.35}
$$

The only nonzero terms in (8.35) are for $A_{t+1}^j \subseteq A_t^i$, where time-$(t+1)$ atoms are obtained by breaking down time-t atoms. Note that in some situations the above formula simplifies to one term, $\widetilde{\mathbb{P}}(Y_{t+1} = Y_{t+1,k'} | Y_t = Y_{t,k}) = \widetilde{\mathbb{P}}(A_{t+1}^j | A_t^i)$ for appropriate particular values of i and j.

Equation (8.35) extends in the obvious manner for a joint (vector) process, say $\mathbf{Y}_t = (Y_t^1, \ldots, Y_t^N), t = 0, 1, \ldots, T$. Any event $\{Y_t^1 = Y_{t,j_1}^1, \ldots, Y_t^N = Y_{t,j_N}^N\}$, where the N-tuple $(Y_{t,j_1}^1, \ldots, Y_{t,j_N}^N)$ is in the support of $\mathbf{Y}_t$, is given by a union of elementary time-t atoms

$$\{Y_t^1 = Y_{t,j_1}^1, \ldots, Y_t^N = Y_{t,j_N}^N\} = \bigcup_{i \in I_t(j_1, \ldots, j_N)} A_t^i,$$

where $\mathbf{Y}_t(A_t^i) = (Y_{t,j_1}^1, \ldots, Y_{t,j_N}^N), \forall i \in I_t(j_1, \ldots, j_N)$. Here, $I_t(j_1, \ldots, j_N)$ is the index set in $1, \ldots, k_t$ such that $\mathbf{Y}_t = (Y_{t,j_1}^1, \ldots, Y_{t,j_N}^N)$. Similarly, we have the event at time $t+1$, $\{Y_{t+1}^1 = Y_{t+1,j_1'}^1, \ldots, Y_{t+1}^N = Y_{t+1,j_N'}^N\} = \bigcup_{j \in I_{t+1}(j_1', \ldots, j_N')} A_{t+1}^j$, with index values $j \in I_{t+1}(j_1', \ldots, j_N')$ such that $\mathbf{Y}_{t+1}(A_{t+1}^j) = (Y_{t+1,j_1'}^1, \ldots, Y_{t+1,j_N'}^N)$. Hence, (8.35) extends as

$$\widetilde{\mathbb{P}}(Y_{t+1}^1 = Y_{t+1,j_1'}^1, \ldots, Y_{t+1}^N = Y_{t+1,j_N'}^N | Y_t^1 = Y_{t,j_1}^1, \ldots, Y_t^N = Y_{t,j_N}^N)$$

$$= \frac{\displaystyle\sum_{i \in I_t(j_1, \ldots, j_N)} \widetilde{\mathbb{P}}(A_t^i) \sum_{j \in I_{t+1}(j_1', \ldots, j_N')} \widetilde{\mathbb{P}}(A_{t+1}^j | A_t^i)}{\displaystyle\sum_{i \in I_t(j_1, \ldots, j_N)} \widetilde{\mathbb{P}}(A_t^i)}. \qquad (8.36)$$

As noted for the scalar case above, in some situations the last expression reduces to a single term $\mathbb{P}(A_{t+1}^j | A_t^i)$, for particular values of i and j. Note that setting $Y_t^i = S_t^i, i = 1, \ldots, N$, gives us the single-time-step risk-neutral transition probabilities of the joint stock price process.

Consider the trinomial tree in the above examples. The risk-neutral transition probabilities $\widetilde{p}_{(t;j_1,j_2) \to (t+1;j_1',j_2')} \equiv \widetilde{\mathbb{P}}(S_{t+1}^1 = S_{t+1,j_1'}^1, S_{t+1}^2 = S_{t+1,j_2'}^2 | S_t^1 = S_{t,j_1}^1, S_t^2 = S_{t,j_2}^2)$ for the two-stock process, are given in Example 8.7. These follow trivially using (8.36) where we set $Y_t^1 \equiv S_t^1, Y_t^2 \equiv S_t^2$. In this case there is only one term in (8.36). In the following example we compute risk-neutral transition probabilities for the individual stock price processes by implementing (8.35).

Example 8.8. Consider the two-period trinomial model with three base assets as given in Examples 8.3 and 8.4. Determine the single-step transition probabilities, under measure $\widetilde{\mathbb{P}} = \widetilde{\mathbb{P}}^{(B)}$, for each stock price process $\{S_t^i\}_{t=0,1,2}, i = 1, 2$.

Solution. For stock #1, we set $Y_t = S_t^1$ and use (8.35) where $S_0^1 = 50$, $S_1^1 \in \{40, 50, 60\}$, $S_2^1 \in \{35, 40, 45, 55, 60, 70\}$ with time $t = 0 \to 1$ nonzero transition probabilities

$$\widetilde{\mathbb{P}}(S_1^1 = 50, | S_0^1 = 50) = \widetilde{\mathbb{P}}(S_1^1 = 50) = \widetilde{\mathbb{P}}(A_1^1 | A_0^1) = \widetilde{\mathbb{P}}(A_1^1) = 1/3,$$

$$\widetilde{\mathbb{P}}(S_1^1 = 40, | S_0^1 = 50) = \widetilde{\mathbb{P}}(S_1^1 = 40) = \widetilde{\mathbb{P}}(A_1^2 | A_0^1) = \widetilde{\mathbb{P}}(A_1^2) = 1/3,$$

$$\widetilde{\mathbb{P}}(S_1^1 = 60, | S_0^1 = 50) = \widetilde{\mathbb{P}}(S_1^1 = 60) = \widetilde{\mathbb{P}}(A_1^3 | A_0^1) = \widetilde{\mathbb{P}}(A_1^3) = 1/3.$$

For $t = 1 \to 2$ the nonzero transition probabilities are:

$$\widetilde{\mathbb{P}}(S_2^1 = 45, |S_1^1 = 50) = \widetilde{\mathbb{P}}(A_2^1|A_1^1) + \widetilde{\mathbb{P}}(A_2^2|A_1^1) = 2/3,$$

$$\widetilde{\mathbb{P}}(S_2^1 = 60, |S_1^1 = 50) = \widetilde{\mathbb{P}}(A_2^3|A_1^1) = 1/3,$$

$$\widetilde{\mathbb{P}}(S_2^1 = 45, |S_1^1 = 40) = \widetilde{\mathbb{P}}(A_2^4|A_1^2) = 2/9,$$

$$\widetilde{\mathbb{P}}(S_2^1 = 35, |S_1^1 = 40) = \widetilde{\mathbb{P}}(A_2^5|A_1^2) = 2/9,$$

$$\widetilde{\mathbb{P}}(S_2^1 = 40, |S_1^1 = 40) = \widetilde{\mathbb{P}}(A_2^6|A_1^2) = 5/9,$$

$$\widetilde{\mathbb{P}}(S_2^1 = 60, |S_1^1 = 60) = \widetilde{\mathbb{P}}(A_2^7|A_1^3) = 2/5,$$

$$\widetilde{\mathbb{P}}(S_2^1 = 55, |S_1^1 = 60) = \widetilde{\mathbb{P}}(A_2^8|A_1^3) = 2/5,$$

$$\widetilde{\mathbb{P}}(S_2^1 = 70, |S_1^1 = 60) = \widetilde{\mathbb{P}}(A_2^9|A_1^3) = 1/5.$$

Note that all conditional probabilities with common condition (same starting value) sum to unity, as required. It also follows in this case that $\{S_t^1\}_{t=0,1,2}$ is a Markov process.

For stock #2, we set $Y_t = S_t^2$ and use (8.35) where $S_0^2 = 100$, $S_1^2 \in \{70, 115\}$, $S_2^2 \in \{40, 70, 95, 100, 105, 120, 130, 135\}$. For $t = 0 \to 1$ the nonzero transition probabilities are:

$$\widetilde{\mathbb{P}}(S_1^2 = 115, |S_0^2 = 100) = \widetilde{\mathbb{P}}(S_1^2 = 115) = \widetilde{\mathbb{P}}(A_1^1|A_0^1) + \widetilde{\mathbb{P}}(A_1^2|A_0^1) = \widetilde{\mathbb{P}}(A_1^1) + \widetilde{\mathbb{P}}(A_1^2) = 2/3$$

$$\widetilde{\mathbb{P}}(S_1^2 = 70, |S_0^2 = 100) = \widetilde{\mathbb{P}}(S_1^2 = 70) = \widetilde{\mathbb{P}}(A_1^3|A_0^1) = \widetilde{\mathbb{P}}(A_1^3) = 1/3.$$

For $t = 1 \to 2$ the nonzero transition probabilities are:

$$\widetilde{\mathbb{P}}(S_2^2 = 130, |S_1^2 = 115) = \frac{\widetilde{\mathbb{P}}(A_1^1) \cdot \widetilde{\mathbb{P}}(A_2^1|A_1^1)}{\widetilde{\mathbb{P}}(A_1^1) + \widetilde{\mathbb{P}}(A_1^2)} = \frac{\frac{1}{3} \cdot \frac{1}{3}}{\frac{2}{3}} = \frac{1}{6},$$

$$\widetilde{\mathbb{P}}(S_2^2 = 120, |S_1^2 = 115) = \frac{\widetilde{\mathbb{P}}(A_1^1) \cdot \widetilde{\mathbb{P}}(A_2^2|A_1^1) + \widetilde{\mathbb{P}}(A_1^2) \cdot \widetilde{\mathbb{P}}(A_2^5|A_1^2)}{\widetilde{\mathbb{P}}(A_1^1) + \widetilde{\mathbb{P}}(A_1^2)} = \frac{\frac{1}{3} \cdot \frac{1}{3} + \frac{1}{3} \cdot \frac{2}{9}}{\frac{2}{3}} = \frac{5}{18},$$

$$\widetilde{\mathbb{P}}(S_2^2 = 95, |S_1^2 = 115) = \frac{\widetilde{\mathbb{P}}(A_1^1) \cdot \widetilde{\mathbb{P}}(A_2^3|A_1^1)}{\widetilde{\mathbb{P}}(A_1^1) + \widetilde{\mathbb{P}}(A_1^2)} = \frac{\frac{1}{3} \cdot \frac{1}{3}}{\frac{2}{3}} = \frac{1}{6},$$

$$\widetilde{\mathbb{P}}(S_2^2 = 135, |S_1^2 = 115) = \frac{\widetilde{\mathbb{P}}(A_1^2) \cdot \widetilde{\mathbb{P}}(A_2^4|A_1^2)}{\widetilde{\mathbb{P}}(A_1^1) + \widetilde{\mathbb{P}}(A_1^2)} = \frac{\frac{1}{3} \cdot \frac{2}{9}}{\frac{2}{3}} = \frac{1}{9},$$

$$\widetilde{\mathbb{P}}(S_2^2 = 105, |S_1^2 = 115) = \frac{\widetilde{\mathbb{P}}(A_1^2) \cdot \widetilde{\mathbb{P}}(A_2^6|A_1^2)}{\widetilde{\mathbb{P}}(A_1^1) + \widetilde{\mathbb{P}}(A_1^2)} = \frac{\frac{1}{3} \cdot \frac{5}{9}}{\frac{2}{3}} = \frac{5}{18},$$

$$\widetilde{\mathbb{P}}(S_2^2 = 100, |S_1^2 = 70) = \widetilde{\mathbb{P}}(A_2^7|A_1^3) = \frac{2}{5},$$

$$\widetilde{\mathbb{P}}(S_2^2 = 40, |S_1^2 = 70) = \widetilde{\mathbb{P}}(A_2^8|A_1^3) = \frac{2}{5},$$

$$\widetilde{\mathbb{P}}(S_2^2 = 70, |S_1^2 = 70) = \widetilde{\mathbb{P}}(A_2^9|A_1^3) = \frac{1}{5}.$$

Again, note that all conditional probabilities with common condition sum to unity.

$\square$

8.3.8   Radon–Nikodym Derivative Process and Change of Numéraire

In the previous sections, the main results were derived using the risk-neutral probability measure $\widetilde{\mathbb{P}} \equiv \widetilde{\mathbb{P}}^{(B)}$, the EMM with numéraire asset $g = B$. Suppose the EMM $\widetilde{\mathbb{P}}^{(B)}$ is known and there exists another asset S with a strictly positive price process. How can we find the EMM $\widehat{\mathbb{P}}^{(S)}$ relative to the numéraire asset $g = S$ without recomputing all risk-neutral probabilities from scratch? As we first learned in Chapter 5, it is the Radon–Nikodym derivative that connects two equivalent probability measures. Recall its definition. Let $\mathbb{P}$ and $\widehat{\mathbb{P}}$ be two probability measures on the same finite sample space Ω so that $\mathbb{P}(\omega) > 0$ and $\widehat{\mathbb{P}}(\omega) > 0$ for all outcomes $\omega \in \Omega$. That is, $\mathbb{P}$ and $\widehat{\mathbb{P}}$ are equivalent probability measures. The Radon–Nikodym derivative of $\widehat{\mathbb{P}}$ w.r.t. $\mathbb{P}$ is the strictly positive random variable, $\varrho \equiv \frac{d\widehat{\mathbb{P}}}{d\mathbb{P}}$, defined by

$$\varrho(\omega) = \frac{\widehat{\mathbb{P}}(\omega)}{\mathbb{P}(\omega)} \text{ for } \omega \in \Omega \,.$$

Note that this random variable has unit expected value under $\mathbb{P}$, $\mathrm{E}[\varrho] = 1$. The mathematical expectation of a random variable X on Ω under $\widehat{\mathbb{P}}$ can also be computed under the probability measure $\mathbb{P}$ with the use of ϱ as follows:

$$\widehat{\mathrm{E}}[X] = \sum_\omega X(\omega)\widehat{\mathbb{P}}(\omega) = \sum_\omega X(\omega) \underbrace{\frac{\widehat{\mathbb{P}}(\omega)}{\mathbb{P}(\omega)}}_{=\varrho(\omega)} \mathbb{P}(\omega) = \sum_\omega \varrho(\omega) X(\omega) \mathbb{P}(\omega) = \mathrm{E}[\varrho\, X] \,. \quad (8.37)$$

In particular, we can calculate the risk-neutral expectation $\widetilde{\mathrm{E}}[X]$ by computing $\mathrm{E}[\varrho\, X]$ under the actual (real-world) probability measure, where ϱ is the Radon–Nikodym derivative of $\widetilde{\mathbb{P}}$ w.r.t. $\mathbb{P}$.

Consider an $\mathcal{F}_t$-measurable random variable X with $0 \leqslant t \leqslant T$. For example, X is a payoff maturing at time n. The computation of expected values of X while switching the probability measure can be simplified further with the use of the *Radon–Nikodym derivative process*, which is defined just below. Moreover, to price derivatives at any intermediate time $t \in \{0, 1, \ldots, T\}$, we need a conditional expectation under $\widehat{\mathbb{P}}$ and the Radon–Nikodym process handles it as well.

Definition 8.3. Let $\mathbb{P}$ and $\widehat{\mathbb{P}}$ be two probability measures on a finite sample space $\Omega \equiv \Omega_T$ so that $\mathbb{P}(\omega) > 0$ and $\widehat{\mathbb{P}}(\omega) > 0$ for every outcome ω (hence, $\mathbb{P}$ and $\widehat{\mathbb{P}}$ are equivalent). Let ϱ be the Radon–Nikodym derivative of $\widehat{\mathbb{P}}$ w.r.t. $\mathbb{P}$ given by $\varrho(\omega) \equiv \frac{d\widehat{\mathbb{P}}}{d\mathbb{P}}(\omega) = \frac{\widehat{\mathbb{P}}(\omega)}{\mathbb{P}(\omega)}$. The Radon–Nikodym derivative process, w.r.t. a given filtration $\{\mathcal{F}_t\}_{t=0,1,\ldots,T}$, is defined as

$$\varrho_t = \mathrm{E}[\varrho|\mathcal{F}_t], \quad t \in \{0, 1, \ldots, T\} \,. \quad (8.38)$$

In particular, $\varrho_T = \mathrm{E}[\varrho|\mathcal{F}_T] = \varrho$ since $\varrho : \Omega_T \to \mathbb{R}$ is a proper $\mathcal{F}_T$-measurable random variable. The Radon–Nikodym derivative process is strictly positive, $\varrho_t > 0$, with unit initial value, $\varrho_0 = \mathrm{E}[\varrho|\mathcal{F}_0] = \mathrm{E}[\varrho] = 1$.

Recall from Chapter 6, for any random variable Y with $\mathrm{E}[|Y|] < \infty$, the process defined by

$$Y_t = \mathrm{E}[Y \mid \mathcal{F}_t], \quad t \in \{0, 1, \ldots, T\},$$

is a (Doob-Lévy) $\mathbb{P}$-martingale with respect to the given filtration. Since $\mathrm{E}[|\varrho|] = \mathrm{E}[\varrho] = 1 < \infty$, the Radon–Nikodym derivative process $\{\varrho_t\}_{0 \leqslant t \leqslant T}$ defined in (8.38) is a $\mathbb{P}$-martingale with respect to the given filtration.

Suppose that X_t is an $\mathcal{F}_t$-measurable random variable for some $t \in \{0, 1, \ldots, T\}$. In this case, equation (8.37) simplifies further. Applying the tower property gives

$$\widehat{\mathrm{E}}[X_t] = \mathrm{E}[\varrho\, X_t] = \mathrm{E}[\mathrm{E}_t[\varrho\, X_t]] = \mathrm{E}[X_t\, \mathrm{E}_t[\varrho]] = \mathrm{E}[X_t\, \varrho_t] \,. \quad (8.39)$$

This formula is a special case of the following formula that relates conditional expectations under two equivalent measures $\mathbb{P}$ and $\widehat{\mathbb{P}}$ with Radon-Nikodym derivative process, ϱ_t, defined in (8.38). That is, if X_t is $\mathcal{F}_t$-measurable then

$$\widehat{\mathrm{E}}[X_t|\mathcal{F}_s] = \varrho_s^{-1}\mathrm{E}[\varrho_t\, X_t|\mathcal{F}_s]\,,\ 0 \leqslant s \leqslant t \leqslant T. \tag{8.40}$$

This is proven by showing that: (i) $Y_s := \varrho_s^{-1}\mathrm{E}[\varrho_t\, X_t|\mathcal{F}_s]$ is $\mathcal{F}_s$-measurable and (ii) $\forall B \in \mathcal{F}_s$, $\widehat{\mathrm{E}}[\mathbb{I}_B\, X_t] = \widehat{\mathrm{E}}[\mathbb{I}_B\, Y_s]$. Property (i) holds trivially since, by definition, both $\mathrm{E}[\varrho_t\, X_t|\mathcal{F}_s]$ and ϱ_s are $\mathcal{F}_s$-measurable and therefore their ratio (with $\varrho_s > 0$) is $\mathcal{F}_s$-measurable. Property (ii) is shown to hold by firstly making use of (8.39) for $\mathcal{F}_s$-measurable $\mathbb{I}_B\, Y_s$, then applying the definition of Y_s, and finally employing the tower property along with the fact that $\mathbb{I}_B$ is $\mathcal{F}_s$-measurable, giving:

$$\widehat{\mathrm{E}}[\mathbb{I}_B\, Y_s] = \mathrm{E}[\mathbb{I}_B\, Y_s\, \varrho_s] = \mathrm{E}[\mathbb{I}_B\, \mathrm{E}[\varrho_t\, X_t|\mathcal{F}_s]] = \mathrm{E}[\mathrm{E}[\mathbb{I}_B\, \varrho_t\, X_t|\mathcal{F}_s]] = \mathrm{E}[\mathbb{I}_B\, \varrho_t\, X_t].$$

Now, note that $\mathbb{I}_B X_t$ is $\mathcal{F}_t$-measurable since $\mathbb{I}_B$ is $\mathcal{F}_t$-measurable ($\mathcal{F}_s$-measurable with $s \leqslant t$) and X_t is $\mathcal{F}_t$-measurable by assumption. Hence, directly applying (8.39) to the $\mathcal{F}_t$-measurable random variable $\mathbb{I}_B X_t$ gives $\widehat{\mathrm{E}}[\mathbb{I}_B X_t] = \mathrm{E}[\varrho_t\, \mathbb{I}_B X_t]$. Hence, property (ii) holds and (8.40) has been proven.

Suppose the σ-algebra $\mathcal{F}_t = \sigma(\mathcal{P}_t)$ is generated from a partition $\mathcal{P}_t$ that consists of atoms $A_t^1, \ldots, A_t^{k_t}$. Hence, an $\mathcal{F}_t$-measurable random variable is constant on the atoms of $\mathcal{P}_t$. Fix an atom $A_t^i \in \mathcal{P}_t$ and consider the random variable $X_t = \mathbb{I}_{A_t^i}$. On the one hand, we have

$$\widehat{\mathrm{E}}[X_t] = \widehat{\mathrm{E}}[\mathbb{I}_{A_t^i}] = \widehat{\mathbb{P}}(A_t^i). \tag{8.41}$$

On the other hand, we have

$$\mathrm{E}[\varrho_t\, X_t] = \sum_{\omega \in \Omega} \varrho_t(\omega)\, \mathbb{I}_{A_t^i}(\omega)\, \mathbb{P}(\omega) = \sum_{\omega \in A_t^i} \varrho_t(\omega)\mathbb{P}(\omega)\,.$$

Since ϱ_t is $\mathcal{F}_t$-measurable, it is constant on A_t^i. Therefore, we have

$$\mathrm{E}[\varrho_t\, X_t] = \varrho_t(A_t^i)\, \mathbb{P}(A_t^i)\,, \tag{8.42}$$

where $\varrho_t(A_t^i)$ is the value of ϱ_t on the atom A_t^i. According to (8.39), the quantities in (8.41) and (8.42) are equal, and hence

$$\varrho_t(A_t^i) = \frac{\widehat{\mathbb{P}}(A_t^i)}{\mathbb{P}(A_t^i)} \tag{8.43}$$

for any $A_t^i \in \mathcal{P}_t$. The partition $\mathcal{P}_t$ consists of k_t atoms. So, the Radon–Nikodym derivative process can be written as a weighted sum of k_t indicator functions:

$$\varrho_t(\omega) = \sum_{i=1}^{k_t} \varrho_t(A_t^i)\, \mathbb{I}_{A_t^i}(\omega) = \sum_{i=1}^{k_t} \frac{\widehat{\mathbb{P}}(A_t^i)}{\mathbb{P}(A_t^i)}\, \mathbb{I}_{A_t^i}(\omega)\,, \quad t = 0, 1, \ldots, T. \tag{8.44}$$

As a random variable we have $\varrho_t = \sum_{i=1}^{k_t} \varrho_t(A_t^i)\, \mathbb{I}_{A_t^i}$, $t = 0, 1, \ldots, T$.

Note that (8.44) also follows as a direct application of (6.21) in Theorem 6.6 of Chapter 6. Indeed, setting $\mathcal{G} = \mathcal{F}_t = \sigma(\mathcal{P}_t) = \sigma(A_t^1, \ldots, A_t^{k_t})$ and $X = \varrho \equiv \frac{\mathrm{d}\widehat{\mathbb{P}}}{\mathrm{d}\mathbb{P}}$ in (6.21) gives

$$\varrho_t \equiv \mathrm{E}[\varrho \mid \mathcal{F}_t] = \sum_{i=1}^{k_t} \mathrm{E}[\varrho \mid A_t^i]\, \mathbb{I}_{A_t^i} = \sum_{i=1}^{k_t} \frac{\mathrm{E}[\varrho \cdot \mathbb{I}_{A_t^i}]}{\mathbb{P}(A_t^i)}\, \mathbb{I}_{A_t^i} = \sum_{i=1}^{k_t} \frac{\widehat{\mathbb{P}}(A_t^i)}{\mathbb{P}(A_t^i)}\, \mathbb{I}_{A_t^i}.$$

The last equality results from (8.37): $\widehat{\mathbb{P}}(A_t^i) = \widehat{\mathrm{E}}[\mathbb{I}_{A_t^i}] = \mathrm{E}[\varrho \cdot \mathbb{I}_{A_t^i}]$.

Example 8.9. Determine the Radon–Nikodym derivative process for the standard binomial tree model with filtration generated by all the independent up/down market moves. Assume the change of measure $\varrho := \frac{d\mathbb{Q}}{d\mathbb{P}}$, with two equivalent probability measures $\mathbb{P} \sim \mathbb{Q}$.

Solution. Let $p = \mathbb{P}(\mathsf{U})$ and $q = \mathbb{Q}(\mathsf{U})$ be the probability of an upward move under probability measures $\mathbb{P}$ and $\mathbb{Q} \equiv \widehat{\mathbb{P}}$, respectively. Each of the 2^t atoms $A_t^i \in \mathcal{P}_t$ is of the form

$$A_t^i = A_{\bar{\omega}_1 \bar{\omega}_2 \ldots \bar{\omega}_t} = \{\omega \in \Omega_T \; : \; \omega_1 = \bar{\omega}_1, \omega_2 = \bar{\omega}_2, \ldots, \omega_t = \bar{\omega}_t\}$$

for some sequence $\bar{\omega}_1, \bar{\omega}_2, \ldots, \bar{\omega}_t \in \{\mathsf{U}, \mathsf{D}\}$. The probability of event $A_t^i \equiv A_{\bar{\omega}_1 \bar{\omega}_2 \ldots \bar{\omega}_t}$ under $\mathbb{P}$ is

$$\mathbb{P}(A_{\bar{\omega}_1 \bar{\omega}_2 \ldots \bar{\omega}_t}) = p^{\#\mathsf{U}(\bar{\omega}_1, \bar{\omega}_2, \ldots, \bar{\omega}_t)}(1-p)^{\#\mathsf{D}(\bar{\omega}_1, \bar{\omega}_2, \ldots, \bar{\omega}_t)} \equiv p^{\mathsf{U}_t(\bar{\omega})}(1-p)^{\mathsf{D}_t(\bar{\omega})}.$$

The probability of the same event computed under the measure $\mathbb{Q}$ is

$$\mathbb{Q}(A_{\bar{\omega}_1 \bar{\omega}_2 \ldots \bar{\omega}_t}) = q^{\#\mathsf{U}(\bar{\omega}_1, \bar{\omega}_2, \ldots, \bar{\omega}_t)}(1-q)^{\#\mathsf{D}(\bar{\omega}_1, \bar{\omega}_2, \ldots, \bar{\omega}_t)} \equiv q^{\mathsf{U}_t(\bar{\omega})}(1-q)^{\mathsf{D}_t(\bar{\omega})}.$$

So, $\varrho(\omega) \equiv \frac{d\mathbb{Q}}{d\mathbb{P}}(\omega) = \frac{\mathbb{Q}(\omega)}{\mathbb{P}(\omega)}$ and the random variable $\varrho_t = \mathrm{E}[\varrho | \mathcal{F}_t]$, $\mathcal{F}_t = \sigma(\mathcal{P}_t)$ is represented as

$$\varrho_t = \sum_{\bar{\omega}_1, \bar{\omega}_2, \ldots, \bar{\omega}_t \in \{\mathsf{U}, \mathsf{D}\}} \frac{\mathbb{Q}(A_{\bar{\omega}_1 \bar{\omega}_2 \ldots \bar{\omega}_t})}{\mathbb{P}(A_{\bar{\omega}_1 \bar{\omega}_2 \ldots \bar{\omega}_t})} \mathbb{I}_{A_{\bar{\omega}_1 \bar{\omega}_2 \ldots \bar{\omega}_t}}, \quad t = 0, 1, \ldots, T.$$

We see that ϱ_t is $\mathcal{F}_t$-measurable, i.e., it is a function of the first t market moves where for any outcome $\omega = \omega_1 \omega_2 \ldots \omega_t \ldots \omega_T$ we have

$$\varrho_t(\omega) = \varrho_t(\omega_1, \omega_2, \ldots, \omega_t) = \left(\frac{q}{p}\right)^{\#\mathsf{U}(\omega_1, \omega_2, \ldots, \omega_t)} \left(\frac{1-q}{1-p}\right)^{\#\mathsf{D}(\omega_1, \omega_2, \ldots, \omega_t)}$$

$$\equiv \left(\frac{q}{p}\right)^{\mathsf{U}_t(\omega)} \left(\frac{1-q}{1-p}\right)^{\mathsf{D}_t(\omega)}.$$

$\square$

The conditioning formula in (8.40) specializes in the case of an *arbitrary* random variable $X : \Omega_T \to \mathbb{R}$. Note that X is necessarily $\mathcal{F}_T$-measurable. Hence, applying (8.40) for $t = T$ and $X_T \equiv X$ gives

$$\widehat{\mathrm{E}}[X | \mathcal{F}_s] = \varrho_s^{-1} \mathrm{E}[\varrho_T X | \mathcal{F}_s] \equiv \varrho_s^{-1} \mathrm{E}[\varrho X | \mathcal{F}_s], \quad 0 \leqslant s \leqslant T, \tag{8.45}$$

where $\varrho_T = \varrho$.

We now derive an analogue of (8.37) for expectations conditional on an event (i.e., an atom). Fix arbitrarily $t \in \{0, 1, \ldots, T\}$. Then, applying the conditional expectation formula (6.21) under the measure $\widehat{\mathbb{P}}$ gives

$$\widehat{\mathrm{E}}[X \mid \mathcal{F}_t] = \sum_{i=1}^{k_t} \widehat{\mathrm{E}}[X \mid A_t^i] \mathbb{I}_{A_t^i}.$$

On the other hand, combining the above identity $\widehat{\mathrm{E}}[X | \mathcal{F}_t] = \varrho_t^{-1} \mathrm{E}[\varrho X | \mathcal{F}_t]$ and (6.21) under measure $\mathbb{P}$ gives

$$\widehat{\mathrm{E}}[X \mid \mathcal{F}_t] = \varrho_t^{-1} \mathrm{E}[\varrho X | \mathcal{F}_t] = \varrho_t^{-1} \sum_{i=1}^{k_t} \widehat{\mathrm{E}}[\varrho X \mid A_t^i] \mathbb{I}_{A_t^i} = \sum_{i=1}^{k_t} \frac{\widehat{\mathrm{E}}[\varrho X \mid A_t^i]}{\varrho_t(A_t^i)} \mathbb{I}_{A_t^i}$$

$$= \sum_{i=1}^{k_t} \widehat{\mathrm{E}}[(\varrho / \varrho_t) X \mid A_t^i] \mathbb{I}_{A_t^i}.$$

The last equality follows since ϱ_t is $\mathcal{F}_t$-measurable, i.e., constant on every atom A_t^i with $\varrho_t^{-1}\mathbb{I}_{A_t^i} = \varrho_t^{-1}(A_t^i)\mathbb{I}_{A_t^i}$. Since the above two expressions for $\widehat{\mathrm{E}}[X \mid \mathcal{F}_t]$ are equivalent, the coefficients multiplying each indicator random variable $\mathbb{I}_{A_t^i}$ must be equal,

$$\widehat{\mathrm{E}}[X \mid A_t^i] = \widehat{\mathrm{E}}[(\varrho/\varrho_t) X \mid A_t^i] = \frac{\mathbb{P}(A_t^i)}{\widehat{\mathbb{P}}(A_t^i)} \widehat{\mathrm{E}}[\varrho X \mid A_t^i] \tag{8.46}$$

where we used (8.43) for $1/\varrho_t(A_t^i)$ in the last expression.

Suppose we deal with two equivalent martingale measures for respective numéraire assets defined by two strictly positive price processes $f = \{f_t\}_{t \geqslant 0}$ and $g = \{g_t\}_{t \geqslant 0}$, where, for example, one numéraire is a risk-free bond and the other is some base stock. Here we assume that numéraire assets pay no dividends. In the case of dividends, we replace $f_t \to f_t/c_t^f$ and $g_t \to g_t/c_t^g$ with respective time-t cumulative after-dividend factors c_t^f and c_t^g, for all values of time t. In this case, we can obtain a simple explicit formula for the Radon–Nikodym derivative process. According to the pricing formula (8.17), we have the following equivalence of conditional expectations under measures $\widetilde{\mathbb{P}}^{(g)}$ and $\widetilde{\mathbb{P}}^{(f)}$:

$$\widetilde{\mathrm{E}}^{(g)}\left[\frac{g_t}{g_T} X \Big| \mathcal{F}_t\right] = \widetilde{\mathrm{E}}^{(f)}\left[\frac{f_t}{f_T} X \Big| \mathcal{F}_t\right] \implies \widetilde{\mathrm{E}}^{(g)}[X|\mathcal{F}_t] = \widetilde{\mathrm{E}}^{(f)}\left[\frac{g_T/g_t}{f_T/f_t} X \Big| \mathcal{F}_t\right] \tag{8.47}$$

for all $t \in \{0, 1, \ldots, T\}$ and any payoff X. On the other hand, applying (8.45) (using time index t in place of s) with $\widehat{\mathbb{P}} = \widetilde{\mathbb{P}}^{(g)}$ and $\mathbb{P} = \widetilde{\mathbb{P}}^{(f)}$ gives

$$\widetilde{\mathrm{E}}^{(g)}[X \mid \mathcal{F}_t] = \widetilde{\mathrm{E}}_t^{(f)}\left[(\varrho_T/\varrho_t) X | \mathcal{F}_t\right] \tag{8.48}$$

where $\{\varrho_t\}_{0 \leqslant t \leqslant T}$ is the Radon–Nikodym derivative process of $\widetilde{\mathbb{P}}^{(g)}$ w.r.t. $\widetilde{\mathbb{P}}^{(f)}$, defined by $\varrho_t := \widetilde{\mathrm{E}}^{(f)}[\varrho \mid \mathcal{F}_t]$, $\varrho \equiv \varrho_T = \frac{\mathrm{d}\widetilde{\mathbb{P}}^{(g)}}{\mathrm{d}\widetilde{\mathbb{P}}^{(f)}}$, $\varrho(\omega) = \frac{\widetilde{\mathbb{P}}^{(g)}(\omega)}{\widetilde{\mathbb{P}}^{(f)}(\omega)}$. Equating (8.47) and (8.48) gives

$$\widetilde{\mathrm{E}}_t^{(f)}\left[(\varrho_T/\varrho_t) X | \mathcal{F}_t\right] = \widetilde{\mathrm{E}}^{(f)}\left[\frac{g_T/g_t}{f_T/f_t} X \Big| \mathcal{F}_t\right] = \widetilde{\mathrm{E}}^{(f)}\left[\frac{(g_T/g_0)/(f_T/f_0)}{(g_t/g_0)/(f_t/f_0)} X \Big| \mathcal{F}_t\right].$$

Let us fix some state $\omega \in \Omega_T$ and take $X = \mathbb{I}_{\{\omega\}}$. The above identity becomes

$$\left(\frac{\varrho_T}{\varrho_t}\right)(\omega) = \left(\frac{(g_T/g_0)/(f_T/f_0)}{(g_t/g_0)/(f_t/f_0)}\right)(\omega).$$

Since, as follows from (5.58), the Radon–Nikodym derivative $\varrho \equiv \varrho_T$ is equal to $\frac{g_T/g_0}{f_T/f_0}$, we have the Radon–Nikodym derivative process explicitly in terms of a ratio of prices of the two numéraire assets at time t and at initial time 0:

$$\varrho_t = \frac{g_t/g_0}{f_t/f_0}, \quad t \in \{0, 1, \ldots, T\}. \tag{8.49}$$

Note that now $\varrho_t(A_t^i) = \frac{\widetilde{\mathbb{P}}^{(g)}(A_t^i)}{\widetilde{\mathbb{P}}^{(f)}(A_t^i)}$. Hence, combining this with (8.49) allows us to express probabilities of atoms under $\widetilde{\mathbb{P}}^{(g)}$ in terms of probabilities under $\widetilde{\mathbb{P}}^{(f)}$:

$$\widetilde{\mathbb{P}}^{(g)}(A_t^i) = \left(\frac{g_t(A_t^i)/g_0}{f_t(A_t^i)/f_0}\right) \widetilde{\mathbb{P}}^{(f)}(A_t^i) \tag{8.50}$$

for every $A_t^i \in \mathcal{P}_t$. In particular, by setting $t = T$ (i.e., every atom is an elementary outcome ω) we obtain the known relation from Chapter 5,

$$\widetilde{\mathbb{P}}^{(g)}(\omega) = \frac{g_T(\omega)/g_0}{f_T(\omega)/f_0} \widetilde{\mathbb{P}}^{(f)}(\omega), \quad \text{for every } \omega \in \Omega_T. \tag{8.51}$$

Also, using (8.50), we can relate the single-time-step transition (conditional) probabilities between two equivalent martingale measures:

$$\widetilde{\mathbb{P}}^{(g)}(A_{t+1}^j | A_t^i) = \frac{\widetilde{\mathbb{P}}^{(g)}(A_{t+1}^j)}{\widetilde{\mathbb{P}}^{(g)}(A_t^i)} = \frac{g_{t+1}(A_{t+1}^j)/g_t(A_t^i)}{f_{t+1}(A_{t+1}^j)/f_t(A_t^i)} \cdot \widetilde{\mathbb{P}}^{(f)}(A_{t+1}^j | A_t^i) \qquad (8.52)$$

where $\widetilde{\mathbb{P}}^{(f)}(A_{t+1}^j | A_t^i) = \frac{\widetilde{\mathbb{P}}^{(f)}(A_{t+1}^j)}{\widetilde{\mathbb{P}}^{(f)}(A_t^i)}$, for all atoms $A_{t+1}^j \subseteq A_t^i$. In particular, assume a numéraire asset as a function of N stock prices, $g_t = g_t(S_t^1, \ldots, S_t^N)$, and take $f_t = B_t$ (bank account value at time t). Take a time-t atom A_t^i in which the time-t values of the stocks and the interest rate are given by $S_t^1 = S_{t,j_1}^1, S_t^2 = S_{t,j_2}^2, \ldots, S_t^N = S_{t,j_N}^N, R_t = R_{t,j}$. Then, g_t evaluates on such an atom to $g_t(A_t^i) = g_t(S_{t,j_1}^1, S_{t,j_2}^2, \ldots, S_{t,j_N}^N)$. Similarly, consider a time-$(t+1)$ atom $A_{t+1}^j \subseteq A_t^i$ in which the time-$(t+1)$ values of the stocks and the interest rate are given by $S_{t+1}^1 = S_{t+1,j_1'}^1, \ldots, S_{t+1}^N = S_{t+1,j_N'}^N, R_{t+1} = R_{t+1,j'}$. Then, g_{t+1} evaluates to $g_{t+1}(A_{t+1}^j) = g_{t+1}(S_{t+1,j_1'}^1, \ldots, S_{t+1,j_N'}^N)$. Hence, applying (8.52) and using $f_t/f_{t+1} = B_t/B_{t+1} = (1 + R_t)^{-1}$, gives us a relationship between the nodal transition probabilities under two EMMs:

$$\widetilde{p}_{(t;j_1,\ldots,j_N,j)\to(t+1;j_1',\ldots,j_N',j')}^{(g)} = \frac{1}{1 + R_{t,j}} \frac{g_{t+1}(S_{t+1,j_1'}^1, \ldots, S_{t+1,j_N'}^N)}{g_t(S_{t,j_1}^1, \ldots, S_{t,j_N}^N)} \widetilde{p}_{(t;j_1,\ldots,j_N,j)\to(t+1;j_1',\ldots,j_N',j')}. \qquad (8.53)$$

Here, the nodal transition probabilities in the measure $\widetilde{\mathbb{P}}^{(g)}$ with numéraire g are denoted by $\widetilde{p}_{(t;j_1,\ldots,j_N,j)\to(t+1;j_1',\ldots,j_N',j')}^{(g)} := \widetilde{\mathbb{P}}^{(g)}(S_{t+1}^1 = S_{t+1,j_1'}^1, \ldots, S_{t+1}^N = S_{t+1,j_N'}^N, R_{t+1} = R_{t+1,j'} \mid S_t^1 = S_{t,j_1}^1, \ldots, S_t^N = S_{t,j_N}^N, R_t = R_{t,j})$ and the transition probabilities in the measure $\widetilde{\mathbb{P}} \equiv \widetilde{\mathbb{P}}^{(B)}$ with numéraire B are denoted by $\widetilde{p}_{(t;j_1,\ldots,j_N,j)\to(t+1;j_1',\ldots,j_N',j')} := \widetilde{\mathbb{P}}(S_{t+1}^1 = S_{t+1,j_1'}^1, \ldots, S_{t+1}^N = S_{t+1,j_N'}^N, R_{t+1} = R_{t+1,j'} \mid S_t^1 = S_{t,j_1}^1, \ldots, S_t^N = S_{t,j_N}^N, R_t = R_{t,j})$.

As we have shown in the previous section, transition probabilities under any probability measure can be computed using (8.35) for a scalar process or (8.36) for a vector process. For any given EMM $\widetilde{\mathbb{P}}^{(g)}$, we can directly use (8.35) or (8.36) where we set $\widetilde{\mathbb{P}} \equiv \widetilde{\mathbb{P}}^{(g)}$. Moreover, employing (8.35) or (8.36), with $\widetilde{\mathbb{P}} \equiv \widetilde{\mathbb{P}}^{(g)}$, in combination with (8.50) and (8.52) allows us to generally compute transition probabilities under $\widetilde{\mathbb{P}}^{(g)}$ in terms of conditional and unconditional $\widetilde{\mathbb{P}}^{(f)}$-measure probabilities associated to the atoms. In particular, (8.35) gives

$$\begin{aligned}
&\widetilde{\mathbb{P}}^{(g)}(Y_{t+1} = Y_{t+1,k'} | Y_t = Y_{t,k}) \\
&= \frac{\sum_{i \in I_t(k)} \widetilde{\mathbb{P}}^{(f)}(A_t^i) \sum_{j \in I_{t+1}(k')} \widetilde{\mathbb{P}}^{(f)}(A_{t+1}^j | A_t^i) \cdot [g_{t+1}(A_{t+1}^j)/f_{t+1}(A_{t+1}^j)]}{\sum_{i \in I_t(k)} \widetilde{\mathbb{P}}^{(f)}(A_t^i) \cdot [g_t(A_t^i)/f_t(A_t^i)]}.
\end{aligned} \qquad (8.54)$$

The obvious analogous formula for the vector case follows from (8.36). Such formulas can be used for arbitrary choices of numéraire asset prices f_t and g_t which are $\mathcal{F}_t$-measurable (i.e., constant on any elementary time-t atom A_t^i for all $t \geq 0$), yet not necessarily $\sigma(Y_t)$-measurable (or $\sigma(\mathbf{Y}_t)$-measurable in the vector case).

In some applications of the risk-neutral pricing formula it may be beneficial to choose a particular stock price as the numéraire asset price. For example, let $g_t = S_t^1$ be the time-t price of stock #1 and let $f_t = B_t$. Let us denote $\widetilde{\mathbb{P}}^{(f)} \equiv \widetilde{\mathbb{P}}^{(B)} \equiv \widetilde{\mathbb{P}}$ and $\widetilde{\mathbb{P}}^{(g)} \equiv \widetilde{\mathbb{P}}^{(S^1)} \equiv \widehat{\mathbb{P}}$. Assume that we already know $\widetilde{\mathbb{P}}(A_t^i)$ and $\widetilde{\mathbb{P}}(A_{t+1}^j | A_t^i)$ for all the atoms. Then, the $\widehat{\mathbb{P}}$-measure probabilities follow directly from (8.50) and (8.52):

$$\widehat{\mathbb{P}}(A_t^i) = \left(\frac{S_t^1(A_t^i)/S_0^1}{B_t(A_t^i)/B_0}\right)\widetilde{\mathbb{P}}(A_t^i) \tag{8.55}$$

and

$$\widehat{\mathbb{P}}(A_{t+1}^j|A_t^i) = \frac{1}{1+R_t(A_t^i)}\frac{S_{t+1}^1(A_{t+1}^j)}{S_t^1(A_t^i)} \cdot \widetilde{\mathbb{P}}(A_{t+1}^j|A_t^i) \tag{8.56}$$

where we used $B_{t+1}(A_{t+1}^j)/B_t(A_t^i) = 1 + R_t(A_t^i)$ since $A_{t+1}^j \subseteq A_t^i$. We can use these within (8.35) which gives

$$\widehat{\mathbb{P}}(Y_{t+1} = Y_{t+1,k'}|Y_t = Y_{t,k}) = \frac{\displaystyle\sum_{i\in I_t(k)}\widehat{\mathbb{P}}(A_t^i)\sum_{j\in I_{t+1}(k')}\widehat{\mathbb{P}}(A_{t+1}^j|A_t^i)}{\displaystyle\sum_{i\in I_t(k)}\widehat{\mathbb{P}}(A_t^i)}$$

$$= \frac{\displaystyle\sum_{i\in I_t(k)}\frac{\widetilde{\mathbb{P}}(A_t^i)}{(1+R_t(A_t^i))B_t(A_t^i)}\sum_{j\in I_{t+1}(k')}S_{t+1}^1(A_{t+1}^j)\cdot\widetilde{\mathbb{P}}(A_{t+1}^j|A_t^i)}{\displaystyle\sum_{i\in I_t(k)}S_t^1(A_t^i)\cdot\frac{\widetilde{\mathbb{P}}(A_t^i)}{B_t(A_t^i)}} \cdot$$

$$\tag{8.57}$$

For the case of constant interest rates $R_t \equiv r$, this formula simplifies to

$$\widehat{\mathbb{P}}(Y_{t+1} = Y_{t+1,k'}|Y_t = Y_{t,k}) = \frac{1}{1+r}\frac{\displaystyle\sum_{i\in I_t(k)}\widetilde{\mathbb{P}}(A_t^i)\sum_{j\in I_{t+1}(k')}S_{t+1}^1(A_{t+1}^j)\cdot\widetilde{\mathbb{P}}(A_{t+1}^j|A_t^i)}{\displaystyle\sum_{i\in I_t(k)}S_t^1(A_t^i)\cdot\widetilde{\mathbb{P}}(A_t^i)}. \tag{8.58}$$

The following example gives a simple application of the above equations.

Example 8.10. Consider again the two-period trinomial model with three base assets as given in Examples 8.3 and 8.4. Determine the single-step transition probabilities under EMM $\widehat{\mathbb{P}} \equiv \widetilde{\mathbb{P}}^{(S^1)}$, for each stock price process $\{S_t^i\}_{t=0,1,2}$, $i = 1, 2$.

Solution. Note that in this case $R_t = r \equiv 0, B_t \equiv 1$. We can directly use (8.58) or we firstly compute the probabilities in (8.55)-(8.56) and then substitute them into the first-line equation in (8.57). We will use the latter approach. By using the known $\widetilde{\mathbb{P}}$-measure probabilities:

$$\widehat{\mathbb{P}}(A_1^1) = \frac{S_1^1(A_1^1)}{S_0^1}\widetilde{\mathbb{P}}(A_1^1) = \frac{50}{50}\cdot\frac{1}{3} = \frac{1}{3},$$

$$\widehat{\mathbb{P}}(A_1^2) = \frac{S_1^1(A_1^2)}{S_0^1}\widetilde{\mathbb{P}}(A_1^2) = \frac{40}{50}\cdot\frac{1}{3} = \frac{4}{15},$$

$$\widehat{\mathbb{P}}(A_1^3) = \frac{S_1^1(A_1^3)}{S_0^1}\widetilde{\mathbb{P}}(A_1^3) = \frac{60}{50}\cdot\frac{1}{3} = \frac{2}{5},$$

$$\widehat{\mathbb{P}}(A_2^1|A_1^1) = \frac{S_2^1(A_2^1)}{S_1^1(A_1^1)}\widetilde{\mathbb{P}}(A_2^1|A_1^1) = \frac{45}{50}\cdot\frac{1}{3} = \frac{3}{10},$$

$$\widehat{\mathbb{P}}(A_2^2|A_1^1) = \frac{S_2^1(A_2^2)}{S_1^1(A_1^1)}\widetilde{\mathbb{P}}(A_2^2|A_1^1) = \frac{45}{50}\cdot\frac{1}{3} = \frac{3}{10},$$

$$\widehat{\mathbb{P}}(A_2^3|A_1^1) = \frac{S_2^1(A_2^3)}{S_1^1(A_1^1)}\widetilde{\mathbb{P}}(A_2^3|A_1^1) = \frac{60}{50}\cdot\frac{1}{3} = \frac{2}{5},$$

$$\widehat{\mathbb{P}}(A_2^4|A_1^2) = \frac{S_2^1(A_2^4)}{S_1^1(A_1^2)}\widetilde{\mathbb{P}}(A_2^4|A_1^2) = \frac{45}{40}\cdot\frac{2}{9} = \frac{1}{4},$$

$$\widehat{\mathbb{P}}(A_2^5|A_1^2) = \frac{S_2^1(A_2^5)}{S_1^1(A_1^2)}\widetilde{\mathbb{P}}(A_2^5|A_1^2) = \frac{35}{40}\cdot\frac{2}{9} = \frac{7}{36},$$

$$\widehat{\mathbb{P}}(A_2^6|A_1^2) = \frac{S_2^1(A_2^6)}{S_1^1(A_1^2)}\widetilde{\mathbb{P}}(A_2^6|A_1^2) = \frac{40}{40}\cdot\frac{5}{9} = \frac{5}{9},$$

$$\widehat{\mathbb{P}}(A_2^7|A_1^3) = \frac{S_2^1(A_2^7)}{S_1^1(A_1^3)}\widetilde{\mathbb{P}}(A_2^7|A_1^3) = \frac{60}{60}\cdot\frac{2}{5} = \frac{2}{5},$$

$$\widehat{\mathbb{P}}(A_2^8|A_1^3) = \frac{S_2^1(A_2^8)}{S_1^1(A_1^3)}\widetilde{\mathbb{P}}(A_2^8|A_1^3) = \frac{55}{60}\cdot\frac{2}{5} = \frac{11}{30},$$

$$\widehat{\mathbb{P}}(A_2^9|A_1^3) = \frac{S_2^1(A_2^9)}{S_1^1(A_1^3)}\widetilde{\mathbb{P}}(A_2^9|A_1^3) = \frac{70}{60}\cdot\frac{1}{5} = \frac{7}{30}.$$

We now simply repeat the steps in Example 8.8 with $\widehat{\mathbb{P}}$-measure probabilities for stock #1:

$$\widehat{\mathbb{P}}(S_1^1 = 50, |S_0^1 = 50) = \widehat{\mathbb{P}}(S_1^1 = 50) = \widehat{\mathbb{P}}(A_1^1) = \frac{1}{3},$$

$$\widehat{\mathbb{P}}(S_1^1 = 40, |S_0^1 = 50) = \widehat{\mathbb{P}}(S_1^1 = 40) = \widehat{\mathbb{P}}(A_1^2) = \frac{4}{15},$$

$$\widehat{\mathbb{P}}(S_1^1 = 60, |S_0^1 = 50) = \widehat{\mathbb{P}}(S_1^1 = 60) = \widehat{\mathbb{P}}(A_1^3) = \frac{2}{5},$$

$$\widehat{\mathbb{P}}(S_2^1 = 45, |S_1^1 = 50) = \widehat{\mathbb{P}}(A_2^1|A_1^1) + \widehat{\mathbb{P}}(A_2^2|A_1^1) = \frac{3}{5},$$

$$\widehat{\mathbb{P}}(S_2^1 = 60, |S_1^1 = 50) = \widehat{\mathbb{P}}(A_2^3|A_1^1) = \frac{2}{5},$$

$$\widehat{\mathbb{P}}(S_2^1 = 45, |S_1^1 = 40) = \widehat{\mathbb{P}}(A_2^4|A_1^2) = \frac{1}{4},$$

$$\widehat{\mathbb{P}}(S_2^1 = 35, |S_1^1 = 40) = \widehat{\mathbb{P}}(A_2^5|A_1^2) = \frac{7}{36},$$

$$\widehat{\mathbb{P}}(S_2^1 = 40, |S_1^1 = 40) = \widehat{\mathbb{P}}(A_2^6|A_1^2) = \frac{5}{9},$$

$$\widehat{\mathbb{P}}(S_2^1 = 60, |S_1^1 = 60) = \widehat{\mathbb{P}}(A_2^7|A_1^3) = \frac{2}{5},$$

$$\widehat{\mathbb{P}}(S_2^1 = 55, |S_1^1 = 60) = \widehat{\mathbb{P}}(A_2^8|A_1^3) = \frac{11}{30},$$

$$\widehat{\mathbb{P}}(S_2^1 = 70, |S_1^1 = 60) = \widehat{\mathbb{P}}(A_2^9|A_1^3) = \frac{7}{30}.$$

For stock #2:

$$\widehat{\mathbb{P}}(S_1^2 = 115, |S_0^2 = 100) = \widehat{\mathbb{P}}(S_1^2 = 115) = \widehat{\mathbb{P}}(A_1^1) + \widehat{\mathbb{P}}(A_1^2) = \frac{3}{5}$$

$$\widehat{\mathbb{P}}(S_1^2 = 70, |S_0^2 = 100) = \widehat{\mathbb{P}}(S_1^2 = 70) = \widehat{\mathbb{P}}(A_1^3) = \frac{2}{5},$$

$$\widehat{\mathbb{P}}(S_2^2 = 130, |S_1^2 = 115) = \frac{\widehat{\mathbb{P}}(A_1^1) \cdot \widehat{\mathbb{P}}(A_2^1|A_1^1)}{\widehat{\mathbb{P}}(A_1^1) + \widehat{\mathbb{P}}(A_1^2)} = \frac{\frac{1}{3} \cdot \frac{3}{10}}{\frac{3}{5}} = \frac{1}{6},$$

$$\widehat{\mathbb{P}}(S_2^2 = 120, |S_1^2 = 115) = \frac{\widehat{\mathbb{P}}(A_1^1) \cdot \widehat{\mathbb{P}}(A_2^2|A_1^1) + \widehat{\mathbb{P}}(A_1^2) \cdot \widehat{\mathbb{P}}(A_2^5|A_1^2)}{\widehat{\mathbb{P}}(A_1^1) + \widehat{\mathbb{P}}(A_1^2)} = \frac{\frac{1}{3} \cdot \frac{3}{10} + \frac{4}{15} \cdot \frac{7}{36}}{\frac{3}{5}} = \frac{1}{6} + \frac{7}{81},$$

$$\widehat{\mathbb{P}}(S_2^2 = 95, |S_1^2 = 115) = \frac{\widehat{\mathbb{P}}(A_1^1) \cdot \widehat{\mathbb{P}}(A_2^3|A_1^1)}{\widehat{\mathbb{P}}(A_1^1) + \widehat{\mathbb{P}}(A_1^2)} = \frac{\frac{1}{3} \cdot \frac{2}{5}}{\frac{3}{5}} = \frac{2}{9},$$

$$\widehat{\mathbb{P}}(S_2^2 = 135, |S_1^2 = 115) = \frac{\widehat{\mathbb{P}}(A_1^2) \cdot \widehat{\mathbb{P}}(A_2^4|A_1^2)}{\widehat{\mathbb{P}}(A_1^1) + \widehat{\mathbb{P}}(A_1^2)} = \frac{\frac{4}{15} \cdot \frac{1}{4}}{\frac{3}{5}} = \frac{1}{9},$$

$$\widehat{\mathbb{P}}(S_2^2 = 105, |S_1^2 = 115) = \frac{\widehat{\mathbb{P}}(A_1^2) \cdot \widehat{\mathbb{P}}(A_2^6|A_1^2)}{\widehat{\mathbb{P}}(A_1^1) + \widehat{\mathbb{P}}(A_1^2)} = \frac{\frac{4}{15} \cdot \frac{5}{9}}{\frac{3}{5}} = \frac{20}{81},$$

$$\widehat{\mathbb{P}}(S_2^2 = 100, |S_1^2 = 70) = \widehat{\mathbb{P}}(A_2^7|A_1^3) = \frac{2}{5},$$

$$\widehat{\mathbb{P}}(S_2^2 = 40, |S_1^2 = 70) = \widehat{\mathbb{P}}(A_2^8|A_1^3) = \frac{11}{30},$$

$$\widehat{\mathbb{P}}(S_2^2 = 70, |S_1^2 = 70) = \widehat{\mathbb{P}}(A_2^9|A_1^3) = \frac{7}{30}.$$

Note that all transition probabilities with common initial condition sum to unity, as required.   □

The problem of pricing multi-stock options can be simplified in cases where the payoff has a certain symmetry. We will demonstrate this by considering a payoff that is a function of only two stock prices where (S_t^1, S_t^2) are assumed to be jointly Markov. The approach can also be employed in higher dimensional multi-stock cases. Consider a given payoff function $\Lambda(x_1, x_2)$ which can be written as a product of one of the spot prices, say x_1, and some given function $\phi(y)$ of the ratio of the spots, $y := x_2/x_1$, i.e., assume the payoff has the form $\Lambda(x_1, x_2) = x_1 \phi(x_2/x_1)$. As an $\mathcal{F}_{T_m}$-measurable random payoff we have

$$V_{T_m} = V_{T_m}(S_{T_m}^1, S_{T_m}^2) = \Lambda(S_{T_m}^1, S_{T_m}^2) = S_{T_m}^1 \phi(Y_{T_m})$$

$Y_{T_m} := S_{T_m}^2/S_{T_m}^1$, $0 \leqslant t \leqslant T_m \leqslant T$. As we now show, it proves convenient to consider the process defined by the ratio, $Y_t := \frac{S_t^2}{S_t^1}$, $t = 0, 1, \dots, T$.

By choosing $g_t = S_t^1$, the risk-neutral pricing formula in (8.31) gives (where $\widehat{\mathbb{E}} \equiv \widehat{\mathbb{E}}^{(S^1)}$)

$$\frac{V_t(S_t^1, S_t^2)}{S_t^1} = \widehat{\mathbb{E}}\left[\frac{V_{t+1}(S_{t+1}^1, S_{t+1}^2)}{S_{t+1}^1}\middle| S_t^1, S_t^2\right] = \widehat{\mathbb{E}}\left[\frac{V_{t+1}(S_{t+1}^1, S_{t+1}^2)}{S_{t+1}^1}\middle| S_t^1, Y_t\right], \qquad (8.59)$$

$t = 0, 1, \ldots, T_{m-1}$. Note that the price V_t can be equally considered as a function of (S_t^1, S_t^2) or (S_t^1, Y_t). At time $t = T_{m-1}$ we have

$$\frac{V_{T_{m-1}}(S_{T_{m-1}}^1, S_{T_{m-1}}^2)}{S_{T_{m-1}}^1} = \widehat{\mathrm{E}}\left[\phi(Y_{T_m}) \big| S_{T_{m-1}}^1, Y_{T_{m-1}}\right],$$

since $\frac{V_{T_m}(S_{T_m}^1, S_{T_m}^2)}{S_{T_m}^1} = \frac{\Lambda(S_{T_m}^1, S_{T_m}^2)}{S_{T_m}^1} = \phi(Y_{T_m})$.

Here is now a crucial assumption that leads to the factorization of the pricing function as a product. In particular, let us *assume that the ratio process* $\{Y_t\}_{t \geqslant 0}$ *is Markov*. Hence, $\widehat{\mathrm{E}}\left[f_{t+1}(Y_{t+1}) \big| S_t^1, Y_t\right] = \widehat{\mathrm{E}}\left[f_{t+1}(Y_{t+1}) \big| Y_t\right] := f_t(Y_t)$, for all $t \geqslant 0$, and therefore $\widehat{\mathrm{E}}[\phi(Y_{T_m}) \big| S_{T_{m-1}}^1, Y_{T_{m-1}}] = \widehat{\mathrm{E}}\left[\phi(Y_{T_m}) \big| Y_{T_{m-1}}\right] := f_{T_{m-1}}(Y_{T_{m-1}})$. Substituing this into the right-hand side of the above relation gives

$$\frac{V_{T_{m-1}}(S_{T_{m-1}}^1, S_{T_{m-1}}^2)}{S_{T_{m-1}}^1} = f_{T_{m-1}}(Y_{T_{m-1}}),$$

i.e., the price at time T_{m-1} is a product, $V_{T_{m-1}}(S_{T_{m-1}}^1, S_{T_{m-1}}^2) = S_{T_{m-1}}^1 f_{T_{m-1}}(Y_{T_{m-1}})$, $Y_{T_{m-1}} = S_{T_{m-1}}^2 / S_{T_{m-1}}^1$. That is, as a function of spot values $S_t^1 = x_1, S_t^2 = x_2$ we have $V_{T_{m-1}}(x_1, x_2) = x_1 f_{T_{m-1}}(y)$, $y = x_2/x_1$. Hence, by the Markov property and repeated use of the backward recurrence formula in (8.59), it readily follows that the $\mathcal{F}_t$-measurable derivative price has the product form $V_t(S_t^1, S_t^2) = S_t^1 f_t(Y_t)$, $Y_t = \frac{S_t^2}{S_t^1}$, with $f_t(Y_t) = \widehat{\mathrm{E}}\left[f_{t+1}(Y_{t+1}) \big| Y_t\right]$, $t = 0, 1, \ldots, T_{m-1}$, and where $f_{T_m}(Y_{T_m}) = \phi(Y_{T_m})$. That is, as a function of (ordinary) spot values $S_t^1 = x_1, S_t^2 = x_2$ we have the time-t pricing function $V_t(x_1, x_2) = x_1 f_t(y)$, $y = x_2/x_1$, with functions $f_t(y)$, $t = 0, 1, \ldots, T$, determined by backward recurrence $f_t(y) = \widehat{\mathrm{E}}\left[f_{t+1}(Y_{t+1}) \big| Y_t = y\right]$, where $f_{T_m}(y) = \phi(y)$. Hence, the pricing procedure is as follows. We have time-t nodes $(t; i, k)$ on the tree of price pairs $(S_t^1, Y_t) = (S_{t,i}^1, Y_{t,k})$. The time-$t$ derivative price, *expressed now as a function of* (S_t^1, Y_t), for each node $(t; i, k)$ is given by

$$V_t(S_{t,i}^1, Y_{t,k}) = S_{t,i}^1 f_t(Y_{t,k}), \quad t = 0, 1, \ldots, T_m, \tag{8.60}$$

where

$$f_t(Y_{t,k}) = \widehat{\mathrm{E}}\left[f_{t+1}(Y_{t+1}) \big| Y_t = Y_{t,k}\right] = \sum_{k'} \widehat{p}_{(t,k) \to (t+1,k')} \cdot f_{t+1}(Y_{t+1,k'}), \tag{8.61}$$

$\widehat{p}_{(t,k) \to (t+1,k')} := \widehat{\mathbb{P}}(Y_{t+1} = Y_{t+1,k'} | Y_t = Y_{t,k})$, for all (t, k) nodes of Y_t, $t = 0, 1, \ldots, T_{m-1}$, and where $f_{T_m}(Y_{T_m,k}) = \phi(Y_{T_m,k})$ for all terminal node values $Y_{T_m,k}$ of Y_{T_m}. The important point here is that the original derivative pricing problem on two stocks is reduced to an effective derivative pricing problem on only one asset with price process $\{Y_t\}_{t=0,1,\ldots,T_m}$ and effective one-dimensional payoff function $\phi(Y_{T_m})$. This symmetry reduction of the pricing problem is hence also referred to as dimensional reduction.

In the following example we implement (8.60)-(8.61) to the same pricing problem already solved in Example 8.7 where we now exploit the apparent symmetry of the payoff.

Example 8.11. Re-consider the derivative pricing problem solved in Example 8.7. Determine all the no-arbitrage derivative prices for the exchange option with payoff $V_T = (S_T^2 - S_T^1)^+$ at maturity $T = 2$ by implementing the risk-neutral pricing formula with stock #1 as numéraire asset.

Solution. The payoff function factors as: $\Lambda(x_1, x_2) = (x_2 - x_1)^+ = x_1(\frac{x_2}{x_1} - 1)^+ \equiv x_1\phi(\frac{x_2}{x_1})$ where $\phi(y) := (y-1)^+$. As a random variable, $\Lambda(S_T^1, S_T^2) = S_T^1\phi(Y_T)$, $Y_T = \frac{S_T^2}{S_T^1}$. In order to correctly implement (8.60)-(8.61) we firstly require the ratio process $Y_t := \frac{S_t^2}{S_t^1}$, $t = 0, 1, 2$ to be Markov and we need the $\widehat{\mathbb{P}}$-measure transition probabilities for this process. The Markov property is easily shown based on the calculations in Example 8.10. The nodes (see Figure 8.4) on the tree of values (S_t^1, Y_t) are $(S_0^1, Y_0) = (50, 2)$ at $t = 0$; $(S_{1,1}^1, Y_{1,1}) = (50, \frac{115}{50})$, $(S_{1,2}^1, Y_{1,2}) = (40, \frac{115}{40})$, $(S_{1,3}^1, Y_{1,3}) = (60, \frac{70}{60})$ at $t = 1$; $(S_{2,1}^1, Y_{2,1}) = (45, \frac{130}{45})$, $(S_{2,2}^1, Y_{2,2}) = (45, \frac{120}{45})$, $(S_{2,3}^1, Y_{2,3}) = (60, \frac{95}{60})$, $(S_{2,4}^1, Y_{2,4}) = (45, \frac{135}{45})$, $(S_{2,5}^1, Y_{2,5}) = (35, \frac{120}{35})$, $(S_{2,6}^1, Y_{2,6}) = (40, \frac{105}{40})$, $(S_{2,7}^1, Y_{2,7}) = (60, \frac{100}{60})$, $(S_{2,8}^1, Y_{2,8}) = (55, \frac{40}{55})$, $(S_{2,9}^1, Y_{2,9}) = (70, 1)$. At maturity we simply have the effective payoff values $f_2(Y_{2,k}) = \phi(Y_{2,k}) = (Y_{2,k} - 1)^+$:

$$f_2(Y_{2,1}) = \left(\frac{130}{45} - 1\right)^+ = \frac{17}{9}, \; f_2(Y_{2,2}) = \left(\frac{120}{45} - 1\right)^+ = \frac{5}{3}, \; f_2(Y_{2,3}) = \left(\frac{95}{60} - 1\right)^+ = \frac{7}{12}$$

$$f_2(Y_{2,4}) = \left(\frac{135}{45} - 1\right)^+ = 2, \; f_2(Y_{2,5}) = \left(\frac{120}{35} - 1\right)^+ = \frac{17}{7}, \; f_2(Y_{2,6}) = \left(\frac{105}{40} - 1\right)^+ = \frac{13}{8},$$

$$f_2(Y_{2,7}) = \left(\frac{5}{3} - 1\right)^+ = \frac{2}{3}, \; f_2(Y_{2,8}) = 0, \; f_2(Y_{2,9}) = 0.$$

The option prices at maturity, given by $V_2(S_{2,i}^1, Y_{2,k}) = S_{2,i}^1 f_2(Y_{2,k})$, obviously correspond to the payoff values: $V_2(45, \frac{130}{45}) = 85$, $V_2(45, \frac{120}{45}) = 75$, $V_2(60, \frac{95}{60}) = 35$, $V_2(45, \frac{135}{45}) = 90$, $V_2(35, \frac{120}{35}) = 85$, $V_2(40, \frac{105}{40}) = 65$, $V_2(60, \frac{100}{60}) = 40$, $V_2(55, \frac{40}{55}) = 0$, $V_2(70, 1) = 0$.

The transition probabilities, computed in Example 8.10, are simply given by

$$\widehat{p}_{(t,k)\to(t+1,k')} = \widehat{\mathbb{P}}(A_{t+1}^{k'}|A_t^k)$$

(note: there is only one term in (8.58) when $Y_t = S_t^2/S_t^1$). Hence, using (8.61) for $t = 1$ and the payoff values:

$$f_1(Y_{1,k}) = \sum_{k'} \widehat{p}_{(1,k)\to(2,k')} \cdot f_2(Y_{2,k'}).$$

For respective values $k = 1, 2, 3$:

$$f_1(Y_{1,1}) = \widehat{p}_{(1,1)\to(2,1)} \cdot f_2(Y_{2,1}) + \widehat{p}_{(1,1)\to(2,2)} \cdot f_2(Y_{2,2}) + \widehat{p}_{(1,1)\to(2,3)} \cdot f_2(Y_{2,3})$$
$$= \frac{3}{10} \cdot \frac{17}{9} + \frac{3}{10} \cdot \frac{5}{3} + \frac{2}{5} \cdot \frac{7}{12} = \frac{13}{10}$$

$$f_1(Y_{1,2}) = \widehat{p}_{(1,2)\to(2,4)} \cdot f_2(Y_{2,4}) + \widehat{p}_{(1,2)\to(2,5)} \cdot f_2(Y_{2,5}) + \widehat{p}_{(1,2)\to(2,6)} \cdot f_2(Y_{2,6})$$
$$= \frac{1}{4} \cdot 2 + + \frac{7}{36} \cdot \frac{17}{7} + \frac{5}{9} \cdot \frac{13}{8} = \frac{135}{72}$$

$$f_1(Y_{1,3}) = \widehat{p}_{(1,3)\to(2,7)} \cdot f_2(Y_{2,7}) + \widehat{p}_{(1,3)\to(2,8)} \cdot f_2(Y_{2,8}) + \widehat{p}_{(1,3)\to(2,9)} \cdot f_2(Y_{2,9})$$
$$= \frac{2}{5} \cdot \frac{2}{3} + \frac{11}{30} \cdot 0 + \frac{7}{30} \cdot 0 = \frac{4}{15}.$$

The time-1 option values, $V_1(S_{1,i}^1, Y_{1,k}) = S_{1,i}^1 f_1(Y_{1,k})$, are:

$$V_1(50, \frac{115}{50}) = 50 \cdot \frac{13}{10} = 65, \; V_1(40, \frac{115}{40}) = 40 \cdot \frac{135}{72} = 75, \; V_1(60, \frac{70}{60}) = 60 \cdot \frac{4}{15} = 16.$$

At time 0:

$$f_0(Y_0) = \widehat{p}_{(0,0)\to(1,1)} \cdot f_1(Y_{1,1}) + \widehat{p}_{(0,0)\to(1,2)} \cdot f_1(Y_{1,2}) + \widehat{p}_{(0,0)\to(1,3)} \cdot f_1(Y_{1,3})$$
$$= \frac{1}{3} \cdot \frac{13}{10} + \frac{4}{15} \cdot \frac{135}{72} + \frac{2}{5} \cdot \frac{4}{15} = \frac{78}{75}.$$

The time-0 option value is: $V_0 \equiv V_0(50, 2) = S_0^1 \cdot f_0(Y_0) = 50 \cdot \frac{78}{75} = 52$. Note that all option prices agree with those in Example 8.7, as required.

$\square$

8.4 More Examples of Discrete-Time Models

8.4.1 Binomial Tree Model with Stochastic Volatility

In the standard binomial model the upward and downward factors, u and d, and the interest rate r were all assumed constant. We now consider a binomial tree model where the sequence of upward and downward factors $\{u_n\}_{n \geqslant 1}$ and $\{d_n\}_{n \geqslant 1}$, respectively, and the interest rates $\{r_n\}_{n \geqslant 1}$ are all stochastic processes adapted to the natural filtration (generated by either the moves or the stock prices at all times). For each time $n \geqslant 1$, the factors u_n and d_n and the interest rate r_n, for the time period $(n-1, n]$, depend on the first $n-1$ market moves, i.e., for a given sequence of moves $\omega_1 \omega_2 \ldots \omega_{n-1}$ the respective random variables u_n, d_n, r_n have values

$$u_n(\omega_1 \omega_2 \ldots \omega_{n-1}), \quad d_n(\omega_1 \omega_2 \ldots \omega_{n-1}), \quad r_n(\omega_1 \omega_2 \ldots \omega_{n-1}).$$

For the first period $(0, 1]$, $(n-1, n]$ with $n = 1$, these quantities are $\mathcal{F}_0$-measurable, i.e., at time zero the initial factors u_1 and d_1 and the initial rate r_1 are known constants. Hence, the processes $\{u_n\}_{n \geqslant 1}$, $\{d_n\}_{n \geqslant 1}$, $\{r_n\}_{n \geqslant 1}$ are all $\mathbb{F}$-predictable. [2] The initial stock price S_0 is positive. Assuming that $0 < d_n(\omega_1 \omega_2 \ldots \omega_{n-1}) \leqslant u_n(\omega_1 \omega_2 \ldots \omega_{n-1})$ holds for every $n \geqslant 1$ and all $\omega_1, \omega_2, \ldots, \omega_{n-1} \in \{\mathsf{D}, \mathsf{U}\}$ guarantees the positiveness of all future stock prices. The dynamics of the stock price at time $n \geqslant 2$ is given by

$$S_n(\omega_1 \omega_2 \ldots \omega_{n-1} \omega_n) = \begin{cases} u_n(\omega_1 \omega_2 \ldots \omega_{n-1}) S_{n-1}(\omega_1 \omega_2 \ldots \omega_{n-1}) & \text{if } \omega_n = \mathsf{U}, \\ d_n(\omega_1 \omega_2 \ldots \omega_{n-1}) S_{n-1}(\omega_1 \omega_2 \ldots \omega_{n-1}) & \text{if } \omega_n = \mathsf{D}. \end{cases} \quad (8.62)$$

At time $n = 1$, we have

$$S_1(\omega_1) = \begin{cases} u_1 S_0 & \text{if } \omega_1 = \mathsf{U}, \\ d_1 S_0 & \text{if } \omega_1 = \mathsf{D}. \end{cases}$$

The bank account process is stochastic and its value at time $n \geqslant 2$ is given by

$$B_n(\omega_1 \omega_2 \ldots \omega_{n-1}) = B_{n-1}(\omega_1 \omega_2 \ldots \omega_{n-2})(1 + r_n(\omega_1 \omega_2 \ldots \omega_{n-1})),$$

with $B_1 = B_0(1 + r_1)$ known at time 0. The initial value B_0 is positive. Typically, we assume that $B_0 = 1$. As already noted, the bank account value B_n depends on $\omega_1 \omega_2 \ldots \omega_{n-1}$ and is independent of ω_n. That is, the process $\{B_n\}_{n \geqslant 0}$ is $\mathbb{F}$-predictable.

The main characteristics of this model are summarized as follows.

- Every outcome $\omega = \omega_1 \omega_2 \ldots \omega_T$ can be considered as a path in a binomial tree, which is not necessarily a recombining one. The recombination of a binomial tree is essential

---

[2] As mentioned previously, by our definition $r_n := \frac{B_n - B_{n-1}}{B_{n-1}}$ is the return within the period $(n-1, n]$, so that r_n is known at time $n - 1$. Similarly, here we are using u_n and d_n to denote the factors for the period $(n-1, n]$. One can instead equivalently and alternatively choose to denote these factors for period $(n-1, n]$ by u_{n-1} and d_{n-1}. In this latter definition one has u_n and d_n as $\mathcal{F}_n$-measurable random variables with given values $u_n(\omega_1 \omega_2 \ldots \omega_n)$ and $d_n(\omega_1 \omega_2 \ldots \omega_n)$ on a given atom $A_{\omega_1 \omega_2 \ldots \omega_n}$.

since the nodes of one time step increase linearly with the number of time steps in a recombining tree. In a nonrecombining tree, in contrast, the nodes increase exponentially, so that the tree can only be built for a few steps, even using modern computers. A two-period binomial tree is recombining iff $u_1 d_2(\mathsf{U}) = d_1 u_2(\mathsf{D})$ holds. A general multi-period binomial tree is recombining iff every two-period sub-tree is recombining. So, we have the following necessary and sufficient condition:

$$u_n(\omega_1 \ldots \omega_{n-1})\, d_{n+1}(\omega_1 \ldots \omega_{n-1}\mathsf{U}) = d_n(\omega_1 \ldots \omega_{n-1})\, u_{n+1}(\omega_1 \ldots \omega_{n-1}\mathsf{D}) \qquad (8.63)$$

for all $n \in \{1, 2, \ldots, T-1\}$ and all $\omega_1, \omega_2, \ldots, \omega_{n-1} \in \{\mathsf{U}, \mathsf{D}\}$. For example, a three-period binomial tree, which contains three two-period sub-trees, is recombining iff the following conditions hold:

$$u_1\, d_2(\mathsf{U}) = d_1\, u_2(\mathsf{D})\,,$$
$$u_2(\mathsf{U})\, d_3(\mathsf{UU}) = d_2(\mathsf{U})\, u_3(\mathsf{UD})\,,$$
$$u_2(\mathsf{D})\, d_3(\mathsf{DU}) = d_2(\mathsf{D})\, u_3(\mathsf{DD})\,.$$

- This model is arbitrage-free iff every single-period binomial sub-model is arbitrage-free. Therefore, there is no arbitrage iff $d_n < 1 + r_n < u_n$:

$$d_n(\omega_1\omega_2 \ldots \omega_{n-1}) < 1 + r_n(\omega_1\omega_2 \ldots \omega_{n-1}) < u_n(\omega_1\omega_2 \ldots \omega_{n-1}) \qquad (8.64)$$

holds for all $n \in \{1, 2, \ldots, T\}$ and all market moves $\omega_1, \omega_2, \ldots, \omega_{n-1} \in \{\mathsf{D}, \mathsf{U}\}$. This condition ensures the existence of an EMM $\widetilde{\mathbb{P}}^{(B)}$ with bank account as numéraire. In what follows we simply denote this by $\widetilde{\mathbb{P}}$.

- To determine the risk-neutral state probabilities $\widetilde{\mathbb{P}}(\omega_1\omega_2 \ldots \omega_T)$ for a binomial model with T periods, we need to compute the risk-neutral probabilities for each single-period sub-model. At time zero, there is only one binomial sub-model and the two risk-neutral probabilities are

$$\widetilde{\mathbb{P}}(\omega_1 = \mathsf{D}) \equiv \tilde{p}_1(\mathsf{D}) = \frac{u_1 - 1 - r_1}{u_1 - d_1}, \quad \widetilde{\mathbb{P}}(\omega_1 = \mathsf{U}) \equiv \tilde{p}_1(\mathsf{U}) = \frac{1 + r_1 - d_1}{u_1 - d_1}. \qquad (8.65)$$

Fix arbitrarily $n \geqslant 1$ and $\omega_1, \omega_2, \ldots, \omega_n \in \{\mathsf{D}, \mathsf{U}\}$. Consider a binomial sub-tree originated from the path $\omega_1, \omega_2, \ldots, \omega_n$ in a binomial tree. The risk-neutral probabilities of the events $\{\omega_{n+1} = \mathsf{D}\}$ and $\{\omega_{n+1} = \mathsf{U}\}$ conditional on the event (atom) $A_{\omega_1\omega_2 \ldots \omega_n}$ are, respectively, given by

$$\widetilde{\mathbb{P}}(\omega_{n+1} = \mathsf{D} \mid A_{\omega_1\omega_2 \ldots \omega_n}) \equiv \tilde{p}_{n+1}(\mathsf{D} \mid \omega_1\omega_2 \ldots \omega_n)$$
$$= \frac{u_{n+1}(\omega_1\omega_2 \ldots \omega_n) - 1 - r_{n+1}(\omega_1\omega_2 \ldots \omega_n)}{u_{n+1}(\omega_1\omega_2 \ldots \omega_n) - d_{n+1}(\omega_1\omega_2 \ldots \omega_n)}, \qquad (8.66)$$

$$\widetilde{\mathbb{P}}(\omega_{n+1} = \mathsf{U} \mid A_{\omega_1\omega_2 \ldots \omega_n}) \equiv \tilde{p}_{n+1}(\mathsf{U} \mid \omega_1\omega_2 \ldots \omega_n)$$
$$= \frac{1 + r_{n+1}(\omega_1\omega_2 \ldots \omega_n) - d_{n+1}(\omega_1\omega_2 \ldots \omega_n)}{u_{n+1}(\omega_1\omega_2 \ldots \omega_n) - d_{n+1}(\omega_1\omega_2 \ldots \omega_n)}. \qquad (8.67)$$

Equations (8.66)-(8.67), together with (8.62), are equivalent to the $\widetilde{\mathbb{P}}$-martingale property of the discounted stock price:

$$S_n(\omega_1 \ldots \omega_n) = (1 + r_{n+1}(\omega_1 \ldots \omega_n))^{-1}\big[\tilde{p}_{n+1}(\mathsf{U}|\omega_1 \ldots \omega_n)S_{n+1}(\omega_1 \ldots \omega_n\mathsf{U})$$
$$+ \tilde{p}_{n+1}(\mathsf{D}|\omega_1 \ldots \omega_n)S_{n+1}(\omega_1 \ldots \omega_n\mathsf{D})\big]. \qquad (8.68)$$

Finally, the risk-neutral state probability $\widetilde{\mathbb{P}}(A_{\omega_1\omega_2...\omega_T}) \equiv \widetilde{\mathbb{P}}(\omega_1\omega_2 \ldots \omega_T)$ of any outcome $\omega = \omega_1\omega_2\ldots\omega_T \in \Omega_T$ can be computed as (see Section 6.3.5 of Chapter 6)

$$\widetilde{\mathbb{P}}(\omega_1\omega_2\ldots\omega_T) = \tilde{p}_1(\omega_1)\tilde{p}_2(\omega_2 \mid \omega_1) \cdots \tilde{p}_T(\omega_T \mid \omega_1\omega_2\ldots\omega_{T-1}).$$

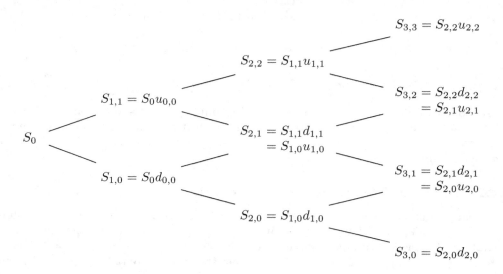

FIGURE 8.5: A schematic representation of a three-period binomial lattice with state-dependent market move factors.

A useful version of a stochastic binomial model is a model with state-dependent returns (see Figure 8.5). Let us recall some basic facts about binomial lattices. In a recombining binomial lattice, all paths with the same number of upward and downward moves lead to the same node. The node (n, m) of a binomial lattice (with integer coordinates n and m so that $0 \leqslant m \leqslant n$) is reached from the root node $(0, 0)$ by making m upward moves and $n - m$ downward moves. There are $\binom{n}{m}$ such paths. The stock price at the node (n, m) is denoted by $S_{n,m}$. Let the factors and the risk-free rate be functions of n and m. That is, we denote the respective factors at each node (n, m) by $d_{n,m}$ and $u_{n,m}$. In this model, the above up and down factors collapse to factors that depend only on the time value n and the total number of up moves m at time n. That is, $u_{n,m} = u_{n+1}(\omega_1\omega_2\ldots\omega_n)$ and $d_{n,m} = d_{n+1}(\omega_1\omega_2\ldots\omega_n)$ for all $\omega_1, \omega_2, \ldots, \omega_n$ s.t. $\#U(\omega_1\omega_2\ldots\omega_n) = m$. In particular, $u_{0,0} = u_1, u_{1,0} = u_2(D), u_{1,1} = u_2(U), u_{2,0} = u_3(DD), u_{2,1} = u_3(DU) = u_3(UD), u_{2,2} = u_3(UU)$ and similarly for the corresponding down factors. In this mdoel, the random up and down factors are functions of the stock price: $u_n = u_n(S_n)$ and $d_n = d_n(S_n)$. Let's further assume that the interest rate for the n-th period $r_{n+1} = R_n(S_n)$ is a function of the stock price with nodal values $R_{n,m} = R_n(S_{n,m})$. The factors have nodal values $d_{n,m} = d_n(S_{n,m})$ and $u_{n,m} = u_n(S_{n,m})$, $n = 0, \ldots, T - 1$, and given initial values $d_0 \equiv d_{0,0}$ and $u_0 \equiv u_{0,0}$.

The recombining condition (8.63) takes the form

$$u_{n,m}\, d_{n+1,m+1} = d_{n,m}\, u_{n+1,m}$$

for all integers n and m with the property $0 \leqslant m \leqslant n \leqslant T - 2$. The stock prices are given by the following iterative formulae:

$$S_{n+1,0} = S_{n,0}d_{n,0},$$
$$S_{n+1,m} = S_{n,m}d_{n,m} = S_{n,m-1}u_{n,m-1} \text{ for } 1 \leqslant m \leqslant n,$$
$$S_{n+1,n+1} = S_{n,n}u_{n,n},$$

for any $n \in \{0, 1, \ldots, T - 1\}$. The risk-neutral transition probabilities in this model are given in terms of the interest rates and the factors (see (8.66) and (8.67)):

$$\tilde{\mathbb{P}}(S_{n+1} = S_{n+1,m+1} | S_n = S_{n,m}) \equiv \tilde{p}_{(n,m) \to (n+1,m+1)} = \frac{u_{n,m} - 1 - R_{n,m}}{u_{n,m} - d_{n,m}},$$

$$\tilde{\mathbb{P}}(S_{n+1} = S_{n+1,m} | S_n = S_{n,m}) \equiv \tilde{p}_{(n,m) \to (n+1,m)} = \frac{1 + R_{n,m} - d_{n,m}}{u_{n,m} - d_{n,m}}.$$

Although we do not discuss here practical examples, a state-dependent binomial tree can be constructed as an approximation of a continuous-time stock price process with time- and state-dependent volatility $\sigma(t, S)$. The tree consistent with a given volatility structure is called the *implied tree*.

8.4.2 Binomial Tree Model for Interest Rates

The goal of this sub-section is to develop a binomial-tree model for stochastic interest rates. This model can be used for pricing (zero-)coupon bonds and other interest rate derivatives. Consider a discrete-time model with the finite state space $|\Omega_T| = 2^T$, natural filtration $\{\mathcal{F}_n\}_{n \in \{0,1,\ldots,T\}}$ generated by the market moves, and risk-free *rate of interest* $\{r_n\}_{n \in \{1,2,\ldots,T\}}$. As done throughout, we are simply denoting time by integer values which may be mapped to actual discrete calendar times for each period. As defined above, the rate r_n is $\mathcal{F}_{n-1}$ measurable for every $n \in \{1, 2, \ldots, T\}$. In particular, the rate r_1 for the first period (from time 0 to time 1) is initially known. We are assuming a binomial model where r_n is $\mathcal{F}_{n-1}$-measurable with constant value $r_n(\omega_1, \ldots, \omega_{n-1})$ on each atom $A_{\omega_1 \ldots \omega_{n-1}}$, for $n \geqslant 2$. In most situations, the rates r_n are all positive for all market outcomes. In general, we only require $r_n > -1$ $(1 + r_n > 0)$, which is assumed to hold for all market outcomes.

The rate r_n is valid for the nth period, i.e., one dollar invested in the bank account at time $n - 1$ grows to $1 + r_n$ dollars at time n. So, P_0 dollars invested at time 0 grows to $P_n = P_0(1 + r_1)(1 + r_2) \cdots (1 + r_n)$ at time n. Since the rates $r_1, r_2, \ldots, r_n$ are all $\mathcal{F}_{n-1}$-measurable, the accumulated random value P_n is $\mathcal{F}_{n-1}$-measurable with given numerical value $P_n(\omega_1, \omega_2, \ldots, \omega_{n-1})$ on a given atom $A_{\omega_1 \ldots \omega_{n-1}}$.

In accordance with the definition in (8.11), the bank account process $\{B_n\}_{n=0,1,\ldots,T}$ is given by

$$B_n = B_0(1 + r_1)(1 + r_2) \cdots (1 + r_n) \text{ for all } n \in \{1, 2, \ldots, T\},$$

with known initial value B_0. We define the discount factor at time $n \geqslant 1$ by

$$D_n := \frac{B_0}{B_n} = \frac{1}{(1 + r_1)(1 + r_2) \cdots (1 + r_n)}, \tag{8.69}$$

and $D_0 = 1$. For any $n < m$ we have the ratio

$$\frac{B_n}{B_m} = \frac{D_m}{D_n} = \frac{1}{(1 + r_{n+1})(1 + r_{n+2}) \cdots (1 + r_m)}. \tag{8.70}$$

The risk-neutral pricing formula (8.17) gives us the time-n value of a time-m payment (payoff) V_m received at time m, $n \leqslant m \leqslant T$ (where V_m is $\mathcal{F}_m$-measurable) is

$$V_n = B_n \tilde{\mathbb{E}}_n \left[\frac{V_m}{B_m} \right] \text{ for } n \in \{0, 1, \ldots, m - 1\}, \tag{8.71}$$

where the risk-neutral expectation is taken under the EMM $\widetilde{\mathbb{P}} \equiv \widetilde{\mathbb{P}}^{(B)}$ with bank account as numéraire asset. Note that here V_n is an $\mathcal{F}_n$-measurable random variable where we are using the convenient shorthand notation $\widetilde{\mathrm{E}}_n[\cdot] \equiv \widetilde{\mathrm{E}}[\cdot|\mathcal{F}_n]$. Therefore the time-$n$ no-arbitrage prices of a zero-coupon bond (ZCB) that pays one dollar at respective maturity times m ($V_m \equiv 1$) are (as $\mathcal{F}_n$-measurable random variables):

$$
\begin{aligned}
Z_{n,m} = B_n \widetilde{\mathrm{E}}_n \left[\frac{1}{B_m} \right] &\equiv \frac{1}{D_n} \widetilde{\mathrm{E}}_n \left[D_m \right] = \widetilde{\mathrm{E}}_n \left[\frac{1}{(1+r_{n+1})(1+r_{n+2})\cdots(1+r_m)} \right] \\
&= \frac{1}{(1+r_{n+1})} \widetilde{\mathrm{E}}_n \left[\frac{1}{(1+r_{n+2})\cdots(1+r_m)} \right],
\end{aligned} \tag{8.72}
$$

for $n = 0, 1, \ldots, m-1$. Note that r_{n+1} is $\mathcal{F}_n$-measurable. At the maturity time m, the price of the ZCB is $Z_{m,m} = V_m = 1$ and for $n = m-1$ we have $Z_{m-1,m} = (1+r_m)^{-1}$.

Equation (8.72) gives $Z_{n,m}$ as $\mathcal{F}_n$-measurable random variables. Given an outcome (atom) $A_{\omega_1 \ldots \omega_n}$, we have all time-$n$, maturity $m > n$, ZCB prices given by (see (6.41) in Chapter 6):

$$
\begin{aligned}
Z_{n,m}(\omega_1 \ldots \omega_n) &= \sum_{\omega_{n+1},\ldots,\omega_m \in \{\mathsf{D},\mathsf{U}\}} \frac{B_n}{B_m}(\omega_1 \ldots \omega_n \omega_{n+1} \ldots \omega_m) \widetilde{p}(\omega_{n+1} \ldots \omega_m | \omega_1 \ldots \omega_n) \\
&= \sum_{\omega_{n+1},\ldots,\omega_m \in \{\mathsf{D},\mathsf{U}\}} \frac{\widetilde{p}(\omega_{n+1} \ldots \omega_m | \omega_1 \ldots \omega_n)}{(1+r_{n+1}(\omega_1 \ldots \omega_n))\cdots(1+r_m(\omega_1 \ldots \omega_{m-1}))} \\
&= \frac{1}{(1+r_{n+1}(\omega_1 \ldots \omega_n))} \sum_{\omega_{n+1},\ldots,\omega_{m-1} \in \{\mathsf{D},\mathsf{U}\}} \frac{\widetilde{p}(\omega_{n+1} \ldots \omega_{m-1} | \omega_1 \ldots \omega_n)}{(1+r_{n+2}(\omega_1 \ldots \omega_{n+1}))\cdots(1+r_m(\omega_1 \ldots \omega_{m-1}))}
\end{aligned}
$$
$$\tag{8.73}$$

where the *risk-neutral conditional probabilities* are given by (see (6.39) in Chapter 6):

$$
\widetilde{p}(\omega_{n+1} \ldots \omega_{m-1} | \omega_1 \ldots \omega_n) := \frac{\widetilde{\mathbb{P}}(A_{\omega_1 \ldots \omega_{m-1}})}{\widetilde{\mathbb{P}}(A_{\omega_1 \ldots \omega_n})} \equiv \frac{\widetilde{\mathbb{P}}(\omega_1 \ldots \omega_{m-1})}{\widetilde{\mathbb{P}}(\omega_1 \ldots \omega_n)}. \tag{8.74}
$$

Note that the last line of (8.73) follows since r_{n+1} is constant on $A_{\omega_1 \ldots \omega_n}$ and the sum over $\omega_m \in \{\mathsf{D}, \mathsf{U}\}$ collapses: $\widetilde{p}(\omega_{n+1} \ldots \omega_{m-1}\mathsf{D}|\omega_1 \ldots \omega_n) + \widetilde{p}(\omega_{n+1} \ldots \omega_{m-1}\mathsf{U}|\omega_1 \ldots \omega_n) = \widetilde{p}(\omega_{n+1} \ldots \omega_{m-1}|\omega_1 \ldots \omega_n)$. Also note that in the case $m = n+1$ the above formula is interpreted with no sum where $Z_{n,n+1}(\omega_1 \ldots \omega_n) = (1 + r_{n+1}(\omega_1 \ldots \omega_n))^{-1}$.

An important special case of (8.73) arises for $n = 0$, giving the time-0 risk-neutral pricing formula for a ZCB maturing at any time $m \geqslant 1$:

$$
Z_{0,m} = \frac{1}{(1+r_1)} \sum_{\omega_1,\ldots,\omega_{m-1} \in \{\mathsf{D},\mathsf{U}\}} \frac{\widetilde{\mathbb{P}}(\omega_1 \ldots \omega_{m-1})}{(1+r_2(\omega_1))\cdots(1+r_m(\omega_1 \ldots \omega_{m-1}))}. \tag{8.75}
$$

Note that $Z_{0,0} = 1$ and for $m = 1$ we simply have $Z_{0,1} = (1+r_1)^{-1}$ with rate r_1 known at time 0. The first nontrivial case is for $m = 2$:

$$
Z_{0,2} = \frac{1}{(1+r_1)} \sum_{\omega_1 \in \{\mathsf{D},\mathsf{U}\}} \frac{\widetilde{\mathbb{P}}(\omega_1)}{(1+r_2(\omega_1))} = \frac{1}{(1+r_1)} \left[\frac{\widetilde{\mathbb{P}}(\mathsf{U})}{1+r_2(\mathsf{U})} + \frac{\widetilde{\mathbb{P}}(\mathsf{D})}{1+r_2(\mathsf{D})} \right], \tag{8.76}
$$

with risk-neutral probabilities $\widetilde{\mathbb{P}}(\mathsf{U})$ and $\widetilde{\mathbb{P}}(\mathsf{D}) = 1 - \widetilde{\mathbb{P}}(\mathsf{U})$.

We can also determine the yield rate $y_{n,m}$, $0 \leqslant n \leqslant m \leqslant T$, which is a fixed rate of interest for the period from time n to m defined so that it is equivalent to the rate of return of the ZCB maturing at time m. Solving the respective time-value equation gives the yield rate:

$$Z_{n,m} = (1 + y_{n,m})^{-(m-n)} \implies y_{n,m} = \left(\frac{1}{Z_{n,m}}\right)^{\frac{1}{(m-n)}} - 1. \tag{8.77}$$

Note that $y_{n,m}$ are $\mathcal{F}_n$-measurable random variables which are known for a given outcome $A_{\omega_1...\omega_n}$.

We can also define other $\mathcal{F}_n$-measurable rates of return which are called *forward rates*. In particular, the forward rate at time $n \leqslant m < T$ for one period $[m, m+1]$ is defined by

$$f_{n,m} := \frac{Z_{n,m}}{Z_{n,m+1}} - 1. \tag{8.78}$$

Note that $Z_{n,m} = (1 + f_{n,m})Z_{n,m+1}$, or $(1 + f_{n,m})^{-1}Z_{n,m} = Z_{n,m+1}$. In the special case when $n = m$ we simply have $f_{m,m} = \frac{1}{Z_{m,m+1}} - 1 = R_m \equiv r_{m+1}$, which is the rate of return on the bank account at current time m. The forward rate $f_{n,m}$ corresponds to the simple rate of return that is locked in at current time n for a future single time period $[m, m+1]$. We can readily see this by considering the following simple trading (investment) strategy. At time n, a trader short sells one unit of an m-maturity ZCB and uses the proceeds (having value $Z_{n,m}$) to purchase $b \equiv \frac{Z_{n,m}}{Z_{n,m+1}}$ units of an $(m+1)$-maturity ZCB with value $Z_{n,m+1}$. At time n this portfolio has zero cost to set up since $-Z_{n,m} + bZ_{n,m+1} = 0$. At time m, the trader covers the short ZCB position by paying one dollar ($Z_{m,m} = 1$). Then, at time $m+1$ the trader receives b dollars from the long position (since $Z_{m+1,m+1} = 1$). Hence, this strategy is equivalent to investing one dollar at time m for a payoff of b dollars at time $m+1$. The simple rate of return over the single period $[m, m+1]$ is therefore $b - 1 = \frac{Z_{n,m}}{Z_{n,m+1}} - 1$. Note that this forward rate must be the agreed rate at time n in order to avoid an obvious arbitrage in the bond market.

The one-period forward rate in (8.78) extends to multi-period forward rates. We define the ($\mathcal{F}_n$-measurable) forward rate at time $n \leqslant m$ for period $[m, m+k]$, $k \geqslant 1$, $m + k \leqslant T$, by

$$f_{n;m,m+k} := \frac{Z_{n,m}}{Z_{n,m+k}} - 1. \tag{8.79}$$

The single-period forward rate is a special case of the k-period rate, i.e., for $k = 1$, $f_{n;m,m+1} \equiv f_{n,m}$. This rate relates the m-maturity ZCB prices with the $(m+k)$-maturity ZCB prices: $Z_{n,m} = (1 + f_{n;m,m+k})Z_{n,m+k}$, or $(1 + f_{n;m,m+k})^{-1}Z_{n,m} = Z_{n,m+k}$. The k-period forward rate is also given in terms of k single-period forward rates. As seen above, (8.78) relates the time-n ZCB prices with adjacent maturities m and $m + 1$ via the single-period forward rate $f_{n,m}$. Hence, applying this relation recursively gives:

$$(1 + f_{n,m})^{-1}Z_{n,m} = Z_{n,m+1},$$
$$(1 + f_{n,m+1})^{-1}(1 + f_{n,m})^{-1}Z_{n,m} = (1 + f_{n,m+1})^{-1}Z_{n,m+1} = Z_{n,m+2}, \ldots,$$
$$\prod_{i=0}^{k-1}(1 + f_{n,m+i})^{-1}Z_{n,m} = Z_{n,m+k} \iff Z_{n,m} = \prod_{i=0}^{k-1}(1 + f_{n,m+i})Z_{n,m+k}.$$

The last line holds for any $k \geqslant 1$, $n \leqslant m$. By comparing this to the relation in (8.79), we have the equivalence between the k-period forward rate and all adjacent single-period forward rates in the time $[m, m+k]$:

$$1 + f_{n;m,m+k} = \prod_{i=0}^{k-1}(1 + f_{n,m+i}) \text{ or } (1 + f_{n;m,m+k})^{-1} = \prod_{i=0}^{k-1}(1 + f_{n,m+i})^{-1}. \tag{8.80}$$

We can relate the single period forward rate $f_{n,m}$ (or equally the difference in the time-n ZCB prices maturing at adjacent times m and $m+1$) to the time-n price of a derivative contract that pays R_m dollars at time $m+1$, where $R_m \equiv r_{m+1}$ is the rate of return on the bank account for the single period $[m, m+1]$. Indeed, defining $V_n := Z_{n,m} - Z_{n,m+1}$, for given $m \geqslant n$, and using the risk-neutral pricing formula in (8.72) for both ZCB prices (and using $\frac{B_{m+1}}{B_m} = (1 + R_m)$) gives

$$
V_n = B_n \widetilde{\mathrm{E}}_n \left[\frac{1}{B_m} \right] - B_n \widetilde{\mathrm{E}}_n \left[\frac{1}{B_{m+1}} \right]
$$

$$
= B_n \widetilde{\mathrm{E}}_n \left[\frac{1}{B_m} - \frac{1}{B_{m+1}} \right] = B_n \widetilde{\mathrm{E}}_n \left[\frac{\frac{B_{m+1}}{B_m} - 1}{B_{m+1}} \right] = B_n \widetilde{\mathrm{E}}_n \left[\frac{R_m}{B_{m+1}} \right] \equiv B_n \widetilde{\mathrm{E}}_n \left[\frac{V_{m+1}}{B_{m+1}} \right],
$$

where $V_{m+1} := R_m$ defines the time-$(m+1)$ payoff of the contract having time-n value V_n. From (8.78), we also have $Z_{n,m} - Z_{n,m+1} = f_{n,m} Z_{n,m+1}$. Hence, $V_n = f_{n,m} Z_{n,m+1}$, i.e.,

$$
f_{n,m} = \frac{B_n}{Z_{n,m+1}} \widetilde{\mathrm{E}}_n \left[\frac{R_m}{B_{m+1}} \right]. \tag{8.81}
$$

The time-0 prices and forward rates follow simply for $n = 0$ (see Exercise 8.7.) The formula in (8.81) can be generalized to the case of k-period forward rates for $k \geqslant 1$. In particular, let $V_{m+k} := R_{m,m+k}$ be the time-$(m+k)$ payoff of a contract, where $R_{m,m+k} \equiv \frac{B_{m+k} - B_m}{B_m}$ is the rate of return on the bank account over the time interval $[m, m+k]$. Then,

$$
f_{n;m,m+k} = \frac{B_n}{Z_{n,m+k}} \widetilde{\mathrm{E}}_n \left[\frac{R_{m,m+k}}{B_{m+k}} \right]. \tag{8.82}
$$

We leave it as an exercise to show this (see Exercise 8.7).

A coupon bond with maturity m can be modeled as a stream of payments $C_1, C_2, \ldots, C_m$ made at the respective times $1, 2, \ldots, m$. For each $k \in \{1, 2, \ldots, m-1\}$, the value C_k is the coupon payment made at time k. The last payment C_m at time m includes the redemption value as well as any coupon due at maturity time m. In the case of the zero-coupon bond maturing at time m, we have $C_1 = C_2 = \cdots = C_{m-1} = 0$ and $C_m = 1$. The no-arbitrage value of a coupon bond (assuming the coupon payments C_n are nonrandom) is equal to the no-arbitrage value of a portfolio of zero-coupon bonds with redemption C_n maturing at time $n = 1, 2, \ldots, m$. Thus, the no-arbitrage initial price $V_0^{(m)}$ of an m-maturity coupon-paying bond is calculated via risk-neutral pricing:

$$
V_0^{(m)} = \sum_{n=1}^m \widetilde{\mathrm{E}}_0 \left[\frac{B_0}{B_n} C_n \right] = \sum_{n=1}^m C_n \widetilde{\mathrm{E}}_0 \left[\frac{B_0}{B_n} \right] = \sum_{n=1}^m C_n Z_{0,n}. \tag{8.83}
$$

In general, the initial no-arbitrage value V_0 of a stream of random (path-dependent) payments $C_1, C_2, \ldots, C_m$, where C_n is $\mathcal{F}_n$-measurable for each $n = 1, 2, \ldots, m$, is a sum of no-arbitrage values of individual payments:

$$
V_0^{(m)} = \sum_{n=1}^m \widetilde{\mathrm{E}}_0 \left[\frac{B_0}{B_n} C_n \right] = \sum_{n=1}^m \widetilde{\mathrm{E}}_0 \left[\frac{C_n}{(1+r_1)(1+r_2)\cdots(1+r_n)} \right]
$$

$$
= \frac{1}{(1+r_1)} \sum_{n=1}^m \sum_{\omega_1, \ldots, \omega_n \in \{\mathrm{D,U}\}} \frac{\widetilde{\mathbb{P}}(\omega_1 \ldots \omega_n) C_n(\omega_1 \ldots \omega_n)}{(1+r_2(\omega_1))\cdots(1+r_n(\omega_1 \ldots \omega_{n-1}))}. \tag{8.84}
$$

For example, if the maturity $m = 2$ the above sum is given explicitly as:

$$V_0^{(2)} = \frac{1}{(1+r_1)}\widetilde{\mathbb{E}}_0[C_1] + \frac{1}{(1+r_1)}\widetilde{\mathbb{E}}_0\left[\frac{C_2}{1+r_2}\right].$$

$$= \frac{1}{(1+r_1)}\sum_{\omega_1 \in \{\mathsf{D},\mathsf{U}\}}\widetilde{\mathbb{P}}(\omega_1)C_1(\omega_1) + \frac{1}{(1+r_1)}\sum_{\omega_1,\omega_2 \in \{\mathsf{D},\mathsf{U}\}}\frac{\widetilde{\mathbb{P}}(\omega_1\omega_2)C_2(\omega_1\omega_2)}{1+r_2(\omega_1)}. \qquad (8.85)$$

In some fixed-income derivatve pricing applications (e.g., see Exercises 8.8, 8.9) each time-n payment C_n is $\mathcal{F}_{n-1}$-measurable, e.g., C_n may be a function of the rate r_n for the period $[n-1, n]$ and C_1 is constant (or given in terms of r_1). Since $\sum_{\omega_n \in \{\mathsf{D},\mathsf{U}\}}\widetilde{\mathbb{P}}(\omega_1 \ldots \omega_{n-1}\omega_n) = \widetilde{\mathbb{P}}(\omega_1 \ldots \omega_{n-1})$, then (8.84) gives

$$V_0^{(m)} = \frac{C_1}{(1+r_1)} + \sum_{n=2}^{m}\sum_{\omega_1,\ldots,\omega_{n-1} \in \{\mathsf{D},\mathsf{U}\}}\frac{\widetilde{\mathbb{P}}(\omega_1 \ldots \omega_{n-1})C_n(\omega_1 \ldots \omega_{n-1})}{(1+r_2(\omega_1))\cdots(1+r_n(\omega_1 \ldots \omega_{n-1}))}. \qquad (8.86)$$

For $m = 3$, (8.86) reads

$$V_0^{(3)} = \frac{C_1}{(1+r_1)} + \frac{1}{(1+r_1)}\sum_{\omega_1 \in \{\mathsf{D},\mathsf{U}\}}\frac{\widetilde{\mathbb{P}}(\omega_1)C_2(\omega_1)}{1+r_2(\omega_1)}$$

$$+ \frac{1}{(1+r_1)}\sum_{\omega_1,\omega_2 \in \{\mathsf{D},\mathsf{U}\}}\frac{\widetilde{\mathbb{P}}(\omega_1\omega_2)C_3(\omega_1\omega_2)}{(1+r_2(\omega_1))(1+r_3(\omega_1\omega_2))}. \qquad (8.87)$$

Example 8.12. Consider a nonrecombining binomial tree model with all single-time-step conditional risk-neutral probabilities and interest rates as given in Figure 8.6. Compute the time-0 price of (i) a zero-coupon bond maturing at time 3 and (ii) a coupon bond with payments $C_1 = C_2 = C_3 = \frac{1}{3}$.

Solution. First, we compute the time-0 prices $Z_{0,m}$ of zero-coupon bonds maturing at times $m = 3, 2, 1$, respectively, by employing (8.75):

$$Z_{0,3} = \frac{1}{(1+r_1)}\sum_{\omega_1,\omega_2 \in \{\mathsf{D},\mathsf{U}\}}\frac{\widetilde{\mathbb{P}}(\omega_1\omega_2)}{(1+r_2(\omega_1))(1+r_3(\omega_1\omega_2))}$$

$$= \frac{1}{(1+r_1)}\left[\frac{\widetilde{\mathbb{P}}(\mathsf{DD})}{(1+r_2(\mathsf{D}))(1+r_3(\mathsf{DD}))} + \frac{\widetilde{\mathbb{P}}(\mathsf{DU})}{(1+r_2(\mathsf{D}))(1+r_3(\mathsf{DU}))}\right.$$

$$+ \frac{\widetilde{\mathbb{P}}(\mathsf{UD})}{(1+r_2(\mathsf{U}))(1+r_3(\mathsf{UD}))} + \left.\frac{\widetilde{\mathbb{P}}(\mathsf{UU})}{(1+r_2(\mathsf{U}))(1+r_3(\mathsf{UU}))}\right]$$

$$= \frac{1}{1.05}\left[\frac{1/5}{1.04 \cdot 1.03} + \frac{3/10}{1.04 \cdot 1.04} + \frac{1/8}{1.06 \cdot 1.05} + \frac{3/8}{1.06 \cdot 1.07}\right]$$

$$\cong 0.86382\,,$$

$$Z_{0,2} = \frac{1}{(1+r_1)}\left[\frac{\widetilde{\mathbb{P}}(\mathsf{D})}{1+r_2(\mathsf{D})} + \frac{\widetilde{\mathbb{P}}(\mathsf{U})}{1+r_2(\mathsf{U})}\right]$$

$$= \frac{1}{1.05}\left[\frac{1/2}{1.04} + \frac{1/2}{1.06}\right] \cong 0.907112\,,$$

$$Z_{0,1} = \frac{1}{1+r_1} = \frac{1}{1.05} \cong 0.952381\,.$$

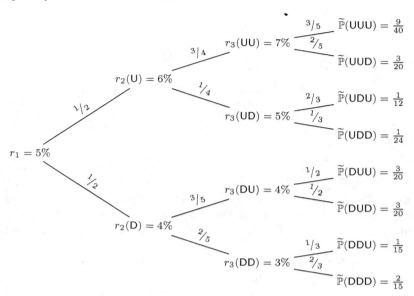

FIGURE 8.6: A three-period binomial model of stochastic interest rates.

In the above calculations, we used the risk-neutral probabilities $\widetilde{\mathbb{P}}(\omega_1\omega_2)$ computed from (conditional) transition probabilities, which are given as weights in Figure 8.6:

$$\widetilde{\mathbb{P}}(DD) = \widetilde{\mathbb{P}}(D)\,\widetilde{p}(D \mid D) = \frac{1}{2} \cdot \frac{2}{5} = \frac{1}{5},$$

$$\widetilde{\mathbb{P}}(DU) = \widetilde{\mathbb{P}}(D)\,\widetilde{p}(U \mid D) = \frac{1}{2} \cdot \frac{3}{5} = \frac{3}{10},$$

$$\widetilde{\mathbb{P}}(UD) = \widetilde{\mathbb{P}}(U)\,\widetilde{p}(D \mid U) = \frac{1}{2} \cdot \frac{1}{4} = \frac{1}{8},$$

$$\widetilde{\mathbb{P}}(UU) = \widetilde{\mathbb{P}}(U)\,\widetilde{p}(U \mid U) = \frac{1}{2} \cdot \frac{3}{4} = \frac{3}{8}.$$

So, the price of the zero-coupon bond maturing at time 3 is $Z_{0,3} \cong 0.868116$. The price of the coupon bond (maturing at time 3) is

$$\frac{1}{3} \cdot Z_{0,1} + \frac{1}{3} \cdot Z_{0,2} + \frac{1}{3} \cdot Z_{0,2} = \frac{0.952381 + 0.907112 + 0.86382}{3} \cong 0.90777. \qquad \square$$

For examples on pricing coupon bonds with random coupon payments see Exercise 8.14.

In Example 8.12, the bond prices are calculated using the risk-neutral probabilities given to us a priori. In practice, we deal with the reverse situation: the market prices of all possible bonds with different maturities are known and then the risk-neutral probability measure $\widetilde{\mathbb{P}}$ needs to be determined (implied) in a manner that is consistent with the prices. Recall that the discounted bond price processes $\{\overline{Z}_{n,m} = Z_{n,m}/B_n\}_{0 \leqslant n \leqslant m}$ are $\widetilde{\mathbb{P}}$-martingales for all $m = 1, 2, \ldots, T$. Hence, using the single-step martingale property $\overline{Z}_{n,m} = \widetilde{\mathbb{E}}_n[\overline{Z}_{n+1,m}]$, for $m \geqslant n + 1$, gives:

$$Z_{n,m}(\omega_1 \ldots \omega_n) = \frac{B_n(\omega_1 \ldots \omega_{n-1})}{B_{n+1}(\omega_1 \ldots \omega_n)} \widetilde{E}_n [Z_{n+1,m}](\omega_1 \ldots \omega_n)$$

$$= \frac{1}{1 + r_{n+1}(\omega_1 \ldots \omega_n)} [Z_{n+1,m}(\omega_1 \ldots \omega_n \mathsf{D}) \widetilde{p}(\mathsf{D}|\omega_1 \ldots \omega_n)$$

$$+ Z_{n+1,m}(\omega_1 \ldots \omega_n \mathsf{U}) \widetilde{p}(\mathsf{U}|\omega_1 \ldots \omega_n)], \qquad (8.88)$$

where $\widetilde{p}(\mathsf{D}|\omega_1 \ldots \omega_n) \equiv \widetilde{\mathbb{P}}(A_{\omega_1 \ldots \omega_n \mathsf{D}} \mid A_{\omega_1 \ldots \omega_n})$ and $\widetilde{p}(\mathsf{U}|\omega_1 \ldots \omega_n) \equiv \widetilde{\mathbb{P}}(A_{\omega_1 \ldots \omega_n \mathsf{U}} \mid A_{\omega_1 \ldots \omega_n})$. Here we used $\frac{B_n(\omega_1 \ldots \omega_{n-1})}{B_{n+1}(\omega_1 \ldots \omega_n)} = [1 + r_{n+1}(\omega_1 \ldots \omega_n)]^{-1}$, which is independent of the value of $\omega_{n+1} \in \{\mathsf{D}, \mathsf{U}\}$. Therefore, the risk-neutral conditional probabilities can be found by solving the following linear 2-by-2 system of equations in the two unknown conditional probabilities, for each outcome $\omega_1, \ldots, \omega_n \in \{\mathsf{D}, \mathsf{U}\}$, and for all times n and m with $1 \leqslant n + 1 < m \leqslant T$:

$$\begin{cases} Z_{n+1,m}(\omega_1 \ldots \omega_n \mathsf{D}) \, \widetilde{p}(\mathsf{D}|\omega_1 \ldots \omega_n) + Z_{n+1,m}(\omega_1 \ldots \omega_n \mathsf{U}) \, \widetilde{p}(\mathsf{U}|\omega_1 \ldots \omega_n) \\ = (1 + r_{n+1}(\omega_1 \ldots \omega_n)) Z_{n,m}(\omega_1 \ldots \omega_n), \\ \widetilde{p}(\mathsf{D}|\omega_1 \ldots \omega_n) + \widetilde{p}(\mathsf{U}|\omega_1 \ldots \omega_n) = 1, \end{cases} \qquad (8.89)$$

where the interest rates $r_{n+1} \equiv R_n$ are given in terms of the ZCB prices for all single periods $[n, n+1]$, i.e., $1 + r_{n+1}(\omega_1 \ldots \omega_n) = (Z_{n,n+1}(\omega_1 \ldots \omega_n))^{-1}$. Note that for $m = n + 1$ the first equation in (8.89) is trivially equivalent to the second since in this case $Z_{n+1,m} = Z_{n+1,n+1} = 1$. Clearly, under the no-arbitrage assumption, all such single-step conditional probabilities must have value between 0 and 1 and have to be independent of maturity m.

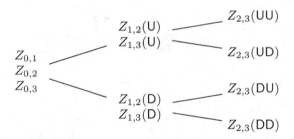

FIGURE 8.7: A binomial tree with ZCB prices.

For example, in the three-period case, we have three zero-coupon bonds with maturities $m = 1$, $m = 2$, and $m = 3$, respectively. The bond prices can be organized in a binomial tree, given in Figure 8.7. Since the face value $Z_{m,m} = 1$, for all $m = 1, 2, 3$, we omit all such nodes in the figure. To determine the two state probabilities $\widetilde{\mathbb{P}}(\mathsf{D})$ and $\widetilde{\mathbb{P}}(\mathsf{U})$, we solve (8.89) for either $m = 2$ or $m = 3$:

$$Z_{1,2}(\mathsf{D})\widetilde{\mathbb{P}}(\mathsf{D}) + Z_{1,2}(\mathsf{U})\widetilde{\mathbb{P}}(\mathsf{U}) = Z_{0,2}/Z_{0,1}$$

or

$$Z_{1,3}(\mathsf{D})\widetilde{\mathbb{P}}(\mathsf{D}) + Z_{1,3}(\mathsf{U})\widetilde{\mathbb{P}}(\mathsf{U}) = Z_{0,3}/Z_{0,1}$$

with $\widetilde{\mathbb{P}}(\mathsf{D}) = 1 - \widetilde{\mathbb{P}}(\mathsf{U})$. The resulting probabilities in either case must be the same, i.e., the result has to be independent of maturity, or otherwise there is an arbitrage opportunity in the market. The general proof of this fact is left as an exercise for the reader. To determine the four conditional probabilities $\widetilde{p}(\omega_2|\omega_1) = \widetilde{\mathbb{P}}(A_{\omega_1\omega_2} \mid A_{\omega_1})$ for $\omega_1, \omega_2 \in \{\mathsf{D}, \mathsf{U}\}$, we solve the following equations:

$$Z_{2,3}(\mathsf{DD})\widetilde{p}(\mathsf{D} \mid \mathsf{D}) + Z_{2,3}(\mathsf{DU})\widetilde{p}(\mathsf{U} \mid \mathsf{D}) = Z_{1,3}(\mathsf{D})/Z_{1,2}(\mathsf{D})$$
$$Z_{2,3}(\mathsf{UD})\widetilde{p}(\mathsf{D} \mid \mathsf{U}) + Z_{2,3}(\mathsf{UU})\widetilde{p}(\mathsf{U} \mid \mathsf{U}) = Z_{1,3}(\mathsf{U})/Z_{1,2}(\mathsf{U}),$$

with $\widetilde{p}(\mathsf{D} \mid \mathsf{D}) + \widetilde{p}(\mathsf{U} \mid \mathsf{D}) = 1$ and $\widetilde{p}(\mathsf{D} \mid \mathsf{U}) + \widetilde{p}(\mathsf{U} \mid \mathsf{U}) = 1$. Since the value of a ZCB at maturity is always one, the above process cannot be used to determine the probabilities $\widetilde{p}(\omega_3|\omega_1\omega_2) = \widetilde{\mathbb{P}}(A_{\omega_1\omega_2\omega_3} \mid A_{\omega_1\omega_2})$ for the last period. In particular, for the last period we simply have $\widetilde{p}(\mathsf{D}|\omega_1\omega_2) + \widetilde{p}(\mathsf{U}|\omega_1\omega_2) = 1$. We remark that the latter probabilities are not required since $Z_{3,3} \equiv 1$.

8.4.3 Interest Rates with the Markov Property

The formulation for pricing interest rate derivatives may be simplified (or recast) in cases where the payoff is a function of the interest rate. In this case we have the Markov property where conditioning on $\mathcal{F}_n$ is reduced to conditioning on the interest rate. In particular, let us conveniently denote the interest rate process by $R_n := r_{n+1}$, $n = 0, 1, \ldots, T$. We are assuming a random time-m payoff $V_m = V_m(R_m)$, $1 \leqslant m \leqslant T$, which is a given time-$m$ function of R_m. We hence consider the atoms in (8.28) at each time $n = 0, 1, \ldots, T$ as given by the sets $A_n^j = \{R_n = R_{n,j}\}$, $j = 1, \ldots, \kappa(n)$, where $\kappa(n)$ is the number of distinct possible values (nodes) of the rate R_n. At initial time $n = 0$ we only have one node with given value $R_0 \equiv r_1$. The random rate R_n, for each n, is constant on the atoms of the partition $\mathcal{P}_n = \{A_n^1, \ldots, A_n^{\kappa(n)}\}$. As well, the prices are constant on these atoms, i.e., the price at each node is given by $V_n(A_n^j) = V_n(R_{n,j}) \equiv V_{n,j}$. Applying (8.28) in this case gives us the backward recurrence pricing equation

$$V_{n,j} = \frac{1}{1 + R_{n,j}} \sum_{k=1}^{\kappa(n+1)} \widetilde{p}_{(n,j)\to(n+1,k)} V_{n+1,k}, \tag{8.90}$$

$j = 0, 1, \ldots, \kappa(n)$, $n = 1, \ldots, m - 1$, with known prices $V_{m,j}$, $j = 1, \ldots, \kappa(m)$, and where

$$\widetilde{p}_{(n,j)\to(n+1,k)} := \widetilde{\mathbb{P}}(R_{n+1} = R_{n+1,k}|R_n = R_{n,j}) \tag{8.91}$$

are the (single-step) risk-neutral transition probabilities for the (Markov chain) process $\{R_n\}_{n=0,1,\ldots,T}$. As an example, for any fixed maturity m, zero-coupon bonds $V_n \equiv Z_{n,m}$ have payoff $V_m = Z_{m,m} = 1$. Hence, the prices are functions of the interest rate. In particular, the time-n prices of ZCBs with time-n rate $R_{n,j}$ and maturing at time $m > n$, denoted by $Z_{n,j;m} \equiv Z_{n,m}(R_{n,j})$, can be computed by applying (8.90):

$$Z_{n,j;m} = \frac{1}{1 + R_{n,j}} \sum_{k=1}^{\kappa(n+1)} \widetilde{p}_{(n,j)\to(n+1,k)} Z_{n+1,k;m}. \tag{8.92}$$

Note that, since $Z_{m,k;m} = 1$ for all k and $\sum_{k=1}^{\kappa(m)} \widetilde{p}_{(m-1,j)\to(m,k)} = 1$, employing (8.92) for $n = m - 1$ gives

$$Z_{m-1,j;m} = \frac{1}{1 + R_{m-1,j}}. \tag{8.93}$$

The conditional probabilities in (8.91) are risk-neutral probabilities for attaining node $(n+1, k)$ at time $n+1$ from node (n, j) at time n. The multi-step tansition probabilities satisfy probability conservation (the Chapman-Kolmogorov relations). For example,

$$\widetilde{p}_{(n,i) \to (n'+1,j)} = \sum_{\ell=1}^{\kappa(n')} \widetilde{p}_{(n,i) \to (n',\ell)} \cdot \widetilde{p}_{(n',\ell) \to (n'+1,j)}$$

for all $n \leqslant n'$, $n' = 1, \ldots, T-1$, where $\widetilde{p}_{(n,i) \to (n,j)} = \delta_{i,j}$ is the trivial zero-step transition probability. In a typical multinomial tree model only some of the transition probabilities in (8.90) are nonzero. For example, in a binomial model there are exactly two nodes emanating from a previous node in one time step, i.e., for each node (n, j) there are two values of k for which $\widetilde{p}_{(n,j) \to (n+1,k)} \neq 0$. For a trinomial model there are exactly three values of k with $\widetilde{p}_{(n,j) \to (n+1,k)} \neq 0$, etc.

Note that the analogue of the formula in (8.73) obtains by iterating the single-step formula in (8.92):

$$Z_{n,j;m} = \frac{1}{1+R_{n,j}} \sum_{j_{n+1}} \cdots \sum_{j_{m-1}} \frac{\widetilde{p}_{(n,j) \to (n+1,j_{n+1})}}{1+R_{n+1,j_{n+1}}} \times \cdots \times \frac{\widetilde{p}_{(m-2,j_{m-2}) \to (m-1,j_{m-1})}}{1+R_{m-1,j_{m-1}}} \quad (8.94)$$

for $m > n+1$ and $Z_{n,n+1}(R_{n,j}) \equiv Z_{n,j;n+1} = (1+R_{n,j})^{-1}$. The sums are over all connecting nodes with nonzero single-step risk-neutral transition probabilities. The analogue of (8.75) is then simply given by setting $n = 0$:

$$Z_{0,m} \equiv Z_{0,0;m} = \frac{1}{1+R_0} \sum_{j_1} \cdots \sum_{j_{m-1}} \frac{\widetilde{p}_{(0,0) \to (1,j_1)}}{1+R_{1,j_1}} \times \cdots \times \frac{\widetilde{p}_{(m-2,j_{m-2}) \to (m-1,j_{m-1})}}{1+R_{m-1,j_{m-1}}} \quad (8.95)$$

for $m \geqslant 2$ and $Z_{0,1} = (1+R_0)^{-1}$ where the initial rate R_0 is a known constant.

To illustrate the application of (8.92) let us reconsider the nonrecombining binomial tree model in Example 8.12. We use Figure 8.6 to obtain the values of the interest rate at each node and the risk-neutral transition probabilities. We have the nodal values

$R_0 \equiv R_{0,0} = 0.05$; $R_{1,1} = r_2(\mathsf{D}) = 0.04$, $R_{1,2} = r_2(\mathsf{U}) = 0.06$;

$R_{2,1} = r_3(\mathsf{DD}) = 0.03$, $R_{2,2} = r_3(\mathsf{DU}) = 0.04$, $R_{2,3} = r_3(\mathsf{UD}) = 0.05$, $R_{2,4} = r_3(\mathsf{UU}) = 0.07$,

where $\kappa(1) = 2$, $\kappa(2) = 4$, $\kappa(3) = 8$ and

$$\widetilde{p}_{(0,0) \to (1,1)} = \widetilde{p}_{(0,0) \to (1,2)} = \frac{1}{2};$$

$$\widetilde{p}_{(1,1) \to (2,1)} = \frac{2}{5}; \ \widetilde{p}_{(1,1) \to (2,2)} = \frac{3}{5}; \ \widetilde{p}_{(1,2) \to (2,3)} = \frac{1}{4}; \ \widetilde{p}_{(1,2) \to (2,4)} = \frac{3}{4}.$$

Let us now compute all ZCB prices with maturity $m = 3$ by backward recurrence. Firstly, we have $Z_{3,3} = 1$, i.e., $Z_{3,j;3} = 1$ for all nodes $(3, j)$. For time $n = 2$ we use (8.93):

$$Z_{2,3}(R_{2,j}) \equiv Z_{2,j;3} = \frac{1}{1+R_{2,j}} \ , \ j = 1, 2.$$

Using the above node values for the rate gives

$$Z_{2,1;3} = \frac{1}{1+R_{2,1}} = \frac{1}{1.03} \cong 0.97087379 \ , \ Z_{2,2;3} = \frac{1}{1+R_{2,2}} = \frac{1}{1.04} \cong 0.96153846,$$

$$Z_{2,3;3} = \frac{1}{1+R_{2,3}} = \frac{1}{1.05} \cong 0.95238095 \ , \ Z_{2,4;3} = \frac{1}{1+R_{2,4}} = \frac{1}{1.07} \cong 0.934579.$$

[Note that we equivalently have: $Z_{2,1;3} = Z_{2,3}(\mathsf{DD})$, $Z_{2,2;3} = Z_{2,3}(\mathsf{DU})$, $Z_{2,3;3} = Z_{2,3}(\mathsf{UD})$, $Z_{2,4;3} = Z_{2,3}(\mathsf{UU})$.]

In turn, using these values within (8.92) for $n = 1$ gives the ZCB prices at time 1, maturing at time 3:

$$Z_{1,3}(R_{1,j}) \equiv Z_{1,j;3} = \frac{1}{1 + R_{1,j}} \sum_{k=1}^{4} \widetilde{p}_{(1,j) \to (2,k)} Z_{2,k;3}, \ j = 1, 2.$$

For either value of j we have only two nonzero terms in the sum, giving

$$Z_{1,1;3} = \frac{1}{1 + R_{1,1}} [\widetilde{p}_{(1,1) \to (2,1)} Z_{2,1;3} + \widetilde{p}_{(1,1) \to (2,2)} Z_{2,2;3}]$$

$$= \frac{1}{1.04} \left[\frac{2}{5} \cdot \frac{1}{1.03} + \frac{3}{5} \cdot \frac{1}{1.04} \right] \cong 0.92815$$

and

$$Z_{1,2;3} = \frac{1}{1 + R_{1,2}} [\widetilde{p}_{(1,2) \to (2,3)} Z_{2,3;3} + \widetilde{p}_{(1,2) \to (2,4)} Z_{2,4;3}]$$

$$= \frac{1}{1.06} \left[\frac{1}{4} \cdot \frac{1}{1.05} + \frac{3}{4} \cdot \frac{1}{1.07} \right] \cong 0.88588.$$

[Note: $Z_{1,1;3} = Z_{1,3}(\mathsf{D})$, $Z_{1,2;3} = Z_{1,3}(\mathsf{U})$.]

Now using (8.92) for $n = 1$ gives the single ZCB price at time 0:

$$Z_{0,3} \equiv Z_{0,0;3} = \frac{1}{1 + R_0} [\widetilde{p}_{(0,0) \to (1,1)} Z_{1,1;3} + \widetilde{p}_{(0,0) \to (1,2)} Z_{1,2;3}]$$

$$= \frac{1}{1.05} \left[\frac{1}{2} \cdot Z_{1,1;3} + \frac{1}{2} \cdot Z_{1,2;3} \right] \cong 0.86382.$$

This, of course, recovers the price we obtained previously. We can apply (8.92) to compute all ZCB prices for maturities $m = 1, 2$. For $n = 1, m = 2$ we apply (8.93):

$$Z_{1,1;2} = \frac{1}{1 + R_{1,1}} = \frac{1}{1.04} \cong 0.961538, \ \ Z_{1,2;2} = \frac{1}{1 + R_{1,2}} = \frac{1}{1.06} \cong 0.943396.$$

[Note: $Z_{1,1;2} = Z_{1,2}(\mathsf{D})$, $Z_{1,2;2} = Z_{1,2}(\mathsf{U})$.] Using these two prices gives

$$Z_{0,2} \equiv Z_{0,0;2} = \frac{1}{1 + R_0} [\widetilde{p}_{(0,0) \to (1,1)} Z_{1,1;2} + \widetilde{p}_{(0,0) \to (1,2)} Z_{1,2;2}] \cong 0.907112.$$

Finally, for $n = 0, m = 1$ we simply have $Z_{0,1} \equiv Z_{0,0;1} = \frac{1}{1 + R_0} = 1/1.05 \cong 0.952381$.

We remark that in a recombining binomial tree we have $(n + 1)$ nodes at each time n. Following our usual convention we denote the interest rate nodes by $R_{n,j}$, $j = 0, 1, \ldots, n$, where j is the total number of upward moves at time n and $R_{0,0} \equiv R_0$ is the initial node. For every node (n, j) the rate can only transition into nodes $(n + 1, j)$ and $(n + 1, j + 1)$. Hence, the backward pricing equations (8.90) and (8.92) are simply

$$V_{n,j} = \frac{1}{1 + R_{n,j}} \left[\widetilde{p}_{(n,j) \to (n+1,j)} V_{n+1,j} + \widetilde{p}_{(n,j) \to (n+1,j+1)} V_{n+1,j+1} \right], \tag{8.96}$$

$j = 0, 1, \ldots, n$, with known $V_{m,k}$, $k = 1, \ldots, m$, and

$$Z_{n,j;m} = \frac{1}{1 + R_{n,j}} \left[\widetilde{p}_{(n,j) \to (n+1,j)} Z_{n+1,j;m} + \widetilde{p}_{(n,j) \to (n+1,j+1)} Z_{n+1,j+1;m} \right], \tag{8.97}$$

$j = 0, 1, \ldots, n$, with $Z_{m,k;m} = 1$, where $\widetilde{p}_{(n,j) \to (n+1,j)} + \widetilde{p}_{(n,j) \to (n+1,j+1)} = 1$. The formula in (8.95) for the initial price of an m-maturity ZCB can be written more explicitly. In particular, for $m = 2$, we have

$$Z_{0,2} = \frac{1}{1 + R_0} \left[\frac{\widetilde{p}_{(0,0) \to (1,0)}}{1 + R_{1,0}} + \frac{\widetilde{p}_{(0,0) \to (1,1)}}{1 + R_{1,1}} \right], \tag{8.98}$$

and for $m = 3$,

$$Z_{0,3} = \frac{1}{1 + R_0} \left[\frac{\widetilde{p}_{(0,0) \to (1,0)}}{1 + R_{1,0}} \left(\frac{\widetilde{p}_{(1,0) \to (2,0)}}{1 + R_{2,0}} + \frac{\widetilde{p}_{(1,0) \to (2,1)}}{1 + R_{2,1}} \right) \right.$$
$$\left. + \frac{\widetilde{p}_{(0,0) \to (1,1)}}{1 + R_{1,1}} \left(\frac{\widetilde{p}_{(1,1) \to (2,1)}}{1 + R_{2,1}} + \frac{\widetilde{p}_{(1,1) \to (2,2)}}{1 + R_{2,2}} \right) \right]. \tag{8.99}$$

Note that (8.98)-(8.99) also follow readily by repeatedly applying (8.97) back to time 0, where $Z_{0,m} \equiv Z_{0,0;m}$.

Some simple examples of recombining binomial tree models with interest rate as a Markov process are those in which the time-n rate is a given function of the number of up moves up to time n, i.e., $R_n = f_n(\mathsf{U}_n)$, where $\mathsf{U}_n(\omega_1 \ldots \omega_n) = \#\mathsf{U}(\omega_1 \ldots \omega_n) = \sum_{i=1}^{n} \mathbb{I}_{\{\omega_i = \mathsf{U}\}}$. In particular, we can model the random rate R_n for the n-th period as a simple linear function of U_n,

$$R_n = a_n + b_n \mathsf{U}_n \tag{8.100}$$

for each time $n \geqslant 1$, and where R_0 is the known initial rate. At each time n we have a pair of parameters a_n, b_n. Note that the term $b_n \mathsf{U}_n$ models the volatility (randomness) of the time-n interest rate whereas a_n corresponds to a nonrandom drift term at time n. In practice, this model can be used to price interest rate derivatives once the model parameters are adjusted (i.e., calibrated) so that the model prices best match the market prices of ZCBs for various maturities. At each time $n \geqslant 1$, the nodes on the recombining binomial tree are given by $R_{n,j} = a_n + b_n \cdot j$, $j = 0, 1, \ldots, n$, since $R_n(\omega_1 \ldots \omega_n) = a_n + b_n \cdot j$ for every sequence $(\omega_1 \ldots \omega_n)$ s.t. $j = \mathsf{U}_n(\omega_1 \ldots \omega_n)$.

The above model is essentially the discrete-time analogue of the well-known Ho-Lee model used in continuous-time single-factor interest rate modelling. A specific version of this model is to assume $b_n \equiv \sigma$, $n \geqslant 1$, where σ is a positive constant volatility parameter. In this case, $R_n = a_n + \sigma \mathsf{U}_n$ with nodes $R_{n,j} = a_n + j\sigma = R_{n,0} + j\sigma$. At each time $n \geqslant 1$, $R_{n,0} \equiv a_n$ is the lowest node and all other adjacent nodes are separated by σ. We can calibrate this simpler version of the Ho-Lee model by matching all market ZCB prices $Z_{0,1}, Z_{0,2}, \ldots, Z_{0,m}$, with respective maturities $t = 1, 2, \ldots, m$, while adjusting the nodes (m unknown parameters) $R_0, R_{1,0} \equiv a_1, R_{2,0} \equiv a_2, \ldots, R_{m-1,0} \equiv a_{m-1}$. Note that σ can be further adjusted. In the general version, the parameters $b_1, \ldots, b_m$ offer extra flexibility at every time step and the binomial tree can therefore be calibrated to match a larger number of fixed-income derivatives.

If the above m market ZCB prices are the only market prices that we require the model to match, then we can simply fix all the risk-neutral transition probabilities and fix a value for σ. In particular, a simple choice is to set all single-step transition probabilities $\widetilde{p}_{(n,j) \to (n+1,j)} = \widetilde{p}_{(n,j) \to (n+1,j+1)} = \frac{1}{2}$. The calibration procedure can be done starting from initial node R_0 which follows trivially from the time-0 market ZCB price with maturity $m = 1$: $Z_{0,1} = (1 + R_0)^{-1} \implies R_0 = \frac{1}{Z_{0,1}} - 1$. The nodes at time $t = 1$ are obtained by using (8.98) with given time-0 market ZCB price with maturity $m = 2$. Writing $R_{1,1} = R_{1,0} + \sigma$, and setting $\widetilde{p}_{(0,0) \to (1,0)} = \widetilde{p}_{(0,0) \to (1,1)} = \frac{1}{2}$, gives an equation in $R_{1,0}$ which can be solved. This hence produces a value for a_1, i.e., the time-1 node values $R_{1,0}, R_{1,1}$ are determined. In turn, by a similar procedure, the time-2 nodal values (or a_2) can then be obtained by matching the time-0 market ZCB price with maturity $m = 3$ via (8.99) and substituting the calculated values for the time-1 nodes. We leave the rest of the details of this calibration procedure as an exercise for the reader (see Exercise 8.15).

Another known example of a simple Markov model is a discrete-time analogue of the so-called Black-Derman-Toy (BDT) model which has been used in continuous-time single-factor interest rate modelling. The interest rate is now assumed to follow the relation

$$R_n = a_n b_n^{\mathsf{U}_n} \tag{8.101}$$

for each time $n \geqslant 1$, and where R_0 is the known initial rate. Note that the logarithm of R_n is linear in U_n. One version is to assume constant parameters $b_n \equiv \sigma$, $n \geqslant 1$. Generally, the binomial tree nodes at time $n \geqslant 1$ are given by $R_{n,j} = R_{n,0}(b_n)^j$, $j = 0, 1, \ldots, n$, with lowest time-n node $R_{n,0} = a_n$ (note: $R_{0,0} \equiv R_0$). The ratio of adjacent nodes at each time $n \geqslant 1$ is fixed by the parameter b_n, i.e., $R_{n,j+1}/R_{n,j} = b_n$ for $j = 0, 1, \ldots, n$ and in the case $b_n = \sigma$, $R_{n,j+1}/R_{n,j} = \sigma$. This model can also be readily calibrated to a set of ZCB prices of various maturities. This model can also be used to price various interest rate derivatives.

Consider now a contract having a stream of payments $C_1, C_2, \ldots, C_m$, $C_n = C_n(R_n)$, $n = 1, 2, \ldots, m$, i.e., each payment C_n is assumed to depend upon only the interest rate for the n-th period. The initial no-arbitrage value $V_0 \equiv V_0(R_0) \equiv V_0(R_{0,0})$ of such a contract is given by repeatedly applying the backward recurrence formula in (8.90), giving

$$V_0 = \frac{1}{1+R_0} \sum_{n=1}^{m} \left(\sum_{j_1} \cdots \sum_{j_{n-1}} \sum_{j_n} \frac{\widetilde{p}_{(0,0)\to(1,j_1)}}{1+R_{1,j_1}} \frac{\widetilde{p}_{(1,j_1)\to(2,j_2)}}{1+R_{2,j_2}} \times \cdots \right.$$
$$\left. \times \frac{\widetilde{p}_{(n-1,j_{n-1})\to(n,j_n)}}{1+R_{n-1,j_{n-1}}} \right) C_n(R_{n,j_n}), \tag{8.102}$$

where the multiple summation is taken over all connecting nodes with nonzero single-step risk-neutral transition probabilities of the interest-rate Markov process. This formula, which can be used for any interest-rate Markov process, is actually a specialization of (8.84). If each payment $C_n = C_n(R_{n-1})$ at time n is assumed to be a function of only the interest rate for the $(n-1)$-th period, then we have

$$V_0 = \frac{1}{1+R_0} \sum_{n=1}^{m} \left(\sum_{j_1} \cdots \sum_{j_{n-1}} \frac{\widetilde{p}_{(0,0)\to(1,j_1)}}{1+R_{1,j_1}} \frac{\widetilde{p}_{(1,j_1)\to(2,j_2)}}{1+R_{2,j_2}} \times \cdots \right.$$
$$\left. \times \frac{\widetilde{p}_{(n-2,j_{n-2})\to(n-1,j_{n-1})}}{1+R_{n-1,j_{n-1}}} \right) C_n(R_{n-1,j_{n-1}}). \tag{8.103}$$

This formula is a specialization of (8.86) for Markovian interest rates. For example, in the case of $m = 3$ future payments (8.103) reads

$$V_0 = \frac{1}{1+R_0} \left[C_1(R_0) + \sum_{j_1} \frac{\widetilde{p}_{(0,0)\to(1,j_1)}}{1+R_{1,j_1}} C_2(R_{1,j_1}) \right.$$
$$\left. + \sum_{j_1} \frac{\widetilde{p}_{(0,0)\to(1,j_1)}}{1+R_{1,j_1}} \sum_{j_2} \frac{\widetilde{p}_{(1,j_1)\to(2,j_2)}}{1+R_{2,j_2}} C_3(R_{2,j_2}) \right] \tag{8.104}$$

We can also consider a path-dependent interest-rate derivative with a single time-m payoff that is dependent on all interest rates up to time m, i.e., $V_m = V_m(R_0, R_1, \ldots, R_m)$.

Risk-neutral pricing gives the time-0 no-arbitrage value $V_0 = V_0(R_0) = V_0(R_{0,0})$ of this derivative as

$$V_0 = \frac{1}{1+R_0} \sum_{j_1} \cdots \sum_{j_{m-1}} \sum_{j_m} \frac{\widetilde{p}_{(0,0)\to(1,j_1)}}{1+R_{1,j_1}} \frac{\widetilde{p}_{(1,j_1)\to(2,j_2)}}{1+R_{2,j_2}} \times \cdots$$

$$\times \frac{\widetilde{p}_{(m-1,j_{m-1})\to(m,j_m)}}{1+R_{m-1,j_{m-1}}} V_m(R_0, R_{1,j_1}, \ldots, R_{m,j_m}). \qquad (8.105)$$

For example, in a binomial tree model with payoff at maturity $m = 2$ assumed as a function $V_2 = V_2(R_0, R_1, R_2)$:

$$V_0 = \frac{1}{1+R_0} \left[\frac{\widetilde{p}_{(0,0)\to(1,0)}}{1+R_{1,0}} \left(\widetilde{p}_{(1,0)\to(2,0)} V_2(R_0, R_{1,0}, R_{2,0}) + \widetilde{p}_{(1,0)\to(2,1)} V_2(R_0, R_{1,0}, R_{2,1}) \right) \right.$$

$$\left. + \frac{\widetilde{p}_{(0,0)\to(1,1)}}{1+R_{1,1}} \left(\widetilde{p}_{(1,1)\to(2,1)} V_2(R_0, R_{1,1}, R_{2,1}) + \widetilde{p}_{(1,1)\to(2,2)} V_2(R_0, R_{1,1}, R_{2,2}) \right) \right].$$
$$(8.106)$$

The analogous explicit formula for maturity $m = 3$, now involving eight terms, also follows simply from (8.105). Applications of such formulas to the pricing of path-dependent interest-rate derivatives are left as exercises at the end of this chapter.

In the simplest non-path-dependent case in which $V_m = V_m(R_m)$ and assuming constant interest rates $R_n \equiv R_0$, the above formula simplifies to a single sum (upon repeated use of the above Chapman-Kolmogorov relation):

$$V_0 = \frac{1}{(1+R_0)^m} \sum_{j_m} \widetilde{p}_{(0,0)\to(m,j_m)} V_m(R_{m,j_m}) \qquad (8.107)$$

where $\widetilde{p}_{(0,0)\to(m,j_m)} := \widetilde{\mathbb{P}}(R_m = R_{m,j_m} | R_0 = R_{0,0})$ is the m-step risk-neutral probability to attain node R_{m,j_m} at time m.

8.4.4   Forward Measures for Interest-Rate Derivative Pricing

We already discussed earlier in this chapter the benefits of employing different numéraires for pricing equity derivatives (stock options). In a previous example, we have employed one of the stocks as numéraire (rather than the bank account) to price a two-stock equity option. The risk-neutral pricing formula in (8.17), and its single-step version in (8.18), allows us to price any financial derivative by selecting any appropriate numéraire g with EMM $\widetilde{\mathbb{P}}^{(g)}$. Any choice of strictly positive domestic asset qualifies as a numéraire. Throughout the previous section, when pricing ZCBs and other interest-rate derivatives, we have exclusively employed the usual EMM $\widetilde{\mathbb{P}} \equiv \widetilde{\mathbb{P}}^{(B)}$ with domestic bank account B as numéraire asset.

An alternate choice of numéraire which can simplify the pricing of some interest-rate derivatives (as well as equity derivatives with stochastic interest rates) is to select a ZCB of a given maturity. In particular, given a maturity $1 \leqslant m \leqslant T$, the so-called m-forward measure is defined as the EMM with numéraire asset price process chosen as the m-maturity ZCB, i.e., $g_n := Z_{n,m}, 0 \leqslant n \leqslant m$. In shorthand notation, we may write $g := Z^{(m)}$ to denote this numéraire asset. We shall denote the EMM $\widetilde{\mathbb{P}}^{(g)}$ by $\widetilde{\mathbb{P}}^{(Z^{(m)})} \equiv \widetilde{\mathbb{P}}^{(m)}$. From (8.49), with $f_n = B_n$ at time n, it follows immediately that the Radon-Nikodym derivative process $\{\varrho_n \equiv \varrho_n^{(m)}\}_{0 \leqslant n \leqslant m}$ is given by:

$$\varrho_n^{(m)} = \frac{g_n/g_0}{B_n/B_0} = \frac{Z_{n,m}}{Z_{0,m}} D_n = \frac{Z_{n,m}}{(1+R_0)(1+R_1)\cdots(1+R_{n-1})} \frac{1}{Z_{0,m}}, \qquad (8.108)$$

where $D_n = B_0/B_n = (1 + R_0)^{-1} \cdots (1 + R_{n-1})^{-1}$ is the discount factor for the first n periods. The change of probability measure $\widetilde{\mathbb{P}} \to \widetilde{\mathbb{P}}^{(m)}$ is given by $\varrho_m^{(m)}$ (setting $n = m$):

$$\frac{\widetilde{\mathbb{P}}^{(m)}(\omega)}{\widetilde{\mathbb{P}}(\omega)} = \varrho_m^{(m)}(\omega) = \frac{D_m(\omega)}{Z_{0,m}}, \, \omega \in \Omega_T. \tag{8.109}$$

Here we used $Z_{m,m} = 1$. Note that $\varrho_n^{(m)}$ is $\mathcal{F}_n$-measurable and $\{\varrho_n^{(m)}\}_{0 \leqslant n \leqslant m}$ is a $\widetilde{\mathbb{P}}$-martingale started at $\varrho_0^{(m)} = 1$ with unit $\widetilde{\mathbb{P}}$-expectation, $\widetilde{\mathbb{E}}[\varrho_n^{(m)}] = 1$. For the trivial case with $m = 1$, we simply have $\varrho_0^{(1)} = \varrho_1^{(1)} = 1$. Note that $\varrho_m^{(m)}$ is $\mathcal{F}_{m-1}$-measurable. In fact, $\varrho_m^{(m)} = \varrho_{m-1}^{(m)}$. This follows since $Z_{m-1,m} = (1 + R_{m-1})^{-1}$ and (8.108), with $n = m - 1$, gives

$$\varrho_{m-1}^{(m)} = \frac{1}{(1 + R_0)(1 + R_1) \cdots (1 + R_{m-1})} \frac{1}{Z_{0,m}} = \varrho_m^{(m)}. \tag{8.110}$$

In particular, let us consider the general binomial model with time-n atoms $A_{\omega_1 \ldots \omega_n} \equiv \omega_1 \ldots \omega_n$. The Radon-Nikodym derivative process on each atom evaluates to

$$\varrho_n^{(m)}(\omega_1 \ldots \omega_n) = \frac{Z_{n,m}(\omega_1 \ldots \omega_n)}{(1 + R_0)(1 + R_1(\omega_1)) \cdots (1 + R_{n-1}(\omega_1 \ldots \omega_{n-1}))} \frac{1}{Z_{0,m}}. \tag{8.111}$$

We therefore have $\varrho_m^{(m)}(\omega_1 \ldots \omega_{m-1}\omega_m) = \varrho_{m-1}^{(m)}(\omega_1 \ldots \omega_{m-1}\omega_m) = \varrho_{m-1}^{(m)}(\omega_1 \ldots \omega_{m-1})$, where

$$\varrho_{m-1}^{(m)}(\omega_1 \ldots \omega_{m-1}) = \frac{1}{(1 + R_0)(1 + R_1(\omega_1)) \cdots (1 + R_{m-1}(\omega_1 \ldots \omega_{m-1}))} \frac{1}{Z_{0,m}}. \tag{8.112}$$

For any outcome $\omega \equiv \omega_1 \ldots \omega_{m-1}\omega_m \ldots \omega_T \in \Omega_T$, the relation in (8.109) allows us to relate its probability under the two equivalent measures $\widetilde{\mathbb{P}}^{(m)}$ and $\widetilde{\mathbb{P}}$. In particular,

$$\widetilde{\mathbb{P}}^{(m)}(\omega) = \varrho_m^{(m)}(\omega_1 \ldots \omega_{m-1}\omega_m)\widetilde{\mathbb{P}}(\omega) = \varrho_{m-1}^{(m)}(\omega_1 \ldots \omega_{m-1})\widetilde{\mathbb{P}}(\omega). \tag{8.113}$$

Note that the ratio $\widetilde{\mathbb{P}}^{(m)}(\omega)/\widetilde{\mathbb{P}}(\omega)$ is constant for all $\omega \in A_{\omega_1 \ldots \omega_{m-1}}$ with the first $m - 1$ market moves fixed. In particular, on any atom $\omega_1 \ldots \omega_{m-1}\omega_m \equiv A_{\omega_1 \ldots \omega_{m-1}\omega_m}$, we have

$$\widetilde{\mathbb{P}}^{(m)}(\omega_1 \ldots \omega_{m-1}\omega_m) = \varrho_{m-1}^{(m)}(\omega_1 \ldots \omega_{m-1})\widetilde{\mathbb{P}}(\omega_1 \ldots \omega_{m-1}\omega_m). \tag{8.114}$$

For example, for $m = 2$ and $m = 3$ we respectively have:

$$\widetilde{\mathbb{P}}^{(2)}(\omega_1\omega_2) = \varrho_1^{(2)}(\omega_1)\widetilde{\mathbb{P}}(\omega_1\omega_2) = \frac{1}{(1 + R_0)(1 + R_1(\omega_1))} \frac{1}{Z_{0,2}}\widetilde{\mathbb{P}}(\omega_1\omega_2)$$

and $\widetilde{\mathbb{P}}^{(3)}(\omega_1\omega_2\omega_3) = \varrho_2^{(3)}(\omega_1\omega_2)\widetilde{\mathbb{P}}(\omega_1\omega_2\omega_3)$, i.e.,

$$\widetilde{\mathbb{P}}^{(3)}(\omega_1\omega_2\omega_3) = \frac{1}{(1 + R_0)(1 + R_1(\omega_1))(1 + R_2(\omega_1\omega_2))} \frac{1}{Z_{0,3}}\widetilde{\mathbb{P}}(\omega_1\omega_2\omega_3). \tag{8.115}$$

Note that in each case, in order to use any of the above equations we need to know the relevant ZCB prices.

Based on the probabilities $\widetilde{\mathbb{P}}^{(m)}(\omega)$, for all outcomes $\omega \equiv \omega_1 \ldots \omega_{m-1}\omega_m \ldots \omega_T \in \Omega_T$, we can compute all conditional probabilities in the $\widetilde{\mathbb{P}}^{(m)}$-measure. For example, consider a binomial model with $m = 3$ with all eight probabilities $\widetilde{\mathbb{P}}^{(3)}(\omega_1\omega_2\omega_3)$, $\omega_1, \omega_2, \omega_3 \in \{\mathsf{D}, \mathsf{U}\}$ being known (e.g., computed using the above formula). Then,

$$\widetilde{p}^{(3)}(\omega_3|\omega_1\omega_2) := \widetilde{\mathbb{P}}^{(3)}(\omega_1\omega_2\omega_3|\omega_1\omega_2) = \frac{\widetilde{\mathbb{P}}^{(3)}(\omega_1\omega_2\omega_3)}{\widetilde{\mathbb{P}}^{(3)}(\omega_1\omega_2)} = \frac{\widetilde{\mathbb{P}}^{(3)}(\omega_1\omega_2\omega_3)}{\widetilde{\mathbb{P}}^{(3)}(\omega_1\omega_2\mathsf{D}) + \widetilde{\mathbb{P}}^{(3)}(\omega_1\omega_2\mathsf{U})} \tag{8.116}$$

gives all eight single-time-step transition probabilities $\widetilde{p}^{(3)}(\mathsf{U}|\omega_1\omega_2)$ and $\widetilde{p}^{(3)}(\mathsf{D}|\omega_1\omega_2) = 1 - \widetilde{p}^{(3)}(\mathsf{U}|\omega_1\omega_2)$, for all four cases of $\omega_1, \omega_2, \in \{\mathsf{D}, \mathsf{U}\}$. Following this,

$$\widetilde{p}^{(3)}(\omega_2|\omega_1) := \widetilde{\mathbb{P}}^{(3)}(\omega_1\omega_2|\omega_1) = \frac{\widetilde{\mathbb{P}}^{(3)}(\omega_1\omega_2)}{\widetilde{\mathbb{P}}^{(3)}(\omega_1)} = \frac{\widetilde{\mathbb{P}}^{(3)}(\omega_1\omega_2)}{\widetilde{\mathbb{P}}^{(3)}(\omega_1\mathsf{D}) + \widetilde{\mathbb{P}}^{(3)}(\omega_1\mathsf{U})} \tag{8.117}$$

gives all four single-time-step transition probabilities $\widetilde{p}^{(3)}(\mathsf{U}|\omega_1)$ and $\widetilde{p}^{(3)}(\mathsf{D}|\omega_1) = 1 - \widetilde{p}^{(3)}(\mathsf{U}|\omega_1)$ for $\omega_1 \in \{\mathsf{D}, \mathsf{U}\}$. Finally, using

$$\widetilde{\mathbb{P}}^{(3)}(\omega_1) = \frac{\widetilde{\mathbb{P}}^{(3)}(\omega_1\omega_2)}{\widetilde{p}^{(3)}(\omega_2|\omega_1)}, \tag{8.118}$$

or simply $\widetilde{\mathbb{P}}^{(3)}(\omega_1) = \widetilde{\mathbb{P}}^{(3)}(\omega_1\mathsf{U}) + \widetilde{\mathbb{P}}^{(3)}(\omega_1\mathsf{D})$, gives $\widetilde{\mathbb{P}}^{(3)}(\mathsf{U})$ and $\widetilde{\mathbb{P}}^{(3)}(\mathsf{D}) = 1 - \widetilde{\mathbb{P}}^{(3)}(\mathsf{U})$. Other conditional probabilities for moves up to time $n = 3$ follow from the above single-time-step probabilities.

Although the above approach can be used in a straightforward (yet tedious) manner to obtain conditional probabilities under $\widetilde{\mathbb{P}}^{(m)}$, we can also determine all the conditional probabilities by exploiting the properties of the change of measure which connects $\widetilde{\mathbb{P}}^{(m)}$-measure conditional probabilities to $\mathbb{P}$-measure conditional probabilities, as follows. By making use of (8.52), with $g_n = Z_{n,m}, f_n = B_n$, we have

$$\widetilde{\mathbb{P}}^{(m)}(A_{n+1}^j|A_n^i) = \frac{\widetilde{\mathbb{P}}^{(m)}(A_{n+1}^j)}{\widetilde{\mathbb{P}}^{(m)}(A_n^i)} = \frac{\varrho_{n+1}^{(m)}(A_{n+1}^j)}{\varrho_n^{(m)}(A_n^i)}\widetilde{\mathbb{P}}(A_{n+1}^j|A_n^i), \tag{8.119}$$

$0 \leqslant n \leqslant m - 1$, for any two atoms $A_{n+1}^j \subseteq A_n^i$ at successive times n and $n + 1$. Specializing this relation to a binomial model with atoms $A_{\omega_1\ldots\omega_n\omega_{n+1}} \subseteq A_{\omega_1\ldots\omega_n}$ gives

$$\widetilde{p}^{(m)}(\omega_{n+1}|\omega_1\ldots\omega_n) = \frac{\varrho_{n+1}^{(m)}(\omega_1\ldots\omega_n\omega_{n+1})}{\varrho_n^{(m)}(\omega_1\ldots\omega_n)}\widetilde{p}(\omega_{n+1}|\omega_1\ldots\omega_n)$$

$$= \frac{Z_{n+1,m}(\omega_1\ldots\omega_n\omega_{n+1})}{Z_{n,m}(\omega_1\ldots\omega_n)(1 + R_n(\omega_1\ldots\omega_n))}\widetilde{p}(\omega_{n+1}|\omega_1\ldots\omega_n), \tag{8.120}$$

where $\widetilde{p}^{(m)}(\omega_{n+1}|\omega_1\ldots\omega_n) := \frac{\widetilde{\mathbb{P}}^{(m)}(\omega_1\ldots\omega_n\omega_{n+1})}{\widetilde{\mathbb{P}}^{(m)}(\omega_1\ldots\omega_n)}$ and $\widetilde{p}(\omega_{n+1}|\omega_1\ldots\omega_n) := \frac{\widetilde{\mathbb{P}}(\omega_1\ldots\omega_n\omega_{n+1})}{\widetilde{\mathbb{P}}(\omega_1\ldots\omega_n)}$ are single-time-step transition probabilities in the respective measures $\widetilde{\mathbb{P}}^{(m)}$ and $\widetilde{\mathbb{P}}$. The last expression in (8.120) follows using (8.111). Note that when $n = m - 1$, the relation in (8.120) gives $\widetilde{p}^{(m)}(\omega_m|\omega_1\ldots\omega_{m-1}) = \widetilde{p}(\omega_m|\omega_1\ldots\omega_{m-1})$ since $\varrho_m^{(m)} = \varrho_{m-1}^{(m)}$ (or $Z_{m,m} = 1$ and $Z_{m-1,m}(1 + R_{m-1}) = 1$). And for the case where $n = 0$ we have the relation between the probabilities for the first move:

$$\widetilde{\mathbb{P}}^{(m)}(\omega_1) = \varrho_1^{(m)}(\omega_1)\widetilde{\mathbb{P}}(\omega_1) = \frac{Z_{1,m}(\omega_1)}{Z_{0,m}(1 + R_0)}\widetilde{\mathbb{P}}(\omega_1). \tag{8.121}$$

Example 8.13. Consider the binomial interest rate model in Figure 8.6 (note: $R_n \equiv r_{n+1}$, $n = 0, 1, 2$). Determine all single-time-step transition probabilities and the probabilities for all outcomes $\omega_1\omega_2\omega_3$ in the forward measure $\widetilde{\mathbb{P}}^{(3)}$.

Solution. Based on the above discussion, there are two basic approaches which we illustrate here.

Method 1. We first compute $\widetilde{\mathbb{P}}^{(3)}(\omega_1\omega_2\omega_3)$ by using the computed ZCB price $Z_{0,3}$, the interest rate values and $\widetilde{\mathbb{P}}(\omega_1\omega_2\omega_3)$ values in Figure 8.6 within (8.115); then (8.116)-(8.118) are used to obtain the single-time-step transition probabilities. In particular, applying (8.115) gives:

$$\widetilde{\mathbb{P}}^{(3)}(\mathsf{UUU}) = \frac{\widetilde{\mathbb{P}}(\mathsf{UUU})\frac{1}{Z_{0,3}}}{(1+R_0)(1+R_1(\mathsf{U}))(1+R_2(\mathsf{UU}))} = \frac{\frac{9}{40}\frac{1}{0.86382}}{(1.05)(1.06)(1.07)} \simeq 0.21872,$$

$$\widetilde{\mathbb{P}}^{(3)}(\mathsf{UUD}) = \frac{\widetilde{\mathbb{P}}(\mathsf{UUD})}{\widetilde{\mathbb{P}}(\mathsf{UUU})}\widetilde{\mathbb{P}}^{(3)}(\mathsf{UUU}) = \left(\frac{3/20}{9/40}\right)0.21872 \simeq 0.14581,$$

$$\widetilde{\mathbb{P}}^{(3)}(\mathsf{UDU}) = \frac{\widetilde{\mathbb{P}}(\mathsf{UDU})\frac{1}{Z_{0,3}}}{(1+R_0)(1+R_1(\mathsf{U}))(1+R_2(\mathsf{UD}))} = \frac{\frac{1}{12}\frac{1}{0.86382}}{(1.05)(1.06)(1.05)} \simeq 0.08255,$$

$$\widetilde{\mathbb{P}}^{(3)}(\mathsf{UDD}) = \frac{\widetilde{\mathbb{P}}(\mathsf{UDD})}{\widetilde{\mathbb{P}}(\mathsf{UDU})}\widetilde{\mathbb{P}}^{(3)}(\mathsf{UDU}) = \left(\frac{1/24}{1/12}\right)0.08255 \simeq 0.04127,$$

$$\widetilde{\mathbb{P}}^{(3)}(\mathsf{DUU}) = \frac{\widetilde{\mathbb{P}}(\mathsf{DUU})\frac{1}{Z_{0,3}}}{(1+R_0)(1+R_1(\mathsf{D}))(1+R_2(\mathsf{DU}))} = \frac{\frac{3}{20}\frac{1}{0.86382}}{(1.05)(1.04)(1.04)} \simeq 0.15290,$$

$$\widetilde{\mathbb{P}}^{(3)}(\mathsf{DUD}) = \frac{\widetilde{\mathbb{P}}(\mathsf{DUD})}{\widetilde{\mathbb{P}}(\mathsf{DUU})}\widetilde{\mathbb{P}}^{(3)}(\mathsf{DUU}) = \widetilde{\mathbb{P}}^{(3)}(\mathsf{DUU}) \simeq 0.15290,$$

$$\widetilde{\mathbb{P}}^{(3)}(\mathsf{DDU}) = \frac{\widetilde{\mathbb{P}}(\mathsf{DDU})\frac{1}{Z_{0,3}}}{(1+R_0)(1+R_1(\mathsf{D}))(1+R_2(\mathsf{DD}))} = \frac{\frac{1}{15}\frac{1}{0.86382}}{(1.05)(1.04)(1.03)} \simeq 0.06862,$$

$$\widetilde{\mathbb{P}}^{(3)}(\mathsf{DDD}) = \frac{\widetilde{\mathbb{P}}(\mathsf{DDD})}{\widetilde{\mathbb{P}}(\mathsf{DDU})}\widetilde{\mathbb{P}}^{(3)}(\mathsf{DDU}) = \left(\frac{2/15}{1/15}\right)0.06862 \simeq 0.13723.$$

Hence,

$$\widetilde{\mathbb{P}}^{(3)}(\mathsf{UU}) = \widetilde{\mathbb{P}}^{(3)}(\mathsf{UUU}) + \widetilde{\mathbb{P}}^{(3)}(\mathsf{UUD}) \simeq 0.36453,$$

$$\widetilde{\mathbb{P}}^{(3)}(\mathsf{UD}) = \widetilde{\mathbb{P}}^{(3)}(\mathsf{UDU}) + \widetilde{\mathbb{P}}^{(3)}(\mathsf{UDD}) \simeq 0.12382,$$

$$\widetilde{\mathbb{P}}^{(3)}(\mathsf{DU}) = \widetilde{\mathbb{P}}^{(3)}(\mathsf{DUU}) + \widetilde{\mathbb{P}}^{(3)}(\mathsf{DUD}) \simeq 0.30580,$$

$$\widetilde{\mathbb{P}}^{(3)}(\mathsf{DD}) = \widetilde{\mathbb{P}}^{(3)}(\mathsf{DDU}) + \widetilde{\mathbb{P}}^{(3)}(\mathsf{DDD}) \simeq 0.20585.$$

Also, $\widetilde{\mathbb{P}}^{(3)}(\mathsf{U}) = \widetilde{\mathbb{P}}^{(3)}(\mathsf{UU}) + \widetilde{\mathbb{P}}^{(3)}(\mathsf{UD}) \simeq 0.48835$ and $\widetilde{\mathbb{P}}^{(3)}(\mathsf{D}) = \widetilde{\mathbb{P}}^{(3)}(\mathsf{DU}) + \widetilde{\mathbb{P}}^{(3)}(\mathsf{DD}) \simeq 0.51165$.

All the single-time-step transition probabilities now follow. Using (8.116):

$$\widetilde{p}^{(3)}(\mathsf{U}|\mathsf{UU}) = \frac{\widetilde{\mathbb{P}}^{(3)}(\mathsf{UUU})}{\widetilde{\mathbb{P}}^{(3)}(\mathsf{UU})} = 0.6, \quad \widetilde{p}^{(3)}(\mathsf{D}|\mathsf{UU}) = 0.4,$$

$$\widetilde{p}^{(3)}(\mathsf{U}|\mathsf{UD}) = \frac{\widetilde{\mathbb{P}}^{(3)}(\mathsf{UDU})}{\widetilde{\mathbb{P}}^{(3)}(\mathsf{UD})} = 0.6666..., \quad \widetilde{p}^{(3)}(\mathsf{D}|\mathsf{UD}) = 0.3333...$$

$$\widetilde{p}^{(3)}(\mathsf{U}|\mathsf{DU}) = \frac{\widetilde{\mathbb{P}}^{(3)}(\mathsf{DUU})}{\widetilde{\mathbb{P}}^{(3)}(\mathsf{DU})} = 0.5, \quad \widetilde{p}^{(3)}(\mathsf{D}|\mathsf{DU}) = 0.5,$$

$$\widetilde{p}^{(3)}(\mathsf{U}|\mathsf{DD}) = \frac{\widetilde{\mathbb{P}}^{(3)}(\mathsf{DDU})}{\widetilde{\mathbb{P}}^{(3)}(\mathsf{DD})} = 0.3333..., \quad \widetilde{p}^{(3)}(\mathsf{D}|\mathsf{DD}) = 0.6666...,.$$

Note that we must have exactly $\widetilde{p}^{(3)}(\omega_3|\omega_1\omega_2) = \widetilde{p}(\omega_3|\omega_1\omega_2)$. Using (8.117):

$$\widetilde{p}^{(3)}(\mathsf{U}|\mathsf{U}) = \frac{\widetilde{\mathbb{P}}^{(3)}(\mathsf{UU})}{\widetilde{\mathbb{P}}^{(3)}(\mathsf{U})} \simeq 0.74645, \quad \widetilde{p}^{(3)}(\mathsf{D}|\mathsf{U}) = \frac{\widetilde{\mathbb{P}}^{(3)}(\mathsf{UD})}{\widetilde{\mathbb{P}}^{(3)}(\mathsf{U})} \simeq 0.25355,$$

$$\widetilde{p}^{(3)}(\mathsf{U}|\mathsf{D}) = \frac{\widetilde{\mathbb{P}}^{(3)}(\mathsf{DU})}{\widetilde{\mathbb{P}}^{(3)}(\mathsf{D})} \simeq 0.59768, \quad \widetilde{p}^{(3)}(\mathsf{D}|\mathsf{D}) = \frac{\widetilde{\mathbb{P}}^{(3)}(\mathsf{DD})}{\widetilde{\mathbb{P}}^{(3)}(\mathsf{D})} \simeq 0.40232.$$

Method 2. We now first compute all single-time-step transition probabilities using (8.120) for $n = 1, 2$ and (8.121). Note that, since $m = 3$, for $n = 2$ we simply have $\widetilde{p}^{(3)}(\omega_3|\omega_1\omega_2) = \widetilde{p}(\omega_3|\omega_1\omega_2)$. Setting $n = 1$ in (8.120) gives:

$$\widetilde{p}^{(3)}(\omega_2|\omega_1) = \frac{Z_{2,3}(\omega_1\omega_2)}{Z_{1,3}(\omega_1)(1 + R_1(\omega_1))}\widetilde{p}(\omega_2|\omega_1).$$

Substituting the previously computed values of $Z_{1,3}(\omega_1)$ and $Z_{2,3}(\omega_1\omega_2)$, and the known probabilities $\widetilde{p}(\omega_2|\omega_1)$, gives:

$$\widetilde{p}^{(3)}(\mathsf{U}|\mathsf{U}) = \frac{Z_{2,3}(\mathsf{UU})}{Z_{1,3}(\mathsf{U})(1 + R_1(\mathsf{U}))}\widetilde{p}(\mathsf{U}|\mathsf{U}) \simeq \frac{0.934579}{0.88588(1.06)}\frac{3}{4} \simeq 0.74645,$$

$\widetilde{p}^{(3)}(\mathsf{D}|\mathsf{U}) \simeq 1 - 0.74645 \simeq 0.25355$, and

$$\widetilde{p}^{(3)}(\mathsf{U}|\mathsf{D}) = \frac{Z_{2,3}(\mathsf{DU})}{Z_{1,3}(\mathsf{D})(1 + R_1(\mathsf{D}))}\widetilde{p}(\mathsf{U}|\mathsf{D}) \simeq \frac{0.96153846}{0.92815(1.04)}\frac{3}{5} \simeq 0.59768,$$

$\widetilde{p}^{(3)}(\mathsf{D}|\mathsf{D}) \simeq 1 - 0.59768 \simeq 0.40232$. Using (8.121) for $\omega_1 = \mathsf{U}$:

$$\widetilde{\mathbb{P}}^{(3)}(\mathsf{U}) = \frac{Z_{1,3}(\mathsf{U})}{Z_{0,3}(1 + R_0)}\widetilde{\mathbb{P}}(\mathsf{U}) = \frac{0.88588}{0.86382(1.05)}\frac{1}{2} \simeq 0.48835,$$

and $\widetilde{\mathbb{P}}^{(3)}(\mathsf{D}) \simeq 1 - 0.48835 \simeq 0.51165$. The probabilities $\widetilde{\mathbb{P}}^{(3)}(\omega_1\omega_2)$ and $\widetilde{\mathbb{P}}^{(3)}(\omega_1\omega_2\omega_3)$ follow by simply multiplying out the above relevant single-time-step transition probabilities:

$$\widetilde{\mathbb{P}}^{(3)}(\omega_1\omega_2) = \widetilde{\mathbb{P}}^{(3)}(\omega_1)\widetilde{p}^{(3)}(\omega_2|\omega_1), \quad \widetilde{\mathbb{P}}^{(3)}(\omega_1\omega_2\omega_3) = \widetilde{\mathbb{P}}^{(3)}(\omega_1\omega_2)\widetilde{p}^{(3)}(\omega_3|\omega_1\omega_2).$$

As required, the two methods give the same results.   □

More generally, we can relate $\widetilde{\mathbb{P}}^{(m)}$-measure conditional probabilities

$$\widetilde{p}^{(m)}(\omega_{n+1}\ldots\omega_k|\omega_1\ldots\omega_n) := \frac{\widetilde{\mathbb{P}}^{(m)}(\omega_1\ldots\omega_n\omega_{n+1}\ldots\omega_k)}{\widetilde{\mathbb{P}}^{(m)}(\omega_1\ldots\omega_n)}$$

to the corresponding $\widetilde{\mathbb{P}}$-measure conditional probabilities

$$\widetilde{p}(\omega_{n+1}\ldots\omega_k|\omega_1\ldots\omega_n) := \frac{\widetilde{\mathbb{P}}(\omega_1\ldots\omega_n\omega_{n+1}\ldots\omega_k)}{\widetilde{\mathbb{P}}(\omega_1\ldots\omega_n)}$$

for $1 \leqslant n < k \leqslant T$. There are three cases, as follows. For $1 \leqslant n < k \leqslant m \leqslant T$:

$$\widetilde{p}^{(m)}(\omega_{n+1}\ldots\omega_k|\omega_1\ldots\omega_n) = \frac{\varrho_k^{(m)}(\omega_1\ldots\omega_k)}{\varrho_n^{(m)}(\omega_1\ldots\omega_n)}\widetilde{p}(\omega_{n+1}\ldots\omega_k|\omega_1\ldots\omega_n). \tag{8.122}$$

For $1 \leqslant n \leqslant m - 1, m \leqslant k \leqslant T$:

$$\widetilde{p}^{(m)}(\omega_{n+1} \ldots \omega_k | \omega_1 \ldots \omega_n) = \frac{\varrho_{m-1}^{(m)}(\omega_1 \ldots \omega_{m-1})}{\varrho_n^{(m)}(\omega_1 \ldots \omega_n)} \widetilde{p}(\omega_{n+1} \ldots \omega_k | \omega_1 \ldots \omega_n). \qquad (8.123)$$

For $0 \leqslant m - 1 \leqslant n < k \leqslant T$:

$$\widetilde{p}^{(m)}(\omega_{n+1} \ldots \omega_k | \omega_1 \ldots \omega_n) = \widetilde{p}(\omega_{n+1} \ldots \omega_k | \omega_1 \ldots \omega_n). \qquad (8.124)$$

The relation in (8.122) is readily proven by iterating (8.120) and using (8.121). In particular, for any $1 \leqslant k \leqslant m$:

$$\begin{aligned}
\widetilde{\mathbb{P}}^{(m)}(\omega_1 \ldots \omega_k) &= \widetilde{\mathbb{P}}^{(m)}(\omega_1) \cdot \widetilde{p}^{(m)}(\omega_2 | \omega_1) \cdots \cdot \widetilde{p}^{(m)}(\omega_k | \omega_1 \ldots \omega_{k-1}) \\
&= \varrho_1^{(m)}(\omega_1) \widetilde{\mathbb{P}}(\omega_1) \cdot \frac{\varrho_2^{(m)}(\omega_1 \omega_2)}{\varrho_1^{(m)}(\omega_1)} \widetilde{p}(\omega_2 | \omega_1) \cdots \cdot \frac{\varrho_k^{(m)}(\omega_1 \ldots \omega_k)}{\varrho_{k-1}^{(m)}(\omega_1 \ldots \omega_{k-1})} \widetilde{p}(\omega_k | \omega_1 \ldots \omega_{k-1}) \\
&= \varrho_k^{(m)}(\omega_1 \ldots \omega_k) \cdot \widetilde{\mathbb{P}}(\omega_1) \cdot \widetilde{p}(\omega_2 | \omega_1) \cdots \cdot \widetilde{p}(\omega_k | \omega_1 \ldots \omega_{k-1}) \\
&= \varrho_k^{(m)}(\omega_1 \ldots \omega_k) \cdot \widetilde{\mathbb{P}}(\omega_1 \ldots \omega_k).
\end{aligned}$$

Hence, repeating this formula for $n < k$ and dividing the two expressions gives (8.122),

$$\frac{\widetilde{\mathbb{P}}^{(m)}(\omega_1 \ldots \omega_k)}{\widetilde{\mathbb{P}}^{(m)}(\omega_1 \ldots \omega_n)} = \frac{\varrho_k^{(m)}(\omega_1 \ldots \omega_k)}{\varrho_k^{(m)}(\omega_1 \ldots \omega_n)} \cdot \frac{\widetilde{\mathbb{P}}(\omega_1 \ldots \omega_k)}{\widetilde{\mathbb{P}}(\omega_1 \ldots \omega_n)}.$$

An alternate proof is to use the fact that $\widetilde{\mathrm{E}}^{(m)}[X] = \widetilde{\mathrm{E}}[\varrho_n^{(m)} X]$ for any $\mathcal{F}_n$-measurable X, where $\mathrm{E}^{(m)}$ is the expectation under measure $\widetilde{\mathbb{P}}^{(m)}$. Fix a sequence of moves $\omega_1 \ldots \omega_n$, $n \leqslant m$. Since $\mathbb{I}_{A_{\omega_1 \ldots \omega_n}}$ is $\mathcal{F}_n$-measurable,

$$\widetilde{\mathbb{P}}^{(m)}(\omega_1 \ldots \omega_n) = \widetilde{\mathrm{E}}^{(m)}\left[\mathbb{I}_{A_{\omega_1 \ldots \omega_n}}\right] = \widetilde{\mathrm{E}}\left[\varrho_n^{(m)} \mathbb{I}_{A_{\omega_1 \ldots \omega_n}}\right] = \varrho_n^{(m)}(\omega_1 \ldots \omega_n) \widetilde{\mathbb{P}}(\omega_1 \ldots \omega_n).$$

Repeating this formula for integer $k \leqslant m, k > n$ and dividing, $\frac{\widetilde{\mathbb{P}}^{(m)}(\omega_1 \ldots \omega_k)}{\widetilde{\mathbb{P}}^{(m)}(\omega_1 \ldots \omega_n)}$, gives (8.122).

Note that (8.124) follows directly from (8.114). In particular,

$$\widetilde{\mathbb{P}}^{(m)}(\omega_1 \ldots \omega_{m-1} \ldots \omega_n) = \varrho_{m-1}^{(m)}(\omega_1 \ldots \omega_{m-1}) \widetilde{\mathbb{P}}(\omega_1 \ldots \omega_{m-1} \ldots \omega_n)$$

for any integer $m - 1 \leqslant n \leqslant T$. Hence, for integers $m - 1 \leqslant n < k \leqslant T$:

$$\frac{\widetilde{\mathbb{P}}^{(m)}(\omega_1 \ldots \omega_k)}{\widetilde{\mathbb{P}}^{(m)}(\omega_1 \ldots \omega_n)} = \frac{\varrho_{m-1}^{(m)}(\omega_1 \ldots \omega_{m-1})}{\varrho_{m-1}^{(m)}(\omega_1 \ldots \omega_{m-1})} \frac{\widetilde{\mathbb{P}}(\omega_1 \ldots \omega_k)}{\widetilde{\mathbb{P}}(\omega_1 \ldots \omega_n)} = \frac{\widetilde{\mathbb{P}}(\omega_1 \ldots \omega_k)}{\widetilde{\mathbb{P}}(\omega_1 \ldots \omega_n)},$$

which is (8.124). By using the above two formulas for $m \leqslant k \leqslant T$ and $1 \leqslant n \leqslant m - 1$, we obtain (8.123):

$$\frac{\widetilde{\mathbb{P}}^{(m)}(\omega_1 \ldots \omega_{m-1} \ldots \omega_k)}{\widetilde{\mathbb{P}}^{(m)}(\omega_1 \ldots \omega_n)} = \frac{\varrho_{m-1}^{(m)}(\omega_1 \ldots \omega_{m-1})}{\varrho_n^{(m)}(\omega_1 \ldots \omega_n)} \frac{\widetilde{\mathbb{P}}(\omega_1 \ldots \omega_{m-1} \ldots \omega_k)}{\widetilde{\mathbb{P}}(\omega_1 \ldots \omega_n)}.$$

Within the m-forward EMM, with conditional expectation $\widetilde{\mathrm{E}}_n^{(m)}[\cdot] \equiv \widetilde{\mathrm{E}}^{(m)}[\cdot | \mathcal{F}_n]$ under measure $\widetilde{\mathbb{P}}^{(m)}$, (8.17) gives the time-$n$ price of a derivative with $\mathcal{F}_m$-measurable payoff V_m as

$$V_n = Z_{n,m} \widetilde{\mathrm{E}}_n^{(m)}[V_m] \qquad (8.125)$$

since $g_m = Z_{m,m} = 1$, and the corresponding single-period recurrence pricing formula is

$$V_n = Z_{n,m} \, \widetilde{\mathrm{E}}_n^{(m)} \left[\frac{V_{n+1}}{Z_{n+1,m}} \right], \tag{8.126}$$

$0 \leqslant n \leqslant m - 1$. Equation (8.125), or equivalently (8.126), is obviously a statement of the fact that any derivative price process divided by the m-maturity ZCB price process is a $\widetilde{\mathbb{P}}^{(m)}$-martingale. In fact, any (domestic) non-dividend paying asset price divided by the m-maturity ZCB price is a $\widetilde{\mathbb{P}}^{(m)}$-martingale. In particular, recall that any (domestic) non-dividend paying stock price process, say $\{S_n\}_{n \geqslant 0}$, has m-*forward price* at time $n \leqslant m$ given by the ratio $F_{n,m}^{(S)} := \frac{S_n}{Z_{n,m}}$. We recall that, in a standard forward contract on the asset, this price corresponds to the delivery price agreed upon at time n for the asset having future value S_m at time m. Hence, the forward price process $\{F_{n,m}^{(S)}\}_{0 \leqslant n \leqslant m}$ is indeed a $\widetilde{\mathbb{P}}^{(m)}$-martingale. Observe that this also follows from (8.125), or (8.126), by considering the special case where the derivative is simply the stock itself, $V_n = S_n$.

For any (domestic) non-dividend paying asset A with price process $\{A_n\}_{n \geqslant 0}$, the m-forward price process of the asset is defined[3] by $F_{n,m}^{(A)} := \frac{A_n}{Z_{n,m}}$, $0 \leqslant n \leqslant m$. Note that at time $n = m$, $F_{m,m}^{(A)} := A_m$. The forward price $F_{n,m}^{(A)}$ corresponds to the particular value of the strike price that makes the m-maturity forward contract on asset A have zero value at time n. Indeed, a forward contract with strike K, maturing at time m, has payoff $V_m = A_m - K$. The time-n value of the forward contract is given by (8.125),

$$V_n = Z_{n,m} \, \widetilde{\mathrm{E}}_n^{(m)}[A_m] - Z_{n,m} K = Z_{n,m} \, \widetilde{\mathrm{E}}_n^{(m)}[A_m] - Z_{n,m} K = A_n - Z_{n,m} K.$$

The last equality follows from the $\widetilde{\mathbb{P}}^{(m)}$-martingale property of the forward price, i.e., $\widetilde{\mathrm{E}}_n^{(m)}[A_m] \equiv \widetilde{\mathrm{E}}_n^{(m)}[F_{m,m}^{(A)}] = F_{n,m}^{(A)}$, hence $Z_{n,m} \widetilde{\mathrm{E}}_n^{(m)}[A_m] = Z_{n,m} F_{n,m}^{(A)} = A_n$. [Note: this also follows directly by using (8.125) for derivative choice $V_n = A_n$.] Now, setting the forward contract price to zero, $V_n = 0$ for all $0 \leqslant n \leqslant m$, gives $K = \frac{A_n}{Z_{n,m}} = F_{n,m}^{(A)}$. We remark that this is even more readily seen by writing $A_m = F_{m,m}^{(A)}$, where $V_m = F_{m,m}^{(A)} - K$. Then, the martingale property $\widetilde{\mathrm{E}}_n^{(m)}[F_{m,m}^{(A)}] = F_{n,m}^{(A)}$ gives the result.

In practice, given m-maturity ZCB prices and the $\widetilde{\mathbb{P}}^{(m)}$-measure transition probabilities on a given tree, we can use (8.126) recursively to compute all the derivative prices on every atom or node of the tree. In the case of a general binomial tree, the no-arbitrage prices at every atom $\omega_1 \ldots \omega_n$ are given by

$$V_n(\omega_1 \ldots \omega_n) = Z_{n,m}(\omega_1 \ldots \omega_n) \, \widetilde{\mathrm{E}}_n^{(m)} \left[\frac{V_{n+1}}{Z_{n+1,m}} \right] (\omega_1 \ldots \omega_n), \tag{8.127}$$

$0 \leqslant n \leqslant m - 1$, where $V_m(\omega_1 \ldots \omega_m)$ is the known payoff. An equivalent useful way to rewrite (8.125) and (8.126) is

$$F_{n,m}^{(V)} = \widetilde{\mathrm{E}}_n^{(m)}\left[F_{n+1,m}^{(V)}\right], \qquad F_{n,m}^{(V)} = \widetilde{\mathrm{E}}_n^{(m)}\left[V_m\right] \tag{8.128}$$

i.e.,

$$F_{n,m}^{(V)}(\omega_1 \ldots \omega_n) = \widetilde{\mathrm{E}}_n^{(m)}\left[F_{n+1,m}^{(V)}\right](\omega_1 \ldots \omega_n), \tag{8.129}$$

---

[3] In the trivial special case that the asset is itself the m-maturity ZCB, i.e., $A_n \equiv Z_{n,m}$, then $F_{n,m}^{(A)} \equiv 1$, $0 \leqslant n \leqslant m$. Moreover, if the derivative has payoff $V_m \equiv Z_{m,m} \equiv 1$ then (8.125)-(8.126) are trivially consistent with $V_n \equiv Z_{n,m}$, $0 \leqslant n \leqslant m$.

where $F_{n,m}^{(V)} := \frac{V_n}{Z_{n,m}}$ is a $\widetilde{\mathbb{P}}^{(m)}$-martingale and is the m-forward price of the derivative at time $n = 0, 1, \ldots, m$, with random payoff $F_{m,m}^{(V)} = V_m$. We can therefore use the single-step recursion pricing formula in (8.129) (equivalently (8.127)) to obtain the derivative prices at every node on the tree.

Beginning at time $m-1$ we have $F_{m-1,m}^{(V)} = \widetilde{\mathbb{E}}_{m-1}^{(m)}[V_m]$, giving $V_{m-1} = Z_{m-1,m}F_{m-1,m}^{(V)}$. At time $m-2$ we have $F_{m-2,m}^{(V)} = \widetilde{\mathbb{E}}_{m-2}^{(m)}[F_{m-1,m}^{(V)}]$, giving $V_{m-2} = Z_{m-2,m}F_{m-2,m}^{(V)}$, and so on until time 0 where $F_{0,m}^{(V)} = \widetilde{\mathbb{E}}_0^{(m)}[F_{1,m}^{(V)}]$, giving $V_0 = Z_{0,m}F_{0,m}^{(V)}$. In this manner we generate the derivative prices at every time-n node $\omega_1 \ldots \omega_n$ using (8.129) and the single-time-step transition probabilities in the m-forward measure:

$$F_{n,m}^{(V)}(\omega_1 \ldots \omega_n) = \tilde{p}^{(m)}(\mathsf{U}|\omega_1 \ldots \omega_n)F_{n+1,m}^{(V)}(\omega_1 \ldots \omega_n \mathsf{U})$$
$$+ \tilde{p}^{(m)}(\mathsf{D}|\omega_1 \ldots \omega_n)F_{n+1,m}^{(V)}(\omega_1 \ldots \omega_n \mathsf{D}) \tag{8.130}$$

where $V_n(\omega_1 \ldots \omega_n) = Z_{n,m}(\omega_1 \ldots \omega_n)F_{n,m}^{(V)}(\omega_1 \ldots \omega_n)$ for all $n = 0, 1, \ldots, m-1$. At maturity m we simply have $F_{m,m}^{(V)}(\omega_1 \ldots \omega_m) = V_m(\omega_1 \ldots \omega_m)$. It is important to note that (8.130) is applicable to any path-dependent European derivative with $\mathcal{F}_m$-measurable payoff V_m. In the special case where the derivative is a stock, $V_n = S_n$, the m-forward prices of the stock at all the nodes are given recursively by the same $\mathbb{P}^{(m)}$-martingale property:

$$F_{n,m}^{(S)}(\omega_1 \ldots \omega_n) = \tilde{p}^{(m)}(\mathsf{U}|\omega_1 \ldots \omega_n)F_{n+1,m}^{(S)}(\omega_1 \ldots \omega_n \mathsf{U})$$
$$+ \tilde{p}^{(m)}(\mathsf{D}|\omega_1 \ldots \omega_n)F_{n+1,m}^{(S)}(\omega_1 \ldots \omega_n \mathsf{D}) \tag{8.131}$$

where $S_n(\omega_1 \ldots \omega_n) = Z_{n,m}(\omega_1 \ldots \omega_n)F_{n,m}^{(S)}(\omega_1 \ldots \omega_n)$, $n = 0, 1, \ldots, m-1$, $F_{m,m}^{(S)} = S_m$.

Consider the specific case where we have a given random payoff V_3 at maturity $m = 3$. Applying (8.128) recursively, for $n = 2, 1, 0$ gives:

$$F_{2,3}^{(V)} = \widetilde{\mathbb{E}}_2^{(3)}[V_3], \qquad V_2 = Z_{2,3}F_{2,3}^{(V)},$$
$$F_{1,3}^{(V)} = \widetilde{\mathbb{E}}_1^{(3)}[F_{2,3}^{(V)}], \quad V_1 = Z_{1,3}F_{1,3}^{(V)},$$
$$F_{0,3}^{(V)} = \widetilde{\mathbb{E}}_0^{(3)}[F_{1,3}^{(V)}] \quad V_0 = Z_{0,3}F_{0,3}^{(V)}.$$

Expressing these equations explicitly within the general binomial model, i.e., the equivalent of (8.130), gives the prices at all nodes:

$$F_{2,3}^{(V)}(\omega_1\omega_2) = \tilde{p}^{(3)}(\mathsf{U}|\omega_1\omega_2)V_3(\omega_1\omega_2\mathsf{U}) + \tilde{p}^{(3)}(\mathsf{D}|\omega_1\omega_2)V_3(\omega_1\omega_2\mathsf{D}),$$
$$\implies V_2(\omega_1\omega_2) = Z_{2,3}(\omega_1\omega_2)F_{2,3}^{(V)}(\omega_1\omega_2),$$
$$F_{1,3}^{(V)}(\omega_1) = \tilde{p}^{(3)}(\mathsf{U}|\omega_1)F_{2,3}^{(V)}(\omega_1\mathsf{U}) + \tilde{p}^{(3)}(\mathsf{D}|\omega_1)F_{2,3}^{(V)}(\omega_1\mathsf{D}),$$
$$\implies V_1(\omega_1) = Z_{1,3}(\omega_1)F_{1,3}^{(V)}(\omega_1),$$
$$F_{0,3}^{(V)} = \tilde{\mathbb{P}}^{(3)}(\mathsf{U})F_{1,3}^{(V)}(\mathsf{U}) + \tilde{\mathbb{P}}^{(3)}(\mathsf{D})F_{1,3}^{(V)}(\mathsf{D}),$$
$$\implies V_0 = Z_{0,3}F_{0,3}^{(V)}.$$

The following example considers the pricing of a European call option and a (path-dependent) lookback option within a non-recombining binomial model with stochastic interest rates and stochastic stock price volatility.

Example 8.14. Consider the three-period binomial model in Figure 8.6. Assume a non-dividend stock price process on this model with time-3 values $S_3(\mathsf{UUU}) = 20$, $S_3(\mathsf{UUD}) = 15$, $S_3(\mathsf{UDU}) = 12$, $S_3(\mathsf{UDD}) = 9$, $S_3(\mathsf{DUU}) = 12$, $S_3(\mathsf{DUD}) = 8$, $S_3(\mathsf{DDU}) = 6$, $S_3(\mathsf{DDD}) = 2$. Determine the derivative prices at all nodes for the following payoffs:

(a) $V_3 = (S_3 - 10)^+$;

(b) $V_3 = (M_3 - 12)^+$, where $M_3 = \max_{0 \leqslant n \leqslant 3} S_n$.

Solution. We shall use the $m = 3$ forward measure $\tilde{\mathbb{P}}^{(3)}$ with recurrence pricing equations as given just above. In order to implement this we need the ZCB prices $Z_{0,3}, Z_{1,3}, Z_{2,3}$ as well as the single-step transition probabilites under $\tilde{\mathbb{P}}^{(3)}$. The transition probabilities are computed in Example 8.13. The ZCB prices were also already computed above:

$$Z_{2,3}(\mathsf{DD}) \cong 0.97087, Z_{2,3}(\mathsf{DU}) \cong 0.96154, Z_{2,3}(\mathsf{UD}) \cong 0.95238, Z_{2,3}(\mathsf{UU}) \cong 0.93458,$$

$$Z_{1,3}(\mathsf{D}) \cong 0.92815, Z_{1,3}(\mathsf{U}) \cong 0.88588, Z_{0,3} \cong 0.86382.$$

(a) The call prices at maturity $m = 3$ are given by the payoffs: $V_3(\mathsf{UUU}) = 10$, $V_3(\mathsf{UUD}) = 5$, $V_3(\mathsf{UDU}) = 2$, $V_3(\mathsf{UDD}) = 0$, $V_3(\mathsf{DUU}) = 2$, $V_3(\mathsf{DUD}) = 0$, $V_3(\mathsf{DDU}) = 0$, $V_3(\mathsf{DDD}) = 0$. Applying the above backward recurrence formulas gives us all the 3-forward prices of the call and the corresponding call prices:

$$F_{2,3}^{(V)}(\mathsf{UU}) = 0.6 \cdot 10 + 0.4 \cdot 5 = 8 \implies V_2(\mathsf{UU}) \cong 0.93458 \cdot 8 \cong 7.477,$$

$$F_{2,3}^{(V)}(\mathsf{UD}) \cong 0.66667 \cdot 2 + 0.33333 \cdot 0 = 1.333$$

$$\implies V_2(\mathsf{UD}) \cong 0.95238 \cdot 1.33334 \cong 1.270,$$

$$F_{2,3}^{(V)}(\mathsf{DU}) \cong 0.5 \cdot 2 + 0.5 \cdot 0 = 1 \implies V_2(\mathsf{DU}) \cong 0.96154 \cdot 1 \cong 0.962,$$

$$F_{2,3}^{(V)}(\mathsf{DD}) = 0.33333 \cdot 0 + 0.66667 \cdot 0 = 0 \implies V_2(\mathsf{DD}) = 0,$$

$$F_{1,3}^{(V)}(\mathsf{U}) \cong 0.74645 \cdot 8 + 0.25355 \cdot 1.33334 \cong 6.310$$

$$\implies V_1(\mathsf{U}) \cong 0.88588 \cdot 6.30967 \cong 5.590,$$

$$F_{1,3}^{(V)}(\mathsf{D}) \cong 0.59768 \cdot 1 + 0.40232 \cdot 0 \cong 0.598 \implies V_1(\mathsf{D}) \cong 0.92815 \cdot 0.59768 \cong 0.555,$$

$$F_{0,3}^{(V)} \cong 0.48835 \cdot 6.30967 + 0.51165 \cdot 0.59768 \cong 3.387$$

$$\implies V_0 \cong 0.86382 \cdot 3.38713 \cong 2.926.$$

(b) We first need the payoff values for all outcomes. Hence, we need the stock prices at times $n = 2, 1, 0$ and these are obtained by the applyng either (8.131) or the fact that $\{S_n/B_n\}_{n \geqslant 0}$ is a $\tilde{\mathbb{P}}$-martingale, i.e., equation (8.68). The reader can check that both sets of equations produce the same stock price values and forward stock prices:

$$S_2(\mathsf{UU}) \cong 16.822, S_2(\mathsf{UD}) \cong 10.476, S_2(\mathsf{DU}) \cong 9.615, S_2(\mathsf{DD}) \cong 3.236,$$

$$S_1(\mathsf{U}) \cong 14.373, S_1(\mathsf{D}) \cong 6.792, S_0 \cong 10.079,$$

and

$$F_{2,3}^{(S)}(\mathsf{UU}) \cong 18.000, F_{2,3}^{(S)}(\mathsf{UD}) \cong 11.000, F_{2,3}^{(S)}(\mathsf{DU}) \cong 10.000, F_{2,3}^{(S)}(\mathsf{DD}) \cong 3.333,$$

$$F_{1,3}^{(S)}(\mathsf{U}) \cong 16.225, F_{1,3}^{(S)}(\mathsf{D}) \cong 7.318, F_{0,3}^{(S)} \cong 11.668.$$

Although not required for the calculation of the derivative prices, we note that the stochastic up and down factors are readily obtained using (8.62):

$$u_3(\text{UU}) = \frac{S_3(\text{UUU})}{S_2(\text{UU})} \cong 1.189, d_3(\text{UU}) = \frac{S_3(\text{UUD})}{S_2(\text{UU})} \cong 0.892,$$

$$u_3(\text{UD}) = \frac{S_3(\text{UDU})}{S_2(\text{UD})} \cong 1.145, d_3(\text{UD}) = \frac{S_3(\text{UDD})}{S_2(\text{UD})} \cong 0.859,$$

$$u_3(\text{DU}) = \frac{S_3(\text{DUU})}{S_2(\text{DU})} \cong 1.248, d_3(\text{DU}) = \frac{S_3(\text{DUD})}{S_2(\text{DU})} \cong 0.832,$$

$$u_3(\text{DD}) = \frac{S_3(\text{DDU})}{S_2(\text{DD})} \cong 1.854, d_3(\text{DD}) = \frac{S_3(\text{DDD})}{S_2(\text{DD})} \cong 0.618,$$

$$u_2(\text{U}) = \frac{S_2(\text{UU})}{S_1(\text{U})} \cong 1.170, d_2(\text{U}) = \frac{S_2(\text{UD})}{S_1(\text{U})} \cong 0.729,$$

$$u_2(\text{D}) = \frac{S_2(\text{DU})}{S_1(\text{D})} \cong 1.416, d_2(\text{D}) = \frac{S_2(\text{DD})}{S_1(\text{D})} \cong 0.476,$$

$$u_1 = \frac{S_1(\text{U})}{S_0} \cong 1.426, d_1 = \frac{S_1(\text{D})}{S_0} \cong 0.674.$$

Using $M_3(\omega_1\omega_2\omega_3) = \max\{S_0, S_1(\omega_1), S_2(\omega_1\omega_2), S_3(\omega_1\omega_2\omega_3)\}$, the payoff values are:

$$V_3(\text{UUU}) = (S_3(\text{UUU}) - 12)^+ = 8, V_3(\text{UUD}) = (S_2(\text{UU}) - 12)^+ \cong 4.822,$$
$$V_3(\text{UDU}) = V_3(\text{UDD}) = (S_1(\text{U}) - 12)^+ \cong 2.373,$$
$$V_3(\text{DUU}) = V_3(\text{DUD}) = V_3(\text{DDU}) = V_3(\text{DDD}) = 0.$$

Applying backward recurrence produces the forward prices of the lookback and the corresponding lookback prices:

$$F_{2,3}^{(V)}(\text{UU}) = 0.6 \cdot 8 + 0.4 \cdot 4.822 \cong 6.729 \implies V_2(\text{UU}) \cong 0.93458 \cdot 6.729 \cong 6.289,$$

$$F_{2,3}^{(V)}(\text{UD}) \cong 0.66667 \cdot 2.373 + 0.33333 \cdot 2.373 = 2.373$$
$$\implies V_2(\text{UD}) \cong 0.95238 \cdot 2.373 \cong 2.260,$$

$$F_{2,3}^{(V)}(\text{DU}) \cong 0.5 \cdot 0 + 0.5 \cdot 0 = 0 \implies V_2(\text{DU}) = 0,$$

$$F_{2,3}^{(V)}(\text{DD}) = 0.33333 \cdot 0 + 0.66667 \cdot 0 = 0 \implies V_2(\text{DD}) = 0,$$

$$F_{1,3}^{(V)}(\text{U}) \cong 0.74645 \cdot 6.729 + 0.25355 \cdot 2.373 \cong 5.625$$
$$\implies V_1(\text{U}) = 0.88588 \cdot 5.625 = 4.983,$$

$$F_{1,3}^{(V)}(\text{D}) \cong 0.59768 \cdot 0 + 0.40232 \cdot 0 = 0 \implies V_1(\text{D}) = 0,$$

$$F_{0,3}^{(V)} \cong 0.48835 \cdot 5.625 + 0.51165 \cdot 0 \cong 2.747$$
$$\implies V_0 \cong 0.86382 \cdot 2.747 \cong 2.373.$$

$\square$

8.5   Exercises

Exercise 8.1. Prove that a multi-period model is complete iff every single-period sub-model is complete (Lemma 8.6).

Exercise 8.2. Consider a binomial model with stochastic volatility and stochastic interest rates presented in Section 8.4.1. Obtain a formula for the number of shares of stock, Δ_t, held at each time $t \in \{0, 1, 2, \ldots, T-1\}$ in a self-financing portfolio strategy that replicates a derivative with a given $\mathcal{F}_T$-measurable payoff V_T, with $\{\mathcal{F}_t\}_{t \geqslant 0}$ as filtration generated by the market up/down moves.

Exercise 8.3. Consider a two-period market model with four scenarios and two base assets: a risk-free asset with the prices $B_0 = 50$, $B_1 = 55$, and $B_2 = 60$ and a risky stock whose prices are given by

Scenario, ω	$S_0(\omega)$	$S_1(\omega)$	$S_2(\omega)$
ω^1	50	60	70
ω^2	50	60	50
ω^3	50	45	50
ω^4	50	45	40

(a) Determine the EMM relative to the risk-free asset as numéraire.

(b) If instead $S_2(\omega^1) = 65$, show how to construct an arbitrage strategy.

Exercise 8.4. Consider the two-period trinomial model in Example 8.4 (see Figure 8.4). For each of the following payoffs, determine the dynamic hedging strategy $(\beta_t, \Delta_t^1, \Delta_t^2)$, $t = 0, 1$, for all nodes in the tree:

(a) $V_2(\Omega) = [10, 20, 30, 40, 50, 60, 70, 80, 90]$;

(b) $V_2 = (S_2^2 - S_2^1)^+$;

(c) $V_2 = \max\{S_2^1, S_2^2\}$.

(d) $V_2 = \min\{S_2^1, S_2^2\}$.

Exercise 8.5. Consider the two-period trinomial model with three base assets as given in Examples 8.3 and 8.4 (see also Figure 8.4).

(a) Construct the Radon–Nikodym process of $\widetilde{\mathbb{P}}^{(S^1)}$ w.r.t. $\widetilde{\mathbb{P}}^{(B)}$: $\{\varrho_t\}_{t \in \{0,1,2\}}$ where $\varrho_t \equiv \left(\frac{d\widetilde{\mathbb{P}}^{(S^1)}}{d\widetilde{\mathbb{P}}^{(B)}}\right)_t$ with filtration generated by the atoms A_t^i (see Figure 8.2).

(b) Determine the probabilities $\widetilde{\mathbb{P}}^{(S^1)}(\omega^i)$ for $i = 1, \ldots, 9$.

Exercise 8.6. Consider the following game. There is a set of three coins labelled 1, 2, and 3. The first coin has probability of 0.4 for heads, the second coin has probability of 0.6 for heads, and the third coin is fair. Your initial wealth is \$1. At each time step you toss the three coins in turn. For each coin $\#i = 1, 2, 3$ that results in heads, your wealth is increased by a factor of $i+1$; otherwise, your wealth is correspondingly decreased by a factor of $i+1$. Let V_n denote the wealth after a total of n time steps. Determine the distribution of V_n for any $n \geqslant 1$. Assume all coin tosses are independent.

Exercise 8.7. (a) For a given integer m with $1 \leqslant m \leqslant T$, consider a contract that pays r_m at time m, where $r_m \equiv R_{m-1}$ is the simple interest rate for the period $[m-1, m]$. Show that the time-zero no-arbitrage price of this contract is equal to $Z_{0,m-1} - Z_{0,m}$, where $Z_{0,k}$ are time-0 zero-coupon bond prices maturing at times $k = 0, 1, \ldots, T$.

(b) Derive equation (8.82).

Exercise 8.8. For a given integer m with $1 \leqslant m \leqslant T$, an m-period (payer's) *interest rate swap* with unit principal is a contract that makes future payments $P_1, P_2, \ldots, P_m$ at times $1, 2, \ldots, m$, respectively, where $P_n = K - r_n$, $n = 1, 2, \ldots, m$, K is a fixed interest rate and $r_n \equiv R_{n-1}$ is the simple interest rate for the period $[n-1, n]$.

(a) Show that the time-zero no-arbitrage value $V_0 \equiv V_0^{\mathrm{Swap}(m)}$ of this swap contract is given by

$$V_0^{\mathrm{Swap}(m)} = K \sum_{n=1}^{m} Z_{0,n} - (1 - Z_{0,m}). \tag{8.132}$$

(b) Compute the time-zero no-arbitrage value V_0 of a three-period swap with rate $K = 4\%$ for the model with interest rates given in Figure 8.6.

(c) For the same interest rate model in part (b), determine the equilibrium swap rate, K_s, for which the present value $V_0 = 0$.

Exercise 8.9. For a given integer m with $1 \leqslant m \leqslant T$, an m-period *interest rate cap* is a contract that makes future payments $C_1, C_2, \ldots, C_m$ at times $1, 2, \ldots, m$, respectively, where $C_n = (r_n - K)^+$ for $n = 1, 2, \ldots, m$, and K is a fixed rate. An m-period *interest rate floor* is a contract that makes future payments $F_1, F_2, \ldots, F_m$ at times $1, 2, \ldots, m$, respectively, where $F_n = (K - r_n)^+$ for $n = 1, 2, \ldots, m$, and K is a fixed rate. Here, $r_n \equiv R_{n-1}$ is the simple interest rate for the period $[n-1, n]$.

(a) Consider an interest rate swap, cap, and floor with the same given period m. Show that the sum of the initial no-arbitrage m-period values of the swap, $V_0^{\mathrm{Swap}(m)}$, and cap, $V_0^{\mathrm{Cap}(m)}$, gives the initial no-arbitrage value of the floor, $V_0^{\mathrm{Floor}(m)}$, i.e.,

$$V_0^{\mathrm{Swap}(m)} + V_0^{\mathrm{Cap}(m)} = V_0^{\mathrm{Floor}(m)}.$$

(b) Determine the initial no-arbitrage values of the three-period floor and cap with $K = 4\%$ for the model with interest rates given in Figure 8.6.

Exercise 8.10. Consider a binomial interest-rate tree model having the following ZCB prices:

$$Z_{0,1} = 0.975, \ Z_{0,2} = 0.825, \ Z_{1,2}(\mathsf{D}) = 0.85, \ Z_{1,2}(\mathsf{U}) = 0.84, \ Z_{1,3}(\mathsf{D}) = 0.8, \ Z_{1,3}(\mathsf{U}) = 0.75,$$

$$Z_{2,3}(\mathsf{DD}) = 0.965, \ Z_{2,3}(\mathsf{DU}) = 0.885, \ Z_{2,3}(\mathsf{UD}) = 0.93, \ Z_{2,3}(\mathsf{UU}) = 0.86.$$

Determine all risk-neutral probabilities $\widetilde{\mathbb{P}}(\omega_1\omega_2)$ under measure $\widetilde{\mathbb{P}} \equiv \widetilde{\mathbb{P}}^{(B)}$.

Exercise 8.11. Consider a general binomial tree model for interest rates. Let the risk-free interest rates and no-arbitrage prices of bonds with all possible maturities be given. Show that the lack of arbitrage implies that the risk-neutral probabilities defined by (8.89) are independent of maturity.

Exercise 8.12. Consider a Markov model with the following interest rates $R_{n,j}$ on the given nodes (n, j) for $n = 0, 1, 2$: $R_{0,0} = R_0 = 2\%, R_{1,1} = 3\%, R_{1,2} = 3.5\%, R_{1,3} = 4\%, R_{2,1} = 2.5\%, R_{2,2} = 2.75\%, R_{2,3} = 3.25\%, R_{2,4} = 3.75\%, R_{2,5} = 4.25\%, R_{2,6} = 4.5\%$. The nonzero risk-neutral transition probabilities in (8.91) (under measure $\widetilde{\mathbb{P}} \equiv \widetilde{\mathbb{P}}^{(B)}$) are as follows:

$$\widetilde{p}_{(0,0)\to(1,1)} = \frac{1}{3}, \ \widetilde{p}_{(0,0)\to(1,2)} = \frac{1}{6}, \ \widetilde{p}_{(0,0)\to(1,3)} = \frac{1}{2};$$

$$\widetilde{p}_{(1,1)\to(2,1)} = \frac{1}{3}, \ \widetilde{p}_{(1,1)\to(2,2)} = \frac{2}{3}, \ \widetilde{p}_{(1,2)\to(2,3)} = \widetilde{p}_{(1,2)\to(2,4)} = \frac{1}{2}$$

$$\widetilde{p}_{(1,3)\to(2,5)} = \frac{1}{4}, \ \widetilde{p}_{(1,3)\to(2,6)} = \frac{3}{4}.$$

(a) Compute the ZCB prices $Z_{n,j;m} = Z_{n,m}(R_{n,j})$ for all nodes (n, j), $0 \leqslant n < m$, $m = 1, 2, 3$.

(b) Compute the time-0 no-arbitrage values of the three-period swap, cap and floor: $V_0^{\text{Swap}(3)}$, $V_0^{\text{Cap}(3)}$ and $V_0^{\text{Floor}(3)}$. Assume a swap rate $K = 4\%$. See Exercises 8.8 and 8.9 for basic formulas and definitions.

Exercise 8.13. Assume an investor is allowed to trade in the bank (money market) account and in zero-coupon bonds of all maturities up to some fixed future time T. In particular, let

$$\Pi_n = \beta_n B_n + \sum_{m=n+1}^{T} \alpha_n^{(m)} Z_{n,m}$$

be the time-n portfolio value of a self-financing portfolio strategy over time $n = 0, 1, \ldots, T - 1$.

(a) Show (argue) that the wealth equation

$$\Pi_{n+1} = \alpha_n^{(n+1)} + \sum_{m=n+2}^{T} \alpha_n^{(m)} Z_{n+1,m} + (1 + R_n)\left(\Pi_n - \sum_{m=n+1}^{T} \alpha_n^{(m)} Z_{n,m}\right)$$

arises, where $R_n \equiv r_{n+1}$ is the simple interest rate for the period $[n, n+1]$.

(b) Assume all the positions β_n, $\alpha_n^{(m)}$ and R_n are $\mathcal{F}_n$-measurable, where $\{\mathcal{F}_n\}_{n\geqslant 0}$ is the filtration generated by the market moves. Show that the discounted portfolio value process $\{D_n \Pi_n\}_{0\leqslant n\leqslant T}$ is a martingale w.r.t. $\{\mathcal{F}_n\}_{n\geqslant 0}$ under the risk-neutral measure $\widetilde{\mathbb{P}} \equiv \widetilde{\mathbb{P}}^{(B)}$ with bank account as numéraire.

Exercise 8.14. Consider the three-period binomial interest rate model in Example 8.12 with interest rates and risk-neutral single-step transition (conditional) probabilites given in Figure 8.6. In each case, compute the no-arbitrage time-0 value, $V_0^{(3)}$, of a coupon-bearing bond with the following time-n path-dependent payments C_n:

(a) $C_1(\mathsf{U}) = 0.01$, $C_1(\mathsf{D}) = 0.02$, $C_2(\mathsf{UU}) = 0.04$, $C_2(\mathsf{UD}) = 0.035$, $C_2(\mathsf{DU}) = 0.03$, $C_2(\mathsf{DD}) = 0.025$, $C_3(\mathsf{UUU}) = 0.05$, $C_3(\mathsf{UUD}) = 0.045$, $C_3(\mathsf{UDU}) = C_3(\mathsf{UDD}) = 0.04$, $C_3(\mathsf{DUU}) = C_3(\mathsf{DUD}) = 0.03$, $C_3(\mathsf{DDU}) = 0.025$, $C_3(\mathsf{DDD}) = 0.02$;

(b) $C_n = 0.01(1 + \mathsf{U}_n)$, where $\mathsf{U}_n(\omega)$ is the number of U's (up moves) in the first n moves in $\omega \in \Omega$.

(c) $C_n = 0.01(1 + \mathsf{D}_n)$, where $\mathsf{D}_n(\omega)$ is the number of D's (down moves) in the first n moves in $\omega \in \Omega$.

Exercise 8.15. Consider the Ho-Lee binomial interest rate model with rates given by equation (8.100) with fixed volatility parameter $b_n \equiv \sigma$, $n \geqslant 1$. Part of the calibration procedure was discussed in Section 8.4.3. Provide the complete steps in calibrating the model to the first m-maturity ZCB market prices $Z_{0,1}, Z_{0,2}, \ldots, Z_{0,m}$.

Exercise 8.16. Consider a three-period Ho-Lee binomial interest rate model with rates given by equation (8.100), where $R_0 = 0.05, a_1 = 0.045, a_2 = 0.04$ and $b_1 = b_2 = 0.01$. Assume all single-step transition probabilities in the risk-neutral measure $\widetilde{\mathbb{P}} \equiv \widetilde{\mathbb{P}}^{(B)}$ are equal to $\frac{1}{2}$.

(a) Compute the ZCB prices $Z_{n,j;m} = Z_{n,m}(R_{n,j})$ for all nodes (n, j), $0 \leqslant n < m$, $m = 1, 2, 3$.

(b) Compute the time-0 no-arbitrage values of the three-period swap, cap and floor: $V_0^{\text{Swap}(3)}$, $V_0^{\text{Cap}(3)}$ and $V_0^{\text{Floor}(3)}$. Assume a swap rate $K = 4\%$. See Exercises 8.8 and 8.9 for basic formulas and definitions.

Exercise 8.17. Consider a BDT binomial interest rate model with rates given by equation (8.101), where $R_0 = 0.02$, $a_n = 0.02(1.1)^{-n}$ and $b_n = 1.21$, $n \geqslant 1$. Assume all single-step transition probabilities in the risk-neutral measure $\widetilde{\mathbb{P}} \equiv \widetilde{\mathbb{P}}^{(B)}$ are equal to $\frac{1}{2}$.

(a) Compute the ZCB prices $Z_{n,j;m} = Z_{n,m}(R_{n,j})$ for all nodes (n, j), $0 \leqslant n < m$, $m = 1, 2, 3$.

(b) Compute the time-0 no-arbitrage value of the derivative having time-2 payoff:

$$V_2 = (K - \min\{R_0, R_1, R_2\})^+ ,$$

with fixed rate $K = 2\%$.

(c) Compute the time-0 no-arbitrage value of the derivative having time-3 payoff:

$$V_3 = (\max\{R_0, R_1, R_2, R_3\} - K)^+ ,$$

with fixed rate $K = 2.2\%$.

(d) Compute the time-0 no-arbitrage value of the derivative having time-3 payoff:

$$V_3 = (K - \min\{R_0, R_1, R_2, R_3\})^+ ,$$

with fixed rate $K = 2\%$.

Exercise 8.18. Show that the k-period forward rate process $\{f_{n,m;m+k}\}_{0\leqslant n\leqslant m}$, defined by (8.79), is a $\widetilde{\mathbb{P}}^{(m+k)}$-martingale, i.e., a martingale under the $(m + k)$-forward measure. Note that for $k = 1$ we have the $\widetilde{\mathbb{P}}^{(m+1)}$-martingale property of the single-period forward rate process $\{f_{n,m}\}_{0\leqslant n\leqslant m}$.

Exercise 8.19. Obtain all the single-period forward rates $\{f_{n,2}\}_{n=0,1,2}$ for the binomial model in Figure 8.6 (see Example 8.14). Then, verify explicitly the $\widetilde{\mathbb{P}}^{(3)}$-martingale property: $\widetilde{\mathbb{E}}_1^{(3)}[f_{2,2}] = f_{1,2}$ and $\widetilde{\mathbb{E}}_0^{(3)}[f_{1,2}] = f_{0,2}$.

A

Elementary Probability Theory

A.1 Probability Space

Probability theory studies regularities and rules arising in experiments that are random and whose results cannot be predicted with certainty in advance. Probability theory gives us tools to model and study such events, to understand how they are connected and how we can make better predictions.

A.1.1 A Sample Space and Events

Consider a random experiment, and let its every possible result be described by one and only one sample outcome. A *sample space* is the set of all outcomes of the experiment. We denote it by the capital Greek letter omega, Ω. Any particular outcome is denoted by the lowercase Greek letter omega, ω. Any subset of a sample space, i.e., any part of the set Ω, including the whole set and an empty set, is called an *event*. Events are denoted by capital Latin letters such as E, F, etc. We say that an event *occurs* if the actual outcome of a random experiment belongs to the event. That is, event E occurs if the outcome ω is an element of E, denoted $\omega \in E$. A sample space is said to be *discrete* if it consists of a finite or countably infinite number of outcomes and *continuous* if it contains an interval (finite or infinite) of real numbers.

The table below provides some examples of experiments, outcomes, and sample spaces.

Experiment	Outcome	Sample Space
toss a coin	head or tail	$\{H, T\}$
roll a die	the upright face number	$\{1, 2, 3, 4, 5, 6\}$
measure the lifetime of a person selected at random	the age (in full years)	$\{0, 1, 2, \ldots\}$
measure the height of a person selected at random	the height	$(0, \infty)$
locate a place around the world selected at random	two coordinates: latitude and longitude	$\mathbb{R}^2$

Example A.1. Toss a coin three times. Describe the sample space Ω and the following events.

(A) Only one head appears.

(B) At most one head appears.

(C) At least one head appears.

(D) At least two heads appear.

(E) No head appears.

Solution. The sample space consists of eight outcomes; each outcome is a sequence of three letters H and T:

$$\Omega = \{TTT, TTH, THT, THH, HTT, HTH, HHT, HHH\}.$$

The first letter in a sequence corresponds to the result of the first tossing, the second letter is for the second tossing result, and so on. The events are the following subsets of Ω:

$$A = \{TTH, THT, HTT\},$$
$$B = \{TTT, TTH, THT, HTT\},$$
$$C = \{TTH, THT, THH, HTT, HTH, HHT, HHH\} = \Omega \setminus \{TTT\},$$
$$D = \{THH, HTH, HHT, HHH\},$$
$$E = \{TTT\}.$$

□

Example A.2. Consider the repeated tossing of a coin until the second time a head appears.

(a) Find the sample space.

(b) Find an event that a coin has been tossed exactly four times.

Solution. We may observe two heads in a row, or a head followed by a tail and then by the other head, or a sequence of tails with one head somewhere in the middle ended by another head. So, the sample space has infinitely many outcomes, but each outcome has the following pattern: $\omega = L_1 L_2 \ldots L_n H$, where $L_1, L_2, \ldots, L_n \in \{T, H\}$ and $L_1, \ldots, L_n$ contain only one H. Thus, the sample space Ω is

$$\{HH, THH, HTH, TTHH, THTH, HTTH, TTTHH, TTHTH, THTTH, HTTTH, \ldots\}.$$

The event that the coin has been tossed four times is $\{TTHH, THTH, HTTH\}$.   □

Since events are simply subsets of a sample space, we can perform set operations on events to create new events. Let A and B be two events of a sample space Ω. That is, $A, B \subset \Omega$.

- The *intersection* of A and B, denoted $A \cap B$ or AB, is a subset of Ω that contains all elements that are in both A and B. We can see that the event $A \cap B$ occurs if and only if both events A and B occur.

- The *union* of A and B, denoted $A \cup B$, contains all elements that are either in A, in B, or in both sets. The event $A \cup B$ occurs if and only if at least one of the events A and B occurs.

- The *difference* of A and B, denoted $A \setminus B$ or $A - B$, contains of all elements of A that are not in B. The event $A \setminus B$ occurs if and only if A occurs but not B.

- The *complement* of A, denoted A^C, or A', or $\bar{A}$, refers to all elements of Ω that are not in A. That is, $A^C = \Omega \setminus A$. The event A^C occurs if and only if A does not occur.

Using the set-theoretic notation, we can write

$$A \cap B = \{x \in \Omega \mid x \in A \text{ and } x \in B\},$$
$$A \cup B = \{x \in \Omega \mid x \in A \text{ or } x \in B\},$$
$$A \setminus B = \{x \in \Omega \mid x \in A \text{ and } x \notin B\},$$
$$A^{\mathsf{C}} = \{x \in \Omega \mid x \notin A\}.$$

For two sets, we can identify three possible situations. Firstly, the sets may have no common elements. Secondly, they may have common and noncommon elements. Thirdly, one set is a subset of the other set. We can describe these relations in terms of events. If event A occurs whenever event B occurs, then B is a subset of A, i.e., $B \subset A$. If events A and B do not occur together, then they have no common elements, i.e., $A \cap B = \emptyset$. In the latter case, we say that A and B are *mutually disjoint* (or *mutually exclusive*) events.

The operations under events and relations between them can be visualized using *Venn diagrams*. Figure A.1 contains sample Venn diagrams for two events. Each event is represented by a circle. The shaded region corresponds to the result of the respective set operation. The outer rectangle represents the sample space Ω.

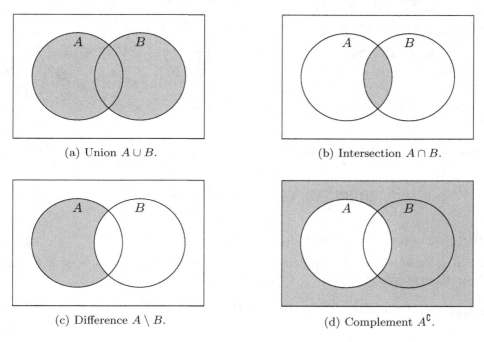

(a) Union $A \cup B$.   (b) Intersection $A \cap B$.

(c) Difference $A \setminus B$.   (d) Complement A^{C}.

FIGURE A.1: Venn diagrams.

Example A.3. Let E, F, G be three events of a common sample space. Find set-theoretic expressions and respective Venn diagrams for the following events:

(a) both E and F occur, but not G;

(b) exactly one of E, F, and G occurs;

(c) at least one of E, F, and G occurs;

(d) at most one of E, F, and G occurs.

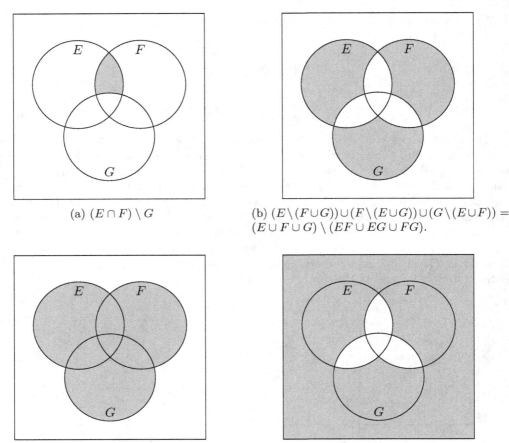

(a) $(E \cap F) \setminus G$

(b) $(E \setminus (F \cup G)) \cup (F \setminus (E \cup G)) \cup (G \setminus (E \cup F)) = (E \cup F \cup G) \setminus (EF \cup EG \cup FG)$.

(c) $E \cup F \cup G$.

(d) $(EF \cup FG \cup EG)^{\complement}$.

FIGURE A.2: Venn diagrams for Example A.3.

Solution. The answers are provided in Figure A.2.   □

When we perform several operations on events, the following laws can be used to simplify set-theoretic expressions.

Commutative Law:
$$A \cup B = B \cup A,$$
$$A \cap B = B \cap A.$$

Associative Law:
$$(A \cup B) \cup C = A \cup (B \cup C),$$
$$(A \cap B) \cap C = A \cap (B \cap C).$$

Distributive Law:
$$(A \cup B) \cap C = (A \cap C) \cup (B \cap C),$$
$$(A \cap B) \cup C = (A \cup C) \cap (B \cup C).$$

DeMorgan's Law:
$$(A \cap B)^{\complement} = A^{\complement} \cup B^{\complement},$$
$$(A \cup B)^{\complement} = A^{\complement} \cap B^{\complement}.$$

A.1.2 Probability

The *probability* of an event is a *chance* of how likely the event occurs, which is measured as the likelihood on a scale of 0 to 1. The probability of an impossible event is zero, the probability of an inevitable (certain) event is one. A rare event has a probability close to zero. A very common event has a probability close to one.

There are three commonly-used interpretations of the concept of probability.

- **Frequentist:** The probability of an event is the long-run proportion of times that the event occurs in independent repetitions (trials) of some random experiment:

$$\mathbb{P}(E) \approx \frac{\text{Number of times } E \text{ occurs}}{\text{Number of trials}}.$$

- **Bayesian:** Probability is interpreted as reasonable expectation representing a state of knowledge or as quantification of a personal belief.

- **Classical:** An event's probability is the ratio of the number of favourable outcomes to the number of possible outcomes in a symmetric experiment where all outcomes are equally likely to happen.

In defining the probability, we follow an axiomatic approach. Consider a random experiment with a sample space Ω. For an event $E \subset \Omega$, we define a number $\mathbb{P}(E)$. The probability of a singleton $\{\omega\}$ is denoted by $\mathbb{P}(\omega)$. The function $\mathbb{P}$, which is defined on subsets of Ω, satisfies the following three axioms known as *Kolmogorov's axioms*.

Axiom 1. (Nonnegativity) $\mathbb{P}(E) \geqslant 0$ for any $E \subset \Omega$.

Axiom 2. (Certainty) $\mathbb{P}(\Omega) = 1$.

Axiom 3. (Countable Additivity) For any countable collection of pairwise disjoint events $E_1, E_2, \ldots$ such that $E_n \cap E_m = \emptyset$ when $n \neq m$, we have

$$\mathbb{P}(E_1 \cup E_2 \cup E_3 \cup \cdots) = \mathbb{P}(E_1) + \mathbb{P}(E_2) + \mathbb{P}(E_3) + \cdots \qquad (A.1)$$

$$\text{or, equivalently,} \quad \mathbb{P}\left(\bigcup_{k \geqslant 1} E_k\right) = \sum_{k \geqslant 1} \mathbb{P}(E_k).$$

The function $\mathbb{P}$ is called a *probability function* or a *probability measure*. It satisfies the following properties that can be derived directly from the probability axioms.

1. $\mathbb{P}(\emptyset) = 0$ where the empty set $\emptyset$ represents a null or impossible event.

2. For any finite collection of disjoint events $E_1, E_2, \ldots, E_n$, we have that

$$\mathbb{P}(E_1 \cup E_2 \cup \cdots \cup E_n) = \sum_{k=1}^{n} \mathbb{P}(E_k).$$

It is a special case of Axiom 3, where all E_i with $i > n$ are empty sets.

3. $\mathbb{P}(A^{\complement}) = 1 - \mathbb{P}(A)$ (*Complementation Rule*).

4. $A \subset B \Rightarrow \mathbb{P}(A) \leqslant \mathbb{P}(B)$ (*Domination Principle*).

5. $\forall A \subset \Omega \quad \mathbb{P}(A) \leqslant 1$.

6. $\mathbb{P}(A \cup B) = \mathbb{P}(A) + \mathbb{P}(B) - \mathbb{P}(A \cap B)$ (*Addition Rule*).

7. $\forall A_1, A_2, \ldots, A_n \subset \Omega$ we have the following *Inclusion-Exclusion Identity*:

$$\mathbb{P}\left(\bigcup_{k=1}^{n} A_k\right) = \sum_{k=1}^{n} \mathbb{P}(A_k) - \sum_{k_1 < k_2} \mathbb{P}(A_{k_1} \cap A_{k_2}) + \cdots$$
$$+ (-1)^{m+1} \sum_{k_1 < \cdots < k_m} \mathbb{P}(A_{k_1} \cap A_{k_2} \cap \cdots \cap A_{k_m}) + \cdots$$
$$+ (-1)^{n+1} \mathbb{P}(A_1 \cap A_2 \cap \cdots \cap A_n). \tag{A.2}$$

Below, we present two approaches to constructing a probability function. The first one is known as *classical probability*, where a sample space has finitely many, equally likely possible outcomes. Let Ω contain $N(\Omega)$ outcomes, and let the probability of any outcome $\omega \in \Omega$ be the same and equal to $\mathbb{P}(\omega) = \dfrac{1}{N(\Omega)}$. For any $E \subset \Omega$, the computation of $\mathbb{P}(E)$ requires the calculation of the number of ways the event E may occur. As defined above, an event occurs if the actual outcome of a random experiment belongs to the event. Since all outcomes are equally likely, the probability of event E should be proportional to the number of outcomes contained in E. Indeed, by Axiom 3, the probability of an event $E \subset \Omega$ is given by

$$\mathbb{P}(E) = \sum_{\omega \in E} \mathbb{P}(\omega) = \sum_{\omega \in E} \frac{1}{N(\Omega)} = \frac{N(E)}{N(\Omega)},$$

where $N(E)$ denotes the number of elements of E.

Here are some examples of random experiments with equally likely outcomes.

- Pick a card at random from a deck of 52 playing cards (where all cards are different). There are 52 equally likely outcomes.

- Take a pair of cards at random from a deck of 52 playing cards. There are $\dfrac{52 \cdot 51}{2} = 1326$ possible pairs (if we ignore the order in which the cards have been selected).

- Toss n coins in order, where $n = 1, 2, \ldots$. Assuming the coins are fair, there are 2^n equally likely outcomes. Each outcome can be represented by a string of n letters from $\{H, T\}$, where the kth letter from the left-hand side is the result of the kth coin tossing for $k = 1, 2, \ldots, n$.

- Roll n balanced six-sided dice in order, where $n = 1, 2, \ldots$. There are 6^n equally likely outcomes.

- Take n balls without replacement from an urn with M distinct balls where $1 \leqslant n \leqslant M$. There are $M(M-1)(M-2) \cdots (M-n+1)$ ordered samples.

In what follows, we only deal with balanced six-sided dice and fair coins.

Example A.4. Two dice are rolled. What is the probability that

(a) both dice will show the same value?

(b) one die shows a 3, and the other shows a 4?

(c) the total is greater than 9?

Solution. Each outcome is a pair of two integers from 1 to 6. In total, there are $6 \times 6 = 36$ equally likely outcomes. The sample space is $\Omega = \{(i, j) : 1 \leqslant i, j \leqslant 6\}$.

(a) The event contains six outcomes:

$$A = \{(1,1), (2,2), (3,3), (4,4), (5,5), (6,6)\}.$$

Its probability is $\mathbb{P}(A) = \dfrac{6}{36} = \dfrac{1}{6}$.

(b) The event contains two outcomes:

$$B = \{(3,4), (4,3)\}.$$

Its probability is $\mathbb{P}(B) = \dfrac{2}{36} = \dfrac{1}{18}$.

(c) The total of two dice is an integer from 2 to 12. Let us list of possible pairs with a sum greater than or equal to 10:

$$C = \{(5,5), (4,6), (6,4), (5,6), (6,5), (6,6)\}.$$

Its probability is $\mathbb{P}(C) = \dfrac{6}{36} = \dfrac{1}{6}$.

$\square$

The second approach to defining a probability function is referred to as *Geometric Probability*. Consider a domain $\Omega \subset \mathbb{R}^n$ with a finite nonzero measure, $\|\Omega\| \in (0, \infty)$. For a linear domain ($n = 1$), the measure is a length. For a planar domain ($n = 2$), the measure is an area. For a solid domain ($n = 3$), the measure is a volume. Select a point $r \in \Omega$ at random. Here, "at random" means that the probability to select a point from any subdomain $A \subset \Omega$ does not depend on the shape or location of A but only depends on its measure $\|A\|$. We say that the point r is distributed *uniformly* in the domain Ω.

An event $A \subset \Omega$ occurs if the selected random point belongs to A. The probability of event A should be proportional to its measure $\|A\|$:

$$\mathbb{P}(A) = K \cdot \|A\| \text{ for some common constant } K$$

Since $1 = \mathbb{P}(\Omega) = K \cdot \|\Omega\|$, we have that $K = 1/\|\Omega\|$. Therefore, the geometric probability measure is defined as $\mathbb{P}(A) = \frac{\|A\|}{\|\Omega\|}$.

Example A.5 (The meeting problem). Romeo and Juliet arranged to meet in a certain place between 2 p.m. and 3 p.m. The person who comes first will wait for the other during 10 min and then leave. What is the probability of meeting these two people if each of them may come at any time (equally likely between 2 p.m. and 3 p.m.) independently on the other one?

Solution. Let R and J be, respectively, the time (in minutes) when Romeo and Juliet arrived at the meeting place after 2 p.m. The pair (R, J) is selected at random in the square $\Omega = [0, 60] \times [0, 60]$ with the area $\|\Omega\| = 3600$. They meet each other if they arrive no more than 10 minutes apart. That is, they meet if $|R - J| \leqslant 10$ holds. The area or the domain

$$\{(R, J) : |R - J| \leqslant 10, \ 0 \leqslant R, J \leqslant 60\}$$

is $60^2 - 50^2 = 1100$. Thus, the probability that the meeting takes place is $\dfrac{1100}{3600} = \dfrac{11}{36}$. $\square$

A.1.3   Probability Space

A probability space combines the sample space Ω and the probability function $\mathbb{P}$ defined on the events of the sample space. First, we consider an elementary case with a finite space $\Omega = \{\omega_1, \omega_2, \ldots, \omega_N\}$, where the probability of each outcome is known. Let $p_i = \mathbb{P}(\omega_i)$ be the probability of ω_i for $i = 1, 2, \ldots, N$. In this case, the probability can be calculated for each event $E \subset \Omega$ as follows. Let $E = \{\omega_{j_1}, \omega_{j_2}, \ldots, \omega_{j_K}\}$ be some event of Ω, where $0 \leqslant K \leqslant N$. If $K = 0$, then $E = \emptyset$; if $K = N$, then $E = \Omega$. Property 2 gives that

$$\mathbb{P}(E) = \sum_{\omega \in E} \mathbb{P}(\omega) = \sum_{\ell=1}^{K} p_{j_\ell}.$$

Thus, we can compute the probability of any event by summing up the probabilities of individual outcomes that constitute the event. The probability $\mathbb{P}$ is a function defined on the power set: $\mathbb{P} : 2^\Omega \to [0, 1]$. We call the pair $(\Omega, \mathbb{P})$ a *finite probability space*.

In a more general situation, we may have infinitely many outcomes, or we may not be able to measure the probability of every possible event. In general, a probability space also includes a collection of events whose likelihood can be measured using the probability function. The triple $(\Omega, \mathcal{F}, \mathbb{P})$ is called a *general probability space* where:

- Ω is a sample space;

- $\mathcal{F}$ is a σ-algebra (sigma-algebra) of events as defined below;

- $\mathbb{P}$ is a probability function defined on $\mathcal{F}$. It satisfies Kolmogorov's axioms.

A σ-*algebra* $\mathcal{F}$ is a special set of events, $\mathcal{F} \subset 2^\Omega$, that satisfies the following properties.

(i) $\Omega \in \mathcal{F}$;

(ii) if $A \in \mathcal{F}$, then $A^{\mathsf{C}} \in \mathcal{F}$;

(iii) if $A_1, A_2, \ldots \in \mathcal{F}$, then $\displaystyle\bigcup_{n \geqslant 1} A_n \in \mathcal{F}$.

The most trivial example of a σ-algebra is the collection $\{\emptyset, \Omega\}$. For any event $E \subset \Omega$, we can construct the following σ-algebra: $\mathcal{F}_E = \{\emptyset, E, E^{\mathsf{C}}, \Omega\}$. It is not difficult to prove the following properties of a σ-algebra $\mathcal{F}$:

(1) $\emptyset \in \mathcal{F}$;

(2) if $A, B \in \mathcal{F}$ then $A \setminus B \in \mathcal{F}$;

(3) if $A_1, A_2, \ldots \in \mathcal{F}$ then $\displaystyle\bigcap_{n \geqslant 1} A_n \in \mathcal{F}$.

You can get acquainted with properties and applications of σ-algebras in Chapter 6.

Example A.6. Find a probability space describing the following situations.

(a) One die is rolled. The outcome of the random experiment is the number of dots facing up. We can only observe whether the number is odd or even.

(b) Two dice are rolled. We can only observe the total number of dots.

Solution.

(a) The sample space is $\Omega = \{1, 2, 3, 4, 5, 6\}$. The event that the number on the upper face is odd is $E = \{1, 3, 5\}$. Its complement $E^C = \{2, 4, 6\}$ is the event that the number is even. The σ-algebra $\mathcal{F} = \{\emptyset, E, E^C, \Omega\}$ describes this situation when we can only observe if the number is odd or even. Clearly, $\mathbb{P}(E) = \mathbb{P}(E^C) = \frac{1}{2}$.

(b) The sample space is $\Omega = \{(i, j) : 1 \leqslant i, j \leqslant 6\}$. If we can only observe the total number of dots, then one of the following 11 events can occur:

$$E_1 = \{(1, 1)\}, \quad E_2 = \{(1, 2), (2, 1)\}, \quad E_3 = \{(1, 3), (2, 2), (3, 1)\},$$
$$E_4 = \{(1, 4), (2, 3), (3, 2), (4, 1)\}, \quad E_5 = \{(1, 5), (2, 4), (3, 3), (4, 2), (5, 1)\},$$
$$E_6 = \{(1, 6), (2, 5), (3, 4), (4, 3), (5, 2), (6, 1)\},$$
$$E_7 = \{(2, 6), (3, 5), (4, 4), (5, 3), (6, 2)\}, \quad E_8 = \{(3, 6), (4, 5), (5, 4), (6, 3)\},$$
$$E_9 = \{(4, 6), (5, 5), (6, 4)\}, \quad E_{10} = \{(5, 6), (6, 5)\}, \quad E_{11} = \{(6, 6)\}.$$

Event E_j occurs if the total number of dots equals $j + 1$ for $j = 1, 2, \ldots, 11$. The σ-algebra $\mathcal{F}$ is generated by the above events. Any element of $\mathcal{F}$ is a union of some selection of the events $E_1, E_2, \ldots, E_{11}$ (including an empty selection that results in an empty set). In total, $\mathcal{F}$ contains $2^{11} = 2048$ events. The probabilities are

$$\mathbb{P}(E_1) = \mathbb{P}(E_{11}) = \frac{1}{36}, \quad \mathbb{P}(E_2) = \mathbb{P}(E_{10}) = \frac{2}{36}, \quad \mathbb{P}(E_3) = \mathbb{P}(E_9) = \frac{3}{36},$$
$$\mathbb{P}(E_4) = \mathbb{P}(E_8) = \frac{4}{36}, \quad \mathbb{P}(E_5) = \mathbb{P}(E_7) = \frac{5}{36}, \quad \mathbb{P}(E_6) = \frac{6}{36}.$$

To find the probability of some event $E \in \mathcal{F}$, you need to find which events from the list $E_1, E_2, \ldots, E_{11}$ constitute the event E and then calculate the sum of probabilities of those events.

$\square$

A.1.4 Counting Techniques and Combinatorial Probabilities

Suppose that two experiments are to be performed. Experiment 1 can result in any one of n_1 possible outcomes. For each outcome of experiment 1, there are n_2 possible outcomes of experiment 2. Then, there are $n_1 \cdot n_2$ possible outcomes of the two experiments in total. This counting rule can be generalized as follows.

Version 1: Multiplication of possibilities. If k experiments are performed such that the first one may result in any one of n_1 possible outcomes, and if for each of these outcomes there are n_2 possible outcomes of experiment 2, and so on, then there are a total of $n_1 n_2 \cdots n_k$ possible outcomes of the k experiments.

Version 2: Multiplication of choices. If sets $A_1, A_2, \ldots, A_k$ contain $n_1, n_2, \ldots, n_k$ elements, respectively, then there are $n_1 n_2 \cdots n_k$ possible ways of choosing first an element of A_1, then an element of A_2, $\ldots$, and, lastly, an element of A_k.

Example A.7. A licence plate number in Ontario consists of four letters followed by three digits. How many distinct licence plates are possible if digits and letters can occur more than once?

Solution. A licence plate number has the following pattern: $L_1 L_2 L_3 L_4 D_1 D_2 D_3$ where L and D denote, respectively, a letter and a digit. There are 26 letters and 10 digits. The total number of distinct plate numbers is $26^4 \cdot 10^3 = 456{,}976{,}000$. $\square$

Consider a typical situation, where a few objects are selected from a finite collection called a *population*. The selection procedure is called *sampling*, and hence the group of objects selected is called a *sample*. Let the population consist of n distinct objects. Suppose that k members of the population are selected. We are interested in the number of possible ways to select (arrange) the k objects. The answer depends on the sampling procedure and how to distinguish different collections. For example, we can consider ordered or unordered samples. In an ordered sample, the selected elements form a sequence (e.g., digits in a phone number, or letters in a word). In an unordered sample, the arrangement of elements is irrelevant (e.g., winning lottery numbers).

Additionally, we can have samples with replacement or without replacement. In the first case, repetition of the same element is allowed (e.g., numbers in a license plate). In the second, repetition is not allowed as in a lottery drawing—once a number has been drawn, it cannot be drawn again.

Let us consider the following three cases:

- an ordered sample with replacement;

- an ordered sample without replacement;

- an unordered sample without replacement.

Suppose we are interested in the number of ways to form an ordered sample with replacement. Every time when an object is selected from a population of size n, there are n possibilities. The object selected is noted and then placed back. Thus, there are n^k ways to select k members of the population one at a time with identity noted, where each member selected is returned to the population for a possible reselection. For example, we have $2^3 = 8$ ways to form a three-digit binary number. Here they are: 000, 001, 010, 011, 100, 101, 110, 111.

Let us find the number of ordered samples without replacement. We start with a population of size n. Once a element has been selected, it is not returned to the population. There are n ways to select the first object. Then, there are $n-1$ ways to select the second object, and so on. For the kth object, we have $n - (k-1)$ candidates. As a result, by the generalized counting rule, there are $n \cdot (n-1) \cdots (n-k+1) = \frac{n!}{(n-k)!}$ possibilities to select k members from the population of n distinct objects with identity noted, where each member selected is not returned to the population. An ordered arrangement of a set of objects is called a *permutation*. The number of different permutations of a subset of k members selected from a population of n distinct objects is denoted $P_{k,n}$. So, we have

$$P_{k,n} = \frac{n!}{(n-k)!}.$$

A special case is when we use all n objects to form an ordered arrangement, i.e., $k = n$. The total number of such arrangements is equal to $n!$. For example, we can list all possible arrangements of letters a, b, c. There are $3! = 6$ of them: abc, acb, bac, bca, cab, cba. By convention, we have $0! = 1$ and hence $P_{n,n} = 1$.

Consider the set $\Omega = \{1, 2, 3, 4\}$. A permutation of this set is a non-repeating list such as 4132. Moreover, 4123 is another permutation of $\{1, 2, 3, 4\}$, but 1124 is not a permutation of Ω. There are $4! = 4 \cdot 3 \cdot 2 \cdot 1 = 24$ permutations of Ω. A two-element permutation of $\{1, 2, 3, 4\}$ is a non-repeating list with two elements such as 41, or 13, or 32. There are $12 = 4 \cdot 3 = \frac{4!}{2!} = P_{2,4}$ such permutations.

Example A.8. A licence plate number in Ontario consists of four letters followed by three digits. How many distinct licence plates are possible if no digit or letter can occur twice?

Solution. If no digit or letter can occur twice, then we have

$$P_{4,26} \cdot P_{3,10} = \frac{26!}{22!} \cdot \frac{10!}{7!} = 258{,}336{,}000$$

distinct licence plate numbers. □

Example A.9. Ten people, including myself and my friend, are randomly arranged in a line. How likely is it that there are exactly three people between me and my friend?

Solution. There are 10! equally likely arrangements. There are $2 \cdot 6$ ways to place two people in a line of 10 and have exactly three people between them. Additionally, there are 8! ways to arrange the other eight people. So, in total, we have $2 \cdot 6 \cdot 8!$ favourable arrangements. Thus, the probability is

$$\frac{2 \cdot 6 \cdot 8!}{10!} = \frac{12}{90} = \frac{2}{15} \cong 13.13\%.$$

□

Lastly, let us find the number of unordered samples without replacement. Select a subset of size k from a set of n elements without noting the order. The total number of such selections is called *the number of combinations*. It is denoted as $\binom{n}{k}$ (read "n choose k") or $C_{k,n}$ and equals

$$\binom{n}{k} = \frac{n!}{k!(n-k)!}.$$

The values $\binom{n}{k}$ are often referred to as *binomial coefficients* thanks to the binomial formula:

$$(x+y)^n = \sum_{k=0}^{n} \binom{n}{k} x^k y^{n-k}.$$

Note the two special cases: $\binom{n}{n} = \binom{n}{0} = 1$ and $\binom{n}{n-1} = \binom{n}{1} = n$. Note also an important version of the binomial formula with $x = p$ and $y = 1 - p$ where $0 \leqslant p \leqslant 1$:

$$\sum_{k=0}^{n} \binom{n}{k} p^k (1-p)^{n-k} = (p + (1-p))^n = 1.$$

In summary, the number $\binom{n}{k}$ represents:

- the number of ways to select k objects out of n given objects (in the sense of unordered samples without replacement);

- the number of k-element subsets of an n-element set.

Example A.10. There are ten books on the bookshelf. Of these, four are mathematics books, three are economics books, two are history books, and one is a language book. The books are arranged so that all the books dealing with the same subject are together on the shelf. How many different arrangements are possible?

Solution. There are 4! ways to arrange four subjects. For a subject with n books, there are $n!$ ways to arrange the books. Thus, there are

$$4! \cdot 4! \cdot 3! \cdot 2! \cdot 1! = 6912$$

different arrangements. □

Example A.11. A committee of size five is to be selected from a group of six humans and nine aliens.

(a) What is the total number of different selections?

(b) How many selections for the committee consisting of three humans and two aliens are possible?

(c) How many different selections are possible if the committee must include at least one alien?

Solution.

$$\text{(a)} \quad \binom{15}{5} = \frac{15!}{5! \, 10!} = 3003.$$

$$\text{(b)} \quad \binom{6}{3} \cdot \binom{9}{2} = \frac{6!}{3! \, 3!} \cdot \frac{9!}{2! \, 7!} = 720.$$

$$\text{(c)} \quad \binom{15}{5} - \binom{6}{5} = 3003 - 6 = 2997.$$

To calculate (c), we subtract the number of committees with humans only from the total number of committees. □

Example A.12. A box contains 10 red socks and six blue socks. What is the probability that two socks chosen at random are of the same colour?

Solution. Here, we deal with an unordered sample without replacement. There are $\binom{16}{2} = 120$ pairs in total. Among those pairs, there are $\binom{6}{2} = 15$ pairs of blue socks and $\binom{10}{2} = 45$ pairs of red socks. Thus, the probability of selecting two socks of the same colour equals

$$\mathbb{P}(\text{two red socks}) + \mathbb{P}(\text{two blue socks}) = \frac{\binom{10}{2}}{\binom{16}{2}} + \frac{\binom{6}{2}}{\binom{16}{2}} = \frac{45 + 15}{120} = \frac{1}{2}.$$

□

Example A.13. A poker hand consists of five cards dealt without noting the order from an ordinary deck of 52 playing cards. Assume that the deck is well-shuffled so that the cards are selected at random. Assume the following hierarchy of card denominations from lowest to highest:

$$2, 3, ..., 10, \text{Jack, Queen, King, Ace.}$$

Compute probabilities of the following combinations:

(A) five cards of the same suit;

(B) five cards of different denominations;

(C) a Jack is the highest card;

(D) one pair;

(E) two pairs;

(F) three of a kind;

(G) four of a kind.

Solution. There are $\binom{52}{5} = 2{,}598{,}960$ different poker hands. The probabilities are:

$$\mathbb{P}(A) = \frac{\binom{4}{1} \cdot \binom{13}{5}}{\binom{52}{5}} = \frac{5148}{2{,}598{,}960} \cong 1.981\%;$$

$$\mathbb{P}(B) = \frac{\binom{13}{5} \cdot \binom{4}{1}^5}{\binom{52}{5}} = \frac{1{,}317{,}888}{2{,}598{,}960} \cong 50.708\%;$$

$$\mathbb{P}(C) = \frac{\binom{4}{1} \cdot \binom{52-16}{4}}{\binom{52}{5}} = \frac{235{,}620}{2{,}598{,}960} \cong 9.066\%;$$

$$\mathbb{P}(D) = \frac{\binom{13}{1} \cdot \binom{4}{2} \cdot \binom{12}{3} \cdot \binom{4}{1}^3}{\binom{52}{5}} = \frac{1{,}098{,}240}{2{,}598{,}960} \cong 42.257\%;$$

$$\mathbb{P}(E) = \frac{\binom{13}{2} \cdot \binom{4}{2}^2 \cdot \binom{52-8}{1}}{\binom{52}{5}} = \frac{123{,}552}{2{,}598{,}960} \cong 4.754\%;$$

$$\mathbb{P}(F) = \frac{\binom{13}{1} \cdot \binom{4}{3} \cdot \binom{12}{2} \cdot \binom{4}{1}^2}{\binom{52}{5}} = \frac{123{,}552}{2{,}598{,}960} \cong 4.754\%;$$

$$\mathbb{P}(G) = \frac{\binom{13}{1} \cdot \binom{52-4}{1}}{\binom{52}{5}} = \frac{624}{2{,}598{,}960} \cong 0.024\%.$$

$\square$

A.1.5 Conditional Probability

We are often interested in calculating probabilities when some partial information concerning the result of the experiment is available. In such situations, the desired probabilities are conditional on the information provided.

Let A and B be events of the same sample space Ω. A *conditional probability* of event A given that event B has already occurred, denoted $\mathbb{P}(A \mid B)$, is calculated by the formula

$$\mathbb{P}(A \mid B) = \frac{\mathbb{P}(A \cap B)}{\mathbb{P}(B)} \quad \text{where} \quad \mathbb{P}(B) \neq 0. \tag{A.3}$$

That is, the conditional probability $\mathbb{P}(A \mid B)$ is only defined when $\mathbb{P}(B) > 0$. We read $\mathbb{P}(A \mid B)$ as "the probability of A given B."

Note that the usual probability of an event, which can be called the *unconditional probability*, is a special case of the conditional probability. Indeed, let $B = \Omega$ in (A.3), then

$$\mathbb{P}(A \mid \Omega) = \frac{\mathbb{P}(A \cap \Omega)}{\mathbb{P}(\Omega)} = \frac{\mathbb{P}(A)}{1} = \mathbb{P}(A).$$

Additionally, the conditional probability function $A \mapsto \mathbb{P}(A \mid B)$ satisfies the axioms of probability. In other words, $\mathbb{P}(\cdot \mid B)$ is a probability function.

Also, note the following properties of conditional probability, which can be proved using the axioms of probability, the definition of conditional probability, and properties of set operations.

(1) $\mathbb{P}(A^{\complement} \mid B) = 1 - \mathbb{P}(A \mid B)$.

(2) $\mathbb{P}(A_1 \cup A_2 \mid B) = \mathbb{P}(A_1 \mid B) + \mathbb{P}(A_2 \mid B)$ whenever $A_1 \cap A_2 = \emptyset$.

(3) If $A \cap B = \emptyset$, then $\mathbb{P}(A \mid B) = 0$.

(4) If $B \subset A$, then $\mathbb{P}(A \mid B) = 1$.

Conditional probabilities are useful for computing probabilities of intersections of multiple events:

$$\mathbb{P}(A \cap B) = \mathbb{P}(A) \cdot \mathbb{P}(B \mid A) \quad \text{if} \quad \mathbb{P}(A) \neq 0, \quad \text{and}$$
$$\mathbb{P}(A \cap B) = \mathbb{P}(B) \cdot \mathbb{P}(A \mid B) \quad \text{if} \quad \mathbb{P}(B) \neq 0.$$

Example A.14. Each student is allowed two attempts to pass an exam. Experience shows that 60% of all students pass on the first try and that, for those who do not, 80% pass on the second try.

(a) What is the probability that a student passes the exam?

(b) If a student passed the exam, what is the probability that he or she passed on the first attempt?

Solution. Introduce three events:

$$E = \{\text{passed the exam}\},$$
$$E_1 = \{\text{passed the exam on the first attempt}\},$$
$$E_2 = \{\text{passed the exam on the second attempt}\}.$$

Clearly, $E_1 \cap E_2 = \emptyset$, and hence $E_2 \subset E_1^{\complement}$ and $E_2 = E_2 \cap E_1^{\complement}$. We have $\mathbb{P}(E_1) = 0.6$ and $\mathbb{P}(E_2 \mid E_1^{\complement}) = 0.8$. Moreover, using properties of probabilities, we find

$$\mathbb{P}(E_2) = \mathbb{P}(E_1^{\complement} \cap E_2) = \mathbb{P}(E_1^{\complement}) \cdot \mathbb{P}(E_2 \mid E_1^{\complement}) = (1 - \mathbb{P}(E_1)) \cdot \mathbb{P}(E_2 \mid E_1^{\complement}) = 0.4 \cdot 0.8 = 0.32.$$

Compute the probabilities:

(a) $\mathbb{P}(E) = \mathbb{P}(E_1 \cup E_2) = \mathbb{P}(E_1) + \mathbb{P}(E_2) - \mathbb{P}(E_1 \cap E_2) = 0.6 + 0.32 - 0 = 0.92.$

(b) $\mathbb{P}(E_1 \mid E) = \dfrac{\mathbb{P}(E_1 \cap E)}{\mathbb{P}(E)} = \dfrac{\mathbb{P}(E_1)}{\mathbb{P}(E)} = \dfrac{0.6}{0.92} = 0.652.$

□

Consider a collection of n events, $A_1, A_2, \ldots, A_n \subset \Omega$, such that $\mathbb{P}(A_1 \cap A_2 \cap \cdots \cap A_n) > 0$. Then, the probability of the intersection is calculated by the *multiplication rule*:

$$\mathbb{P}(A_1 \cap A_2 \cap \cdots \cap A_n) = \mathbb{P}(A_1) \cdot \mathbb{P}(A_2 \mid A_1) \cdot \mathbb{P}(A_3 \mid A_1 \cap A_2) \cdots$$
$$\mathbb{P}(A_n \mid A_1 \cap A_2 \cap \cdots \cap A_{n-1}). \tag{A.4}$$

Example A.15. Suppose that an ordinary deck of 52 cards (which contains four Aces) is randomly divided into four hands of 13 cards each. We are interested in finding the probability that each hand has an Ace. Let E_i be the event that the ith hand has exactly one Ace. Determine the probability $\mathbb{P}(E_1 E_2 E_3 E_4)$ by using the multiplication rule.

Solution. Using the multiplication rule (A.4) with $n = 4$ gives

$$\mathbb{P}(E_1 E_2 E_3 E_4) = \mathbb{P}(E_1) \cdot \mathbb{P}(E_2 \mid E_1) \cdot \mathbb{P}(E_3 \mid E_1 E_2) \cdot \mathbb{P}(E_4 \mid E_1 E_2 E_3).$$

We have,

$$\mathbb{P}(E_1) = \frac{\binom{4}{1} \cdot \binom{52-4}{12}}{\binom{52}{13}} = \frac{\binom{4}{1} \cdot \binom{48}{12}}{\binom{52}{13}} = \frac{9139}{20{,}825},$$

$$\mathbb{P}(E_2 \mid E_1) = \frac{\binom{3}{1} \cdot \binom{52-13-3}{12}}{\binom{52-13}{13}} = \frac{\binom{3}{1} \cdot \binom{36}{12}}{\binom{39}{13}} = \frac{325}{703},$$

$$\mathbb{P}(E_3 \mid E_1 E_2) = \frac{\binom{2}{1} \cdot \binom{52-26-2}{12}}{\binom{52-26}{13}} = \frac{\binom{2}{1} \cdot \binom{24}{12}}{\binom{26}{13}} = \frac{13}{25}.$$

Since, $\mathbb{P}(E_4 \mid E_1 E_2 E_3) = 1$, we have $\mathbb{P}(E_1 E_2 E_3 E_4) = \frac{9139}{20{,}825} \cdot \frac{325}{703} \cdot \frac{13}{25} = \frac{2197}{20{,}825} \cong 10.55\%.$ $\square$

A.1.6 Law of Total Probability and Bayes' Formula

Very often, a sample space can be represented as a union of mutually exclusive events, say, $B_1, B_2, \ldots, B_n$, which represent results of some experiment. Such partitioning of a sample space may simplify the problem of computing $\mathbb{P}(A)$ for an event A by calculating *prior* probabilities $\mathbb{P}(B_j)$ and conditional probabilities $\mathbb{P}(A \mid B_j)$. The rule used to calculate $\mathbb{P}(A)$ is known as the *law of total probability*. Additionally, given $\mathbb{P}(A)$, we can calculate a *posterior* probability $\mathbb{P}(B_j \mid A)$ using the formula for the conditional probability.

A collection of events $B_1, B_2, \ldots, B_n$ such that $\bigcup_{k=1}^{n} B_k = \Omega$ is said to be *exhaustive*. The events $\{B_k\}$ are usually called *hypotheses* and from their definition it follows that at least one of them must occur when the random experiment is performed. An exhaustive collection of mutually exclusive events is called a *partition*. For a partition $B_1, B_2, \ldots, B_n$, we have $\mathbb{P}(B_1) + \mathbb{P}(B_2) + \cdots + \mathbb{P}(B_n) = 1$. Here are some examples.

- For any event B such that $B \neq \emptyset$ and $B \neq \Omega$, the sets B and $B^{\complement}$ form a partition.

- Pick a card from a deck and note its suit. We have a partition with four events:

$$B_1 = \{\text{spades}\}, \quad B_2 = \{\text{hearts}\}, \quad B_3 = \{\text{diamonds}\}, \quad B_4 = \{\text{clubs}\}.$$

- Roll two dice and compute the sum of values. We have a partition with 11 events:

$$B_j = \{\text{the sum equals } j + 1\}, \quad j = 1, 2, \ldots, 11.$$

- Toss a coin three times and count the number of heads. An outcome is a sequence of three letters from $\{H, T\}$, where H denotes a head and T denotes a tail. We have a partition with four events:

$$\{HHH\}, \quad \{HHT, HTH, THH\}, \quad \{HTT, THT, TTH\}, \quad \{TTT\}.$$

Theorem A.1 (The Law of Total Probability). *Let $B_1, B_2, \ldots, B_n$ be a partition of a sample space Ω; let all $\mathbb{P}(B_k) > 0$. The probability of any event $A \subset \Omega$ can be calculated as follows:*

$$\mathbb{P}(A) = \mathbb{P}(A \cap B_1) + \mathbb{P}(A \cap B_2) + \cdots + \mathbb{P}(A \cap B_n) \tag{A.5}$$
$$= \mathbb{P}(A \mid B_1)\mathbb{P}(B_1) + \mathbb{P}(A \mid B_2)\mathbb{P}(B_2) + \cdots + \mathbb{P}(A \mid B_n)\mathbb{P}(B_n). \tag{A.6}$$

From the definition of conditional probability,

$$\mathbb{P}(A \cap B) = \mathbb{P}(A \mid B)\mathbb{P}(B) = \mathbb{P}(B \cap A) = \mathbb{P}(B \mid A)\mathbb{P}(A).$$

Thus, we can write

$$\mathbb{P}(B \mid A) = \frac{\mathbb{P}(A \mid B)\,\mathbb{P}(B)}{\mathbb{P}(A)}.$$

From the law of total probability, we have

$$\mathbb{P}(B \mid A) = \frac{\mathbb{P}(A \mid B)\,\mathbb{P}(B)}{\mathbb{P}(A \mid B)\,\mathbb{P}(B) + \mathbb{P}(A \mid B^{\complement})\,\mathbb{P}(B^{\complement})}.$$

Example A.16. In answering a question on a multiple-choice test, a student either knows the right answer or he guesses. Let p be the probability that the student knows the answer and $1 - p$ the probability that he guesses. Assume that a student who guesses at the answer will be correct with probability $1/m$, where m is the number of multiple choice alternatives. What is the conditional probability that the student knew the answer to a question given that he answered the question correctly?

Solution. Define the events: $A = \{$answered correctly$\}$ and $B = \{$knew the answer$\}$. Thus, $B^{\complement} = \{$guessed the answer$\}$. We have $\mathbb{P}(B) = p$ and $\mathbb{P}(A \mid B^{\complement}) = \frac{1}{m}$.

$$\mathbb{P}(B \mid A) = \frac{\mathbb{P}(A \mid B)\mathbb{P}(B)}{\mathbb{P}(A \mid B)\mathbb{P}(B) + \mathbb{P}(A \mid B^{\complement})\mathbb{P}(B^{\complement})} = \frac{p}{p + (1 - p)/m}.$$

$\square$

In general, if $\mathbb{P}(A)$ in the denominator of $\mathbb{P}(B \mid A) = \frac{\mathbb{P}(A|B)\mathbb{P}(B)}{\mathbb{P}(A)}$ is calculated using the law of total probability for multiple events, we obtain the following general result known as *Bayes' Theorem.*

Theorem A.2 (Bayes' Theorem). *If $B_1, B_2, \ldots, B_n$ are n mutually exclusive and exhaustive events, and A is any event such that $\mathbb{P}(A) > 0$, then*

$$\mathbb{P}(B_j \mid A) = \frac{\mathbb{P}(A \mid B_j)\mathbb{P}(B_j)}{\mathbb{P}(A \mid B_1)\mathbb{P}(B_1) + \mathbb{P}(A \mid B_2)\mathbb{P}(B_2) + \cdots + \mathbb{P}(A \mid B_n)\mathbb{P}(B_n)} \tag{A.7}$$

for $j = 1, 2, \ldots, n$.

Bayes' theorem provides a formula for finding the probability that the "effect" A was "caused" by the event B_j.

Example A.17. Suppose that an insurance company classifies people into one of three classes: *good risk, average risk,* and *bad risk.* The company's records indicate that the probabilities that good-, average-, and bad-risk persons will be involved in an accident over a one-year span are, respectively, 0.05, 0.15, and 0.30. If 20% of the population is a good risk, 50% an average risk, and 30% a bad risk, what proportion of people have accidents in a fixed year? If a policyholder had no accidents in 2020, what is the probability that he or she is a good risk? is an average risk? is a bad risk?

Solution. Let G, A, B denote the events that a policyholder selected at random is, respectively, a good-, average-, or bad-risk person. Let D denote the event that an accident happened to the policyholder in a given year. We have that

$$\mathbb{P}(G) = 0.2, \quad \mathbb{P}(A) = 0.3, \quad \mathbb{P}(B) = 0.5,$$
$$\mathbb{P}(D \mid G) = 0.05, \quad \mathbb{P}(D \mid A) = 0.15, \quad \mathbb{P}(D \mid B) = 0.3.$$

By the total probability formula,

$$\mathbb{P}(D) = \mathbb{P}(D \mid G)\mathbb{P}(G) + \mathbb{P}(D \mid A)\mathbb{P}(A) + \mathbb{P}(D \mid B)\mathbb{P}(B)$$
$$= 0.05 \cdot 0.2 + 0.15 \cdot 0.3 + 0.3 \cdot 0.5 = 20.5\%.$$

Thus, we have

$$\mathbb{P}(G \mid D^{C}) = \frac{\mathbb{P}(G) \cdot \mathbb{P}(D^{C} \mid G)}{\mathbb{P}(D^{C})} = \frac{\mathbb{P}(G) \cdot (1 - \mathbb{P}(D \mid G))}{1 - \mathbb{P}(D)} = \frac{0.2 \cdot (1 - 0.05)}{1 - 0.205} \cong 23.90\%,$$

$$\mathbb{P}(A \mid D^{C}) = \frac{\mathbb{P}(A) \cdot \mathbb{P}(D^{C} \mid A)}{\mathbb{P}(D^{C})} = \frac{\mathbb{P}(A) \cdot (1 - \mathbb{P}(D \mid A))}{1 - \mathbb{P}(D)} = \frac{0.3 \cdot (1 - 0.15)}{1 - 0.205} \cong 32.08\%,$$

$$\mathbb{P}(B \mid D^{C}) = 1 - \mathbb{P}(G \mid D^{C}) - \mathbb{P}(A \mid D^{C}) \cong 44.02\%.$$

$\square$

A.1.7 Independence of Events

In some cases, the occurrence of one event does not affect the probability that the other event occurs. Two events A and B are said to be *independent* if one of the following equivalent statements holds:

- $\mathbb{P}(A \mid B) = \mathbb{P}(A)$ given that $\mathbb{P}(B) > 0$;

- $\mathbb{P}(B \mid A) = \mathbb{P}(B)$ given that $\mathbb{P}(A) > 0$;

- $\mathbb{P}(AB) = \mathbb{P}(A)\mathbb{P}(B)$.

Clearly, if events A and B are independent, then so are: A and B^{C}; A^{C} and B; and A^{C} and B^{C}. Two events that are not independent are said to be *dependent*.

If A and B are events with positive probabilities, exactly one of three things must be true:

1. $\mathbb{P}(A \mid B) > \mathbb{P}(A)$. The occurrence of B makes A more likely. The events A and B are said to be *positively correlated*. **Example:** Toss a coin twice, let $A = \{$two heads$\}$, $B = \{$head on first$\}$.

2. $\mathbb{P}(A \mid B) < \mathbb{P}(A)$. The occurrence of B makes A less likely. The events A and B are said to be *negatively correlated*. **Example:** Toss a coin twice, let $A = \{$at least one head$\}$, $B = \{$tail on first$\}$.

3. $\mathbb{P}(A \mid B) = \mathbb{P}(A)$. The occurrence of B does not influence the likelihood of A. The events A and B are independent. **Example:** Toss a coin twice, let $A = \{$head on second$\}$, $B = \{$head on first$\}$.

Example A.18. Three coins are tossed in order. Verify whether the following two events are independent or not:

$$A = \{\text{the first toss results in head}\} \quad \text{and} \quad B = \{\text{there is at least one tail}\}.$$

Solution. There are eight equally likely outcomes, each of which can be represented by a sequence $R_1 R_2 R_3$, where $R_i \in \{T, H\}$ is the result of the ith coin toss. Calculate probabilities of events A, B, AB:

$$\mathbb{P}(A) = \mathbb{P}(\{HTT, HTH, HHT, HHH\}) = \frac{1}{2},$$

$$\mathbb{P}(B) = 1 - \mathbb{P}(B^{\mathbb{C}}) = 1 - \mathbb{P}(\{HHH\}) = \frac{7}{8},$$

$$\mathbb{P}(AB) = \mathbb{P}(A) - \mathbb{P}(AB^{\mathbb{C}}) = \mathbb{P}(A) - \mathbb{P}(\{HHH\}) = \frac{1}{2} - \frac{1}{8} = \frac{3}{8}.$$

Since, $\mathbb{P}(AB) = \frac{3}{8} \neq \frac{7}{16} = \mathbb{P}(A)\mathbb{P}(B)$, the events A and B are dependent. □

For more than two events, there are several ways to define the notion of independence. Three events A, B, and C are said to be *pairwise independent* if all of the following hold:

$$\mathbb{P}(AB) = \mathbb{P}(A)\,\mathbb{P}(B), \quad \mathbb{P}(BC) = \mathbb{P}(B)\,\mathbb{P}(C), \quad \mathbb{P}(AC) = \mathbb{P}(A)\,\mathbb{P}(C). \qquad \text{(A.8)}$$

Three events A, B, and C are said to be *mutually independent* if in addition to (A.8) the following holds:

$$\mathbb{P}(ABC) = \mathbb{P}(A)\,\mathbb{P}(B)\,\mathbb{P}(C).$$

Example A.19. Suppose that two dice are rolled, one black and one grey. Let

$$A = \{\text{the black die comes up even}\},$$
$$B = \{\text{the grey die comes up even}\},$$
$$C = \{\text{the sum of the dies is even}\}.$$

Show that A, B, and C are pairwise independent events but are not mutually independent. That is, the pairwise independence does not imply the mutual independence.

Solution. The sample space consists of 36 equally likely pairs (i, j) where $i, j \in \{1, 2, \ldots, 6\}$ are the outcomes of the black and grey dice, respectively. First, let us find the probabilities of individual events:

$$\mathbb{P}(A) = \mathbb{P}(B) = \frac{3 \cdot 6}{36} = \frac{1}{2}, \quad \mathbb{P}(C) = \sum_{k \text{ is odd}} \mathbb{P}(\text{sum} = k) = \frac{1 + 3 + 5 + 5 + 3 + 1}{36} = \frac{1}{2}.$$

The probabilities of pairwise intersections are

$$\mathbb{P}(AB) = \mathbb{P}(AC) = \mathbb{P}(BC) = \mathbb{P}\big(\{(2, 2), (2, 4), (2, 6), (4, 2), (4, 4), (4, 6), (6, 2), (6, 4), (6, 6)\}\big)$$
$$= \frac{9}{36} = \frac{1}{4}.$$

Here, we use the fact that if the sum of two numbers is even and one summand is even, then the other is even as well. Since, $\mathbb{P}(AB) = \mathbb{P}(A)\mathbb{P}(B)$, $\mathbb{P}(AC) = \mathbb{P}(A)\mathbb{P}(C)$, and $\mathbb{P}(BC) = \mathbb{P}(B)\mathbb{P}(C)$, we have the pairwise independence of A, B, and C. The probability of the triple intersection is

$$\mathbb{P}(ABC) = \mathbb{P}\big(\{(2, 2), (2, 4), (2, 6), (4, 2), (4, 4), (4, 6), (6, 2), (6, 4), (6, 6)\}\big) = \frac{1}{4}.$$

Since $\mathbb{P}(ABC) \neq \mathbb{P}(A)\,\mathbb{P}(B)\,\mathbb{P}(C)$, the events A, B, and C are not mutually independent. □

The events $E_1, E_2, \ldots, E_n$ are said to be *mutually independent* if and only if for any collection of these events, $E_{i_1}, E_{i_2}, \ldots, E_{i_k}$, we have

$$\mathbb{P}(E_{i_1} E_{i_2} \ldots E_{i_k}) = \mathbb{P}(E_{i_1})\,\mathbb{P}(E_{i_2}) \cdots \mathbb{P}(E_{i_k}).$$

A.2   Univariate Probability Distributions

A.2.1   Random Variables

When a random experiment is performed, we may be interested in some property of the outcome as opposed to the actual outcome itself. For instance, when dice are rolled, we are often interested in the sum of the dice' facing-up values and are not concerned about the individual number on each die. In flipping a coin, we may be interested in the total number of heads that occur and do not care at all about the actual head-tail sequence that results. These numerical quantities whose values are determined by the outcome of a random experiment are known as random variables.

A *random variable* X on a sample space Ω is a rule that assigns a numerical value $X(\omega)$ to each outcome $\omega \in \Omega$. In other words, a random variable is a function $X : \Omega \to \mathbb{R}$ that maps the sample space Ω into the set of real numbers $\mathbb{R} = (-\infty, \infty)$. The *range*, D_X, of a random variable is the collection of all possible values the random variable takes. That is, $D_X = X(\Omega) = \{x \in \mathbb{R} \mid X(\omega) = x \text{ for some } \omega \in \Omega\}$.

We use uppercase letters for random variables (e.g., X, Y, Z), and lower case letters (e.g., x, y, z) to denote particular values a random variable can take. To indicate that random variable X takes value x, we write $X = x$. The expression $\{X = x\}$ refers to the set of all elements in Ω assigned the value x by the random variable X, i.e., it is the set $\{\omega \in \Omega \mid X(\omega) = x\} = X^{-1}(x)$. Additionally, for $a, b \in \mathbb{R}$, we write

$$\{X \leqslant a\} = \{\omega \in \Omega \mid X(\omega) \leqslant a\},$$
$$\{a < X \leqslant b\} = \{\omega \in \Omega \mid a < X(\omega) \leqslant b\} = \{a < X\} \cap \{X \leqslant b\}.$$

Since a random variable refers to elements in Ω, we can compute the probability that random variable X takes a value x by finding all respective outcomes and then computing their total probability:

$$\mathbb{P}(X = x) = \mathbb{P}\left(\{\omega \in \Omega \mid X(\omega) = x\}\right)$$

Similarly, for $a, b \in \mathbb{R}$, we denote:

$$\mathbb{P}(X \leqslant a) = \mathbb{P}\left(\{\omega \in \Omega \mid X(\omega) \leqslant a\}\right),$$
$$\mathbb{P}(X > a) = \mathbb{P}\left(\{\omega \in \Omega \mid X(\omega) > a\}\right) = 1 - \mathbb{P}(X \leqslant a),$$
$$\mathbb{P}(a \leqslant X \leqslant b) = \mathbb{P}\left(\{\omega \in \Omega \mid a \leqslant X(\omega) \leqslant b\}\right) = \mathbb{P}(X \leqslant b) - \mathbb{P}(X < a).$$

In general, for a countable sample space Ω and any $A \subset \mathbb{R}$, we have

$$\mathbb{P}(X \in A) = \sum_{\omega \in X^{-1}(A)} \mathbb{P}(\omega), \text{ where } X^{-1}(A) = \{\omega \in \Omega \mid X(\omega) \in A\}.$$

The simplest example of a random variable is a Bernoulli variate, which is an indicator function of some event. The only possible values of a Bernoulli random variable are 0 and 1. Let E be an event of the sample space Ω, i.e., $E \subset \Omega$. Define the *indicator function* $\mathbb{I}_E$ of event E as follows: $\mathbb{I}_E = 1$ if the event E occurs, and $\mathbb{I}_E = 0$ if the event E does not occur. That is, formally,

$$\mathbb{I}_E : \Omega \to \{0, 1\}, \quad \mathbb{I}_E(\omega) = \begin{cases} 0 & \text{if } \omega \notin E, \\ 1 & \text{if } \omega \in E. \end{cases}$$

The probability distribution of $\mathbb{I}_E$ is as follows:

$$\mathbb{P}(\mathbb{I}_E = 0) = \mathbb{P}(\{\omega \mid \mathbb{I}_E(\omega) = 0\}) = \mathbb{P}(E^{\complement}) = 1 - \mathbb{P}(E),$$
$$\mathbb{P}(\mathbb{I}_E = 1) = \mathbb{P}(\{\omega \mid \mathbb{I}_E(\omega) = 1\}) = \mathbb{P}(E).$$

Example A.20. Roll two dice. Let X be the sum of two facing-up values. X can take any integer value between 2 and 12 with the respective probabilities:

$$\mathbb{P}(X = 2) = \mathbb{P}\big(\{(1,1)\}\big) = \frac{1}{36},$$

$$\mathbb{P}(X = 3) = \mathbb{P}\big(\{(1,2),(2,1)\}\big) = \frac{2}{36},$$

$$\mathbb{P}(X = 4) = \mathbb{P}\big(\{(1,3),(2,2),(3,1)\}\big) = \frac{3}{36},$$

$$\mathbb{P}(X = 5) = \mathbb{P}\big(\{(1,4),(2,3),(3,2),(4,1)\}\big) = \frac{4}{36},$$

$$\mathbb{P}(X = 6) = \mathbb{P}\big(\{(1,5),(2,4),(3,3),(4,2),(5,1)\}\big) = \frac{5}{36},$$

$$\mathbb{P}(X = 7) = \mathbb{P}\big(\{(1,6),(2,5),(3,4),(4,3),(5,2),(6,1)\}\big) = \frac{6}{36},$$

$$\mathbb{P}(X = 8) = \mathbb{P}\big(\{(2,6),(3,5),(4,4),(5,3),(6,2)\}\big) = \frac{5}{36},$$

$$\mathbb{P}(X = 9) = \mathbb{P}\big(\{(3,6),(4,5),(5,4),(6,3)\}\big) = \frac{4}{36},$$

$$\mathbb{P}(X = 10) = \mathbb{P}\big(\{(4,6),(5,5),(6,4)\}\big) = \frac{3}{36},$$

$$\mathbb{P}(X = 11) = \mathbb{P}\big(\{(5,6),(6,5)\}\big) = \frac{2}{36},$$

$$\mathbb{P}(X = 12) = \mathbb{P}\big(\{(6,6)\}\big) = \frac{1}{36}.$$

Find (a) $\mathbb{P}(X > 2)$; (b) $\mathbb{P}(2 \leqslant X < 5)$.

Solution.

(a) $\mathbb{P}(X > 2) = 1 - \mathbb{P}(X \leqslant 2) = 1 - \mathbb{P}(X = 2) = 1 - \frac{1}{36} = \frac{35}{36}.$

(b) $\mathbb{P}(2 \leqslant X < 5) = \mathbb{P}(X = 2) + \mathbb{P}(X = 3) + \mathbb{P}(X = 4) = \frac{1+2+3}{36} = \frac{1}{6}.$

$\square$

There are two main types of random variables: discrete and continuous. A *discrete random variable* has a finite or countably infinite range, i.e., there exists a countable set $D \subset \mathbb{R}$ such that $\mathbb{P}(X \in D) = 1$. Therefore, all possible values of a discrete random variable can be listed in a sequence in which there is a first element, a second element, and so on. Obviously, if the sample space Ω is countable, then any random variable defined on Ω is discrete. Moreover, we can write any random variable X defined on a countable set Ω as a weighted sum of indicator functions:

$$X(\omega) = \sum_{\omega_j \in \Omega} x_j \, \mathbb{I}_{\{\omega_j\}}(\omega), \quad \text{where} \quad x_j = X(\omega_j).$$

In the above sum, for any outcome ω, exactly one term is nonzero.

Here are some examples of discrete random variables.

- Draw a poker hand of five cards from a deck. Any hand is an outcome ω from the sample space Ω of all hands. Let $X \in \{0, 1, 2\}$ be the number of pairs.

- Throw one dart at a dartboard. The sample space Ω consists of all possible locations of impact. This is uncountable since it includes all points on the board. However, the score X takes its values in the finite set of integers from 1 and 60.

- Toss n coins. An outcome can be represented by a sequence of n letters from $\{H, T\}$. Clearly, we have that $N(\Omega) = 2^n$. Let $X(\omega)$ be equal to the number of heads in ω. The range D_X of X is $\{0, 1, \ldots, n\}$.

A random variable X is said to be *continuous* if both of the following apply.

(i) The range D_X consists of all numbers in a single interval on the number line (possibly with infinite endpoints) or all numbers in a disjoint union of such intervals (e.g., $D_X = [0, 1]$, or $D_X = [0, \infty)$, or $D_X = [-1, 0] \cup [1, 4]$).

(ii) No possible value of the variable has positive probability, that is, $\mathbb{P}(X = c) = 0$ for any c.

Here are some examples of continuous random variables.

- Select a person at random. The height (or the weight) of the selected person is a continuous random variable with its range within $(0, \infty)$.

- Throw one dart at a dartboard. The distance from the point of impact to the centre of the board is a continuous random variable with the range $[0, R)$ where $R > 0$ is a radius of the board.

A.2.2 Cumulative Distribution Function

A probability distribution can be characterized by its *cumulative distribution function (CDF)*. The CDF F_X of a random variable X is defined by

$$F_X(x) = \mathbb{P}(X \leqslant x) \quad \text{for any real number } x.$$

That is, $F_X(x)$ is the probability that X takes a value that is *less than or equal to x*.

Here are some properties of cumulative distribution functions. Let X be a random variable with CDF F.

1. $F(x) \leqslant F(y)$ whenever $x \leqslant y$ (i.e., a CDF is a non-decreasing function).

2. For any x, we have $0 \leqslant F(x) \leqslant 1$. Moreover,

$$\lim_{x \to -\infty} F(x) = 0 \quad \text{and} \quad \lim_{x \to \infty} F(x) = 1.$$

3. Since $\lim_{x \to a^+} \mathbb{P}(X \leqslant x) = \mathbb{P}(X \leqslant a)$, the CDF F is a right-continuous function. That is, $F(a) = F(a^+) = \lim_{x \to a^+} F(x)$.

4. $\mathbb{P}(X = a) = F(a) - \lim_{x \to a^-} F(x) = F(a) - F(a^-)$.

5. F is continuous at $x = a$ iff $\mathbb{P}(X = a) = 0$. Vice versa, $\mathbb{P}(X = a) > 0$ iff F_X is discontinuous at $x = a$, and the size of a jump at a equals the mass probability $\mathbb{P}(X = a)$.

6. For any two numbers $a \leqslant b$, we have

$$\mathbb{P}(a \leqslant X \leqslant b) = F(b) - F(a^-),$$
$$\mathbb{P}(a < X \leqslant b) = F(b) - F(a),$$
$$\mathbb{P}(a \leqslant X < b) = F(b^-) - F(a^-),$$
$$\mathbb{P}(a < X < b) = F(b^-) - F(a).$$

7. Assume that F is a strictly monotonic, continuous function on the range D_X. Then, $\mathbb{P}(F(X) \leqslant x) = x$ holds for all $x \in (0,1)$.

We can construct a new distribution by taking a mixture of two distributions. Let F_1 and F_2 be two CDFs, then $F(x) = a\, F_1(x) + bF_2(x)$ with $a, b \geqslant 0$ such that $a + b = 1$, is again a CDF, i.e., F is a non-decreasing, right-continuous function such that $F(-\infty) = 0$ and $F(+\infty) = 1$.

A.2.3   Discrete Probability Distributions

The *probability distribution* of a random variable X is a description of the probabilities associated with all possible values of X. The distribution of a discrete random variable can be specified by listing all possible values of X along with the probability of each value. The *probability mass function* (PMF) p_X of a discrete random variable X is a nonnegative function such that

$$p_X(x) = \mathbb{P}(X = x) = \mathbb{P}\big(\{\omega \in \Omega \mid X(\omega) = x\}\big) \quad \text{for any } x \in \mathbb{R}.$$

Let the range of X be a finite or countably infinite set $D_X = \{x_1, x_2, \ldots, x_m, \ldots\}$. Then, the following properties hold:

(i) $p_X(x) > 0$, if $x = x_k$, $k = 1, 2, \ldots, m, \ldots$, and otherwise $p_X(x) = 0$;

(ii) $\displaystyle\sum_{x \in D_X} p_X(x) = \sum_{k \geqslant 1} p_X(x_k) = 1.$

Any probability related to a discrete probability distribution can be calculated using the *Fundamental Probability Formula*. For any subset $A \subset \mathbb{R}$, we have

$$\mathbb{P}(X \in A) = \sum_{x \in A} p_X(x) = \sum_{x \in A \cap D_X} p_X(x).$$

At the beginning of this section, we introduced the Bernoulli distribution, which is the simplest example of a discrete probability distribution. Another example is a discrete uniform random variable that has a finite range with equally likely outcomes.

A random variable X is said to have a *discrete uniform* distribution on a finite, non-empty set S with m elements, denoted $X \sim Unif(S)$, if $p_X(x) = \frac{1}{m}$ for $x \in S$, and $p_X(x) = 0$ otherwise. To compute probabilities for this distribution, we use the formula

$$\mathbb{P}(X \in A) = \sum_{x \in A \cap S} p_X(x) = \frac{N(A \cap S)}{m}, \quad \text{for any } A \subset \mathbb{R},$$

where $N(A)$ denotes the number of elements of A.

A function of a random variable is again a random variable. We can find the PMF of a new distribution as follows. Let X be a discrete random variable, and g be a real-valued function defined on the range D_X of X. The PMF of the random variable $Y = g(X)$ is

$$p_Y(y) = \sum_{x \in g^{-1}(y)} p_X(x)$$

for y in the range $D_Y = g(D_X)$ of Y, and $p_Y(y) = 0$ otherwise. If g is a one-to-one function defined on the range of X, then the PMF of $Y = g(X)$ is

$$p_Y(y) = p_X(g^{-1}(y))$$

for y in the range of Y; otherwise, $p_Y(y) = 0$.

To illustrate these new concepts, we solve a few problems.

Example A.21. Two dice are tossed. Let X denote the total of the two faces.

(a) Find the PMF of X as a closed-form expression.

(b) Find the PMF of $Y = |X - 7|$.

Solution. The mass probabilities $\mathbb{P}(X = x)$ have been given in Example A.20. Since, $p_X(x) = \mathbb{P}(X = x) \neq 0$ for $x \in \{2, 3, \ldots, 12\}$, the PMF can be written as a piecewise function:

$$p_X(x) = \begin{cases} \frac{\min\{x-1, 13-x\}}{36} & x = 2, 3, \ldots, 12, \\ 0 & \text{otherwise.} \end{cases}$$

Note that $\min\{a, b\} = \frac{a+b-|a-b|}{2}$, and hence $p_X(x) = \frac{6-|7-x|}{36}$ for $x = 2, 3, \ldots, 12$.

The random variable $Y = |X-7|$ has the range $\{0, 1, \ldots, 5\}$. Since, $\mathbb{P}(Y = 0) = \mathbb{P}(X = 7)$ and $\mathbb{P}(Y = k) = \mathbb{P}(X = 7 + k) + \mathbb{P}(X = 7 - k)$ for $k = 1, 2, 3, 4, 5$, the PMF of Y is

$$p_Y(y) = \begin{cases} \frac{1}{6} & y = 0, \\ \frac{6-y}{18} & y = 1, 2, \ldots, 5, \\ 0 & \text{otherwise.} \end{cases}$$

$\square$

Example A.22. The PMF of X is

$$p(x) = \begin{cases} C \cdot |x| & x = -3, -2, -1, 1, 2, 3, \\ 0 & \text{otherwise,} \end{cases}$$

where C is some positive constant.

(a) Find the constant C.

(b) Find the PMFs of $Y := X^2$ and $Z := (X)^+ \equiv \max\{X, 0\}$.

Solution.

(a) The sum of values of a PMF is one. Thus,

$$1 = \sum_x p(x) = C \cdot (|-3| + |-2| + |-1| + |1| + |2| + |3|) = 12 \cdot C \implies C = \frac{1}{12}.$$

(b) The range of $Y := X^2$ is $\{1, 4, 9\}$. The values of p_Y are

$$p_Y(1) = \mathbb{P}(Y = 1) = \mathbb{P}(X = -1 \text{ or } X = 1) = p(-1) + p(1) = \frac{1}{6},$$

$$p_Y(4) = \mathbb{P}(Y = 4) = \mathbb{P}(X = -2 \text{ or } X = 2) = p(-2) + p(2) = \frac{1}{3},$$

$$p_Y(9) = \mathbb{P}(Y = 9) = \mathbb{P}(X = -3 \text{ or } X = 3) = p(-3) + p(3) = \frac{1}{2}.$$

The range of $Z := \max\{X, 0\}$ is $\{0, 1, 2, 3\}$. The values of p_Z are

$$p_Z(0) = \mathbb{P}(Z = 0) = \mathbb{P}(X \leqslant 0) = p(-3) + p(-2) + p(-1) = \frac{1}{2},$$

$$p_Z(k) = \mathbb{P}(Z = k) = \mathbb{P}(X = k) = p(k) = \frac{k}{12} \text{ for } k = 1, 2, 3.$$

$\square$

The CDF F_X of a discrete random variable X can be expressed in terms of the PMF p_X as follows

$$F_X(y) = \mathbb{P}(X \leqslant y) = \sum_{x \in D_X \,:\, x \leqslant y} p_X(x).$$

If possible values of X form an increasing sequence $x_1 < x_2 < x_3 < \cdots$, then the distribution function F_X is a piecewise constant function given by

$$F_X(y) = \sum_{\ell=1}^{k} p_X(x_\ell) \text{ for } y \in [x_k, x_{k+1}), \; k \geqslant 1 \; \text{ and } \; F_X(y) = 0 \text{ for } y < x_1.$$

The CDF F_X is constant on the intervals $[x_k, x_{k+1})$, and then it has a jump of size $p_X(x_k)$ at x_k for each $k \geqslant 1$.

Example A.23. Consider a discrete random variable X such that

$$\mathbb{P}(X = -1) = \frac{1}{2}, \quad \mathbb{P}(X = 0) = \frac{1}{4}, \quad \text{and} \; \mathbb{P}(X = 1) = \frac{1}{4}.$$

Find and plot its CDF $F_X(x)$.

Solution. X is a discrete random variable. Thus, the CDF $F_X(x)$ is a piecewise constant function that changes its value at the points $x = -1, 0, 1$ only. We have that

$$\mathbb{P}(X < -1) = 0, \quad \mathbb{P}(X \leqslant -1) = \frac{1}{2}, \quad \mathbb{P}(X \leqslant 0) = \frac{1}{2} + \frac{1}{4} = \frac{3}{4}, \quad \mathbb{P}(X \leqslant 1) = \frac{1}{2} + \frac{1}{4} + \frac{1}{4} = 1.$$

So, the CDF is

$$F_X(x) = \begin{cases} 0 & x < -1, \\ 0.5 & -1 \leqslant x < 0, \\ 0.75 & 0 \leqslant x < 1, \\ 1.0 & 1 \leqslant x. \end{cases}$$

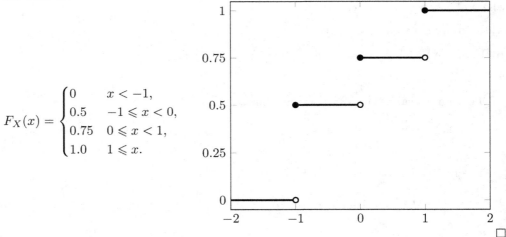

Example A.24. Determine the PMF of a random variable X from its CDF

$$F_X(x) = \begin{cases} 0 & x < -2, \\ 0.2 & -2 \leqslant x < 0, \\ 0.7 & 0 \leqslant x < 2, \\ 1.0 & 2 \leqslant x. \end{cases}$$

Solution. The mass probabilities are equal to jump sizes of the CDF:

$$p_X(x) = \mathbb{P}(X = x) = F_X(x) - F_X(x^-).$$

Thus,

$$p_X(-2) = F_X(-2) - F_X(-2^-) = 0.2 - 0 = 0.2,$$
$$p_X(0) = F_X(0) - F_X(0^-) = 0.7 - 0.2 = 0.5,$$
$$p_X(2) = F_X(2) - F_X(2^-) = 1 - 0.7 = 0.3.$$

□

A.2.4 Continuous Probability Distributions

X is said to be a *continuous* random variable if for any $x \in \mathbb{R}$ the mass probability $\mathbb{P}(X = x)$ is zero. For example, select a point in a continuous domain at random. Each coordinate is a continuous random variable.

Recall that for any CDF we have $F(x^+) = F(x) = \mathbb{P}(X = x) + F(x^-)$. Since any continuous random variable X has zero mass probabilities, i.e., $\mathbb{P}(X = x) = 0$ for any x, we have that $F_X(x^+) = F_X(x^-)$. In other words, the CDF of a continuous random variable is a *continuous* function. As a result, for a continuous random variable X, we have

$$\mathbb{P}(a \leqslant X \leqslant b) = \mathbb{P}(a < X \leqslant b) = \mathbb{P}(a \leqslant X < b) = \mathbb{P}(a < X < b) = F_X(b) - F_X(a)$$

for any $a \leqslant b$. Recall that the CDF for a discrete distribution is a piecewise constant function with a countable number of discontinuities.

Example A.25. Select U at random in $[0, 1]$. Find and plot the CDF $F_U(x)$.

Solution. Using the geometric probability, we can obtain the CDF of U:

$$F_U(x) = \mathbb{P}(U \in [0, x]) = \frac{x}{1} = x \quad \text{for any } x \in [0, 1].$$

Thus,

$$F_U(x) = \begin{cases} 0 & x < 0, \\ x & 0 \leqslant x \leqslant 1, \\ 1 & 1 < x. \end{cases}$$

□

Example A.26. Consider a discrete random variable X such that

$$\mathbb{P}(X = -1) = \frac{1}{2}, \quad \mathbb{P}(X = 0) = \frac{1}{4}, \quad \text{and} \quad \mathbb{P}(X = 1) = \frac{1}{4}.$$

Let U be selected at random in $[0, 1]$. Define Y as follows. Toss a coin. If it results in a head, then $Y = X$. Otherwise, $Y = U$. Use the total probability formula to find the CDF of Y.

Solution. Let $C \in \{H, T\}$ be the result of the coin toss. Find the CDF of Y by conditioning on the value of C:

$$F_Y(y) = \mathbb{P}(Y \leqslant y) = \mathbb{P}(Y \leqslant y \mid C = H) \cdot \mathbb{P}(C = H) + \mathbb{P}(Y \leqslant y \mid C = T) \cdot \mathbb{P}(C = T)$$

$$= \frac{1}{2} \cdot \mathbb{P}(X \leqslant y) + \frac{1}{2} \cdot \mathbb{P}(U \leqslant y) = \frac{1}{2} F_X(y) + \frac{1}{2} F_U(y).$$

Combining CDFs F_X and F_U found in Examples A.23 and A.25, respectively, gives

$$F_Y(y) = \frac{1}{2} \cdot \begin{cases} 0 & y < -1, \\ 0.5 & -1 \leqslant y < 0, \\ 0.75 & 0 \leqslant y < 1, \\ 1.0 & 1 \leqslant y, \end{cases} + \frac{1}{2} \cdot \begin{cases} 0 & y < 0, \\ y & 0 \leqslant y < 1, \\ 1 & 1 \leqslant y. \end{cases}$$

Thus, the CDF of Y is

$$F_Y(y) = \begin{cases} 0 & y < -1, \\ 0.25 & -1 \leqslant y < 0 \\ 0.375 + \frac{y}{2} & 0 \leqslant y \leqslant 1, \\ 1 & 1 < y. \end{cases}$$

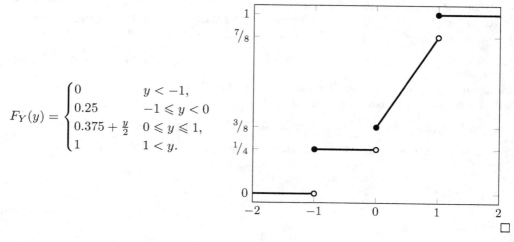

Example A.27. A dart is tossed at a circular board of radius 9 cm at random. Let R be the distance from the bullseye.

(a) Show that the CDF of R is

$$F(r) = \begin{cases} 0 & \text{if } r < 0, \\ \left(\frac{r}{9}\right)^2 & \text{if } 0 \leqslant r \leqslant 9, \\ 1 & \text{if } r > 9. \end{cases} \tag{A.9}$$

(b) Show that

$$\mathbb{P}(a < R \leqslant b) = \mathbb{P}(R \leqslant b) - \mathbb{P}(R \leqslant a) = \frac{b^2}{81} - \frac{a^2}{81} = \int_a^b \frac{2r}{81} \, dr$$

for any $0 \leqslant a \leqslant b \leqslant 9$.

Solution. Using the definition of the geometric probability, we have

$$\mathbb{P}(R \leqslant r) = \frac{\text{area of the disk of radius } r}{\text{area of the disk of radius } 9} = \frac{\pi r^2}{\pi 9^2} = \left(\frac{r}{9}\right)^2.$$

Clearly, $\mathbb{P}(R < 0) = \mathbb{P}(R > 9) = 0$ and $\mathbb{P}(0 \leqslant R \leqslant 9) = 1$. Thus, the CDF of R is given by (A.9). For any $0 \leqslant a < b \leqslant 9$, we have

$$\mathbb{P}(R \leqslant b) - \mathbb{P}(R \leqslant a) = F(b) - F(a) = \frac{b^2}{81} - \frac{a^2}{81}.$$

Since $F(r) = \frac{r^2}{81}$ is an antiderivative of $f(r) = \frac{2r}{81}$, by the Fundamental Theorem of Calculus, $F(b) - F(a) = \int_a^b \frac{2r}{81} \, dr$. $\qquad\square$

Consider the density of a loading on a long, thin beam. For any point x along the beam, the density can be described by a nonnegative function (measured in weight/length). Intervals with large loadings correspond to large values of the function. The total loading between points a and b is determined as the integral of the density function from a to b.

Similarly, a *probability density function (PDF)* $f_X(x)$ can be used to describe the probability distribution of a continuous random variable X, whose range is an interval of the real line. If a subinterval $(x - \epsilon/2, x + \epsilon/2)$ is likely to contain a value for X, the probability $\mathbb{P}(X \in (x - \epsilon/2, x + \epsilon/2)) \approx \epsilon f_X(x)$ is large, and this corresponds to large values for $f_X(x)$.

The probability that X is between a and b is determined as the integral of $f_X(x)$ from a to b. For example, the function $f(r) = \frac{2r}{81}$ in Example A.27 is a PDF on the interval $[0, 9]$.

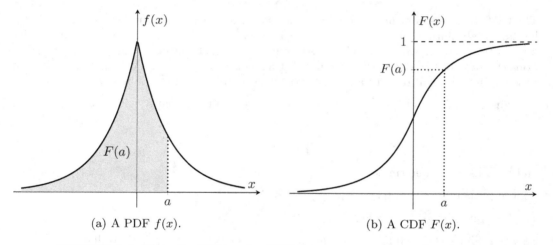

(a) A PDF $f(x)$. (b) A CDF $F(x)$.

FIGURE A.3: Plot of a PDF and a CDF for some continuous distribution.

A *probability density function (PDF)* is any function f such that

$$\text{(i)} \ f(x) \geqslant 0 \text{ for all } x; \qquad \text{(ii)} \int_{-\infty}^{\infty} f(x) \, dx = 1.$$

Here are some examples:

- $f(x) = e^{-x} \mathbb{I}_{[0,\infty)}(x)$, i.e., $f(x)$ is nonzero for nonnegative x;

- $f(x) = 0.5 e^{-|x|} \mathbb{I}_{(-\infty,0)}(x) + 0.5 e^{-x} \mathbb{I}_{[0,\infty)}(x)$ for $x \in \mathbb{R}$;

- $f(x) = \dfrac{1}{b-a} \mathbb{I}_{[a,b]}(x)$, i.e., $f(x)$ is nonzero for $x \in [a, b]$.

Example A.28. How can the function $f(x) = e^{-2x} \mathbb{I}_{[0,\infty)}(x)$ be modified to become a PDF?

Solution. Clearly, $f(x) \geqslant 0$ for all $x \in \mathbb{R}$. Let us find the integral of f on the real line:

$$\int_{-\infty}^{\infty} e^{-2x} \mathbb{I}_{[0,\infty)}(x) \, dx = \int_{0}^{\infty} e^{-2x} \, dx = \frac{1}{2}.$$

If we divide f by $\frac{1}{2}$, then the resulting function is still nonnegative, but its full integral is one. Thus, $f(x) = 2e^{-2x} \mathbb{I}_{[0,\infty)}(x)$ is a PDF. $\qquad\square$

A random variable X is said to be *absolutely continuous* if there exists a PDF f such that

$$\mathbb{P}(a \leqslant X \leqslant b) = \int_{a}^{b} f(x) \, dx$$

for any $a, b \in \mathbb{R}$ with $a \leqslant b$. That is, for an absolutely continuous random variable, the probability is equal to an area bounded by the lines $x = a$, $x = b$, $y = 0$, and $y = f(x)$. The probability that X is contained in an interval of length ε around the point a is approximately $\varepsilon f(a)$. We see that $f(a)$ is a measure of how likely it is that the random variable will be near a. However, the value $f(a)$ is not a probability that $X = a$ (recall that $\mathbb{P}(X = a) = 0$ for any continuous random variable X).

If X is an absolutely continuous random variable with PDF f, its CDF is given by

$$F(x) = \mathbb{P}(X \leqslant x) = \int_{-\infty}^{x} f(y) \, dy \quad \text{for any } x \in \mathbb{R}.$$

The CDF $F(x)$ is a continuous function of x thanks to the Fundamental Theorem of Calculus. So, an absolutely continuous random variable (with a PDF) has a continuous probability distribution (with a continuous CDF). Although there exist continuous probability distributions that do not have PDFs, we will not distinguish these two cases. We will omit the word "absolutely" and will refer to all such distributions and random variables as continuous.

Example A.29. Find the CDF of a continuous random variable X with the PDF

$$f(x) = \begin{cases} 0 & \text{for } x < 0, \\ 2e^{-2x} & \text{for } x \geqslant 0. \end{cases}$$

and then use it to determine

(a) the probabilities: (i) $\mathbb{P}(1 < X)$, (ii) $\mathbb{P}(1 < X < \ln 2)$;

(b) the value of x such that $\mathbb{P}(X < x) = 0.1$.

Solution. Since $f(x) = 0$ for $x < 0$, the range of X is $[0, \infty)$. For $x \geqslant 0$, we have

$$F(x) = \int_{-\infty}^{x} f(y) \, dy = \int_{0}^{x} 2e^{-2y} \, dy = 1 - e^{-2x}.$$

For $x < 0$, we have $F(x) = 0$.

(a) Compute the probabilities:

    (i) $\mathbb{P}(1 < X) = 1 - \mathbb{P}(X \leqslant 1) = 1 - F(1) = 1 - (1 - e^{-2 \cdot 1}) = e^{-2}$;

    (ii) $\mathbb{P}(1 < X < \ln 2) = F(\ln 2) - F(1) = (1 - e^{-2 \ln 2}) - (1 - e^{-2 \cdot 1}) = e^{-2} - 2^{-2} = e^{-2} - \frac{1}{4}$,

        where we used the fact that $e^{-2 \ln 2} = \left(e^{\ln 2}\right)^{-2} = 2^{-2} = \frac{1}{4}$.

(b) To find the value of x such that $\mathbb{P}(X < x) = 0.1$, solve $F(x) = 0.1$ for x:

$$1 - e^{-2x} = 0.1 \iff e^{-2x} = 0.9 \iff -2x = \ln 0.9 \iff x = -\frac{\ln 0.9}{2}.$$

$\qquad\square$

A.3   Mathematical Expectations and Other Moments

When a random experiment has been repeated multiple times, we can observe that the average of outcomes converges to some limiting value. For example, when we flip a fair coin, the long-run count of heads is approximately half of the total number of tosses. So, the average proportion of heads is close to one half, and the difference between the actual frequency and $\frac{1}{2}$ is becoming smaller as the number of tosses increases. This limiting value of the average is called the *mathematical expectation*.

A.3.1   Mathematical Expectation of a Discrete Random Variable

Let $\hat{x}_1, \hat{x}_2, \ldots, \hat{x}_n$ be observations of a discrete random variable X with the PMF $p_X(x) = p_k$ for $x = x_k$ with $k = 1, 2, \ldots, m$. Calculate the average value

$$\bar{x} = \frac{\hat{x}_1 + \hat{x}_2 + \cdots + \hat{x}_n}{n}$$

and form m groups to obtain

$$\bar{x} = \frac{1}{n} \left(\underbrace{\hat{x}_{i_1} + \cdots + \hat{x}_{i_{n_1}}}_{\hat{x}=x_1} + \cdots + \underbrace{\hat{x}_{j_1} + \cdots + \hat{x}_{j_{n_m}}}_{\hat{x}=x_m} \right),$$

where the kth group has $n_k \geqslant 0$ elements. In total, there are $n_1 + n_2 + \cdots + n_m = n$ terms. Hence,

$$\bar{x} = \frac{1}{n} \left(n_1 \cdot x_1 + n_2 \cdot x_2 + \cdots + n_m \cdot x_m \right).$$

Using the frequentist interpretation of probability, we can write $\dfrac{n_k}{n} \approx \mathbb{P}(X = x_k) = p_X(x_k)$. As a result, we obtain

$$\bar{x} \approx \sum_{k=1}^{m} x_k p_X(x_k). \tag{A.10}$$

Surprisingly, the expression in the right-hand side of the above formula depends on the PMF and not on the observed values $\hat{x}_i$.

The *expected value* of a discrete random variable X, denoted $\mathrm{E}[X]$ or μ_X, is defined by

$$\mathrm{E}[X] = \sum_{x \in D_X} x \cdot p_X(x), \tag{A.11}$$

where D_X and p_X are, respectively, the range and the PMF of X. That is, $\mathrm{E}[X]$ is an average of all possible values weighted by respective probabilities. As we can see from (A.10) and (A.11), if an experiment is repeated a large number of times, then the sample average $\bar{x}$ is approximately equal to $\mathrm{E}[X]$. We will demonstrate later in the Law of Large Numbers that the sample average converges to the expected value as the sample size n goes to infinity.

The concept of expectation is analogous to the physical concept of the centre of gravity for a distribution of weights. Consider a discrete random variable X with mass points x_k and mass probabilities $p_k = p_X(x_k) > 0$ with $k = 1, 2, \ldots, m$. Imagine a rod in which weights with mass p_k are located at the respective points x_k. The point at which the rod would be balanced is known as the centre of gravity. Its location is given by the mathematical expectation $\mathrm{E}[X]$.

Let us consider a few simple examples.

- The mathematical expectation of a constant variable $X \equiv C$ is equal to the constant:

$$E[C] = C.$$

- Consider the indicator of an event, $\mathbb{I}_E$, which equals one with probability $p = \mathbb{P}(E)$ and zero with probability $1 - p = \mathbb{P}(E^{\complement})$. The expected value is

$$E[\mathbb{I}_E] = 1 \cdot p + 0 \cdot (1 - p) = p.$$

- Let X be a discrete uniform random variable over a set of m distinct points $\{x_k\}_{1 \leqslant k \leqslant m}$. The expected value of X is given by the average of the mass points:

$$E[X] = \sum_{k=1}^{m} x_k \frac{1}{m} = \frac{1}{m} \sum_{k=1}^{m} x_k.$$

- The mathematical expectation of random variable X defined on a finite sample space Ω is given by

$$E[X] = \sum_{x \in \Omega} X(\omega)\mathbb{P}(\omega).$$

To compute the mathematical expectation of a random function $g = g(X)$ of a discrete random variable X, we can use one of the following two approaches. First, we can find the range and the PMF of the new variate $Y = g(X)$ and then apply (A.11):

$$E[Y] = \sum_{y \in D_Y} y \cdot p_Y(y). \tag{A.12}$$

Second, we can use the following formula, which is equivalent to (A.12):

$$E[g(X)] = \sum_{x \in D_X} g(x) p_X(x). \tag{A.13}$$

Example A.30. Suppose X has the PMF $p_X(x) = 1/3$ for $x = -1, 0, 1$.

(a) Compute $E[X]$.

(b) Compute $E[X^2]$ using both (A.12) and (A.13).

(c) What do your answers imply about $(E[X])^2$ and $E[X^2]$?

Solution.

(a) $E[X] = (-1) \cdot p_X(-1) + 0 \cdot p_X(0) + 1 \cdot p_X(1) = -\frac{1}{3} + \frac{1}{3} = 0.$

(b) First, find the PMF of $Y = X^2$ and use (A.12) to calculate $E[Y] = E[X^2]$. Since $(-1)^2 = 1^2 = 1$, the range of Y is $D_Y = \{0, 1\}$. That is, Y is a Bernoulli variable with $p = \mathbb{P}(Y = 1) = \mathbb{P}(X = 1 \text{ or } X = -1) = \frac{2}{3}$. Thus, $E[Y] = p = \frac{2}{3}$. Second, use (A.13):

$$E[X^2] = \sum_{x} x^2 \cdot p_X(x) = (-1)^2 \cdot p_X(-1) + 0^2 \cdot p_X(0) + 1^2 \cdot p_X(1) = \frac{2}{3}.$$

(c) As we can see, $(E[X])^2 \neq E[X^2]$, in general.

□

Here is a list of important properties of the mathematical expectation.

1. In general, for any random variable X and function g for which $E[g(X)]$ is defined and finite, we have

$$E[g(X)] \neq g(E[X]).$$

The exception is a linear function $g(x) = ax + b$ with constant a and b, for which we have $E[aX + b] = aE[X] + b$.

2. For any two random variables X and Y and two constants a and b, we have

$$E[aX + bY] = aE[X] + bE[Y].$$

We can generalize this property. For any collection of random variables $X_1, X_2, \ldots, X_\ell$ and functions $g_1, g_2, \ldots, g_\ell$ for which all $E[g_k(X_k)]$ are defined and finite, we have

$$E[g_1(X_1) + g_2(X_2) + \cdots + g_\ell(X_\ell)] = E[g_1(X_1)] + E[g_2(X_2)] + \cdots + E[g_\ell(X_\ell)].$$

That is, the expected value is a *linear functional*.

3. If $X \geqslant 0$ a.s., then $E[X] \geqslant 0$. If $X \leqslant Y$ a.s., then $E[X] \leqslant E[Y]$. In particular, $|E[X]| \leqslant E[|X|]$ since $-|X| \leqslant X \leqslant |X|$. Here, "a.s." stands for *almost surely*, i.e., it happens with probability 1.

Example A.31.

(a) If the mean of $3Y - 7$ is equal to -4, what is the mean of Y?

(b) If $E[Z] = 2$ and $E[Z^2] = 8$, what is the value of $E[(2 + 4Z)^2]$?

Solution.

(a) We have $-4 = E[3Y - 7] = 3E[Y] - 7$. Solve it for $E[Y]$ to obtain $E[Y] = 1$.

(b) $E[(2 + 4Z)^2] = E[4 + 16Z + 16Z^2] = 4 + 16E[Z] + 16E[Z^2] = 4 + 16 \cdot 2 + 16 \cdot 8 = 164$.

$\square$

Example A.32. For a power-law discrete distribution, the PMF is proportional to a power function. That is,

$$p(x) = c\,x^{-k}, \quad \text{for } x = 1, 2, 3, \ldots$$

with exponent $k > 1$ and some scaling constant $c > 0$. Let a random variable X follow the power law. For what values of k is $E[X^n]$ with $n \geqslant 1$ finite?

Solution. To find the constant c, we use the fact that a sum of all mass probabilities of a discrete random variable is one. Thus,

$$1 = \sum_{x \geqslant 1} c\,x^{-k} \implies c = \Big(\sum_{x \geqslant 1} x^{-k}\Big)^{-1}.$$

The distribution is defined if the series $\sum_{x \geqslant 1} x^{-k}$ converges. Similarly, the expectation $E[X^n]$ is finite if the series $\sum_{x \geqslant 1} x^{-k+n}$ converges. By the integral test for convergence, this series converges iff the integral $\int_1^\infty x^{-k+n}\,dx$ is finite. This integral converges iff $k - n > 1$ or, equivalently, $k > n + 1$. Thus, the distribution is well-defined if $k > 1$. Its expectation is finite if $k > 2$.

$\square$

A.3.2   Variance and Other Moments

Expected values of powers of X are called *moments about 0* or simply *moments*.

- $E[X]$ is the first moment.

- $E[X^2]$ is the second moment.

- $E[X^k]$ is the kth moment.

Here and below, $k = 1, 2, 3, \dots$. Expected values of powers of $X - \mu$ with $\mu \equiv E[X]$ are called *moments about the mean* or *central moments*.

- $E[X - \mu]$ is the first central moment. Clearly, $E[X - \mu] = E[X] - \mu = 0$.

- $E[(X - \mu)^2]$ is the second central moment. It is also known as the variance of X.

- $E[(X - \mu)^k]$ is the kth central moment.

Let X be a discrete random variable with PMF p, range D, and mean μ.

- The *kth moment of X* is calculated as

$$E[X^k] = \sum_{x \in D} x^k p(x).$$

- The *kth central moment of X* is calculated as

$$E[(X - \mu)^k] = \sum_{x \in D} (x - \mu)^k p(x).$$

The *variance* of X, denoted $\mathrm{Var}(X)$ or σ_X^2, is the second central moment. It is given by

$$\mathrm{Var}(X) = E\left[(X - \mu)^2\right] = \sum_{x \in D} (x - \mu)^2 \, p(x). \qquad (A.14)$$

The *standard deviation* of X is $\sigma_X = \sqrt{\mathrm{Var}(X)}$. In general, the variance increases in magnitude with the likelihood of extreme values. The variance measures how spread out the distribution is. In finance, the standard deviation is used as a measure of risk.

Lastly, define the kth *standardized moment* as $E\left[\left(\frac{X - \mu_X}{\sigma_X}\right)^k\right]$. Here, $Y = \frac{X - \mu_X}{\sigma_X}$ is called the *standardization* of random variable X. Clearly, the expected value of Y is zero, and its variance is one.

Example A.33. Consider the following two games of chance.

Game A: You win \$2 with probability 2/3 and lose \$1 with probability 1/3.

Game B: You win \$1002 with probability 2/3 and lose \$2001 with probability 1/3.

(a) Find and compare the expected winnings for the games.

(b) Find and compare the variances of winnings for the games.

Solution.

(a) Let A and B denote the winnings for games A and B, respectively. Calculate the average values:

$$E[A] = \$2 \cdot \frac{2}{3} + (-\$1) \cdot \frac{1}{3} = \$1,$$

$$E[B] = \$1002 \cdot \frac{2}{3} + (-\$2001) \cdot \frac{1}{3} = \$1.$$

(b) Find the variances of winnings for the games:

$$\text{Var}(A) = (2-1)^2 \cdot \frac{2}{3} + (-1-1)^2 \cdot \frac{1}{3} = 2,$$

$$\text{Var}(B) = (1002-1)^2 \cdot \frac{2}{3} + (-2001-1)^2 \cdot \frac{1}{3} = 2,003,002.$$

Although the games have the same average winnings, the standard deviation for game B is significantly larger than that for game A:

$$\sigma_A = \$1.41 \ll \$1415.63 = \sigma_B.$$

□

Example A.34. Let X denote the outcome of throwing a die. Find its mean and variance.

Solution. The random variable X has a discrete uniform distribution over the set $D_X = \{1, 2, 3, 4, 5, 6\}$. Therefore, $E[X] = \frac{1}{6} \sum_{x=1}^{6} x = \frac{21}{6} = 3.5$. The variance is

$$\text{Var}(X) = \sum_{x=1}^{6} \frac{(x-3.5)^2}{6} = \frac{(-2.5)^2 + (-1.5)^2 + (-0.5)^2 + (0.5)^2 + (1.5)^2 + (2.5)^2}{6}$$

$$= \frac{8.75}{3} \cong 2.9167.$$

□

In addition to equation (A.14), we can also use another formula where the variance is expressed as a difference of the second moment and the squared mean. Let X be a discrete random variable with PMF p and range D. Then, the variance $\text{Var}(X)$ can be computed as follows:

$$\text{Var}(X) = E[X^2] - (E[X])^2 = \sum_{x \in D} x^2 \, p(x) - \left(\sum_{x \in D} x \, p(x) \right)^2.$$

Using the above formula, it is not difficult to prove that

$$\text{Var}(aX + b) = a^2 \, \text{Var}(X)$$

for any constants a and b. That is, $\sigma_{aX+b} = |a| \sigma_X$. Moreover, for any constant, $\text{Var}(C) = 0$, since $E[C^2] = E[C]^2 = C^2$.

Example A.35. The returns on three stocks A, B, and C are random variables with the following PMFs

r	-5%	1%	7%
$\mathbb{P}(R_A = r)$	0.25	0.50	0.25
$\mathbb{P}(R_B = r)$	0.1	0.8	0.1

r	-10%	1%	12%
$\mathbb{P}(R_C = r)$	0.1	0.8	0.1

(a) Show that the expected return on all three stocks is the same and equal to 1%.

(b) Compute the standard deviation of the return for each stock.

(c) Given the variances of the returns, which stock would a risk-averse investor prefer to invest in?

Solution. Find the expected return $\mu = \mathrm{E}[R]$ for each stock:

$$\mu_A = (-5) \cdot 0.25 + 1 \cdot 0.5 + 7 \cdot 0.25 = 1\%,$$
$$\mu_B = (-5) \cdot 0.1 + 1 \cdot 0.8 + 7 \cdot 0.1 = 1\%,$$
$$\mu_C = (-10) \cdot 0.1 + 1 \cdot 0.8 + 12 \cdot 0.1 = 1\%.$$

Find the standard deviation for each return:

$$\sigma_A = \sqrt{(-5)^2 \cdot 0.25 + (1)^2 \cdot 0.5 + (7)^2 \cdot 0.25 - 1^2} \cong 4.924\%,$$
$$\sigma_B = (\sqrt{(-5)^2 \cdot 0.1 + (1)^2 \cdot 0.8 + (7)^2 \cdot 0.1 - 1^2} \cong 2.683\%,$$
$$\sigma_C = (\sqrt{(-10)^2 \cdot 0.1 + (1)^2 \cdot 0.8 + (12)^2 \cdot 0.1 - 1^2} \cong 4.919\%.$$

The risk associated with each of these stocks can be gauged by the variance of return: the higher the variance—the larger the risk. A risk-averse investor would prefer to invest in stock B with the lowest variance of return. $\square$

A.3.3 Mean, Variance, and Median of a Continuous Random Variable

The mean and the variance of a continuous random variable are defined similarly as it was done for discrete random variables. Integration replaces summation in the definitions. Let X be a continuous random variable with probability density function $f(x)$. The *mean* or *expected value* of X, denoted μ_X or $\mathrm{E}[X]$, is

$$\mathrm{E}[X] = \int_{-\infty}^{\infty} x\, f(x)\, \mathrm{d}x$$

The *variance* of X, denoted $\mathrm{Var}(X)$ or σ_X^2, is

$$\mathrm{Var}(X) = \int_{-\infty}^{\infty} (x - \mu_X)^2\, f(x)\, \mathrm{d}x = \int_{-\infty}^{\infty} x^2\, f(x)\, \mathrm{d}x - \mu_X^2.$$

The standard deviation of X is defined as a square root of the variance: $\sigma_X = \sqrt{\mathrm{Var}(X)}$. If X is a continuous random variable with PDF f, and h is some function, then

$$\mathrm{E}[h(X)] = \int_{-\infty}^{\infty} h(x)\, f(x)\, \mathrm{d}x$$

provided that $\mathrm{E}[|h(X)|] = \int_{-\infty}^{\infty} |h(x)|\, f(x)\, \mathrm{d}x < \infty$. In particular,

$$\mathrm{E}[X^n] = \int_{-\infty}^{\infty} x^n\, f(x)\, \mathrm{d}x,$$

$$\mathrm{E}[(X - \mu)^n] = \int_{-\infty}^{\infty} (x - \mu)^n\, f(x)\, \mathrm{d}x, \quad n = 1, 2, 3 \ldots$$

Example A.36. The lifetime (in hours) of electronic tubes is a random variable having a probability density function given by

$$f(x) = a^2 x e^{-ax}, \quad x \geqslant 0; \quad \text{where } a > 0.$$

Compute the expected lifetime of a tube.

Solution. Use the integration by parts to find the expected lifetime μ:

$$\mu = \int_0^\infty a^2 x^2 e^{-ax}\,\mathrm{d}x = -ax^2 e^{-ax}\Big]_0^\infty + \int_0^\infty 2axe^{-ax}\,\mathrm{d}x$$

$$= -2xe^{-ax}\Big]_0^\infty + \int_0^\infty 2e^{-ax}\,\mathrm{d}x = -\frac{2}{a}e^{-ax}\Big]_0^\infty = \frac{2}{a}.$$

$\square$

Example A.37. The density function of X is given by $f(x) = \begin{cases} a + bx^2 & 0 \leqslant x \leqslant 1, \\ 0 & \text{otherwise.} \end{cases}$

If $\mathrm{E}[X] = \frac{3}{5}$, find a and b.

Solution. Solve the system with two equations to find a and b:

$$\begin{cases} \int_0^1 (a + bx^2)\,\mathrm{d}x = 1 \\ \int_0^1 x(a + bx^2)\,\mathrm{d}x = \frac{3}{5} \end{cases} \iff \begin{cases} 3a + b = 3 \\ 2a + b = 2.4 \end{cases} \iff \begin{cases} a = 0.6 \\ b = 1.2 \end{cases}$$

Clearly, $0.6 + 1.2x^2 \geqslant 0$ for all $x \in [0, 1]$, so $f(x) = (0.6 + 1.2x^2)\mathbb{I}_{[0,1]}(x)$ satisfies the definition of a PDF. $\square$

Let X be a continuous random variable. The $100p$th *percentile* of X is the smallest number x_p for which

$$\mathbb{P}(X \leqslant x_p) = p, \quad \text{where } p \in (0, 1).$$

That is, x_p is a threshold value below which random draws would fall $100p\%$ of the time. The percentile x_p solves the equation $F_X(x_p) = p$ with the CDF F_X of X. The function that maps p to the percentile x_p, is called a *quantile function*. If the CDF F is continuous and strictly monotonic, then the quantile function is given by the inverse CDF F^{-1}. In finance, a quantile function for a profit-and-loss distribution is used to calculate the Value-at-Risk (VaR), which is a commonly-used risk measure.

Let X be a continuous random variable with CDF F, then its *median* is defined as the value m for which $F(m) = \frac{1}{2}$. That is, the median is the 50th percentile of X. The median m minimizes the expected value of the absolute error. That is, $\mathrm{E}|X - c|$ attains its minimum value when c is equal to the median of X.

Example A.38. The time (in seconds) between consecutive hits to a website can be modelled as a continuous random variable T with the PDF $f(t) = 2e^{-2t}\mathbb{I}_{[0,\infty)}(t)$.

(a) Compute $\mathbb{P}(T = 1)$.

(b) Compute $\mathbb{P}(\text{within a millisecond of one second}) = \mathbb{P}(0.999 \leqslant T \leqslant 1.001)$ first exactly and then approximately.

(c) Determine the 50th and 90th percentiles of T.

Solution.

(a) Since T is a continuous random variable, $\mathbb{P}(T = 1) = 0$.

(b) To compute the probability, we use

$$\mathbb{P}(a \leqslant T \leqslant b) = \int_a^b 2e^{-2t}\,\mathrm{d}t = e^{-2t}\Big]_a^b = e^{-2b} - e^{-2a} \quad\text{or}\quad \int_a^b 2e^{-2t}\,\mathrm{d}t \approx e^{-2\left(\frac{a+b}{2}\right)}\cdot(b-a).$$

The exact value is $e^{-1.998} - e^{-2.002} \cong 0.054134\%$. The approximate value is actually almost the same as the exact one: $2e^{-2}\cdot 0.002 \cong 0.054134\%$. The reason is that the rectangle approximation for definite integrals works well for small integration intervals.

(c) The CDF of T is $F(x) = 1 - e^{-2x}$ for $x \geqslant 0$ and $F(x) = 0$ for $x < 0$. Solve $F(x) = p$ for $p \in (0, 1)$:

$$F(x) = p \iff 1 - e^{-2x} = p \iff x = -\frac{\ln(1-p)}{2}.$$

For $p = 0.5$, we have $x_{0.5} = -\frac{\ln 0.5}{2} \cong 0.3466$. For $p = 0.9$, we have $x_{0.9} = -\frac{\ln 0.1}{2} \cong 1.1513$.

□

A.3.4   Moment Generating Functions

The *moment generation function (MGF)* of a random variable X, denoted $M_X(t)$, is

$$M_X(t) = \mathrm{E}[e^{tX}]. \tag{A.15}$$

For a discrete random variable, it is given by

$$M_X(t) = \sum_{x \in D_X} e^{tx} p_X(x). \tag{A.16}$$

For a continuous random variable, we have

$$M_X(t) = \int_{-\infty}^{\infty} e^{tx} f_X(x)\,\mathrm{d}x. \tag{A.17}$$

The MGF is defined for all values of t such that $\mathrm{E}[e^{tX}]$ is finite. Note that $M_X(t) \geqslant 0$ for all t in the domain of the MGF. Additionally, $M_X(0) = \mathrm{E}[e^{0X}] = \mathrm{E}[1] = 1$.

Example A.39. Suppose that X has the PMF $p(x) = \frac{1}{6}\left(\frac{5}{6}\right)^x$ for $x = 0, 1, 2, \ldots$ Find the MGF of X and state its domain.

Solution. The MGF of X is

$$M_X(t) = \sum_{x=0}^{\infty} e^{tx} \cdot \frac{1}{6}\left(\frac{5}{6}\right)^x = \frac{1}{6}\sum_{x=0}^{\infty}\left(\frac{5}{6}\cdot e^t\right)^x = \frac{1}{6}\cdot\frac{1}{1 - 5e^t/6} = \frac{1}{6 - 5e^t}.$$

The series in the above expression converges iff $\left|\frac{5}{6}e^t\right| < 1$, i.e., iff $t < \ln(6/5)$. Thus, the MGF $M_X(t)$ is defined for $t \in (-\infty, \ln(6/5))$.

□

Example A.40. Consider an indicator random variable X, which is zero with probability $1/3$ and one with probability $2/3$.

(a) Compute $M(t) = \mathrm{E}[e^{tX}]$ for any $t \in \mathbb{R}$.

(b) Compute $M'(t)$ and $M''(t)$ and evaluate them at $t = 0$.

Solution. The MGF of X is

$$M(t) = e^{t \cdot 0} \cdot \frac{1}{3} + e^{t \cdot 1} \cdot \frac{2}{3} = \frac{1}{3} + \frac{2e^t}{3}.$$

Differentiate $M(t)$ at zero:

$$M'(0) = \left. \frac{2e^t}{3} \right|_{t=0} = \frac{2}{3} \quad \text{and} \quad M''(0) = \left. \frac{2e^t}{3} \right|_{t=0} = \frac{2}{3}.$$

$\square$

The main feature of an MGF, which explains its name, is the moment generation property.

Theorem A.3. *Let X be a random variable whose MGF $M_X(t)$ exists for all t with $|t| < \epsilon$ for some $\epsilon > 0$. Then, all moments of X exist and*

$$\mathrm{E}[X^k] = M_X^{(k)}(0) \quad \text{for} \quad k = 1, 2, 3, \ldots$$

Other two important properties of MGFs are as follows.

(1) If for two random variables X and Y there exists $\epsilon > 0$ such that $M_X(t) = M_Y(t)$ for all t with $|t| < \epsilon$, then X and Y have the *same probability distribution*. In other words, the MGF (if it exists) uniquely defines the probability distribution.

(2) Let X be a random variable with MGF $M_X(t)$ and a and b are constants, then

$$M_{a+bX}(t) = e^{at} M_X(bt)$$

Example A.41. Suppose that the MGF of a random variable Z is $M(t) = e^{t^2/2}$.

(a) Find the mean and variance of Z.

(b) Find the moment generating function of $X = 2 - 3Z$.

(c) Compute the mean and variance of $Y = e^Z$.

Solution.

(a) Differentiate $M(t)$ at zero:

$$M'(0) = \left. te^{t^2/2} \right|_{t=0} = 0 \implies \mathrm{E}[Z] = 0,$$

$$M''(0) = \left. (1 + t)e^{t^2/2} \right|_{t=0} = 1 \implies \mathrm{Var}(Z) = 1 - 0 = 1.$$

(b) Using the property that $M_{a+bZ}(t) = e^{at} M_Z(bt)$, we obtain

$$M_X(t) = M_{2-3Z}(t) = e^{2t} M(-3t) = e^{2t} e^{(-3t)^2/2} = e^{2t+9t^2/2}.$$

(c) We have $E[e^Z] = M(1)$ and $E[e^{2Z}] = M(2)$. Thus,

$$E[Y] = E[e^Z] = M(1) = e^{1^2/2} = \sqrt{e} \cong 1.6487,$$

$$\text{Var}(Y) = E[e^{2Z}] - E[e^Z]^2 = M(2) - M(1)^2 = e^{2^2/2} - \left(e^{1/2}\right)^2 = e^2 - e \cong 4.6708.$$

$\square$

Example A.42. Given $M_X(t) = 0.2 + 0.3e^t + 0.5e^{3t}$, find $p_X(x)$, $E[X]$, and $\text{Var}(X)$.

Solution. If a discrete random variable has a finite range $D = \{x_1, x_2, \ldots, x_m\}$, then its MGF is

$$M(t) = \sum_{k=1}^{m} e^{tx_k} \cdot p_X(x_k).$$

Compare this expression with the one for M_X, we can conclude that X has a three-point discrete distribution with mass points $x_1 = 0$, $x_2 = 1$, $x_3 = 3$ and respective mass probabilities $p_X(x_1) = 0.2$, $p_X(x_2) = 0.3$, $p_X(x_3) = 0.5$.

The expected value and the variance can be now calculated using the PMF or by differentiating the MGF:

$$E[X] = \sum_{j=1}^{3} x_j \cdot p_X(x_j) = 0 \cdot 0.2 + 1 \cdot 0.3 + 3 \cdot 0.5 = 1.8$$

$$= M_X'(0) = 0.3e^0 + 1.5e^0 = 1.8;$$

$$\text{Var}(X) = \sum_{j=1}^{3} (x_j - E[X])^2 \cdot p_X(x_j) = (-1.8)^2 \cdot 0.2 + (-0.8)^2 \cdot 0.3 + (1.2)^2 \cdot 0.5 = 1.56$$

$$= E[X^2] - E[X]^2 = M_X''(0) - (M_X'(0))^2 = 0.3e^0 + 4.5e^0 - 1.8^2 = 1.56.$$

$\square$

A.4   Discrete and Continuous Probability Distributions

A.4.1   Bernoulli Trials

Consider the following random experiments.

- Flip a coin 10 times. Count the number of heads observed.

- A machine tool produces 1% defective parts. Find the number of defective parts in the next 25 parts produced.

- Of all packages transmitted through a digital transmission channel, 10% are received in error. Identify the number of packages in error in the next five packages transmitted.

- A multiple choice test contains 10 questions, each with four choices, and you guess at each question. Find the number of questions answered correctly.

Each of these random experiments can be thought of as consisting of a series of *repeated, independent*, random trials: 10 flips of the coin in experiment 1, the production of 25 parts in experiment 2, and so forth. Each trial results in one of the two exclusive outcomes.

A series of random experiments is called *Bernoulli trials* if:

(i) There are only two possible outcomes for each trial, which are labelled as *success* and *failure*, respectively.

(ii) The probability of success, denoted as p, is the same for each trial. Clearly, the failure probability equal to $1 - p$ is the same for all trials. That is,

$$\mathbb{P}(\text{"success"}) = p, \quad \mathbb{P}(\text{"failure"}) = 1 - p.$$

(iii) The outcomes from different trials are independent of one another.

Here are other examples of Bernoulli trials.

- Tossing a coin. *Success* is that a head (tail) is faced up. For a balanced coin, $p = 0.5$.

- Playing a roulette. *Success* is that red comes up. For a fair roulette, $p = \frac{18}{38}$.

- Sampling with replacement from a population that contains objects of two types (one is labelled as success). The probability p is equal to the proportion of "successful" objects.

- Roll a die, we have success if the result does not exceed 2. For a balanced die, we have $p = \frac{1}{3}$.

If at least one condition listed above is violated, we deal with a series of non-Bernoulli trials. Here are some examples of such series:

- tossing nonidentical coins;

- rolling nonidentical unbalanced dice;

- sampling without replacement from a finite population.

For such series of identically distributed and independent trials with a binary result, we are often interested in the number of successes in the next n trials, which is a discrete random variable. We can also construct other probability distributions from Bernoulli trials by counting, for example, the number of failures happened before the first success (or the kth success with $k = 1, 2, 3, \dots$).

A.4.2 Bernoulli Distribution

Consider a single Bernoulli trial with the probability of success $p \in (0, 1)$. Let $X = 1$ if the outcome is a success, and $X = 0$ when it is a failure. A random variable X is said to be a *Bernoulli random variable*. The mass probabilities are $\mathbb{P}(X = 1) = p$ and $\mathbb{P}(X = 0) = 1 - p$. The PMF is $p_X(x) = p^x(1-p)^{1-x}$ with $x \in \{0, 1\}$. A Bernoulli random variable X is an indicator function of the success.

Example A.43. Let X be a Bernoulli random variable with probability of success p. Find its MGF $M(t)$; compute the derivatives of $M(t)$ and evaluate them at $t = 0$ to find μ_X and σ_X^2.

Solution. The MGF is $M(t) = e^{t \cdot 0} \cdot (1-p) + e^{t \cdot 1} \cdot p = 1 - p + pe^t$. Differentiate $M(t)$ at zero:

$$M'(0) = pe^t \Big]_{t=0} = p \implies \mu_X = p,$$

$$M''(0) = pe^t \Big]_{t=0} = p \implies \sigma_X^2 = p - p^2 = p(1-p).$$

$\square$

A.4.3   Binomial Distribution

Consider a series of n Bernoulli trials. Represent a success in each trial by the letter s and a failure by the letter f. The sample space Ω consists of n-tuples involving s and f:

$$\Omega = \{r_1 r_2 \cdots r_n \mid r_i \in \{s, f\}\}.$$

A representative outcome is $\underbrace{ssffsf \ldots fs}_{n}$, where the letter in the ith position refers to the result of the ith trial. Clearly, since there are n trials with two possible results in each trial, the sample space Ω consists of 2^n different outcomes. Let us find the probability of a particular outcome. First, rearrange the trial results:

$$\underbrace{ssffsf \ldots fs}_{n} \to \underbrace{ss \ldots s}_{x} \underbrace{ff \ldots f}_{n-x}.$$

Suppose that in this series there are x *successes* and $n - x$ *failures*, where x can be any integer between 0 and n. Thus, the probability of this outcome is

$$\underbrace{pp \ldots p}_{x} \underbrace{qq \ldots q}_{n-x} = p^x q^{n-x} \quad \text{where} \quad q = 1 - p.$$

The probability only depends on the number of successes rather on their locations in the series.

Let X count the number of successes. Every possible outcome included in the event $\{X = x\}$ is a rearrangement of x successes and $n - x$ failures; it is assigned the same probability. The number of distinct arrangements of x successes in a series of n trials is

$$\binom{n}{x} = \frac{n!}{x!(n-x)!}.$$

Thus, for given n and p, the probability of the event $\{X = x\}$ is

$$\mathbb{P}(X = x) = \binom{n}{x} p^x (1-p)^{n-x}$$

for $x = 0, 1, 2, \ldots, n$, and $\mathbb{P}(X = x) = 0$ otherwise.

If X represents the number of successes that occur in n Bernoulli trials, then X is said to be a *binomial random variable* with parameters n (the number of trials) and p (the probability of success). We write $X \sim Bin(n, p)$. The probability mass function of the binomial distribution with parameters n and p is

$$b(x; n, p) = \binom{n}{x} p^x (1-p)^{n-x}, \quad x = 0, 1, \ldots, n.$$

This distribution is called the binomial distribution because the values of the PMF are terms of the binomial expansion

$$1 = (p + (1-p))^n = \sum_{x=0}^{n} \binom{n}{x} p^x (1-p)^{n-x}.$$

Example A.44. A coin is tossed six times. It is a sequence of Bernoulli trials where a head is success. The number of heads follows the binomial distribution with $n = 6$ and $p = 1/2$. Find the probability that

(a) exactly two heads occur;

(b) at least four heads occur;

(c) no heads occur.

Solution. The PMF of X is $p_X(x) = \binom{6}{x} \cdot \left(\frac{1}{2}\right)^x \cdot \left(1 - \frac{1}{2}\right)^{6-x} = \frac{1}{64} \cdot \binom{6}{x}$ with $x = 0, 1, \ldots, 6$.

(a) $\mathbb{P}(X = 2) = p_X(2) = \frac{1}{64} \cdot \binom{6}{2} = \frac{1}{64} \cdot \frac{6 \cdot 5}{2} = \frac{15}{64}$.

(b) $\mathbb{P}(X \geqslant 4) = p_X(4) + p_X(5) + p_X(6) = \frac{1}{64} \cdot \left[\binom{6}{4} + \binom{6}{5} + \binom{6}{6}\right] = \frac{11}{32}$.

(c) $\mathbb{P}(X = 0) = p_X(0) = \frac{1}{64} \cdot \binom{6}{0} = \frac{1}{64}$.

$\square$

In a series of n Bernoulli trials, let the Bernoulli random variable B_i represent the result of the ith trial with $i = 1, 2, \ldots, n$. The number X of successes in the series of n Bernoulli trials is given by the formula

$$X = B_1 + B_2 + \cdots + B_n.$$

The Bernoulli random variables B_i are said to be *mutually independent* since the events $\{B_i = 1\}$ are mutually independent. So, a binomial random variable $X \sim Bin(n, p)$ can be expressed as the sum of n independent Bernoulli random variables with the common probability of success p. The expected value of X is then given by

$$E[X] = E[B_1 + B_2 + \cdots + B_n] = E[B_1] + E[B_2] + \cdots + E[B_n] = nE[B_1] = np.$$

Alternatively, we can find the mean and other moments including the variance by differentiating the moment generating function (MGF). First, let us find the MGF of $Bin(n, p)$:

$$M(t) = \sum_{k=0}^{n} e^{tk} \binom{k}{n} p^k (1-p)^{n-k} = \sum_{k=0}^{n} \binom{k}{n} (pe^t)^k (1-p)^{n-k} = (1 - p + pe^t)^n.$$

Second, we differentiate $M(t)$ and find the first two moments. The mathematical expectation is

$$E[X] = M'(0) = npe^t(1 - p + pe^t)^{n-1}\Big|_{t=0} = np.$$

The variance is

$$\text{Var}(X) = E[X^2] - E[X]^2 = M''(0) - (np)^2 = n(n-1)p^2 + np - n^2p^2 = np(1-p).$$

A.4.4   Hypergeometric Distribution

Suppose that a day's production contains a few manufactured parts that do not conform to customer requirements. Several parts are selected at random, *without replacement*, from the batch. Let us count the number of nonconforming parts in the sample. This experiment is fundamentally different from the examples with Bernoulli trials since the trials are *not independent*. Note that if each selected unit were replaced before the next selection, the trials would be independent, and the probability of a nonconforming part on each trial would not be changing from trial to trial. Then, the number of nonconforming parts in the sample would be a binomial random variable.

Let a set of N objects contain

K objects classified as successes $(K \leqslant N)$,

$N - K$ objects classified as failures.

A sample of size n objects is selected randomly (*without replacement*) from the N objects, where $n \leqslant N$. Let the random variable X denote the number of successes in the sample. In this case, the random variable X is said have a *hypergeometric distribution*, denoted $HyperGeom(N, K, n)$, with the PMF

$$p(x; N, K, n) = \frac{\binom{K}{x}\binom{N-K}{n-x}}{\binom{N}{n}}, \quad \begin{array}{l} \text{where } x = 0, 1, \ldots, n, \\ 0 \leqslant n \leqslant N, \\ n - x \leqslant N - K, \\ \text{and } x \leqslant K. \end{array}$$

There are at most K successes, thus $p(x) = 0$ if $x > K$. There are at most $N - K$ failures, thus $p(x) = 0$ if $n - x > N - K$.

Note that if n is small relative to N, the hypergeometric distribution can be approximated by the binomial distribution $Bin(n, p)$ with $p = K/N$.

Example A.45. A province runs a lottery in which 6 numbers are randomly selected between 1 and 40, without replacement. A player chooses 6 numbers before the province's sample is selected.

(a) What is the probability that all six numbers chosen by a player match the six numbers in the province's sample?

(b) What is the probability that at least four of the six numbers chosen by a player appear in the province's sample?

Solution. The number of correct lottery numbers is $X \sim HyperGeom(40, 6, 6)$.

(a) $\mathbb{P}(X = 6) = \dfrac{\binom{6}{6} \cdot \binom{34}{0}}{\binom{40}{6}} = \dfrac{1}{3,838,380} \cong 0.0000261\%.$

(b) $\mathbb{P}(X \geqslant 4) = \dfrac{\binom{6}{4} \cdot \binom{34}{2} + \binom{6}{5} \cdot \binom{34}{1} + \binom{6}{6} \cdot \binom{34}{0}}{\binom{40}{6}} = \dfrac{8415 + 204 + 1}{3,838,380} \cong 0.2246\%.$

□

A hypergeometric random variable $X \sim HyperGeom(N, K, n)$ can be expressed as a sum of identically distributed, *dependent* Bernoulli random variables: $X = B_1 + B_2 + \cdots + B_K$, where $B_m = 1$ if the mth "success" was selected, and $B_m = 0$ otherwise. It is not difficult to show that $B_m = 1$ with probability $\frac{n}{N}$. We can find the expected value of X by using the linearity of the mathematical expectation:

$$\mathrm{E}[X] = \mathrm{E}[B_1] + \mathrm{E}[B_2] + \cdots + \mathrm{E}[B_K] = K\mathrm{E}[B_1] = \frac{nK}{N}.$$

A.4.5 Geometric Distribution

Consider an infinite series of Bernoulli trials. Now we ask ourselves the question "how many *failures* are there before the first success?" If p is the probability of success, then to achieve the first success at the $(x+1)$st trial, we must observe x failures in the first x trials. The probability of this event is

$$\mathbb{P}(\underbrace{f f \cdots f}_{x} s) = (1-p)^x p \quad \text{with } x = 0, 1, 2, \ldots.$$

The random variable X that counts the number of failures until the first success is said to have a *geometric distribution* with parameter $p \in (0,1)$, denoted $X \sim Geom(p)$. The probability mass function is

$$p(x; p) = (1-p)^x p \quad \text{for } x = 0, 1, 2, \ldots$$

Note that

$$\sum_{x=0}^{\infty} p(x; p) = p \sum_{x=0}^{\infty} (1-p)^x = \frac{p}{1-(1-p)} = 1,$$

since the probabilities form a geometric series.

We can also ask the question: "how many *trials* are required until the first success?" The random variable X that counts the number of *trials* until the first success has the probability mass function

$$p(x; p) = (1-p)^{x-1} p \quad \text{for } x = 1, 2, 3, \ldots$$

It is an alternative definition of the geometric probability distribution.

Example A.46. An urn contains N white and M black balls. Balls are selected at random, one at a time until a black ball is drawn. Assuming the sampling with replacement, find the probability that

(a) exactly n draws are needed;

(b) at least k draws are needed.

Solution. Define the selection of a black ball as success. The probability of success is $p = \frac{M}{N+M}$. Let X denote the number of white balls selected until the first black one. Then, $X \sim Geom(p)$.

(a) The probability that exactly $n \geq 1$ draws are needed equals

$$\mathbb{P}(X + 1 = n) = \mathbb{P}(X = n - 1) = (1-p)^{n-1} p.$$

(b) At least k draws are needed with probability

$$\mathbb{P}(X + 1 \geq k) = \mathbb{P}(X = k - 1) + \mathbb{P}(X = k) + \mathbb{P}(X = k + 1) + \cdots$$

$$= \sum_{x=k-1}^{\infty} (1-p)^x p = (1-p)^{k-1} \sum_{j=0}^{\infty} (1-p)^j p = (1-p)^{k-1}.$$

$\square$

The result of part (b) of the above exercise yields a formula for the *tail probability*. Let $X \sim Geom(p)$. The probability of having at least k failures before the first success is given by

$$\mathbb{P}(X \geq k) = (1-p)^k, \quad k = 0, 1, 2, \ldots \tag{A.18}$$

The formula (A.18) can be used to derive two properties. First, we can express the expected value in terms of tails probabilities and then compute the mean by evaluating a geometric series. Second, it allows for establishing the *memoryless property* of a geometric random variable.

Theorem A.4. *Consider a random variable X that takes nonnegative integer values, then its expected value is given by*

$$E[X] = \sum_{k=1}^{\infty} \mathbb{P}(X \geqslant k).$$

For a geometric random variable $X \sim Geom(p)$ with tail probabilities $\mathbb{P}(X \geqslant k) = (1-p)^k$, we have

$$E[X] = \sum_{k=1}^{\infty} (1-p)^k = \frac{1-p}{p}.$$

The event $\{X \geqslant k\}$ can be viewed as a long run of failures. The *memoryless property* of the geometric distribution states that a long run of failure does not increase or decrease the likelihood of a future success. Specifically, we have

$$\mathbb{P}(X \geqslant k + n \mid X \geqslant k) = \mathbb{P}(\text{next } n \text{ trials are failures} \mid \text{first } k \text{ trials are failures})$$
$$= \frac{\mathbb{P}(X \geqslant k + n)}{\mathbb{P}(X \geqslant k)} = (1-p)^n = \mathbb{P}(X \geqslant n).$$

Similarly, we can prove that $\mathbb{P}(X = k + n \mid X \geqslant k) = \mathbb{P}(X = n)$, as is shown in the example below.

Example A.47. A player buys one ticket each week. The probability of winning a prize is p. Compare the following two probabilities.

(a) Win a prize in the second week

(b) Given that the player did not win a prize during last year (52 weeks), find the (conditional) probability that it takes exactly two additional weeks for the player to win a prize.

Solution. Let X count the number of "failed" weeks until the first prize has been won. Then, $X \sim Geom(p)$.

(a) $\mathbb{P}(X = 1) = p \cdot (1 - p)$.

(b) $\mathbb{P}(X = 53 \mid X \geqslant 52) = \frac{\mathbb{P}(X=53; X \geqslant 52)}{\mathbb{P}(X \geqslant 52)} = \frac{\mathbb{P}(X=53)}{\mathbb{P}(X \geqslant 52)} = p \cdot (1 - p)$.

$\square$

A.4.6   Negative Binomial Distribution

Consider Bernoulli trials with success probability p. Let X equal the number of failures that precede the rth success. Let us calculate $\mathbb{P}(X = n)$ for $n = 0, 1, 2, \ldots$ We have a series of $n + r$ trials: $\omega = \underbrace{xx \cdots xs}_{n+r \text{ trials}}$, where the first $n + r - 1$ trials result in $r - 1$ success and n failures. There are $\binom{n+r-1}{r-1}$ ways to arrange $r - 1$ successes among $n + r - 1$ trials. Therefore, for the number X of failures that precede the rth success, we have

$$\mathbb{P}(X = n) = \binom{n + r - 1}{r - 1} p^r (1 - p)^n, \quad n = 0, 1, 2, \ldots$$

The random variable X is said to have the *negative binomial distribution* with parameters r and p, denoted $X \sim NBin(r, p)$. Notice that if $r = 1$, then

$$\mathbb{P}(X = n) = \binom{n}{0} p^1 (1-p)^n = p(1-p)^n.$$

That is, $NBin(1, p) = Geom(p)$.

We can also ask how many *trials* are required until the rth success. Consider a random variable X that counts the number of *trials* required to achieve r successes. In other words, X is the trial number when the rth success happens. The probability mass function of X is

$$p(n) = \binom{n-1}{r-1} p^r (1-p)^{n-r}, \quad n = r, r+1, r+2, \ldots$$

Example A.48. What is the probability that r successes occur before m failures in a series of Bernoulli trials?

Solution. Let $X \sim NBin(r, p)$. The probability in question is

$$\mathbb{P}(X < m) = \sum_{n=0}^{m-1} \binom{n+r-1}{r-1} \cdot p^r \cdot (1-p)^n.$$

$\square$

A negative binomial random variable $X \sim NBin(r, p)$ can be expressed as a sum of r independent geometric random variables with common parameter p. Therefore, the expected value of X is equal to

$$E[X] = \frac{(1-p)r}{p}.$$

Alternatively, to find the mean and variance of X, we can differentiate its MGF:

$$M_X(t) = \frac{p^r}{\left(1 - e^t(1-p)\right)^r} \implies \mu_X = M_X'(0) = \frac{(1-p)r}{p}, \quad \sigma_X^2 = M_X''(0) - \mu_X^2 = \frac{(1-p)r}{p^2}.$$

A.4.7 Poisson Distribution

Consider a series of n Bernoulli trials with the probability of success p. Let $X \sim Bin(n, p)$. Suppose that n is getting larger, and p is getting smaller, such that $np = \lambda$ remains constant. For $k \ll n$, we have

$$\begin{aligned}
\mathbb{P}(X = k) &= \frac{n!}{k!(n-k)!} p^k (1-p)^{n-k} \\
&= \frac{n \cdot (n-1) \cdots (n-k+1)}{n \cdot n \cdots n} \frac{1}{k!} (np)^k (1 - \frac{np}{n})^{n-k} \\
&\approx \frac{\lambda^k}{k!} e^{-\lambda}
\end{aligned}$$

since $\frac{n-i}{n} \approx 1$ for $i = 0, 1, \ldots, k-1$ and $(1 - \lambda/n)^n \approx e^{-\lambda}$ for large values of n. Therefore, $\mathbb{P}(X = k) \approx e^{-\lambda} \frac{\lambda^k}{k!}$, $k \ll n$. The approximation is excellent, provided that $np^2 \ll 1$.

A random variable X is said to have the *Poisson* distribution with parameter $\lambda > 0$, denoted $X \sim Pois(\lambda)$, if its PMF is of the form

$$p(x) = e^{-\lambda} \frac{\lambda^x}{x!}, \quad x = 0, 1, 2, \ldots$$

and $p(x) = 0$ otherwise. The Poisson distribution is a good probabilistic model when we count some rare events occurring during a fixed period (e.g., earthquakes, defaults, phone calls, etc.)

Let us find the MGF $M_X(t)$ of $X \sim Pois(\lambda)$ and then use it to find μ_X and σ_X^2:

$$M_X(t) = \sum_{k=0}^{\infty} e^{tk} \frac{\lambda^k}{k!} e^{-\lambda} = e^{-\lambda} \sum_{k=0}^{\infty} \frac{(\lambda e^t)^k}{k!} = e^{-\lambda} e^{\lambda e^t} = e^{\lambda(e^t - 1)}.$$

Differentiate $M_X(t)$ at $t = 0$ to find the first two moments of X:

$$\mathrm{E}[X] = M_X'(0) = e^{\lambda(e^t - 1)} \lambda e^t \Big]_{t=0} = \lambda,$$

$$\mathrm{E}[X^2] = M_X''(0) = e^{\lambda(e^t - 1)} (\lambda^2 + \lambda) e^t \Big]_{t=0} = \lambda^2 + \lambda.$$

Thus, $\mu_X = \lambda$ and $\sigma_X^2 = \lambda$.

A.4.8 Continuous Uniform Distribution

Suppose that the range of a random variable X is a finite interval $[a, b] \subset \mathbb{R}$. Additionally, X lies equally likely anywhere in $[a, b]$. That is, $\mathbb{P}(c \leqslant X \leqslant d) = \frac{d-c}{b-a}$ for any c and d such that $a \leqslant c \leqslant d \leqslant b$. The CDF of X is

$$F_X(x) = \mathbb{P}(X \leqslant x) = \mathbb{P}(a \leqslant X \leqslant x) = \frac{x - a}{b - a} \quad \text{for} \quad a \leqslant x \leqslant b.$$

The PDF is then

$$f_X(x) = \frac{dF_X(x)}{dx} = \left(\frac{1}{b - a} \right) \mathbb{I}_{[a,b]}(x).$$

The random variable X is said to have a continuous uniform distribution on the interval $[a, b]$, denoted $X \sim Unif(a, b)$.

The distribution $Unif(a, b)$ can be expressed in terms of $Unif(0, 1)$ and vice versa. Indeed, let $U \sim Unif(0, 1)$. Define $X \equiv a + (b - a)U$. Then $X \sim Unif(a, b)$ On the other hand, for $X \sim Unif(a, b)$, we can define $U \equiv \frac{X-a}{b-a}$. Then, $U \sim Unif(0, 1)$.

The continuous uniform distribution can be used to model random variables having non-uniform distributions. Consider a continuous random variable X with CDF F. Assume that the CDF F is a strictly increasing function, and hence the inverse function F^{-1} exists. Then, we can prove that $\mathbb{P}(F(X) \leqslant x) = x$ for all $x \in (0, 1)$. That is, $F(X) \sim Unif(0, 1)$. Let $U \sim Unif(0, 1)$. The random variable $F^{-1}(U)$ has the same probability distribution as that of X.

A.4.9 Exponential Distribution

Fix arbitrarily $n \in \mathbb{N}$. Suppose that Bernoulli trials (with the probability of success p) are performed at times $\frac{1}{n}, \frac{2}{n}, \ldots$. Assume that $np = \lambda$, as $n \to \infty$. On average, we observe λT successes in any interval $[0, T]$. Let X be the time of the first success. We have that $nX \sim Geom(p)$. For a real number $x > 0$, compute the tail probability $\mathbb{P}(X \geqslant x)$:

$$\mathbb{P}(X \geqslant x) = \mathbb{P}(nX \geqslant nx) = \mathbb{P}(nX \geqslant \lfloor nx \rfloor) = (1 - p)^{\lfloor nx \rfloor}.$$

In the limiting case, as $n \to \infty$, we have

$$\mathbb{P}(X \geqslant x) = \left(1 - \frac{\lambda}{n} \right)^{\frac{n}{\lambda} \frac{\lambda \lfloor nx \rfloor}{n}} \to e^{-\lambda x}$$

since $\frac{\lfloor nx \rfloor}{n} \to x$.

If a continuous random variable X, which takes positive real values, has the tail probability $\mathbb{P}(X \geqslant x) = e^{-\lambda x}$ with $\lambda > 0$, then its CDF is

$$F(x) = \mathbb{P}(X \leqslant x) = 1 - \mathbb{P}(X > x) = 1 - e^{-\lambda x}, \quad x \geqslant 0.$$

The PDF is then

$$f(x) = F'(x) = \lambda e^{-\lambda x} \text{ for } x \geqslant 0, \quad \text{and} \quad f(x) = 0 \text{ for } x < 0.$$

Such a random variable is called an *exponential* random variable with parameter $\lambda > 0$. It is denoted as $X \sim Exp(\lambda)$.

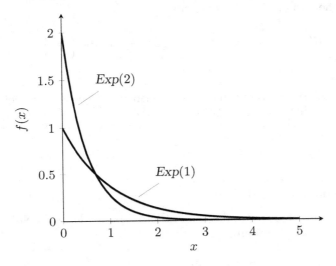

FIGURE A.4: Exponential PDFs.

The exponential distribution is used in engineering, natural and social sciences, including finance and economics. It is often used as the distribution for a time between successive occurrences of some rare event, or for a time required to complete a certain task. The following phenomena can be modelled using the exponential distribution:

- the lifetime of a person or a product;

- the time between two successive hits on a website;

- the time until a corporation defaults on its obligations;

- the service time at a hospital emergency room.

The exponential distribution can be viewed as a continuous analogue of the geometric distribution. In particular, like a geometric random variable, the exponential variate has the memoryless property. Let $X \sim Exp(\lambda)$. Then, for any positive real numbers t and s,

$$\mathbb{P}(X \geqslant t + s \mid X \geqslant t) = \mathbb{P}(X \geqslant s).$$

Example A.49. Compute the mean, the variance, and the MGF of $X \sim Exp(\lambda)$.

Solution. First, find the MGF of X:

$$M_X(t) = \int_0^\infty e^{tx} \lambda e^{-\lambda x}\, dx = \int_0^\infty \lambda e^{-(\lambda - t)x}\, dx = \frac{\lambda}{\lambda - t},$$

where $t < \lambda$. Differentiate the MGF:

$$M_X'(t) = \frac{\lambda}{(\lambda - t)^2}, \quad M_X''(t) = \frac{2\lambda}{(\lambda - t)^3}.$$

The mean and the variance of X are, respectively,

$$\mu_X = M_X'(0) = \frac{1}{\lambda} \quad \text{and} \quad \sigma_X^2 = M_X''(0) - (M_X'(0))^2 = \frac{2}{\lambda^2} - \frac{1}{\lambda^2} = \frac{1}{\lambda^2}.$$

$\square$

The exponential distribution can also be described by another parameter θ, which relates to λ as $\theta = 1/\lambda$. In this case, the PDF of X is written as

$$f(x; \theta) = \begin{cases} \frac{1}{\theta} e^{-x/\theta} & x \geq 0, \\ 0 & \text{otherwise.} \end{cases}$$

A.4.10   Normal Distribution

Consider a binomial random variable $X \sim Bin(n, p)$ with large n. Note that the PMF is bell-shaped and admits the approximation

$$p(x) = \binom{n}{x} p^x (1 - p)^{n-x} \qquad (x = 0, 1, \ldots, n)$$

$$\approx \frac{1}{\sqrt{2\pi}\sqrt{npq}} \exp\left(-\frac{(x - np)^2}{2npq}\right) \qquad (q = 1 - p)$$

Denote $\sigma^2 = npq$ and $\mu = np$ and then rewrite this approximation:

$$f(x) = \frac{1}{\sqrt{2\pi}\sigma} \exp\left(-\frac{(x - \mu)^2}{2\sigma^2}\right). \tag{A.19}$$

The function f is a probability density function on $(-\infty, \infty)$.

We say that a continuous random variable X that has the PDF in (A.19) is a *normal random variable* with parameters μ and σ^2, denoted $X \sim Norm(\mu, \sigma^2)$. We can show that

$$E[X] = \mu \quad \text{and} \quad Var(X) = \sigma^2.$$

Set $Z = \dfrac{X - \mu}{\sigma}$; then $Z \sim Norm(0, 1)$, and it is called a *standard normal* variable. The PDF of Z is

$$n(x) = \frac{1}{\sqrt{2\pi}} e^{-x^2/2}, \quad -\infty < x < \infty.$$

Denote $\mathcal{N}(z)$ the standard normal CDF given by

$$\mathcal{N}(z) = \int_{-\infty}^z n(x)\, dx = \int_{-\infty}^z \frac{1}{\sqrt{2\pi}} e^{-x^2/2}\, dx.$$

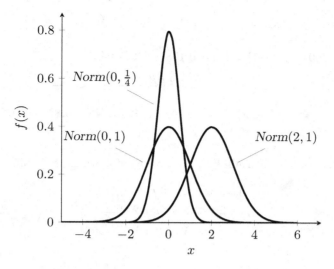

FIGURE A.5: Normal PDFs.

We can express the CDF F and PDF f of a nonstandard normal variable $X \sim Norm(\mu, \sigma^2)$ in terms of the standard normal CDF $\mathcal{N}$ and PDF n:

$$f(x) = \frac{1}{\sigma} \frac{1}{\sqrt{2\pi}} \exp\left(-\frac{1}{2}\left(\frac{x-\mu}{\sigma}\right)^2\right) = \frac{1}{\sigma} n\left(\frac{x-\mu}{\sigma}\right),$$

$$F(x) = \mathbb{P}(X \leqslant x) = \mathbb{P}\left(\frac{X-\mu}{\sigma} \leqslant \frac{x-\mu}{\sigma}\right) = \mathbb{P}\left(Z \leqslant \frac{x-\mu}{\sigma}\right) = \mathcal{N}\left(\frac{x-\mu}{\sigma}\right).$$

When we transform a random variable by first subtracting its expected value and then dividing the difference obtained by the standard deviation, it is called a *standardization*. Suppose X has mean μ and standard deviation $\sigma > 0$. Then, the standardization of X is the random variable $Z = \dfrac{X - \mu}{\sigma}$.

Example A.50. An exam is regarded as being good if the test scores can be approximated by a normal distribution. Suppose that the instructor uses the test scores to estimate μ and σ^2, and then assign the letter grade A to those whose test score is greater than $\mu + \sigma$, B if the score is between μ and $\mu + \sigma$, C if it is between $\mu - \sigma$ and μ, D if it is between $\mu - 2\sigma$ and $\mu - \sigma$. What percentage of participants received grade A? grade B? grade C? grade D?

Solution. Let $X \sim Norm(\mu, \sigma^2)$ be the test score of a participant selected at random. We have the following probabilities to receive grades A, B, C, and D:

$$\mathbb{P}(A) = \mathbb{P}(X > \mu + \sigma) = \mathbb{P}\left(\frac{X-\mu}{\sigma} > \frac{\mu+\sigma-\mu}{\sigma}\right)$$

$$= 1 - \mathcal{N}(1) \cong 1 - 0.8413 = 15.87\%,$$

$$\mathbb{P}(B) = \mathbb{P}(\mu < X < \mu + \sigma) = \mathbb{P}\left(\frac{\mu-\mu}{\sigma} < \frac{X-\mu}{\sigma} < \frac{\mu+\sigma-\mu}{\sigma}\right)$$

$$= \mathcal{N}(1) - \mathcal{N}(0) \cong 0.8413 - 0.5 = 34.13\%,$$

$$\mathbb{P}(C) = \mathbb{P}(\mu - \sigma < X < \mu) = \mathbb{P}\left(\frac{\mu - \sigma - \mu}{\sigma} < \frac{X - \mu}{\sigma} < \frac{\mu - \mu}{\sigma}\right)$$
$$= \mathcal{N}(0) - \mathcal{N}(-1) \cong 0.5 - 0.1587 = 34.13\%,$$
$$\mathbb{P}(D) = \mathbb{P}(\mu - 2\sigma < X < \mu - \sigma) = \mathbb{P}\left(\frac{\mu - 2\sigma - \mu}{\sigma} < \frac{X - \mu}{\sigma} < \frac{\mu - \sigma - \mu}{\sigma}\right)$$
$$= \mathcal{N}(-1) - \mathcal{N}(-2) = \mathcal{N}(2) - \mathcal{N}(1) \cong 0.9772 - 0.8413 = 13.59\%.$$

□

For any normal random variable X, we have the following probabilities:

$$\mathbb{P}(\mu - \sigma < X < \mu + \sigma) \cong 68.3\%,$$
$$\mathbb{P}(\mu - 2\sigma < X < \mu + 2\sigma) \cong 95.5\%,$$
$$\mathbb{P}(\mu - 3\sigma < X < \mu + 3\sigma) \cong 99.7\%.$$

For many applications in statistics, we will need the values on the measurement axis that capture small tail areas under the standard normal curve. Let $Z \sim Norm(0,1)$. The value z_α denotes the value for which $\mathbb{P}(Z > z_\alpha) = \alpha$ or, equivalently, $\mathcal{N}(z_\alpha) = 1 - \alpha$. That is, z_α is the $100(1-\alpha)$th percentile of the standard normal distribution. The values z_α are usually referred to as *z critical values*.

TABLE A.1: Standard normal percentiles and critical values.

Percentile	90	95	97.5	99	99.5
α (tail area)	0.1	0.05	0.025	0.01	0.005
z_α	1.28	1.645	1.96	2.33	2.58

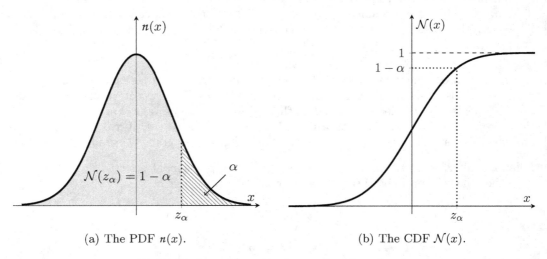

(a) The PDF $n(x)$.   (b) The CDF $\mathcal{N}(x)$.

FIGURE A.6: Critical value z_α for $Norm(0,1)$.

Example A.51. Find (a) $z_{0.01}$ and (b) $z_{0.05}$. Use them to find 95th and 99th percentiles for $Norm(\mu, \sigma^2)$.

Solution. Let $X \sim Norm(\mu, \sigma^2)$, and then $Z = \frac{X-\mu}{\sigma} \sim Norm(0,1)$.

(a) $\mathbb{P}(Z \geqslant z_{0.01}) = 0.01 \iff \mathcal{N}(z_{0.01}) = 99\%$. Thus, $z_{0.01} \cong 2.33$.

(b) $\mathbb{P}(Z \geqslant z_{0.05}) = 0.05 \iff \mathcal{N}(z_{0.01}) = 95\%$. Thus, $z_{0.05} \cong 1.645$.

To find the $(100p)$th percentile, x_p, for $Norm(\mu, \sigma^2)$, solve $\mathbb{P}(X \leqslant x) = p$ for x:

$$\mathbb{P}(X \leqslant x) = p \iff \mathbb{P}\left(Z \leqslant \frac{x - \mu}{\sigma}\right) = p \iff \frac{x - \mu}{\sigma} = z_{1-p} \iff x = \mu + \sigma\, z_{1-p}.$$

Therefore, $x_{95} = \mu + \sigma\, z_{0.05} = \mu + 1.645\,\sigma$ and $x_{99} = \mu + \sigma\, z_{0.01} = \mu + 2.33\,\sigma$. $\square$

Example A.52.

(a) Show that the MGF of a standard normal random variable is $M(t) = e^{t^2/2}$.

(b) Using the MGF of Z, compute the MGF of $X = \mu + \sigma Z \sim Norm(\mu, \sigma^2)$.

(c) Compute the mean and the variance of $S = e^X$ with $X \sim Norm(0.07, 0.25)$.

Solution.

(a) The MGF of a standard normal random distribution is

$$\mathrm{E}[e^{tX}] = \int_{-\infty}^{\infty} e^{tx}\, n(x)\, dx = \int_{-\infty}^{\infty} \frac{1}{\sqrt{2\pi}} e^{tx}\, e^{-x^2/2}\, dx$$
$$= \int_{-\infty}^{\infty} e^{t^2/2} \frac{1}{\sqrt{2\pi}} e^{-(x-t)^2/2}\, dx = e^{t^2/2}.$$

where t is any real number.

(b) Using the property that $M_{\mu+\sigma Z}(t) = e^{\mu t} M_Z(\sigma t)$, we have

$$M_X(t) = e^{\mu t} e^{(\sigma t)^2/2} = e^{\mu t + \sigma^2 t^2/2}.$$

(c) Note that $\mathrm{E}[S^a] = \mathrm{E}[e^{aX}] = M_X(a)$ for any $a \in \mathbb{R}$. Therefore,

$$\mathrm{E}[S] = \mathrm{E}[e^X] = M_X(1) = e^{0.07 + 0.25^2/2} \cong 1.10655,$$
$$\mathrm{Var}(S) = \mathrm{E}[e^{2X}] - \mathrm{E}[e^X]^2 = M_X(2) - M_X(1)^2$$
$$= e^{0.07\cdot 2 + (0.25\cdot 2)^2/2} - e^{0.2025} \cong 0.67202.$$

$\square$

Example A.53. The continuously-compounded return on a stock is normally distributed with an average of 7% and a standard deviation (volatility) of 25% (both are annual quantities). Compute the probability that

(a) the return is between 4% and 10%;

(b) the return is negative;

(c) the return is more than 20%.

How do the answers change if volatility is 50%?

Solution. The return R has the distribution $Norm(0.07, 0.25^2)$. Thus, we can write $R = 0.07 + 0.25Z$ where $Z \sim Norm(0, 1)$.

(a) $\mathbb{P}(4\% < R < 10\%) = \mathbb{P}(|Z| < 0.12) = \mathcal{N}(0.12) - \mathcal{N}(-0.12) = 2\mathcal{N}(0.12) - 1 \cong 9.55\%$.

(b) $\mathbb{P}(R < 0) = \mathbb{P}(Z < -0.28) = \mathcal{N}(-0.28) = 1 - \mathcal{N}(0.28) \cong 38.97\%$.

(c) $\mathbb{P}(R > 20\%) = \mathbb{P}(Z > 0.52) = 1 - \mathcal{N}(0.52) \cong 30.15\%$.

In general, for $R \sim Norm(\mu, \sigma^2)$, an interval probability of the form $\mathbb{P}(a < R < b)$ with fixed a and b decreases as σ increases. One-sided probabilities $\mathbb{P}(R < a)$ and $\mathbb{P}(R > a)$ increase with the growth of σ. So, if volatility changes from 25% to 50%, the probability in (a) decreases, whereas the probabilities in (b) and (c) increase. □

A.4.11 Gamma Distribution

The *gamma function*, denoted $\Gamma(x)$, is defined by

$$\Gamma(t) = \int_0^\infty x^{t-1} e^{-x} \, dx, \quad t > 0.$$

It has the following properties:

1. $\Gamma(t + 1) = t\Gamma(t)$;

2. $\Gamma(1) = 1$;

3. $\Gamma(n + 1) = n!$.

The gamma function is a *generalization of factorials* to all positive real numbers.

Fix arbitrarily $n \in \mathbb{N}$. Suppose that Bernoulli trials (with the probability of success p) are performed at times $\frac{1}{n}, \frac{2}{n}, \ldots$. Assume that $np = \lambda$, as $n \to \infty$. Let X be the time of the rth success. We have that $nX \equiv Y \sim NBin(r, p = \frac{\lambda}{n})$. For $x > 0$, compute $\mathbb{P}(X > x) = \mathbb{P}(nX > nx) = \mathbb{P}(nX > \lfloor nx \rfloor)$. Observe that for $Y \sim NBin(r, p)$ and $Z \sim Bin(k, p)$ we have

$$\mathbb{P}(Y > k) = \mathbb{P}(Z \leqslant r - 1) = \sum_{j=0}^{r-1} \binom{k}{j} p^j (1 - p)^{k-j}$$

$$\approx \sum_{j=0}^{r-1} e^{-kp} \frac{(kp)^j}{j!}$$

Let us use this approximation for $k = \lfloor nx \rfloor$. In the limiting case, as $n \to \infty$, we have

$$\mathbb{P}(X > x) \to e^{-\lambda x} \sum_{j=0}^{r-1} \frac{(\lambda x)^j}{j!}, \quad \text{since} \quad \frac{\lfloor nx \rfloor}{n} \to x \text{ and } np = \lambda.$$

Consider a continuous random variable X that takes positive real values and has the CDF

$$F(x) = \mathbb{P}(X \leqslant x) = 1 - \mathbb{P}(X > x) = 1 - e^{-\lambda x} \sum_{j=0}^{r-1} \frac{(\lambda x)^j}{j!}, \quad x > 0.$$

We can find its PDF by differentiating the distribution function:

$$f(x) = F'(x) = \frac{\lambda^r}{\Gamma(r)} x^{r-1} e^{-\lambda x}, \quad x > 0.$$

A random variable is said the have the *gamma probability distribution* with parameters $\lambda > 0$ and $\alpha > 0$, denoted $Gamma(\alpha, \lambda)$, if its PDF is

$$f(x) = \frac{\lambda^\alpha}{\Gamma(\alpha)} x^{\alpha-1} e^{-\lambda x}, \quad x > 0. \tag{A.20}$$

The parameter α is called the *shape parameter*, and λ is called the *rate parameter*. The exponential distribution is a special case of the gamma distribution when $\alpha = 1$: $Gamma(\alpha = 1, \lambda) = Exp(\lambda)$. For $\alpha = \nu/2$, where $\nu = 1, 2, 3, \ldots$, this distribution is also known as the *chi-square distribution* with ν degrees of freedom.

A.4.12 Transformation of Continuous Random Variables

Let X be a continuous random variable with PDF f_X. Introduce a function g defined on the range of X. How can we find the PDF f_Y of $Y = g(x)$ and express it in terms of f_X? A solution to this problem consists of the following three steps.

1. Identify the range of Y.

2. Determine F_Y in terms of F_X.

3. Differentiate F_Y to find f_Y and use that $F' = f$.

First, let us consider the case with a linear transformation. That is, find the distribution of $Y = a + bX$, where a and $b \neq 0$ are real constants. Let the range of X be an interval (c, d). The range of Y is the interval $(a + bc, a + bd)$ if $b > 0$ or $(a + bd, a + bc)$ if $b < 0$. Consider the case with $b > 0$ and find the CDF of Y:

$$F_Y(y) = \mathbb{P}(Y \leqslant y) = \mathbb{P}(a + bX \leqslant y) = \mathbb{P}\left(X \leqslant \frac{y-a}{b}\right) = F_X\left(\frac{y-a}{b}\right).$$

Therefore, the PDF of Y is

$$f_Y(y) = F_Y'(y) = \frac{1}{b} F_X'\left(\frac{y-a}{b}\right) = \frac{1}{b} f_X\left(\frac{y-a}{b}\right).$$

Suppose X has a normal distribution $Norm(\mu, \sigma^2)$. We can identify the distribution of $Y = a + bX$, with $b > 0$:

$$f_Y(y) = \frac{1}{b} f_{Norm(\mu,\sigma^2)}\left(\frac{y-a}{b}\right) = \frac{1}{b\sigma} n\left(\frac{y-a-b\mu}{b\sigma}\right) = f_{Norm(a+b\mu,(b\sigma)^2)}(x).$$

So, Y is again a normal random variable with mean $a + b\mu$ and variance $(b\sigma)^2$.

Secondly, consider a more general case with a monotonic, differentiable function g. If g is *monotonically increasing*, then

$$F_Y(y) = \mathbb{P}(g(X) \leqslant y) = \mathbb{P}(X \leqslant g^{-1}(y)) = F_X(g^{-1}(y))$$

$$\implies f_Y(y) = F_X'(g^{-1}(y)) \cdot \frac{d}{dy} g^{-1}(y) = \frac{1}{g'(x)} f_X(x), \text{ where } x = g^{-1}(y).$$

If g is a *monotonically decreasing* function, then

$$F_Y(y) = \mathbb{P}(g(X) \leqslant y) = \mathbb{P}(X \geqslant g^{-1}(y)) = 1 - F_X(g^{-1}(y))$$

$$\implies f_Y(y) = -F_X'(g^{-1}(y)) \cdot \frac{d}{dy} g^{-1}(y) = -\frac{1}{g'(x)} f_X(x) = \frac{1}{|g'(x)|} f_X(x), \text{ where } x = g^{-1}(y).$$

Let Y have a normal probability distribution $Norm(\mu, \sigma^2)$. Then, the distribution of $X = \exp(Y)$ is called the *lognormal distribution*. Since Y takes its values anywhere in $\mathbb{R}$, the range of X is the interval $(0, \infty)$. The cumulative distribution function of X is

$$F(x) = \mathbb{P}(X \leqslant x) = \mathbb{P}(Y \leqslant \ln(x)) = \mathbb{P}\left(Z \leqslant \frac{\ln(x) - \mu}{\sigma}\right) = \mathcal{N}\left(\frac{\ln(x) - \mu}{\sigma}\right)$$

where $x > 0$, and $F(x) = 0$ for $x \leqslant 0$. Since $Y = \ln(X)$, and $\ln'(x) = \frac{1}{x}$, the probability density function for X is given by

$$f(x) = \frac{1}{x}\frac{1}{\sigma}n\left(\frac{\ln(x) - \mu}{\sigma}\right) = \frac{1}{x\sigma\sqrt{2\pi}}\exp\left(-\frac{(\ln x - \mu)^2}{2\sigma^2}\right), \quad 0 < x < \infty.$$

Using the normal MGF (see Example A.52), we can find the mean and the variance of a lognormal distribution:

$$\mathrm{E}[X] = \mathrm{E}[e^Y] = e^{\mu + \sigma^2/2},$$

$$\mathrm{Var}(X) = \mathrm{E}[e^{2Y}] - \mathrm{E}[e^Y]^2 = e^{2\mu + \sigma^2}\left(e^{\sigma^2} - 1\right).$$

As you can see in Chapters 2 and 4, the lognormal distribution plays an important role in mathematical finance.

Example A.54. Suppose that $X \sim Norm(\mu, \sigma^2)$ is the continuously-compounded (random) return on a stock over one year. Find the CDF and the PDF of the one-year future stock price $S = S_0 e^X$, where $S_0 > 0$ is today's stock price, which is known. This is a standard model for stock returns and share prices.

Solution. Write the stock price as an exponential of a normal random variable:

$$S = e^{\ln S_0 + X} = e^{\ln S_0 + \mu + \sigma Z},$$

where $Z \sim Norm(0, 1)$. Thus, S is a lognormal random variable, and its CDF is

$$F_S(s) = \mathbb{P}(e^{\ln S_0 + \mu + \sigma Z} \leqslant s) = \mathcal{N}\left(\frac{\ln(s/S_0) - \mu}{\sigma}\right), \quad s > 0.$$

Differentiate F_S to obtain the PDF:

$$f_S(s) = \frac{\mathrm{d}}{\mathrm{d}s}F_S(s) = \frac{1}{s\sigma}n\left(\frac{\ln(s/S_0) - \mu}{\sigma}\right), \quad s > 0.$$

$\square$

A.5 A Joint Probability Distribution

The world is not univariate. Many quantities are related to or depend on other quantities. If X and Y are discrete random variables, their *joint probability distribution* is a description of the set of points (x, y) in the joint range of (X, Y) along with the probability of each point. The joint probability distribution of *two random variables* is referred to as a *bivariate probability distribution*.

The *joint PMF* of the discrete random variables X and Y, denoted $p_{X,Y}(x, y)$ or $p(x, y)$, satisfies the following two properties:

(1) $p(x, y) = \mathbb{P}(X = x; Y = y) \geqslant 0$,

(2) $\sum_x \sum_y p(x, y) = 1$.

For any set A consisting of pairs of (x, y) values, the probability that (X, Y) lies in A is obtained by summing the values of the joint PMF over all pairs in A:

$$\mathbb{P}[(X, Y) \in A] = \sum_{(x,y)\in A} \sum p(x, y).$$

Example A.55. Roll two dice. Let V_1 and V_2 denote the face values which are turned up. Find the joint probability distribution of the maximum $X = \min\{V_1, V_2\}$ and the minimum $Y = \max\{V_1, V_2\}$ of the face values.

Solution. Since $\mathbb{P}(V_1 = i, V_2 = j) = \mathbb{P}(V_1 = i)\mathbb{P}(V_2 = j) = \frac{1}{36}$ for $1 \leqslant i, j \leqslant 6$, the joint PDF of V_1 and V_2 is

$$p(u, v) = \begin{cases} \frac{1}{36}, & u, v \in \{1, 2, 3, 4, 5, 6\}, \\ 0, & \text{otherwise.} \end{cases}$$

Clearly, $\mathbb{P}(X \leqslant Y) = 1$; hence, the joint PMF $p_{X,Y}(x, y)$ is zero if $x > y$. For $x, y \in \{1, 2, \ldots, 6\}$ with $x \leqslant y$, we have the following two cases:

$$p_{X,Y}(x, x) = \mathbb{P}(V_1 = V_2 = x) = \frac{1}{36},$$

$$p_{X,Y}(x, y) = \mathbb{P}(V_1 = x; V_2 = y) + \mathbb{P}(V_1 = y; V_2 = x) = \frac{2}{36} \text{ if } x \neq y.$$

So, we have

$$p_{X,Y}(x, y) = \begin{cases} 1/36 & 1 \leqslant x = y \leqslant 6, \\ 2/36 & 1 \leqslant x < y \leqslant 6, \\ 0 & \text{otherwise.} \end{cases}$$

$\square$

It is important to distinguish between the *joint probability distribution* of X and Y and the *probability distribution of each variable* individually. The individual probability distribution of a random variable is referred to as its *marginal probability distribution*.

If X and Y are discrete random variables with joint PMF $p(x, y)$, then the marginal probability mass functions of X and Y are, respectively,

$$p_X(x) = \mathbb{P}(X = x) = \sum_{y \in R_x} p(x, y) \text{ and } p_Y(y) = \mathbb{P}(Y = y) = \sum_{x \in R_y} p(x, y),$$

where R_x denotes the set of all values of Y in the range of (X, Y) for which $X = x$, and R_y denotes the set of all values of X in the range of (X, Y) for which $Y = y$.

The mean and the variance of a single random variable X can be calculated using the joint PMF p or the marginal PMF p_X:

$$E[X] = \mu_X = \sum_x x\, p_X(x) = \sum_x x \left(\sum_{y \in R_x} p(x,y) \right)$$

$$= \sum_x \sum_{y \in R_x} x\, p(x,y) = \sum_{x,y} x\, p(x,y),$$

$$\mathrm{Var}(X) = \sigma_X^2 = \sum_x (x - \mu_X)^2 p_X(x) = \sum_x (x - \mu_X)^2 \left(\sum_{y \in R_x} p(x,y) \right)$$

$$= \sum_x \sum_{y \in R_x} (x - \mu_X)^2 p(x,y) = \sum_{x,y} (x - \mu_X)^2 p(x,y).$$

Example A.56. Let V_1 and V_2 denote the face values which are turned up on two dice. Find the marginal PMF of $X = \min\{V_1, V_2\}$ and $Y = \max\{V_1, V_2\}$. Use the marginal PMFs to find μ_X and μ_Y.

Solution. The joint PMF $p_{X,Y}(x,y)$ can be represented in a table form, where columns and rows contain mass probabilities with fixed $Y = y$ and fixed $X = x$, respectively. The marginal PMFs $p_X(x)$ and $p_Y(y)$ can be found by calculating row and column sums, respectively.

$x \backslash y$	1	2	3	4	5	6	$p_X(x)$
1	$\frac{1}{36}$	$\frac{2}{36}$	$\frac{2}{36}$	$\frac{2}{36}$	$\frac{2}{36}$	$\frac{2}{36}$	$\frac{11}{36}$
2	0	$\frac{1}{36}$	$\frac{2}{36}$	$\frac{2}{36}$	$\frac{2}{36}$	$\frac{2}{36}$	$\frac{9}{36}$
3	0	0	$\frac{1}{36}$	$\frac{2}{36}$	$\frac{2}{36}$	$\frac{2}{36}$	$\frac{7}{36}$
4	0	0	0	$\frac{1}{36}$	$\frac{2}{36}$	$\frac{2}{36}$	$\frac{5}{36}$
5	0	0	0	0	$\frac{1}{36}$	$\frac{2}{36}$	$\frac{3}{36}$
6	0	0	0	0	0	$\frac{1}{36}$	$\frac{1}{36}$
$p_Y(y)$	$\frac{1}{36}$	$\frac{3}{36}$	$\frac{5}{36}$	$\frac{7}{36}$	$\frac{9}{36}$	$\frac{11}{36}$	

Compute the expected values of X and Y:

$$E[X] = 1 \cdot \frac{11}{36} + 2 \cdot \frac{9}{36} + 3 \cdot \frac{7}{36} + 4 \cdot \frac{5}{36} + 5 \cdot \frac{3}{36} + 6 \cdot \frac{1}{36} = \frac{91}{36} \cong 2.528,$$

$$E[Y] = 1 \cdot \frac{1}{36} + 2 \cdot \frac{3}{36} + 3 \cdot \frac{5}{36} + 4 \cdot \frac{7}{36} + 5 \cdot \frac{9}{36} + 6 \cdot \frac{11}{36} = \frac{161}{36} \cong 4.472.$$

Let us verify the above values using the fact that $X + Y = V_1 + V_2$:

$$E[X] + E[Y] = E[X + Y] = E[V_1 + V_2] = E[V_1] + E[V_2] = 2E[V_1] = 7 = \frac{161 + 91}{36}.$$

$\square$

Given a joint PMF $p(x,y)$ of discrete random variables X and Y, the *conditional PMF* of Y given $X = x$ is defined as

$$p_{Y|X}(y \mid x) = \mathbb{P}(Y = y \mid X = x) = \frac{\mathbb{P}(Y = y; X = x)}{\mathbb{P}(X = x)} = \frac{p(x,y)}{p_X(x)} \quad \text{for} \quad p_X(x) > 0.$$

Similarly, the *conditional PMF* of X given $Y = y$ is defined as

$$p_{X|Y}(x \mid y) = \frac{p(x, y)}{p_Y(y)} \quad \text{for} \quad p_Y(y) > 0.$$

The conditional probability mass function $p_{Y|X}$ provides the conditional probabilities for the values y of Y in the set R_x. A conditional PDF has the following properties:

(1) $p_{Y|X}(y \mid x) \geqslant 0$,

(2) $\sum_{y \in R_x} p_{Y|X}(y \mid x) = 1$

for all x and y. That is, a conditional PDF satisfy the definition of a regular, unconditional PDF.

Example A.57. Let V_1 and V_2 denote the face values which are turned up on two dice. Find the conditional PMF of $X = \min\{V_1, V_2\}$ given $Y = \max\{V_1, V_2\} = 3$.

Solution. Given that $Y = 3$, X can only take values in $\{1, 2, 3\}$. The conditional PMF is:

$$p_{X|Y}(x|3) = \frac{p_{X,Y}(x, 3)}{p_Y(3)} = \frac{p_{X,Y}(x, 3)}{5/36} = \begin{cases} 2/5 & \text{if } x = 1 \text{ or } x = 2, \\ 1/5 & \text{if } x = 3, \\ 0 & \text{otherwise.} \end{cases}$$

$\square$

A.5.1 Bivariate Continuous Probability Distributions

The joint (bivariate) probability density function of two continuous random variables X and Y, denoted by $f_{X,Y}(x, y)$ or simply by $f(x, y)$, satisfies

(1) $f(x, y) \geqslant 0$,

(2) $\int_{-\infty}^{\infty} \int_{-\infty}^{\infty} f(x, y) \, dx \, dy = 1$.

For any region R of the two-dimensional plane, we have

$$\mathbb{P}((X, Y) \in R) = \iint_R f(x, y) \, dx \, dy.$$

The double integral of the joint PDF f over a region R provides the probability that (X, Y) assumes a value in R. This integral can be interpreted as the volume under the surface $z = f(x, y)$ over the region R.

Let X and Y be continuous random variables with a joint PDF $f(x, y)$. The *marginal probability density functions* of X and Y are, respectively,

$$f_X(x) = \int_{-\infty}^{\infty} f(x, y) \, dy \quad \text{and} \quad f_Y(y) = \int_{-\infty}^{\infty} f(x, y) \, dx.$$

The *conditional probability density function* of X given $Y = y$ is

$$f_{X|Y}(x \mid y) = \frac{f(x, y)}{f_Y(y)} \quad \text{for} \quad f_Y(y) > 0.$$

Correspondingly, the *conditional probability density function* of Y given $X = x$ is

$$f_{Y|X}(y \mid x) = \frac{f(x, y)}{f_X(x)} \quad \text{for} \quad f_X(x) > 0.$$

Example A.58. Suppose that X and Y have a joint density given by

$$f(x,y) = \begin{cases} C & \text{for } 0 \leqslant x, y \leqslant A, \\ 0 & \text{otherwise,} \end{cases}$$

for some constants $C, A \in (0, \infty)$.

(a) Find C such that f is a valid PMF.

(b) Calculate $\mathbb{P}(X < Y)$.

(c) Find the marginal density of X and then use it to compute $\mathrm{E}[X]$.

Solution.

(a) The double integral of f should be equal to one:

$$1 = \int_{-\infty}^{\infty} \int_{-\infty}^{\infty} f(x,y)\, dx\, dy = \int_{0}^{A} \int_{0}^{A} C\, dx\, dy = C \cdot A^2 \implies C = 1/A^2.$$

(b) $\mathbb{P}(X < Y) = \iint\limits_{0 \leqslant x < y \leqslant A} C\, dx\, dy = \dfrac{A^2}{2A^2} = \dfrac{1}{2}.$

(c) For any $x \in [0, A]$, we have

$$f_X(x) = \int_0^A C\, dy = \frac{1}{A}.$$

Thus, $\mathrm{E}[X] = \displaystyle\int_0^A \frac{x}{A}\, dx = \frac{A}{2}.$

$\square$

A.5.2   Independence of Random Variables

To compute the expected value of a function of two random variables, we can use one of the following two formulae:

$$E[h(X,Y)] = \begin{cases} \displaystyle\sum_x \sum_y h(x,y)\, p(x,y) \\ \qquad \text{if } X \text{ and } Y \text{ are discrete random variables with a joint PMF } p; \\[1em] \displaystyle\int_{-\infty}^{\infty} \int_{-\infty}^{\infty} h(x,y)\, f(x,y)\, dx\, dy \\ \qquad \text{if } X \text{ and } Y \text{ are continuous random variables with a joint PDF } f. \end{cases}$$

When two or more random variables are defined on the same probability space, it is useful to describe how they vary together. A common measure of the relationship between two random variables is the *covariance*. The *covariance* between the random variables X and Y with respective expectations μ_X and μ_Y is denoted by $\mathrm{Cov}(X,Y)$ or $\sigma_{X,Y}$ and is calculated as

$$\mathrm{Cov}(X,Y) = \sigma_{X,Y} = \mathrm{E}\big[(X - \mu_X)(Y - \mu_Y)\big]. \tag{A.21}$$

It is the *sign* of covariance as opposed to the *magnitude* that is most important. The sign of $(X - \mu_X)(Y - \mu_Y)$ is positive iff both variables are above their averages, or both are below the averages. It is negative iff one is above its average, while the other is below.

Therefore, a positive covariance indicates a tendency for X and Y to move together. A negative covariance indicates a tendency for X and Y to move in opposite directions. The above statements refer to what happens "on average."

In addition to (A.21), we can use the following shortcut formula to compute the covariance:

$$\text{Cov}(X, Y) = \text{E}[XY] - \text{E}[X]\,\text{E}[Y]. \tag{A.22}$$

Example A.59. Let V_1 and V_2 denote the face values which are turned up on two dice. Find the covariance $\text{Cov}(X, Y)$ between $X = \min\{V_1, V_2\}$ and $Y = \max\{V_1, V_2\}$.

Solution. In Example A.56, we found that $\text{E}[X] = \frac{91}{36}$ and $\text{E}[Y] = \frac{161}{36}$. To apply the shortcut formula for $\text{Cov}(X, Y)$, we only need to find $\text{E}[XY]$:

$$\text{E}[XY] = \sum_{x=1}^{6}\sum_{y=x}^{6} x\,y\,p_{X,Y}(x, y) = \frac{1}{36} \cdot \sum_{x=1}^{6} x^2 + \frac{2}{36} \cdot \sum_{x=1}^{5}\sum_{y=x+1}^{6} xy = \frac{49}{4}.$$

Thus, $\text{Cov}(X, Y) = \frac{49}{4} - \frac{91}{36} \cdot \frac{161}{36} \cong 0.9452.$ □

Here are some properties of the covariance. Let X, Y, Z be random variables, and a, b, c be constants.

(1) $\text{Cov}(X, X) = \text{Var}(X)$.

(2) $\text{Var}(aX + bY) = a^2\,\text{Var}(X) + b^2\,\text{Var}(Y) + 2ab\,\text{Cov}(X, Y)$.

(3) $\text{Cov}(aX + bY, cZ) = ac\,\text{Cov}(X, Z) + bc\,\text{Cov}(Y, Z)$.

Example A.60. Let X_1 and X_2 be quantitative and verbal scores on one exam, and let Y_1 and Y_2 be corresponding scores on another exam. If $\text{Cov}(X_1, Y_1) = 5$, $\text{Cov}(X_1, Y_2) = 1$, $\text{Cov}(X_2, Y_1) = 2$, and $\text{Cov}(X_2, Y_2) = 8$, what is the covariance between the two total scores $X_1 + X_2$ and $Y_1 + Y_2$?

Solution.

$$\text{Cov}(X_1 + X_2, Y_1 + Y_2) = \text{Cov}(X_1, Y_1) + \text{Cov}(X_1, Y_2) + \text{Cov}(X_2, Y_1) + \text{Cov}(X_2, Y_2)$$
$$= 5 + 1 + 2 + 8 = 16.$$

□

The *correlation* between random variables X and Y, denoted $\text{Corr}(X, Y)$ or $\rho_{X,Y}$, is given by

$$\rho_{X,Y} = \frac{\text{Cov}(X, Y)}{\sqrt{\text{Var}(X)}\sqrt{\text{Var}(Y)}} = \frac{\sigma_{X,Y}}{\sigma_X \sigma_Y}. \tag{A.23}$$

Random variables with nonzero correlation are said to be *correlated*. Random variables with zero correlation are said to be *uncorrelated*. Note that correlation is a unitless quantity that measures *the strength of the linear association* between random variables, and its sign indicates the direction of the relationship.

Let us list important properties of correlation.

- For any two random variables X and Y, the magnitude of correlation is bounded by one:
$$-1 \leqslant \rho_{X,Y} \leqslant 1.$$

- If a and c are either both positive or both negative, and b and d are any constants, then
$$\text{Corr}(aX + b, cY + d) = \text{Corr}(X, Y)$$

- The correlation $\rho_{X,Y}$ equals -1 or $+1$, iff $Y = aX + b$ for some constants $a \neq 0$ and b. That is, $\rho_{X,Y} = \pm 1$ iff X and Y are *linearly related*.

- If X and Y are independent, then $\rho_{X,Y} = 0$, but $\rho_{X,Y} = 0$ does not imply the independence of X and Y (unless both X and Y are normally distributed).

We say that two random variables X and Y are *independent* if any one of the following conditions is true

(1) For any two sets A and B in the ranges of X and Y, respectively, we have
$$\mathbb{P}(X \in A; Y \in B) = \mathbb{P}(X \in A) \cdot \mathbb{P}(Y \in B).$$

(2) $f_{Y|X}(y \mid x) = f_Y(y)$ for all x and y with $f_Y(y) > 0$.

(3) $f_{X|Y}(x \mid y) = f_X(x)$ for all x and y with $f_X(x) > 0$.

(4) $f_{X,Y}(x,y) = f_X(x)f_Y(y)$ for all x and y.

If the above is not satisfied for all pairs (x, y), then X and Y are said to be *dependent*. Here, f denotes a PDF (if we deal with continuous random variables) or a PMF (if we deal with discrete random variables).

Example A.61. Let $X_1, X_2, \ldots, X_n$ be i.i.d. *Unif*(0, 1) random variables. Find the CDF and the PDF of $M = \max\{X_1, X_2, \ldots, X_n\}$.

Solution. The range of each X_i is the interval $[0, 1]$. Thus, the range of M is $[0, 1]$. Since $X_1, X_2, \ldots, X_n$ are independent random variables, then $\{X_i \leqslant x\}$, $i = 1, 2, \ldots, n$ are independent events. Thus, for any $x \in [0, 1]$,

$$F_M(x) = \mathbb{P}(M \leqslant x) = \mathbb{P}(X_1 \leqslant x, X_2 \leqslant x, \ldots, X_n \leqslant x)$$
$$= \mathbb{P}(X_1 \leqslant x)\mathbb{P}(X_2 \leqslant x) \cdots \mathbb{P}(X_n \leqslant x) = F_{X_1}(x)F_{X_2}(x) \cdots F_{X_n}(x) = x^n.$$

Differentiate $F_M(x)$ to get $f_M(x) = nx^{n-1}$ for $x \in [0, 1]$. As a result,

$$F_M(x) = \begin{cases} 0 & x < 0 \\ x^n & 0 \leqslant x \leqslant 1 \\ 1 & x > 1 \end{cases} \quad \text{and} \quad f_M(x) = nx^{n-1}\,\mathbb{I}_{[0,1]}(x).$$

$\square$

If X and Y are independent random variables then

$$\mathrm{E}[h_1(X) \cdot h_2(Y)] = \mathrm{E}[h_1(X)] \cdot \mathrm{E}[h_2(Y)]$$

for any functions h_1 and h_2 (assuming the respective expected values exist). Particularly, if X and Y are independent random variables, then $\mathrm{E}[XY] = \mathrm{E}[X] \cdot \mathrm{E}[Y]$. Therefore, for independent random variables X and Y, we have

$$\sigma_{X,Y} = \rho_{X,Y} = 0.$$

In general, $\sigma_{X,Y} = \rho_{X,Y} = 0$ does not imply that X and Y are independent. The bivariate normal distribution is an exception. If X and Y are normal random variables and $\sigma_{X,Y} = \rho_{X,Y} = 0$, then X and Y are independent. In summary, if X and Y are independent, then $\mathrm{Cov}(X,Y) = 0$. However, $\mathrm{Cov}(X,Y) = 0$ does not imply that X and Y are independent.

Example A.62. Consider $p_X(x) = 1/3, x = -1, 0, 1$ and define $Y = X^2$. Show that $\mathrm{Cov}(X,Y) = 0$, but X and Y are not independent.

Solution. We have

$$\mathrm{E}[X] = -1 \cdot \frac{1}{3} + 0 \cdot \frac{1}{3} + 1 \cdot \frac{1}{3} = 0,$$

$$\mathrm{E}[Y] = (-1)^2 \cdot \frac{1}{3} + 0^2 \cdot \frac{1}{3} + 1^2 \cdot \frac{1}{3} = \frac{2}{3},$$

$$\mathrm{E}[XY] = \mathrm{E}[X^3] = (-1)^3 \cdot \frac{1}{3} + 0^3 \cdot \frac{1}{3} + 1^3 \cdot \frac{1}{3} = 0.$$

Thus, $\mathrm{Cov}(X,Y) = 0$. However, X and Y are dependent since $Y = X^2$. For example,

$$\mathbb{P}(X = 1; \, Y = 1) = \mathbb{P}(X = 1) = \frac{1}{3} \neq \frac{2}{9} = \frac{1}{3} \cdot \frac{2}{3} = \mathbb{P}(X = 1) \cdot \mathbb{P}(Y = 1).$$

$\square$

A.6 Limit Theorems

There are two fundamental theorems in the probability theory: the Central Limit Theorem (CLT) and the Law of Large Numbers (LLN). Both theorems refer to the limiting distribution of a sum of random variables as the number of terms increases.

A.6.1 Chebyshev's Theorem

The standard deviation σ measures the variation of a probability distribution. It reflects the concentration of probability in the neighborhood of the mean value. If σ is large, then the probability of getting values farther away from the mean is high. Chebyshev's inequality allows us to estimate the probability of observing a random variable far away from its mathematical expectation.

Theorem A.5 (Chebyshev's Inequality). *For any random variable X with mean μ and variance σ^2, we have*

$$P(|X - \mu| \geqslant k\sigma) \leqslant \frac{1}{k^2}. \tag{A.24}$$

That is, the probability that a random variable differs from its mean by at least k standard deviations is less than or equal to $1/k^2$.

Notice that the rule is useful only for $k > 1$. The proof is based on Markov's inequality which states that for any nonnegative random variable Y with a finite expectation,

$$\mathbb{P}(Y \geqslant t) \leqslant \mathrm{E}[Y]/t \ \text{ for all } \ t > 0. \qquad (A.25)$$

Letting $Y = (X - \mu_X)^2$ and $t = k^2 \sigma_X^2$ in (A.25) proves (A.24).

The so-called three-σ rule is a special case of Chebyshev's theorem. The rule says that most of observations in any data set lie within three standard deviations to either side of the mean regardless of the actual distribution. According to (A.24), we have

$$\mathbb{P}(|X - \mu| \geqslant 3\sigma) \leqslant \frac{1}{9} \ \text{ or, equalivalently, } \ \mathbb{P}(|X - \mu| < 3\sigma) \geqslant 88.89\%.$$

Another corollary of Chebyshev's inequality is that $X = \mathrm{E}[X]$ with probability one if $\mathrm{Var}(X) = 0$. Since, the variance of a constant variable is zero, we can conclude that $\mathrm{Var}(X) = 0$ iff X is a constant.

The importance of Markov's and Chebyshev's inequalities is that they allow us to derive bounds on probabilities when only the mean or both the mean and variance are known.

Example A.63. Show that for one million flips of a balance coin, the probability that the proportion of heads falls between 0.495 and 0.505 is at least 99%.

Solution. Let X denote the number of heads in 10^6 flips. Since $X \sim Bin(n = 10^6, p = \frac{1}{2})$, we have

$$\mu_X = \frac{10^6}{2} = 500{,}000 \ \text{ and } \ \sigma_X^2 = \frac{10^6}{4} = 250{,}000.$$

Therefore,

$$\mathbb{P}\left(0.495 \leqslant \frac{X}{10^6} \leqslant 0.505\right) = \mathbb{P}(495{,}000 \leqslant X \leqslant 505{,}000) = \mathbb{P}(-5000 \leqslant X - 500{,}000 \leqslant 5000)$$

$$= \mathbb{P}(|X - \mu_X| \leqslant 5000) = 1 - \mathbb{P}\left(|X - \mu_X| > \frac{5000 \sigma_X}{\sigma_X}\right)$$

$$\geqslant 1 - \frac{1}{(5000/\sigma_X)^2} = 1 - \frac{\sigma_X^2}{5000^2} = 1 - \frac{1}{100} = 99\%.$$

$\square$

A.6.2   Sum of Random Variables

Let us recall the basic properties of the mathematical expectation and the variance of a sum of random variables. Let X_i have mean μ_i and variance σ_i^2 for $i = 1, 2, \ldots, n$. Then,

$$\mathrm{E}[a_1 X_1 + a_2 X_2 + \cdots + a_n X_n] = a_1 \mathrm{E}[X_1] + a_2 \mathrm{E}[X_2] + \cdots + a_n \mathrm{E}[X_n] = \sum_{i=1}^{n} a_i \mu_i.$$

If the random variables are independent, then

$$\mathrm{Var}[a_1 X_1 + a_2 X_2 + \cdots + a_n X_n] = a_1^2 \, \mathrm{Var}[X_1] + \mathrm{Var}[X_2] + \cdots + a_n^2 \, \mathrm{Var}[X_n] = \sum_{i=1}^{n} a_i^2 \sigma_i^2.$$

To identify the distribution of a sum of random variables, we can use two approaches. Firstly, we can use the moment generating functions. Secondly, it is often possible to find the density function of the sum in closed form by integrating the joint PDF.

Let $X_1, X_2, \ldots, X_n$ be *independent* variables with MGFs $M_{X_1}(t), M_{X_2}(t), \ldots, M_{X_n}(t)$, respectively. Define $Y = X_1 + X_2 + \cdots + X_n$. Then,

$$M_Y(t) = M_{X_1}(t) M_{X_2}(t) \cdots M_{X_n}(t).$$

That is, the MGF of a *sum* of independent random variables is the *product* of the individual MGFs. Once we have obtained the MGF of Y, we can find its distribution.

It is interesting to know when a sum preserves the probability distribution. Suppose that $X_1, X_2, \ldots, X_n$ are independent random variables from the same family of probability distributions. Under what condition does the distribution of $Y = X_1 + X_2 + \cdots + X_n$ belong to the same distribution family? Here are results for selected probability distributions.

Binomial: If $X_i \sim Bin(m_i, p)$, then $Y \sim Bin(m_1 + m_2 + \cdots + m_n, p)$.

Poisson: If $X_i \sim Pois(\lambda_i)$, then $Y \sim Pois(\lambda_1 + \lambda_2 + \cdots + \lambda_n)$.

Normal: If $X_i \sim Norm(\mu_i, \sigma_i^2)$, then $Y \sim Norm\left(\sum_{i=1}^n \mu_i, \sum_{i=1}^n \sigma_i^2\right)$.

Gamma: If $X_i \sim Gamma(\alpha_i, \beta)$, then $Y \sim Gamma(\alpha_1 + \alpha_2 + \cdots + \alpha_n, \beta)$.

A.6.3 Sample Mean and Limit Theorems

A *statistic* is any quantity whose value can be calculated from a data sample. Before obtaining the data, there is uncertainty as to what value of any particular statistic will result. A statistic is a random variable, denoted by an uppercase letter. A lowercase letter is used to denote the calculated or observed value of the statistic.

The *sample mean* is denoted by $\overline{X}$, and the calculated mean from a sample is $\bar{x}$. The *sample standard deviation* is S, and the calculated value from a sample is denoted by s.

Random variables $X_1, X_2, \ldots, X_n$ are said to form a *random sample* of size n, if they are all mutually independent and identically distributed (i.i.d.). That is, every X_i has the same probability distribution. Let $X_1, X_2, \ldots, X_n$ be a random sample from a distribution with $E[X_i] = \mu$ and $Var(X_i) = \sigma^2$. The sample total $T_o = X_1 + X_2 + \cdots + X_n$ has the mean $E[T_o] = n\mu$ and the variance $Var(T_o) = n\sigma^2$. The sample mean $\overline{X} = T_o/n = (X_1 + X_2 + \cdots + X_n)/n$ has the mean $E[\overline{X}] = E[T_o]/n = \mu$ and the variance $Var(\overline{X}) = Var(T_o)/n^2 = \sigma^2/n$.

The Central Limit Theorem (CLT) allows us to find the limiting distribution of the sample mean. As $n \to \infty$, the standardized versions of $\overline{X}$ and T_o have the standard normal distribution. That is,

$$\lim_{n \to \infty} \mathbb{P}\left(\frac{\overline{X} - \mu}{\sigma/\sqrt{n}} \leq z\right) = \mathbb{P}(Z \leq z) = \mathcal{N}(z)$$

and

$$\lim_{n \to \infty} \mathbb{P}\left(\frac{T_o - n\mu}{\sqrt{n}\sigma} \leq z\right) = \mathbb{P}(Z \leq z) = \mathcal{N}(z),$$

where $Z \sim Norm(0, 1)$, and $\mathcal{N}(z)$ is a standard normal CDF.

Provided the sample size n is large enough, we can pretend as though the standardized T_o and $\overline{X}$ have a standard normal distribution to compute probabilities. For large n, we have

- $T_o \sim Norm(n\mu, n\sigma^2)$ (approximately);

- $\overline{X} \sim Norm(\mu, \sigma^2/n)$ (approximately).

Why is the CLT important? No matter what distribution you sample from, if you average enough observations, the probability distribution of the average is approximately normal. As a rule of thumb, the normal approximation can safely be used for $n \geqslant 30$.

Example A.64. 100 dice are tossed. Use the CLT to estimate the probability that their sum exceeds 370.

Solution. Each $X_i \sim Unif\{1, 2, 3, 4, 5, 6\}$, and hence $\mu = \mathrm{E}[X_i] = 7/2 = 3.5$ and $\sigma^2 = \mathrm{Var}(X_i) = 91/6 - 49/4 = 35/12 \cong 2.91667$. Let T_o denote the sum of X_i's. Using the CLT, we obtain

$$\mathbb{P}(T_o > 370) = \mathbb{P}\left(\frac{T_o/100 - \mu}{\sqrt{\sigma^2/100}} > \frac{370/100 - \mu}{\sqrt{\sigma^2/100}}\right)$$

$$\approx \mathbb{P}\left(Z \geqslant \frac{3.7 - 3.5}{0.1708}\right) = \mathbb{P}(Z \geqslant 1.1711) = 1 - \mathcal{N}(1.171) \cong 43.21\%.$$

□

Let $X_1, X_2, \ldots$, be i.i.d. random variables with common mean μ and variance σ^2. The *sample variance* S^2 is

$$S^2 = \frac{1}{n-1} \sum_{i=1}^{n} (X_i - \overline{X})^2.$$

One can show that $\mathrm{E}[S^2] = \sigma^2$.

A.6.4   The Law of Large Numbers

We define the expected value of a random variable as a long-range average of n independent sample values of X:

$$\bar{x} = \frac{x_1 + x_2 + \cdots + x_n}{n} \approx \mathrm{E}[X] \quad \text{for } n \gg 1.$$

Now we establish a mathematically precise version of this relation as a theorem known as the *Law of Large Numbers* (LLN).

Let $X_1, X_2, \ldots$ be i.i.d. random variables with finite mean $\mathrm{E}[X]$. Then,

$$\overline{X} = \frac{X_1 + \cdots + X_n}{n} \to E[X], \quad \text{as } n \to \infty.$$

Two questions need to be answered.

1. What type of convergence is it?

2. What are the conditions for the convergence to hold?

Theorem A.6 (The Weak Law of Large Numbers (LLN)). *Let $X_1, X_2, \ldots$ be i.i.d. random variables with common mean μ and finite variance σ^2. Then,*

$$\forall \varepsilon > 0 \quad \lim_{n \to \infty} \mathbb{P}\left(\left|\frac{X_1 + \cdots + X_n}{n} - \mu\right| < \varepsilon\right) = 1.$$

A corollary of the above theorem is the Bernoulli Weak Law of Large Numbers.

Theorem A.7. *Let E be an event associated with a random experiment such that $\mathbb{P}(E)$ remains constant. Consider an infinite series of independent repetitions of the random experiment, and let $N(E)$ denote the number of times the event E occurs in the first N repetitions. Then,*

$$\forall \varepsilon > 0 \quad \lim_{N \to \infty} \mathbb{P}\left(\left|\frac{N(E)}{N} - \mathbb{P}(E)\right| < \epsilon\right) = 1.$$

As we can see, the Bernoulli LLN confirms the frequentist interpretation of the probability function.

Let us list the important types of convergence.

Convergence in Probability: Random variables $\{X_n\}_{n \geqslant 1}$ converges in *probability* to the random variable X, as $n \to \infty$, denoted $X_n \xrightarrow{p} X$, if

$$\forall \epsilon > 0 \quad \mathbb{P}(|X_n - X| > \epsilon) \to 0, \quad \text{as } n \to \infty.$$

Theorem A.6 tells us that

$$\overline{X}_n = \frac{X_1 + \cdots + X_n}{n} \xrightarrow{p} \mu, \quad \text{as } n \to \infty.$$

Convergence in Distribution: Random variables $\{X_n\}_{n \geqslant 1}$ converges in *distribution* to the random variable X, as $n \to \infty$, denoted $X_n \xrightarrow{d} X$, if

$$F_{X_n}(x) \to F_X(x), \quad \text{as } n \to \infty, \quad \forall x \in \mathbb{R} \text{ such that } F_X \text{ is continuous at } x.$$

Convergence Almost Surely (strong): Random variables $\{X_n\}_{n \geqslant 1}$ converges *almost surely (a.s.)* to the random variable X, as $n \to \infty$, denoted $X_n \xrightarrow{a.s.} X$, if

$$\mathbb{P}(\{\omega \mid X_n(\omega) \to X(\omega), \quad \text{as } n \to \infty\}) = 1.$$

Theorem A.8 (The Weak LLN). *Let $X_1, X_2, \ldots$ be i.i.d. random variables with finite mean μ. Then,*

$$\overline{X}_n = \frac{X_1 + \cdots + X_n}{n} \xrightarrow{p} \mu, \quad \text{as } n \to \infty.$$

Theorem A.9 (The Strong LLN). *Let $X_1, X_2, \ldots$ be i.i.d. random variables with finite mean μ. Then,*

$$\overline{X}_n = \frac{X_1 + \cdots + X_n}{n} \xrightarrow{a.s.} \mu, \quad \text{as } n \to \infty.$$

B

Answers and Hints to Exercises

B.1 Chapter 1

1. (a) Use $1 + r_{[0,2]} = (1 + r_1)(1 + r_2)$. (b) Express $1 + r_{[0,k+1]}$ in terms of $r_{[0,k]}$ and r_{k+1}.

2. (a) 8.2118%. (b) $4220.26. (c) 2.6656%.

3. Use that $R_{[t,s]} = (1 + i)^{(s-t)}$ for annual compounding.

4. Compute the derivatives $V'(t)$ and $V''(t)$.

5. (a) 10.25%, 10.4713%, 10.5171%. (b) 9.7618%, 9.5609%, 9.5310%.

6. (a) IRR $= (1 + i^{(m)}/m)^m - 1$. (b) IRR $= (1 + rt)^{1/t} - 1$.

7. Use Proposition 1.1.

8. (a) 4.9595 years. (b) 4.7071 years. (c) 4.6210 years.

9. $t = \dfrac{\ln n}{\ln(1 + i)}$.

10. For annual rates, we have $i_a = 7.1\% < i_c \cong 7.1436\% < i_b \cong 7.1859\%$

11. (a) $y(t) = 0.035$ for $t \in [0, 2]$, $y(t) = 0.02 + 0.03/t$ for $t \in [2, 5]$, $y(t) = 0.0275 - 0.0075/t$ for $t \in [5, 8]$. (b) $4269.57. (c) $4600.00. (d) $3600.

12. The only positive solution is $i = -3/4 + \sqrt{265}/20 \cong 6.3941\%$.

13. It is better to purchase the equipment by instalments, since the present value of the instalments is $198,036.37.

14. NPV$_A$ = $2023.03 and NPV$_B$ = $4356.83. Since NPV$_B$ > NPV$_A$, proceed with project B.

15. (a) $y(T) = 0.01 + \frac{0.04\sqrt{(T)}}{3} + 0.0005T$. (b) $2385.21. (c) $4611.75. (d) $-$477.30.

16. (a) $71.18. (b) $ 723.87. (c) 2.579 years.

18. $576.61.

19. $149,482.24.

20. The NPV equals $-$1,330.74. Since it is negative, then the company should not proceed with the purchase.

21. $y(T) = b + \frac{(a-b)\ln(1+T)}{T}$ and $Z(0, T) = (1 + T)^{(b-a)}e^{-bT}$.

22. (a) The graph needs to (i) be decreasing and convex, (ii) have the horizontal axis as a horizontal asymptote, (iii) have a vertical intercept at $(0, 1100)$, and (iv) pass through the point $(.02, 1000)$. (b) The graph needs to (i) be decreasing and convex, (ii) have the vertical axis as a vertical asymptote, (iii) have a horizontal intercept at $(1100, 0)$, and (iv) pass through the point $(1000, .02)$.

23. (a) $y(T) = 0.03 + 0.01T - \frac{0.01}{3}T^2$; the graph needs to a concave and have an inverted shape. (b) $Z(0, T) = \exp(-0.03T - 0.01T^2 + \frac{0.01}{3}T^3)$. (c) \$2807.22.

24. (a) The price $P = \min\{\$1112.47, \$1118.68, \$1119.40\} = \$1112.47$ guarantees the yield rate $i^{(2)}$ of 7% or higher. (b) By the method of averages: $i^{(2)} \cong 6.86\%$. Interpolation between $i^{(2)} = 6\%$ and $i^{(2)} = 7\%$ gives $i^{(2)} \cong 6.853\%$.

25. \$981.14.

26. (a) \$11,942.45. (b) 8.021%.

27. We have $r_A = 2.5\% > r_C = 2\%$ and $r_B = 5\% > r_D = 4.5\%$. On the other hand, $r_{\text{Jack}} = 3.75\% < r_{\text{Tom}} = 4\%$.

28. The increase associated with a drop in yield by 1%.

29. Calculate the bond value for the yield $i^{(2)} = 5\%$ and compare it with the actual price.

30. Calculate the bond value for the yield $i^{(2)} = 7\%$ and compare it with the actual price.

31. 1%, 1.5%, and 4%.

32. 1.9438% on the one-year bond, 3.0008% on the two-year bond.

33. (a) 6.4539%. (b) 9.3560%.

34. (a) $i^{(2)} \cong 10.42\%$. (b) Interpolation between $i^{(2)} = 10\%$ and $i^{(2)} = 11\%$ gives $i^{(2)} \cong 10.54\%$.

35. (a) $y^{(2)} \cong 4.00\%$. (b) After three iterations, the bracket is $[a(3), b(3)] = [3.875\%, 4\%]$. The midpoint is $c(3) = 3.9375\%$.

36. (a) $y^{(2)} \cong 4.1095\%$. (b) After one iterations, $y^{(2)} \cong 4.0790\%$.

37. (a) $y^{(2)} \cong 3.679\%$. (b) Overestimate. (c) It becomes $(3.65\%, 4.2\%)$. (d) $y^{(2)} \cong 3.637314\%$.

38. (a) \$4347.22, 1.89270 years. (b) \$4344.78, 1.89266 years. (c) \$4342.30, 1.89263 years.

39. \$1520.21, 1.94194 years.

40. 10.0333 years.

41. They are the same.

42. (a) 4.3550 years. (b) Individual changes are 0.105%, 0.48%, 0.765%, 1.305%; portfolio change is weighted average of individual changes, approximately 0.653254%. (c) 0.653254%.

43. (a) $P_A = \$3630.75$, $P_B = \$2475.13$, $P_C = \$3163.75$, $P_D = \$3446.04$. (b) \$12,715.67, 18.8815 years. (c) \$240.09, 1.8881%.

44. (a) \$28,293.00, 5.1101 years. (b) $-\$699.12$, -2.471%.

45. (a) \$188,847.21. (b) \$165,871.13 in bond A, \$22,976.07 in bond B. (c) 177.600 shares of bond A, 14.977 shares of bond B.

47. C, B, D, A.

B.2   Chapter 2

1. (a) $x = -\frac{100}{3}$ and $y = \frac{400}{9}$. (b) The expected rate of return is 8%.

2. $\{x \leqslant 0, 12x + 11y \geqslant 0\} \cup \{x \geqslant 0, 9x + 11y \geqslant 0\}$.

3. Any portfolio (x, y) with $x = 100 - y$ and $0 \leqslant y \leqslant 30$.

4. $\Pi_0 = \$1560$; $\Pi_T = \$1843.20$ if the stock price increases, $\Pi_T = \$1363.20$ if the stock price decreases. (b) $r \cong 68.62\%$ if the stock price increases, $r \cong -52.75\%$ if the stock price decreases.

5. $\Pi_0 = \$1440$; $\Pi_T = \$1497.60$ if the stock price increases, $\Pi_T = \$1400.40$ if the stock price decreases. (b) $\$1424.70$. (c) -3.2%.

6. $S_1 \in \{105, 180\}$, $S_2 \in \{73.5, 126, 216\}$, $S_3 \in \{51.45, 88.2, 151.2, 259.2\}$.

7. The probability distribution of S_4 is

S_4	190.33	147.08	113.65	87.82	67.86
p_{S_4}	.39%	4.69%	21.09%	42.19%	31.64%

$\mathbb{P}(S_4 \geqslant S_0) = 5.08\%$.

8. $\mathrm{E}[S_T] = S_0(up + d(1-p))^T$.

10. (a) $u = \frac{1}{d} = \exp\left(\frac{0.2}{\sqrt{40}}\right) \cong 1.0321281$; $p \cong 51.9764\%$. (b) $\mathrm{E}[r_1] \cong 0.175025\%$. (c) $\mathrm{E}[S_{40}] = S_0(1 + \mathrm{E}[r_1])^{40} \cong 75.071729$.

11. (a) $u_N = \dfrac{1}{d_N} = \exp(\sqrt{\sigma^2 \delta_N + \mu^2 \delta_N^2})$, $p_N = \dfrac{1}{2} + \dfrac{\mu\sqrt{\delta_N}}{2\sqrt{\sigma^2 + \mu^2 \delta_N}}$.
    (b) $d_N = \exp(\mu\delta_N - \sigma\sqrt{\delta_N})$, $u_N = \exp(\mu\delta_N + \sigma\sqrt{\delta_N})$.

12. $\mathrm{Var}[X] = e^{a^2 + 2b}(e^{a^2} - 1)$, $\mathrm{E}[X] = e^{\frac{a^2}{2} + b}$.

13. (a) 122.1402758. (b) 58.8468%. (c) 49.5816%.

15. At time 0, take a short position in V and a long position in W. The cash proceeds are $V_0 - W_0 > 0$.

16. (a) two stock shares and $-205/177 \cong -1.1582$ bond units. (b) $\$66.27$. (c) $\$29$.

17. (a) $r_S^- = -7.143\%$, $r_S^+ = 17.857\%$, $r_B = 20.85\%$. (b) Yes, since $r_S^+ < r_B$. Short x stock shares and buy y bonds; (x, y) with $x < 0$ and $y = 7x/3$ is an arbitrage portfolio.

18. (a) $-\frac{2}{3}$ stock shares and $\frac{125}{96}$ bond units; $\Pi_0 = \$31.77080$. (b) $\mathrm{E}[\Pi_T] = \$56.00$; the weights are $w_{\text{stock}} = -104.92\%$ and $w_{\text{bond}} = 204.92\%$. (c) $r_S^\pm = \pm 30\%$, $r_B = 28\%$. No arbitrage.

19. Arbitrage exists since the bond return is the less than the stock return in the down state. Buy $x > 0$ stock shares and borrow $\$130x$ by short selling $\frac{13x}{7}$ bonds.

20. Arbitrage exists since the bond return is the greater than the stock return in the up state. Buy $y > 0$ bonds and borrow $\$150y$ by short selling y stock shares.

21. (a) 83.2727%. (b) \$63.45. (c) Buy 20 claims. The arbitrage profit is \$442 if the stock price increases and \$127 if the stock price decreases.

22. $x = \dfrac{\text{Cov}(S_T, C_T)}{\text{Var}(S_T)} = \dfrac{C^+ - C^-}{S^+ - S^-}$ and $y = \dfrac{C_0 - x S_0}{B_0}$.

23. (a) No. (b) Yes, the replicating portfolio is $(x, y) = (-\frac{1}{10}, \frac{1}{5})$.

24. (a) $Y > 110$. (b) $X < 110 < Y$. (c) $(x, y) = (-1, 10)$ is an arbitrage portfolio. (d) $(x, y) = (1, -10)$ is an arbitrage portfolio.

25. (a) $\tilde{p} = \frac{2}{3} \cong 66.6667\%$ and $1 - \tilde{p} = \frac{1}{3}$.

26. (a) $\tilde{p} = \frac{13}{15} \cong 86.6666\%$. (b) The risk-neutral probability distribution of S_3 is

S_3	199.65	145.2	105.6	76.8
p_{S_3}	0.6510	0.3004	0.0462	0.0024

  (c) 34.90%

27. (a) 67%. (b) 6%. (c) \$83.37.

28. (a) $\tilde{p}_1 = \frac{1}{3}$ and $\tilde{p}_2 = \frac{2}{3}$.

29. (a) The replicating portfolio is $(x, y) = (\frac{2}{3}, -\frac{60}{11})$; its initial value is $\frac{400}{33} \cong 12.12$. (b) $C_0 = \frac{400}{33} \cong 12.12$.

30. (a) The values of S_4 are 2, 8, 32, 128, and 512. (b) The real-world probability distributions of S_4 is

S_4	2	8	32	128	512
p_{S_4}	$\frac{16}{625}$	$\frac{96}{625}$	$\frac{216}{625}$	$\frac{216}{625}$	$\frac{81}{625}$

The risk-neutral probability distributions of S_4 is

S_4	2	8	32	128	512
p_{S_4}	$\frac{1}{81}$	$\frac{8}{81}$	$\frac{24}{81}$	$\frac{32}{81}$	$\frac{16}{81}$

  (c) $E[S_4] = 122.9312$ and $\widetilde{E}[S_4] = 162$. (d) $\mathbb{P}(S_4 > S_0) = 47.52\%$ and $\widetilde{\mathbb{P}}(S_4 > S_0) = \frac{16}{27} \cong 59.26\%$.

31. (a) Show that a zero portfolio is the only portfolio that satisfies $\Pi_0 = 0$ and $\Pi_T(\omega^i) \geqslant 0$ with $i = 1, 2, 3$. (b) The general solution is $\tilde{p}_1 = \tilde{p}_3 = t$ and $\tilde{p}_2 = 1 - 2t$ with $t \in (0, 0.5)$.

33. (a) The set of possible portfolios is a union of the following two sets:

$$\left\{ (x, y) : x \geqslant 2/3, \ y \geqslant -\frac{90}{11}x \right\} \quad \text{and} \quad \left\{ (x, y) : x \leqslant 2/3, \ y \geqslant \frac{20 - 120x}{11} \right\}.$$

  (b) $\max_{(x,y)} \Pi_0^{(x,y)}$ is infinite; $\min_{(x,y)} \Pi_0^{(x,y)} = \frac{400}{33}$ is attained at the point $(x, y) = (\frac{2}{3}, -\frac{60}{11})$.

34. (a) An arbitrage strategy: $x_0 = y_0 = 0$, $x_1(\omega^1) = x_1(\omega^2) = 0$, $y_1(\omega^1) = y_1(\omega^2) = 0$, $x_1(\omega^3) = x_1(\omega^4) = -t$, $y_1(\omega^3) = y_1(\omega^4) = 9t/11$ with $t > 0$. (b) No.

35. No.

B.3   Chapter 3

3. $E[u(W)] = 6.25$ and $u(E[W]) \cong 6.9462$.

4. (a) Since $E[u(V_1)] = \frac{1}{2}$, $E[u(V_2)] = \frac{1}{2} + \frac{1}{2e^2} \cong 0.56767$, and $E[u(V_2)] = \frac{2}{e} - \frac{1}{e^2} \cong 0.60042$, she should choose investment 3. (b) Since $E[V_1] = E[V_2] = E[V_3] = 1$, a risk-indifferent investor is indifferent between these three investments.

5. (a) $a = \max\{\frac{2p-(1+r)}{1-r}, 0\}$. (b) $a = 0$ if $p \leqslant \frac{1+r}{2}$, $a = 1$ if $\frac{(1+r)e^W}{1-r+(1+r)e^W} \leqslant p$, otherwise $a = \frac{5}{W} \ln\left(\frac{p(1-r)}{(1+r)(1-p)}\right)$.

6. (a) $\frac{8300}{3} \cong 2766.6667$. (b) 1.0183012. (c) 4.6417.

7. (a) 7.843775. (b) 7.852217; it is less than the result of (a) since the utility function is concave down. (c) \$2549.81.

8. (a) 4.680509. (c) \$2246.29.

9. (a) 2.63%; he should buy the ticket. (b) $\lfloor 18.8177 \times 100 \rfloor / 100 = \$18.81$.

10. (a) $A_u = a$. Thus, $\pi \approx \frac{1}{2}A_u \operatorname{Var}(V) = \frac{a}{2}V_0^2\sigma^2$ and $C \approx V_0(1+\mu) - \frac{a}{2}V_0^2\sigma^2$. (b) π and C are the same as those in (a).

11. (a) $\alpha \geqslant 0$.

12. (a) \$105.27. (b) \$111.15. (c) \$111.11.

13. (a) 0.6247; not since his current utility is 0.6321. (b) \$18.01. (c) $X \geqslant \$6.53$.

14. 62.1499%

16. (b) 1. (c) $\frac{1}{2}$. (d) $\frac{1}{2}$.

20. 2.

21. $x = \frac{4}{9}(10\mu - 1)$ and $y = 10 - x$.

22. (a) $\mu_1 = 9.5\%$, $\mu_2 = 4.5\%$, $\sigma_1 = 3.5\%$, $\sigma_2 \cong 5.22\%$, $\rho \cong -97.151\%$. (b) $(w_1, w_2) = (0.6, 0.4)$, $\mu_{\mathrm{mv}} = 7.5\%$, $\sigma_{\mathrm{mv}} = 0.5\%$.

23. (a) $\mu(w) = 0.1 - 0.05w$, $\sigma^2(w) = 0.0225w^2$. (b) (i) $w_1 = -5.5556\%$, $w_2 = 105.5556\%$; (ii) -2.22 stock shares and 33.776 bonds; (iii) 10.2775%; (iv) 0.8325%.

24. (a) $E[V_T] = 1050 - 50x$, $\operatorname{Var}[V_T] = (19x^2 - 24x + 9) \cdot 10^4$. (b) $(x_1, x_2) = (\frac{23}{38}, \frac{15}{38})$.

25. (a) 3.79%. (b) 1.5119%.

26. (a) $w_1 = -125\%$, $w_2 = 225\%$, $\mu = 7\%$. (b) $w_1 \cong 35.7143\%$, $w_2 \cong 64.2857\%$, $\mu \cong 13.4286\%$.

27. (a) $w_1 = 57.0492\%$, $w_2 = 42.9508\%$. (b) $w_1 = 57.3770\%$, $w_2 = 42.6230\%$. (c) For (a), $\mu = 12.1475\%$ and $\sigma = 11.0891\%$; for (b), $\mu = 12.1311\%$ and $\sigma = 11.0883\%$. The portfolio in (a) converges to the portfolio in (b) as $V_0 \to \infty$.

28. (a) $\mu = 23\%$, $\sigma \cong 43.2666\%$. (b) $w_1 = -75\%$, $w_2 = 125\%$. (c) $w_1 \cong 123.5294\%$, $w_2 \cong -23.5294\%$.

29. $w_1 = 60\%$, $w_2 = 40\%$; $\mu = 14.6\%$, $\sigma = 18.2147\%$.

31. (a) $w_1 \cong 45.9391\%$, $w_2 \cong 54.0609\%$. (b) $x_1 \cong 11.485$, $x_2 \cong 10.136$. (c) $\mu \cong 13.5127\%$, $\sigma \cong 6.8008\%$.

32. (a) $\mu(x) = 0.01x + 0.11$, $\sigma^2(x) = 0.12360x^2 - 0.13200x + 0.0484$. (b) $w_1 \cong 53.3981\%$, $w_2 \cong 46.6019\%$. (c) $\mu \cong 11.5340\%$, $\sigma = 11.4705\%$. (d) $w_1 \cong 54.5146\%$, $w_2 \cong 45.4854\%$. (e) $\mu \cong 11.5451\%$, $\sigma = 11.4772\%$.

33. (a) 7.2956%. (b) The points are $(0.139, 0.077)$, $(0.040884, 0.045818)$, $(0.044, 0.042)$. (c) Should look like a segment of a hyperbola passing through the given points; should include part of the upper branch and only a bit of the lower branch.

34. (a) -2.5315%. (b) The points are $(0.996, -0.38)$, $(0.088198, -0.025315)$, $(0.0213, 0.044)$. (c) Should look like a segment of a hyperbola passing through the given points; should include part of the upper branch and only a bit of the lower branch.

35. (a) 4.1034% and 6.73495%. (b) 3.232% and -1.9605%. (c) $\alpha \geqslant 0.0040265$.

36. (a) It is a parabola opening down, roots at 0.1321 and 1.3350, maximum attained at 0.7335, 73.35% in bonds. (b) It is a parabola opening down, roots at -3.2375, 1.8710, maximum attained at -0.6831, 100% in equities. (e) $\alpha \leqslant 0.001767$. (f) 0.8909. (g) identically zero for $\alpha \in [0, 0.001767]$, then concave and increasing for $\alpha > 0.001767$ (not differentiable at $\alpha = 0.001767$) with a horizontal asymptote at 0.8909.

37. (a) It is a parabola opening up with one real root at $w = 0$. (b) It is a parabola opening down; it passes through $(0, r)$, has one negative root and one positive root. (c) Graph is increasing at $w = 0$, therefore expected utility can always be increased by putting some money in the risky asset. Alternatively, maximum of objective function occurs at $w = (\mu - r)/2\lambda\sigma^2$, which is strictly positive.

38. (a) $\mu > 0.03125$. (b) $\mu > 0.065$. (c) $\sigma < 0.3162$. (d) $\sigma < 0.4472$.

40. (b) $x \in (-0.9761355821, 0.6761355821)$.

41. (a) $\begin{bmatrix} 0.04 & -0.02 & 0.009 \\ -0.02 & 0.0625 & 0.02625 \\ 0.009 & 0.02625 & 0.0225 \end{bmatrix}$. (b) $\sigma_3 > 0.0981203$.

42. (a) $\mathbf{w}_{mv}^\top \cong [.8455, .7366, -.5820]$. (b) $\mathbf{w}_{meu}^\top \cong [1.1077, 1.0624, -1.1701]$. (c) $\mu_{mv} \cong 5.3186\%$, $\sigma_{mv} \cong 11.7680\%$; $\mu_{meu} \cong 6.2325\%$, $\sigma_{meu} \cong 12.1500\%$.

43. (a) $\mathbf{w}_{\mathcal{M}}^\top \cong [1.9398, 2.0962, -3.0360]$; $\mu_{\mathcal{M}} \cong 9.1321\%$, $\sigma_{\mathcal{M}} \cong 17.2517\%$. (b) $\mu = 6.2793$, $\sigma = 10.3510\%$. (c) 31.7868%; $\mu = 6.8651\%$.

B.4 Chapter 4

1. (a) $\$80.98$. (b) $\$9.89$. (c) An arbitrage exists. The profit on the maturity date is $\$91.22 - \$80.98 - \$5e^{0.07 \cdot 0.5} \cong \$5.09$.

2. $\$269.53 \cdot (1 + 0.0065 \cdot 0.5) \cong \$270.41$ per 100 pounds.

3. 17.57 cents per pound.

4. $9687.28.

5. (b) Must exceed $384.19.

6. (a) $42.17. (b) $39.14. (c) The forward price in one month would be $39.50; the contract value is $0.36. (d) The forward price in three months would be $38.29; the contract value is −$0.84.

7. (a) $4811.73. (b) The forward price in two months would be $4652.53; the contract value is −$153.98.

8. $7325.25.

9. (a) $510.64. (b) The forward price in one month would be $550.14; the contract value is $37.95.

10. (a) $198.02. (b) The forward price on June 1 would be $262.03; the contract value is $61.09. (c) Kate should enter into a short forward contract with delivery on January 1 next year and the delivery price of $262.03. The profit is $262.03 − $198.02 = $64.01. (d) Kate should enter into a short forward contract with delivery on January 1 next year and the delivery price of $156.64. The loss is $198.02 − $156.64 = $41.38.

13. It will increase. The change in value is approximately $\delta V = V(r + \delta r) - V(r) \approx \frac{\delta V}{\delta r} \delta r = KT \delta r e^{-rT}$.

14. (b) Arbitrage profit of $F - S_0 e^{r_b T}$ in T-year dollars. (b) Arbitrage profit of $S_0 e^{r_\ell T} - F$ in T-year dollars.

15. (a) $F \geqslant S_0 e^{r_\ell T}$. (b) $F \leqslant S_0 e^{r_b T}$

16. (a) Buy one barrel, borrow $100, enter into a short forward; in six months, the arbitrage profit is $103 − $102.53 = $0.47. (b) Sell short one barrel, invest $100, enter into a long forward; in six months, the arbitrage profit is $101.51 − $100 = $0.51.

17. Enter into a long forward; sell one coupon bond for $9620; to pay two coupons out, buy an annuity that gives $400 in six months and $400 in one year for $749.07, invest $8870.93. The arbitrage profit is $9687.28 − $9200 = $487.28 in one year time.

18. (a) $105.13. (b) $7.47.

19. (b) Enter into a short forward with delivery time in eight months and delivery price $1525.21. (c) Enter into a short forward with delivery time in eight months and delivery price $915.13.

20. (a) $525.57. (b) Sell short one share for $500, invest the proceed without risk, enter into a long forward contract; the arbitrage profit is $15.57. (c) Borrow $500, buy on share, enter into a short forward contract; the arbitrage profit is $34.43.

27. (a) $5.71. (b) $13.02.

28. (a) −$3.02. (b) −$3.59.

29. $71.18.

30. (a) $4.26. (b) $4.83.

31. $0.28.

32. (a) $7.13. (b) −$0.37.

33. (a) $31. (b) $42.

34. (a) $2. (b) $3. (c) $57.

35. (a) $9.61/ (b) $1.43. (c) $1.29. (d) $1.72.

36. (a) 0.0129 shares. (b) The portfolio value is $(1 − 0.0129)S_T + (175 − S_T)^+$. (d) For $S_T < 172.7713$.

38. (a) $7.61. (b) Form the portfolio with one long call, one short put, one short stock, and a deposit with the three-month future value of $195. Its total value is −$4.39 today with zero obligations in the future. (c) Form the portfolio with one long put, one short call, one long stock, and a loan with the three-month future value of $195. Its total value is −$2.61 today with zero obligations in the future.

41. The replicating portfolio consists of −0.2 stock shares and 21.9048 bonds. The initial price of the put is $1.90.

42. The replicating portfolio consists of one stock shares and K bonds.

43. $C(K = 190) = 10$, $C(K = 200) = \frac{20}{3}$, $C(K = 220) = 0$.

44. (a) The replicating portfolio consists of −0.2 stock shares and 0.1896 bonds. (b) $0.44.

45. (a) The portfolio replicating the call payoff consists of $1/3$ stock shares and −13.6364 bonds. The portfolio replicating the put payoff consists of −2/3 stock shares and 36.3636 bonds. (b) $C_0 = \$3.03$. $P_0 = \$3.03$. (c) $2.25.

46. $0.83.

50. (a) $C_0 = S_0 \mathcal{N}(d_+)$, $P_0 = S_0 \mathcal{N}(-d_+)$. (b) $C_0 + P_0 = S_0$.

51. (a) 0. (b) ∞. (c) $(S − K)^+$. (d) $\left(Se^{-q(T-t)} − Ke^{-r(T-t)}\right)^+$. (e) $Se^{-q(T-t)}$.

52. (a) The delta-gamma neutral portfolio consists of 31.59 shares of the stock, 575.19 call options with an expiry of 120 days and −500 call options with an expiry of 90 days. Its current values is $2213.71. After five days, the relative change in value is (i) −0.01%, (ii) 1.42%, (iii) 7.86%. (b) The delta-rho neutral portfolio consists of 71.34 shares of the stock, 484.79 call options with an expiry of 120 days and −500 call options with an expiry of 90 days. Its current values is $5024.47. After 5 days, the relative change in value is (i) 6.96%, (ii) 0.34%, (iii) 6.63%.

53. (b) One long put struck at K_1 and one long call struck at K_2. (c) Its no-arbitrage value is a sum of the call and put Black–Scholes values. (d) The delta is a sum of the call and put Black–Scholes deltas.

54. (a) $\frac{\partial C_t}{\partial K} + \frac{\partial P_t}{\partial K} = -1$. (b) $\frac{\partial C_T}{\partial K} = -\mathbb{I}_{\{S>K\}}$, $\frac{\partial P_T}{\partial K} = \mathbb{I}_{\{S<K\}}$.

56. (a) $C_0 = e^{-rT} \mathcal{N}(d_-)$, $P_0 = e^{-rT} \mathcal{N}(-d_-)$.
(b) $C_0 + P_0 = e^{-rT}$. (c) $\Delta_C = e^{-rT} \frac{n(d_-)}{S_0 \sigma \sqrt{T}}$, $\Delta_P = -e^{-rT} \frac{n(d_-)}{S_0 \sigma \sqrt{T}}$.

57. (a) $V_0 = e^{-rT} \left(\mathcal{N}(d_-(S_0/K_1), T) − \mathcal{N}(d_-(S_0/K_2), T)\right)$.
(b) $\Delta_V = e^{-rT} \left(\mathcal{N}(d_-(S_0/K_1), T) − \mathcal{N}(d_-(S_0/K_2), T)\right)$.

58. (a) \$681.87. (b) \$40.47. (c) An out-of-the-money option is more sensitive. (d) An one-month option is more sensitive. (e) The XYZ call more valuable (worth \$114.19). The same is true for puts (\$107.61 for the XYZ at-the-money put versus \$33.90 for the ABC at-the-money put).

B.5   Chapter 5

1. (a) $\beta^+ = -\frac{d}{u-d}$, $\Delta^+ = \frac{1}{(u-d)S_0}$; $\beta^- = \frac{u}{u-d}$, $\Delta^- = -\frac{1}{(u-d)S_0}$.

(b) $\pi_0(\mathcal{E}^+) = S_0\Delta^+ + B_0\beta^+ = \frac{1}{1+r}\frac{1+r-d}{u-d} \equiv \frac{1}{1+r}\tilde{p}$,

$\pi_0(\mathcal{E}^-) = S_0\Delta^- + B_0\beta^- = \frac{1}{1+r}\frac{u-(1+r)}{u-d} \equiv \frac{1}{1+r}(1-\tilde{p})$.

(c) $[\beta_X, \Delta_X] = \left[\frac{x^- u - x^+ d}{u-d}, \frac{x^+ - x^-}{(u-d)S_0}\right]$.

(d) $\pi_0(X) = x^+\pi_0(\mathcal{E}^+) + x^-\pi_0(\mathcal{E}^-) = \frac{1}{1+r}(\tilde{p}x^+ + (1-\tilde{p})x^-)$.

(e) For $X = \max(S_T, K)$: (i) $\frac{K}{S_0} \le d$ gives $\pi_0(X) = S_0$, (ii) $\frac{K}{S_0} \ge u$ gives $\pi_0(X) = \frac{K}{1+r}$, (iii) $d < \frac{K}{S_0} < u$ gives $\pi_0(X) = \frac{1}{1+r}(\tilde{p}S_0 u + (1-\tilde{p})K)$. For $X = \max(S_T - K, 0)$, note that $\max(S_T - K, 0) = \max(S_T, K) - K$, hence $\pi_0(X) = \pi_0(\max(S_T, K)) - \frac{K}{1+r}$.

2. Hedging portfolio value is $\Pi_t = \Delta S_t - C_t$ for $t \in \{0, T\}$ with variance

$$\text{Var}(\Pi_T) = \left[(C^- - C^+) + (S^+ - S^-)\Delta\right]^2 p(1-p).$$

The variance attains its minimal value of zero when $\Delta = \frac{C^+ - C^-}{S^+ - S^-}$.

3. (a) $\Pi_T(\omega) = \Delta S_T(\omega) - 1.03(100\Delta - 11) - (S_T(\omega) - 95)^+$. Hence, $\Pi_T(\omega_1) = 22\Delta - 18.67$, $\Pi_T(\omega_2) = -3\Delta + 6.33$, $\Pi_T(\omega_3) = -18\Delta + 11.33$. (b) A perfect hedge requires $\Pi_T(\omega_j) = 0$, for all $j = 1, 2, 3$. This yields no solution for Δ, i.e., there is no perfect hedge.

4. (b) The replicating portfolio is $\theta = [\frac{-4}{105}, \frac{1}{2}]$. Hence, **X** is attainable. (c) State price vectors are $\mathbf{\Psi}(t) = [\frac{25}{63} - \frac{2}{3}t, t, \frac{5}{9} - \frac{1}{3}t]^T$, where $t \in (0, \frac{25}{42})$ is the no-arbitrage condition. (d) Bid-ask spread for call option price is $(\frac{25}{21}, \frac{100}{63}) \sim (1.1905, 1.5873)$. (e) The new model is arbitrage-free since $C_0 = \frac{3}{2} \in (\frac{25}{21}, \frac{100}{63})$. [Alternatively, the new 3-by-3 matrix leads to a strictly positive state price vector.] (f) Solving the martingale equations gives the EMM as a probability vector $[\tilde{p}_1, \tilde{p}_2, \tilde{p}_3] = [\frac{7}{2}t - 1, -\frac{9}{2}t + 2, t]$, $t \in (\frac{2}{7}, \frac{4}{9})$.

5. (a) Market is complete iff $R_1^1 R_2^2 - R_2^1 R_1^2 \ne 0$. (b) $\tilde{p} = \frac{(R_2^1 - R_2^2)R_1^1}{R_2^1 R_1^2 - R_1^1 R_2^2}$ and $1 - \tilde{p} = \frac{(R_2^2 - R_1^2)R_2^1}{R_2^1 R_1^2 - R_1^1 R_2^2}$.

(c) $\varphi_1 = \frac{x_1 R_2^2 - x_2 R_1^2}{S_0^1(R_1^1 R_2^2 - R_2^1 R_1^2)}$, $\varphi_2 = \frac{x_2 R_1^1 - x_1 R_2^1}{S_0^2(R_1^1 R_2^2 - R_2^1 R_1^2)}$. The initial fair price of payoff X is

$$\pi_0(X) = \varphi_1 S_0^1 + \varphi_2 S_0^2 = \frac{x_1(R_2^2 - R_2^1) + x_2(R_1^1 - R_1^2)}{R_1^1 R_2^2 - R_2^1 R_1^2}.$$

7. (b) $S_T^1 = 2S_T^2 - S_T^3$, e.g., S_T^1 is a redundant security.

8. (a) Show that the model admits arbitrage. (b) Hint: form a portfolio with zero initial value and strictly positive terminal value.

9. A portfolio with one share of stock and $-\frac{K}{B_T}$ units of a bond replicates the payoff. The initial value of this portfolio is

$$F_0 = 1 \cdot S_0 + \left(-\frac{K}{B_T}\right) \cdot B_0 = S_0 - \frac{K}{1+r},$$

where r is the risk-free rate of interest.

10. (a) $\varphi_A = 0.4$, $\varphi_B = 0.6$. (b) $\varphi = [0.4, 0.6]$; $\Pi_0[\varphi] = 64$. There is an arbitrage opportunity, e.g., $\varphi_A = 0.4$, $\varphi_B = 0.6$, and $\varphi_C = -1$. (c) State price vector $[\Psi_1, \Psi_2] = [\frac{8}{10}, \frac{3}{10}]$ is strictly positive. Risk-neutral probabilities are $\tilde{p}_1 = \frac{8}{11}$, $\tilde{p}_2 = \frac{3}{11}$.

11. (a) Three base assets: zero-coupon bond denoted by S^1, the stock denoted by S^2, and the call option denoted by S^3, where

$$\mathbf{S}_0 = \begin{bmatrix} 10/11 \\ 45 \\ 10 \end{bmatrix} \quad \text{and} \quad \mathbf{D} = \begin{bmatrix} 1 & 1 & 1 \\ 60 & 50 & 30 \\ 20 & 10 & 0 \end{bmatrix}.$$

(b)

$$\varphi_1 = \left[3, -\frac{1}{10}, \frac{1}{5}\right], \quad \varphi_2 = \left[-6, \frac{1}{5}, -\frac{3}{10}\right], \quad \varphi_3 = \left[4, -\frac{1}{10}, \frac{1}{10}\right].$$

The risk-neutral probabilities are $\tilde{p}_1 = \frac{1}{4}$, $\tilde{p}_2 = \frac{3}{5}$, $\tilde{p}_3 = \frac{3}{20}$.
(c) $\varphi = [40, -1, 1]$. No-arbitrage price $P_0 = \Pi_0^\varphi = \frac{15}{11}$.
(d) Show that it gives the same price; arbitrage-free model.

12. (a) Show that $\text{rank}(\mathbf{D}) < 3$. We have $x = \frac{12 - 2Z}{R}$ and $y = \frac{Z}{4} - 1$ s.t. $S_T^2 = xB_T + yS_T^1$.
(b) Show that $\det(\mathbf{D}) \neq 0$. (c) No arbitrage iff $\frac{48}{5} < Z \leqslant 12$ and $R \geqslant 1$.
(d) $Z = \frac{21}{2} = 10.5$ and $R = \frac{12}{11}$; $\Psi_1 = \frac{2}{3}$ and $\Psi_2 = \frac{1}{4}$. The risk-neutral probabilities are $\tilde{p}_1 = R\Psi_1 = \frac{8}{11}$ and $\tilde{p}_2 = R\Psi_2 = \frac{3}{11}$.

13. (a) Yes. (b) Model is incomplete. (c) Payoff is attainable. Replicating portfolios are $\varphi = [x, \frac{1}{5} - x, 0]$, $x \in \mathbb{R}$.

14. (a) Show that Ψ solves $\mathbf{D}\Psi = \mathbf{S}_0$. (b) Yes, since Ψ is strictly positive. (c) $r = \frac{2}{47} \approx 4.255\%$. (d) $[\tilde{p}_1, \tilde{p}_2, \tilde{p}_3] = [\frac{34}{47}, \frac{2}{47}, \frac{11}{47}]$. (e) $\pi_0 = \frac{48}{49}$

15. (b) The unique state-price vector is $\Psi = [\frac{1}{4}, \frac{1}{4}, \frac{3}{20}, \frac{1}{4}]^\top$; risk-neutral probabilities are $\tilde{p}_1 = \frac{5}{18}$, $\tilde{p}_2 = \frac{5}{18}$, $\tilde{p}_3 = \frac{3}{18}$, $\tilde{p}_4 = \frac{5}{18}$. (c) The initial put price is $\pi_0 = \frac{21}{20} = 1.05$. (d) The initial call price is $\pi_0 = \frac{1}{4}$. (e) The initial price is $\pi_0 = 1.2$.

16. (a) Consider the return on the stock $r_S(\omega)$ and show whether $\min_\omega r_S(\omega) < r_B < \max_\omega r_S(\omega)$, where r_B is the return on the bond. The model is incomplete.
(b) $\Psi = [\Psi_1, \Psi_2, \Psi_3]^\top = [\frac{5}{14} - \frac{3}{4}t, t, \frac{25}{42} - \frac{1}{4}t]$, where $0 < t < \frac{10}{21}$. Market model is arbitrage-free and admits a family of state-price vectors since model is incomplete. (c) Thus the bid-ask spread is $(\frac{20}{21}, \frac{25}{21})$. (d) Show that a portfolio that is long the (zero-value) call is an arbitrage.

17. (a) Show that there are no strictly positive state price vectors Ψ. Thus, the model admits arbitrage. You can construct an arbitrage portfolio $\phi = [x, y, z]$ s.t. $\Pi_0^\phi = 0$ and $\Pi_T^\phi(\omega) \geqslant 0$ with $\Pi_T^\phi(\omega^j) > 0$ for some j.
(b) In this case the general solution is: $\Psi_1 = \frac{1}{18} + \frac{1}{3}t$, $\Psi_2 = \frac{4}{9} - \frac{4}{3}t$, $\Psi_3 = t$, $\Psi_4 = \frac{1}{2}$, $t \in \mathbb{R}$. All Ψ_i, $i = 1, 2, 3, 4$ are strictly positive iff $0 < t < \frac{1}{3}$. For example, let $t = \frac{1}{6}$, then $\Psi = [1/9, 2/9, 1/6, 1/2] \gg 0$. Hence, the model is arbitrage-free.

18. The initial fair price of j-th AD security, $\pi_0(\mathcal{E}^j) = \Psi_j$. We have the bid-ask spreads:

$$\pi_0(\mathcal{E}^1) \in \left(\frac{(1+r-m)^+}{(1+r)(u-m)}, \frac{1+r-d}{(1+r)(u-d)} \right),$$

$$\pi_0(\mathcal{E}^2) \in \left(0, \frac{u-1-r}{(1+r)(u-m)} \right) \quad \text{if } m \leqslant 1+r,$$

$$\pi_0(\mathcal{E}^2) \in \left(0, \frac{1+r-d}{(1+r)(m-d)} \right) \quad \text{if } m \geqslant 1+r,$$

$$\pi_0(\mathcal{E}^3) \in \left(\frac{(m-1-r)^+}{(1+r)(m-d)}, \frac{u-1-r}{(1+r)(u-d)} \right).$$

19. (a) $\boldsymbol{\Psi}$ is given by

$$\Psi_1 = -\frac{5}{11} + \frac{1}{2}x + 2y, \quad \Psi_2 = \frac{15}{11} - \frac{3}{2}x - 3yy, \quad \Psi_3 = xy, \quad \Psi_4 = y,$$

$x, y \in \mathbb{R}$. Hence, $\boldsymbol{\Psi}$ is strictly positive iff $x \in \left(0, \frac{10}{11}\right)$ and $y \in \left(\frac{5}{22} - \frac{x}{4}, \frac{5}{11} - \frac{x}{2} \right)$.

(b) $P_0 \in \left(0, \frac{15}{11}\right)$.

20. (a)

$$\mathbf{S}_0 = \begin{bmatrix} 10 \\ 10 \end{bmatrix}, \quad \mathbf{D} = \begin{bmatrix} 8 & 14 \\ 12 & 12 \end{bmatrix}.$$

(b) Since rank $\mathbf{D} = 2 = M$ (or $\det \mathbf{D} \neq 0$), the market is complete. Use the geometric method to show explicitly that $\mathbf{S}_0$ lies inside the cone generated by the two column vectors of $\mathbf{D}$. Or, explicitly, show $\min \frac{S_T^1(\omega)}{S_0^2} < \frac{S_T^2(\omega)}{S_0^1} < \max \frac{S_T^1(\omega)}{S_0^1}$ Hence, the market is arbitrage-free. (c) $p_1 = \frac{2}{3}$, $p_2 = \frac{1}{3}$. (d) Portfolio positions are not finite unless we impose no shorting. For no shorting we have the maximizing portfolio positions $\varphi_1 = 0$, $\varphi_2 = 10$.

21. (a) Show strictly positive $\boldsymbol{\Psi}$ for $s \in \left(\frac{5}{33}, \frac{5}{9} \right)$.

(b) $B_0 = \frac{25}{24} + \frac{s}{8}$. This price is not unique since the bond payoff is not attainable. Bid-ask spread is $B_0 \in \left(\frac{35}{33}, \frac{10}{9} \right) = (1.06061, 1.11111)$.

(c) The call option initial price $C_0 = -\frac{5}{12} + \frac{43}{4}s$ is not unique since it's payoff is not attainable. Bid-ask spread is $C_0 \in \left(\frac{40}{33}, \frac{50}{9} \right) = (1.2121, 5.5556)$.

22. (a) $N = M = 2$. The model is complete with no redundant base assets, i.e., show $\det(\mathbf{D}) \neq 0$. (b) $\tilde{p}_1^{(1)} = \frac{1}{4}$, $\tilde{p}_2^{(1)} = \frac{3}{4}$. (c) $\tilde{p}_1^{(2)} = \frac{2}{3}$, $\tilde{p}_2^{(2)} = \frac{1}{3}$.

(d) For the numeraire asset S^1, we have $(\tilde{p}_1, \tilde{p}_2) = \left(\frac{1}{4}, \frac{3}{4} \right)$. For the numeraire asset S^2, we have $(\tilde{p}_1, \tilde{p}_2) = \left(\frac{2}{3}, \frac{1}{3} \right)$. Show that both give the same price, $C_0 = 2.5$. Note: The price does not depend on the choice of the numeraire. This must be the case since the claim is attainable and no arbitrage implies the Law of One Price.

23. (a)

$$\mathbf{D} = \begin{bmatrix} 1 & 1 & 1 \\ 60 & 50 & 30 \end{bmatrix}.$$

Number of states is greater than number of base assets.

(b) State price is

$$\Psi_1 = \frac{13}{22} - \frac{2}{3}t, \quad \Psi_2 = t, \quad \Psi_3 = \frac{7}{22} - \frac{1}{3}t,$$

for $t \in (0, 39/44)$. The market is arbitrage-free.

(c) The initial price is a function of parameter t: $C_0(t) = \frac{130}{11} - \frac{10}{3}t$. The bid-ask spread is $\left(\frac{195}{22}, \frac{130}{11}\right) \cong (8.86, 11.82)$.

(d) The new payoff matrix and initial price vector are, respectively,

$$\mathbf{D} = \begin{bmatrix} 1 & 1 & 1 \\ 60 & 50 & 30 \\ 20 & 10 & 0 \end{bmatrix} \quad \text{and} \quad \mathbf{S}_0 = \begin{bmatrix} 10/11 \\ 45 \\ 10 \end{bmatrix}.$$

This model is complete since $\det(\mathbf{D}) \neq 0$ and arbitrage-free since the call price is within the bid-ask spread: $10 \in (8.86, 11.82)$.

(e) Using part (c), the unique state-price vector and unique initial prices of the AD securities are:

$$\pi_0(\mathcal{E}^1) = \Psi_1 = \frac{5}{22}, \quad \pi_0(\mathcal{E}^2) = \Psi_2 = \frac{6}{11}, \quad \pi_0(\mathcal{E}^3) = \Psi_3 = \frac{3}{22}.$$

We have $r = 0.1$ and the risk-neutral probabilities are $\tilde{p}_1 = \frac{1}{4}$, $\tilde{p}_2 = \frac{3}{5}$, $\tilde{p}_3 = \frac{3}{20}$.

(f) $P_0 = \frac{15}{11}$.

24. (a) Solving $\mathbf{D}\Psi = \mathbf{S}_0$ gives a unique strictly positive solution $\Psi = \left[\frac{3}{7}, \frac{1}{7}, \frac{4}{7}\right]^\top$. The market is arbitrage-free.

(b) $\Psi_1 + \Psi_2 + \Psi_3 = \frac{8}{7} = (1+r)^{-1}$, hence $r = -\frac{1}{8}$. Replicating portfolio for $[1,1,1]$ is $\phi = \left[\frac{5}{7} - 2t, \frac{1}{7}, \frac{2}{7}, t\right]$, $t \in \mathbb{R}$.

(c) The initial no-arbitrage price of the call is $C_0 = \frac{1}{7} \cong 0.1429$.

(d) The initial no-arbitrage price of the put is $P_0 = \frac{29}{28} \cong 1.0357$.

25. (a) The market is complete iff $c \neq 5$.

(b) $\tilde{p}_1 = \frac{27.5-c}{2c-10}$ and $\tilde{p}_2 = 1 - \tilde{p}_1$, where $\tilde{p}_1 \in (0,1) \implies c \in (12.5, 27.5)$.

(c) The market is arbitrage-free iff $c \in (12.5, 27.5)$.

27. (a) Hint: use $q_j = \frac{g_T(\omega^j)/g_0}{B_T/B_0} \tilde{p}_j$, where $\tilde{p}_j = \widetilde{\mathbb{P}}^{(B)}(\omega^j)$, $j = 1, 2$.

(c) Use $S_- < (1+r)S_0 < S_+$.

(d) Use risk-neutral pricing: $\pi_0(\mathcal{E}^j) = g_0 \widetilde{\mathbb{E}}^{(g)}\left[\frac{\mathcal{E}^j}{g_T}\right]$, $j = 1, 2$.

28. (a) The conditions can be written as a system of three linear equations:

$$\begin{cases} \frac{q_1}{S_+} + \frac{q_2}{S_0} + \frac{q_3}{S_-} = \frac{1}{S_0(1+r)} \\ q_1 + q_2 + q_3 = 1 \\ \left(1 - \frac{S_0}{S_+}\right) q_1 = \frac{C_0}{S_0} \end{cases}$$

(b) $[q_1, q_2, q_3] = \left[\frac{1}{2}, \frac{1}{3}, \frac{1}{6}\right]$.

(c) $C_0 = \frac{40}{3} \cong 13.3333$.

29. (a) $\Psi_1 = \frac{22}{63}$ and $\Psi_2 = \frac{38}{63}$; risk-neutral probabilities are $\tilde{p}_1 = \frac{11}{30}$ and $\tilde{p}_2 = \frac{19}{30}$. The initial no-arbitrage price of the third base asset is $S_0^3 = \frac{220}{63} \cong 3.4921$

(b) $\varphi_1 = \frac{-20c_1+80c_2}{63}$, $\varphi_2 = \frac{c_1-c_2}{15}$; $\Pi_0^\varphi = \frac{22c_1+38c_2}{63}$.

(c) $[c_1, c_2] = [V_T(\omega^1), V_T(\omega^2)] = [20, 5]$. Using (b), $V_0 = \frac{22\cdot20+38\cdot5}{63} = 10$.

30. (a) $\mathbf{D} = \begin{bmatrix} S_+ & S_0 & S_- \\ S_+ - S_- & S_0 - S_- & 0 \\ S_+ - S_0 & 0 & 0 \end{bmatrix}$.

(b) $\det(\mathbf{D}) = -S_-(S_0 - S_-)(S_+ - S_0) < 0$. Since it is nonzero, the 3-by-3 model is complete and has no redundant securities.

(c) $\boldsymbol{\varphi}^{(1)} = [0, 0, \frac{1}{S_+ - S_0}]$, $\boldsymbol{\varphi}^{(2)} = [0, \frac{1}{S_0 - S_-}, -\frac{S_+ - S_-}{(S_+ - S_0)(S_0 - S_-)}]$,
$\boldsymbol{\varphi}^{(3)} = [\frac{1}{S_-}, -\frac{S_0}{S_-(S_0 - S_-)}, \frac{1}{S_0 - S_-}]$.

(d) $C_T^1(\omega) \geqslant C_T^2(\omega)$ for all $\omega \in \Omega$ and $C_T^1(\omega^j) > C_T^2(\omega^j)$ for $j = 1, 2$, hence $C_0^1 > C_0^2$ by the law of one price. [Note: this inequality also follows since the initial prices are given by their respective discounted risk-neutral expected values where (as random variables) $C_T^1 - C_T^2 > 0$. This implies $C_0^1 = \pi_0(C_T^1) > \pi_0(C_T^2) = C_0^2$.

(e) The bid-ask spread is $C_0^2 \in (0, \frac{1}{2})$.

31. (a) $\varphi_X/\varphi_Y = -3$. (b) $\hat{p}_1 = \frac{1}{9}$, $\hat{p}_2 = \frac{2}{9}$, $\hat{p}_3 = \frac{2}{3}$.

32. (a) $\Delta_{\mathrm{opt}} = \frac{\mathrm{Cov}(C_T, S_T)}{\mathrm{Var}(S_T)}$. (b) Show that $\mathrm{Var}(\Pi_T)\Big|_{\Delta = \Delta_{\mathrm{opt}}} = \mathrm{Var}(C_T)\left(1 - \rho^2\right)$ with correlation coefficient $\rho := \mathrm{Corr}(C_T, S_T)$. We have $|\rho| < 1$ for uncorrelated C_T and S_T and $\rho = \pm 1$ if C_T and S_T are linearly dependent.

(c) The optimal Δ's for the three submodels are

$$\Delta_1 = \frac{C^u - C^d}{S^u - S^d}, \quad \Delta_2 = \frac{C^u - C^m}{S^u - S^m}, \quad \Delta_3 = \frac{C^m - C^d}{S^m - S^d},$$

where $S^u = S_0 u$, $S^m = S_0 m$, and $S^d = S_0 d$. Computing $\mathrm{Var}(S_T)$ and $\mathrm{Cov}(C_T, S_T)$, for the trinomial model, and dividing gives the optimal Δ:

$$\begin{aligned}
\Delta_{\mathrm{opt}} &= \frac{\mathrm{Cov}(C_T, S_T)}{\mathrm{Var}(S_T)} \\
&= \frac{(C^u - C^d)(S^u - S^d)p_u p_d + (C^u - C^m)(S^u - S^m)p_u p_m + (C^m - C^d)(S^m - S^d)p_m p_d}{(S^u - S^d)^2 p_u p_d + (S^u - S^m)^2 p_u p_m + (S^m - S^d)^2 p_m p_d} \\
&= \left(\frac{(S^u - S^d)^2 p_u p_d}{(S^u - S^d)^2 p_u p_d + (S^u - S^m)^2 p_u p_m + (S^m - S^d)^2 p_m p_d}\right)\left(\frac{C^u - C^d}{S^u - S^d}\right) \\
&\quad + \left(\frac{(S^u - S^m)^2 p_u p_m}{(S^u - S^d)^2 p_u p_d + (S^u - S^m)^2 p_u p_m + (S^m - S^d)^2 p_m p_d}\right)\left(\frac{C^u - C^m}{S^u - S^m}\right) \\
&\quad + \left(\frac{(S^m - S^d)^2 p_m p_d}{(S^u - S^d)^2 p_u p_d + (S^u - S^m)^2 p_u p_m + (S^m - S^d)^2 p_m p_d}\right)\left(\frac{C^m - C^d}{S^m - S^d}\right) \\
&\equiv a_1 \Delta_1 + a_2 \Delta_2 + a_3 \Delta_3.
\end{aligned}$$

33. Denote $S_T^{ij} = S_T^i(\omega^j)$ and $C_T^j = C_T(\omega^j)$ for $i = 1, 2, \ldots, N$ and $j = 1, 2, \ldots, M$. The terminal value of the hedging portfolio φ is

$$\Pi_T = -C_T + \sum_{i=1}^{N} \varphi_i S_T^i.$$

It may be convenient to define $\mu_C \equiv \mathrm{E}[C_T]$ and $\mu_S^i \equiv \mathrm{E}[S_T^i]$. Calculate the variance $\mathrm{Var}(\Pi_T) = \mathrm{E}\left[(\Pi_T - \mathrm{E}[\Pi_t])^2\right]$ and then set the derivatives w.r.t. the positions in the base assets, φ_k, $k = 1, 2, \ldots, N$, to zero. Show that this leads to a system of linear equations in the portfolio positions which takes the required matrix form $\varphi \Sigma = \mathbf{B}$.

B.6   Chapter 6

1. The events and their respective probabilities are as follows:

   (a) $E = \{HHH, HHT, THH, THT\}$, $\mathbb{P}(E) = \mathbb{P}(H) = p$,

   (b) $E = \{THH, THT, HTH, TTH\}$, $\mathbb{P}(E) = \mathbb{P}(THH) + \mathbb{P}(THT) + \mathbb{P}(HTH) + \mathbb{P}(TTH) = 2p(1-p)$,

   (c) $E = \{TTT\}$, $\mathbb{P}(E) = \mathbb{P}(TTT) = (1-p)^3$.

2. Make use of the fact that $\omega \in A \cap B \iff (\omega \in A$ and $\omega \in B)$ and $\omega \in A \cup B \iff (\omega \in A$ or $\omega \in B)$.

3. Use the properties in question 1.

4. (a) $S_0 = 4$; $S_1 \in \{2, 8\}$; $S_2 \in \{1, 4, 16\}$; $S_3 \in \{1/2, 2, 8, 32\}$. (b) $S_3 = S_0 u^{U_3} d^{3-U_3}$ has range $\{S_{3,0} = 1/2, S_{3,1} = 2, S_{3,2} = 8, S_{3,3} = 32\}$, where

$$\mathbb{P}(S_3 = S_{3,k}) = \binom{3}{k} p^k (1-p)^{3-k} = \binom{3}{k}\left(\frac{1}{4}\right)^k\left(\frac{3}{4}\right)^{3-k}, \quad k = 0, 1, 2, 3,$$

and $\mathrm{E}[S_3] = \frac{343}{128} \simeq 2.6797$.

   (c) Hint: Consider $X := (S_1 S_2 S_3)^{1/3}$, where $S_n = S_0 u^{U_n} d^{n-U_n}$, $n = 1, 2, 3$. The range of X is the set of values

$$\{x_k = S_0 u^{k/3} d^{2-k/3} \,;\, k = 0, 1, 2, 3, 4, 5, 6\} = \{1, \sqrt[3]{4}, \sqrt[3]{16}, 4, 4\sqrt[3]{4}, 8\sqrt[3]{2}, 16\}.$$

The PMF of X is the set of values: $\mathbb{P}(X = x_0) \equiv \mathbb{P}(X = 1)$ and

$$\mathbb{P}(X = 1) = \left(\frac{3}{4}\right)^3 = \frac{27}{64}; \ \mathbb{P}(X = \sqrt[3]{4}) = \frac{1}{4}\left(\frac{3}{4}\right)^2 = \frac{9}{64}; \ \mathbb{P}(X = \sqrt[3]{16}) = \frac{1}{4}\left(\frac{3}{4}\right)^2 = \frac{9}{64};$$

$$\mathbb{P}(X = 4) = \frac{1}{4}\left(\frac{3}{4}\right)^2 + \frac{3}{4}\left(\frac{1}{4}\right)^2 = \frac{12}{64}; \ \mathbb{P}(X = 4\sqrt[3]{4}) = \frac{3}{4}\left(\frac{1}{4}\right)^2 = \frac{3}{64};$$

$$\mathbb{P}(X = 8\sqrt[3]{2}) = \frac{3}{4}\left(\frac{1}{4}\right)^2 = \frac{3}{64}; \ \mathbb{P}(X = 16) = \left(\frac{1}{4}\right)^3 = \frac{1}{64}.$$

Hence $\mathrm{E}[(S_1 S_2 S_3)^{1/3}] \equiv \mathrm{E}[X] \simeq 2.7696$.

5. (a) $\{\mathcal{T} \leqslant 3\} = A_U \cup \{DUU\} = \{UUU, UUD, UDU, UDD, DUU\}$; $\mathbb{P}(\mathcal{T} \leqslant 3) = \frac{19}{64}$.

   (b) By part (a), $\{\mathcal{T} = 3\} = \{DUU\}$; $\mathbb{P}(\mathcal{T} = 3) = \frac{3}{64}$.

   (c) $\{\mathcal{T} > 3\} = \{\mathcal{T} \leqslant 3\}^{\complement} = \Omega_3 \backslash \{\mathcal{T} \leqslant 3\} = \{DUD, DDU, DDD\}$; $\mathbb{P}(\mathcal{T} > 3) = \frac{45}{64}$.

6. (a) $\{\mathcal{T}_b^+ = 0\} = \{\mathcal{T}_b^+ = 1\} = \{\mathcal{T}_b^+ = 3\} = \{\mathcal{T}_b^+ = 5\} = \emptyset$, $\{\mathcal{T}_b^+ = 2\} = A_{UU}$;

   $\{\mathcal{T}_b^+ = 4\} = A_{UDUU} \cup A_{DUUU}$;

   $\{\mathcal{T}_b^+ = 6\} = \{UDDUUU, UDUDUU, DDUUUU, DUDUUU, DUUDUU\}$.

   By definition, $\{\mathcal{T}_b^+ = \infty\} \equiv \{\mathcal{T}_b^+ > 6\} = \{\mathcal{T}_b^+ \leqslant 6\}^{\complement} = \Omega_6 \backslash \{A_{UU} \cup A_{UDUU} \cup A_{DUUU} \cup \{UDDUUU, UDUDUU, DDUUUU, DUDUUU, DUUDUU\}\}$.

   (b) $\mathbb{P}(\mathcal{T}_b^+ = 0) = \mathbb{P}(\mathcal{T}_b^+ = 1) = \mathbb{P}(\mathcal{T}_b^+ = 3) = \mathbb{P}(\mathcal{T}_b^+ = 5) = 0$; $\mathbb{P}(\mathcal{T}_b^+ = 2) = \frac{1}{16}$;

$\mathbb{P}(\mathcal{T}_b^+ = 4) = 2\frac{3}{4}\left(\frac{1}{4}\right)^3 = \frac{3}{128}$; $\mathbb{P}(\mathcal{T}_b^+ = 6) = 5\left(\frac{3}{4}\right)^2\left(\frac{1}{4}\right)^4 = \frac{45}{4096}$; $\mathbb{P}(\mathcal{T}_b^+ = \infty) = 1 - \frac{397}{4096} = \frac{3699}{4096}$.

(c) For $t = 2$: $\widehat{S}_t = \widehat{S}_2 \in \{S_{2,0}, S_{2,1}, u^2\} = \{u^{-2}, 1, u^2\}$ with PMF values:

$$\mathbb{P}(\widehat{S}_2 = u^{-2}) = \frac{9}{16}, \quad \mathbb{P}(\widehat{S}_2 = 1) = \frac{3}{8}, \quad \mathbb{P}(\widehat{S}_2 = u^2) = \frac{1}{16}.$$

For $t = 3$: $\widehat{S}_3 \in \{S_{3,0}, S_{3,1}, S_{3,2}, u^2\} = \{u^{-3}, u^{-1}, u, u^2\}$ with PMF values:

$$\mathbb{P}(\widehat{S}_3 = u^{-3}) = \frac{27}{64}, \quad \mathbb{P}(\widehat{S}_3 = u^{-1}) = \frac{27}{64}, \quad \mathbb{P}(\widehat{S}_3 = u) = \frac{3}{32}, \quad \mathbb{P}(\widehat{S}_3 = u^2) = \frac{1}{16}.$$

For $t = 4$: $\widehat{S}_4 \in \{S_{4,0}, S_{4,1}, S_{4,2}, u^2\} = \{u^{-4}, u^{-2}, 1, u^2\}$ with PMF values:

$$\mathbb{P}(\widehat{S}_4 = u^{-4}) = \frac{81}{256}, \quad \mathbb{P}(\widehat{S}_4 = u^{-2}) = \frac{27}{64}, \quad \mathbb{P}(\widehat{S}_4 = 1) = \frac{45}{256}, \quad \mathbb{P}(\widehat{S}_4 = u^2) = \frac{22}{256}.$$

For $t = 5$: $\widehat{S}_5 \in \{S_{5,0}, S_{5,1}, S_{5,2}, S_{5,3}, u^2\} = \{u^{-5}, u^{-3}, u^{-1}, u, u^2\}$ with PMF values:

$$\mathbb{P}(\widehat{S}_5 = u^{-5}) = \frac{243}{1024}, \quad \mathbb{P}(\widehat{S}_5 = u^{-3}) = \frac{405}{1024}, \quad \mathbb{P}(\widehat{S}_5 = u^{-1}) = \frac{243}{1024},$$
$$\mathbb{P}(\widehat{S}_5 = u) = \frac{45}{1024}, \quad \mathbb{P}(\widehat{S}_5 = u^2) = \frac{88}{1024}.$$

For $t = 6$: $\widehat{S}_6 \in \{S_{6,0}, S_{6,1}, S_{6,2}, S_{6,3}, u^2\} = \{u^{-6}, u^{-4}, u^{-2}, 1, u^2\}$ with PMF values:

$$\mathbb{P}(\widehat{S}_6 = u^{-6}) = \frac{729}{4096}, \quad \mathbb{P}(\widehat{S}_6 = u^{-4}) = \frac{1458}{4096}, \quad \mathbb{P}(\widehat{S}_6 = u^{-2}) = \frac{1134}{4096},$$
$$\mathbb{P}(\widehat{S}_6 = 1) = \frac{378}{4096}, \quad \mathbb{P}(\widehat{S}_6 = u^2) = \frac{397}{4096}.$$

7. (a) Define $E_1 := \{(i,j) : i \in \{2,4,6\}, 1 \leqslant j \leqslant 6\}$ and $E_2 := \{(i,j) : j \in \{2,4,6\}, 1 \leqslant i \leqslant 6\}$. Then, we have four atoms: $A_{EE} = E_1 \cap E_2$, $A_{EO} = E_1 \cap E_2^{\mathsf{C}}$, $A_{OE} = E_1^{\mathsf{C}} \cap E_2$ and $A_{OO} = E_1^{\mathsf{C}} \cap E_2^{\mathsf{C}}$, where $\mathcal{P} = \{A_{EE}, A_{EO}, A_{OE}, A_{OO}\}$.

(b) $\mathcal{P} = \{E, E^{\mathsf{C}}\}$, where

$$E = \{(i,j) : i = j, 1 \leqslant i, j \leqslant 6\} \equiv \{(1,1), (2,2), (3,3), (4,4), (5,5), (6,6)\}.$$

(c) $\mathcal{P} = \{E, E^{\mathsf{C}}\}$, where $E = \{(i,j) : |i - j| = 1, 1 \leqslant i, j \leqslant 6\}$, or equivalently

$$E = \{(1,2), (2,1), (2,3), (3,2), (3,4), (4,3)(4,5), (5,4), (5,6), (6,5)\}.$$

8. (a) The partition $\mathcal{P}(\mathsf{D}_3)$ has four atoms:

$$A_0 = \{\mathsf{D}_3 = 0\} = \{\mathsf{UUU}\}, \quad A_1 = \{\mathsf{D}_3 = 1\} = \{\mathsf{DUU}, \mathsf{UDU}, \mathsf{UUD}\},$$
$$A_2 = \{\mathsf{D}_3 = 2\} = \{\mathsf{DDU}, \mathsf{DUD}, \mathsf{UDD}\}, \quad A_3 = \{\mathsf{D}_3 = 3\} = \{\mathsf{DDD}\}.$$

Hence, $\sigma(\mathsf{D}_3) = \sigma(\mathcal{P}(\mathsf{D}_3)) = 2^{\mathcal{P}(\mathsf{D}_3)}$:

$$\sigma(\mathsf{D}_3) = \{\emptyset, A_0, A_1, A_2, A_3, A_0^{\mathsf{C}}, A_1^{\mathsf{C}}, A_2^{\mathsf{C}}, A_3^{\mathsf{C}}, A_0 \cup A_1, A_0 \cup A_2, A_0 \cup A_3,$$
$$(A_0 \cup A_1)^{\mathsf{C}}, (A_0 \cup A_2)^{\mathsf{C}}, (A_0 \cup A_3)^{\mathsf{C}}, \Omega_3\}.$$

Note: we can also write this set using $(A_0 \cup A_1)^{\complement} = A_2 \cup A_3$, $(A_0 \cup A_2)^{\complement} = A_1 \cup A_3$, $(A_0 \cup A_3)^{\complement} = A_1 \cup A_2$.

(b) We have

$$X_1 = \ln[u^{B_2} d^{1-B_2}] = (\ln u)\, \mathbb{I}_{\{B_2=1\}} + (\ln d)\, \mathbb{I}_{\{B_2=0\}} = (\ln u)\, \mathbb{I}_{\{\omega_2=U\}} + (\ln d)\, \mathbb{I}_{\{\omega_2=D\}},$$

where B_2 is the Bernoulli random variable for the second market move. Similarly,

$$X_2 = \ln[u^{B_3} d^{1-B_3}] = (\ln u)\, \mathbb{I}_{\{B_3=1\}} + (\ln d)\, \mathbb{I}_{\{B_3=0\}} = (\ln u)\, \mathbb{I}_{\{\omega_3=U\}} + (\ln d)\, \mathbb{I}_{\{\omega_3=D\}},$$

where B_3 is the Bernoulli random variable for the third market move. Here we assume $d < u$. Hence, The corresponding 4 atoms are

$$A_1 = A_{*UU} \equiv \{\omega_2 = U, \omega_3 = U\} = \{UUU, DUU\}$$

$$A_2 = A_{*UD} \equiv \{\omega_2 = U, \omega_3 = D\} = \{UUD, DUD\}$$

$$A_3 = A_{*DU} \equiv \{\omega_2 = D, \omega_3 = U\} = \{UDU, DDU\}$$

$$A_4 = A_{*DD} \equiv \{\omega_2 = D, \omega_3 = D\} = \{UDD, DDD\}$$

with σ-algebra $\sigma(X_1, X_2) = \sigma(\{A_1, A_2, A_3, A_4\}) = 2^{\mathcal{P}}$.

9. (a) $n = 11$. (b) unbounded ($n = \infty$). (c) $n = 100$. (d) $n = 0$.

10. $X \equiv S_1 \cdot S_2 \cdot S_3 \in \{0, 1, 2, 3, 4, 5, 6\}$, i.e., $\sigma(X)$ is generated by seven atoms. But $\sigma(S_1, S_2, S_3) = \sigma(\omega_1, \omega_2, \omega_3)$ is generated by eight atoms. i.e., $\sigma(S_1, S_2, S_3) \neq \sigma(S_1 \cdot S_2 \cdot S_3)$.

For the product $S_1 \cdot S_2$ we have $\sigma(S_1 \cdot S_2) = \sigma(\omega_1, \omega_2) = \sigma(S_1, S_2)$.

11. Show that $\sigma(M_1) = \sigma(\{A_U, A_D\})$, $\sigma(M_1 + M_2) \equiv \sigma(\{A_{DD}, A_{DU}, A_{UD}, A_{UU}\})$. Hence, $\sigma(M_1) \subset \sigma(M_1 + M_2)$. However, then show that

$$\sigma(M_1 + M_2 + M_3) = \sigma(\{UUU\}, \{UUD\}, \{UDU\}, \{UDD, DUU\}, \{DUD\}, \{DDU\}, \{DDD\}).$$

That is, not all unions of atoms in $\sigma(M_1 + M_2)$ are contained in $\sigma(M_1 + M_2 + M_3)$, e.g., $A_{DU} = \{DUU, DUD\} \notin \sigma(M_1 + M_2 + M_3)$. Hence, $\sigma(M_1 + M_2) \not\subseteq \sigma(M_1 + M_2 + M_3)$.

12. Hint: use the fact that $\sigma(M_1, \ldots, M_n) = \sigma(X_1, \ldots, X_n) = \sigma(\omega_1, \ldots, \omega_n) \equiv \mathcal{F}_n$.

13. (a) $\mathcal{P}(S_2) = \mathcal{P}(\{S_2 = \frac{1}{4}\}, \{S_2 = 1\}, \{S_2 = 4\}) \equiv \mathcal{P}(A_{DD}, A_{DU} \cup A_{UD}, A_{UU})$.

(b) $E[S_3 \mid S_2] \equiv E[S_3 \mid \sigma(S_2)] = \frac{5}{16}\mathbb{I}_{\{S_2=\frac{1}{4}\}} + \frac{5}{4}\mathbb{I}_{\{S_2=1\}} + 5\mathbb{I}_{\{S_2=4\}}$.

14. (a) Take the logarithm of S_n in terms of U_n, where $u = \frac{1}{d} = e$.

(b) $E[L_n] = n(2p - 1)$.

(c) p^{N-1}.

(d) We have $\{\tau = 0\} = \{\tau = 2\} = \emptyset$, $\{\tau = 1\} = A_U$, $\{\tau = 3\} = A_{DUU}$ and $\{\tau > 3\} \equiv \{\tau = \infty\} = \{\tau \leqslant 3\}^{\complement} = (A_U \cup A_{DUU})^{\complement} = A_{DUD} \cup A_{DDU} \cup A_{DDD}$. Hence, there are three atoms with partition $\mathcal{P}(\tau) = \{A_U, A_{DUU}, A_{DUD} \cup A_{DDU} \cup A_{DDD}\}$.

15. (a) $E[f(S_{n+1}) \mid D_n] = pf(uS_n) + (1-p)f(dS_n)$.

(b) $E[S_1 S_2 \cdots S_n S_{n+1} \mid \mathcal{F}_n] = S_1 S_2 \cdots S_n^2(pu + (1-p)d)$.

(c) $E[S_n \mid U_n \geqslant n - 1] = \frac{E[S_n \mathbb{I}_{\{U_n \geqslant n-1\}}]}{\mathbb{P}(U_n \geqslant n-1)} = S_0 u^{n-1} \frac{n(1-p)d + pu}{n(1-p)+p}$.

16. Letting $B = \{$at least two U's$\}$, $E[D_4 \mid B] = \frac{E[D_4 \mathbb{I}_B]}{\mathbb{P}(B)} = \frac{4(1-p)(3-2p)}{3p^2 - 8p + 6}$.

17. Hint: use $B = \bigcup_{m=1}^{M} (B \cap A_m)$.

19. (a) $E[D_N \mid \mathcal{F}] = D_N$.

(b) Denote $X \equiv U b D$. Use the binomial distribution of U_N to obtain an expression for $E[X \mid U_N = n] = \frac{E[X \, \mathbb{I}_{\{U_N=n\}}]}{\mathbb{P}(U_N=n)}$, $n = 0, 1, \ldots, N$. Finally,

$$E[X \mid U_N] \equiv E[U b D \mid \mathcal{F}] = \sum_{n=1}^{N} \left\{ \frac{1}{\binom{N}{n}} \sum_{k=1}^{n} k \binom{N-k-1}{n-k} \right\} \mathbb{I}_{\{U_N=n\}}.$$

20. $E_n[U_m] = U_n + (m-n)p$, if $n < m$, and $E_n[U_m] = U_m$, if $n \geqslant m$. $E_n[D_m] = D_m$, if $n \geqslant m$, and $E_n[D_m] = D_n + (m-n)(1-p)$, if $n < m$.

21. (a) $(S_n)^2 (u^2 p + d^2(1-p))^{m-n}$. (b) $p = \frac{1-d^2}{u^2-d^2}$.

22. (a) $S_3 \in \{S_{3,k} = S_0 u^k d^{3-k}; k = 0, 1, 2, 3\}$ with PMF $\mathbb{P}(S_3 = S_{3,k}) = \frac{1}{8}\binom{3}{k}$.

(b) $|\sigma(S_3)| = 16$.

(c) $\{S_3 > S_1\} = \{UUU, DUU\}$ and $\{S_3 = S_1\} = \{UUD, UDU, DDU, DUD\}$.

(d) $\mathbb{P}(S_3 \geqslant S_1) = p(2-p)$.

(e) $\mathbb{P}(S_3 > S_1 \mid S_1 = uS_0) = p^2$.

23. (a) $M_1 \in \{1, 2\}$, $M_2 \in \{1, 2, 4\}$, $M_3 \in \{1, 2, 4, 8\}$.

(b) $\mathbb{P}(M_3 = 1) = 1 - p - p^2 + p^3$, $\mathbb{P}(M_3 = 2) = p(1 - p^2)$, $\mathbb{P}(M_3 = 4) = p^2(1-p)$, $\mathbb{P}(M_3 = 8) = p^3$.

(c) $E[M_2 \mid \sigma(S_1)] = 2(1+p)\mathbb{I}_{A_U} + \mathbb{I}_{A_D}$; $E[M_3 \mid \sigma(S_1)] = 2(1 + p + 2p^2)\mathbb{I}_{A_U} + (1 + 2p^2)\mathbb{I}_{A_D}$.

(d) Show that: $\mathcal{P}_0 = \{\Omega_3\}$, $\mathcal{P}_2 = \{A_{UU}, A_{DU}, A_{DD}\}$ and $\mathcal{P}_3 = \{\{UUU\}, \{UUD\}, A_{UD} \cup \{DUU\}, A_{DD} \cup \{DUD\}\}$.

Explain why $\{\mathcal{F}_n := \sigma(M_n)\}_{0 \leqslant n \leqslant 3}$ is not a filtration.

24. Hint: $\{X_n\}_{n=0,1,\ldots}$ is a Doob-Lévy martingale and $\mathcal{T}_\ell$ is a stopping time w.r.t. the given filtration.

25-28, 30-31, 33-35. See martingale examples of Section 6.3.6.

36. $p = \frac{1+r-d}{u-d}$, where we require $d < 1 + r < u$.

37. (a) Show or argue that $\{\mathcal{T} \leqslant t\} \in \mathcal{F}_t$, for any $t \in \{0, 1, \ldots\}$, where $\mathcal{F}_t = \sigma(S_0, S_1, \ldots, S_t)$. (b) Show or argue that $\{\mathcal{T} \leqslant t\} \in \mathcal{F}_t$, for any $t \in \{0, 1, \ldots\}$. (c) Show or argue that $\{\mathcal{T} \leqslant t\} \in \mathcal{F}_t$, for any $t \in \{0, 1, \ldots\}$. (d) Show or argue that $\{\mathcal{T} \leqslant t\} \in \mathcal{F}_t$, for any $t \in \{0, 1, \ldots\}$.

38. Fix any $t \geqslant 0$ and show that $\{\mathcal{T} \wedge m \leqslant t\} \in \mathcal{F}_t$. Hint: It may be useful to consider the identity $\{X \wedge Y \leqslant t\} = \{X \leqslant t\} \cup \{Y \leqslant t\}$ and the basic closure properties of subsets of $\mathcal{F}_t$.

39. Fix any $t \geqslant 0$ and show that $\{\mathcal{T}_1 \wedge \mathcal{T}_2 \leqslant t\} \in \mathcal{F}_t$ and $\{\mathcal{T}_1 \vee \mathcal{T}_2 \leqslant t\} \in \mathcal{F}_t$. A similar hint as in the previous question may be used here.

40. (b) See martingale examples of Section 6.3.6.

(c) $\mathbb{P}(M_\mathcal{T} = a) = \frac{1-(q/p)^b}{(q/p)^a-(q/p)^b}$.

(d) $E[\mathcal{T}] = \frac{b}{p-q} - \frac{b-a}{p-q}\frac{1-(p/q)^b}{1-(p/q)^{b-a}}$.

B.7   Chapter 7

1. (a) $\mathbb{P}(S_t = S_{t,n}, \ S_v = S_{v,m}) = \binom{v}{m}\binom{t-v}{n-m}p^n(1-p)^{t-n}$.

    (b) $\mathbb{P}(S_t = S_{t,n} \mid S_v = S_{v,m}) = \binom{t-v}{n-m}p^{n-m}(1-p)^{(t-v)-(n-m)}$.

    Note: the conditional probability is zero for $n - m < 0$ or $n - m > t - v$, as implied by the definition of $\binom{t-v}{n-m}$.

2. (a) $\mathbb{P}(S_3 = 93.15|S_1 = 115) = (1-p)^2$; $\mathbb{P}(S_3 = 119.025|S_1 = 115) = 2p(1-p)$;
    $\mathbb{P}(S_3 = 152.0875|S_1 = 115) = p^2$.

    (b) $\mathbb{P}(S_3 = 72.9 \mid S_1 = 90) = (1-p)^2$; $\mathbb{P}(S_3 = 93.15 \mid S_1 = 90) = 2p(1-p)$;
    $\mathbb{P}(S_3 = 119.025 \mid S_1 = 90) = p^2$.

3. (a) $P_{3,0} = (10-1)^+ = 9$, $P_{3,1} = (10-3)^+ = 7$, $P_{3,2} = (10-9)^+ = 1$, $P_{3,3} = (10-27)^+ = 0$; $P_{2,0} = 6$, $P_{2,1} = 2$, $P_{2,2} = 0.2$, $P_{1,0} = 2.4$, $P_{1,1} = 0.52$, $P_{0,0} = 0.792$.

    (b) $\Delta_{0,0} = -0.235$; $\Delta_{1,0} = -1$; $\Delta_{1,1} = -0.15$; $\Delta_{2,0} = -1$; $\Delta_{2,1} = -1$; $\Delta_{2,2} \cong -0.0556$.

5. Hint: make use of induction on the wealth equation, where $\Pi_0 = V_0$.

6. $V_0 = \sum_{t=n}^{T}(1+r)^{-t}\binom{t-1}{n-1}\tilde{p}^n(1-\tilde{p})^{t-n}$, where $\tilde{p} = (1+r-d)/(u-d)$.

7. (a) $V_t(S,G) = \dfrac{\tilde{p}V_{t+1}\left(Su,(G^{t+1}Su)^{\frac{1}{t+2}}\right)+(1-\tilde{p})V_{t+1}\left(Sd,(G^{t+1}Sd)^{\frac{1}{t+2}}\right)}{1+r}$,

    for $t = 0,1,\ldots,T-1$, where $\tilde{p} = (1+r-d)/(u-d)$ and $V_T(S,G) = (G-K)^+$.

    (b) $\Delta_t(S,G) = \dfrac{V_{t+1}\left(Su,(G^{t+1}Su)^{\frac{1}{t+2}}\right)-V_{t+1}\left(Sd,(G^{t+1}Sd)^{\frac{1}{t+2}}\right)}{S(u-d)}$

    for $t = 0,1,\ldots,T-1$.

8. (a) $V_3(\mathrm{UUU}) = 6.69694$, $V_3(\mathrm{UUD}) = 3.16726$, $V_3(\mathrm{UDU}) = 0.485281$, and $V_3(\omega) = 0$ for all other outcomes $\omega \in \Omega_3$. For $t = 0,1,2$:

    $$V_2(\mathrm{UU}) = 4.651616, \ V_2(\mathrm{UD}) = 0.291169, \ V_2(\mathrm{DU}) = 0, \ V_2(\mathrm{DD}) = 0,$$
    $$V_1(\mathrm{U}) = 2.8492, \ V_1(\mathrm{D}) = 0, \ V_0 = 1.70952.$$

    (b) $\Delta_2(\mathrm{UU}) = 0.1961$, $\Delta_2(\mathrm{UD}) = 0.081$, $\Delta_2(\mathrm{DU}) = 0$, $\Delta_2(\mathrm{DD}) = 0$, $\Delta_1(\mathrm{U}) = 0.3637$, $\Delta_1(\mathrm{D}) = 0$, $\Delta_0 = 0.3564$.

9. (a) $\widetilde{\mathbb{E}}_0[A_T] = \dfrac{S_0}{T+1}\dfrac{(1+r)^{T+1}-1}{r}$.

    (b) $C_0 - P_0 = \dfrac{(1+r)-(1+r)^{-T}}{r(T+1)}S_0 - (1+r)^{-T}K$.

10. (a) Show that $\{[S_t, M_t]\}_{t \geqslant 0}$ is Markovian. Hence, argue that

    $$(1+r)^{-(T-t)}\widetilde{\mathbb{E}}_t\left[(M_T-K)^+\right] = C_t(S_t, M_t),$$
    $$(1+r)^{-(T-t)}\widetilde{\mathbb{E}}_t\left[(K-M_T)^+\right] = P_t(S_t, M_t),$$

    for all $t = 0,1,\ldots,T$.

(b) The no-arbitrage prices of the lookback call and put options are, respectively,

$$C_0 = (1+r)^{-3}\Big((S_0u^3 - K)^+\tilde{p}^3 + (S_0u^2 - K)^+\tilde{p}^2(1-\tilde{p})$$

$$+ (S_0u - K)^+(2\tilde{p}^2(1-\tilde{p}) + \tilde{p}(1-\tilde{p})^2) + (S_0 - K)^+(2\tilde{p}(1-\tilde{p})^2 + (1-\tilde{p})^3)\Big),$$

$$P_0 = (1+r)^{-3}\Big((K - S_0u^3)^+\tilde{p}^3 + (K - S_0u^2)^+\tilde{p}^2(1-\tilde{p})$$

$$+ (K - S_0u)^+(2\tilde{p}^2(1-\tilde{p}) + \tilde{p}(1-\tilde{p})^2) + (K - S_0)^+(2\tilde{p}(1-\tilde{p})^2 + (1-\tilde{p})^3)\Big),$$

where $\tilde{p} = \frac{1+r-d}{u-d}$.

11. (a) Hint: use (7.4) and the fact that $q_{t+1}Y_{t+1}$ is strictly positive.

(b) Show that $\widetilde{E}_t^{(B)}\left[\frac{S_{t+1}}{c_{t+1}B_{t+1}}\right] = \frac{S_t}{c_tB_t}$. You can use $\widetilde{E}_t^{(B)}[Y_{t+1}] = 1 + r$.

(c) Show that $\widetilde{E}_t^{(S)}\left[\frac{c_{t+1}B_{t+1}}{S_{t+1}}\right] = \frac{c_tB_t}{S_t}$.

(d) Make use of (7.4).

(e) Make use of (7.16) to show that $\widetilde{E}_t\left[\frac{\Pi_{t+1}}{B_{t+1}}\right] = \frac{\Pi_t}{B_t}$.

(f) Make use of (7.16) to show that $\widetilde{E}_t^{(S)}\left[\frac{c_{t+1}\Pi_{t+1}}{S_{t+1}}\right] = \frac{c_t\Pi_t}{S_t}$.

12. $\alpha = \frac{1+r-d}{u-d} \cdot \frac{u}{1+r}$, $\beta = 1 - \alpha = \frac{u-(1+r)}{u-d} \cdot \frac{d}{1+r}$.

13. (a)

$$V_t(\omega_1,\ldots,\omega_t) = \begin{cases} \dfrac{\tilde{p}^{\#U(\bar{\omega}_{t+1}\ldots\bar{\omega}_T)} \cdot (1-\tilde{p})^{\#D(\bar{\omega}_{t+1}\ldots\bar{\omega}_T)}}{(1+r)^{T-t}} & \text{if } \omega_1 = \bar{\omega}_1, \ldots, \omega_T = \bar{\omega}_T \\ 0 & \text{otherwise.} \end{cases}$$

(b) The delta hedge process is given by

$$\Delta_t(\omega_1,\ldots,\omega_t) = \begin{cases} \dfrac{V_{t+1}(\bar{\omega}_1\ldots\bar{\omega}_t\mathsf{U})}{S_t(\bar{\omega}_1\ldots\bar{\omega}_t)(u-d)} & \text{if } \omega_k = \bar{\omega}_k \text{ for } k = 1,2\ldots,t, \ \bar{\omega}_{t+1} = \mathsf{U} \\ -\dfrac{V_{t+1}(\bar{\omega}_1\ldots\bar{\omega}_t\mathsf{D})}{S_t(\bar{\omega}_1\ldots\bar{\omega}_t)(u-d)} & \text{if } \omega_k = \bar{\omega}_k \text{ for } k = 1,2\ldots,t, \ \bar{\omega}_{t+1} = \mathsf{D} \\ 0 & \text{otherwise.} \end{cases}$$

15. (a) $V_t(S_t) = (1+r)^{-(T-t)}K\mathcal{B}(m_{T-t}; T-t, \tilde{p}) + S_t(1 - \mathcal{B}(m_{T-t}; T-t, \tilde{q}))$

where $\tilde{q} = \frac{u}{1+r}\tilde{p}$, $\tilde{p} = \frac{1+r-d}{u-d}$ and

$$m_{T-t} \equiv m_{T-t}(S_t, K) = \left\lceil \frac{\left|\ln\left(\frac{K}{S_t}\right) - (T-t)\ln d\right|}{\ln\left(\frac{u}{d}\right)} \right\rceil,$$

for $K \in [S_td^{T-t}, S_tu^{T-t}]$; $m_{T-t} \equiv -\infty$ for $K < S_td^{T-t}$; $m_{T-t} \equiv T-t$ for $K > S_tu^{T-t}$.

(b) $V_t^b(S_t) = S_t\mathcal{B}(m_{T-t}; T-t, \tilde{q}) + K(1+r)^{-(T-t)}(1 - \mathcal{B}(m_{T-t}; T-t, \tilde{p}))$

with $\tilde{q}, \tilde{p}, m_{T-t}$ given in (a).

16. (a) $V_t(S_t) = S_t[1 - \mathcal{B}(m_{T-t}; T-t, \tilde{q})]$ where $\tilde{q} = \frac{u}{1+r}\tilde{p} = \frac{u}{1+r}\frac{1+r-d}{u-d}$ and m_{T-t} given in 15.(a).

(b)

$$V_t(S_t) = P_t^E(S_t; K_2) - P_t^E(S_t; K_1)$$

$$= \frac{K_2}{(1+r)^{T-t}} \mathcal{B}(m_{T-t}(S_t, K_2); T-t, \tilde{p}) - S_t \, \mathcal{B}(m_{T-t}(S_t, K_2); T-t, \tilde{q})$$

$$- \frac{K_1}{(1+r)^{T-t}} \mathcal{B}(m_{T-t}(S_t, K_1); T-t, \tilde{p}) + S_t \, \mathcal{B}(m_{T-t}(S_t, K_1); T-t, \tilde{q}),$$

with $\tilde{q}$, $\tilde{p}$ as in (a); $m_{T-t}(S_t, K_1)$ and $m_{T-t}(S_t, K_2)$ given by $m_{T-t}(S_t, K)$ in 15.(a) with respective strikes $K = K_1$ and $K = K_2$.

(c)

$$V_t(S_t) = P_t^E(S_t; K_1) + C_t^E(S_t; K_2)$$

$$= \frac{K_1}{(1+r)^{T-t}} \mathcal{B}(m_{T-t}(S_t, K_1); T-t, \tilde{p}) - S_t \, \mathcal{B}(m_{T-t}(S_t, K_1); T-t, \tilde{q})$$

$$+ S_t \left(1 - \mathcal{B}(m_{T-t}(S_t, K_2); T-t, \tilde{q})\right)$$

$$- \frac{K_2}{(1+r)^{T-t}} \left(1 - \mathcal{B}(m_{T-t}(S_t, K_2); T-t, \tilde{p})\right),$$

with $\tilde{q}$, $\tilde{p}$, $m_{T-t}(S_t, K_1)$ and $m_{T-t}(S_t, K_2)$ as in (b).

17. Hint: make use of the no-arbitrage condition $d < 1 + r < u$.

18. (a) $C_0^E \cong 0.2977$.

(b) In this case, you may argue that $C_t^A = C_t^E$ for all $t = 0, 1, 2, 3$. i.e., $C_0^A = C_0^E \cong 0.2977$.

19. (a) $C_0^E \cong 1.8965$. (b) $C_0^A \cong 5.1871$.

20. The prices $V_t(S)$ of the American straddle are: $V_3(1) = 7$, $V_3(3) = 5$, $V_3(9) = 1$, $V_3(27) = 19$; $V_2(2) = 6$, $V_2(6) = 2$, $V_2(18) = \frac{58}{5}$, $V_1(4) = 4$, $V_1(12) = \frac{184}{25}$; $V_0(8) = \frac{652}{125}$. The consumption values $C_t(S)$ are: $C_2(2) = 1.6$, $C_2(6) = 0.4$, $C_2(18) = 0$; $C_1(4) = 1.6$, $C_1(12) = 0$; $C_0(8) = 0$. The delta hedge positions $\Delta_t(S)$ are: $\Delta_2(2) = -1$, $\Delta_2(6) = -0.6$, $\Delta_2(18) = 1$; $\Delta_1(4) = -1$, $\Delta_1(12) = 0.8$; $\Delta_0(8) = 0.42$. This American straddle should be exercised early at $t = 1$ when $S_1 = 4$, or at time $t = 2$ when $S_2 \in \{2, 6\}$, or at time $t = 3$ when the option is in the money. That is, the optimal exercise rule, τ^*, is as follows:

$$\tau^*(\omega_1\omega_2\omega_3) = \begin{cases} 1 & \text{if } \omega_1 = \mathsf{D} \\ 2 & \text{if } \omega_1\omega_2 = \mathsf{UD} \\ 3 & \text{if } \omega_1\omega_2 = \mathsf{UU} \end{cases}$$

21. Note: $\Lambda_t = \Lambda(S_t, A_t) = (4 - A_t)^+$ is nonzero only at the node $(S_3, A_3) = (1, \frac{15}{4})$, with value $\Lambda(1, \frac{15}{4}) = 0.25$. Hence, it is not optimal to exercise the option early so its value is equal to the value of the corresponding European-style Asian option with maturity $T = 3$ and payoff $\Lambda_3 = (4 - A_3)^+$. Therefore, the consumption process is zero. The optimal stopping rule is given by

$$\tau^*(\omega) = \begin{cases} 3 & \text{if } \omega = \mathsf{DDD} \\ \infty & \text{otherwise} \end{cases}$$

By backward recurrence, with $V_3(1, \frac{15}{4}) = 0.25$ and $V_3(S_3, A_3) = 0$ for all other nodes (S_3, A_3). We have the following prices at time $t = 2, 1, 0$:

$$V_2(2, \frac{14}{3}) = 0.05, \ V_2(6, 6) = 0, \ V_2(6, \frac{26}{3}) = 0, \ V_2(18, \frac{38}{3}) = 0,$$
$$V_1(4, 6) = 0.01, \ V_1(12, 10) = 0,$$
$$V_0 = V_0(8, 8) = 0.002.$$

The delta positions, $\Delta_t(S, A)$, $t = 2, 1, 0$, are:

$$\Delta_2(2, \frac{14}{3}) = -0.125, \ \Delta_2(6, 6) = 0, \ \Delta_2(6, \frac{26}{3}) = 0, \ \Delta_2(18, \frac{38}{3}) = 0,$$
$$\Delta_1(4, 6) = -0.0125, \ \Delta_1(12, 10) = 0,$$
$$\Delta_0 = \Delta_0(8, 8) = -0.00125.$$

22. Hint: Use the optimal exercise policy for the derivative in the previous example.

23. (a) The American prices $V_t(S_t)$ are:

$V_3(1) = 3, \ V_3(3) = 1, \ V_3(9) = 0, \ V_3(27) = 15,$
$V_2(2) = 2, \ V_2(6) = 0.2, \ V_2(18) = 9, \ V_1(4) = 0.52, \ V_1(12) = 5.44, \ V_0 \equiv V_0(8) = 3.368.$

The only nonzero consumption value is $C_2(2) = 0.8$.

(b) $(\Delta_0, \beta_0) = (0.615, -3.03125)$.

(c) The optimal exercise rule is

$$\tau^*(\omega) = \begin{cases} 2 & \text{if } \omega \in A_{DD} \\ 3 & \text{if } \omega \in \{UUU, UDD, DUD\} \end{cases}$$

and otherwise $\tau^* = \infty$.

(d) Make use of part (c).

24. (a) The American prices $V_t(S_t)$ are:

$V_3(2.53125) = 0, \ V_3(4.21875) = 0, \ V_3(7.03125) = 1.53125, \ V_3(11.71875) = 0,$
$V_2(3.375) = 0, \ V_2(5.625) = 1.1484375, \ V_2(9.375) = 1.125,$
$V_1(4.5) = 0.861328125, \ V_1(7.5) = 2, \ V_0 \equiv V_0(6) = 1.57177734.$

The nonzero consumption values are: $C_2(9.375) = 0.997395833, \ C_1(7.5) = 1.06054688$.

(b) $(\Delta_0, \beta_0) = (0.379557293, -1.21921877)$.

(c) The optimal exercise rule is

$$\tau^*(\omega) = \begin{cases} 1 & \text{if } \omega \in A_U \\ 3 & \text{if } \omega \in \{DUU\} \end{cases}$$

and otherwise $\tau^* = \infty$.

(d) Make use of part (c).

25. For $t = 1, 2, 3$:

$$V_{t-1}(S, A) = \max\left\{(S - A)^+, \frac{3}{5}V_t\left(\frac{3}{2}S, \frac{tA + \frac{3}{2}S}{t+1}\right) + \frac{1}{5}V_t\left(\frac{1}{2}S, \frac{tA + \frac{1}{2}S}{t+1}\right)\right\}$$

where $V_3(S, A) = (S - A)^+$. The nodes (S_t, A_t), $t = 0, 1, 2, 3$, are as in Exercise 21. At expiry we have: $V_3(1, \frac{15}{4}) = V_3(3, \frac{17}{4}) = V_3(3, \frac{21}{4}) = V_3(3, \frac{29}{4}) = 0$, $V_3(9, \frac{25}{4}) = \frac{11}{4}$, $V_3(9, \frac{35}{4}) = \frac{1}{4}$, $V_3(9, \frac{47}{4}) = V_3(27, \frac{65}{4}) = 0$. For $t = 2, 1, 0$:

$$V_2(2, \frac{14}{3}) = 0, \ V_2(6, 6) = 1.65, \ V_2(6, \frac{26}{3}) = 0.15, \ V_2(18, \frac{38}{3}) = \frac{16}{3} \approx 5.333,$$

$$V_1(4, 6) = 0.99, \ V_1(12, 10) = 3.23,$$

$$V_0 = V_0(8, 8) = 2.136.$$

(b) For $t = 1, 2, 3$: $\Delta_{t-1}(S, A) = \dfrac{V_t\left(\frac{3}{2}S, \frac{tA + \frac{3}{2}S}{t+1}\right) - V_t\left(\frac{1}{2}S, \frac{tA + \frac{1}{2}S}{t+1}\right)}{S}$.

(c) The deltas $\Delta_t = \Delta_t(S_t, A_t)$ are

$$\Delta_2(2, \frac{14}{3}) = 0, \ \Delta_2(6, 6) = \frac{11}{24} \approx 0.45833, \ \Delta_2(6, \frac{26}{3}) = \frac{1}{24} \approx 0.041667,$$

$$\Delta_2(18, \frac{38}{3}) = 0, \ \Delta_1(4, 6) = 0.4125, \ \Delta_1(12, 10) \approx 0.43194,$$

$$\Delta_0 = \Delta_0(8, 8) = 0.28,$$

and $\beta_t B_t = \beta_t(S_t, A_t)B_t = V_t(S_t, A_t) - \Delta_t(S_t, A_t)S_t$ values are

$$\beta_2(2, \frac{14}{3})B_2 = 0, \ \beta_2(6, 6)B_2 = -1.1, \ \beta_2(6, \frac{26}{3})B_2 = -0.1,$$

$$\beta_2(18, \frac{38}{3})B_2 = \frac{16}{3} \approx 5.33333, \ \beta_1(4, 6)B_1 = -0.66, \ \beta_1(12, 10)B_1 \approx -1.95333,$$

$$\beta_0 B_0 = -0.104.$$

B.8   Chapter 8

2. On a given time-t outcome $\omega_1 \dots \omega_t$:

$$\Delta_t(\omega_1\omega_2 \dots \omega_t) = \frac{V_{t+1}(\omega_1\omega_2 \dots \omega_t\mathsf{U}) - V_{t+1}(\omega_1\omega_2 \dots \omega_t\mathsf{D})}{S_{t+1}(\omega_1\omega_2 \dots \omega_t\mathsf{U}) - S_{t+1}(\omega_1\omega_2 \dots \omega_t\mathsf{D})}$$

$$= \frac{V_{t+1}(\omega_1\omega_2 \dots \omega_t\mathsf{U}) - V_{t+1}(\omega_1\omega_2 \dots \omega_t\mathsf{D})}{(u_{t+1}(\omega_1\omega_2 \dots \omega_t) - d_{t+1}(\omega_1\omega_2 \dots \omega_t))S_t(\omega_1\omega_2 \dots \omega_t)}$$

for $t = 1, \dots, T - 1$ and $\Delta_0 = \frac{V_1(\mathsf{U}) - V_1(\mathsf{D})}{(u_1 - d_1)S_0}$.

3. (a) We can label the time-1 atoms as $A_1^1 = \{\omega^1, \omega^2\}$, $A_1^2 = \{\omega^3, \omega^4\}$ and at time-2: $A_2^1 = \{\omega^1\}$, $A_2^2 = \{\omega^2\}$, $A_2^3 = \{\omega^3\}$, $A_2^4 = \{\omega^4\}$. By the single-step martingale property: $\widetilde{\mathbb{P}}(A_1^1|A_0) \equiv \widetilde{\mathbb{P}}(A_1^1) = \frac{2}{3}$, $\widetilde{\mathbb{P}}(A_1^2|A_0) \equiv \widetilde{\mathbb{P}}(A_1^2) = \frac{1}{3}$; $\widetilde{\mathbb{P}}(A_2^1|A_1^1) = \frac{17}{22}$, $\widetilde{\mathbb{P}}(A_2^2|A_1^1) = \frac{5}{22}$, $\widetilde{\mathbb{P}}(A_2^3|A_1^2) = \frac{10}{11}$, $\widetilde{\mathbb{P}}(A_2^4|A_1^2) = \frac{1}{11}$. Hence, $\widetilde{\mathbb{P}}(\omega^1) = \frac{17}{33}$, $\widetilde{\mathbb{P}}(\omega^2) = \frac{5}{33}$, $\widetilde{\mathbb{P}}(\omega^3) = \frac{10}{33}$, $\widetilde{\mathbb{P}}(\omega^4) = \frac{1}{33}$.

(b) We now have the 2×2 single-period sub-model with (initial time-1) stock price $S_1 = S_1(\omega^1) = 60$ and time-2 stock prices $(S_2(\omega^1), S_2(\omega^2)) = (65, 50)$. Note: arbitrage exists since $S_2(\omega^1) < \frac{B_2}{B_1} S_1(\omega^1) \simeq 65.4545$. At time-1 with $S_1 = 60$ (at node A_1^1) we form a portfolio with positions $(\beta_1, \Delta_1) = (y, -y\frac{B_1}{S_1}) = (y, -\frac{11}{12}y)$, for any $y > 0$. Hence, this portfolio has zero time-1 value, $\Pi_1 = \beta_1 B_1 + \Delta_1 S_1 = 0$, and strictly positive time-2 value: $\Pi_1(\omega^1) = y[60 - \frac{11}{12} \cdot 65] > 0$ and $\Pi_1(\omega^2) = y[60 - \frac{11}{12} \cdot 50] > 0$.

4. In each case, apply (8.29) and solve the system of three equations in three unknowns $(\beta_t, \Delta_t^1, \Delta_t^2) \equiv (\beta_t(A_t), \Delta_t^1(A_t), \Delta_t^2(A_t))$ at each time-t node for $t = 1, 0$.

(a) At node A_1^1: $(\beta_1, \Delta_1^1, \Delta_1^2) = (185, -1, -1)$. At A_1^2: $(\beta_1, \Delta_1^1, \Delta_1^2) = (130, 0, -\frac{2}{3})$. At A_1^3: $(\beta_1, \Delta_1^1, \Delta_1^2) = (\frac{38}{3}, \frac{6}{5}, -\frac{4}{15})$.
At the initial node A_0: $(\beta_0, \Delta_0^1, \Delta_0^2) = (\frac{10,514}{27}, -\frac{10}{3}, -\frac{238}{135})$.

(b) At node A_1^1: $(\beta_1, \Delta_1^1, \Delta_1^2) = (0, -1, 1)$. At A_1^2: $(\beta_1, \Delta_1^1, \Delta_1^2) = (0, -1, 1)$. At A_1^3: $(\beta_1, \Delta_1^1, \Delta_1^2) = (56, -\frac{8}{5}, \frac{4}{5})$. At the initial node A_0: $(\beta_0, \Delta_0^1, \Delta_0^2) = (\frac{46}{3}, -1, \frac{13}{15})$.

Hint: For part (c) and (d) you can combine a simple replication argument and the result in part (b).

(c) At A_1^1: $(\beta_1, \Delta_1^1, \Delta_1^2) = (0, 0, 1)$. At A_1^2: $(\beta_1, \Delta_1^1, \Delta_1^2) = (0, 0, 1)$. At A_1^3: $(\beta_1, \Delta_1^1, \Delta_1^2) = (56, -\frac{3}{5}, \frac{4}{5})$. At the initial node A_0: $(\beta_0, \Delta_0^1, \Delta_0^2) = (\frac{46}{3}, 0, \frac{13}{15})$.

(d) At node A_1^1: $(\beta_1, \Delta_1^1, \Delta_1^2) = (0, 1, 0)$. At A_1^2: $(\beta_1, \Delta_1^1, \Delta_1^2) = (0, 1, 0)$. At A_1^3: $(\beta_1, \Delta_1^1, \Delta_1^2) = (-56, \frac{8}{5}, \frac{1}{5})$. At the initial node A_0: $(\beta_0, \Delta_0^1, \Delta_0^2) = (-\frac{46}{3}, 1, \frac{2}{15})$.

5. (a) $\varrho_t = \frac{S_t^1/S_0^1}{B_t/B_0} = \frac{S_t^1}{S_0^1} = \frac{S_t^1}{50}$, $t = 0, 1, 2$. At time $t = 0$, $\varrho_0 \equiv 1$. For $t = 1$, $\varrho_1(A_1^1) = 1$, $\varrho_1(A_1^2) = \frac{4}{5}$, $\varrho_1(A_1^3) = \frac{6}{5}$. For $t = 2$, $\varrho_2(A_2^1) = \varrho_2(A_2^2) = \varrho_2(A_2^4) = \frac{9}{10}$, $\varrho_2(A_2^3) = \varrho_2(A_2^7) = \frac{6}{5}$, $\varrho_2(A_2^5) = \frac{7}{10}$, $\varrho_2(A_2^6) = \frac{4}{5}$, $\varrho_2(A_2^8) = \frac{11}{10}$, $\varrho_2(A_2^9) = \frac{7}{5}$.

(b) Denote $\widehat{\mathbb{P}} \equiv \widehat{\mathbb{P}}^{(S^1)}$ and $\widetilde{\mathbb{P}} \equiv \widetilde{\mathbb{P}}^{(B)}$. We have $\widehat{\mathbb{P}}(\omega^i) = \varrho_2(\omega^i)\widetilde{\mathbb{P}}(\omega^i)$, with probabilities $\widehat{\mathbb{P}}(\omega^1) = \widehat{\mathbb{P}}(\omega^2) = \frac{1}{10}$, $\widehat{\mathbb{P}}(\omega^3) = \frac{2}{15}$, $\widehat{\mathbb{P}}(\omega^4) = \frac{1}{15}$, $\widehat{\mathbb{P}}(\omega^5) = \frac{7}{135}$, $\widehat{\mathbb{P}}(\omega^6) = \frac{4}{27}$, $\widehat{\mathbb{P}}(\omega^7) = \frac{12}{75}$, $\widehat{\mathbb{P}}(\omega^8) = \frac{11}{75}$, $\widehat{\mathbb{P}}(\omega^9) = \frac{7}{75}$.

6. $V_n \in \{2^{2k_1-n}3^{2k_2-n}4^{2k_3-n} \equiv \frac{2^{2k_1}3^{2k_2}4^{2k_3}}{(4!)^n}, \; 0 \leqslant k_1, k_2, k_3 \leqslant n\}$.

$\mathbb{P}\left(V_n = 2^{2k_1-n}3^{2k_2-n}4^{2k_3-n}\right) = \binom{n}{k_1}\binom{n}{k_2}\binom{n}{k_3}p_1^{k_1}p_2^{k_2}p_3^{k_3}(1-p_1)^{n-k_1}(1-p_2)^{n-k_2}(1-p_3)^{n-k_3}$ where $p_1 = 0.4, p_2 = 0.6, p_3 = 0.5$.

7. (a) This follows immediately by considering the time-0 case of (8.81). [Remark: The result can also be re-derived by showing that $D_{m-1} - D_m = \frac{B_0}{B_m}r_m$ and then taking the expectation $\widetilde{\mathbb{E}}_0[\;]$ of both sides.] (b) Use similar steps as in the derivation of (8.81).

8. (a) Hint: use the risk-neutral pricing formula in (8.84), with $C_n = K - r_n$, and use the result in Exercise 8.7. In the end, collapse the term with the telescoping sum.

(b) $V_0 \equiv V_0^{\text{Swap}(3)} \simeq -0.02725$.

(c) Generally: $K_s = \frac{1 - Z_{0,m}}{\sum_{n=1}^m Z_{0,n}}$; $K_s \simeq 5.0005\%$.

9. (a) Hint: make use of risk-neutral pricing formula in (8.86) and the identity $(K - r_n)^+ - (r_n - K)^+ = K - r_n$.

(b) Hint: appropriately apply (8.87), giving $V_0^{\text{Cap}(3)} \simeq 0.0290247$; $V_0^{\text{Floor}(3)} \simeq 0.00178$.

10. $\widetilde{\mathbb{P}}(\text{DD}) \simeq 0.43213, \widetilde{\mathbb{P}}(\text{DU}) \simeq 0.18326, \widetilde{\mathbb{P}}(\text{UD}) \simeq 0.18053, \widetilde{\mathbb{P}}(\text{UU}) \simeq 0.20408$

12. (a) $Z_{2,1;3} \simeq 0.9756098$, $Z_{2,2;3} \simeq 0.9732360$, $Z_{2,3;3} \simeq 0.9685230$, $Z_{2,4;3} \simeq 0.9638554$, $Z_{2,5;3} \simeq 0.9592326$, $Z_{2,6;3} \simeq 0.9569378$, $Z_{1,1;3} \simeq 0.945658$, $Z_{1,2;3} \simeq 0.933516$, $Z_{1,3;3} \simeq 0.920684$, $Z_{0,0;3} \equiv Z_{0,3} = 0.91289$, $Z_{1,1;2} \simeq 0.97087$, $Z_{1,2;2} \simeq 0.96618$, $Z_{1,3;2} \simeq 0.96154$, $Z_{0,0;2} \equiv Z_{0,2} = 0.94649$, $Z_{0,0;1} \equiv Z_{0,1} = 0.98039$.

(b) $V_0^{\mathrm{Swap}(3)} = 0.02648$, $V_0^{\mathrm{Cap}(3)} = 0.001974$, and $V_0^{\mathrm{Floor}(3)} = 0.028454$.

13. (a) Hint: use the fact that the money market portion grows at rate R_n within period $[n, n+1]$ and note that $Z_{k,m} = 0$ for all values $k > m$ and that $Z_{k,k} = 1$ for all k.

(b) Hint: to show $\widetilde{\mathrm{E}}_n[D_{n+1}\Pi_{n+1}] = D_n\Pi_n$ use the r.h.s. of the wealth equation in part (a) within the expectation. Then, make use of the martingale property of $\{D_n Z_{n,m}\}_{0 \leqslant n \leqslant m}$ (for any fixed m), equation (8.72), the identity $Z_{n,n+1} = (1 + R_n)^{-1}$, and the fact that $\alpha_n^{(m)}$, R_n, $Z_{n,m}$ and Π_n are $\mathcal{F}_n$-measurable.

14. (a) $V_0^{(3)} \simeq 0.0792$; (b) $V_0^{(3)} \simeq 0.0597$; (b) $V_0^{(3)} \simeq 0.0527$.

16. (a) $Z_{0,1} \simeq 0.9524$, $Z_{0,2} \simeq 0.9071$, $Z_{0,3} \simeq 0.8639$, $Z_{1,0;2} = 0.9569$, $Z_{1,1;2} = 0.9479$, $Z_{2,0;3} = 0.9615$, $Z_{2,1;3} = 0.9524$, $Z_{2,2;3} = 0.9434$, $Z_{1,0;3} = 0.9158$, $Z_{1,1;3} = 0.8985$.

(b) Price the cap by using the formula in (8.104) for the binomial case. The swap is readily priced using (8.132) and the floor price then follows by symmetry. The respective prices are: $V_0^{\mathrm{Swap}(3)} = -0.02715$, $V_0^{\mathrm{Cap}(3)} = 0.02715$ and $V_0^{\mathrm{Floor}(3)} = 0$.

17. (a) $Z_{0,1} \simeq 0.9804$, $Z_{0,2} \simeq 0.9601$, $Z_{0,3} \simeq 0.9390$, $Z_{1,0;2} = 0.9821$, $Z_{1,1;2} = 0.9764$, $Z_{2,0;3} = 0.9837$, $Z_{2,1;3} = 0.9785$, $Z_{2,2;3} = 0.9716$, $Z_{1,0;3} = 0.9636$, $Z_{1,1;3} = 0.9520$.

(b) $V_0 = 0.00127$; (c) $V_0 = 0.00375$; (d) $V_0 = 0.00143$.

C

Glossary of Symbols and Abbreviations

$a_{\overline{n}|j}$	discount factor for n payments and rate j
$a(t)$	accumulation function over a time period of length t
$A_{\omega_1\ldots\omega_n}$	atom (event) corresponding to the sequence of first n moves in a binomial tree
$B(t)$ or B_t	value of a risk-free security (e.g., a bond or a bank account) at time t
$\mathscr{b}(x; n, p)$	probability mass function for the binomial law $Bin(n, p)$
$\mathcal{B}(x; n, p)$	(cumulative) probability distribution function for the binomial law $Bin(n, p)$
$Bin(n, p)$	binomial probability distribution with number of trials n and success probability p
C^A	American call
C^E	European call
CDF	cumulative (probability) distribution function
$\mathrm{Corr}(X, Y)$	correlation coefficient of X and Y
$\mathrm{Cov}(X, Y)$	covariance of X and Y
CRR	Cox–Ross–Rubinstein
D	downward move in a binomial tree
div	dividend
$d(t)$	discounting function over a time period of length t
$\frac{d\mathbb{P}}{d\mathbb{Q}}$	Radon–Nikodym derivative of $\mathbb{P}$ w.r.t. $\mathbb{Q}$
EMM	equivalent martingale measure
$Exp(\lambda)$	exponential probability distribution with rate λ
$\mathrm{E}[X]$	mathematical expectation of X
$\widetilde{\mathrm{E}}[X]$	risk-neutral mathematical expectation of X
$\mathrm{E}[X \mid \mathcal{F}]$	mathematical expectation of X conditional on a σ-algebra $\mathcal{F}$
$\mathrm{E}_t[X]$	mathematical expectation of X conditional on $\mathcal{F}_t$
$\widetilde{\mathrm{E}}^{(g)}[X]$	mathematical expectation of X w.r.t. the probability measure $\mathbb{P}^{(g)}$

$\widetilde{\mathrm{E}}_t^{(g)}[X]$	mathematical expectation of X conditional on $\mathcal{F}_t$ w.r.t. the probability measure $\widetilde{\mathbb{P}}^{(g)}$
$\mathrm{E}_{t,x}[X]$	mathematical expectation of X conditional on an underlying process having value x at time t
$\mathrm{E}_{t,\mathbf{x}}[X]$	mathematical expectation of X conditional on a underlying vector process having value $\mathbf{x}$ at time t
$\mathcal{F}_t$	σ-algebra generated by information available up to time t
$\mathbb{F} \equiv \{\mathcal{F}_t\}_{0 \leqslant t \leqslant T}$	filtration up to time T
$\mathcal{F}_t^X$	σ-algebra generated by a process X up to time t
$\mathbb{F}^X \equiv \{\mathcal{F}_t^X\}_{0 \leqslant t \leqslant T}$	natural filtration for a process X up to time T
$f_{n,m}$	forward rate at time n for interval $[m, m+1]$ in a binomial tree
$f_{n;m,m+k}$	forward rate at time n for interval $[m, m+k]$ in a binomial tree
$f_X,\ f_D$	probability density function (of random variable X or probability distribution D)
$F_X,\ F_D$	(cumulative) distribution function (of random variable X or probability distribution D)
$Gamma(\kappa, \lambda)$	gamma probability distribution with shape parameter κ and rate parameter λ
$i^{(m)}$	nominal interest rate compounded at frequency m
$\mathbb{I}_A$	indicator of event (or set) A
iff	if and only if
i.i.d.	independent and identically distributed
K	strike price
$\Lambda(\cdot)$	payoff function
m_t^X	minimum over $[0, t]$ of the process X
M_t^X	maximum over $[0, t]$ of the process X
$n(x)$	probability density function for a standard normal law
$\mathcal{N}(x)$	(cumulative) probability distribution function for a standard normal law
$Norm(\mu, \sigma^2)$	normal probability distribution with mean μ and variance σ^2
NPV	net present value
ω	outcome or scenario (element of a state space)
ω_n	n-th move in a binomial tree
Ω	state space

P	present value, or principal, or purchase price
$\mathbb{P}$	probability measure
$\widetilde{\mathbb{P}} \equiv \widetilde{\mathbb{P}}^{(B)}$	risk-neutral probability measure with bank account as numéraire
$\widetilde{\mathbb{P}}^{(g)}$	risk-neutral probability measure (or EMM) with asset g as numéraire
$\mathbb{P}(A)$	probability of event A
$\mathbb{P}(A \mid B)$	probability of event A conditional on event B
$\mathcal{P}_t$	partition of Ω generated by information available at time t
$\mathcal{P}(X)$	partition of Ω generated by random variable X
$\mathcal{P}(X_1, \ldots, X_n)$	partition of Ω generated by random variables $X_1, \ldots, X_n$
P_A	present (discounted) value of an annuity
P^A	American put
P^E	European put
$\Pi(t)$ or Π_t	portfolio value at time t
$\overline{\Pi}(t)$ or $\overline{\Pi}_t$	discounted portfolio value at time t
PDF	probability density function
$Pois(\lambda)$	Poisson probability distribution with rate $\lambda > 0$
ϱ	Radon–Nikodym derivative
ϱ_t	Radon–Nikodym derivative process at time t
ρ	correlation coefficient
r	interest rate
$r(t)$	instantaneous interest rate at time t
$r_{[t_1, t_2]}$	rate of return for time interval $[t_1, t_2]$
$R_{[t_1, t_2]}$	total return for time interval $[t_1, t_2]$
$R_m \equiv r_{m+1}$	rate of return for single time period $[m, m+1]$
$\mathbb{R}$	set of real numbers $(-\infty, \infty)$
$\mathbb{R}_+$	set of nonnegative real numbers $[0, \infty)$
$\sigma(X)$	σ-algebra generated by random variable X
$\sigma(\{X_\lambda\})$	σ-algebra generated by a collection $\{X_\lambda\}$
$S(t)$ or S_t	price of a risky asset (e.g., a stock) at time t
$S_i(t)$ or S_t^i	price of the ith risky asset at time t
T	maturity time; expiry time; exercise time

$\mathcal{T}_b^X$	first passage time of a process X to level b
$\mathcal{T}_{(a,b)}^X$	first exit time of a process X from the interval (a,b)
U	upward move in a binomial tree
$Unif(a,b)$	uniform probability distribution on an interval (a,b)
$V(t)$	(accumulated) value function at time t
$v(\tau,S)$	derivative pricing function of time to maturity τ and spot S
$V(t,S)$ or $V_t(S)$	derivative pricing function of calendar time t and spot S
$\overline{V}(t,S)$ or $\overline{V}_t(S)$	discounted derivative pricing function
V_A	future (accumulated) value of an annuity
$Var(X)$	variance of X
VaR	Value at Risk
w.r.t.	with respect to
$y(\tau)$	yield rate for time to maturity τ
$y(t,T)$	yield rate at time t for maturity T
$y_{n,m}$	yield rate in a binomial tree from time n to m
$Z(t,T)$	zero-coupon bond price at time t for maturity T
$Z_{n,m}$	zero-coupon bond value in a binomial tree at time n with maturity m
ZCB	zero-coupon bond

D

Greek Alphabet

A, α alpha

B, β beta

Γ, γ gamma

Δ, δ delta

E, ϵ, ε epsilon

Z, ζ zeta

H, η eta

Θ, θ, ϑ theta

I, ι iota

K, κ kappa

Λ, λ lambda

M, μ mu

N, ν nu

Ξ, ξ xi

O, o omicron

Π, π pi

R, ρ, ϱ rho

Σ, σ sigma

T, τ tau

Υ, υ upsilon

Φ, ϕ, φ phi

X, χ chi

Ψ, ψ psi

Ω, ω omega

References

Theory of Probability and Stochastic Processes

D. Applebaum. *Lévy Processes and Stochastic Calculus*. Cambridge University Press, 2009.

K.B. Athreya and S.N. Lahiri. *Measure Theory and Probability Theory*. Springer Texts in Statistics. Springer-Verlag, 2006.

Z. Brzeźniak and T. Zastawniak. *Basic Stochastic Processes: A Course Through Exercices*. Springer-Verlag, 1999.

M. Capinski and P.E. Kopp. *Measure, Integral and Probability*. Springer Undergraduate Mathematics Series. Springer-Verlag, 2004.

R. Cont and P. Tankov. *Financial Modelling with Jump Processes*. Chapman & Hall/CRC Financial Mathematics Series. Taylor & Francis, 2004.

W. Feller. *An Introduction to Probability Theory and Its Applications*, volume 1. John Wiley & Sons, 1971a.

W. Feller. *An Introduction to Probability Theory and Its Applications*, volume 2. John Wiley & Sons, 1971b.

A. Gut. *Probability: A Graduate Course*. Springer-Verlag, 2005.

A. Gut. *An Intermediate Course in Probability*. Springer-Verlag, 2009.

R.V. Hogg and E.A. Tanis. *Probability and Statistical Inference*. Pearson/Prentice Hall, 8th edition, 2010.

M. Jeanblanc, M. Yor, and M. Chesney. *Mathematical Methods for Financial Markets*. Springer Finance. Springer-Verlag, 2009.

I. Karatzas and S.E. Shreve. *Brownian Motion and Stochastic Calculus*. Graduate Texts in Mathematics. Springer New York, 1991.

S. Karlin and H.M. Taylor. *A First Course in Stochastic Processes*. Academic Press, 1975.

S. Karlin and H.M. Taylor. *A Second Course in Stochastic Processes*. Academic Press, 1981.

F.C. Klebaner. *Introduction to Stochastic Calculus with Applications*. Imperial College Press, 2005.

H.H. Kuo. *Introduction to Stochastic Integration*. Springer-Verlag, 2006.

B.K. Øksendal. *Stochastic Differential Equations: An Introduction with Applications*. Springer-Verlag, 2010.

M.M. Rao and R.J. Swift. *Probability Theory with Applications.* Mathematics and Its Applications. Springer-Verlag, 2006.

A.N. Shiryaev. *Probability.* Graduate Texts in Mathematics. Springer-Verlag, 1996.

Introduction to Mathematics of Finance

R. Brown, S. Kopp, and P. Zima. *Mathematics of Finance.* McGraw-Hill Ryerson Limited, 7th edition, 2011.

J.R. Buchanan. *An Undergraduate Introduction to Financial Mathematics.* World Scientific, 2008.

M. Capiński and T. Zastawniak. *Mathematics for Finance: An Introduction to Financial Engineering.* Springer Undergraduate Mathematics Series. Springer-Verlag, 2003.

M. Davis, L. Bachelier, A. Etheridge, and P.A. Samuelson. *Louis Bachelier's Theory of Speculation: The Origins of Modern Finance.* Princeton University Press, 2011.

D. Lovelock, M. Mendel, and A.L. Wright. *An Introduction to the Mathematics of Money: Saving and Investing.* Springer-Verlag, 2007.

T. Mikosch. *Elementary Stochastic Calculus: with Finance in View.* World Scientific, 1998.

S. Roman. *Introduction to the Mathematics of Finance: From Risk Management to Options Pricing.* Undergraduate Texts in Mathematics. Springer-Verlag, 2004.

S.M. Ross. *An Elementary Introduction to Mathematical Finance.* Cambridge University Press, 2011.

P. Wilmott, S. Howison, and J. Dewynne. *The Mathematics of Financial Derivatives: A Student Introduction.* Cambridge University Press, 1995.

Mathematics of Finance (Discrete-Time)

M. Capiński and E. Kopp. *Discrete Models of Financial Markets.* Mastering Mathematical Finance. Cambridge University Press, 2012.

H. Föllmer and A. Schied. *Stochastic Finance: An Introduction in Discrete Time.* De Gruyter Textbook Series. De Gruyter, 2011.

P.K. Medina and S. Merino. *Mathematical Finance and Probability. A Discrete Introduction.* Birkhauser, 2004.

S.R. Pliska. *Introduction to Mathematical Finance: Discrete Time Models.* John Wiley & Sons, 1997.

S.E. Shreve. *Stochastic Calculus for Finance I: The Binomial Asset Pricing Model.* Springer Finance. Springer-Verlag, 2012.

Mathematics of Finance (Continuous-Time)

C. Albanese and G. Campolieti. *Advanced Derivatives Pricing and Risk Management: Theory, Tools and Hands-on Programming Application.* Academic Press Advanced Finance Series. Elsevier Academic Press, 2005.

M. Avellaneda and P. Laurence. *Quantitative Modeling of Derivative Securities: From Theory To Practice.* Chapman & Hall/CRC, 1999.

K. Back. *A Course in Derivative Securities: Introduction to Theory and Computation.* Springer Finance. Springer-Verlag, 2005.

M. Baxter and A. Rennie. *Financial Calculus: An Introduction to Derivative Pricing.* Cambridge University Press, 1996.

D. Brigo and F. Mercurio. *Interest Rate Models – Theory and Practice: With Smile, Inflation and Credit.* Springer Finance. Springer-Verlag, 2007.

A.J.G. Cairns. *Interest Rate Models: An Introduction.* Princeton University Press, 2004.

R.-A. Dana and M. Jeanblanc. *Financial Markets in Continuous Time.* Springer Finance. Springer-Verlag, 2003.

R.J. Elliott and P.E. Kopp. *Mathematics of Financial Markets.* Springer-Verlag, 2005.

A. Etheridge. *A Course in Financial Calculus.* Cambridge University Press, 2002.

J.-P. Fouque, G. Papanicolaou, and K.R. Sircar. *Derivatives in Financial Markets with Stochastic Volatility.* Cambridge University Press, 2000.

Y.K. Kwok. *Mathematical Models of Financial Derivatives.* Springer Finance. Springer-Verlag, 2008.

A.L. Lewis. *Option Valuation under Stochastic Volatility: with Mathematica Code.* Finance Press, 2000.

A.N. Shiryaev. *Essentials of Stochastic Finance: Facts, Models, Theory.* Advanced Series on Statistical Science & Applied Probability. World Scientific, 1999.

S.E. Shreve. *Stochastic Calculus for Finance II: Continuous-Time Models.* Springer Finance. Springer-Verlag, 2010.

P. Wilmott. *Derivatives: the Theory and Practice of Financial Engineering.* Wiley Frontiers in Finance Series. John Wiley & Sons, 1998.

Computational Methods

G. Fusai and A. Roncoroni. *Implementing Models in Quantitative Finance: Methods and Cases.* Springer Finance. Springer-Verlag, 2007.

P. Glasserman. *Monte Carlo Methods in Financial Engineering.* Applications of Mathematics: Stochastic Modelling and Applied Probability. Springer-Verlag, 2004.

A. Hirsa. *Computational Methods in Finance.* Chapman & Hall/CRC Financial Mathematics Series. Taylor & Francis, 2012.

P.E. Kloeden and E. Platen. *Numerical Solution of Stochastic Differential Equations.* Applications of Mathematics: Stochastic Modelling and Applied Probability. Springer-Verlag, 1992.

R. Korn, E. Korn, and G. Kroisandt. *Monte Carlo Methods and Models in Finance and Insurance.* Chapman & Hall/CRC Financial Mathematics Series. Taylor & Francis, 2010.

D.L. McLeish. *Monte Carlo Simulation and Finance.* Wiley Finance. John Wiley & Sons, 2011.

E. Platen and N. Bruti-Liberati. *Numerical Solution of Stochastic Differential Equations with Jumps in Finance.* Stochastic Modelling and Applied Probability. Springer-Verlag, 2010.

E. Platen and D. Heath. *A Benchmark Approach to Quantitative Finance.* Springer Finance. Springer-Verlag, 2006.

D. Tavella. *Quantitative Methods in Derivatives Pricing: An Introduction to Computational Finance.* Wiley Finance. John Wiley & Sons, 2003.

Financial Economics

S.L. Allen. *Financial Risk Management: A Practitioner's Guide to Managing Market and Credit Risk.* Wiley Finance. John Wiley & Sons, 2012.

J.P. Danthine and J.B. Donaldson. *Intermediate Financial Theory.* Academic Press Advanced Finance Series. Elsevier Academic Press, 2005.

J. Gatheral. *The Volatility Surface: A Practitioner's Guide.* John Wiley & Sons, 2011.

J.C. Hull. *Options, Futures, and Other Derivatives.* Pearson Education, 8th edition, 2011.

D.G. Luenberger. *Investment Science.* Oxford University Press, 1997.

R.L. McDonald. *Derivatives Markets.* Addison-Wesley series in Finance. Addison Wesley, 2003.

H.H. Panjer and P.P. Boyle. *Financial Economics: with Applications to Investments, Insurance, and Pensions.* Actuarial Foundation, 1998.

Index

Printed in the United States
by Baker & Taylor Publisher Services